Joseph Heller

Nature's Biodiversity

The Search, for Beginnings and for Order

Springer

Joseph Heller
Evolution, Systematics and Ecology
The Hebrew University of Jerusalem
Jerusalem, Israel

ISBN 978-3-031-89792-4 ISBN 978-3-031-89790-0 (eBook)
https://doi.org/10.1007/978-3-031-89790-0

This Springer imprint is published by the registered company Springer Nature Switzerland AG
The registered company address is: Gewerbestrasse 11, 6330 Cham, Switzerland

Nature's Biodiversity

To Chava, whom I love so much;
And in memory of my beloved
brother, Michael

Acknowledgements

Netta Kasher took upon herself the graphic editing the illustrations, carefully and with skilful perfectionism restoring the special charm of those ancient illustrations, which were related and printed hundreds of years ago.

I thank Naomi Sivan, my research assistant for many years who has assisted me in so many different aspects of this book. Her cooperation and companionship are priceless.

I could not have written this book without the warm, devoted and sympathetic assistance of Chava Nowerstern and Haya Herr-Meyer, of the Edelstein Collection, in the National Library of Israel. They patiently helped me find many sources of centuries-old books, gave me permission to use illustrations from these books and translated ancient Latin, Spanish and German texts for me. The policy of the National Library of Israel is that they should not be given credit for citing their books, and I respect their policy. The staff at the Harman Library of the Hebrew University also generously opened up their hidden shelves for me, laden with treasures of ancient books.

I am grateful to Esther Lachman for being the best sort of scientific editor I could have wished. Closely reading the manuscript, she made many helpful comments and corrections, eliminating numerous inconsistencies, redundancies and stylistic infelicities in the process, with many constructive ideas to improve, both each illustration and the overall image of the book.

Tuvia Kurz has created, with a master's hand, top-quality illustrations. I am deeply grateful to him for the meticulous way in which he has drawn and painted many of the figures.

I owe a great debt of gratitude to Zeev Arad, of the Technion in Haifa, to Dieter Ebert, of the University of Basel, to Frida Ben Ami, of the University of Tel Aviv, to Ariel Chipman and to Ronen Kadmon of the Hebrew University, who read drafts of some chapters, pointed out errors and omissions and made numerous constructive suggestions. My friends at the Department of Evolution, Ecology and Behaviour at the Hebrew University have very patiently put up with my continued presence in the Department, long past my official retirement.

Last but not least I thank Éva Lőrinczi, my editor at Springer, for the many ways she found to improve the text.

All these people have contributed, to the quality of this book. I am fortunate to have worked with such friends for so long. I thank them, all.

Jerusalem, 2024 Joseph Heller

Introduction

Man is a classifying animal by nature. Our urge to classify the environment is part of our biology; it is a basic instinct like hunger, like fear or love. Like every animal, we need classifying thought to distinguish food from what is not food, a predator from a non-predator, a possible mate from an unlikely one. With the multitude of creatures being so diverse, we would not have survived had we not distinguished between what is edible and what is poisonous, who is a predator and from whom we need not flee. Some measure of classification occurs in every animal, but humans differ in that they express their classifications through names. Language gives us a far greater ability to distinguish, in nature, the details of objects surrounding us. "Scorpion", and thanks to language not merely "scorpion" but yellow scorpion, black scorpion, thick-tailed scorpion, lethal scorpion, sharp-ribbed scorpion—details of scorpion diversity that, through language, we can refine in our classification. Far from merely identifying and naming, however, classifications are also theories about the basis of order in nature—whether we should classify whales with fishes or with mammals, with seals or with dolphins.

This book deals with different theories concerning biodiversity as they have developed from ancient days down to our days, in an attempt to show how we have arrived at our present state of knowledge about biodiversity. It deals with the search for overall patterns, in the grand diversity of life, both with theories that well-established facts have proven true and with theories that we have found to be wrong, and that have faded away almost completely. For somewhat similar reasons, beyond real animals, I deal with some fabulous ones as well: they were very real to many people, throughout many ages. I myself have devoted my professional career to the study of various aspects of diversity in one group, the molluscs.

Students of biodiversity have sought, throughout the ages, answers to the question how living creatures first came to be, each suited so impressively well to its specific environment. Indeed, beyond the living creatures themselves they have sought to understand how the entire world came to be out of some amorphous, unorganized matter. Some have suggested creation by a supernatural conscious entity; others have offered ideas involving one physics-based element or another. I shall also present ideas as to how the entire world, with its incredibly complex creatures, came to be.

Plants and animals are living entities, and whatever their diversity, this calls for a definition of Life. Early philosophers used the terms Soul and Spirit to describe the difference between a lifeless body and a living one, and they understood the soul as something within the body. Later philosophers considered the soul as a divine entity that enters the body. What is a soul and do animals have souls? Do souls migrate, from man to animal and back again? These questions were such a central part of the overall view of nature, from times of antiquity onwards, that we shall, throughout the book, touch also upon views concerning the soul.

Animal reproduction posed a problem broadly comparable to that of the creation of the world: the appearance of order from nothing, of form from matter. How does an embryo develop? From a very early age in history and onwards, students of diversity have realized that the generation of new life is a major component in the entity of different kinds of plants and animals. They included, in their texts and as part of their overall view of nature, various aspects of the differences in habits of courtship, mating, conception and reproduction, in different animal groups. We shall follow the ways in which, through the centuries, students of diversity have sought to understand, both sexual mating and whether living creatures can arise from non-living matter such as dust, with no parent being involved. Students of diversity have also wondered about the significance of diverse modes of embryological development, and I present these too, as part of these naturalists' views.

Diversity means and requires diverse functioning bodies. Medicine has led our understanding of the way in which the vertebrate body functions. The study of diversity also involves functional comparative morphology but in general, we shall touch upon these aspects only briefly and where relevant, so as not to extend over too many fields. I do present, however, the basic knowledge of (human) morphology in ancient times, and the great advance of morphology and anatomy during the Medieval and the Renaissance. It captures the spirit of the period more than other branches of biology, in that it at first sought to glorify and restore Greek and Roman observations, later to break away from them, and finally to develop original views. For similar reasons, I present also the great advance in Botany, during the Renaissance.

My approach to the history of the study of diversity is thus a broad one that links classification together with reproduction, embryology, behaviour and morphology, biological disciplines between which the boundaries are often so vague that no single discipline can develop in total isolation from another.

To put this study in perspective I present a long timescale, working through the history of the western world over the past 2300 years, up to Darwin and beyond. In the century and a half since Darwin, we have witnessed a surge in many aspects of the study of diversity, which has led knowledge and theory to split into multiple disciplines. These include the refining of classifying techniques, the rise of new disciplines such as genetics, evolution, ecology, biogeography and behaviour (species-specific courting and mating aspects). The historic treatment of all these new disciplines is far too broad for any one book. In describing the research of biodiversity during the post-Darwin era, I therefore focus on just two disciplines, Systematics and Biogeography. These two final chapters are not up-to-date summaries of everything in modern Systematics or in modern Biogeography; they are for readers interested in historic perspectives.

This book aims to expand the intellectual horizons of biologists, at all levels, in that it concerns such aspects of diversity as classification, functional morphology, reproduction, embryonic development and evolution, as they developed through history. Some historians of biology have preferred to single out specific problems and discuss them throughout the centuries (e.g., Mayr 1982). The advantage of this approach is in that biology is characterized by the longevity of its scientific problems, which persist from generation to generation and from century to century. This approach traces a problem or idea back to its beginning, and follows up its fate and ramifications, from its historic beginning, either to its solution or to present times. My approach differs in that it presents, together, many aspects of the study of biodiversity of a given period. By doing so I seek to provide the reader with a better feeling of the spirit of the age and of the intellectual milieu in which naturalists of different ages were active—each discovery and each theory is better assessed when viewed in its contemporary setting. My approach often reflects the horizons of the naturalists themselves, many of whom studied both straightforward classification of biodiversity and embryology, morphology and modes of reproduction, seemingly distinct academic fields that often merge into one another. In general, I have attempted not to separate the discoveries and theories from the biographies of the personae who contributed to developing them, as the history of the study of biodiversity is inseparable from those who have made it.

Beyond western culture, I originally intended to present the history of the study of biodiversity in other cultures as well, of China and of India. China's oldest written records preserve records of animals, reports of animal activity and descriptions of the use of animals in socio-economic, religious and ritual practice. However, ancient Chinese writings did not classify animal diversity in depth, and none of them described animal life and morphology. Motives underlying animal classification in China were not zoological and did not perceive the world as a purely physical reality, to be analysed as a biological system. They were contained in a larger project to explain a ritual order, or the structure of the cosmos as a whole. For readers seeking to broaden their horizons on Chinese approaches to animals I warmly recommend *Animals Through Chinese History* (Sterckx et al. 2019). In the ancient Indian culture too, I could not find much beyond a very coarse classification of animal and plant biodiversity. I focus, therefore, on western culture.

I have been generous with illustrations, because the book aims at both biologists, historians and the general reader, and because no one can be familiar with every single animal that, over the centuries, one naturalist or another has chosen to discuss. I also provide portraits of the naturalists themselves, both because they deserve it and to illustrate, through them and their costumes, the age in which they were living (for some of the ancient Greeks I present their coins, instead). Some illustrations are from ancient books, which—I cannot resist—I love so much. Tuvia Kurz has carried out many other illustrations, and I am deeply grateful to him for the meticulous way in which he has drawn and painted many of the figures, bringing into it the fascination of animals with rich and accurate detail down to the last bristle.

References

Mayr, E. 1982. *The growth of biological thought. Diversity, evolution and inheritance.* Belknap: Harvard University Press.

Sterckx, R., M. Siebert, and D. Schäfer, editors. 2019. *Animals through chinese history: Earliest times to 1911.* New York: Cambridge University Press.

Contents

Chapter 1
Ancient Days

Abstract Early Greek philosophers discussed what matter is, and how man and animals came about. Some suggested that Man first developed within an egg case, from which he hatched and emerged onto land. Others claimed that animals originated from primordial moisture. Another theory proposed that creatures formed under a cosmic cycle of eternal change, growth, and decay. At first, only limbs formed, later they came together; but only creatures with homogeneous limbs, as found in the animals of today, have survived. Others suggested that the world consists of eternal, invisibly minute atoms, moving about in an infinite void and forming perishable combinations. Animals are classified into those with blood, versus those without red blood.

Plato believed that a world of ideas exists beyond this real world, which compares to a huge well-organized animal, and that entities of this world are mere shadows of eternal concepts. In our world, each element consists of geometrically unique atoms; and some entities are included within each other. To classify and define them, one should start with the broadest entity, and narrow the possibilities by dividing it in two, by the presence/absence of a single trait (dichotomy). We thus obtain a hierarchy, of small groups in relation to more inclusive ones.

Aristotle believed that entities of this world are not mere shadows of eternal concepts. He claimed that we can understand the basic principles of the world by sensing it, by reasoning, or by generalizing. The universe is timeless. It consists of a series of globes, one inside the other with an unmovable Earth at the centre, consisting of the four elements and four qualities. Furthermore, it consists of matter and form, and movement expresses the dominance of form over matter. Earth, the "underworld", forever changes, by growing, generating, and perishing. The heavens, from the moon and beyond, form the "upper world": a series of constantly rotating transparent spheres, one surrounding the other and containing the unchanging stars, which consist of a fifth, never-changing element. Above these is a starless heaven in which there is no matter, only form. Nothing exists beyond it. The world is thus limited, finite, and closed. All spheres touch each other, and they transmit movement of the uppermost sphere, down to Earth. The unchanging eternal, constantly repeating movement of the heavens is thus responsible for everything that exists and changes on Earth.

J. Heller, *Nature's Biodiversity*, https://doi.org/10.1007/978-3-031-89790-0_1

Aristotle criticized Plato's abstract classification by single trait dichotomies, since his method might split up natural groups and, on the other hand, create meaningless ones. Our world consists of real groups and real sub-groups, so a correct classification of animals should begin by applying straightforward intuition and clustering of animals into large, generally accepted groups, that we clearly define. Within each intuitive group, sub-groups usually differ merely by quantifiable traits, of more and of less.

Species are what they are because they belong to groups that are more inclusive. In his terminology of grouping, Aristotle used the terms *Genos* to describe any group above species level and *Eidos* to represent the subgroup of any *genos*. We should classify animals by syllogisms: If A belongs to B, and B belongs to C, then A necessarily belongs to C. The outcome is a formal definition of what a kind is. We need not state all features of a kind, just its position in the structure of kinds, with a definition that separates it from other kinds. Comprehensive intuitive classification and subsequent divisions are thus counterparts of the intrinsic order in animal diversity.

Aristotle developed a rather complete, full-blown classification of the animal world. All animals have a heart and a digestive tract. Some animals have no blood (our invertebrates), others have blood (vertebrates). Invertebrates divide into four major groups: those with soft bodies, soft shells, hard shells, and terrestrial invertebrates. Two minor groups, sea anemones and sponges, link animals to plants. Aquatic vertebrates are all fishes, but sharks and rays differ from bony fishes, and so do dolphins and whales. Among birds, Aristotle appears to have recognized six subgroups. Among terrestrial quadrupeds, some lay eggs (our reptiles and amphibians), and others bear live young (mammals, including man). Most mythological animals are fantasies. Aristotle thus set up a complete taxonomic system of very many animal groups and created a comprehensive and broadly correct classification. He erred in placing sea urchins and sea squirts with snails and bivalves, starfishes close to sea anemones, and whales and dolphins with bony fish and sharks. For a first-ever attempt at classifying the entire animal world, his classification broadly lives up to the expectations of his theory and method. The number Four, which had a special significance in the ancient world, may perhaps have inadvertently seeped into parts of his animal classification.

Aristotle thought that earthly mortals imitate heavenly immortals by generation. The diverse generative abilities of animals form a ladder, in which terrestrial mammals are at the top as they nourish their embryos via placentas and, after their birth, they suckle their babies. Birds come next because they lay two-coloured eggs that are perfect, in that they have a hard shell. They feed their offspring, but only for a short time. Reptiles lay imperfect eggs in that they have a soft shell; they do not feed their hatchlings. Fishes and cuttlefish lay imperfect eggs in that they lack a shell; hatchlings resemble adults. Flies (and other insects) lay imperfect eggs in that they lack a shell; furthermore, the hatchlings differ considerably from adults. Snails, bivalves and sea urchins generate spontaneously from rotting matter, their diversity requiring species-specific temperatures. Spontaneously generated species lack the continuum of generations, produced in sexual animals.

Eggs need heat to develop into embryos. This heat may come from the sun, but many eggs need also the warmth of their brooding parent bird. Animals consist of matter and form. To explain generation, namely life forming from nothing, Aristotle distinguished between potentiality (ability) and actuality. Development is the realization of a potential form. The male, who activates matter and bestows future form on the offspring, is more divine than the female, who merely contributes matter. The soul makes matter and forms into a living entity by activating nutrition, growth, reproduction, movement, sensation, and understanding. Life occurs at three levels: nutrition, growth and generation, common to plants, animals and humans; movement, sensing, and response to stimuli, restricted to animals and man; and rational abilities, restricted to man. The embryo sequentially actualizes the corresponding vegetative, sensitive, and rational souls. The soul acts via the spirit. During embryo formation, the character of a higher taxonomic group forms before that of a lower group: the embryo forms characters that make it an animal, not a plant, before characters that make it a man, not a horse. The embryo's sex depends on the father's heat. Maleness is a dominant character, beneath which femininity is a hidden potential. If the male character is strong, the offspring will be male, if weak, female. Environmental factors also determine sex. Aristotle also believed in spontaneous generation.

Coming slightly after Aristotle, Theophrastus proposed a general classification of plant diversity into trees, shrubs, under-shrubs and herbs. He believed that some plants change their kind as they grow, and he believed in spontaneous generation. In emphasizing applied aspects, he described the many different uses of wood and resins of different trees.

Epicurus believed that the world consists of eternal, unchanging atoms, and of void. The world is unlimited in size, with an unlimited number of atoms, and an unlimited amount of void. All things are conglomerations of atoms, all events are atom collisions, that result in ever-changing conglomerations. Our present world is a temporary agglomeration that eventually will fall apart. Atomic collisions may form organs and, if an organ can fulfil a function, then function sets in. Form, however, was created before function. The soul too consists of atoms and void, and it disintegrates when the body dies. In Alexandria, the Greek rulers built a museum—a storehouse of texts and a school of philosophy, the first-ever state-supported institute of research.

In Rome, Pliny's military and administrative experience may have shaped some of his interpretations of animal behaviour. Bee's daily life resembles the daily routine in a military camp. Moral aspects in the behaviour of two animals reflect traditional, unwritten codes of Roman social norms. Elephants think rationally, understand human language, obey orders, and remember duties. They are honest, wise and just, have reverence for the sun, moon and their king, and conduct purification rituals. Being modest, they mate only in private; adultery is unknown. Good-tempered, peaceful and gentle, they enjoy marks of honour, yet hate shame and disgrace. Lions are noble and merciful. When enraged they turn their fury on men rather than women and attack children only when extremely hungry. Otherwise, they are enraged only at the disrespect of those who dare challenge them. When pursuing they charge by leaps, when retreating they do so in slow dignity. All this suggests that Roman values and customs are part of the world's grand order of nature.

Pliny further wrote that the universe is an eternal being, the Sun being its soul and mind. Earth, at its centre, consists of four elements. Heavens belong to the gods but Earth, with its nature, belongs to men. Nature creates everything, including plants and animals, solely for the benefit of man, to exploit it. Nature is rational, so well-ordered that even tiny animals display bodily perfection; and so powerful that it can change animals, from one kind to another. Animal remedies are part of this grand Nature-for-Man plan. Pliny classified medicines by their animal origin: remedies from Man, farm animals, wild animals, and plants.

Pliny classified animals into those on land, in the water, in the air, and insects, thereby mixing habitat and natural groups, and somewhat reflecting the four elements. In his terrestrial animals, there was not much order, beyond clustering all domestic kinds in one sequence. Among aquatic ones he described mammals, then fishes, and then invertebrates, the latter in increasing order of their armour: without shells, crusty, and hard shells. He believed that animals on land have aquatic versions: certain aquatic animals resemble sheep, horses, and bulls. Among aerial animals, Pliny clustered birds of augury. His "plumage birds", included the peacock, sacred to the goddess Juno; the farmyard cock, who tells the future; and the goose, sacred to Romans. Some insects dwell on the ground but others fly (through fire). Bees, alone among insects, serve man with their honey and wax. Like man, they toil, construct hives, and have a government with collective leaders. Drones, lacking stings, serve the true bees, who order them about and drive them out to work. Hatched offspring resemble maggots but the king has wings, from the start; a retinue of peers escorts him everywhere. Pliny's animal world contained several fabulous animals (phoenix, unicorn, and sphinx).

Dioscorides, a Roman physician, wrote of plants and medications, in a medicinal order: aromatic oils and plants that produce them, animals and agricultural plants from which medications are prepared, roots, seeds and various plants, and grapevine.

Galen was a Roman physician. As law prohibited cadaver dissection, he dissected animals, assuming they resembled humans. He mastered the approach of controlled experiments, to test functions of an organ. The body operates through a natural soul (liver), for appetite and pleasure; a vital soul (heart) that governs emotions; and a rational soul (brain) that governs reason, memory, and motion. These souls have different vessels: veins, arteries, and nerves. Disease results from an imbalance of the four body humours, each associated with qualities of one of the four elements. Concerning reproduction, semen derives from the body's humours, drawn from all organs to the testicles, complete and with souls. Females produce imperfect semen, which has generation powers upon mixing with male semen, during copulation; a fleshy embryo develops. During intercourse, if male sperm falls on the right side, the baby will be male; on the left side—female. If male semen is more abundant, the child will look like the father; if the mother's is, then like the mother.

If in the semen, one humour is more abundant than the others are, the child will follow that humour and be sanguine, choleric, phlegmatic or melancholic; and not like its parents.

1.1 Greece: A Dawn

1.1.1 Introduction

The study of biodiversity is the study of the variety of living things and, as a discipline, it is as old as mankind—no specific date in human history would be too early to begin the history of the classification of diversity. Nevertheless, a history of the approach to diversity has to start at some point. One such point could be with the first prominent biologist, Aristotle. However, to place Aristotle in the context of the intellectual environment in which he acted and developed his biological thought, we must go back two to three centuries before his time, to the dawn of Greek philosophy. Philosophy first evolved in the small independent Greek city-states of Ionia, (central coastal Anatolia in Turkey of today) from about 650 BC onwards.

"Philosophy" in Greek means "love of wisdom" and the ancient Greeks used the term in a broad sense, to cover what we now consider both sciences and liberal arts. Philosophers regularly divided their subjects into three parts: logic, ethics and physics. Logic included the study of language as well as the study of thought and argument. Ethics included moral and political theorizing, and it embraced what would now fall under sociology and ethnography. Physics was the study of nature and the phenomena of the natural world.

Aristotle, according to this triple definition, would have termed those early philosophers who studied the nature of the universe centuries before him the *phusikoi*, and their activity *phusiologia*: they were students of nature and their subject was the study of nature. The word *phusis* derives from a verb meaning "to grow", and this term introduces a clear distinction between the natural and the artificial world (in Greek, *tekhne*), between things that have grown and things that have been made. Chairs and ploughs are artefacts made by designers; they have no nature, for they do not grow. Trees, wheat, horses and snakes, on the other hand, are not artefacts but natural objects—they grow. The word *nature* designates the sum of the natural objects and natural events; in this sense to discourse "On Nature" is to talk about the totality of the natural world—the entire universe. In another important sense, when early Greek philosophers inquired into nature, they were inquiring into the nature of things. Any natural object—that grows, rather than being manufactured—has a nature of its own. Its nature is an intrinsic, essential feature—it is not accidental or coincidental.

The first ancient Greek philosophers speculated about how the physical cosmos emerged, what it was composed of and what gave it order. Many people the world over believe in creation legends. They explain the origin of the entire universe including the sun, the trees, the animals and the humans, by some mysterious supernatural creature, deity, or a god; a Creator that once upon a time had the power to create the universe, and still (in some legends) has the power to govern it. Many people also believe that the whole world is breathing and alive and that spirits, souls, or gods inhabit even the rocks, the mountains and the heavens, which are part of a nature in which everything is alive, active and creative.

Some early Greek philosophers believed in an eternal and unchanging master plan of nature and they sought, by concentrated rational thought, to decipher the basic principles of this master plan of the universe. They strove to explain the motions of the heavenly bodies, the periodicity of meteorological events, the nature of different minerals and the birth, nourishment, growth and death of plants and animals. To them, the world was orderly and logical, with its various components arranged in a coherent system. Some of them searched for a minimal amount of basic laws that could explain a maximal number of phenomena. An extreme example is a certain Anaximanes of Miletus (mid-sixth century BCE) who claimed that everything in our universe could be explained by one single element, air, which is constantly in motion and which occurs in two physical states, dense or thin, depending on whether the environment is hot or cold. These philosophers thus sought to understand the world in terms that were, to them, not religious. Whether or not gods exist, a small group of internal, fundamental laws governs the world, not mythological deities with capricious desires. The vast majority of the early Greek philosophers also believed in a static world, in which the only major changes are the cycles of seasons and the history of Man. They had no concept of evolution or ongoing changes in the surrounding plants and animals in nature.

Many of these philosophers put their thoughts down in writing, some in verse others in prose, some in one single text and others in numerous books. Of all these texts, not even one survived to our day. All our knowledge regarding the theories of early Greek philosophers (unlike our knowledge of Plato and Aristotle) is indirect, from quotations of later Greek philosophers or of Christian scholars in the fourth to sixth centuries. These quotations are often in the style of "Alpha says that Beta told him that Gamma once said". The theories of early Greek philosophers thus reach us through the words of philosophers who lived much later, between two hundred and a thousand years after they originally wrote them down. Naturally, some of these second-hand texts are unreliable, or conflicting. It is from these second-hand texts that we must reconstruct the original views of early Greek philosophers. We have second-hand fragments that mention the theories of 17 early Greek philosophers, who focused mainly on basic problems of the substance of the universe, and very little on plants and animals (Barnes 1987). I briefly present three of them, Anaximander, Empedoceles and Democrites, before going on to Plato.

1.1.2 Life Emerged from the Sea: Anaximander

The Greek philosopher Anaximander (ca. 610–540 BCE) lived in the Greek city-state of Miletus (Fig. 1.1). We know very little about his life. He seems to have travelled a lot, led a mission that founded a colony on the coast of the Black Sea and to have erected a sundial in Sparta. He also developed a theory concerning the manner in which the universe is organized and a biological theory on the origin of Man.

Anaximander suggested the existence of a "Boundless" (*Apeiron* in Greek): An eternal, ageless, undifferentiated entity that has no beginning and no end, which is the origin of all things that exist, including the heavens and the earth (Barnes 1987; Kocandrle and Couprie 2017). He thus presented a concept of an open universe (unlike the later concept of Aristotle, of a closed universe covered by a celestial dome). Anaximander's Boundless added two distinct features to the traditional Greek concept of the divinity of the immortal Homeric gods of his day: it is an impersonal entity, and it is not only immortal but also not born. By making the Boundless the source of all things, Anaximander's theory, on its high level of abstraction, perceives ultimate reality in something that is non-perceptible, unlike earth, air, fire and water, which we perceive through our senses. This Boundless is the underlying cause of everything, the origin of all existing things, the source from which existing things derive their existence and to which they return upon their destruction. Thus, the ultimate cause of things is some invisible and limitless physical substance, which is capable of morphing into all the physical things that we see. The *Apeiron* is something so vague that it is neither everything nor nothing, neither organized nor disorganized, neither complete nor a set of parts. It is something far beyond our understanding (Barnes 1987). Anaximander's *Apeiron* seems similar to the "formless and void" Earth of the Creation as written in *Genesis*.

Anaximander also proposed that the celestial bodies lie behind one another—there is depth in the universe. This idea is remarkable because we would naturally think of the fixed stars as a dome-like structure into which the celestial bodies are set, all at the same distance. He erred in that he placed the celestial bodies in the

Fig. 1.1 A coin from Miletos, Ionia, before 575 BC, the days of Anmaximander. On one side is a reclining lion, its head turned back; on the other side is a central oblong depression containing a standing fox, flanked by two squares one of which contains a stag's head. (Figure by Tuvia Kurz)

wrong order: he suggested that the stars were nearest to the Earth, followed by the moon, with the sun farthest away. He further claimed that all celestial bodies make full circles and pass beneath the Earth, both the stars around the polar, as he could have observed, and the more southern stars, that disappear behind the horizon. The celestial bodies are like chariot wheels. The rims of these wheels consist of an opaque vapour that is hollow, and filled with fire. Tubular channels, or holes, penetrate the opaque vapour, and it is through these openings, in the vapours enveloping the wheels, that the fire shines through; and this is how the heavenly bodies appear to us. This is what we see as the sun, the moon and the stars. Sometimes these holes are blocked, and hence eclipses occur. Indeed, the moon regularly waxes and wanes according to whether the channels are regularly blocked or open, and this is what accounts for the phases of the moon. The Earth floats free in the centre of the universe, not supported by anything, but resting where it is. The top of the Earth is flat and circular, as suggested by the horizon. Its overall shape is cylindrical, like a stone pillar or a column drum, and we live on top of this stone pillar. In short, Anaximander developed a cosmology (Barnes 1987; Kocandrle and Couprie 2017).

The Greek concept of the universe is expressed in the word *kosmos*, hence our "cosmos" and "cosmology". The noun *kosmos* derives from a word that means "to order", "to arrange" and "to marshal"—it is used by Homer to describe the Greek generals marshalling their troops for battle. Moreover, it is a beautiful arrangement: the word *kosmos* in ordinary Greek meant not only ordering but also adornment (hence the English word "cosmetic"), something which beautifies and is pleasant to contemplate. The cosmos is the universe, the totality of things; but it is also the ordered universe, the aesthetic, elegant universe—and hence, in principle, it must be explicable.

Anaximander also wrote about biology. In his opinion, animal life originated from a primordial moisture that covered the earth, in some distant age before the sun had dried it up (Barnes 1987; Couprie and Pott 2005). Humans, however, originated from animals different from the humans of today. He knew this because other animals can look after themselves soon after being born, while only humans require a long period of nursing; that is why they would not have survived if they had originally been like the humans of today. At this point, Anaximander seems to have drawn on observations on shark reproduction (Figs. 1.2, 1.3 and 1.4):

> […] the first animals were born in moisture, surrounded by prickly barks. As they grew older, they emerged onto drier parts, the barks burst, and, for a short time, they lived a different kind of life. […] from hot water and earth there arose fish, or animals very like fish, that humans grew in them, and that the embryos were retained inside up to puberty whereupon the fish-like animals burst and men and women emerged already able to look after themselves. […] at first men came into being inside fish and were nourished there—like sharks—only emerging and taking to the land when they were able to look after themselves (Barnes 1987).

Shark reproductive strategy involves the birth of a small number of young, sometimes only one or two. In some Mediterranean sharks, the female places the large, fertilized eggs inside a special leathery egg case ("mermaid's purse") of collagen, which is permeable to water and oxygen (Fig. 1.2). The parents do not guard these

Fig. 1.2 A mermaid purse with a developing shark inside, attached to its yolk. (Tuvia Kurz)

purses. Each case has four elongated tendrils, which entangle in seaweeds or boulders on the seabed, anchoring the case and preventing it from being swept to sea. The embryo develops inside the purse until fully developed, and eventually, a miniature shark, already capable of catching its first prey, cuts its way out. One Mediterranean case-producing shark is the catshark *Sciliorhynus canicula*, another one is the rabbit shark *Chimaera monstrosa*, usually found in deep, but sometimes also in shallow waters.

In other sharks, the embryo develops inside the female uterus, and the female gives birth to live young. In the Mediterranean dog shark (Fig. 1.4), the large, fertilized eggs continue to develop after reaching the uterus, independently from one another, at the expense of the yolk sac, which provides their nourishment until just before birth, some 2 years after fertilization. In hammerheads too (Sphyrnidae, Fig. 1.5; four species in the Mediterranean), the female gives birth to live young. At first, the developing embryos feed on a yolk sac. Later, when the supply of yolk is exhausted, the depleted yolk sac transforms into a placenta and an umbilical cord, which contains blood vessels. Nutrients and oxygen pass directly from the mother's bloodstream to the embryo, and waste products from the embryo to the mother, for elimination (Mojetta 2004).

Fig. 1.3 The offspring of some sharks live protected inside the mother's body, feeding on their own yolk. (Tuvia Kurz)

Fig. 1.4 Dog shark, *Squallus acanthias.* (Tuvia Kurz)

Reproduction and embryo development are even more specialized in the Mediterranean bull shark *Eugomphodus taurus*. The embryos fight one another from this early stage, feeding on one another until only the two strongest embryos remain, one in each uterus (Mojetta 2004).

The ancient Greeks would have seen all these Mediterranean sharks, and the many various modes of shark reproduction could perhaps have inspired Anaximander's ideas regarding the origins of humans. From the few second-hand statements remaining from his texts, I suggest the following reconstruction of Anaximander's views on the subject of human origins: Originally, both fish and men were born in the sea from the mixture of two elements, water and earth. The first humans somewhat resembled some kind of shark as they were born into egg

Fig. 1.5 Hammerhead shark, *Sphyrna zygaena.* (Tuvia Kurz)

Fig. 1.6 Dolphin, *Delphinus delphis.* (Tuvia Kurz)

cases with horn-like tendrils. As they grew older, the case burst, and they emerged on land. At some later stage, men and women resembled other kinds of sharks as they came into being inside those fish, or animals very like fish, that, similarly to a viviparous shark, nourished them inside the female body. They only emerged and moved on to land when they were able to care for themselves. In reconstructing Anaximander's theory, I suggest that the animal that was similar to a live-bearing shark, but more mammal-like, could perhaps have been a dolphin (Fig. 1.6). The next step could have been a siren (Sirenia) because like humans, the mammary glands of sirens are between the front flippers/legs. Furthermore, like humans, they cannot hold their breath as long as other marine-adapted mammals such as dolphins and whales (they can submerge for only up to 10 min); and when resting, they may "tail stand" (in an appropriate depth) with their heads out of the water (Fig. 1.7). Whether or not my reconstruction is true, the theory of Man originating from fish led Anaximander to a moral conclusion: because fish were once the fathers and mothers of men, he urges us not to eat them (Barnes 1987).

Fig. 1.7 Dugong, *Dugong dugong*. (Tuvia Kurz)

Plutarch (46–119 C), a Greek priest at the Temple of Appolo in Delphi, was one of the first to suggest that Anaximander had live-bearing sharks in mind when constructing his theory, which suggests a marine origin of man. Alternatively, however, man could have originated in freshwater habitats. Links between river mud and theories on the origin of life appear already in ancient Egypt, where life depended on the annual flooding of the Nile, and in Babylonia, where life depended on the flooding of the Euphrates and the Tigris. Creatures living both in rivers and nearby mud might have inspired Anaximander's ideas on man's conquest of land. In caddisflies (Trichoptera), the aquatic larva builds itself a protective case of plant material or pebbles that somewhat resembles pieces of "prickly bark" and within which it dwells for several months. Eventually, it leaves its case, crawls onto dry land, moults and flies away (Fig. 1.8). The caddisfly lifecycle would have been easy to observe.

Anaximander's opinion that animal life originated from a primordial moisture that covered the earth in some distant age before the sun dried it up, would accompany Western thought for a thousand years to come, up to the sixth century CE; and it would turn up again, much later, in the eighteenth and nineteenth centuries.

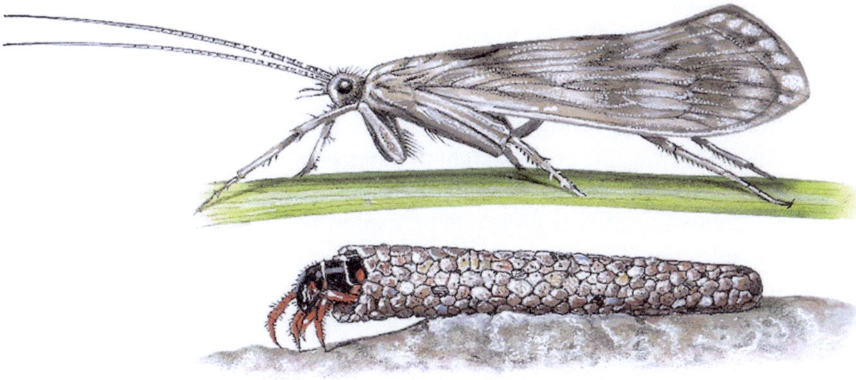

Fig. 1.8 Caddisfly, larva in case and adult. (Tuvia Kurz)

Fig. 1.9 Akragas, Empedoceles' times: a coin with crab. (Tuvia Kurz)

1.1.3 Love, Strife and Animals Created from Floating Limbs: Empedocles

From the distant days of Anaximander, we move forward a century, to the time in history when a Greek soldier ran 40 km from Marathon to Athens to announce the Greek victory, and when the Pantheon of Athens was built. At that time, Empedocles (ca 495–435 BCE) lived at Akragas (Sicily), then a leading city of *Magna Graecia*, with about half a million inhabitants (Fig. 1.9). A wealthy man, Empedocles liked to wear purple robes, a golden girdle, bronze sandals and a Delphic laurel wreath; he also liked to keep a train of boy attendants and to provide dowries for many of the girls of Akragas. Legend has it that he died by throwing himself into the volcano of Mount Etna, believing that he was immortal and that he would come back as a god after being consumed by the fire.

Empedocles wrote in verse. One of his poems, *On Nature*, was concerned with the formation and structure of the physical world, and the formation of the animals and plants in it. In a way, the poem sought to replace the traditional Greek gods with four lifeless, impersonal elements and two cosmic forces. Empedocles claimed that the world is composed of four elements of matter, which he termed "roots": fire, aither, earth and water (Fig. 1.10). Whereas Aristotle would later consider the term aither as referring to air, Empedoceles may well have had in mind a mysterious entity that is bright, rare, energetic and glorious, integral perhaps to some fiery entities of the heavenly bodies (Shaw 2014). It may be a "root" mixed throughout the universe, which brings life by mingling with the other elements. Aither would thus be the spark of life that permeates Empedocles' cosmos.

These four roots are both material and spiritual. Empedocles' choice of the term "roots", which Plato later called "elements", stresses their potential for creation and growth. "Elements" (Greek *stoicheia*, Latin *elementa*) is the word for the letters of the alphabet and is a metaphor for the ability of the elements of matter to form different types of materials by changing positions, just as a limited number of letters form all sorts of different words. The four elements were able to create all things, including living creatures, by being mixed in different proportions. However, each of the elements retained its own characteristics in the mixture, and each was eternal and unchanging. The different proportions in which these four indestructible and unchangeable elements combine with each other produce differences in structure and the properties of matter. The stone contains a high amount of earth, whereas a lamb contains more water and fire, and is thus both soft and a living creature. Nothing new comes or can come into being; the non-existent cannot generate anything; the only change that can occur is a change in the juxtaposition of elements with elements. This theory was an important development in scientific thinking because it suggested that some matter, even if it looked as consisting of pure

Fig. 1.10 The four elements and four qualities of Empedocles

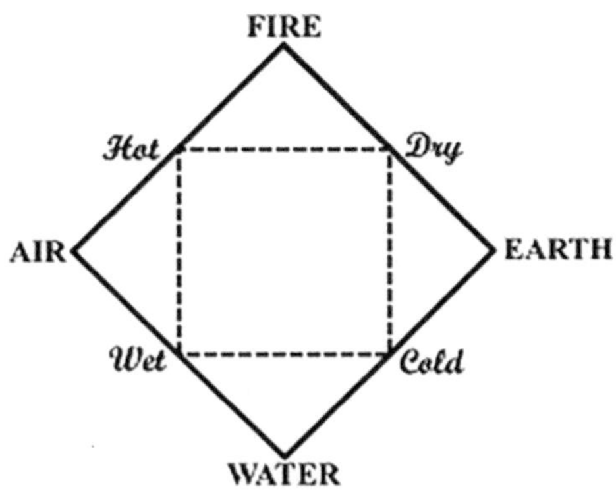

material, might actually consist of a combination of the four core elements (Kingsley et al. 2020).

The four elements are eternal and unalterable, and as change is the consequence of their mixture and separation, Empedocles supposed the existence of moving powers that bring about mixture and separation. As the basis for change, he suggested two opposing eternal forces that act upon these four elements, forces that operate in creative versus destructive ways, to combine all things in the world as well as to separate them. He personified these two forces as Love (*Philia*)—a force of attraction and combination—and Strife (*Neikos*)—a force of repulsion and separation. When Love is completely dominant, it draws all the elements very tightly together into a sphere in which they are completely intermingled, motionless and indistinguishable from each other. No matter can exist in this extreme state, and no life is possible. Then, as Love's power gradually weakens and Strife's power begins to grow, it gradually separates out the four elements from the sphere into distinct entities of fire, air, earth and water. At a certain point of a harmonious balance between Love and Strife, the world (*kosmos*) was created: the mass of the Earth is at its centre, water surrounds the Earth, air forms the next layer and, from the fire at the periphery, the sun came into being as a distinct entity. Within this world, from the mixture of elements in different proportions, various forms of mortal animal life arose. However, once Strife went beyond this point of a harmonious balance between itself and Love and achieved total domination, the elements separated so completely that the world fell apart and destroyed all life. Empedocles assumed a cyclical universe whereby the elements return and prepare the formation of the sphere for the next period of the universe (Barnes 1987).

At this point, another cycle begins. Love increases in power again and draws the elements together until, at a certain point of a harmonious balance between itself and Strife, the elements are once again intermingled, and as this happens, a world is again created and mortal life is reborn. Eventually, Love achieves full and total dominance again, and both the cosmos and animals perish. The cosmic cycle goes on in this manner, endlessly repeating itself in continual alternation in an endless cosmic cycle in which the influences of Love and Strife endlessly wax and wane in dominance, neither force ever wholly escaping the imposition of the other. The constant interplay of these elements, alternately attracting and repelling each other, formed the universe in which we live today.

How does Life come to be? Empedocles presented a comprehensive theory for the origin of all plant and animal life, including humans. His theory of generation posited two processes that work to produce the life we see around us today: "the coming together and the unfolding of birth" (more strictly "un-leafing", since the metaphor originates from the leaves of plants). In his theory of animal creation, at first, only individual limbs and organs were produced from the Earth, and they wandered around aimlessly:

> Here sprang up many faces without necks, naked arms wandered about unattached to shoulders, and eyes strayed alone, begging for foreheads (I cite from Barnes 1987).

Then, under the combined power of Love, the limbs came together and joined. However, these chance combinations were random, and the result included, besides normal humans and normal oxen, all sorts of wild and random hybrid combinations, such as double-headed creatures, hermaphrodites, ox-faced man creatures and man-faced ox-creatures:

> Many creatures grew double-headed, double-chested—many man-faced oxen arose, while others again sprang forth as ox-headed offspring of man—creatures mixed partly from male, partly from female form and fitted with shadowy limbs (Barnes 1987).

Eventually, Empedocles suggested, those creatures unfit for life, because chance had wrongly assembled their disparate parts, died out. They went extinct either immediately or because they were unable to breed. Only the creatures that chance had assembled their limbs correctly survived, as the animals we see today. In other words, a great burst of birth in the beginning produced the entire variety of nature, which extinctions then whittled down to the creatures we see today.

Empedocles adds another theory concerning the origins of life. He described "shoots" of men and women arising from the earth, drawn up by the element of fire as it separates out from the other elements, under the power of increasing Strife (Fig. 1.11):

> Come now, hear how [the element of] fire, as it separated, raised up the nocturnal shoots of men and pitiable women: it is no erring nor ignorant tale. Whole-nature shapes first sprang up from the earth, having a portion of both water and heat. These [the element of] fire sent up, wishing to come to its like: they did not yet display the desirable form of limbs nor voice, which is the part proper to men (Barnes 1987).

As his choice of the word "shoots" indicates, these were not yet fully articulated people with distinct limbs but "whole-nature shapes" that "did not as yet display the desirable shape of limbs, nor voice or language native to man". As Strife increased in power and harmony with Love was achieved, these "shoots" gradually became fully articulated with distinct limbs and features, just as plant buds develop into flowers. He thus accounted for human origins using a botanical analogy, with humans as biological products of the Earth itself. Thus, Empedocles proposed two quite different theories for the creation of life: one by the joining of limbs, the other similar to the formation of flowers.

In conclusion, Empedocles explained diversity and changes in the world as a combination and separation of the four elements. Each of the four had its particular quality, but these qualities alone were not sufficient to create the animal world. Besides the interaction of fire, air, earth and water, there had to be other forces at work in order to explain how the world we live in, and its creatures, come to be. Life forms only when competing principles of Love and Strife reach a harmonious balance (Kingsley et al. 2020).

Empedocles, like Pythagoras before him, believed in the transmigration of souls between humans and animals, that souls can be reincarnated between humans, animals and even plants. Accordingly, he followed a vegetarian lifestyle and warned his fellows to abstain from bloodshed and from eating flesh. The traditional practice

Fig. 1.11 Empedocles: the genesis of Man. (Tuvia Kurz)

of animal sacrifice was a fundamental basis of the Greek religion, and those attending the sacrifice ate the flesh of the sacrificed animals. Empedocles, however, urged his readers to abstain from meat since all souls reincarnate, as various sorts of animals; killing and eating animals is, in reality, cannibalism. This applied even to Empedocles himself, "For already I have at some time been a boy and a girl and a bush and a bird and a silent fish in the sea". Empedocles pictured a time in the past, a sort of golden age, when universal harmony existed. In those ideal days, man would respect the gods by painting animal figures, and by offering only perfumes of subtle fragrance, distilled myrrh, sweet-smelling frankincense and golden honey. The unspeakable slaughter of bulls did not yet drench the altar. In those days, "all creatures, both animals and birds, were tame and gentle to men and bright was the flame of their friendship". Many of Empedocles' ideas would surface again two thousand years after his time, in early modern Europe and in the Age of Enlightenment.

1.1.4 Atoms: Democritus

Empedocles's theory was problematic in that no matter how many times you break down (say) a stone, the resulting pieces never resemble any of the four core elements. We now move half a century ahead, to Democritus (460–370 BCE, Fig. 1.12) from Abdera (northeastern Greece), "the laughing philosopher" because of his emphasis on the value of cheerfulness. Democritus founded the atomist theory: Everything in the universe is composed of small indivisible particles, which move about in an infinite void.

One key issue in Greek philosophy was the problem of change versus constancy: how can things change over time, yet remain constant? Not only are there many kinds of things in the world, but each one is subject to change. Water comes and goes, yet it remains water; animals are born and then mature, grow old, die and decay, yet the next generation will resemble its ancestors, which decomposed in the past. Individual things in nature undergo endless changes, yet something about them enables them to retain their identity. This suggests that many properties of materials must derive from something internal, that will itself never decay, something that stores the same inherent, indivisible properties for eternity, enabling some intrinsic order in the world: the internal principles of nature are sufficient to explain its structure and its history. Why has everything in the world not yet decayed, and how can exactly the same materials, plants and animals be recreated, again and again? One solution to explain how indivisible properties can be conveyed in a way not easily visible to human senses was to assume the existence of atoms (Barnes 1987; Kingsley et al. 2020).

Fig. 1.12 Democritus and a coin of his time, from Abdera, presenting a phoenix. (Tuvia Kurz)

Atomism was an attempt to understand this contrast between change versus constancy. It made change possible by suggesting that there were multiple unchanging material beings, which persisted and merely rearranged themselves to form the changing world of appearances. These unchanging beings are indivisible particles, the atoms. Two fundamentally different kinds of realities compose the natural world: atoms and void. Atoms (from the Greek *atomon*, "indivisible") are solid beings, so small that they escape our senses. Infinite in number and suspended in an infinite void, they move about, and upon colliding, they either repel one another or combine into clusters. Clusters form because of a "like-to-like" tendency, which exists in nature: internal preferences cause animals of a kind to attract one another and thus to cluster, "doves with doves, cranes with cranes". Similarly, atoms of a kind choose to cluster by size and shape. Like-to-like also occurs because of external forces applied on the being in the same manner that different kinds of seeds, when whirled in a sieve, cluster "lentils with lentils, barley with barley, wheat with wheat". Similarly, "oval pebbles are forced into the same place as oval pebbles, and round pebbles as round pebbles by the motion of the waves". Whether because of internal attraction or external forces, like-to-like explained how an ordered arrangement could arise automatically as a by-product of the random collisions of beings in motion. The disorderly motion of individual atoms could produce an orderly cosmos in which atoms are not just randomly scattered, but cluster to form masses of distinct types (Barnes 1987; Kingsley et al. 2020).

Atoms attach to each other and remain together until a stronger external force overtakes them, shakes them and scatters them apart. In this manner, generation and dissolution occur, not only of animals but also of worlds. The atoms are unchangeable, un-generated and indestructible and can only change their positions. While the atoms are eternal, the objects composed of them are not.

We now proceed from the question of "Why" to that of "How". Democritus suggested that the atoms adhere to one another by means of convex and concave surfaces, or by tiny hooks and barbs on their surfaces, which entangle with one another, thus accounting for the cohesiveness of some compounds. The atoms of different substances differ from each other by their shape, their size and the arrangement of their parts—there is an infinite variety of atoms. Iron atoms are solid and strong with hooks that lock them together; water atoms are smooth and slippery; salt atoms are sharp and pointed; air atoms are light. The combination of atom properties is sufficient to account for the multitude of differences between the objects in the world. Differences between objects are three in number—shape, order and position (Barnes 1987; Kingsley et al. 2020). Democritus' atoms are nearer to our modern concept of molecules than to the atoms of modern science.

Atoms and voids are eternal, and they are all that exists. Everything that we sense is merely different combinations of atoms, and what comes into life and eventually dies is merely the way these atoms rearrange. Principles underlying atom motion are also eternal, and consequently, motion itself is eternal. The basic reality is therefore eternal in that it consists only of atoms, their motion and the void. What does change is the varying rearrangement of atoms, and this results in the resultant diversity of life that we observe.

Democritus may have been the first biologist to propose a method for the classification of different types of animals. He divided all living creatures into two groups: animals that had red blood [our vertebrates] and those that did not [our invertebrates]. He may also have been the originator of an ancient theory about the historical development of human communities. Unlike the earlier Hesiodic view that the human past was a golden age compared to which the present day is a decline, Democritus suggested that human life was originally like that of animals. He described the gradual development of human communities for purposes of mutual aid, and the origin of language, crafts and agriculture. Democritus wrote many books that dealt with biology in a broad sense: *On the Nature of Man, On Flesh, On Mind, On the Senses, On Flavours, On Colours, On Causes concerned with Seeds, Plants and Fruits* and On *Causes concerned with Animals*. None of these have survived.

1.1.5 Ideals Versus Reality and the World's Organization: Plato

One generation after Democritus, Plato (428–348 BCE, Fig. 1.13; I have read the translation of his works by Liebes 1955–1979) was born into a family deeply involved in the political life of Athens. As a youth, he was a devoted follower of Socrates, but later developed his own line of philosophical thought. Nevertheless, Socrates' ideas continued to influence Plato's thought to such an extent that I find it sometimes difficult to differentiate between Plato's reports of Socrates' theories and his own. During the Peloponnesian War between Athens and Sparta, he was recruited for 5 years of military service, and when the war ended, he joined Athens' 30 Tyrants oligarchy. A year later, the Athenians restored democracy, and Plato hoped to enter politics. However, after the execution of Socrates, he left Athens and travelled to Egypt, Sicily and Italy, only eventually returning to Athens. Then another war broke

Fig. 1.13 Plato, and a coin from Athens, from his time, presenting an owl. (Tuvia Kurz)

out, and Plato was again recruited, and this time he was perhaps decorated for bravery in battle. At the age of 40, he founded a school of philosophy and sciences, in which he hoped to train young men who would eventually become statesmen and would improve the political leadership of the Greek city-states (the school was termed "Academy" because it was on land which had belonged to some obscure person, Akademos). Later in life, Plato became involved in the politics of the city-state of Syracuse (Sicily) and was sold into slavery, but was ransomed and sent back to Athens. Plato was thus a widely travelled man and an experienced soldier, who had close contacts with many leading men of affairs, in his own city and elsewhere. He was also a philosopher.

In his philosophy, Plato sought to classify entities by defining those conditions which are necessary and sufficient to typify each entity, and only it. Some of the infinite entities in our world are included in each other, like a series of Russian babushkas. To classify and define these entities, while focusing on the smallest one that does not contain any additional entity within it, Plato suggested starting with the broadest entity in the series, and then gradually narrowing the possibilities within the vast space of the large entity, by dividing it in two, according to the presence or absence of any single trait. He next divided the two sub-groups obtained in this manner according to another single trait, and so on until he reached the trait of the smallest group. Such an operation is termed a dichotomy (*diaeresis* in Greek). This very simple logical procedure provided Plato with a rigorous method of defining entities by a succession of dichotomous divisions. Furthermore, in this manner, he obtained a hierarchy of the smallest group in relation to the more inclusive entities to which it is related: by looking through the hierarchy, we get the definition of the smallest entity. In principle, there is no end to this process of conceptual subdivision. As a method of identifying an entity, it ends at the point at which further divisions no longer differentiate between the entities that belong to it.

In an example of dichotomy from everyday life in Greece of those days, Plato (in his text *Sophist*) first classifies and then defines "anglers". He begins with a very broad entity, artisanship.

This can be divided into productive (husbandry, manufacturing) versus acquisitive artisanship (learning, hunting, trading).

Acquisition divides into what we acquire by exchange, versus what we acquire by conquest.

Conquest divides into what we conquer by open force (fighting) versus what we conquer by secret force (hunting).

Hunting can be divided into the pursuit of lifeless prey (e.g., diving for corals and sponges, which the Greeks thought were lifeless) versus the pursuit of live prey.

Hunting of live prey divides into hunting land animals and water animals. Aquatic hunting divides into fowling and fishing.

Fishing divides into fishing by enclosures (twig baskets, casting nets, nooses and the like) and fishing by striking.

Fishing by striking divides into fishing with a three-pronged spear, which strikes any part of the fish body, and fishing by hook, which strikes only the mouth and lips.

In this manner, by going through a series of dichotomies from the general to the particular, we arrive at a definition of the entity “angler”: it is an acquisitive craftsmanship involved in aquatic fish hunting, that captures its target with a hook that grasps the inside of the mouth. In comparison, “net hunting” is an acquisitive artisanship that involves (aquatic) hunting of fish captured by an enclosure. These two entities, angling and net fishing, are contained within the more general entity “fishing”, which in turn is contained within an even more general entity, “aquatic hunting”, and so on.

Plato did not offer examples of how to use dichotomy in classifying and defining plants or animals. If, for example, we were to take his dichotomous method and use it to first classify and then define a squirrel, then we would first divide all animals into marine versus terrestrial. We would then divide the terrestrial animals into those possessing legs versus those without legs; those with legs into two-legged versus four-legged; the four-legged into those covered by fur versus those covered by scales; the furry ones into those with fingers versus those with hoofs; and so on. Applying these dichotomies, we would define the entity “squirrel” as a four-legged, furry, fingered terrestrial animal.

Plato believed that understanding comes through concentrated thought, not from observation (“from the senses, we shall not gain any information or truth, only pleasures of the eye”). He had a special interest in geometry: his observation that a triangle, no matter what combination of angles it has, is always a triangle, discontinuously different from a quadrangle or any other polygon, became the basis for his essentialism (Mayr 1982). In line with this geometric approach, he argued that the world consists, not of four but of five elements: earth, water, fire, air—and the skies; and he also suggested that each of these elements consists of tiny, geometrically unique atoms: The atom of fire is a four-sided pyramid, that of earth a six-sided cube, air is eight-sided, water twenty-sided and heavens twelve-sided (Fig. 1.14).

Beyond these five elements and all the real objects and creatures in our ever-changing world, and beyond the processes of growth and decay of which we are aware through our senses, Plato believed that there is also an unchangeable reality, an eternal, perfect, abstract model of each group of objects, a model which Plato termed an idea (*eide*). Every object in the world is merely a fleeting, imperfect, approximate expression of an eternal, perfect, unchanging idea, an abstract archetype that we can only perceive by reason. It is a timeless reality independent of any object, which bears the same relationship to the things of the sensory world that,

Fig. 1.14 The atoms of Plato. (Tuvia Kurz)

say, a perfect geometric figure, such as a sphere, bears to an imperfect approximation, such as an orange (Rubenstein 2003). The real world is merely like a bad copy of a masterpiece; the world of abstract ideas is more real than the objects themselves. There are thus two kinds of worlds: the earthly, apparent, ever-changing, always-becoming, imperfect world, which we observe with our five senses; and an eternal, unseen, unchanging world of timeless perfect forms and ideas, way beyond our senses, which we grasp by our intellect. Each *eides* is constant and discontinuous with other *eides*. This concept applies also to manmade objects, in that each artefact is only an external expression of an eternal *eide*. A carpenter has in mind an idea of a bed, a model that he himself does not create and does not shape, but it is there and it is eternal. Then there is the specific bed made by that specific carpenter, which we see now, but which he can dismantle tomorrow. The bed idea is discontinuous with that of a chair and a table.

As applied to animals, this theory of ideas suggests that every mortal animal of a species is only the external expression of an ideal eternal idea of that species; every individual horse, whether big or small, brown grey white or black, is only an imperfect representation of some perfect, eternal idea of a horse. When one of his critics commented that 'I see particular horses, but not the quality of "horseness", Plato replied, "That is because you have eyes but no intelligence". Plato's *eide* was termed *essence* by scholars of the High Middle Ages, who followed him in suggesting that variation is the imperfect manifestation of an underlying essence. Plato's essentialism, with its emphasis on discontinuity between ideas, constancy and typical values ("typology"), would dominate the thought of the Western world for two millennia.

Now, if all we see are no more than transient expressions of eternal principles, there is not much point in studying these ever-changing objects we see around us—they are interesting only to the extent of what they indicate about the nature of the constant, unchanging, super-real, eternal idea. Plato was not deeply interested in the nearby surrounding nature, in plants and animals.

He was, however, interested in the creation of the universe. He proposed (in *Timaeus*) that in the beginning there was chaos, a state of non-balance in which the elements (fire, air, earth and water) were in a shapeless state, mixed together and in constant motion. Creation was the act of imposing a rational and wise order on this disordered chaos, by an intelligent, good, reasoning and eternal creating God (Mayr 1982; Plato used the term Demiurge, *Demiourgos* in Greek, a craftsman-like entity).

According to Plato, the universe is composed of all the air, fire, earth and water that exist—there are no elemental remnants beyond it. This is so, both the universe should be as complete as possible, and that there should be only one single universe, with no remnants from which an additional universe could form. The universe is comparable to a huge, gigantic animal, the shape of which is that of a completely smooth sphere. It has no eyes because outside it nothing has remained to be seen, no ears because beyond it nothing has remained that can produce sounds, no mouth because nothing has remained beyond it to be eaten, no hands because nothing has remained to be held and no legs because there is nothing to stand upon. Hence, the universe is a giant animal in the shape of a huge, perfect sphere, a completely smooth rolling ball. The soul of the world both surrounds it and fills it as a misty cloud, governing all the world's actions. The universe, in short, is a living creature, with a body, a soul and with reason. Man is also composed of body, soul and reason; furthermore, Plato argued (*Philebus*) that the elements in man's body derive from elements in the body of the cosmos. Whatever man has, the cosmos must also have: "Whence can a human body have received a soul, if the body of the universe does not possess a soul?" However, Man is different because when he dies, his body decomposes back to the elements. There is thus a difference between the universe (macrocosm) and the human body (microcosm), but also a similarity, in that Man (microcosm), reflects in his structure the whole world (macrocosm). The whole world is a web of hidden affinities between Man and his external world, between the heavens and Earth, between living and non-living entities. Plato's arguments on these matters call to mind the biblical text in which "God created Man in his own image".

Once formed, the world filled up with four groups of creatures, some eternal and others mortal (*Timaeus*). There are: the celestial mythological gods, created from high-quality matter and then scattered over the sky as stars, to adorn it; and there are the mortals. These include the winged animals, which fly through the air. Then come the terrestrial mortal creatures that dwell on the land. Finally, there are the aquatic mortals that dwell in the water. At first, the creating God made the souls of these mortals from low-quality archaic matter, which he divided into as many souls as there were stars. Next, he assigned each soul to a different star, which saw to its upbringing. Finally, the mythological gods took the four elements and created the bodies of the mortals from them, which were sent, with their souls, down to earth. At first, only men were created. Those who lived righteously would return to their star after death and live on it happily. Those who did not live righteously would return to earth as inferior creatures. Men who came to this world and were cowards were reincarnated as women. Men who came to this world and were frivolous and fickle, and thought that to reach heaven all they needed was to fly, grew feathers and were reincarnated as birds. Men who came to this world and were ignorant, lacked philosophical thought and stopped using their minds were reincarnated as creatures with elongated arms that reached the ground, and heads elongated to such an extent that their soul was crushed, and they returned as four-legged beasts. Men who came to this world but were the most stupid of all were reincarnated as creatures dwelling in the deep water, and that did not even breathe fresh air, since mud was their element of breath, and these are the fish and the bivalves. "These are the laws according to which all animal types change one into another, today as from ancient times"—this is how biodiversity was created and how it is maintained. (In *Phaedrus,* Plato states that crickets were also once humans. Upon hearing the Muses singing, they were so charmed and enchanted that they forgot to eat and to drink and therefore died. Eventually, they became crickets, but received a gift of honour from the Muses: they could sing continuously from birth to death without the need for food and drink.)

Plato believed that in Man, the first organ to be created was the head, in which the lofty eternal soul lodges. Its shape is close to that of a perfect sphere. The torso and limbs were created to save the head from the inconvenience of rolling over rough ground. Another inferior mortal soul dwells inside the torso. In it is the heart, which is both the organ of courage and the fountain of blood. The lungs cool the heart and calm it down when it heats up in situations such as rage. The diaphragm separates the heart from the intestines. The most important organ below the diaphragm is the liver, in which there is a third soul, which creates dreams during sleep; these can predict the future.

In conclusion, the early Greek philosophers that preceded Aristotle contributed the very basic concepts of biological thought: what the cosmos is and how it came to be; what the Earth is and how it is composed; how Man and animals came to be and the differences between them. Somewhat like sharks, Man at first developed within an egg case from which he eventually hatched and emerged onto land. At a later stage, he developed within the body of a life-bearing mother. Early Greek philosophers also invented the four-element theory of matter (earth, air, fire and water), and proposed a theory for the formation of plants, animals and humans, in a world under a cosmic cycle of eternal change, growth and decay, because of two cosmic forces engaged in an eternal battle for supremacy. Some of them believed in the transmigration of souls. Others suggested that the universe consists of numerous invisibly minute entities that are eternal, the atoms, which constantly move about in an infinite void. They also suggested that humans originally lived like animals; gradually, they formed communities, and eventually, they developed language, crafts and agriculture. Some believed in a world of ideas that exists beyond this real world, and that the world is comparable to a huge, well-organized animal. Early Greek philosophers also developed the dichotomous method to classify entities and to define them. While their contribution to knowledge was often mistaken, they were the ones who laid the foundations of biological thought, and the world of biological thought was entered by Aristotle.

1.2 Greece: Aristotle and Later

1.2.1 The World—Was It Ever Created?

When the octopus (Fig. 1.15) seeks prey, it changes its colour to blend with the colour of nearby stones. Neat and thrifty in its habits, after eating all the edible parts of its prey, it discards the shells of snails and crabs and the skeletons of little fishes outside its lair. It has an ink pouch with a terminal outlet to the ventral surface of its body, and when disturbed and frightened, it discharges this ink to make the surrounding water black and turbid, thus putting an opaque shield in front of its body. After which, new ink accumulates in the pouch. All cephalopods have this characteristic ink, but in the cuttlefish, it is more abundant than in the rest, because it lives near the shore where enemies are abundant, and because it has no other means of protection, whereas the octopus has its long, twining feet to use in its defence. The calamary, another cephalopod, lives far out at sea, and this distance from the shore

Fig. 1.15 Octopus. (Tuvia Kurz)

gives it protection. The octopus also ventures onto dry land, preferably on rough ground. When mating, males and females intertwine their arms and bring their suckers into mutual contact. One of the male's arms, with particularly large suckers, functions as a penis that penetrates the female. After copulation, the female spawns many eggs, which she glues in clusters to the sides of her lair. She broods them until they hatch, some 50 days later, and the little octopi creep out, like so many tiny spiders. The detailed form of their limbs is not yet discernible, and most of these young are so small and helpless that they perish. Soon after eggs are laid, both parents stop feeding. They submit passively when tossed about by waves, get covered with slime, become leathery, and are also so feeble that they are easily preyed upon by even small fishes and are dragged out of their dens with ease. The parent octopus does not live out the year.

This meticulous description of the natural history of the octopus (Fig. 1.15), mostly accurate to this very day (we know now, of course, that octopuses do not venture onto land), is assembled with slight modifications from two of Aristotle's books. It represents his outstanding ability to both observe the natural history of animals and to describe his observations in minute detail; it also illustrates the great advances that he contributed to biology. The loving detail with which he describes the octopus, as well as other marine animals, indicates that much of his information is original, or at least was obtained personally from fishermen. Unlike Plato, Aristotle believed that we should gain knowledge from observation—he was probably the first to conceive the idea of detailed scientific research of nature, for its own sake.

We might also consider the following text on the classification and anatomy of snakes by Aristotle (Fig. 1.16): Of all animals, the serpents are most similar to lizards—we need only imagine lizards lengthening, and having their legs removed. Scales coat the body of a serpent, like that of a lizard, and a snake's internal organs are identical to those of lizards except that, owing to the narrowness and length of its body, the viscera are narrow and elongated. In both serpent and lizard, the tongue is thin, long, forked and can extend greatly. The serpent's heart, situated close to the pharynx, is small and kidney-shaped. The single lung is very long, and the liver is long and simple; the spleen is short and round, as in lizards. In water snakes, the gallbladder is next to the liver; in other snakes, it is usually at the side of the gut. The teeth are saw-like. There are some 30 ribs. Some say that if you pick out a serpent's

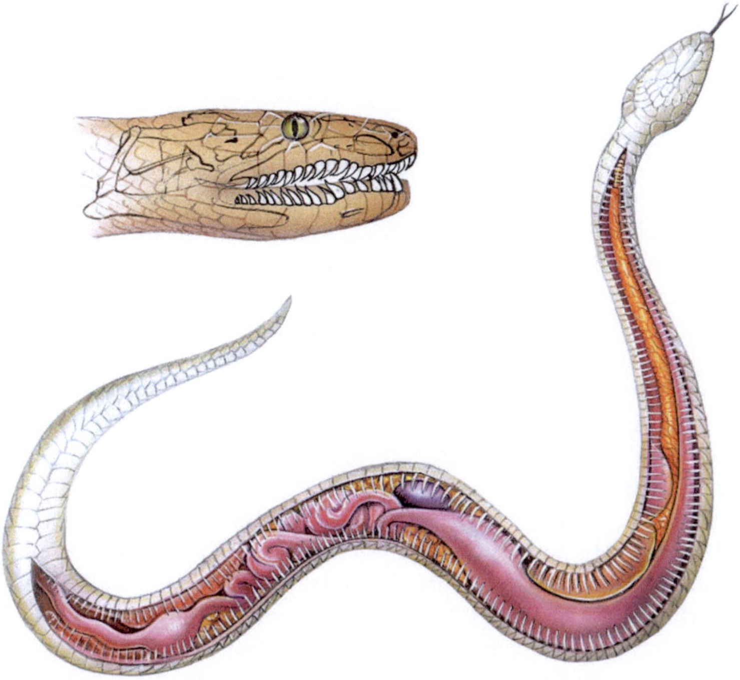

Fig. 1.16 Snake dissection. (Tuvia Kurz)

eyes, they will grow again. Furthermore, the tails of both lizards and serpents will grow again if we cut them off.

This text, too, is reasonably accurate to this very day (serpent's eyes do not grow again when picked out, and snake tails do not regenerate as do those of geckos). It is a sign of Aristotle's outstanding abilities that he conducted detailed anatomical dissections that enabled him to arrange animal groups into a reasonably correct classification, with snakes close to lizards. Aristotle, who dissected about 50 different animal species, was perhaps the first to appreciate the great importance of painstaking and detailed internal morphological investigations. By his time, people surely knew well the inner organs of mammals, due both to animal husbandry and the custom of animal sacrifices. No one, however, sacrificed snakes. To dissect a snake would require intellectual curiosity in biology as well as the determination to translate this curiosity into active dissection and detailed description. If I were to choose one single moment in history in which biological research was born, I would choose that moment when Aristotle dissected a snake to observe its anatomy.

Aristotle was also very observant of animal behaviour. He noted, for example, that when a male partridge approaches the female, he often makes her keep silent, to avoid having to battle other males who might have heard him. This is the first written reference to sneaker males: small males that live on the edges of a dominant male's territory (also termed satellite males). Sneaker males are not very conspicuous, and they do not attempt to attract females with mating displays. Rather, they ambush a female whom a dominant male had drawn and, if she consents, they mate with her secretly. As another example, Aristotle noted that if a man chances upon a young brood of partridges and tries to catch them, the mother bird rolls in front of him and feigns injury. The man thinks he is on the point of catching her every moment and follows her, and so she draws him on and on to a distance until every one of her broods has had time to escape. She then flies off and later returns to the nest and calls her young back, a very intelligent behaviour indeed. He further noted that pigeons, like man, are peculiar among birds in that they do not tilt their head back when drinking. This is indeed so, as most other birds take a mouthful of water and then tip their head back to send the water down their gullet. He further wrote that in temples, where cocks are set apart from hens, they all tread over any newcomer, peck its head and "treat him with every possible outrage". This description by Aristotle is the first record of the pecking order, a system of hierarchical social organization that refers to pecking as a measure of dominance and leadership rank in chickens.

Aristotle (384–322 BCE, Fig. 1.17) came from Stagira, on the northern edge of Classical Greece. His father was a physician to the King of Macedonia, and his mother came from a wealthy family on the island of Euboea. At 18, he moved to Athens (perhaps to enter Plato's Academy), where he remained for nearly 20 years. When 37, he travelled to the northeastern Aegean, first to Assos on the coast of Asia Minor and then to the nearby island of Lesbos. When almost 40, he was invited by Philip of Macedon to tutor his son Alexander, and during that time he also tutored two other future kings, Ptolemy and Cassander. With the assassination of King Philip, his son Alexander became king, and left Greece to conquer Persia and become Alexander the Great. Aristotle then returned to Athens (under the patronage of Alexander?) and, at the age of 47, established a school named the Lyceum where he taught for the next 12 years. Alexander's sudden death kindled anti-Macedonian sentiment in Athens, and a local high religious official denounced Aristotle for not honouring the gods. Aristotle fled Athens and died soon after, on his mother's island of Euboea.

During the years he ran the Lyceum, Aristotle wrote many treatises, which were probably working drafts for his lectures and not meant for publication. Roman scholars compiled these drafts, some two centuries after his death. Aristotle wrote about zoology, anatomy, embryology, geography, geology, meteorology, astronomy and physics. In addition to science, he wrote on aesthetics, ethics, government, logic, metaphysics, politics, economics, psychology, rhetoric, theology, education, foreign customs, literature and poetry. His works in biology represent the first-ever comprehensive attempt to observe and describe life in a scientific way. His most important zoological texts are *History of Animals* (I have read the Thompson 1910 translation), *Parts of Animals* (translated by Peck 1983 and by Balme 2001), *Generation of Animals* (translated by Platt 1912), *Physics* (Translated by Reeve 2018), *Metaphysics* (translated by Rosen 1966) and *Meteorology* (Barri 1974). They are the earliest-ever textbooks on animal biology and the written point of origin of basic biological thought.

Fig. 1.17 Aristotle. (Tuvia Kurz)

We recall that Plato, a generation before Aristotle, had viewed the physical world as a mere shadow of a perfect spiritual world completely separate from our concrete world, a world that contains eternal and unchanging ideas of all the different types of things that exist in our real world, ideas that we can reach only through reflection and reason, not through the senses. Plato considered each entity surrounding us as an imperfect copy of an abstract *eidos*. Aristotle, on the other hand, saw reality in physical objects of the physical universe, objects that we can grasp through our senses, through experience and through generalizations. Things of this world are not mere shadows or approximations of eternal concepts. The universe is a living, integrated place whose basic principles we can understand from the data presented by impressions of our senses, by reasoning and by generalizing. The *eidos* exists but is not abstract; it is an existing entity within each individual, it is the species as we see it. Everything in nature that the senses can perceive is both real and knowable, and direct observation, combined with reason and logic, are the most reliable guides to truth. Aristotle, in dividing the universe into eternal versus perishable regions, directed his philosophy of biology at a scientific understanding of the perishable compounds, their matter and their form. True, the eternal level of the heavens is nobler, in cases, it is even divine, but the perishable—plants and animals—are easier to know, since they live all around us (Randall 1960; Barri 1974; Ross 1977; Lear 1988; Lennox 2001).

Before delving into Aristotle's texts, we should remember that Aristotle never edited or authorized the texts that we read today, 2300 years after his time. His original texts were copied time and again, translated from one language to another, commented upon and edited and are bound to have at least some distortions. One obvious case is where (*History of Animals* 9), in the midst of a text on marine animals including various marine snails, the text says that "Wherever snails are present you may be sure there is neither pig nor partridge in the neighbourhood; for both pig and partridge eat up the snails". Clearly, someone after Aristotle, very unaware of the fact that there are snails also in the sea, not only on the land, where they live together with pigs and partridges, was confusing marine snails with land snails; otherwise, this statement, in the context of marine animals, makes no sense. With this cautionary note, we can now turn to the texts.

Aristotle's approach to biology rested upon a sharp separation between two types of questions: questions of "what" (Greek *hoti*) and questions of "why" (Greek *dioti*, Randall 1960). "What" questions demand a detailed, meticulous definition and description of the phenomenon under study and "Why" questions require the explanation of facts. There is purpose in Nature, and the functioning of animal parts, for example, should be explained by examining how they contribute to the functioning of the organism as a whole. Philosophy was, for Aristotle, the knowledge of the whys, of the *dioti*'s—why things are as we observe them to be; and why they must be so. *History of Animals* is largely a "what" book, in that it records observations concerning different animals, and records the main facts of animal life (in Greek, *History* means the search for knowledge, of information). It is a huge and rich store of knowledge built up by careful observations by himself and, surely, by others. Aristotle investigated differences between animals in particular body parts, in ways of life and types of activity and in specific characters.

Thus, in describing the neck and its organs, he noted that those animal groups that have lungs (mammals, lizards, snakes and birds) also have a neck that contains two noteworthy organs, the windpipe (larynx) and food-pipe (gullet, oesophagus). Fishes, however, lack a lung, and they lack a neck, windpipe and food-pipe. Aristotle next asked *why* mammals, reptiles and birds have a neck with a windpipe and a gullet. He replied in his *Parts of Animals*, largely a "why" book, in which he explained the functional morphology of the neck: animals with lungs must breathe, and the windpipe must be there to transfer air, to and from the lungs. The neck exists so as to envelope the windpipe and protect it. Unlike the windpipe, the food-pipe is not necessary, as it exerts no action upon the food; there is really no reason why the stomach should not be next to the mouth. However, once the windpipe is there, it must exist to bridge the gap between the mouth and the stomach. The windpipe, because it is in front of the oesophagus, might suffer from accidentally swallowing food that would damage it. To cope with this, in (terrestrial) mammals we have the epiglottis (birds having a larynx). The epiglottis rises during breathing and falls during eating, thereby preventing food particles from slipping into the windpipe. In spite of this measure, should a food particle mistakenly enter the windpipe, choking and coughing ensue. While food is passing over the epiglottis, seldom does a particle of it slip into the windpipe. The epiglottis is thus a contrivance of nature to remedy the problematic position of the windpipe in front of the oesophagus, a position that is the result of necessity. Since the lung surrounds the heart, which is in the front, it follows that the windpipe must lie in front of the oesophagus, as it leads to the lung and heart. It is a universal law that, as regards above and below, front and back, right and left, the nobler and more honourable organ is invariably placed uppermost, in front, and on the right, rather than in the opposite positions, unless some more important object stands in the way. Accordingly, in all animals but man, the heart is at the centre of the pectoral region; in man, it is a little to the left, to counterbalance the chilliness of that side. This is so because in man, as compared with other animals, the left side is colder than the right side. Thus, an animal that must have one functional organ (a windpipe) must sometimes have another organ (a food-pipe), even if the latter lacks function. In this manner, all parts of each animal collaborate for the benefit of the whole and are intelligently organized for that same purpose. *Parts of Animals* thus aimed at developing a theory from the facts recorded in *History of Animals* (Randall 1960; Barri 1974; Ross 1977; Lear 1988; Lennox 2001); Aristotle, not knowing the real nature of respiration, believed that fishes, since they lack lungs, do not breath and he regarded the lungs of terrestrial vertebrates as a device to cool body heat by means of the inspired air.

Another example of *hoti* versus *dioti* is Aristotle's record of the development of the chick embryo in the *History of Animals*. On the third or fourth day after an egg has been laid, a tiny pulsating heart appears, and from it, two blood vessels eventually arise and head towards the surrounding membranes. Only later do we see the body of the embryo, the head with its large eyes and other organs. This is an accurate "*what*" description of the chick embryo development. "*Why*" is the heart the first organ to develop? *Parts of Animals* gives the answer: Aristotle believed that the heart is the main source of internal heat for the entire embryo, which throughout its entire development is independent of its parents. Further, it is the site for the production of blood. When the developing embryo departs from both the male and the female contributions, it must manage its life on its own, like a son who builds a house far away from his father. The heart is the source of development, from which the later development of the body proceeds. Aristotle thus proposed a functional answer to the question of *why* the heart is the first organ in embryonic development (he erred in believing that the heart is the organ that produces blood and that this blood then forms the other organs).

Aristotle's distinction between "what" and "why" questions goes together with the study of animal morphology, to this day, under the terms of "form" and "function". In a way, his distinction is also with us from another aspect: wherever possible, we tend to strictly separate "results", the hard "what" facts, from "discussion", which may contain a "why" element of speculation. In later generations, Aristotle's "why" questions merged and developed into teleology (from Greek *telos*, meaning end or purpose), the study of Nature by attempting to describe things in terms of their purpose or goal. Natural teleology contends that natural entities have intrinsic purposes. For instance, an acorn's intrinsic *telos* is to become a fully-grown oak tree. Thus, to explain something is to determine its benefit, its purpose, its *telos* or "reason for which".

The world—was it ever created? How does it work? Reaching far beyond biology, Aristotle's intellectual horizons carried him to the levels of cosmic theory. Unlike Plato, who believed in creation as an action of making order in chaos, Aristotle believed that, since something must always come from something, there must always have been something before, and therefore, the universe must be eternal. The world is timeless—it was always there and always will be, both uncreated and indestructible. The cosmos consists of matter and form, and Aristotle suggested that the tool by which form expresses its dominance over matter is movement, the factor that causes one form to change into another. Everything that happens in the universe is only movement (Dynamics might be a better term than Movement). The eternal stars possess the perfection of all movements, the circular one, because eternal movement takes place only in circles, whereas the four earthly elements move up and down. The universe consists of two parts, the terrestrial and the celestial regions. In the realm of Earth, all bodies consist of combinations of the four elements, and they are perishable. In the region beyond the Moon, however, the heavenly bodies consist of a fifth, eternal element (*ether*; *quintessence* is a later term), a pure element that, being imperishable, is subject neither to generation nor to destruction. Earth rests at the centre of the entire universe that revolves around it in circular movements, and Life, with its generation and decay, inhabits it alone. Earth has the shape of a huge sphere that remains at rest because of its huge size. Beyond Earth and the heavens, there is a spiritual space which Mankind cannot fathom. Aristotle thus opposed the view of Plato that the divine celestial bodies consist mostly of fire.

The world beyond the moon is a series of transparent globes, one inside the other. The stars do not fall down, because they are mounted, in a series of solid globes (Fig. 1.18; elsewhere they are termed "spheres", "wheels", "skies", "heavens"). Earth is at the centre of these globes, it consists of the four elements (fire, water, earth and air), and of the four qualities of hot, cold, moist and dry. The element of earth is at the core, beyond it is the sphere of the element of water (oceans and seas), outside of which is the sphere of air (atmosphere) and beyond it the sphere of fire. This series of four spheres forms the world of the four elements. Had it not been for the external (and eternal) force of movement, the four elements would have settled down as a series of four globes with earth in the centre, surrounded by a sphere of water, then a sphere of air and beyond that, one of fire. However, movement constantly stirs and mixes them. Thus, we do not always encounter the elements in their pure state; we sometimes see them in a mixture, such as water mixed with earth (*Meteorology* 1: 4, as translated by Webster 1923). Every object, whether a metal, plant, animal or human, consists of a mixture of these four elements. Each of these elements tends to move vertically; earth and water move downward and air and fire move upward. All movement in the underworld stems from this desire of each of these elements to return to its natural place. In a piece of burning wood, the fire element in it will rise upwards, since it strives to return to its natural place in the

Fig. 1.18 Aristotle: The World is a series of crystal spheres one inside another. (Tuvia Kurz)

outermost sphere. If we throw a stone upwards, it will fall down because the earth's element inside it tries to return to its natural place in the innermost sphere of the earth. Of the four element qualities, two, hot and cold, are active and they determine, conjoin and change things; dry and moist are passive elements, and do so but to a lesser extent. Natural changes, in plants and animals, are the work of these powers.

Putrefaction is the destruction of the peculiar and natural heat by the heat of the environment. Everything on Earth is liable to putrefy, except fire. Everything that putrefies grows drier and ends by becoming earth or dung. Putrefaction takes place less in cold than in hot seasons, but not in boiling hot environments, less in flowing environments than in motionless ones; and less readily in great quantities of a thing than in little quantities. Some animals generate putrefying bodies because the secreted heat organizes the particles secreted with it. This (outdated) theory, which we today term spontaneous generation, claims that some living creatures arise directly from non-living matter, with no parent being involved. Aristotle believed that spontaneous generation varies with the nature of the environment. In the sea, sand gives rise to scallops, slime to oysters and intertidal rock crevices to give rise to limpets.

The concoction is a process in which the natural and proper heat of an object perfects corresponding qualities. When a concoction takes place, we say that a thing is being perfected and is coming to be itself. The proper heat of a thing is what sets up this perfecting. Things that undergo concoction become more compact, thicker, hotter and drier. Fruit ripening is a concoction of the nutriment in fruit. Since concoction is a sort of perfecting, the process of ripening is perfect when the seeds in the fruit become able to reproduce that fruit; this is what "ripening" means, as applied to fruit.

Aristotle also suggested that environments might change repeatedly, sometimes in a cyclic manner, yet sometimes not. Water, for instance, could evaporate from the sea and rain down again, and rivers might come into existence and then dry up; what was sea may become dry land by river silting and what was land may become sea. The interior of the earth causes such changes because it cyclically grows, then decays and perishes "like the bodies of plants and animals"; and it is these perishable entities, the plants and animals inhabiting the sub-lunar world, which form the rich diversity of nature. Earth is forever changing, but sometimes the changes are so gradual and over such immense periods as compared with our length of life that they are difficult to observe.

Above Earth and its four elements, and as we have just mentioned, from the moon and beyond, the spheres in the sky form the world of the heavenly stars: a series of spheres, one surrounding the other and constantly rotating. The shape of the heavens is spherical. The stars differ from Earth in that they never change. As they do not participate in the vertical movement of the four elements, Aristotle concluded that they consist of a completely different, fifth element that, he claimed, never changes and never moves. Accordingly, Aristotle divided the universe into the sub-lunar world, which is constantly changing, and the upper world beyond the moon, the heavens, which is eternal, exempt from decay and from generation, and never changing. Comets do appear occasionally and unpredictably, but Aristotle considered them as atmospheric events of the sub-lunar world.

The moon is within the lowermost of the transparent heavenly spheres, and above it come the spheres of Mercury, Venus, Sun, Mars, Jupiter and Saturn. Above Saturn is a sphere inlaid with fixed stars, above which comes the ninth and last sphere, a starless, abstract heaven in which there is no matter but merely form; this is the cyclical wheel, which in turning around completes its circle every single day and night. Nothing exists beyond this outermost wheel: no further spheres, no vacuum, no dimensions of space or time.

Thus, Aristotle's world is a limited, finite, closed world. All the crystal heavens touch each other, each is in contact with the one inside and the one outside it, and thus the movement of the ninth, cyclic sphere is transmitted through them until it is transmitted to the Earth. The last heaven in this series is the lunar sphere, which influences the cycles of tides in the sea. Thus, the unchanging eternal, constantly repeating movement of the heavens is responsible for everything that changes beneath the moon. Many centuries after Aristotle, Christian theologians would welcome his cosmic theories, and furthermore, they would view the eternal heavens as pure and the earth as corrupt and doomed. Two millennia later, Robert Fludd would elaborate on Aristotle's order of the celestial elements, but with the sun in the middle.

1.2.2 Classifying Animal Diversity: Theory

Aristotle, beyond being a devoted naturalist, was also a logician, and beyond his interest in classifying the vast diversity of animals, he sought to categorize all forms of life. Somewhat like Plato, he viewed each group of animals as an entity that we must define, in terms of when almost all its members have the same set of distinguishing features; for example, all birds have feathers, wings and beaks and can fly. However, every large group also splits into several smaller sub-groups, which in turn split further and so on, down to the smallest group among animals, the species. Aristotle was against applying, to animal classification, Plato's method, of taking *any* set of different dichotomies based on *any* set of single features, to arrive at the definition of a given entity (group). A random single trait that we might choose for dividing a large group might be irrelevant to this group, and therefore, it might split up groups that are obviously natural. For example, in a *living on land or living in water* dichotomy, the natural entity Birds, of which some are terrestrial whereas others are aquatic, would appear in both branches of the division, and we would thus get two groups of Birds (Wightman 1953; Fig. 1.19). Similarly, a dichotomy of *winged or not winged* would split the glow-worms, which are both; and a dichotomy of *wild or domestic* would separate horse, bull, dog, pig, goat and sheep from similar animals of the same kind in the wild. We might thus break up natural groups, the members of which have more than one feature in common. A random dichotomy might also lead to groupings that are meaningless in terms of reality. A dichotomy of *two legs or four legs* would lump Man together with Birds, a meaningless group, as the legs in these groups are completely different. As a further objection, in a series of dichotomies, the consecutive divisions may be on different levels, and therefore, they would not really be narrowing down the previous divisions. For example, in a series of two dichotomies, the first step could divide animals into *those in water or those on land* and the second step could divide each of the two resulting groups into, for example, *legs or no legs* (Fig. 1.19). However, having legs (or no legs) is not a trait of water (or land), and therefore, this second division does not narrow down the entities of *living in water or living on land.* Aristotle also argued that a dichotomic classification of animals demands bifurcations that are strict opposites, like *White or Black*, or *Straight or Bent.*

Aristotle further claimed (*Parts of Animals*) that a classification of organisms in accord with the canons of logic cannot be made, because there is no selection of a single feature that permits serial subordination of genera and species. Each of the different kinds of animals is to be defined only by the consideration of many different features, none of which by itself is adequate for essential definition (Sloan 1972).

For these reasons, Aristotle opposed using Plato's method of dichotomies based on any set of features for classifying animals. He claimed that the process of splitting and sub-splitting must narrow down the options by using only relevant features. A correct approach that would narrow down the entities achieved by the first division could be the division of those on land into *burrow-dwellers or surface dwellers;* and of those in water into *swimmers or crawlers* (Fig. 1.19). Among those

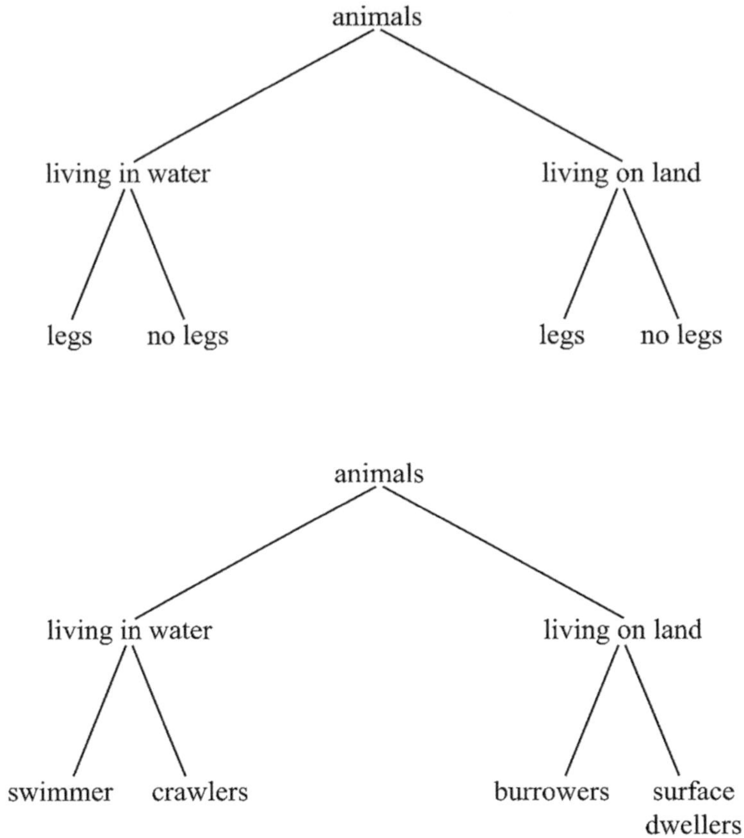

Fig. 1.19 Dichotomies as considered by Aristotle: A meaningless dichotomy (above), and a meaningful one (below)

endowed with feet, as a further example, we must not differ between those with feathers versus those that are featherless, as the presence/absence of feathers is not an outcome of footness. We may split, however, those endowed with feet, into those that are cloven-footed versus those not cloven, as these are features of the foot, cloven-footedness being a form of footedness. In this manner, we continue until we reach the species, that group among which there are no differences. The last differentiating feature will be the substance and definition of that last group.

Large animal groups are those that we instinctively recognize, and they differ from each other in the presence or absence of several major features. Birds and Fishes, for example, two large groups, differ in that birds have feathers, beaks and lungs, whereas fishes have scales, spines and gills. Other large groups may differ by having hands in one versus claws in the other, or fingernails in one versus hoofs in the other. A correct method of classifying animals is to begin by applying straightforward intuition, yours and that of others, and using common knowledge, to cluster animals into such large and generally accepted groups as mammals, birds, lizards, fishes and so on. The next step should be to define each of these large, intuitive groups by a clear-cut set of several traits. We may divide one group into several subgroups at the same time, such as dividing the bloodless animals (our invertebrates) into insects, cephalopods, crustaceans and snails with bivalves; and blood-possessing animals (our vertebrates) into fishes, reptiles, birds, quadrupeds and man.

Within each large group, with its unique set of traits, the various sub-groups usually differ by mere quantifiable differences ("the more and the less"). These would include minor variations in colour and shape, many or few of a given character, large or small, smooth or rough, long or short—the difference in degree. Subgroups are what they are because they belong to groups that are more inclusive. In his grouping terminology, Aristotle used only two terms. One, *Genos* is a flexible term, which he used to describe any group above the species level, even if it encompasses most of the animal world: the *genos* of Birds, the *genos* of Fishes, of Whales and so on. The other term, *Eidos* (originally coined by Plato), is also a flexible term, and it represents the subgroup of any *genos*. The raptors are an *eidos* within the *genos* Birds, the eagle is an *eidos* within the genos Raptors, and the golden eagle is an *eidos* within the *genos* Eagles. This is different from their modern equivalents, *genus* and *species* (the latter term derived from Latin), which in modern taxonomy represent the two lowest ranks within the biological grouping hierarchy.

Aristotle claimed that in classifying animals into groups and subgroups, it is through the definitions that we come to know a group's essence (he did not use "essence", a Latin term that came into use later, during the Medieval). Each entity has many features, some of which are relevant to classification in that they contribute to that entity being what it is. We bind up the essential features involved in a group being what it is, and an individual belonging to one group or another, into its "what". As an example, the essential feature of Humans is that they are rational animals. Now, take an individual (Aristotle, in his example, took Socrates). When we say that the essence of Socrates is to be human, we are saying that he has the entire set of features necessary to be a human, from a general, living nature down to a specific human one. Socrates belongs to a group (humans) unique in its rational abilities. This group belongs to a larger one, that of furry, lactating animals (our mammals), which belongs to a still larger group, those with red blood (our vertebrates). Vertebrates belong to an even larger group, (animals), the members of which have five senses and the ability to move. Further essential features assign him to living beings, including plants: his abilities of growth, reproduction and death. Man is mortal and this belongs to his "what", to what makes him what he is, it is an essential feature of being a living entity, without it, no man would be a man; were he immortal he might be a god or a mineral, but he would not be human. All these are essential features of (our example of) Socrates, and they define him: he is a subgroup of a subgroup of a subgroup of a subgroup. From What to Why, if we seek to know why men are mortal, they are so because men are animals, and all animals are mortal. Man belongs to that entity that possesses that feature, and hence they all exhibit that feature: it is essential to being a human that men are mortal.

This is proven by a syllogism: If A belongs to B, and B belongs to C, then A necessarily belongs to C. If all men are mortal, and all animals are mortal, then man necessarily belongs to the animals. Aristotle's syllogism is, for him, an instrument of logical proof. The syllogism will operate in a world consisting of both "kinds" of things, *eide*; and of much more inclusive kinds, *gene*. This is a world of real species and real genera, not of abstract Platonic ideas. It is a world in which individuals are what they are because they are of a certain kind, to a certain species. This species is what it is because it, in turn, belongs to a more inclusive kind, to a certain genus. In this real world, classification is concerned with what a given species is, and what are the differences distinguish it from all other species in the same genus. The outcome of such an enquiry will be a formal definition, a statement of what a thing is, what is its "essence" or "nature". It is not necessary in such a definition to state all features of a kind, not even all essential features, in order to state "what the kind of thing really is". It is sufficient to state its position in the structure of kinds: to state the kind to which that species belongs, its genus, and what marks it off as a distinctive species. All the features of Man come either from the fact that "man is a mammal"—this is his genus—or that he possesses the power of rational thought—this is his distinctive trait, distinguishing him from all other animals. Therefore, we will state these two traits, which are man's "essence", in his definition, "Man is an animal with the power of rational thought" (more traditionally, this runs "Man is a rational animal"). A classification of animals must start with large groups, each having its unique set of traits. It then proceeds through the subgroups (when present) to the species, and not conversely, from the species level upwards, via subgroups to large groups.

In this manner, Aristotle laid down coherent theoretical foundations on how to construct a comprehensive classification system for animals. Classification and division are counterparts of the intrinsic order in nature. Each group consists both of the concrete individuals that form it and of its essence, its general formula. Individuals are subject to destruction and are capable of generation, but there is no destruction of the formula.

Beyond these essences of groups, particular individuals within a group may possess other features that are not essential to that group—eye colour or hair colour among humans, for example—features that occur but not always and not for necessity (what we, today, term intraspecific variation): a man is a man, no matter what his eye colour. Aristotle considered such features as "accidental" additions to the essential features, additions that we should ignore when classifying. It is luck when events happen by accident or spontaneously, in contrast to reason and nature, which dominate the essential. (Accident—is such as if someone, in digging a pit for a plant, has found a treasure). In structuring a classification, we must not split by accidental qualities, say by subdividing those endowed with feet into those that are white and those that are black. Classification is the demonstration of essential features, as stated in definitions; or putting it from the other side, definition is the grasp of the essence, which marks one species off from another.

We can best illustrate Aristotle's search for order among and within animal groups in his treatment of bird diversity. Birds are a group that differs from all other animal groups in that their members have feathers, and a bony beak (Aristotle was probably unaware that tortoises also have bony beaks). Within this group and in response to the demands of habitat and environment, the greater or lesser development of organs that occur in all birds determines the various subgroups. All birds have two legs, but in some of them the legs are long, whereas in others they are short; all have a neck, but in some it is long, and in others short. Furthermore, variation between organs may correlate, say between neck and leg, in that long-legged birds have long necks, and short-legged birds have short necks. This is so because for a bird perched on long legs, a short neck would be of no use in collecting food from the ground, and a long neck would be useless if the legs were short. Aristotle went on to explain that a long neck is disadvantageous in that it is weak, and as carnivorous birds depend on strength for their subsistence, birds that have talons never have long necks. Beaks also vary, being straight in those who use them for straightforward grabbing, and hooked in those who tear away at raw flesh. Hooked beaks are advantageous also in fighting, and raptors must use violence to get their food, from the bodies of other animals. In some marsh-dwelling herbivorous birds, the beak is broad and flat, to dig, crop and pull up plants; in others it is elongated, to reach food from below the water surface. In some aquatic birds, whose feet are webbed, the neck acts as a fishing rod, and the beak represents the line and hook. As another correlation, and moving from legs and necks to wings and weight, Aristotle noted that flight by wing extension is in a bird's essence. Birds with talons that live on flesh are well adapted for flight in that their wings are large and strong, yet their body is small since consumed nutriment goes mainly to wing production and maintenance. Ground-dwelling (poor flying) frugivorous birds, on the other hand, have bulky, heavy bodies that are not built for strong flight. In some of these heavy-bodied birds, the legs have spurs as weapons. Spurs and talons never co-exist in the same bird, for nature never makes anything superfluous. If a bird can fly and has talons, it has no use for spurs, as these are weapons for fighting on the ground. Ground-dwelling kinds, on the other hand, would find talons useless and actually injurious, as they would stick into the ground and interfere with their walking and running. This is why birds with talons walk so badly. Foot shape is also in harmony with the bird's mode of life. Birds living on the water, where wings are useless, require swimming feet and indeed, their feet resemble the oars of a boat or the fins of a fish. Marsh-dwelling birds, on the other hand, have very long legs and their feet lack full webbing because they are predominantly not swimmers; their legs and toes are elongated because they live on a ground that gives way under the foot. In these birds, the nutriment that should go to their tail feathers goes to their legs, to increase their size; and, therefore, these birds, when in flight, use their legs as a tail, that stretches out behind them.

Beyond the more or less, having the same organ but positioned in different places also contributes to diversity. All birds have four toes (leaving aside the ostrich, with its cloven hoof), and in almost all of them, three toes point forward and the fourth points backwards, as a steadying heel. In the wryneck, however, two toes point forward and two backwards. Aristotle suggested that this is so because the wryneck's body does not incline forward as much as that of other birds. He classified the ostrich as intermediate, between birds and mammals: It resembles a bird in being bipedal and in having feathers, but differs from birds in being unable to soar and in its feathers, which resemble hair and are useless for flight. It resembles mammals also in having upper eyelashes consisting of (what he thought were) hairs, and in having cloven hoofs, not toes. Aristotle suggested that these peculiarities arise from the ostrich's bulk, which is that of a mammal rather than that of a bird, and prevent it from flying.

In classifying birds as a distinct large group made for flying, Aristotle noted that another large group, fishes, are made for swimming, and accordingly they have no legs, feet, hands or wings. Gills and breathing water are one major feature distinguishing fishes from lung-possessing, air-breathing birds. Other features are their fins rather than limbs, and scales rather than feathers. Turning to diversity within fishes, the tail resembles a fin in some, is short and fleshy in others, and is spinous and elongated in cartilaginous flat fish such as the torpedo and trygon, in which nutriments that should have gone to the tail have been diverted to increase the breadth of the body. A "bony lid" (the operculum) covers the gills in most fishes, but it is absent in sharks and rays. Diversity within the group of fishes also concerns the mouth, which is at the very front tip of the body in some, whereas in others (dolphins, sharks, rays) it is on the under surface; these later turn on their back to take food. In this, nature's purpose is to prevent these fish from giving way to their gluttonous ravening after food. Had they been able to seize their prey more easily than they do, they would have perished from over-repletion. In those in which the mouth is not on the under surface, there is diversity in the extent to which it can open. It gapes widely in carnivorous kinds that have a strong mouth and sharp teeth; in others, it is at the point of a small, tapering snout.

Aristotle blundered in classifying dolphins and whales as fish (Romero 2012). Observing that they have lungs rather than gills, he thought that this was because, in general, animals of large size require an excess of heat to facilitate their motion, so nature had therefore placed a lung within their blood-heated body. In classifying them, he wrote that these creatures are both land and water animals, since by inhaling air they resemble land animals, whereas by having no feet and by deriving their food from the sea, they resemble aquatic animals. Seals, too, are between land and water animals: as water animals, they have feet; and as land animals, they have fins and sharp, interfitting teeth. Along broadly similar lines also bats are halfway creatures, but between ground and flying mammals.

1.2.3 Classifying Animal Diversity: Practice

Readers of this book might perhaps question whether people of over two thousand years ago could have differed among the different species of, say, octopi. In answer, we go beyond Greece and beyond Euroasia, and we also go into the field of vernacular names, as they reflect the ability of a culture to distinguish among animals. In Hawaii, the name *he'e mauli* refers to the blue octopus *Octopus cyanea* whereas *he'e puloa* refers to the octopus *Callistoctopus ornatus,* and *mu he'e* to squid; *Leho kolea* refers to the cowry *Cypraea maculifera*, whereas *Leho ahi* refers to the cowry *C. mauritiana*; *Pupu o ha'upu* refers to the topshell *Trochus intextus* versus *Kashelelani* to the top shell *Leptothyra verruca*; and *Momi* to the dove snail *Euplica varians* versus *Laiki* to the dove snail *Graphicomassa maragarita* (data from the Internet). Likewise, also in eastern Africa, local languages differ even among the many small species of mammals (Swynnerton 1946). We have no reason to doubt the abilities of the ancient Greeks to distinguish among species.

What, then, was Aristotle's zoological horizon? Beginning at a level beneath animals, he described plants as living beings in that they grow, and generate via seeds and fruit; however, they get their food from the earth by means of their roots; and this food has already been concocted when taken in. The earth and its heat thus serve them, instead of a stomach. He then separated animals from plants in claiming that, though they too grow and generate young, they also possess a heart, a digestive system and sense perception.

Next, within the animal world and following Democritus who had lived a generation before him, he differentiated between those two large groups, "non-blooded" and "blooded" animals (*Anhaima* and *Enhaima*), corresponding to our modern distinction between invertebrates and vertebrates. The invertebrates are animals that do not breathe, in addition to not having blood. In the following pages, I present Aristotle's classification in some detail, to show that he aimed at a rather complete, full-blown classification of the animal world.

Non-blooded Animals (Invertebrates)

Invertebrate diversity classifies into four major groups: those with soft bodies (*Malakya*), with soft shells (*Malakostraka*), with hard shells (*Ostrakoderma*) and terrestrial invertebrates (*Entoma*). There are also two smaller groups, which bridge the gap between animals and plants: sea anemones (*Acalephae*) and sponges. Almost every invertebrate in Aristotle's world falls into one of these four large or two small groups, some of which he further classified, into subgroups (Pellegrin 1986).

In these four large groups, he placed cardinal emphasis on the calcareous skeleton. In crabs, lobsters, snails and bivalves this skeleton is external, holding the living, fleshy parts together and protecting them from injury. In addition, as these animals are bloodless and have, therefore, only a little warmth of themselves, the shell surrounds the body "as a bed-warmer does the embers", keeping the heat inside. Aristotle added that in this, these invertebrates resemble another animal group, the tortoises. In cephalopods and in insects there is no external skeleton distinctly separate from the rest of the body. Cuttlefish and calamari have an internal skeleton; octopi have none. In insects, there is no distinction between soft and hard parts. The whole body is hard, a hardness, however, being more flesh-like than bony, yet harder than flesh. This makes the insect body less liable to get broken up (Pellegrin 1986).

Soft-Bodied Non-blooded Sea Animals (Fig. 1.20) The *Malakya* ("soft-bodied", this group corresponds to our group of cephalopods, of which most recent representatives are soft on their outside) are a large group in which the hard parts, when

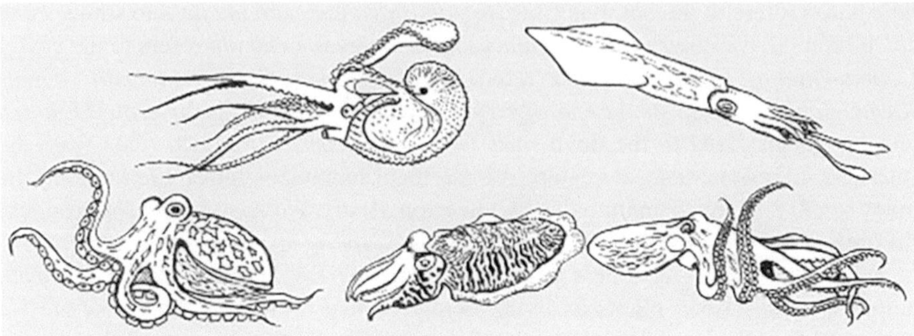

Fig. 1.20 Aristotle's grouping of invertebrates: Malakya. (Tuvia Kurz)

present (as in cuttlefish), are inside the fleshy parts, and in this respect, they resemble the red-blooded animals. Their body consists of eight arms, each usually with a double row of suckers. Cuttlefish and squid also have two long tentacles, each with a double row of suckers, with which they catch their food and bring it to their mouth. In stormy weather, they also "cling by them to a rock and sway about in the rough water like ships lying at anchor". The head is continuous with the arms, they have a mantle sac that contains the internal parts and a gland from which they discharge ink when stressed (*HA* 4:1, 2, 16).

This main group of "soft-bodied" is subdivided into two. One subgroup consists of octopuses, with a small body and long arms, which can both swim and crawl and which lack a hard internal part. The other subgroup consists of the cuttlefish and squid that have a large body and short arms and move about only by swimming. Their set of ten arms is useless for crawling and for clinging to rocks during storms, as eight of them are small and weak; the remaining two are very big, and specialize in catching prey. This second subgroup has an internal hard element, the "sepium", which varies in shape from one species to another. Each of these two subgroups further divides into species.

We are fortunate in that Aristotle's brief descriptions allow us to identify extant cephalopod species in today's marine fauna of Greece. The octopuses of Greece divide into (what we now term) the common octopus *Octopus vulgaris,* distinguished by its large size; the musky octopus *Eledone moschata,* which he distinguished by its long arms, each with only a single row of suckers; and the peculiar paper nautilus *Argonauta argo,* a pelagic octopus distinguished by a unique brood chamber, which resembles the valves of bivalves. The cuttlefish and squid also divide into three: the cuttlefish *Sepia officinalis;* the sagittal squid *Todarodes sagittatus,* which is bigger; and the common squid *Loligo vulgaris,* which has a longer fin (Voultsiadou and Vafidis 2007).

Beyond classification, Aristotle described how cephalopods copulate, the male possessing a penis-like organ that he inserts into the female, the couple swimming forwards and backwards. He also described *Argonauta*, as a peculiar octopus in that it rises from the depths of the sea and sails the waves. He erroneously believed that it has an inverted, empty shell in which it dwells, and which serves as a boat. He noticed that in this pelagic octopus, a thin membrane extends between two of its arms, resembling the (much thicker) webs on the feet of aquatic birds. This suggested to him that *Argonauta* uses this membrane as a sail when the wind blows and that it extends two other arms, as rudders. If alarmed it fills its shell with seawater and sinks. He believed that it generates by spontaneous generation, not by sexual intercourse, as do other octopi.

Soft-Shelled Non-blooded Sea Animals (Fig. 1.21) The *Malakostraka* ("soft-shelled animals", a general name for what we now call crustaceans, mostly Decapoda) are a major group of bloodless animals that differ from the "soft-bodied" by their hard part, the shell, which is outside the fleshy part (rather than inside it, as in some *Malakya*). This shell is rather crusty and consequently, it breaks by crushing, rather than by shattering. All crustaceans can both crawl and swim, and accordingly, they have many feet. Their mouth is equipped with two "teeth" (what we today call mandibles).

Aristotle divided the crustaceans into four subgroups, two of lobsters (*Karaboi* and *Astakoi*), one of shrimps, prawns and mantis-shrimps that lack claws (*Karides*) and one of crabs (*Karkinoi*), distinguished from lobsters and prawns by having a rounded body and lacking a tail. The crabs and lobsters, Aristotle wrote, resemble each other in possessing claws, which serve as hands for seizing and holding objects; their right claw is invariably larger and stronger, "for it is natural to every animal in active operations to use the parts on its right side in preference to those on its left". This seems to be so in many other animal groups that possess defensive or offensive weapons, such as tusks, teeth, horns and spurs. In lobsters, chance determines which claw shall be the larger one. This is so because, belonging as they do to a group in which this is a constant character, they must have claws, but they have the larger claw in this indeterminate way because they use their claws not for their natural purpose, but for locomotion. Lobsters have a tail while members of the crab subgroup do not, because lobsters swim about and a tail is therefore of use to them, as

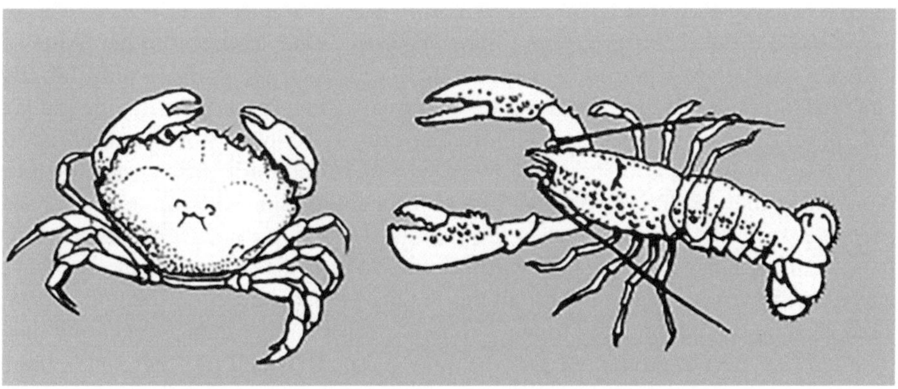

Fig. 1.21 Aristotle's grouping of invertebrates: Malakostraka. (Tuvia Kurz)

it serves for propulsion like the blade of an oar, whereas a tail would be of no use to crabs, who habitually creep into holes and crevices. Some crabs, however, have their hindermost feet flattened out and resembling fins or oar-blades, to help the animal in swimming.

Some of these subgroups further divide into numerous species, which differ from one another in shape and size. As with the cephalopods, we are fortunate that Aristotle's brief descriptions often enable identification with crustaceans of the extant Greek aquatic fauna. Among the *Astakoi* he described the common lobster *Hommarus gammarus* with its long and slender antennas, long and pointed rostrum, smooth carapace and massive pincers. In the *Karaboi* he described the slipper lobster *Scyllarus arctus* with its peculiar second antenna. Among crabs, he noted the ghost crab *Ocypode cursor* ("in the neighbourhood of Phoenice there are found on the beach certain crabs that are nicknamed the 'horsemen', because they run with such speed that it is difficult to overtake them") and the freshwater crab *Potamon fluviatile* (Voultsiadou and Vafidis 2007). These are a few examples, of many.

Hard-Shelled Non-blooded Sea Animals (Fig. 1.22) The *Ostracoderma* ("hard-shelled animals") are a group of bloodless animals that resemble crustaceans in that their hard parts are outside the fleshy part, but differ from them in that the hard parts consist of a chalky shell, rather than a crusty one.

Consequently, Aristotle wrote, we shatter this harder part, rather than crush it. In "hard-shelled" animals the head faces downwards as in plants, because they take in their nourishment from below, just as plants do with their roots. Some "hard-shelled" animals resemble plants in that they too are motionless (mussels, sea-squirts); others move about, though only a little (snails). Many are motionless, or nearly so. Nature, however, has provided them with protection in the hardness of their shell, which may have one valve, or two valves, or spiral, as in whelks, or globular, as in sea urchins. All hard-shelled animals generate spontaneously from within the substratum on which they live, whether it be mud, sand, or rocks. Aristotle was aware that the snails, while resembling crabs and lobsters in having an external skeleton, resemble octopus and squid in their general body plan, with the digestive system bending forwards.

Aristotle divided this group into five subgroups. One subgroup of hard-shelled animals consists of those with a spiral shell, to which the body is only loosely attached (this sub-group corresponds to what we consider most of the marine snails). All spiral-shaped animals creep, and they always do so away from the shell apex. They have an operculum ("cover") to protect their body, for were it not for this, Aristotle argues, colliding with external objects could injure their foot. Their head protrudes from the shell and, when the snail is alarmed, it draws it in. Some snails have small, delicate and yet sharp teeth. Some, such as the purple murex, also have a trunk with which they bore through the shells of other snails, on which they prey. They feed by extruding their "tongue" that may reach longer than a finger and that, in some (the purple-producing genera), may pierce the shell of other snails upon which they prey. Aristotle was here referring to the long proboscis of predatory snails. In dividing this subgroup of spiral shells into species, Aristotle's text

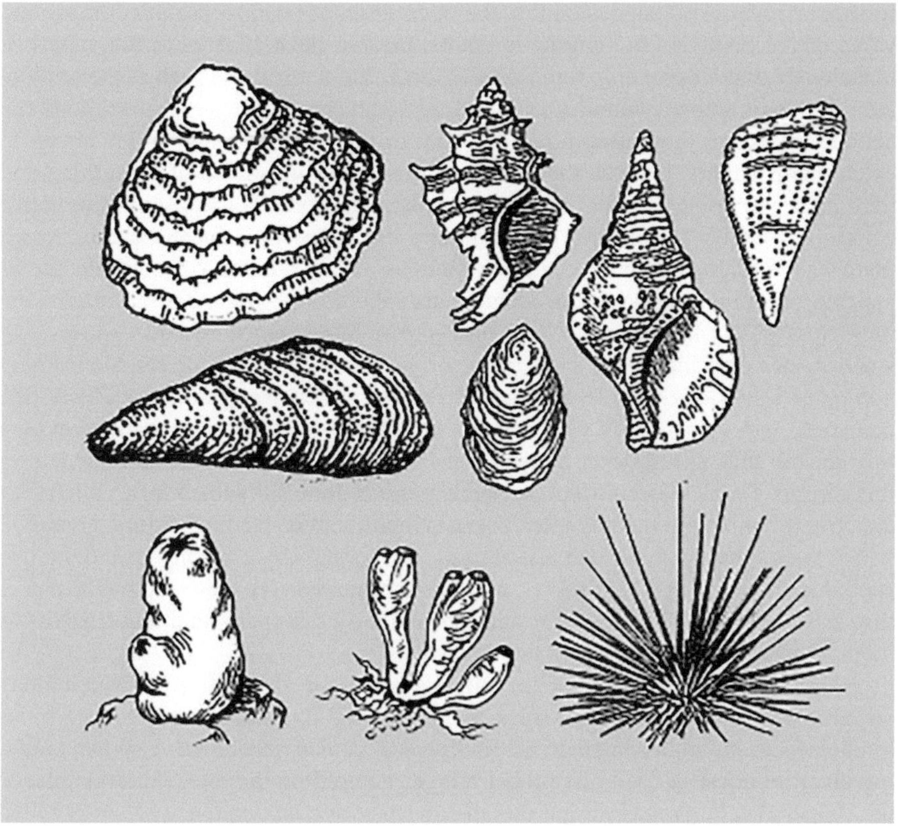

Fig. 1.22 Aristotle's grouping of invertebrates: Ostracoderma. (Tuvia Kurz)

mentions what we identify as the nerite *Nerita*, the purple dye murexes *Bolinus brandaris, Hexaplex trunculus* and *Stramonita haemastoma*, *Strombus* or perhaps *Cerithium* and also a few additional species which we cannot identify. Aristotle also wrote about hermit crabs (Paguroidea): they originate in mud and make their way into an empty shell in which they live, carrying the shell about. When they grow too large for one shell, they leave it and move into a larger shell. I find much innocent charm in Aristotle's words, that the hermit crab *Karkinium* is intermediate between crustaceans and snails, because in its morphology it resembles crawfish, but in its habit of introducing itself into a shell and living there, it resembles snails; it thus partakes of the characters of both groups. He further commented that the hermit crab differs from snails in that it easily slips out of its shell, rather than adhering to it.

A second subgroup of "hard-shelled" is the bivalves, which are protected by being closed between two "shells" (valves) which may gape, as in scallops and mussels, in which the valves are united on one side only (the hinge) and they open and shut on the other; or they may be fused on both sides, as in razor-shells. Aristotle further (erroneously) suggested that the operculum of snails represents the second valve of the bivalves. In contrast to snails, bivalve flesh adheres to the valves so tenaciously that we can only remove it by force. Some bivalves, such as the scallop, are capable of active swimming, and indeed some people suggest that scallops can actually fly, since they often jump right out of the traps that anglers lay about, to catch them. Others, such as the pen-shell *Pinna*, are incapable of swimming and they attach to some external object. In Aristotle's texts we can, with reasonable certainty, identify the scallops (Pectinidae; his term *Kteis* means comb, which matches the Latin *Pecten*; he describes them as having a radial-ribbed shell, and as capable of swimming). We can identify also the edible oyster *Ostrea edulis*, the mussel *Mytilus galloprovincialis* ("they grow, like onions, by side shoots, the smaller ones growing on the sides of the original ones"), the noble pen-shell *Pinna nobilis* and the razor shell (Solenidae). Aristotle writes that the bivalve *Pinna* (the pen shell), grows out of its silky threads and that it contains in its mantle cavity a tiny shrimp that, if removed, the pen shell will die; the bivalve actually generates this shrimp. Today we know that penshells produce long silky threads by which they attach to the substratum. They often host a symbiotic filter-feeding shrimp, *Pontonia*, which warns the bivalve when it feels a threat, perhaps by pinching it; and the pen shell then closes shut. The bivalves, scallops and mussels are protected by the power they have of closing their valves; and the snails by the operculum that transforms them, as it were, from univalves into bivalves.

A third subgroup of hard-shelled animals is the univalves (limpets, which today are placed among the marine snails), which powerfully attach themselves to rocks by their foot; and in this manner, the rock serves as their second valve. When crawling about in quest of food, the limpet relaxes its hold on the rock. Aristotle placed the limpet *Patella caerulea*, and the abalone *Haliotis tuberculata* (which is called the ear-of-the-sea) in this group.

The most protected of all hard-shelled subgroups is the fourth one, that of the sea urchins. Their globular shell, covered by sharp spines, completely encloses the body. Sea urchins also use their spines as feet, thus moving from place to place. Here Aristotle mentioned what we consider the edible stony sea urchin *Paracentrotus lividus* and its habit of always having something fixed onto its spines, the species *Cidaris cidaris*, the heart urchin *Brissus unicolor* and the violet urchin *Sphaerechinus granularis*, which is the largest sea urchin in the Greek seas (Voultsiadou and Vafidis 2007). In relying on a single feature, shell hardness, for classification, Aristotle erred in placing the sea urchins and sea squirts together with snails and bivalves. He further erred in suggesting that what people consider as their female gonad is merely fat, of importance to the sea urchin's survival, as it cannot cope with cold temperatures. These "ova" grow considerably during the full moon, when the nights are warmer, owing to moonlight. In some sea urchins the ova are very small, in others they are large because they are of a warmer constitution, and therefore able to concoct their food more thoroughly. Furthermore, those of a warmer constitution tend, because of their warmth, to move more about, thus making more expeditions in search of food than those that are stationary. As evidence of this, they always have things sticking to their spines, for they use their spines as feet.

The final sub-group of hard-shelled animals is the *Tethya* (our tunicates), which resemble plants in that they always live attached to something, and in that they lack that basic feature of animals, the ability to move. They differ from most other animals in that they have no organs of movement and no sensory organs. They differ (though slightly) from plants in that they consist of flesh. A thin transverse partition runs across the middle of the body, which has two openings, one through which the tunicate takes in fluid matter for nutrition and another through which it discharges the surplus of unused 'juice'. Aristotle seems to have been distinguishing between the white sea-squirt *Phalusia mammilata* and the red sea-squirt *Halocynthia papillosa* (Voultsiadou and Vafidis 2007). Again, in relying on a single feature, shell hardness, for classification, Aristotle erred in placing tunicates together with cephalopods, marine snails and bivalves.

Sponges are a small major group, which resemble plants in that they attach to rocks throughout their life, and when detached, they die. According to Aristotle, they develop spontaneously, namely they generate from non-living matter and they get their food from slime. The sponge possesses sensation and as evidence for this, if the sponge is aware of an attempt to pluck it from its place of attachment, it draws itself together, to such an extent that it is difficult to detach. Sponges can regenerate, so that "when the sponge is broken off, a new one grows up again from the remaining portion and takes the place of the old one". It also makes contractile movements in stormy weather, to tighten its hold. Sponge species which we can recognize from Aristotle's descriptions include the black ircinia *Sarcotragus muscarum*, the honeycomb *Hippospongia communis*, the bathing sponge *Spongia officinalis,* the elephant ear *Spongia agaricina* and the leather sponge *Spongia zimocca* (Voultsiadou 2007. Sponges were of importance to Greek warriors, who placed them beneath helmets and thigh-pieces, to soften blows).

Slightly different from the sponges are the sea cucumbers on the one hand and the jellyfish on the other. The latter are free and unattached yet they have no feeling, their life being simply that of a plant separated from the ground.

In an overview paragraph of the stationary ostracoderm, the sea anemones and the sponges, Aristotle commented that this is how nature passes in a continuous sequence from "lifeless objects" to animals, interposing between them beings that live and yet are not animals. Indeed, in many cases scarcely any difference seems to exist between two neighbouring groups.

Non-Blooded Land Animals The *Entoma* are a group of bloodless animals that dwell on land. They all have many feet. The elongated individuals have more feet, with the millipedes having the greatest number. The *Entoma*'s entire body is hard, both back and belly, with no distinction between soft and hard parts. However, Aristotle wrote that their hardness is more flesh-like than bone, and more bone-like than flesh, "so as to make their body less liable to get broken into pieces". Insects resemble lobsters and crabs in their tendency to have the internal organs arranged in a straight line. In this, they also resemble red-blooded animals from which they differ, however, in those the external organs enable locomotion.

The body of these animals is inspected, and the number of their feet corresponds to the number of insections. All *Entoma* have eyes. In most of them, no other sense organ is discernible, but in some, there is a tongue, with which they taste and imbibe their food (we may think of a butterfly's proboscis). This tongue corresponds to a similar organ in snails, the trunk of the purple shell. Aristotle placed all terrestrial arthropods, including insects (beetles, bees and wasps, flies and gnats, cicadas, locusts, fleas), scorpions, centipedes, millipedes, isopods and probably also earthworms in his large group of *Entoma*. Insects all have many feet, to compensate for their natural slowness and frigidity, and enable them greater activity when moving about. Those that, as the millipedes, have exceptionally long bodies and are therefore the most liable to refrigeration, have the greatest number of feet. Their number of feet corresponds to that of their insections. Insects eat little, because of their cold temperament rather than because of their tiny size. In the cicadas, the mouth and tongue fuse into a single pipe through which, as through a root, the insect sucks up the fluids on which it lives.

The division of the insect body into segments is a result of their having several internal organs that repeat themselves, rather than one single organ. This essential character of insects also makes them close to plants, and Aristotle claimed that insects still live when cut into pieces, like plants. However, the cut sections of an insect survive for only a limited time, whereas those of plants can persist and even attain the perfect form of the whole plant so that we can obtain two or more plants from a single plant. Being insected also enables many insects to roll themselves up when in danger, to protect their soft abdomen (here, Aristotle was probably thinking of isopods). Others that cannot do so, such as the *Kanthari* (Cantharidae, certain beetles), draw their segments up into the insected spaces and thus they harden their bodies (*PA* 4: 6). We can feel this (playing dead) behaviour by putting our finger on one such insect. The touch frightens the insect, its body becomes hard and it remains motionless. Some insects have also another means of defence namely a sting, in the front or rear end of the body.

Some insects have weak feet, which they compensate for by their ability to fly. Insects do not use tail steerage when flying, as do birds and their wings lack quill structures. Of such flying insects, the bees make long expeditions in search of food; their light-weighing body has four wings, two on either side. As they are strong in their hind parts, their sting is at the posterior end of the body. Other flying insects, the flies and gnats, are so small and light that they require only two wings to support them in flight. Their sting is in the front because their weakness does not enable them to strike efficiently with the hind of the body. Another group of insects, the beetles, have heavy bodies and stationary habits. They have protective sheaths for their wings because their inactive behaviour exposes their wings to much greater risks than those of the more active bees and flies.

In flies and gnats, the sting is in the front of the body, associated with the mouth, because their weakness does not allow them to strike efficiently with the rear part of their body. Now of course, it is better that each organ should serve a single function: that there should be one organ to serve as a weapon, which can then be very sharp, and a separate one to serve as a tongue, which can then be of spongy texture and suited for absorbing food. Therefore, whenever nature is able to provide two separate organs for two separate functions without one hampering the other, it does so, and it is only when this is impossible, that it uses one organ for several functions. In bees and wasps, therefore, the sting is at the rear. It is internal and delicate because were the sting external it would soon be damaged; and were it thicker, as in scorpions, its weight would encumber their flight.

If we now examine the legs of insects, we find that the anterior pair of legs is in some cases longer than the others are, so they may wipe away any foreign matter that might lodge on the insect's eyes and obstruct its sight. Indeed, flies and bees often use their forelegs to clean themselves. Of the other legs, the hind pair is bigger than the middle pair, so that the insect, when it takes flight, may spring more easily from the ground. This difference in leg size is even more marked in insects that leap, such as locusts and fleas.

In writing on bees, Aristotle many times used the phrase "they say that", suggesting that he relied on the experience of beekeepers rather than on his own observations, and he often added aspects of beekeeping. He wrote, for example, that some beekeepers recognize their own bees in the fields by sprinkling them with flour, beforehand. He also wrote that bees use material from flowers and trees to build a comb; and if a comb collapses, they set it up again, putting props beneath it to enable a pass beneath. They build cells for themselves, for the king and for the drones.

When flying in search of nectar each bee, during each of its flights, settles only upon flowers of one plant species, not on flowers of different kinds ("it flies from violet to violet"). With its forefeet it collects food from flowers, transfers it to the middle pair of legs and from there to its hind legs; and, thus loaded, it flies home. In strong winds, it carries a stone, as a balance. Not all individuals of a hive seek food from flowers. Others bring water or clean and erect cells. The elder bees work in the hives, younger ones go out in the fields. Of those that work in the hive, some attend to the honey, others to the grubs, to the beebread, to building the comb, or to carrying water to the cells and mixing it with honey. Small bees are more industrious than large ones, and as a result, their wings become worn at the edges and their colour black; the bright and shiny bees are "idle, like women". Bees drive out all individuals who are idle or wasteful, and they eject the drones if honey is short. If a bee dies in the hive, other bees carry its corpse out. Early in the morning, they are silent, until one bee arouses them by humming two or three times, and then they all fly off to work. When they return there is some disturbance at first, which gradually decreases until one of them flies round with a humming noise, as if warning them to sleep; they then all become silent.

The drone is larger than the bees, has no sting, and is stupid. As long as the king is alive, the bees produce drones in a separate place. When he is dead, they produce them in their own cells. A new king ventures out of the hive only if accompanied by a swarm of bees that gathers around him, making a peculiar noise 2–3 days beforehand. If a swarm wanders to a distance without its king, it will retrace its steps and return until it finds him. When unable to fly the swarm carries him; and if he perishes, the whole swarm dies with him. If the king that has been left at home emerges from the hive and follows the swarm, they kill him.

Wasps predate bees, as do the titmouse, swallow and bee-eater. Frogs in marshes also prey upon them, when they come for water, and toads kill bees by blowing into the entrance of the hive and then catching them as they fly out. Bees fight such enemies by stinging them, and a swarm can sting and kill animals as large as a horse; the single bee, however, perishes once it has stung anything, for it cannot withdraw its sting from the wound without tearing its own entrails. Bees fight also among themselves, not when at a distance from their hives but when near home.

Aristotle believed that most insects generate by producing worms. Unaware of the insect egg, he thought that from the worm an animal is produced; the whole worm grows and becomes an articulated animal. He observed that some of his "insects" (harvestmen, spiders, locusts, grasshoppers) generate from similar animals, but others do not originate in animals of the same species. Their production is spontaneous, in that they sprout from dew that falls upon plants, or from rotten mud and in dung, from wood, or from animal hair, flesh or excrements. Butterflies generate from caterpillars, and these generate in the leaves of green plants. They are tiny at first, later they grow into little worms that become small caterpillars that take food, that grow and then become motionless and change their form to a pupa (a chrysalis) that does not feed. It has a hard covering but moves when touched. After a short time, a winged animal bursts the covering and flies away—a butterfly. This mode of generation is likewise in all creatures that originate in worms, including those that produce worms after sexual intercourse.

There are three kinds of worms, the flat worms, the round worms and those called ascarides. [Today *Ascaris* is the name of a parasitic nematode worm up to 50 cm long that infects human intestines.] From these creatures, nothing is produced, but the broad worm is attached to the intestine and produces something like the seed of the colocynth. It originates in the mud of wells, and in running waters that flow over an earthy bottom. Other worms originate in the dung of oxen and asses. Flies originate in dung.

Scorpions bring forth many egg-like maggots, which they incubate. When the young ones grow to a certain stage ("are perfect") they drive out and destroy their parents. Harvestmen deposit their ova in a thick basket that they weave and incubate. The young, when they are grown, surround their parent in a circle and kill her. They often seize the male in the same manner if they can catch him, for he assists the female in incubation. Bugs proceed from the moisture that collects on the bodies of animals, and lice from the flesh of other creatures. All lice originate in the animals that are infested with them. There are also other minute animals, some of which occur in wool, and in woollen goods; as those moths that are generated in greatest abundance when the wool is dusty. This worm also occurs in garments. Others also live inside books, and resemble those that occur in old honeycombs; some are like scorpions that have no tails and are very small. Insects, overall, occur in everything.

There is a little worm (called *xylophthorus*), extraordinary in that its head projects beyond its case whereas its feet are at the other end, as in other worms. The rest of the body is contained in a case made of a substance like a spider's web, with dry material on the outside so that it appears to walk about with this attached to it. These creatures live attached to their cases as a snail is to its shell; the whole of the case is joined to the worm and it does not fall out of it but is drawn out of it as if they were joined together. If a person pulls off the case, the creature dies, being as helpless as a snail without its shell. As time advances, this grub becomes a chrysalis, like a caterpillar, and lies without motion; the nature of the winged creature that emerges has never been ascertained.

Blooded Animals (Vertebrates)

After he dealt with the major group of "non-blooded" invertebrates, Aristotle moved on to the other major group, the "blooded" animals (our vertebrates), animals that breathe and have blood. The vertebrate body consists of four liquids, blood, bone marrow, milk and sperm. Bone marrow occurs in the hollow bones of all vertebrates. Milk is found in only one group of blooded animals (mammals), and usually only before birth and during the suckling of the young. Sanguineous animals eject sperm, which contributes to generation (we shall discuss this point, further on). Blood is the only liquid that permeates the entire frame of living animals at all times, as long as life lasts. Aristotle believed that it forms in the heart, from a very early embryonic stage onwards, and that in addition to the heart it occurs in the veins. He further believed that different types of blood differ in their degrees of thinness or thickness, of clearness or turbidity, of coldness or heat; and this is so whether we compare blood from different parts of the same individual or blood of different animals. In the individual, differences distinguish the blood of the upper and the lower halves of the body. The thicker and hotter the blood is, the more the animal is both strong and courageous; the thinner and colder, the more it is sensitive and intelligent. Similar trends occur also in the fluid analogous to blood, and this explains how it is that bees are more intelligent than many blood-containing animals; and that, among the sanguineous animals, those most intelligent are the ones whose blood is thin and cold. The noblest of all are those animals whose blood is hot, thin and clear, for they possess both courage and intelligence. Accordingly, the upper parts are superior in these respects to the lower, the male is superior to the female, and the right side to the left. All blooded animals have a head, that consists mainly of the brain and of major senses. All those possessing a lung have also a neck, which exists to defend the windpipe and the oesophagus. In most animals, this neck is flexible and contains several vertebrae, but in wolves and lions, it contains only a single bone.

Marine Blooded Animals (Fig. 1.23)

Aristotle placed all the vertebrate aquatic representatives in one large group, the fishes. However, he noted that sharks and rays differ from bony fishes in four main traits: they lack an operculum to cover the gill slits, they have rough skin that lacks scales, they tend to bear live young, and they have a cartilaginous skeleton. He mistakenly believed that gill-breathing fishes lack testicles. We, today, can recognize many of the fishes he described.

Fig. 1.23 Aristotle's grouping of vertebrates: Fishes. (Tuvia Kurz)

We now come to what seems to me as Aristotle's major blunder in classification: He erred in classifying the major vertebrate groups by their habitat. Some vertebrates are all-terrestrial, others are all-aquatic. His marine cetaceans thus fell into his aquatic groups. In his Cetacea he included the dolphins (probably a combination of the striped dolphin *Stenella coeruleoalba*, common dolphin *Delphinus delphis*, and bottlenose dolphin *Tursiops truncates*); harbour porpoise *Phocoena phocoena*, "similar to dolphins but smaller and found in the Black Sea"; and the fin whale *Balaenoptera physalus*, a common species in the Mediterranean at that time. Dolphins and whales, though resembling fish in that they have fins rather than feet, come to the surface to breathe air through a blowhole, and they have lungs, like other air-breathing creatures; they breathe air, have a windpipe and lungs and a blowhole through which they spout water. They differ from fish in that they lack gills. In addition, they have wombs and deliver their offspring fully developed (Greek *delphis*, meaning dolphin; *delphys*, meaning womb); and they produce milk to feed their offspring, instead of laying eggs like fishes. Also, the "bones" (= skeleton) of cetaceans resemble those of terrestrial, viviparous quadrupeds more than those of fish. They have ears and can hear. Further, "the [whale] has no teeth but does have hair that resembles hog bristle". Still, he classified the whales among the fishes (Romero 2012). When hunting they will "plunge into the deep, calculate how long they can stay down, and then shoot to the surface like arrows, flying out into the air, sometimes clearing the masts of boats […] when caught underwater in nets they drown. […] Sleeping dolphins do snore, however". In this description, of dolphins leaping out of the water, we might see a relationship with Homer's Hymn to Apollo in which Apollo takes the form of a dolphin and, just as Aristotle describes, bursts out of the water and leaps aboard a ship; Apollo then hijacks it from the amazed sailors by commanding the wind to blow it wherever he likes.

A recent study of Aristotle's texts yielded 109 individual fish taxa (of which 13 had multiple names) and 12 terms that corresponded, whether to fish developmental stages or to their behavioural patterns (Ganias et al. 2017). Fifty-four fishes could be identified to species or genus level, 24 fishes to higher categories (family, order), but 27 fishes could only be classified as Actinopterygii or Elasmobranchii; four could not be classified at all. Among the most easily recognizable fishes were *Lophius* spp., *Conger conger*, *Muraena helena, Anguilla anguilla* and *Mugil cephalus*. The latter two and the tunas (*Thunnus* spp.) were the most frequently recorded fishes. Aristotle's texts were not limited to fishes of human interest, he also mentioned species that fascinated him in their peculiar biology, from body structure and function, via reproduction and development, to feeding habits, migrations and diseases (Ganias et al. 2017).

Birds Birds have internal testicles and produce no milk. Aristotle is vague on the division of birds into subgroups. He seems to have recognized six bird subgroups: crows, pigeons (common pigeons, ringdoves, turtledoves), those with hooked talons (diurnal and nocturnal raptors), weak fliers (partridges, pheasants and grouse), web-toed groups (pelicans, cormorants, ducks, gulls, some rails) and long-legged, non-webbed birds along riverbanks and in marshes (storks, herons and bitterns, cranes). There are no subgroupings for small birds and, apparently, Aristotle did not consider them a group. Throughout his text, he mentioned some 90 species of birds. As a classifying character, he noted the mode in which birds bathe: Some wash themselves by wallowing in dust and others in water, and some neither in dust nor in water. Weak fliers that live on the ground wallow in dust, such as the domestic fowl, partridge, grouse, lark, and peacock. Birds that have straight talons and live along the banks of rivers, in marshes or in the sea bathe themselves. Some bathe both in water and in dust, such as pigeons and sparrows. Most birds with hooked talons do not dust and do not bathe at all. Aristotle's classification of bathing habits reappears 18 centuries later, during the Renaissance, as we shall see further on.

Aristotle mentions “bird zephyria”, eggs that the wind had impregnated and other eggs impregnated by the touch of a hand. He rejected an earlier Greek philosopher (Herodotus, around 400 BCE) who had claimed that vultures come from a part of the earth that is invisible to us and consequently, we see them, suddenly and in great numbers, only when they follow the path of an army. Aristotle insisted that the vulture builds its nest in high, inaccessible rocks, wherefore we rarely see its nest and young.

From Aristotle’s description of the natural history of the cuckoo, we get a glimpse of what he thought about the possibility of one kind of bird changing into another. He wrote that the cuckoo appears for a short time in summer and disappears in winter. Because the hawk that it resembles disappears when the cuckoo arrives, people say that it is a changed hawk. Indeed, one can see very few hawks of any sort during the period in which the cuckoo sings. However, Aristotle noted that the hawk has crooked talons but the cuckoo has not, and the shape of its head is different. It resembles a hawk in colour, but whereas the markings on the hawk form lines, the cuckoo is spotted. True, in size and manner of flight it resembles a small kind of hawk, which disappears during the season in which the cuckoo comes. However, people have seen both at the same time and furthermore, hawks devour cuckoos, something never done by birds of the same kind. Some say that no one has ever seen the young of the cuckoo. This is because it lays its eggs in the nests of birds of other kinds, having first devoured their eggs; when the cuckoo hatches, the host brings it up. The cuckoo makes no nest. It lays its eggs in the nests of other birds such as the sparrow, lark and greenfinch. It lays one egg, upon which it does not sit, but the bird in whose nest it lays both hatches the egg and nurses the young. When the young cuckoo grows, it throws out of the nest the other fledgelings, which thus perish. Some people say that the foster mother bird kills them, and feeds the young cuckoo with them; for the beauty of the young cuckoo makes her despise her own offspring. Others say that the old cuckoo comes and devours the young of the other bird. Others say that the great size of the young cuckoo enables it to seize the food that the foster mother brings to the nest so that the rest starve and perish. Others say that the cuckoo, being the stronger bird, kills its foster brothers. The cuckoo is very cowardly, and when even small birds peck it, it flies away from them. To conclude, Aristotle stated that the cuckoo does not change its species.

Quadrupeds Among terrestrial quadrupeds, Aristotle distinguished between those that lay eggs (oviparous) and those that bear live young (viviparous). The legs of oviparous quadrupeds, which include the crocodile, lizard and tortoise (corresponding to our reptiles), are placed on the side when the animal is in motion, the skin is covered in scales, the testicles are always internal and, except for tortoises, they lack a urinary bladder. The oviparous quadrupeds also include snakes, which resemble long footless lizards. Serpents in general are oviparous; the viper, an exceptional case, is viviparous. Aristotle noted that in this character the taxonomic position of the viper among other snakes, lizards, tortoises and crocodiles, resembles that of viviparous sharks among the usually egg-laying fishes.

Aristotle further noticed that the body of viviparous quadrupeds (our terrestrial mammals) contains more blood than the body of oviparous ones. Hair covers their skin, they bear their young alive, and the females have milk in their breasts, whereas oviparous animals, such as fishes, reptiles and birds, have neither breasts nor milk. The testicles are usually external (but internal in some cases, such as the hedgehog and elephant) and they all possess a urinary bladder. However, Aristotle noted that the dolphin, porpoise and whale, which he placed in the fishes, also bear live young and have milk. Strangely, he mentioned only two subgroups of terrestrial mammals. One is the equids, "those who have a bushy mane and a bushy tail: horse, donkey, mule, the jennet and those that in Syria are named half-donkeys" (we, today, identify these as *Equus hemionus,* the onager). Another subgroup includes the monkeys and apes (our primates), which resemble man in the structure of their face, their teeth, the position of the teats, hands, fingers and nails and female genitals; apes further resemble man in having no tail. However, they also resemble the viviparous quadrupeds in general, in their hairy body, their posture on all fours and the lack of a hip; monkeys also resemble other mammals in having a tail. In line with this similarity to other mammals, Aristotle also noted that the baboon resembles a dog in its face and its large, powerful canine teeth. Beyond these two sub-groups, the many mammalian species are not divided into sub-groups or, in his words, there are "many species, but under no common appellation". Elsewhere, he seems to have recognized carnivores as a group ("all creatures are at enmity with the carnivores, and the carnivores with all the rest, for they all subsist on living creatures"; *History of Animals*). I cannot explain why he failed to describe such rather obvious mammal groups as rodents and ruminants. He considered bats as intermediate between mammals and birds, as is also the ostrich (*Parts of Animals*). In a similar manner, he considered dolphins and whales as intermediate between fish and mammals.

Ever so correctly, Aristotle included Man among the terrestrial mammals, even though he is a biped, not a quadruped. He wrote that of all animals, Man alone stands erect, in accordance with his godlike nature and essence, for it is the function of the god-like to think and to be wise. This would be a difficult task, were it under the burden of a heavy body, pressing down from above and obstructing by its weight the motions of the intellect and of the general sense. This is why no other animal is as intelligent as Man is. Furthermore, among men themselves, in comparing children with adults, or short with tall adults, we find that the former is less intelligent. This is so because their psychical principle is corporeal, and much impeded in its motions.

Standing thus erect, Man has no need for legs in front, and instead of them, nature has endowed him with arms and hands, a consequence of his superior intelligence. Whereas an earlier philosopher (Anaxagoras, c. 500–428 BCE) had argued that the possession of hands was the cause of Man being the most intelligent animal, Aristotle claimed that Man's endowment with hands was the consequence rather than the cause of his superior intelligence. Hands are instruments or organs, and nature, in distributing organs, gives each organ to such animal that can make use of it. In his words, it is wiser to take a person who is already a flute player and give him a flute, than to take one who possesses a flute and teach him the art of flute-playing.

Man owes his hands to his superior intelligence, rather than owing his superior intelligence to his hands. Hands gave Man a wider diversity of modes of defence. Animals have each just one mode of defence, which they can never change but numerous modes of defence are open to Man and he may choose among them at will, adopting one weapon or another, as he pleases. The hand is thus either talon, hoof, or horn, whatever Man chooses, and so too it is spear, and sword, and any other weapon—all from his hand's power, of grasping and holding. This question, as to whether form comes before function or vice versa, will turn up again 23 centuries later, in Paris.

Aristotle further noted that in quadrupeds the chest must be narrow because the front legs extend forwards during locomotion (rather than sideways, as in most reptiles). Because of this narrowness, the quadruped mammae are never on the chest. In Man, the arms are at the sides of the body, not beneath it, and consequently, the chest can be broad, with ample space on it; it is also fleshy, to protect the heart beneath. The mammae in the male, being an additional fleshy substance, are on his chest, side by side. In the female, nature has given the mammae an additional function, as a store-place of nutriment for the offspring. In animals other than man, the mammae are not between the forelegs, as they would interfere with locomotion; they therefore occur elsewhere. In those animals that produce just a few offspring at birth, there are only two mammae, near the thighs; the elephant is an exception in that its mammae are between the forelimbs. In those producing large litters, the mammae are either numerous and on the sides of the belly, as in swine and dogs; or there are only two, in the centre of the belly, as in the lion. The lioness has a poor supply of milk because being a flesh-eater, it gets food at rare intervals and such nourishment that it obtains goes to the maintenance of her body.

Moving on from the mammae to the feet, Aristotle wrote that their function is to enable us to stand securely. The foot's extremity divides into toes because, in an undivided foot, disease of any one part would spread to the whole organ, whereas in one split into separate digits, disease would occur to a lesser extent. Man's toes are short, so they would be less liable to injury. For these reasons, the feet in man are many-toed, and each digit is of no great length. The toes have nails because such projecting parts are weak and therefore require special protection.

Aristotle also sought differences between Man and beast that are not morphological. In describing the essential behavioural characters of various animals, he wrote that the stag is intelligent and timid, the snake is mean and treacherous, the lion noble and courageous, the fox crafty and mischievous, the goose cautious and watchful, the peacock jealous and self-conceited—but of all animals, Man alone is capable of deliberation. He also noted that many animals have memory and are capable of instruction, but no other creature except Man can recall the past at will. Elsewhere he noted that of all animals, Man alone has the ability to learn to make equal use of both hands. Man has the largest brain in proportion to his size, and it is larger in men than in women. This is because the region of the heart and of the lung is hotter and richer in blood in Man than in any other animal; and in men than in women. This also explains why man, alone of animals, stands erect.

Aristotle complemented his classification of animals with impressive correlations. He wrote that viviparous quadrupeds always have fur, and egg-laying quadrupeds have scales. Likewise, "I have never seen a single-hoofed creature with a pair of horns"; animals with well-developed canines do not have horns; and horns do not occur in small mammals. No animal with small pointed teeth has tusks; one single horn occurs in animals with a solid hoof and not in those with a cloven one; two horns occur in animals with cloven hoofs; and birds with talons never possess spurs.

Aristotle was fully aware of the occurrence, in nature, of hybrids. He believed that they occur mainly in Africa rather than in Asia and Europe and that this results in Africa possessing a highly diverse fauna. In Africa, which is hot and dry, thirst brings many different animals together at drinking places, where they copulate and produce young, if their periods of gestation happen to be the same and their size is not very different. The thirst makes them gentle to each other, this results in hybrids and has even brought about the proverb that Libya is always producing something new.

There are several fabulous animals in Greek mythology, mainly the mixed offspring of very different creatures. The fearsome *Chimaera* in Homer's Iliad was part lion, part goat and part snake. Coming from Mount Chimaera (perhaps Yanartaş in southern Turkey, where methane and other gases emerge from the rock and burn) it could run faster than any human and breathe flames of such scorching heat that they would melt any bronze armour. The *Centaur* of Thessaly (modern Greece) had the upper body of a human and the lower body and legs of a horse and dwelled in the mountains of Thessaly; and in Crete, there was the *Minotaur*, a man with the head and tail of a bull. There were the *Sirens*, half woman half bird, who played music so tempting that it allured sailors to run their ships onto rocky shores, where the ships would smash and the sailors drown. *Cecrops* of Attica, the legendary founder of Athens and who taught the Athenians the basics of civilized life (reading and writing, navigation, marriage and ceremonial burial) had the top half of a man and the bottom half of a snake. *Triton*, who lived on the bottom of the sea, had the upper body of a human and the tailed, lower lower part of a fish. There was also *Sphinx*, with the head of a woman, the body of a lion and the wings of a bird, who devoured those who failed to answer her riddle.

Aristotle, in his *Generation of Animals,* dismissed such myths because two such parents have such widely different gestation periods as to make the birth of offspring impossible. Accordingly, in all of Aristotle's descriptions, I could find only one reference to what we would consider a completely extraordinary animal and even it, with cautious reservation on his behalf. In stating that no mammal has double rows of teeth, he noted that "if we are to believe Ctesias" then there is, however, such an animal. The Indian wild beast called the "*Martichoras*" has three rows of teeth in both the upper and lower jaw; it is as big as a lion, as hairy and its feet resemble those of the lion. It resembles man in its face and ears; its eyes are blue, and its colour is vermilion; its tail is like that of a scorpion; it has a sting in the tail and can shoot the spines attached to its tail. Its voice is somewhere between that of a pan-pipe and a trumpet; it can run as swiftly as a deer and is a savage man-eater. (Ctesias was a late-fifth century BCE Greek physician and historian, who lived for a time in Persia and wrote on the wonders of India. He also wrote that the *Martichora* has, in addition to stingers on each side of its tail, an additional stinger on the crown of its head; that it can shoot its stingers to a distance of one hundred feet; that no creature survives its sting except the elephant; and, that fresh stingers grow, to replace those shot while fighting (Fig. 1.24)).

In evaluating Aristotle's classification of animals, we note that he referred to most of the major groups. They are almost all there—octopus, squid, lobster, prawn, crab, snail, bivalve, sea urchin, sea squirt, sea star, sea anemone, sponge, sharks, rays, bony fishes, whales, dolphins, worms and insects. He correctly placed the octopus, cuttlefish and squid into one group; he correctly placed lobsters, crayfish, prawns and crabs into another; he correctly placed snails and bivalves together; and he correctly placed the sea anemones and sponges in separate groups. Aristotle erred in placing the sea urchins and sea squirts together with snails and bivalves, mistakenly placed starfishes close to the sea anemones and classified whales and dolphins together with bony fish and sharks. However, we must remember that Aristotle possessed no evolutionary thinking.

Most of Aristotle's zoology referred to the fauna that he observed in Greece and its surrounding sea. When writing about animals from foreign countries he obviously relied on descriptions by others; and sometimes we can follow this, such as when he copied their mistakes. He described the crocodile in the words of Herodotus, less than a century earlier (495–425 BCE), as an animal that moves its upper rather than lower jaw, and this is a statement that whoever saw a crocodile would not make. He described oxen in Phrygia (an ancient Anatolian kingdom) that can move

Fig. 1.24 *Martichora.* (Tuvia Kurz)

their horns, as they do their ears; and in Persia, the female foetus of the mice are pregnant already in the uterus of their parent. Aristotle also never saw a lion, otherwise, he would not have written that its bones contain virtually no marrow; and that they are so hard that, when we strike them, fire is kindled, as when striking stones; and that it has only one bone in the neck, but no vertebræ. He idealized the lion as an ideal King of the Animals: When hunting, he is very cruel; but when not hungry he is gentle, playful and affectionate towards those animals that grew up with him, and to which he is accustomed. When hunted, he never retreats in panic; and if compelled to yield to his hunters, he retreats slowly and leisurely. If a person aims a blow at him without hitting him, the lion will rush upon him and seize him, but will not do him any injury, nor tear him with his claws; he merely shakes and frightens him and then leaves him. Lions enter towns and attack people only when they grow old; for old age renders them unable to hunt, from worn-away teeth. In the midst of all these myths, Aristotle wrote a very intriguing sentence: "There are two kinds of lions. One of these has a round body and more curly hair and is more cowardly. The other is longer, has straight hair and is more courageous". This might *perhaps* not be a myth. It might perhaps be evidence that in his days there was another species of lion, one with a round body and more curly hair, which has since gone extinct.

In his classification Aristotle created four main groups of invertebrates and four vertebrates; in his malacostracans, he created four subgroups, five ostracoderms (four if we merge his univalves into the snails) and four subgroups of reptiles. Furthermore, by and large, Aristotle's animal classification contains four taxonomic ranks, as can be seen in the two examples below:

- Invertebrates
  - Soft animals
    - Octopuses
      - Musky octopus
- Vertebrates
  - Oviparous quadrupeds
    - Snakes
      - Viper

Far beyond biology also, in probing into the essence of created objects, Aristotle noted that every created object has four factors: the matter it consists of, its shape, who made it and the reason he made it. Could it be that symbolic, mystic elements were trickling into Aristotle's biological thought? The number Four had special significance in the ancient world. There were four Greek classical elements (fire, air, water, earth), four qualities (hot, cold, dry, wet), four humours of the body (blood, yellow bile, black bile, phlegm), four seasons, four cardinal directions and 4 years between each Olympiad. The number four is also significant in Jewish symbolism, such as the four rivers in the Garden of Eden, four ritual species of the biblical festival of Tabernacles, four archangels and, on the evening of Passover, four cups of wine to drink, four questions to be asked and four sons to be dealt with. Four is also of symbolic significance in Buddhism and Hinduism, and similarly, but much later, in Christianity: four canonical gospels and four horsemen of the Apocalypse. To conclude, the number four could perhaps have seeped into parts of Aristotle's classification of animals.

Classification is, by one possible definition, the creation of a hierarchical system of groups and subgroups into which, in a separate process, we place different specific animals. Viewed in this context, Aristotle clearly set up a complete taxonomic system of very many of the groups and subgroups in the animal world, thereby creating a comprehensive classification. Furthermore, his groups are well-defined and well-described. He did not assign each and every genus and species he knew of to its group. Surely, he would have been aware of winkles, which live on every rocky beach; and of ladybugs, which live in every field. However, his wide range of groups, covering the entire Animal Kingdom of his day, would have enabled every species and genus to fall into its (reasonably) correct slot. Generally, for a first-ever attempt at classifying the entire animal world, Aristotle's classification is very good.

1.2.4 Animal Diversity: The Ladder of Nature

We have seen that Aristotle, in seeking patterns in the animal world, recognized that all groups divide into sub-groups, with only the group at the bottom, that of species, not dividing any further. Another of Aristotle's major steps in this context was that he determined the major animal groups: mammals, birds, reptiles, insects and so on. We now come to a third major step of Aristotle in his search for order and pattern in the diversity of nature. He suggested that the entire natural world forms an orderly, ladder-like pattern, a "Great Ladder of Nature" (*scala naturae* in Latin). Factors such as the presence of blood, mode of reproduction, the degree of development reached by the offspring at the time of extrusion from the parent's body, and the degree of parental care, determine the rung of each group on the ladder. All these, in turn, depend on the degree of vital warmth of the parent. Impressed by the part played by heat in the hatching of eggs, Aristotle concluded that warmth was the dominant agent in all embryonic development. Warmth was for him also the major agent in the production of change, even in the inorganic world. Beyond warmth, other factors of the environment such as moisture and dryness may also influence modes of generation. Aristotle carefully noted that the mode of generation corresponds with morphological features only generally, thus for example man, a biped, is live-bearing but birds, another group of bipeds, lay eggs; as do lizards that are quadrupeds and snakes that are legless.

Starting from the very bottom of the Great Ladder of Nature, the lowermost rung consists of lifeless objects—minerals, rocks and stones. All these lack the ability to grow and to generate, and they lack a soul. The soul is better than the body and therefore all lifeless beings, which have no soul, are lower than all the living, which have one.

Above the lifeless come the plants, living creatures as they feed through their roots, grow and generate; yet lacking the ability to move and lacking sense perception.

Animals are more diverse in form than plants. Above the plants come several rungs of animals that, in addition to being able to grow and generate, have two typical traits: they (mostly) can move about, and they can sense the environment and respond to it; for example, when touched, they respond by contracting. As sensation advances, the animal mode of life gains pleasure, both in sexual intercourse and in the rearing of young. Some animals leave their offspring soon after birth (or egg-laying) with no further contact; but those that are more intelligent, and possess more memory, treat their offspring in a more civilized manner.

Among the diverse animal groups, Aristotle believed that the greatest difference occurs between animals with red blood, versus those without red blood. Heat is a force in nature that enables animals to be more independent and to reach a greater size, and high heat occurs only in those animals that have red blood. Lowermost among the animals Aristotle therefore placed the mega-group of all invertebrates, imperfect in that they lack blood, the substance that forms heat and of which, so Aristotle believed, the viscera consist. The lowermost rung of "bloodless" animals consists of various groups that lack one or another of the typical traits of a perfect animal. This includes the sponges, that resemble plants in that they lack the ability to move, and like plants, they live only while growing on something; when pulled off they die. Yet they possess sensitivity, in that when touched they contract. Here comes also the jellyfish, which drift in the sea, with no ability to actively move.

At a higher rung of blood-lacking animals, Aristotle placed his group of marine invertebrates with hard external shells: snails, bivalves, sea urchins and tunicates, his *Ostracoderms* (a reminder: snails, bivalves, sea urchins, sea squirts). They reproduce asexually, whether by the young simply budding off from the parents as do plants, or by spontaneous generation from such lifeless matter as rotting, non-living debris, from mud or slime (today we know that many of them reproduce via a larval planktonic stage that eventually settles and undergoes metamorphosis to adult form. Aristotle could not have been aware of the planktonic larva, as it is microscopic). He suggested that ostracoderms generate spontaneously—directly from rotting, non-living debris—from an earthy element that is not hereditary. To create the form of an ostracoderm, this temperature-dependent earthy matter requires vital heat, which is species-specific. Interactions with matter occur randomly: In a certain place, the air or sea coincidentally reaches a certain vital heat, the required rotting matter is present by chance—and a new individual spontaneously comes into being. The spontaneously formed ostracoderm is not able to generate in the manner of sexually reproducing animals, as no offspring-producing father had formed it. Therefore, spontaneously produced animals are imperfect creatures; therefore also, their appearance is patchy and ephemeral in time and space and they perish when the environment changes. They lack the continuum of generations, found in sexual animals (Balme 1962). As concerns mobility, ostracoderm either do not move, or they move but only very little and with difficulty. Many of them resemble plants in that they live by clinging to something else, and in generating spontaneously they have no sex any more than (Aristotle thought that) plants have. Indeed, ostracoderm and plants are different groups that complement one another, in that there are many ostracoderm in the sea and only a few on land (terrestrial snails) and there are many plants on land and only a few in the sea (algae). Tunicates and sea anemones also resemble plants in that they are sedentary, yet they resemble animals in being fleshy. Some plants, such as the onions, generate by budding clusters, and ostracoderms generate in a similar manner. As evidence that all ostracoderms form spontaneously Aristotle noted that they come into being on the sides of boats when frothy mud putrefies (he was probably thinking of fouling barnacles). Similarly, oysters too can appear from nothing. Thus, once when a naval armament cast anchor at Rhodes, the sailors threw a number of clay vessels out into the sea; after some time, when mud had collected around them, oysters were found inside them. (Balme (1962) suggested that for Aristotle, in this case, the female is presented by the water and earth which the climate concocts into the right condition; and the male principle by that part which is enclosed and causes motion.) In some ostracoderm, there are structures that seem to be eggs, but these are only a sign of the generally good condition of the animal (say, an oyster), just like fat in the blood-possessing groups, and they contribute nothing to generation. Many of the bivalves lack the ability to move, to such an extent that it is difficult to determine whether they are animal or vegetable: The oyster *Ostrea* cements itself to boulders and resembles plants in its sessile, motionless, mode of life. The fan mussel *Pinna* dwells in fine sands and secretes long, delicate, golden byssus threads, which secure it by sticking onto stones that are in the sand. In this manner, to use Aristotle's words, the fan mussel *Pinna* "is rooted to a particular spot". (We today know that the fan mussel usually buries itself almost completely, with only the rear margins of its valves protruding a little above the surface, to filter plankton from the water. If a passing predator bites off these protruding valves, the fan mussel responds by rapidly shrinking downwards, deep into the lower parts of its valves. When the danger has passed, it extends upwards again and repairs its damaged valves by secreting new shell matter. Fan mussels resemble plants also in that after being "grazed to its roots" in this manner, they regenerate and grow rapidly once again.) In addition, the mussel *Mytilus* secretes strong byssus rope-like threads, and it too is largely sedentary on rocks, on the wave-beaten shores of the intertidal. Furthermore, among mussels, Aristotle wrote that small individuals grow alongside large individuals. The immoveable bivalves live by absorbing food from a fluid that percolates through the sea and infiltrates their body (*HA* 8: 3). The sea anemone, with its mouth in the upper centre of its body (*Actinia;* Aristotle's term: *Acelaphe*), feeds upon any small fish that may cross its way. Aristotle felt that this mouth resembles that of an oyster, which also has a mouth that is above (Aristotle may here have been thinking of the incurrent aperture of some bivalve). He argued that the sea anemone resembles the internal part of the oyster and further, it makes use of the rock, as the oyster does of its shell. To his explanations as to why bivalves resemble plants, Aristotle added that in all of them, "as in plants, the head is down below; this is because they uptake their food from below, as plants take it up by their roots". Turning from bivalves to such marine snails as the whelks and purplefish, Aristotle claimed that they spawn "combs" but not real semen (he failed to see that these masses consist of a multitude of fertilized eggs). Alongside one such spontaneously generated whelk, a number of smaller individuals bud off because of concoction: when something encloses water and warm air, a frothy bubble arises, from which living creatures may form. In this manner ostracoderms may form and, as there is plenty of earthy matter in the sea, it hardens around them and creates a protective shell. Land and freshwater ostracoderm include only snails, which are not as diverse as marine ones, because the nature of seawater is more "plastic" than earth; because it contains more matter, and because it is warmer than freshwater. Though fleshy as animals, whatever sensation ostracoderms have, is very dull. Snails are the only group of ostracoderms that have been seen copulating—but "it has never been sufficiently observed whether their generation is the result of copulation or not". At this point, Aristotle, very uncharacteristically, seems to have been putting concept before fact. He had observed that generation is the result of copulation in all other sexually reproducing animals but he believed all ostracoderms generated asexually, and he would not allow the fact that snails copulated to spoil this concept. Among the invertebrates, to conclude, those that lack vitality and movement and in which spontaneous generation is abundant are lowermost: sponges, sea squirts, bivalves and snails are beneath insects, octopus and squid.

Above his ostracoderm come the insects and relatives (his *Entoma*), many of which are "complete" animals because they grow, generate sexually, feed, move about and sense the environment. At this point, Aristotle introduced another scale of animal perfection: In general, the less the offspring of an animal group resemble the parent, the less it is "perfect". Many insects produce offspring that do not resemble their parents; they generate spontaneously, forming worm-like creatures. Aristotle used the term *Scolex*, ancient Greek *skolex* "worm". Today this term refers to the structure at the front end of a tapeworm (Cestoda), a ribbon-like parasitic worm of the mammal digestive tract. The adult tapeworm, which may attain a length of 15 m, has a pear-shaped scolex—a grasping knob armed with hooks and suckers by which it attaches itself to a host; a short, unsegmented neck with high regenerative capacity; and a string of hundreds of segments, which bud off from the neck. Tapeworms absorb nutrients directly from the host's gut, as they have no alimentary system. Mature segments are full of eggs and they leave the host into the environment, where they infect other organisms. Humans are subject to infection by tapeworms if they eat undercooked meat; and tapeworm prevalence may, today, exceed 12%, in some populations (Eom and Rim 2001). The seemingly out-of-nothing appearance of a tapeworm in a digestive tract could perhaps have suggested, to Aristotle, spontaneous generation. He suggested that some insects and ostracoderms do not develop out of a part of something (as do animals from eggs)—they grow out of nothing, like a scolex.

An egg, something undeveloped that has not yet acquired its own form, differs from a scolex in that only part of it develops into the embryo, the rest being nutriment for it, but the whole of a scolex develops into the whole of the young animal. Aristotle claimed that flies produce worms (he was referring to what we call maggots); these eventually become "eggs" (what we call pupae) and only from these "eggs" do the adults finally emerge. Maggots differ considerably from the adult flies that gave birth to them, and they themselves are not able to give birth to offspring that resemble the parent fly. Aristotle here erred in not identifying the real insect eggs, from which maggots emerge, and in thinking that pupae were eggs. He described the grub as turning into an egg (namely, a pupa). Aristotle added that these insects resemble plants also in that after we slice them up, they can survive. However, whereas two plants can develop from one single plant cut into two, this is not so among insects. This statement makes sense if we understand Aristotle's "insect" as referring to earthworms as well: If cut behind the clitellum, which contains the reproductive elements, the head of an earthworm may survive and regenerate its tail, but the tail will not grow a new head and it will die. Aristotle also claimed that in some insects, the offspring, in general, resemble the parent: locusts, cicads, spiders, wasps and ants.

Aristotle further claimed that in general, those animals that move about actively, whether by swimming, flying or walking, also generate sexually, whether they are red- blooded or non-blooded animals. Accordingly, as uppermost among the bloodless groups he placed the fast-swimming cephalopods *Malakia* and fast-running crustaceans *Malakostraka*, both of which generate sexually; yet they lay imperfect eggs.

Higher up the ladder, and above all bloodless animals, come all blood-containing groups (our vertebrates) in which egglaying versus livebearing, combined with the degree of development reached by the offspring at the time of extrusion from the parent's body, determines the rung of the group. The lowermost of Aristotle's vertebrate rungs contained his fishes, which are lower than other vertebrates as they lack a lung (Aristotle, therefore, believed that they do not breathe), as they lay imperfect eggs in that they lack a hard shell, as these eggs grow after being laid and because they have only one colour. Fish eggs complete their growth outside the mother's body because they are so very numerous that they must be spawned quickly, even if small, so as to make room for more eggs in the mother's body. Aristotle wrote that some small fishes originate not from ova or semen, they appear spontaneously in mud, in sand or in sea foam. The loach is derived from another fish, the small and inferior goby. These originate both in the sea and in rivers where putrid matter is abundant. Some people assert that in fishes, the female conceives by swallowing the semen of the male and even fishermen repeat this simple tale. Aristotle commented that these people are unaware that conception in the mouth is impossible because the mouth leads to the intestine, not to the uterus. This point is interesting because mouth brooding (not mouth-fertilization) does indeed occur in certain cichlid fishes, but these do not reach Greece. In the Levant in these fish, after courtship, the female spawns, the male fertilizes the eggs and then he collects them in his mouth, brooding them there until they hatch. Furthermore, among the maternal mouth-brooding cichlids it is quite common for the male to fertilize the eggs only once they are in the female's mouth. In all cases, mouth brooding protects the eggs until they hatch and the fry becomes free swimming. Observations in the Levant on cichlid mouth brooding might perhaps have reached Greece. Eels perplexed Aristotle, because they live in freshwater, but return to spawn in the ocean, in the Sargasso Sea. As European anglers had never caught young eels, their life cycle was a mystery for people of the Ancient World. Aristotle speculated that they are generated in the sea, along riverbanks and in ponds, places in which the heat causes much putridity. They generate neither from sexual intercourse nor from ova but from earthworms that, he believed, form out of mud, growing from the "guts of wet soil" rather than through sexual reproduction. Only in the eighteenth century did naturalists demonstrate that eels are a kind of fish.

Among Fishes, Aristotle faced two major difficulties in constructing his ladder of nature. He generalized that most fishes lay "imperfect" eggs, but among the cartilaginous fishes, many produce live offspring rather than eggs and thus do not fit into the general scheme. Aristotle explained their production of live young as being due to a deficiency of heat. Sharks are "of a cold nature" and therefore they produce eggs that are not hot enough to harden the outer surface of the egg into a shell; they must therefore retain their eggs for protection within the parent's body until they develop into live young. Aristotle also found it difficult to place dolphins and whales on this ladder. He assigned them to the Fishes because they resemble fish in their overall shape, in lacking fur, in lacking well-defined feet and in having internal testes. Yet, he noted, they resemble terrestrial mammals in being warm-blooded, live-bearing, lactating and that they breathe air by lungs.

Mentioning these two difficulties but putting them aside, Aristotle argued that frogs are higher than fishes as they have lungs, yet like fishes, they reproduce by laying eggs that are incomplete as they lack a hard shell, and because the hatchlings (tadpoles) grow in the water and differ considerably from the adult.

A still higher rung is that of the oviparous quadrupeds (our reptiles), whose egg is more "perfect" than that of frogs, as it is laid within a somewhat hard, though flexible shell. The reptile egg is higher than the egg of fishes in that it does not grow after being laid. Air passes through the shell and into the body of the developing embryo, evidence that respiration exists for the creation of a spirit *pneuma,* a mystic vital force essential for life among the blooded animals who live out of water. Without respiration, there would be no way for air to heat up while cooling the heart and transforming into spirit. Aristotle was aware that one kind within the oviparous quadrupeds, the vipers, are viviparous, and thus do not fit into his general scheme; this is so, he claimed, because they are "of a cold nature".

Next, come the Birds, which lay an egg that is "perfect", in that it has a hard shell; accordingly, the egg does not grow in bulk after being laid. The eggs of all birds are alike in that they have a hard shell, and that inside, they are two-coloured: white on the periphery, and yellow in the centre. The shell is made of earthy material, in micro-habitats where heat squeezes out the moisture; consequently, for this to happen, sites in which birds lay eggs must be hot. Birds are involved in procuring food for their young, but after having reared them just briefly, the adults leave them to manage on their own. Ground-dwelling birds lay large clutches because they can more easily obtain their food, compared to birds that are more on the wing and find their food with difficulty, such as predatory birds. Birds that lay many eggs have a shorter lifespan, just like certain plants that bear too many fruits, and that wither away after their fruit has ripened, for lack of food for themselves; they consume all their reserves on making fruit and then become exhausted. Aristotle's text on birds suggests that he was aware of what we, in modern times, consider a K-strategy versus an R-strategy in reproduction.

Among red-blooded animals, those with lungs are hotter than those that do not, since the purpose of lungs, according to Aristotle, is to moderate excessive heat. The highest group of red-blooded animals are the terrestrial mammals, as they have enough vital heat to produce offspring that resemble the parents. Since the embryonic terrestrial mammal is already potentially an animal but still an imperfect one, it must obtain its nourishment from elsewhere. It accordingly makes use of the uterus and of the mother, as a plant does of the earth, to get nourishment until it reaches perfection to the point of now being an animal potentially locomotive. The female provides matter, the male the form and the principle of motion. Terrestrial mammals form a placenta, and their young are born alive as creatures that perfectly resemble their parents (*GA* 2: 1). The young feed on their mother's milk, and further, many mammals keep their offspring with them for a long period even after they stop lactating. (As he had not detected the mammalian ovum, Aristotle regarded the embryo as the direct product of copulation).

Humans are at the pinnacle of all terrestrial mammals, as they alone have the unique ability of reasonable thought, they alone possess speech, and, of all living beings, they alone partake of the divine, or at any rate partake of it in a fuller measure than the rest. Heat tends to make the body erect; thus, terrestrial mammals are more erect than reptiles, and humans are the most erect of all animals. Men are on a higher rung than Women because they have heat that is more vital. Aristotle, in these ideas, was broadly following the Greek physician Hippocrates (c.460–c.370 BCE), who believed in an association between temperature and the soul.

Beyond Mankind, Aristotle believed that the eternal, heavenly entities in our universe are better than the mortal entities, as they contain no evil. They serve as standards for whatever is good in our world. In the lower, sub-lunar world, mortals seek to imitate the heavenly immortals ("reach for the stars"), and one way of achieving this is by generation, because by doing so the species (*eidos*) gains eternal existence. Each individual, because of their ability to create similar descendants, is a link in an infinite chain of individuals that have always existed and always will exist, and this continuum of an infinitely repeating process ensures eternity. Individuals are mortal, yet the species lives on forever. In this manner, mortal creatures can share in the good traits of the heavenly world, which extends above the moon. This is why there shall always be species of men, of animals and of plants.

Aristotle's eternal, orderly Great Ladder of Nature, thus organized all things in the natural world, both living and non-living, in a chain that represents their degree of perfection—a series of stages in which modes of generation entwine with the presence of blood, with the ability to sense the environment, to move about, to think rationally, and to speak. Aristotle did not consider this Chain of Being in any context of evolutionary change. Species were, he believed, timeless in origin; they would never die out and never change—they are permanent, fixed and immutable. The notion of the gradual change of a species, in any modern sense, did not occur to him. Aristotle's Great Ladder of Nature, his *scala natura,* was thus:

Eternal	**Stars** **Red blood animals**	
Viviparous, reason, speech	Humankind:	Man
		Woman
Viviparous, much parental care	Land mammals	
Oviparous, egg perfect, little parental care	Birds	
Oviparous, egg almost perfect	Reptiles[1]	
	Fishes: dolphin, whale[2]	
	Fishes: sharks & rays[3]	
	Fishes	
	Bloodless animals	
Oviparous, highly mobile, sexual generation	Malacia (cephalopods)	
Oviparous, highly mobile, sexual generation	Malacostraca (crustacea)	
Vermiparous	Insects[4]	
Spontaneous generation, limited movement, dull sensation	Ostracoderma (molluscs etc)	
Spontaneous generation, no movement, little sensation	Zoophytes	
Spontaneous generation, no movement, no sensation	**Plants**	
No generation, no growth	**Inanimate matter**	

[1]Vipers, however, brood eggs inside the mother's body, from which they hatch viviparously.
[2]Marine mammals, however, are live-bearing and lactate.
[3]Some cartilaginous fishes brood eggs inside the mother's body, from which they hatch viviparously.
[4]Some insects generate spontaneously.

For Aristotle, animal classification, inserted into the Great Ladder of Nature, was thus evidence for a grand order in the world. This idea of his, of a *scala naturae*, would accompany zoological thought for over two thousand years.

1.2.5 Life Is Born: Male, Female, Generation and Embryonic Development

A block of marble matter has the potential to assume whatever form a sculptor gives it, a seed has the potential to grow into a large living tree, and a tiny embryo has the potential to grow into a huge elephant. Every creature strives to fulfil itself, to realize its inherent potential (in his language, to "actualize"). Actualization, the principle of development from an original state, exists within all plants and animals. It is a process whereby an embryo becomes a baby, who becomes a child, who becomes an adult, who becomes a grandfather. This intrinsic actualization involves the development of Form. Once a creature has reached maturity, its form will no longer be a potentiality; it will exist as a fully-fledged actuality. The growth of an organism is thus a process directed toward an end (*telos*), and the mature functioning organism is the fully actualized form. In a species, the actuality of all individuals is identical Matter, is indefinite and, in itself, it lacks order; form organizes it. Everything that is produced is something produced from something, and by something. Therefore, there is a first mover, one that already exists actually: The world is eternal (*Metaphysics* 9: 8).

During generation (and from here on, I am relying mainly on Aristotle's *Generation of Animals*), each parent supplies a different element. Aristotle defined a male as that which generates in another, and a female as that which generates in itself. He elaborated that these terms sometimes also apply to the entire world, by naming Earth's *mother* as being female, and calling the Sun *father* as being a male and initiating generation. The influence of the Sun triggers plant generation on Earth, and thus the Sun generates in another, whereas Earth generates in itself. Next, Aristotle stated that the female provides the Matter from which the offspring will develop. By itself, this matter is usually passive and not capable of causing a change in form. In mammals, the Matter from which the potential embryo forms concocts in the menstrual fluid (the existence of the mammalian ovum was then unknown; it was discovered only some twenty-one centuries ago). Beyond mammals, Aristotle suggested that Matter occurs in the egg. The male, in his semen, produces the force that activates this Matter and bestows its future Form on the offspring; in addition, he bestows sensory ability on the embryo. Semen concocts from the residue of blood. The semen of every species operates at a specific vital heat—too much heat or too little, and the developing embryo will be deformed. However, Form cannot exist without Matter. Hence, an animal is a concrete entity made up of both Matter and Form. The semen does not become part of the female secretion, but the Form of the product comes from it by means of the semen, which nature uses as a tool much like a carpenter and his saw, or a sculptor's chisel. Aristotle thus viewed Matter as pure potentiality, with new parts developing by a gradual shift from potentiality to actuality. He thought of Matter and Form in terms of a seed germinating on a strip of earth: The earth provides matter for the growing plant, whereas the seed determines its form, which is unique compared to another germinating seed of a different species, germinating on that same strip of earth. As evidence for the broad occurrence, in nature, of activating semen versus passive matter, Aristotle noted that mammalian females who do not menstruate are not fertile. Menstruating females, also, are fertile only when the semen of the male enables the development of shape in the embryo. Then, some birds lay eggs even before they have copulated, but these are never fertile (he named them "wind eggs"). Finally, the male fish spreads its sperm over eggs laid by the female, but only those eggs touched by the male semen become fertile and develop into embryos. The female always provides the material, the male is that which fashions it. Also, while the body is from the female, the soul is from the male.

In animals, males and females are separate. This is so because the male, with his capability of activating the process of embryogenesis, of determining the embryo's shape and of transferring sensory ability to the offspring, is a more divine and better creature than the female, who merely contributes passive matter, and whose blood is of lesser quality. The two sexes persist as separate entities because He, who is better and more divine, should well separate from She, who is less good. The male-female distinction is a global principle and occurs in nature not only in reproductive processes. In curdling milk, for example, the milk is the passive female matter, and the fig juice added to it is the active male element, which initiates the change. The active non-contributing male versus the passive matter-contributing female is also relevant to human artisanship.

As described by Aristotle, when a carpenter makes a new bed, the fact that the carpenter does not become part of the bed's matter parallels the fact that in nature, the male does not become part of the matter of his offspring. The carpenter designs and builds the shape of the bed, but no part of him, or his tools, will become part of the bed—they merely transmit the transition of creation, from creator to created. As to the wood, it will not move itself—the carpenter's art must act on it. Likewise, for a new generation, the semen must act on the menstrual blood, and the seeds must act on the earth. In the early stages of embryonic development, nature only traces out the organs in outline; later on, it works in the details (Wilson 1995). Aristotle believed that insects do not emit sperm; they therefore remain united in copulation for a very long time, until the male forms the embryo.

However, matter and form are not enough—shaped matter is not a living animal, indeed, a dead corpse has the same matter and shape as a living body, and yet it is not a man, and similarly, the bronze sculpture of a hand is not a hand, except in name. The function that an organ performs is what determines it; an eye, for instance, determines when it can see. When a thing cannot do so, it is that organ only in name, like a dead eye or one made of stone. In Aristotle's biology, two mysterious vital forces, the Soul (in Greek *psyche*, Latin *anima*) and Spirit (*pneuma*), are necessary to make matter and form into a living entity. The Soul, comparable to the element of which the stars consist, is a set of activities that distinguish living creatures from non-living ones. It is a mysterious vital principle that enables such activities as nutrition, growth, reproduction, movement, sensation and knowing. Aristotle believed that the Soul resides in the heart, a point that we shall soon elaborate. Not all living organisms have and perform the full set of activities of the soul, and accordingly, Aristotle distinguished between three main levels of being alive. The first is the level of life common to all living organisms, namely to plants, animals and humans: the activities of nutrition, growth and generation. Above this basic level comes a second level of life, restricted to animals, which includes, in addition to the first level, the abilities of movement, sensing and effective response to stimuli. It develops later than the nutritive soul. If this sensitive soul is not present, an animal will be no better than a corpse. It is the male who has the power of making the sensitive soul, and therefore, a female cannot generate an animal from herself alone. The female is merely a mutilated male, and her catamenia are impure semen, in that they do not contain the principle of the soul.

Birds may lay wind-eggs (sterile eggs), however, and this is evidence that the female can generate unaided, up to a certain point. Females, therefore, do have some soul—the low, nutritive soul that exists in animals and plants alike; and, in addition, a weak, sensitive soul. If there is any group of animals that is all-female and has no separate males from it, then the females may generate young without copulation. Aristotle observed no such case, but he noted that in the fish *erythrinos,* only females have been seen. (We, today, identify Aristotle's *erythrinos* with the common pandora *Pagellus erythrinus* (Sparidae), common in the Mediterranean. Today we know that it is a sequential hermaphrodite, being female for the first 2 years of its life and male during its third year.) Aristotle further elaborated that eels and some kinds of mullet are neither male nor female.

Above these activities of animal life, humans alone have also a third, even higher kind of soul, which, in addition to the plant and animal souls, is also capable of conscious planning, and of deliberately rational action. This "higher" activity cannot take place without the lower activities. Aristotle believed that each creature has a single soul, which is capable of three levels of function, depending on the group to which it belongs. He did not think of the soul in terms of later Christianity, as an immortal power that returns to God after death. (In the centuries after Aristotle, philosophers would call these three Aristotelian levels of life the vegetative, sensitive, and intellectual souls.)

The soul acts via the Spirit *pneuma*. The male semen is a foamy broth consisting of water and Spirit. The Spirit consists of a special matter that is beyond the four elements that make up everything on Earth. It is analogous to the *quintessence,* the divine fifth element, which the stars consist of. The innate heat of the Spirit has the ability to generate and, in this, it resembles the Sun, which is also composed of *quintessence,* and that generates, whereas Fire, the sub-lunar element, cannot. The presence of innate heat is what essentially defines living things. It is responsible for all physiological activities, from nutrition to cognition. The higher up an animal is in the scale of nature, the more heat it possesses and the more honourable its soul. After copulation, the Spirit does not become part of the embryo's matter, and yet it forms part of the embryo in that it enters all its tissues as an entity around which matter envelopes and organizes itself, thereby creating both embryonic shape and the material basis for the soul. The Spirit is the immediate instrument of the Soul, and it is through it that the Soul expresses itself. It serves as the vehicle of the innate heat; the channel that conveys and transmits sensations; and the instrument by which, during movement, the Soul moves limbs. Every faculty of the soul communicates and actualizes in the body through the activity of the *pneuma*.

Aristotle considered the heart as the key organ in animal embryonic development because his observations led him to believe that it is the first organ to form, in blooded animals. Furthermore, of all organs, life in the heart fails last of all. It is, therefore, the main organ governing nutrition, sensation, and other faculties. Furthermore, for Aristotle, the heart was also the seat of unified sensing, and since it is sensing that distinguishes the animal from the plant, this organ must take form first. Our perceptions of sight, sound, and smell form a unified perceptual awareness of the world. We do not have several distinct types of consciousness, one of touch, another of sight, another of sound, another of smell. Rather, one unified organ is the seat of this consciousness. We would expect to find this organ located near the centre of the animal, for it would then be almost equally close to every part of the body. Turning from this rational inference to empirical investigation, we do indeed find such an organ—the heart. Since sensations require heat, we would expect that the primary source of sensation would also have to be the source of heat. This would also have to be the source of blood, for it is from the source of heat that blood derives its warmth and fluidity. Studies in embryonic development provide further evidence of the primary importance of the heart: it is the first organ to develop. Beyond this empirical observation, Aristotle added theoretical reasoning regarding why the heart is the first organ to develop. Just as a young animal that has left its parents must be able to manage for itself, so the embryo must control its own development, once sufficiently separated from its parents to establish itself as a distinct entity. Thus, in birds and in viviparous quadrupeds, the heart appears first. The heart is the origin of growth, which is the first activity and the very beginning of the embryo. The heart also regulates the orderly differentiation of body parts; hence, it must develop first.

Aristotle held that actualization of the potential form of a plant or animal proceeds stepwise during development. For example, the human embryo would successively actualize the functions of his hierarchy of souls—vegetative, animal or sensitive, and rational—reflecting the hierarchy of the order of natural things, first plants, then animals, and finally man. Accordingly, the human embryo first passes through the state of a plant, rooted in its mother's womb by the umbilical cord. Later, the animal soul comes into action, and the embryo acquires sensitivity and the capacity for movement. The rational soul comes into action only after birth, which is why the young baby resembles animals in its behaviour. Aristotle further noted that during embryo formation, the character of a higher taxonomic group (what we would rank as a class, for example), is assumed before that of a lower group (what we would rank as only a family, genus or species), Thus, an embryo first becomes a terrestrial mammal, and only then a horse (or a human). In his example, the embryo assumes the character that makes it an animal, not a plant, before it assumes the specific character that makes it a man or a horse or some other animal. Therefore, it must acquire its sensitive soul, which marks it as an animal, before it is specialized as some specific animal. For example (and using our modern terminology), an early stage of a chick embryo is recognizable as a member of the phylum Vertebrata, but not of any particular vertebrate sub-group; later, as structural specialization continues, it is recognizable as a member of the class Aves, and finally as a member of a particular species, *Gallus domesticus*. This statement of Aristotle comes remarkably close to the nineteenth-century Von Baer's Law of embryology, that structures formed early in development are more widely distributed among groups of organisms than structures arising later in development.

Aristotle was aware of the fact that in the order of embryonic development of red-blooded animals, the front part of the body is first marked out; later on, the rear part also reaches full size. At first, each organ appears only in outline and only later does it acquire its colour, softness or hardness. He suggested that in this, the developing embryo resembles a painter producing a work of art, since painters too sketch the animal in outline first, and only later do they put in the colours. Because the source of the sensations is in the heart, this is the first organ to form in the developing embryo, with blood vessels terminating in the brain.

In nature, animals have their season and age for coition; usually, this is the spring, but some copulate in the autumn and winter. Mankind differs in that he is ready to copulate at all seasons, and so are many other animals that associate with man and are pregnant for only a short time (hog, dog, chicken). This is because of the greater warmth in domestic environments and the better food, which is available throughout the year.

Organs form from nutriment. The "most honourable" organs form from the purest and fully concocted nutriment, others from inferior nutriment and from leftover residues. Nature resembles a good householder in that it does not throw away anything from which it is possible to make something useful. Now, in a household, the best food goes to the free men; the inferior, and the residue of the best, goes to the slaves; and the worst goes to the household animals. Similarly, in a developing embryo, flesh and sense organs form from the purest nutriment, but bones, sinews, hair, nails and hoofs form from the residues; hence, these are last to assume their form, for they must wait till there is some residue to spare. In general, there are two kinds of nutriment: one goes to the formation of organs, the other to organ growth.

Aristotle closely studied the embryonic chick, and I present his observations in some detail in this paragraph to give an idea of what the first-ever recorded study of embryogenesis was like. Three days after egg laying, the yolk in the egg comes into being and rises towards the egg's sharp end. It will be here that the heart will appear, as a speck of blood, in the white of the egg. From this sharp end, the chick will eventually hatch. The heart beats as though endowed with life, and veins and arteries with blood inside them branch out from it, as do also nerves—all channels of communication thus originate in the heart. A membrane now envelops the yolk. A bit afterwards, the head is clearly distinguished, and in it, the eyes, considerably protruding. Each of the two blood ducts that lead from the heart next branch, one proceeding towards the integument and the other, like a navel-cord, towards the yolk. Aristotle believed that the life-element of the chick is in the egg white, while nourishment comes from the yolk, through the navel-cord. By the tenth day, the embryonic chick and all its parts are distinctly visible, though the head is still larger than the rest of its body, and the eyes are larger than the head. By this time, other internal organs are also visible. By the twentieth day, the chick is already covered with down, and if the egg is opened and the chick is touched at this time, it moves and chirps. After the twentieth day, the chick breaks the shell and hatches.

Aristotle, in addition to his interest in embryonic development, was also interested in heredity. He sought to explain what determines whether the offspring will be male or female, and why some offspring are similar to one of their parents, or to some distant ancestor. An offspring's sex is determined at the time of conception, and this depends on the male's heart, which creates vital heat. If the amount of heat is sufficient, then the embryo will become a male. Beyond such internal sex-determining factors, there are also external, environmental ones. More males are born among farmyard animals if copulation takes place when the northern dry winds blow, than when the southern, moist ones do. Some shepherds say that even if the animals merely look towards the south or the north, while copulating, this will cause cold or heat, and influence the embryo's sex. The waxing or waning of the moon and the season of the year also influence the sex of the offspring. Yet another factor that determines sex is a specific hereditary entity in the semen of the father. Within this entity, maleness is the strong, dominant character; "beneath it" is the female character, as a hidden potential. If the male character in this entity is strong, then the offspring will be a male; but if it is weak, then the female character surfaces, and then the offspring will be a female.

Another specific hereditary entity in the semen determines resemblance to the father or to a remote ancestor. Within this second entity, resemblance to the father is the strong, dominant character; beneath it is a gradual, continuous series of resemblances to the grandfather, the great-grandfather and so on. If the resemblance to the father is strong, the offspring will resemble the father; the weaker it is, the more the resemblance decreases. If the hereditary entity for resemblance to the father is very weak in the semen, the offspring will resemble the mother. Hence, male offspring usually resemble the father, and female offspring, the mother (in other characters as well as sex). The two hereditary entities, of sex and resemblance, operate independently. Delving further into questions of heredity, Aristotle observed that in general, offspring resemble their parents more than they resemble more remote ancestors; they resemble remote ancestors more than they do any chance persons; and they resemble chance persons more than individuals of some other species. However, offspring also differ from their parents. This, he suggested, is because nature seeks to reproduce the parent in the offspring very exactly, but it fails in different degrees.

Other characteristics that, according to Aristotle, exhibit considerable variation between individuals, especially in man, are eye colour and hair. The eyes in many new-borns are blue but later darken, whereas in others they remain blue throughout life. This suggested to him that the presence of blue eyes in adults is a kind of weakness, in the hereditary entity that determines eye colour.

Some of the variation in human characters is age-related; for example, dark hair, which turns grey with age due to a weakening of the fluid in the brain and a lack of capacity to concoct properly because of a heat deficiency. Further variation in human characters may result from quantitative variation in the four classical qualities (hot, cold, dry, moist). Hair texture is determined by the moisture within the brain, which happens to correspond with environmental moisture: Straight hair develops in northern countries where there is much moisture both in people's brains and in the environment, whereas curly hair develops in Africa, where both people's brains and the surrounding air are dry.

To put these ideas of Aristotle in historical context, his contemporary, Hippocrates of Kos (c. 460–c. 370 BCE), in discussing heredity versus environment and their influences on health, wrote that all parts and functions of the body contribute to heredity, via the semen. Healthy "seed" comes from healthy parts, and diseased seed from diseased parts. Consequently, bald parents have mostly bald children, grey-eyed parents have grey-eyed children, squinting parents have squinting children, and so on. Environment, however, may influence heredity, and as a case, he mentioned the Longheads. Among these people, as soon as a child is born and when his skull is still soft and tender, the adults used to remodel its head with their hands, and force it to increase in length by applying bandages that distorted the roundness of the head and increased its length. As time went on, the process became natural, so the custom no longer exercised compulsion: nature had reinforced custom (in other words, this would be the inheritance of acquired traits. Coming from ancient to modern times, the Mangbetu, an originally Sudanic ethnic tribe living today in Congo, traditionally used to wrap their babies' heads tightly with cloth to give them this distinctive appearance. This practice began dying out in the 1950s).

Aristotle also believed in spontaneous generation, mainly of the bivalves. He placed them as intermediate between plants and animals. As plants, they have no sexes, and one does not generate in another; as animals, they do not bear fruit from themselves. They are formed and generated from a liquid and an earthly concretion. Plants reproduce from seed, from slips, or by budding off alongside, as do onions. Likewise, in mussels, small individuals always grow alongside the original individual, and Aristotle considered this as evidence of spontaneous generation. Rotting matter, concocted with rainwater, is the matter from which spontaneously reproducing animals and plants come into being. They do so in damp earth because there is water in earth, and air in water, and air contains the all-important vital heat. Under these conditions, a frothy, life-containing bubble may arise. In the sea, the earthy matter is present in large quantities, and it is compacted into a shell surrounding the frothy part. In this manner, bivalves form spontaneously on the side of boats when frothy mud putrefies. In places where previously nothing of the kind had existed, a kind of oyster may come into being when the spot turns muddy through want of water. Thus, once a naval armament cast anchor at Rhodes, a number of clay vessels were thrown out into the sea, and after some time, when mud had collected around them, oysters were found in them. As to snails, though people have observed them uniting, they have not sufficiently observed whether snail generation is the result of this union. Aristotle generalized that stationary animals, which resemble plants in that they live attached to some substrate without moving about (whether by walking, swimming or flying), lack the male-female mode of sexual reproduction. Additional spontaneously-generating invertebrates include clothes-moths (our Tineidae) that generate from wool and linen, mites (Acaria) that generate from wood, and fig-wasps (Agaonidae) from figs.

Vertebrates almost never reproduce spontaneously. The product of unfertilized eggs by birds and fishes might resemble spontaneous generation, but Aristotle was careful to insist that such eggs can never develop into an animal without the action of the male: Wherever there is both male and female in a species, the female cannot generate by herself. Only two kinds of fish appear to have no males but consist solely of females, and they regularly generate spontaneously. These fish are the mullet and the eel, and Aristotle suggested that the eel forms from mud.

Several zoological phenomena may have contributed to Aristotle's mistaken belief in spontaneous generation. One is that many marine animals reproduce by transparent planktonic larvae of about a tenth of a millimetre, too tiny to be observed by the naked eye. These larvae settle and metamorphose into tiny adult forms that could well have seemed to have formed out of nothing. Internal parasites, when initially infecting humans and farmyard animals, are far too tiny for the human eye to see, and they too would have seemed to have come out of nowhere. Another phenomenon would be eels, riverine fishes that reproduce in distant seas; upon maturing, they swim all the way up Greek rivers to reproduce, and then they die. Aristotle, not seeing any juvenile eels, could have gotten the impression that adult eels come out of nowhere, thus leading him to believe in spontaneous generation.

Aristotle was involved in many aspects of the animal world: in animal diversity, in the structure and function of the animal body, in the manner in which animals reproduce and in the way they behave in their natural environment. He proposed a system to classify animal diversity, and he further classified the diversity in animal reproduction modes, from bearing live animals, via hard eggshells, to soft shells and further, to spontaneous generation. He distinguished between the functions of the female, who provides the embryo with matter, and the male, who activates the female matter, bestows its future form on the embryo, and provides it with sensory ability. He was the first comprehensive study of biology. Aristotle, the beginning of written biological thought, was a giant, especially when we view him in the context of his times. People of his times believed that salamanders extinguish fire by walking over it; goats breathe through their ears; winged serpents live in Africa; serpents in Libya are so large that they jump into the sea to overturn passing warships; swans, when ending their life, fly out to sea, sing a mournful song, after which they die; and that the goatsucker, a nocturnal bird, sucks goats, resulting in the udder giving no more milk and the goat becoming blind.

Aristotle also erred. He believed that the hen partridge becomes impregnated from the wind that blows from the direction of the cock, or from just hearing his voice, or from his breath, or from him flying over her. He also believed that long-lived persons have one or two lines that extend through the whole palm of the hand, whereas short-lived persons have two lines that do not extend through the whole hand. He brought into Western culture the lore that the lion is liberal, noble, and generous, whereas the fox is cunning and ill-disposed. Also, in his text *On Dreams,* he suggested that mirrors can be stained by the gaze of a menstruating woman.

Yet he also reached many beautiful generalizations concerning the animal kingdom:

- Many terrestrial animals procure their food from the water, but no aquatic or marine animal finds its food on land (we, today, may think of the hippo as an exception).
- Some animals at first inhabit the water, but afterwards change into a different form, and live out of the water (mosquitoes).
- There is no animal that has only wings, in the manner that fish have only fins, all winged animals can also walk.
- No bird with hooked claws is gregarious.
- Only one sense, touch, is common to all animals.
- Every living creature is furnished with moisture and will die if deprived of this moisture.
- Bloodless animals [our invertebrates], but for the cephalopods, are smaller than those which have blood.

- Bloodless animals of the sea reach larger sizes than those on land and in freshwater.
- All creatures, whether they have four feet or many, move on the long axis of their bodies.
- Animals with two horns also have cloven feet and do not possess teeth in both jaws.
- In deer, the horns are solid throughout, and they cast them; in other cloven-footed animals, the horns are hollow.
- No very small animal has horns.
- No animal that has horns also has front teeth in both jaws.
- All high-flying birds have four toes: three forwards, and one backwards; a few birds have two toes forwards and two backwards.
- In terrestrial mammals, males are generally larger than females; in reptiles, females are larger than males.
- In most animals, the voice of the female and of the young is more acute than that of the male.
- All animals with hairs on the body have lashes on the eyelids; birds, and animals with scale-like plates, being hairless, have none. The ostrich forms an exception; though a bird, it has eyelashes.
- Males are stronger than females, and it is in them that weapons occur, exclusively or more fully developed. Stags have horns while does have none, cocks have spurs, females none. Females sometimes possess weapons, but they are less elaborate.
- No animal with many fingers has horns.

Beyond biology, his beautiful opening sentence of his *Metaphysics* that "All men, by nature, desire to know" holds the very essence of all sciences.

1.2.6 Does Chance Govern Our World? Many Other Worlds?

The period after Aristotle's death saw, in the study of biodiversity, the beginning of botany. Theophrastus (ca 371–ca 287 BCE) was a student at the Academy during the closing years of Plato's life, a close friend of Aristotle for most of his life, and his successor as head of the Lyceum. He focused mainly on plants and less on animals, in contrast to Aristotle.

In his *Enquiry into Plants*, he discussed such applied subjects as the effect of climate on tree growth, budding and bearing of fruit, the number of crops per year, and differences between wild and cultivated trees. He proposed a general classification of the vegetable kingdom into trees, shrubs, under-shrubs and herbs (Wightman 1953):

A tree is a plant that springs from the root with a single stem, having knots and several branches, and it cannot easily be uprooted; for instance, olive, fig, or vine.

A shrub is a plant that rises from the root with many branches; for instance, a bramble, Christ's thorn.
An under-shrub is a plant that rises from the root with many stems as well as many branches; for instance, savory-rue.
A herb is a plant that comes up from the root with its leaves, has no main stem, and the seed is borne on the stem; for instance, corn and pot herbs.

Theophrastus emphasized, however, that we should accept these definitions only generally, since in some plants our definitions may overlap; and some, when under cultivation, become different from their essential nature. The mallow may grow tall and become tree-like in merely 6 months, to such an extent that men shall use it as a walking stick. On the other hand, the myrtle turns into a shrub unless pruned. Neither the apple nor the pomegranate nor the pear would seem to be a tree of a single stem, nor indeed any of the trees that have side stems from the roots, but they acquire the character of a tree when the other stems are removed. In some cases, indeed, perhaps we should classify simply by size, or by robustness or longevity. For these reasons, then, we must not make too precise definitions.

Theophrastus suggested that further classification should follow such distinctions as cultivated versus wild, flowering versus flowerless, fruit-bearing versus fruitless and deciduous versus evergreen. To complicate things, many of these traits may vary with geography and with climate. As concerning the distinction between deciduous and evergreen, in the district of Elephantine, vines and figs do not lose their leaves. Thus, the distinction between wild and cultivated seems to be due simply to cultivation, since any plant may be either wild or cultivated. Among undershrubs, he classified some as spiny versus others that are spineless (thistle, eryngo and safflower, versus marjoram, savoury, sage, horehound and balm). He claimed that some plants change their kind as they grow, thus bergamot will turn into mint, and wheat into darnel. Theophrastus believed in reproduction also by spontaneous generation, and he was always concerned with giving it a regular physical cause, such as the action of heat, moisture, air, or corruption (Balme 1962). As to animals, he rejected the belief that frogs and snails come from rain.

Beyond these very broad aspects of plant classification, I am too far from botany to present Theophrastus' contributions to any scientific aspects of botany; for this, I recommend *Flora Unvieled* (Taiz and Taiz 2017). In the coming paragraphs, I present only his keen interest in applied aspects of botany, in the many different uses of wood from different trees. I briefly mention them here to revive the daily life of the Greeks in those ancient days. Wood suitable for shipbuilding can be yielded from pine, silver fir, fir, cedar, and in some cases also from bay, myrtle and beech. Dagger handles and wooden cups can be made of the dark, close-grained wood of the terebinth tree [*Pistacia palaestina*], coming from Syria. Door pivots and sockets, which must be straight, are best when made from elm wood, which does not warp. The light, soft wood of palms is suitable for carving images. Holm oak makes the best charcoal in that it lasts longest, wherefore it is used in silver mines, for the smelting of the ore. Better charcoal comes from trees growing on sunny, dry, north-facing slopes than from those growing on shady, damp, south-facing ones. In iron mines, people use charcoal made of sweet chestnut. Smiths require charcoal of fir rather than of oak as it burns up better into a flame, and is less apt to smoulder. Fire-sticks take fire quicker and better in a northern than in a southern wind. Prickly cedar serves to make statues.

From wood and its many uses, via the various qualities of resins coming from different trees, Theophrastus went on to the medicinal uses of plants. Dittany is useful in childbirth, wolfsbane can be made into poison, and hemlock is a poison that brings a painless death. *Strychnos* causes madness, oleander root in wine makes people gentle and cheerful, and birthwort is good for snakebites. The kind that produces madness, which some call thryoron and some peritton [perhaps the thorn apple *Datura stramonium*] has a white hollow root about a cubit long. One gives three-twentieths of an ounce of this root if the patient is to become merely sportive and to think himself a fine fellow. One gives twice this dose if he is to go mad outright and have delusions; thrice the dose, if he is to be permanently insane; and four times the dose, if the man is to be killed. From here, Theophrastus went on to give a general list of the plants used for perfumes.

As a beautiful piece of folklore, Theophrastus presented statements made by druggists and herb-diggers, which contain many exaggerations. In cutting some roots, one should stand windwards, and in cutting certain plants, one should first anoint oneself with oil since one's body will swell up if one stands the other way. One must stand windward while gathering the fruit of the wild rose; otherwise, there is danger to the eyes. One should gather some roots at night, others by day, and the honeysuckle before the sun strikes them. Peony should be dug up at night because if a man does it in the daytime and is observed by a woodpecker while he is gathering the fruit, he risks the loss of his eyesight; and, if he is cutting the root at that time, he gets a prolapsed rectum. While cutting feverwort, one must beware of the buzzard-hawk, if one wishes to come off unhurt. The fruit of the all-heal is useful against spells and magic arts, in cases of miscarriage, for disorders of the bladder, for the ears, and to strengthen the voice. Its root is used in childbirth and for diseases of women. Its flower is useful in making the iris perfume because of its fragrance. When cutting the all-heal, one should put in the ground, in its place, an offering of all kinds of fruits and a cake. When one is cutting gladwyn, one should put in its place, to pay for it, cakes of meal from spring-sown wheat, and one should cut it with a two-edged sword, first making a circle around it three times; and the piece first cut must be held up in the air, while the rest is being cut. The leaves of the mandrake, when used with a meal, are useful for wounds. One should draw three circles around the mandrake with a sword, and cut it with one's face towards the west; at cutting a second piece one should dance around the plant and say as many things as possible about the mysteries of love (like the direction given about cummin, that one should utter curses at the time of sowing). The root and fruit of hellebore are useful as a purge, and the black kind of hellebore makes wine diuretic. Men also purify horses and sheep with it, at the same time chanting an incantation. Hellebore makes the head heavy, so people cannot go on digging it up for long; therefore, they first eat garlic and take a draught of neat wine therewith. One should also draw a circle around the black hellebore and cut it while facing the east and saying prayers, and one should look out for an eagle, both on the right and on the left; for if the eagle should come near, one may die within a year. From medicines to amulets and charms, folks say that tripolion is useful for every good purpose, wherefore they dig it up by night, camping on the spot. They also say that the snap-dragon produces fair fame, so he who anoints himself with it wins fair fame. The same result follows if he crowns himself with the flower of a plant that Theophrastus named gold-flower, sprinkling over it with a vessel of unfired gold. People use the gold flower in wine against snakebites and for burns.

Epicurus (341–271 B.C.E., Fig. 1.25) was the son of an Athenian citizen whom the authorities had transferred to the island of Samos, as part of the Athenian colonization policy. As a youth, he studied the philosophies of Plato and Democritus and became acquainted with the atom theory. During the civil unrest that followed the death of Alexander, the Athenians were expelled from Samos, and Epicurus moved to Kolophon (on what is now the coast of Turkey). At the age of 32, he began teaching philosophy, first on the island of Lesbos and then on the nearby mainland. At 34, he headed to Athens, where he founded the "Garden", an institute that combined a philosophical community with a school, where he remained for the rest of his life.

Epicurus believed that the world is composed of two basic elements: undifferentiated matter in the form of minute, discrete, solid and indivisible particles, below the threshold of perception; and void, the counterpart of matter, the space where there is no matter (Spiegel 1981). He named the particles "atoms" (in Greek, *atom*

Fig. 1.25 Epicurus. (Tuvia Kurz)

means "the uncuttable") and noted that there is no void within an atom. He explained all natural phenomena in atomic terms—all macroscopic things are conglomerations of atoms, and all events that happen are merely the results of the collisions, rebounding, and entanglements of atoms. Atoms in infinite numbers are continuously moving through space at high speeds, and their chance collisions result in certain combinations and sometimes in other combinations that make the world what it is, at any given moment. Since this ceaseless clashing of an atom with an atom is forever giving rise to new combinations, the life of the universe continues to perpetuate itself inexhaustibly. Things do not arise out of nothing, and they do not vanish into nothing, and therefore it is inconceivable that the atoms had ever come into existence.

Atoms and voids have always existed, and they are eternal and unchanging. However, things consisting of atoms constantly change and decay. This view, of everything being just moving atoms constantly changing the things they form, led Epicurus to his concept of the creation of the world. Our particular world is only a temporary agglomeration of atoms, and furthermore, it is only one of an infinite number of endless worlds, which come into existence and then fall apart and dissolve. In this manner, the whole world alters through time. Nothing remains forever what it was; everything is on the move, and everything undergoes transformation. As one thing withers, decays and dies, another emerges and grows. Broadly speaking, Epicurus inherited this scheme from Democritus. However, Epicurus also argued that the world is unlimited in size, and therefore there must be an unlimited number of atoms, and an unlimited amount of void, since without a limitless amount of void the infinite number of atoms would be unable to move. Epicurus also modified Democritus' theory by suggesting that atoms have weight. When Democritus had explained atomic motion as the result of previous atomic collisions plus the inertia of atoms, Aristotle argued that Democritus did not explain why atoms move, rather than remaining motionless. Epicurus apparently responded to this criticism by claiming that atoms have a natural motion of direction because of their weight, forever vertically downwards, even though there is no bottom to the universe.

Furthermore, in addition to the regular tendency of atoms to move downward, Epicurus suggested that occasionally, and at random times, the atoms move laterally. This lateral movement causes atomic collisions, which form macroscopic bodies that may cluster into organs. If by chance an organ forms, which could fulfil a certain function, then the function sets in. In this manner, Form is created before Function begins: first, the eye is formed, and then it begins to see. In general, Epicurus viewed the world as structured randomly by atoms on the move, which happen to collide. This was very different from Aristotle, who reasoned about each component of the world, about each organ of each creature in nature's vast diversity of animals. In centuries to come, as we shall see further in this book, the question of chance versus purpose as the force structuring the world would continue to occupy Western thought (Spiegel 1981).

Epicurus' atomic theory of the world served as a basis from which he launched his ethical and moral outlook on life. The Gods dwell somewhere out there in space where they eat, drink, and discuss philosophical matters, yet they, too, consist of atoms. They do not try to administer the world, do not trouble themselves with its miseries and are not interested in us.

Having established the physical basis of the world, Epicurus proceeded to explain the nature of the soul that, he argued, also consists of atoms and void. Its atoms are particularly fine, and through them, we experience sensations, pain and pleasure. The soul is mortal, and when the body dies, it too perishes and disintegrates. There can be, consequently, no transmission of souls to other bodies, and no punishment of the soul after death, as Plato had claimed (and as Christianity would claim, some centuries ahead).

The teachings of Epicurus spread throughout the Greek world, and later they spread to Rome as well. One Roman convert to Epicurus' doctrine was Lucretius, another was Marcus Aurelius (Spiegel 1981); we shall meet them later on. In Judaism, followers of Epicurus are mentioned, in the *Mishnah,* as those who shall not reach resurrection, because their philosophy contradicts Divine Providence.

For many centuries, Greece was the centre of intellectual activity in the Western world. After the disintegration of Alexander's brief empire, intellectual activity crossed the Mediterranean to Alexandria, the all-Hellenic, Greek-speaking city, which Alexander had founded in Egypt. The Macedonian general Ptolemy became king of Egypt, declared himself Pharaoh, and with noteworthy modesty added the title Soter ("saviour") to his name. Under the wealthy Ptolemy dynasty, Alexandria soon surpassed Athens as the cultural centre of the Greek world. Fortunately for science, Ptolemy I Soter (367 BC–283 BC) was a former student of Aristotle and was deeply saturated with his ideas. It was he, or perhaps his son Ptolemy II "Philadelphus" (309–246 BCE, Fig. 1.26), who sought to set up an encyclopaedic storehouse of knowledge. Accordingly, he built a *"mousaeion"*, a "hall of the muses", the goddesses of wisdom and poetry, in Alexandria (this is the source of the term for the modern museum). The Alexandria museum was a storehouse of texts, namely a library, and a school of philosophy. Scholars from any country in the world could come to this "Institution of the Muses" where they would receive scholarships, meals and lodging to pursue their research—the museum was supported by

Fig. 1.26 Ptolmey II Philadelphus. (Tuvia Kurz)

the patronage of the royal family. Indeed, the Museum of Alexandria was the first ever state-supported institute of research.

A high priest of the muses was appointed as the religious head of the museum. Academic authority was in the hands of the chief librarian, whose duty was to purchase every book he could lay his hands on. If an owner refused to part with his book, a battery of scribes and copyists in the museum stood ready to copy it and then hand the copy to the owner while they deposited the original in the library. In addition to Greek books, scribes would translate foreign texts into Greek from various languages, including Assyrian, Persian, Hebrew, and Hindi. It was in line with this policy that Ptolemy II ordered 70 of the Jews of Alexandria to translate the Bible into Greek, thus creating the famous Septuagint. It translated several biblical terms into names of fabulous, mythological creatures, including the unicorn and the basilisk, and we shall return to this point later. The library consisted of some 700,000 books and was in the royal sector of the city, in a grand marble building with a roofed walkway, an arcade of seats, and a communal dining room where scholars routinely ate and shared ideas. The complex contained private study rooms, residential quarters, lecture halls, and theatres. More than a thousand scholars lived in the museum at any given time. The museum paid staff members (and scholars), exempted them from taxes, and supplied them with free servants.

The museum brought together the best scholars of the Hellenistic world. The staff included such authorities of the day as Euclid (geometry), Archimedes (engineering), Eratosthenes (who measured the size of the earth and came within 50 miles of its true diameter), Hipparchus (who made the first star map and catalogue), and Herophilus and Erasistratus, who were leading anatomists. For a generation or so, during the reigns of Ptolemy I and Ptolemy II, there was a blaze of knowledge and discovery in Alexandria that the world was not to see for another 18 centuries. Four faculties carried out the teaching: literature, mathematics, astronomy and medicine. Medicine focused mainly on functional human anatomy. Staff conducted dissections of human corpses and of criminals sentenced to have a limb chopped off, in a special room devoted to the study of anatomy. Whereas Aristotle still believed that the site of the intellect is in the heart and that the brain functions in cooling the blood by secreting a special liquid, scholars in Alexandria already considered the brain as the central organ of the nervous system (and its fourth chamber as the site of the intellect). They suggested that every organ receives three vessels: an artery, a vein and a nerve (which they believed was hollow like the other two) that contained blood of two types, two *pneuma*. Air taken in by the lungs is changed in the left ventricle of the heart to the first *pneuma*, a "spirit of life", which is carried to all parts of the body. It then becomes a second *pneuma*, the "spirit of the soul", which spreads throughout all parts of the body from the brain through the nervous system (Franklin 1961).

In establishing the museum, the early, enlightened kings of the Ptolemy dynasty founded a higher-education institute like nothing seen before in the ancient world—an institute, which allowed systematic gathering, creation and dissemination of knowledge. The museum and library marked the true beginning of what would become, over the centuries, our universities and other institutes of higher education.

Eventually, feeble, corrupt kings replaced the enlightened, educated, generous kings of the early Ptolemaic dynasty; they neglected matters of the museum just as they neglected their other responsibilities, and the museum began deteriorating. Julius Caesar burned parts of the library when he laid siege on Alexandria. Later, a fire further destroyed that section of Alexandria, which included the museum, probably on the order of Emperor Aurelian (272 CE). Somehow, the museum survived another few decades until a rioting Christian mob led by the bishop of Alexandria sacked and burnt it, the greatest manuscript collection of ancient times, around 391 CE.

Ptolemy Philadelphus most probably also founded one of the earliest large-scale zoos, one that housed many animals from foreign countries (Bodson 1998). Ancient Greece harboured many introduced species from Africa and Asia. These included monkeys, cheetahs, domestic cats, camels, tigers, elephants, cocks, guineafowl, pheasants and peacocks. Eventually, some of these became acclimatized (domestic cat, cock, guineafowl, pheasant, peacock), whereas others entered the Greek tradition through ancient records: antelope, gazelle, hippopotamus, hyaena, rhinoceros, dromedary, panther and giraffe. Information on African and Asian animals had reached Greece already in the Bronze Age, when carved objects were imported from Egypt and Asia, and wall paintings depicted such foreign animals as baboons, antelopes, monkeys and felines. Cheetahs, from Africa or Asia, were painted on Greek vases in the mid-sixth century, domestic cats imported from Egypt were represented on monuments and vase paintings as early as 500 BCE, pheasants were bred and reared by the mid-fifth century, cocks arrived in Greece during the Bronze Age. Shells from the Red Sea, ostrich eggs, hippopotamus teeth and antelope horns were all excavated from the seventh-century strata of the Samian sanctuary of the goddess Hera. Herodotus (c. 484–c. 425 BC), who travelled through Egypt, wrote of the crocodile that it was known to the Ionians, who had extended to it their name for "lizard": *crocodeilos*. His chapter on the crocodile covers its zoology (anatomy, environment, breeding behaviour, cooperation with the plover); anthropology (its sacred status to the dwellers of Thebes and Elephantine); and techniques of crocodile hunting, in areas where it was not sacred. An early Greek description of a giraffe comes from the second century BCE: "In the country of the Trogodytes, there is also found the animal which the Greeks call camelopard, an animal that, like its name, has in a certain sense a composite nature. For it has the spotted coat of a leopard and is the size of a camel, and it is very fast, and its neck is so long that it obtains its food from the tops of trees". The peacock probably arrived in Greece in the second half of the fifth century, where its history in mainland Greece is linked to an Athenian aristocrat by the name of Pyrilampes, who was Plato's maternal great-uncle and father-in-law, Pericles' close friend, and ambassador to Asian rulers. Pyrilampes brought the birds from Persia to Athens, where he initiated their breeding. He succeeded so well that he was suspected of using his birds to suborn Athenian ladies for Pericles. By the end of the fifth century, the peacock was still uncommon and all the more admired, since its natural beauty, intensified by the attractiveness of its oriental origin, excited public interest. Both men and women came from all over the country to pay a visit (for a fee) to the peacock aviary, maintained by Pyrilampes' son Demos. Many people were eager to buy eggs and to start breeding peacocks on their own, even though this was an expensive endeavour, due to the permanent security guards who had to be housed in the breeding area to protect the birds from being stolen. By the end of the fifth century, a pair of peacocks was valued at 1000 drachmas (the average daily wage of a worker was 1 drachma). By the fourth century, however, peacocks had become common birds and their cost had decreased. Despite the bird's bad reputation for so-called pride and vanity, it continued to be the object of admiration and a meaningful sign of wealth and prestige. The peacock's beauty surpassed that of all other foreign animals, and the spreading of its long feathers was a spectacle the ancient Greeks never stopped marvelling at (Bodson 1998).

On special occasions, foreign animals were offered to kings as tribute. Alexander the Great, while on his way to India, was presented by local kings and chiefs with Indigenous breeds of domestic animals and wild species: dromedaries, Indian elephants, Indian hounds with immense physical strength and bold fighting abilities, Indian horses, sheep and cattle. Furthermore, Indian tribes gave him lions and tigers, thereby giving him symbolic value of power, strength, courage, and endurance. Coming now back to Ptolemy II Philadelphus, he was much famed for the interest (perhaps influenced by Alexander) he took in animals. This was extensively evidenced in his elephant hunting, in the private collection of animals he housed in the royal garden and outbuildings, and the great animal parades that took place publicly, exhibiting hundreds of domestic and wild animals gathered from all over the known world. One of his most extraordinary animals was a giant python *Python sebae,* 13 metres long, trapped in the upper Nile valley and taken to Alexandria, where it was exhibited to the public. All these exotic animals had to be maintained somehow, sometimes for years, and this suggests that Ptolemy Philadelphus kept a large zoo (Bodson 1998).

1.3 Rome: Mainly Pliny

1.3.1 Nature's Purpose

Eventually, Rome became the capital of the ancient world. It was here, in Rome, that one of the earliest encyclopaedias on the natural history of the world was written. Gaius Plinius Secundus (23–79 CE), better known by his Anglicised name Pliny the Elder, fulfilled a succession of military duties as a commander in the infantry, cavalry (Fig. 1.27) and navy and, in addition, civilian duties as governor of several imperial provinces of the Roman Empire. All these took him far and wide into Germany, Spain, Africa, Italy, and probably to France and Belgium as well. In these countries, besides his busy career as a commander and as an imperial administrator, he devoted his time both to observing nature and to scholarly activities: He studied and investigated local geography, nature, peoples and their handicrafts. A number of notes on Rome's society in his encyclopaedia, such as the spectacle of the Empress Lollia Paulina (third wife of Caligula) encrusted with 40,000,000 sesterces-worth of pearls and emeralds at a dinner, suggest that he often moved about in court circles; as do his occasional unfavourable comments on Emperor Nero. He rose to prominence under the powerful patronage of Emperor Vespasian. Pliny met his death when he came too close to the toxic fumes pouring out of the erupting Vesuvius while commanding a fleet of ships, dispatched to help refugees fleeing from Pompeii and Herculaneum.

Pliny, with his broad-minded curiosity about nature, was both a collector of books and a prolific writer. The only book that has survived, of the many he wrote, is *Inquiries into the Natural World* (*Historia Naturalis;* I am closely following Rackham's translation). He adored knowledge, and *Historia Naturalis* contained all the knowledge he could gather. In his words, the book is "an assemblage of twenty thousand facts gathered from books written by one hundred authors" (actually, his lists cover over four hundred authors). A man of remarkably broad horizons, his one-man encyclopaedia encompasses astronomy, geography, botany, medicine, agriculture, zoology, geology, mineralogy and many other fields. Indeed, *Historia Naturalis* purported to cover the entire field of ancient knowledge, based on the best authorities then available. Its sources included some material from his personal memories, but mainly material from other sources, which he collected on an ongoing basis by using one slave as a reader and another as a writer, to take down the notes he dictated, whether while taking his bath, while dining, or while resting. During the winters, he would even furnish the writing slave with gloves and long sleeves, so his hand would not stiffen with cold and slow down his writing (Beagon 2002, 2014).

Fig. 1.27 Part of a piece of a harness bearing an inscription formed from punched dots: PLINIO PRAEF EQ; i.e., Plinio praefecto equitum, "Pliny, prefect of cavalry". This piece may have been issued to every man in Pliny's unit. The figure is the bust of the emperor. Circa 50 CE. Found in Germany, this horse badge is now in the British Museum (presented here with permission). (Tuvia Kurz)

Pliny considered the world as an immense eternal being, never created and that will never perish, with the Sun as its soul and mind. Earth is at the centre of the world, its shape is that of a perfect sphere, it revolves around itself once in 24 h, and consists of the four elements. Just as the sky belongs to the gods, so Earth belongs to men. Earth is always beneficial to Man, and nothing that she generates is against him. Nature is the great benefactor, on whose works we must look with wonder and gratitude. Nothing in nature is born without some deep inner cause, and it is our task to try to discover the beneficial purpose for which each thing has been created. It is Nature that puts healing powers, even in plants that men spurn, like those with thorns. Such is Nature's providence that she makes them prickly and nasty to look at, in order to put off browsing animals and birds and preserve them as remedies for Man.

Nature, with its many plants and animals, is so highly organized that all events that occur fit within a coherent, well-structured scheme. Even the smallest of animals, such as fleas, display body perfection. Their organs of sense are in the very best place on the body for them to function, as are their wings, legs and feet; and with what genius did Nature provide them with a sharp weapon, pointed both for piercing the skin, and tube-shaped for sucking. Using "Nature" as a term of such divine significance that it could read as a synonym of an adored god, he wrote of the "divine power of Nature, which is diffused through everything"; and of "the world of Nature, or in other words – Life". Pliny also presented Nature as the balancing force in the animal world, which prevents any one animal, which happens to have some exceptionally advantageous traits, from spreading excessively. Regarding snakes, for example, on one hand, Nature has bestowed poison on the snake, but on the other, dim eyesight, and war to death with the mongoose. Nature, the most glorious of all sovereigns on Earth, is so powerful that she can change some animals from one kind to another.

Nature, thoughtfully designed for humans to use, deliberately supplies everything that men need. Perhaps not all men, but definitely the Romans. Nature is there to serve man, and our task is to discover the beneficial purpose of each element. Accordingly, Pliny described components of nature not just for themselves, but also in the context of their role in Roman life, and how the Roman can apply these components to his service. He wrote that some things may appear to be useless, but it is only a matter of time before people will discover their uses. The centrality of man was a major aspect of not only Greek and Roman philosophy but also of the biblical outlook on life: God had placed Adam to rule the world (*Genesis*). This suggests that either Pliny was aware of the first chapter in the Old Testament, written centuries before him or that this was a widespread belief in the ancient world.

As the purpose of Nature is to benefit man, then the ways man uses Nature are an essential part of the subject matter. Pliny, the educated but practical Roman statesman-scholar, sought to direct man towards the best use of Nature. His chapters on plants focused on their medicinal value, those on minerals described their uses in architecture, sculpture and jewellery, those on animals mentioned, where relevant, their many uses in domestic households, and for human benefit in general. For example, he elaborated on marine snails as the source of the precious purple dye, for the textiles of the rich, and on marine bivalves as the source of precious pearls. Pliny thus combined Roman practicality with a comprehensiveness of scope, and these two aspects are prominent features in his *Historia Naturalis*. In line with this applied aspect, he also noted when each animal was first presented in Rome. By doing so, he familiarized his reader with a Roman awareness of the animal, as if telling him that the specific animal he is reading about is not a mythical creature, it is something actually observable, and thus the text is relevant.

Nature is the sum of all its parts, and knowing these parts is an end in itself. In presenting his ambitious project of describing everything that exists, Pliny took his broad outlook of the world mainly from Aristotle, who so heavily influenced him that he openly stated that he cites the authority of Aristotle and that he intends to follow his lead on various subjects. Indeed, in zoological parts of his encyclopaedia, he takes much of the information, nearly word-by-word, from Aristotle's works (even to the point of repeating Aristotle's error that the upper jaw of crocodiles is mobile). He apparently intended to update Aristotle's facts and to bring him to his Roman readership; he did not intend to rethink or build upon the philosophical foundations established by Aristotle or to advance any other theoretical thought. He just wished to expand upon the facts, to extend his curiosity, to learn how vast and extensive natural history could be. He completely ignored Epicurus' view that the world consists of atoms. As to the possibility that all things consist that had collided by chance, Pliny responded that nature is not blind, and not accidental; it displays a purpose, in making everything adapted to man's needs. Those who claim that there are a countless number of worlds are mad.

As nature exists solely for the benefit of man, whose task is to exploit the many resources nature has so generously put at his disposal, one aspect of the approach was that plants and animals offer medicines. Pliny believed that in every habitat, in deserts as in forests, Mother Nature has distributed remedies for healing us. Both animals and remedies are part of that one, big, Nature-for-Man plan.

The beetle we call the "Spanish fly", *cantharis* in Pliny's terms, is just one of Pliny's many medicines of animal origin, and I here describe it in some detail, to give the reader a taste of Roman medicine. *Cantharis* beetles were sometimes so numerous that swarms of them devastated crops (Scarborough 1997). Pliny elaborated the one effective measure against such invasions: have a menstruating woman walk through the fields, and then "caterpillars, worms, beetles, and other vermin will fall to the ground". To Pliny, this illustrated the "incredible power of women", especially when menstruating, a folk tradition linked with magic and witchcraft in many early cultures. To prepare a cantharis salve, one should put the beetles into an earthen pot not lined with pitch (as a normal wine jug would be), its mouth closed with a piece of cloth, and hang it over boiling vinegar and salt; after the steam has suffocated the beetles, they can be stored. Before applying the salve, the skin must first be prepared with soda, since the application is caustic, and one must take care not to cause deep sores; one should apply the salve with the addition of gall, mice dung, hellebore, and pepper. Prepared cantharis was very valuable, and Pliny tells of Cato the Younger (95–46 BCE) being brought to trial for auctioning off a king's royal store of the drug for 60,000 sesterces. Pliny recognized *cantharis* as medically both harmful and beneficial. He wrote that *cantharis* in a drink is poisonous and causes sharp pains in the bladder, and adds by way of illustration that a friend of Emperor Nero died when a doctor gave him a drink of cantharis for treatment of "leeches". He went on to say that *cantharis* mixed with grape juice and sheep or goat suet is helpful against skin diseases when applied externally. Furthermore, *cantharis* cauterizes the flesh and forms scabs and hence is very efficacious for leprous sores. It is also a diuretic, and Hippocrates used it for dropsy. *Cantharis* was also a beneficial antidote against the touch of a salamander, which the Romans believed to be very poisonous. In addition to its poisonous effects, medicine derived from *cantharis* irritates the genitals as it passes through the body, resulting in increased blood flow that can mimic the engorgement occurring during sexual arousal. Pliny did not place it on his list of aphrodisiacs, but later Roman sources claimed that the Roman empress Livia (58 BCE–29 CE) drugged members of the imperial family with cantharidin through their food, and used the committed sexual indiscretions that followed to blackmail them. (Fifteen centuries later, Madame de Montespan would slip Spanish fly into the food of Louis XIV, to secure his lust for her.) Nowadays, cantharidin is prepared from dried specimens of the beetle *Lytta vesicatoria* (Meloidae; Fig. 1.28), and it serves in modern veterinary and human medicine, for the treatment of benign epithelial growths, including warts (Scarborough 1977; Prasad 2012).

Pliny classified medicines for man mainly according to their animal origin. There are remedies derived from Man himself, and from farm animals (camel, cattle, goat, sheep, horse, pig, goose), whether from their bodies or their products (milk, cheese, butter, cream, fat, grease, suet, bone marrow, gall, blood). There are remedies from terrestrial animals (elephant, lion, camel, hyena, crocodile, chameleon, skink, hippo, lynx), as from the aquatic ones (sponge, electric ray, sea hare, corals, beaver, tortoise, frog, crab, oyster), indeed water itself has many healing qualities.

Not only Man. Pliny believed that many animals know how to treat their ailments by using medicines to cure themselves. The hippopotamus, when it has overeaten, comes ashore to places of recently cut rushes, seeks an extremely sharp stalk, and squeezes its body down onto it, wounding a vein in its leg and thus letting blood, unburden its body, which would otherwise be susceptible to disease; having finished, the hippo dresses the wound with mud. Stags, when wounded by an arrow,

Fig. 1.28 Spanish fly *Lytta vesicatoria*. (Tuvia Kurz)

feed upon dittany and thereby eject the arrow; and when bitten by a spider, they cure themselves by eating crabs. Hinds use hartwort as a purge, for an easier delivery. Bears, having accidentally swallowed the fruit of mandrake, lick up ants, and loosen their bowels upon awakening from winter dormancy; they devour a plant, the wake-robin. Elephants, having accidentally swallowed a chameleon, feed on wild olives as a remedy. Weasels, when wounded by mice they have hunted, cure themselves with rue. Panthers cure themselves from the poison that hunters put into bait by feeding on human excrement. Ibises use their curved beak to purge themselves. Swallows use Celandine to cure the sore eyes of their chicks (*Celandine* comes from the Greek *chelidṓn*, "swallow". Ancient writers claimed that this flower blooms when the swallows come, and fades when they leave.) Storks, when sick, drug themselves with marjoram. Snakes, when emerging from hibernation, have obscure sight. They anoint and revive their eyes by rubbing themselves against a fennel plant. Certain large snakes quench their spring nausea with the juice of wild lettuce. Lizards, when bitten by snakes, heal themselves with a special herb; tortoises eat ox-grass to restore strength after snakebites. Indeed, so very many animals have made so many medicinal discoveries that these remedies are bound to be useful for man as well. Medicine was thus a factor that increased interest in animal diversity.

Beyond presenting the many uses of animals in medicine, Pliny also discussed applied aspects of the breeding and exploitation of farmyard animals. Geese should be overfed to make their liver grow; the removed liver should be soaked in milk and sweetened with honey, and one should pluck goose feathers twice a year. One should feed chickens with foodstuffs soaked in milk, and manage sitting hens by specific rules. One should also exploit birds for their song. Indeed, he reported, a beautifully singing nightingale once fetched 600,000 sesterces, as a present to Emperor Claudius' consort Agrippina (the worth, in those days, of about 1,200,000 loaves of bread, or 3,000,000 litres of olive oil, or 1,800,000 litres of beer). Pliny discussed also other aspects of the animal economy, noting that an ass was once purchased for 400,000 sesterces (about 3200 pounds of gold), to serve for breeding mules.

The approach that Nature has created everything for the benefit of Man may, sometimes, be problematic: if this is so, why does Nature also produce such evil things as poisonous animals and herbs? Pliny's response was that poison is, in fact, beneficial to man and out of compassion for us, as it provides us with a means of a quick, painless and easy suicide. A kind thought.

Pliny was one of the most influential biologists in antiquity, second only to Aristotle. His massive compilation of knowledge on the natural world was widely read, from antiquity through the Medieval Ages and into the Renaissance, as a vital source of useful scientific knowledge. Physicians edited the medical sections of his *Natural History* to produce the *Medicina Plinii* (Pliny's Medicine), which first appeared in the fourth century with a selection of medical recipes, and was repeatedly copied in centuries to come. His importance as a practical source of scientific information began to decline only in the fifteenth century. In 1492, a professor of medicine at Ferrara compared Pliny's text with that of Dioscorides, uncovering mistakes that Pliny had made in his understanding of Greek sources, mistakes that could have disastrous consequences for those who followed his prescriptions. Who was Dioscorides?

Pedanius Dioscorides (40–90 CE) was a Roman physician, pharmacist and botanist of Greek descent, from Asia Minor. As a military physician travelling the vast empire with the Roman legions, he was constantly on the lookout for medicinal materials. He wrote a voluminous treatise (in Greek), *Periylis iatrikis*, usually known by the name of its Latin translation, *de Materia Medica*, "on Medicinal Materials", which was an encyclopaedia of plants, and of medications which could be extracted from them. It described some six hundred plants, a few animals and minerals, and about one thousand medications. The book consisted of five volumes, which do not correspond to any modern classification:

I. Aromatic oils and plants that produce them: cardamom, nard, valerian, cinnamon, pine resin, quince, apple, peach, apricot, lemon and pear.
II. Animals from which medications are concocted: sea urchins, sea horses, bivalves, crabs, scorpions, rays, vipers, cuttlefish and others; and agricultural plants from which medications are prepared: cereals, beetroot, asparagus, garlic and onion.
III-IV. Roots, seeds and various plants.
V. The vine and its products: wine, grapes, raisins; exceptionally strong medications formed from plants; metals such as zinc and iron oxides.

Dioscorides presented each type of plant by its description, its general uses (such as for food) and its uses in medicine. For example, he noted that one kind of poppy has "a long small head and white seeds" and is used for baking bread, whereas another species has "a head turning downwards" and a third species "its head is longer". From here, he moved to pharmacology: The flower causes sleepiness, and physicians use it to treat skin inflammations, coughs, pains in the ear, and eye inflammations. In small doses, it relieves pain, helps in digestion, and encourages sleep; in large doses, it may cause pain, exhaustion, and even death. He then listed various names of poppies familiar to the Romans, and one known to the Egyptians. *De Materica Medica* was thus a treatise in applied botany, a pharmacological encyclopaedia; it provided simple descriptions of many kinds of plants and the medicinal uses of each.

Back to Pliny, in our modern times, not all students of the history of biology admire him. Sirks and Zirkle wrote (1964) that he was credulous, uncritical, almost incapable of discriminating between reasonable and unreasonable, or between important and trivial. They further claimed he was superstitious, sophomoric and dogmatic, that his zoology was fanciful and founded on hearsay, and that he endorsed practically every traveller's tale including queer and impossible monsters resulting from promiscuous hybridization that supposedly routinely occurred in faraway countries. Frank Goodyear (in Wallace-Hadrill 1990) complained that he was catastrophically indiscriminate, a narrow-minded popularizer, a sceptic infected by traditional sentiment, an aspirant to style who can hardly frame a coherent sentence, one who instead of adopting the plain and sober style appropriate to his theme, succumbed to his lust for decorative details which he added to his text to make it more attractive.

This criticism is somewhat unjust. In writing *Historia Naturalis*, Pliny did not blindly accept every single fact noted down by previous scholars. He rejected as false some of the facts he presented, occasionally he presented contrasting opinions without concluding and sometimes, when an opinion on record contradicted his own personal opinion, he said so. His text on bees illustrates these points. He wrote that bees obtain beebread from fig trees; a certain Menecrates says it is obtained from a flower, "but he is the only authority to make this statement". He wrote that wax is made from flowers of all plants except for two specific herbs, but it is a mistake to say that esparto grass is also an exception, because a great deal of the honey obtained in Spain tastes like that plant. Regarding bee reproduction, Pliny wrote that most authorities suggest that the offspring form in the bees' mouth, by blending blossoms of reed and olive. Others, however, think they form by copulation with a single male, the king, who is of exceptional size so as not to grow weary. The rest of the bees accompany him, "as do women accompanying a husband". Pliny added that the production of drones refutes this second view; for why should the same act of union engender standard bees on one hand and drones on the other? He noted that the first opinion seems nearer to the truth, but it does not explain how reproduction via the mouth could form large individuals. Regarding the sting, Pliny commented that certain persons think that once the bees have planted their sting, they die, while others believe that this only occurs if the bee drives the sting so far that some of the gut follows it. These latter bees, however, become drones and do not make honey; it is as though they have lost their power and confidence. Turning to other animals, Pliny declared the fable of men turning into wolves and then back to themselves again as false. He wrote (Book 8) that beavers practice self-amputation of the male organs when beset by danger, as they know that people hunt them for the sake of their secretion; later, however (Book 28), he wrote that a certain Sextius, a very careful inquirer into medical subjects, denies that beavers ever bite off their own testicles when captured. Pliny was thus careful and critical in handling the zoological data available to him. This style of presenting the entire range of opinions and sometimes deciding which is the one to follow, which is not, and when we cannot decide between alternative opinions, occurs in another scholarly treatise of the Roman period, the *Mishnah*. The *Mishnah* is a written collection (third century CE) of the Jewish oral traditions of law, of different rabbis who lived over a period of four to five centuries. Mostly in Hebrew, parts of it are Aramaic. Pliny wrote in this scholarly style of his day. Yes, he made mistakes.

1.3.2 Animal Diversity: Pliny's Classification

Pliny classified all animals into four groups: those on land, in the water, in the air, and insects. He thereby mixed habitat and what we, in modern eyes, would consider a broad natural group. His groups also reflected the four classic elements of earth, water, air and fire, which is obvious in the first three groups. The fourth can be reached by default, whether because of the lightweight of insects, or because of his belief that some insects are indeed, directly generated by fire. He wrote that in the copper foundries of Cyprus, in the middle of the fire, there flies a creature the size of a fly with wings and four legs, a creature that lives when inside the fire but dies when leaving it.

Pliny classified the wild terrestrial vertebrates, with not much "natural" order (Fig. 1.29). The order in which his series of terrestrial animals was presented reflects only loose associations, resulting from some common character or other, which forms a short chain of animals, after which he moved on to another short chain. Thus, after describing elephants and how large snakes prey on them, he moves on to the next group—that of giant snakes. After the camel, which walks by alternately moving both legs on each side, comes the giraffe, which also has this mode of locomotion. After the amphibious beaver comes the amphibious toad, then the amphibious seal. After the porcupine, which he believed hibernated, comes the hibernating bear, then the hibernating marmot, and then the hibernating hedgehog. All domestic mammals are clustered in one sequence: dog, horse, ass, mule (for work) cow, sheep, goat, and pig (for meat and milk). Oddly, his list of domestic animals does not include the cat.

Among aquatic animals, Pliny followed Aristotle's natural groups (Fig. 1.30). At first, he described (what we term) marine mammals (whale, dolphin, seal), then the fishes, in which he separated sharks because of their cartilaginous skeleton. Then come the marine invertebrates, in increasing order of their armour, all in Latin (not Greek) terms. First those without shells, having soft bodies (*"Mollia"*, Aristotle's

Fig. 1.29 Pliny's land vertebrates. (Tuvia Kurz)

Malakya, our cephalopods); then those with a crusty shell (*"Crustacea"*, Aristotle's *Malakostraka*, our crustaceans); and finally, those with hard, calcareous shells (*"Testacea"*, Aristotle's *Ostracoderma*, broadly, our snails, bivalves, and sea urchins). Within the crustaceans, he correctly placed together the prawn, shrimp, lobster, crayfish and crabs; he correctly noted that they moult and that the right claw is larger than the left. Like Aristotle, he erred in grouping sea urchins together with snails and bivalves into "those with hard calcareous shells". He correctly placed scallops, murexes, whelks, venus-shells, mussels and other genera together; he accurately noted that the paddock *Pholas*, which bores into very hard rock, is luminescent, and that a tiny crab dwells inside the pen mussel *Pinna*. Nature's "love of sport", a term he used to describe the great variety in the colour and shape of marine snails and bivalves, bewildered Pliny. He singled out what he considered the Venus shell that sails like a ship which, projecting its hollow portion and setting it to catch the wind, rides the surface of the water: "The scallop gives a leap and soars out of the water, and it also uses its own shell as a boat". Did this description by Pliny inspire Botticelli in his Birth of Venus?

Pliny confirmed (*Natural History* IX) the opinion common in his days, that "everything born in any department of nature exists also in the sea, as well as a number never found elsewhere". In the water, there are creatures resembling sheep and others with the heads of horses, asses, and bulls, animals that emerge from the water to the shore to graze, and then return to the water (was he thinking of hippopotamuses?). In addition to animals, the sea contains various objects found on land

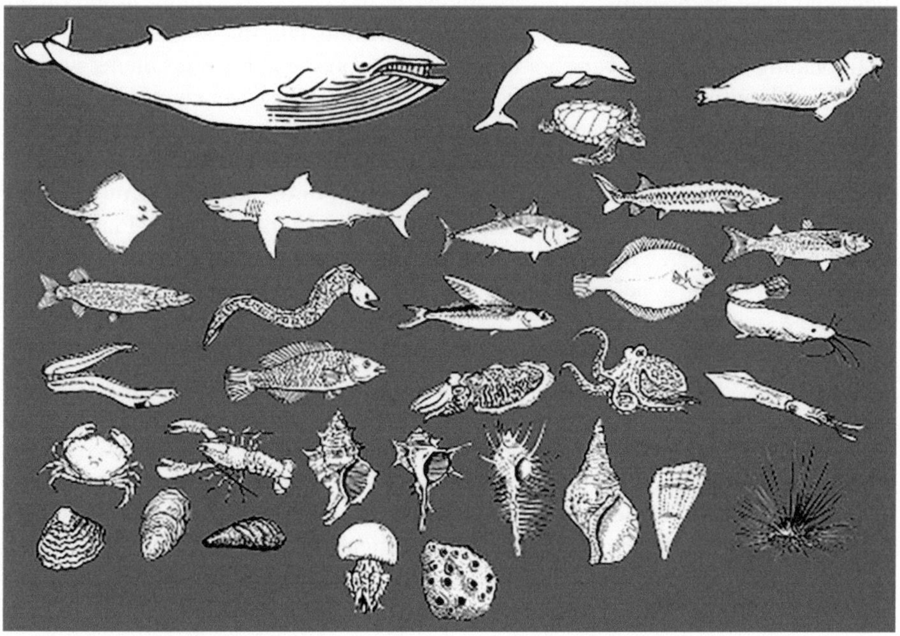

Fig. 1.30 Pliny's aquatic animals. (Tuvia Kurz)

such as swords (swordfish), saws (sawfish) and cucumbers (cucumber-fish), the latter resembling a real cucumber both in colour and scent; which makes it less surprising that in cockle-shells *cocleis*, which are so tiny, there are horse heads.

Some of Pliny's descriptions, though completely incorrect, have been part of Western culture for almost 20 centuries. *Argonauta* is an octopus that dwells near the sea surface. The female forms a nursery for her fertilized eggs by broadening the first arm on each side into a broad membrane. These arms adhere to each other and secrete a chalky container resembling a shell, between them, which the female carries outside her body. The female lays threads of eggs and attaches them to the inner sides of this chalky container where they develop, while she herself sits at the opening of the nursery. Far into the nineteenth century, people still believed that *Argonauta* used its nursery as a boat in which it sailed the oceans, with its two thin arms as sails. This belief dates back to Pliny. What we call *Argonauta*, he named *Nautilus*. He wrote that *Nautilus* twists its two forearms back and spreads a thin membrane between them, which serves as a sail in the breeze, while it uses its other arms underneath as oars, steering with its tail between them like a rudder. Elsewhere, he described *Nautilus* again as a sailing ship: It is a shell with a keel like a boat and a curved stern and bow, which a creature resembling a cuttlefish secretes. In calm weather, the carrier shell strikes the water by dipping its "flippers" like oars, but if the breeze is right, these same flippers are extended to serve as a rudder, and the curves of the shell are spread to the breeze. Another of Pliny's incorrect descriptions that became part of Western culture for almost 20 centuries is of Man fighting giant squids and octopuses. According to Pliny, it is the most savage of animals, killing men who have been shipwrecked, or diving in the water, by coiling around them with its sucker-cups and tearing them asunder with its multiple sucker-cups.

In dealing with flying vertebrates (Fig. 1.31), one of Pliny's problems was the birds' two intuitive outliers: the ostrich, which has feathers yet cannot fly, and the bat, which can fly but has no feathers. Should these two "intermediary creatures" be included within the flying animals, the birds? Pliny coped with this problem by placing the ostrich at the very beginning of his list of birds, separated from all other "standard" birds by a lengthy description of the mythical phoenix. Placed here, within but well separated, he stated that its wings merely assist in running, and he added that it "almost belongs to the group of beasts" in that its talons actually resemble a stag's hooves. Likewise, he placed the bat at the very end of his chapter on birds, just before a series of general chapters concerning generation in such other groups as crocodiles, vipers, man and other terrestrial vertebrates. Pliny correctly stated that the bat is the only viviparous flier, that it has membranes for wings rather than feathers, that it nourishes its young with milk from teats, and that it has a single hipbone, like mammals. In this manner, Pliny coped with the divergence from a bird's characteristic habitat, versus convergence to a bird's characteristic habitat.

Having dealt with the ostrich and the mythical phoenix (we shall return to it later), Pliny now came to the birds with which the Romans were familiar. Throughout his survey of birds, Pliny never ceased to put Man in the centre: birds are interesting insofar as they serve man. He declared that he accepted Aristotle's suggestion to classify birds by the morphology of their feet, into birds of prey with hooked talons,

Fig. 1.31 Pliny's vertebrates of the air. (Tuvia Kurz)

and waterfowl with webbed feet. In his list, the birds following the phoenix might, at first glance, indeed present Aristotle's "birds of prey": First the diurnal raptors (eagle, vulture, hawk, kite), last the nocturnal raptors (owls); and between these two groups the cuckoo, which he believed, following the tale mentioned by Aristotle, seasonally changes into a hawk. However, Pliny also inserted diurnal and nocturnal raptors between the crow and raven, and he placed the woodpecker immediately after the nocturnal raptors.

Now, these are almost precisely the birds serving in Roman augury: the eagle, vulture, hawk, crow, raven, owl, woodpecker and hen. Augury was the practice of taking signs from nature and interpreting them to determine the will of the gods: whether they approve or disapprove of a proposed action, such as a wedding, a war, or the erection of a temple. Romans believed that certain birds, in special signs such as their mode and direction of flight, bring the opinion of the gods to man. Certain birds, termed the *Alites*, served as signs in augury by their flight. These were mainly eagles, hawks, vultures, and osprey; and they also included the woodpecker, ever since those distant days when the king of Latium gave his name to this bird (a witch had changed Picus, father of Latinus, into a woodpecker, because he had slighted her love). Other birds, termed the *Oscines,* bring the word of the gods by their calls. These were the raven, owl, woodpecker and hen. The persistent croak of a crow predicted bad luck, as did some special croaks of a raven. The screech of an eagle owl was an extremely bad omen that one should definitely not ignore. Thus, during the reign of Emperor Claudius, 1 day in the year 43 CE, because an eagle owl entered a shrine on the Capitol, the Romans had the whole city purified. The way farmyard fowls ate grain also provided omens: If they ate greedily, letting grain drop on the ground in a certain manner, the Romans considered this a good sign. The Romans' heavy reliance on augury emerges even from the legendary founding of Rome: When the twins Romulus and Remus could not agree on the site on which to build their future city, they let birds make the decision. Romulus saw 12 vultures settle on the Palatine hill, Remus saw only six alight upon Aventine—and the site of the city was thus determined. (Incidentally, Romulus decided on the site after he had murdered Remus…) This mystic religious approach to birds, rooted so deeply in Roman culture, seems to have shaped Pliny's outlook regarding the ranking of birds: Man is at the centre, and birds offer signs for his use. Pliny first presented the birds of augury. He was aware of Aristotle's classification of birds by the morphology of their feet, birds of prey with hooked talons, and waterfowl with webbed feet, but he ignored it.

His second group consisted of two subgroups, "songbirds" and "plumage birds", and he started with the latter, which are also larger. Among the plumage birds, first on his list was the peacock, sacred to the goddess Juno as the bird that draws her chariot. Then came the farmyard cock, another bird that tells the future, in this case by the manner in which the chicks feed. It was followed by the goose, sacred to the Romans because, in 390 BCE, when the Gauls had taken Rome, the cackling of the geese in the temple of Juno awakened the Romans; this was just in time to save the Capitol from the enemy that was storming it. For this reason, feeding the sacred geese in the Capitol was a public duty of the Roman censors. At this point, the text leaves the issue of bird classification and wades off into issues of avian natural history. The first of these is seasonal movement, whether by long-distance migration (as in crane, stork, geese, swan, quail, blackbird, thrush, starling), by short-distance local migration to nearby places (swallow), or by no migration at all, the bird merely retreating into some hiding place for a few days (turtle dove, pigeon, wood pigeon, night owls). Next came a note about birds found in some regions but not in others (jay, magpie), then a text on changes in bird plumage and song (blackbird, thrush, nightingale, wheatear, oriole, ibis) and then one on nesting and reproduction. This, in short, was the extent of Pliny's classification (and natural history) of birds.

In describing Insects (actually all terrestrial arthropods, including spiders and scorpions) Pliny followed Aristotle, describing them as minute animals with incisions that encircle their body, which do not breathe, and do not have blood, sinews, bones, cartilage, fat or flesh, and which have no internal organs but for the intestine. However, they possess the senses of sight, touch, taste and smell, and some of them also hear. Some insects are ground-dwellers while others are fliers. Some flying insects have two wings, while others have four. Those with a sting in the belly (bees) have four wings, but those with a weapon in the mouth (flies) never have more than two wings. Insect's wings, when torn off, never grow again. Pliny complained that we marvel at elephants' huge shoulders that carry castles, at bulls' powerful necks and the manner in which they fiercely toss their heads, at the rapacity of tigers and the manes of lions, whereas actually, Nature best presents itself in her entirety in her smallest creatures, the insects. There is no further classification; Pliny presents the insect groups one after another in no specific order. He mixed considerable natural history in his description of insects. He wrote most of his text in a matter-of-fact style, but when it came to ants, he enthusiastically remarked about their industry and diligence in their work. Ants, ever so wisely, nibble seeds before they store them away so that they will not germinate in the earth; they split large seeds to get them into their nests, and when the seeds have been dampened by the rain, they bring them out and dry them.

> Since they bring their burdens together from opposite directions and are unknown to one another, certain days are assigned for the market so that they become acquainted. How they flock together on these occasions! How busily they converse, so to speak, with those they meet and press them with questions! We see rocks worn by their passage and a path made by their labours, so that nobody may doubt how much can be accomplished in any matter by even a trifling amount of assiduity!

Of all the insects described by Pliny, bees aroused the most admiration in him, since 'they alone of this group have been created for the sake of man' in that they collect honey and make wax (Fig. 1.32). He believed honey comes from the air and that in spring the leaves of trees are bedewed with honey, whether as the respiration of the sky, or a sort of saliva of the stars or a moisture of the air purging itself. In addition to collecting honey bees also toil, construct hives, have a government, have individual enterprises and collective leaders, and have a system of manners that outstrips that of all other animals. Bewildered, Pliny said "Nature is so mighty a power that out of what is almost a tiny ghost of an animal she has created something incomparable! What sinews and muscles can we match with such efficacy and industry as that of the bees? [...] What men can we rank in rationality with these insects, which unquestionably excel mankind in this, that they recognise only the common interest?"

Observing that drones have no stings, he suggested that this was so because they are imperfect bees, the last made, the incomplete product of bees that are exhausted and discharged from service. They are a late brood and function as servants of the true bees, who consequently order them about and drive them out to work, punishing laggards without mercy. In addition to work, the drones serve the bees also when breeding, as they contribute much to the warmth of the hive. Regarding bee generation, Pliny noted that no one had ever observed their sexual intercourse. According to Pliny, most authorities believed that the bees blended various blossoms in their mouth, and this is where the offspring formed. Others believed it was by copulation with a single exceptionally sized male, the king in each swarm, whom other bees accompany, as women accompanying a husband. However, the production of drones

Fig. 1.32 The honeybee: worker, drone and queen. (Tuvia Kurz)

refutes this view, for why does the same act of union engender some perfect offspring, and others imperfect? Whatever the mode of generation, bees brood their eggs like brooding hens. The hatched offspring resemble white maggots, but the king is honey-coloured and has wings, from the start. The brood matures in about 6 weeks, and they get to work with their mothers under some sort of tuition. A retinue of peers escorts the youthful king. The production of several future kings begins but afterwards, by unanimous vote, they kill the worst of them to prevent the division of the forces. Many of Pliny's erroneous views on bees and their natural history would turn up 15 centuries later, during the late Renaissance and early modern Europe, as we shall see in further parts of this book.

Besides bees, Pliny also said that the ichneumon wasps kill spiders, carry them to their nests, and by incubating them, produce their own kind. (Indeed, we know today that Ichneumonidae injects their eggs into their host's body; after hatching, the larva eats its host alive. The most common hosts are larvae or pupae of butterflies, beetles and flies, but some species also parasitize spiders.) Then, Pliny wrote, there are also nasty "insects", such as scorpions. They are as poisonous as snakes but worse, as they inflict upon the victim a torturing, lingering death lasting 3 days. Their sting is always fatal to girls and, in general, to women, but it is fatal to men only if they are stung in the morning. The poison of a scorpion in Scythia is so strong it will kill even pigs, which are normally immune to such poisons if they plunge into the water. The ashes of the creatures themselves, drunk in wine, are a cure for a human victim. Pliny notes that some scorpions possess wings, perhaps having in mind the scorpion flies Panorpidae in which males have enlarged genitals that resemble the stinger of a scorpion.

Pliny believed that in some instances, one type of animal could change into another one: the dried-up bodies of dead crabs transform into living scorpions, and hawks seasonally change into cuckoos. He followed Aristotle, who had written that the eagle lays three eggs and hatches two of them; as the young ones grow, the mother wearies of feeding them and expels one of them from the nest; a bird of another kind then rears it. Indeed, all birds of prey, when their brood has grown to the point of being able to fly, beat their offspring and expel them from the nest. So much for Aristotle.

Coming back to Pliny, he wrote that the lammergeyer is fond of its young, for whom it fetches food to its nest. A kind bird, it rears both its own young and those of the eagle; when the eagle ejects its young from the nest, this bird catches them as they fall, and feeds them. In a very strange and not yet deciphered text, Pliny writes: eagle hybrids grow to become ospreys, *haliaëtus*, the "sea eagle" (Fig. 1.33):

> The *haliaëtus* does not have its own genus but is the result of crossbreeding with a wide range of other eagles. The offspring that result has its genus among the *ossifraga* genus, from which the smaller vultures are begotten, and from these the large vultures, which do not breed at all. Some people add a genus of eagle which they call the bearded eagle, but which the Tuscans call *ossifraga*.

These in turn breed large vultures named *ossifraga* ("the bone-breakers", perhaps the bearded vulture *Gyptaeus barbatus*, which feeds on bone marrow (Fig. 1.34). It cracks bones by carrying them in flight and then dropping them from a height onto rocks below, thus smashing them to pieces and exposing the nutritious marrow; hence its name *ossifraga*, the "bone-breaker".)

This idea, that eagles could give birth to a hybrid which gives birth to a pseudo-eagle species which in turn gives rise to a small vulture which gives rise to a big vulture which gives rise to a bearded vulture, is perplexing, to say the least. Pliny attributed this only to the eagle (and other diurnal raptors), the birds so admired and respected by Romans, the messengers of Jupiter. This concept of an eagle of one type being brought up by an eagle of another type might perhaps somehow remotely relate to the Roman custom of child adoption. In ancient Rome, adoption was common. It was neither secretive nor shameful, nor was the adopted boy expected to cut ties to his original family. In Roman law, the power to give children for adoption was one of the recognized powers of the *pater familias*. One famous adopted persona was Augustus Caesar. He was born Gaius Octavius, adopted by his great-uncle Julius Caesar and given the name Gaius Julius Caesar Octavianus. Pliny himself adopted his nephew. The story that in nature one type of eagle adopts another may perhaps reflect this custom.

Fig. 1.33 Osprey, *Pandion haliaetus*. (Tuvia Kurz)

Fig. 1.34 Bearded vulture. (Tuvia Kurz)

Pliny observed that Nature's power is so strong that she can crossbreed different animal groups. It is easy to cross domestic animals with the wild variety: The offspring of such unions are termed "hybrid" (*hybridas*), meaning half wild. Here and there, Pliny referred to such interbreeding, which he might have observed, as between dog and wolf, which he thought of as different species (we today regard them as one). He also presented cases of interbreeding beyond his personal observations. According to Pliny, a lioness crossed with a hyena gives birth to the mythical *corocotta*; the moray eel leaves the sea and crawls onto land to copulate with a snake; the skate [*squatinum*] crosses with the ray [*raiam*], resulting in an offspring named *Rhinobates*. He also related how an eel crosses with a viper. In a somewhat favourable or at least understandable style, Pliny referred to the lion as having such a very strong sexual passion that he also sought out females of other kinds. In his lust for indiscriminately mating with females, sometimes violently and by force, a lion often mates with a pard [*pardi*, unidentified; a cheetah?], concentrating his entire strength on her "chastisement". As a result, she keeps her distance when accompanying him. The offspring is then termed a *leopard*, namely, of a lion and a pard (Sirks and Zirkle 1964). Pliny did not omit the unfortunate consequences of human-animal crosses either and stated (book VI) that a woman once gave birth to an elephant, another once gave birth to a snake, and that in Thessaly a creature was once born, half man, half horse.

The Greeks and the Romans thought that animals in distant countries might lead a weird way of life. Herodotus of the fifth century BCE believed that in north India, giant ants ("bigger than foxes but smaller than dogs") dig up gold-enriched sand, in the process of excavating their burrows. The Indians devise elaborate gold-gathering strategies to steal this gold and, to elude pursuit by these ultra-swift ants, they snatch the sand while the ants are taking a siesta underground, during the heat of the day. Julius Caesar, of the first century BCE, described the *alces* (elk) as having legs without joints, and this prevents it from lying down or from getting up if it falls. It sleeps while leaning against trees, and hunters capture it by loosening the roots, or sawing partially, through favourite elk resting-posts. When the unsuspecting animal composes itself for rest, the tree gives way, with tree and sleeper collapsing to the ground and into the hands of the hunters. By Pliny's time, the elk with no joints had acquired, in addition to its idiosyncratic legs, an upper lip so huge that it had to walk backwards, while grazing, so as not to trip over it.

Beyond these real animals with exaggerated or distorted natural history, Pliny's world also contained fabulous, mythological ones. Far from being just his own personal ideas, belief in fabulous animals was widespread throughout the ancient world, both centuries before him and centuries after. In this section, I present, somewhat in length, the very broad extent to which belief in fabulous animals was widespread in the ancient world, and I begin with the phoenix and the half-mouse, which are also mentioned by non-Roman literature of those times. "Phoenix" is the Greek term for the Egyptian "Bennu", an ancient Egyptian deity that represented the soul of Ra, and was linked with the sun and with rebirth. It may perhaps have been a heron of one species or another; perhaps also, the bird that people identified with Bennu shifted from one species to another, over the centuries. In Egyptian mythology, Bennu flew over the waters of Nun, which existed before creation, landing on a rock and issuing a call that determined the essence of creation. It renewed itself like the sun, but over very long periods. The Egyptians worshipped Bennu at Heliopolis, where the priests taught that Bennu lives for five hundred years, and it then builds its own funerary pyre and sets it alight. The newly born offspring of the previous phoenix rises from the ashes of this fire, carries the ashes to Heliopolis, and deposits them on the temple altar. The Egyptian Bennu was, most probably, the mythological source for the Greek descriptions of the phoenix, in which they likened it to an eagle with red and gold plumage, reminiscent of the sun or flames.

Pliny wrote about the phoenix of Arabia, a mysterious sacred bird, dedicated to the sun. There is never but a single phoenix in the entire world; it is the size of an eagle, has beautiful colours, never feeds, and it lives 660 years. When growing old, it builds itself a nest of cinnamon and frankincense twigs, upon which it lies—and dies. After many days, a worm emerges from its bone marrow, starts growing, and becomes a beautiful young bird that carries the remains of its mother to a faraway island. Here, it conducts a burial ceremony, whereupon the stars in heaven reset in their places, and a new cycle of time begins. Somewhat in line with this, Pliny went on to relate that in Sestos (Gallipoli), there once was a maiden who brought up a young eagle by hand. When the maiden died and her funeral pyre was set afire, the eagle flew into it and it too burned to ashes.

Two centuries after Pliny, the phoenix myth was very much alive and evolving, and Lactantius (250–325 CE, advisor to the Roman emperor Constantine) described it as dwelling inside the sun, in an enchanted grove of happiness. After one thousand years of life, the bird descends to this world, where death reigns. It seeks a secret grove, in a desert, and here it chooses a lofty palm, which reaches up so high that no serpent and no bird of prey can reach it. It builds a nest of many spices on this palm, lies down and dies, whereupon a fire bursts out of its dead body, blazes and burns it to ashes. These ashes unite into a seed, which becomes an egg in which a milky coloured worm arises, at first without limbs but then developing into a perfect creature. It bursts out of the egg like a caterpillar, and feeds on heavenly nectar until it grows fully, after which it embalms the remains of its own previous body. Before returning, with this embalmed body, to its enchanted garden in the sun, it presents itself so that mortals (such as the people of Egypt) can admire it in all its royal splendour. It has a multi-colour plumage of red, yellow and purple ('her tail is extended, varied with yellow metal, in the spots of which mingled purple blushes'), green and rose, with a shining crown on its head. Then it flies into the sun, attended by a crowd of birds that accompany it, on part of its journey. The phoenix is thus born from itself, whether female or male, or neither, or both. Death is Life to it; for it to gain eternal life, it must first die. The sun and its bird are therefore a symbol of the fiery renewal of life after death; the sun dies every night and shines every morning.

The phoenix was also very much present in Jewish lore. The Septuagint (second century BCE) translated into Greek the bird mentioned in Psalms 50 ("*ziz saddai*") into "birds of heaven"; thus, the sentence should read "I know every bird in the mountains, and the birds of the heaven [=phoenix] are mine". Four hundred years later and a hundred years after Pliny, Jewish traditions of the second century CE placed the phoenix near the sun. In the Baruch Apocalypse, Baruch reaches the fifth heaven, where the sun goes forth, and an angel shows him a four-wheel chariot drawn by 40 angels, upon which sits a man wearing a fiery crown. A bird of enormous size, "as large as nine mountains", runs along before the sun, appointed by God to be the guardian of the world. It accompanies the sun and, by spreading its wings, absorbs the sun's fiery rays to prevent all living creatures from burning to death. This bird, named Phoenix, feeds on the manna of heaven and the dew of the earth. The mythical bird sacred to an Egyptian cult of the sun thus metamorphosed, in Judaism, into God's minister, who regulates a hostile sun. Some four hundred years later, Jewish traditions of the early medieval period, in the fifth-sixth centuries, placed the phoenix in the Garden of Eden (*Genesis Rabbah* 19). When Adam and Eve ate from the forbidden fruit, they also offered it to the animals in the garden. Almost all the animals ate from the forbidden fruit, and God therefore punished them to become mortal. One animal, the phoenix bird, did not eat from the forbidden fruit, and thus maintained its eternal life—a fulfilment of God's original plan of the world, which humanity and the other animals had spoiled. The phoenix myth was thus re-contextualized, as a theological explanation of an ancient crux: why also animals have to die, if mortality came to the world because of Adam and Eve's first sin. Death derives from Man's disobedience of God, yet the phoenix remains outside the framework of real time and regular history. Also in this text, the term *chol,* mentioned in in the Book of *Job* (29:18: "then I thought I shall die with my nest and shall multiply my days as the *chol"*) is translated as referring to the phoenix, thus reading then I thought I shall die with my nest and shall multiply my days as the *phoenix*. Some six centuries later, during the High Medieval, Bahya ben Asher of Spain (1255–1350) wrote that the eagle is the king of the birds, and that it flies higher than any other bird. When reaching the age of ten, it flies even higher and approaches the sun. Here it is so hot that it drops itself into the sea, where it loses its old set of feathers and then grows a new one. This is the eagle's custom every 10 years, until it reaches the age of one hundred. It now flies even higher, burns to death and falls into the sea. The fabulous phoenix was thus, in one way or another, very much alive in the minds of people for centuries upon centuries. (In another interpretation, of Wazana 2009, Ziz was a cosmic bird, its legs embedded in the foundations of the earth and its head reaching heaven. Together with two other primaeval giant creatures, Leviathan, the water monster, and Behemoth, the terrestrial beast, Ziz personified the third realm—sky, and together the three represented the entire living environment: water, land and air. Ziz thus reflects Anzu, the lion-faced roaring eagle of ancient Mesopotamian religion, which also had cosmic dimensions.)

To me, the phoenix, definitely as described by Lactantius and perhaps also in its earlier versions, closely resembles a mixture between an impressive raptor and a skin beetle. The eastern imperial eagle *Aquila heliaca* (Fig. 1.35; notice the species Latin name) is a large bird of prey that nests in isolated trees, and Aristotle had written that the eagle is a long-lived bird that flies very high, to have a more extensive view; and because of its high flight, people claim it to be the only bird that resembles the gods. Skin beetles (Dermestidae, up to 12 mm) scavenge on dry animal material, such as skin, hair, feathers, dead insects and natural fibres, and they occur in animal carcasses and in nests. In my opinion, the eagle and the skin beetle could perhaps have been the inspiration for the phoenix. So far as to Bennu, the phoenix.

Pliny wrote that after the flooding Nile recedes, "water mice are found with the work of generative water and earth uncompleted—they are already mice in a part of the body, but the most recent part of their structure is still of earth". Pliny took this description from the Greek historian Diodorus Siculus, who wrote, about a century

Fig. 1.35 Inspirations for the phoenix? An eastern imperial eagle and a dermestid. (Tuvia Kurz)

earlier, that in Egypt the soil at certain times generates mice in large numbers (Bazaapoulou-Kyrkaniddou 2001). Some of these mice are fully formed as far as the breast and front feet are concerned, and are able to move, while the rest of the body is unformed, consisting of only a clod of earth. Many people in the ancient world knew of the half-flesh, half-earth mouse. A passage in the *Mishnah* reads that "A mouse that is half flesh and half earth: if one touches the fleshy portion, he is unclean; if the earthy portion, he is clean". The *Talmud*, compiled in the fifth century, some four centuries after Pliny, in discussing the raising of the righteous dead (resurrection), copes with the problem of how the earth could recreate the flesh of those buried, who, as we know, disintegrates into the earth. One of the sages replied that a man could resurrect from the earth in the way that a new man is created from a drop of semen, which is merely liquid. As evidence from nature, that life can be directly created from earth and water, he added (from Pliny?) that a person that goes in the plain after the rains could see a mouse that is half flesh and half earth, and after some time becomes all flesh—which means that earth can create living flesh. Furthermore, a person who goes to the mountains before the rains will not see any snails, but immediately after the rain, the mountain is full of snails. Hence, rainwater and earth have created snails, and this teaches us that the earth can resurrect the dead. In the High Medieval Era, Maimonides (1135–1204) wrote of the spontaneous generation of the mouse from the dust of the earth, so that it is part flesh and part of mud, yet its entire body is in motion, and this is a well-known fact. Countless persons have told him that they have seen it. This myth brings to mind voles, those rodents with five to ten litters per year, the young reaching maturity in a month. Their populations grow very large within a short time, especially after the rains and when the vegetation begins sprouting, one pair producing a hundred more voles in a year. It brings to mind also those land snails that aestivate by burrowing in the ground, and emerging to the surface with the rains (such as the genus *Sphincterochila*). The natural history of these animals may have inspired the belief in the story of the half-flesh, half-earth mice, which emerge seasonally. Indeed, the Hebrew medieval term for spontaneously generated organisms is "species that are born of putrefaction" (*minim ha-noladim meha-'ipush;* Gaziel 2012).

The fiercest mythological creature was the unicorn, with the body of a horse, the head of a stag, the feet of an elephant, the tail of a pig, and a single black horn 1 m long on the middle of its forehead. No one could catch it alive. The earliest unicorn I am aware of is from a ninth-century BCE archaeological site in Syria, at the springs of the Khabur River (Tell Halaf), which was the capital of an Aramaean principality. A basalt relief from the palace of an Aramaean prince presents what may well be a unicorn (Fig. 1.36).

For the ancient Greeks, the unicorn *Hippoi Monokerata*, was a magnificent horse, completely snow-white, with a single, coloured horn arising from its forehead. They sometimes referred to it as a unicorn-ass, *Onoi Monokerata*. The Greek unicorn myth may well have begun with the description by a certain Greek physician, Ctesias, who served in the court of Darius II, king of Persia, early in the fourth century BCE. Upon returning to Greece, after spending several years in Persia, Ctesias wrote that in India there are wild asses the size of a horse and even larger (McCrinde 1882). Their body is white, their head red, their eyes blue, and on their forehead is a single 50 cm long horn. The base of the horn is white, its middle section black, and its pointed tip is red. People who drink from a goblet made of this horn become immune to stomach pains, epilepsy and even poison. Unicorns are so strong and swift that no other animal can catch up with them (McCrinde 1882). Ctesias' Persian myth about unicorns in India might perhaps relate in some way to some of the Indus Valley seals, of the Bronze Age Indus civilization, of between 2500 and 1900 BCE. These seals generally include an animal representation together with a handful of glyphs, and the animals represented are often bulls, each with a single horn (Fig. 1.37).

Some four centuries later, this tale was absorbed into Roman culture, with variations. Pliny wrote that the mightiest animal of all is the unicorn *Monoceratem*, which has the body of a horse, a head like a deer, feet like an elephant and the tail of a boar. It has a prominent horn on its forehead that is three feet long, and no one can capture it. Philostratus (first-second century CE) repeated these descriptions, and added that because of the outstanding qualities of the horn, only kings were permitted to capture a unicorn and to possess a goblet made of the horn.

Fig. 1.36 Unicorn Khabur River (Tell Halaf), Pergamon Museum, Berlin. (Tuvia Kurz)

The unicorn myth would probably have sunk into oblivion, were it not for the Septuagint translation of the biblical Hebrew term *re'em*, which appears seven times in the bible. The Septuagint translated this term into *monokeras*—thereby securing the unicorn's status as a biblical creature created by God, and anyone who denied the existence of unicorns defied the existence of God. We shall return to the unicorn later in this book, in the part concerning the Dark Ages, to follow the development of its legend.

Whereas the phoenix was a mythical bird and the unicorn a mythical mammal, the basilisk was a mythical reptile—a snake native of Cyrenaica, 30 cm long and with a bright white mark on its head like a sort of crown. Pliny writes that when crawling, it does not move its body forward in manifold coils like other snakes, but advances with the middle section of each coil raised high. It defeats other snakes with its hiss; its breath scorches grass and bursts rocks. Its effect on other animals is so disastrous that once, when a man on horseback killed a basilisk with a spear, the killing power rose from within the basilisk, climbed up the spear and killed both rider and horse. Pliny further noted that the basilisk hisses so loudly that all other serpents flee from it.

Fig. 1.37 A single-horned bull, as in a seal of the Bronze Age Indus civilisation. (Tuvia Kurz)

Other mythic animals Pliny described, whose body was a combination of very different animals, included the Yale, of faraway India. The size of a hippopotamus, with an elephant's tail and a wild boar's jaws, the Yale can move its 45 cm long horns forwards and backwards, like donkey ears. (One thousand four hundred years later, in the Renaissance, Lady Margaret Beaufort, mother of Henry VII, used the mythical Yale in her heraldry.)

Pliny's inventory of mythological animals also included the *Sphynx* with udders on its breast. Then there was *Leucrocota*: The size of an ass, with a stag's haunches, the tail, breast and neck of a lion, a badger's head, a mouth opening back to the ears and ridges of bone in place of rows of teeth; this animal could imitate the voice of human beings. Pliny also described how an eel crossed with a viper and how a lioness hybridized with a panther. Nor did he omit the unfortunate consequences of human-animal crosses, stating that a woman once gave birth to an elephant, another once gave birth to a snake, and that in Thessaly a creature that was half man, half horse was once born. Pliny accepted as real all these mythical animals, alongside Aristotle's Martichore that had a triple row of teeth, the face and ears of a human being, grey eyes, a blood-red colour, a lion's body, and could inflict stings with its tail like a scorpion.

Altogether, Pliny's world of animals did not include too many mythical animals: only about a dozen of his 180-odd animals are imaginary (Beagon 2014). This is considerably more than the few mentioned by Aristotle five centuries earlier, but not too bad in view of the knowledge available in his days. This rather reasonable ratio of fabulous to real animals worsens considerably in Pliny's Chaps. 5, 6 and 7, on humans, where he reported on numerous monstrous human races. Some have an anomalous body that combines human with animal parts, such as centaurs, people with feet like horses, and people with dogs' heads. There were also people with different sexual organs on the same body, and others with feet turned backwards and eight toes on each foot. There were people lacking mouths, or heads, or an eye, or nostrils; or they had only one foot, which they used as a parasol; and people with two pupils in each eye, and people with feet like leather thongs, who crawled instead of walking. He also mentioned the pygmies, who deviated completely from the Roman standard of size and reached a height of only 70 cm, and other people with feet 45 m long, and a tribe consisting of women only, and another tribe consisting of cannibals, who ate human flesh. There were also the Pandae, who live 200 years, and have white hair in their youth that grows black in old age, and a people of women only, who conceive at the age of five and do not live beyond the age of eight. In almost all these cases, Pliny carefully mentioned the source of his knowledge, several of his authors having lived quite a few centuries before his days (Ctesias ca. 400 BCE, Eudoxus ca. 380 BCE, Aristotle 350 BCE, Megasthenes ca. 320 BCE, Agatharchides ca. 150 BCE, Cicero 70 BCE and several others). This suggests that he was quoting them more because a detailed review should contain all data, and not necessarily to take responsibility for the words of these authors. He also compiled an endless list of perfectly normal human tribes, in comparison to which his list of only about 20 human monsters is not that long.

1.3.3 Animal Behaviour: Military, Moral, and More Aspects

Pliny's many years of service as infantry and cavalry commander, and then as governor, may have affected his biological thought, and may have influenced the interpretation both of his direct observations of animal behaviour and of information gleaned from other writers. Soldiers in the Roman armies spent only a fraction of their lives in battle. They spent most of their time on routine military duties such as training, guarding, patrolling, and maintenance of camp and equipment. Some of Pliny's descriptions of the daily life of bees are reminiscent of daily life in a military base. He writes that a guard stands at the gates of the hive, as in the gates of a military camp. At dawn, one bee wakes up all the others with a double or triple buzz, as a sort of bugle-call; then they fly forth in a formation. Inside the hive, some bees build, others polish, others bring up material, and some prepare food from what those coming from the outside bring in. This is so because bees do not feed separately, so that there shall be no inequality of work, or food, or time. They mark the idleness of any who are slack and chastise them, and may even punish them with death. When evening approaches, one bee flies around as though giving the order to take repose, with the same loud buzz with which she woke them, and thus, in the manner of a military camp, they all suddenly become quiet. In constructing the hive, they build homes for the commonality first, and then for the kings. The drones have no stings, and are thus imperfect bees, they are the incomplete product of those that are exhausted and now discharged from service. The perfect bees order them about, and drive them out to the works, punishing laggards without mercy.

Pliny's career as a provincial governor also seems to bear on his biological thought, particularly in his interpretation of the daily life of bees. They build large and splendid separate palaces at the bottom of the hive, for those who are to be their rulers. They produce several kings, but afterwards, by unanimous vote, they kill the worst of them so that they would not split the forces. The king does not have a sting; he does not need one because the commons surround him in marvellous obedience. When he goes in procession, the whole swarm accompanies him, masses around him and protects him by not allowing anyone to see him. During the rest of the time, while the "people" are engaged in labour, he himself does the circuit of the works inside, urging the workers on, while he alone is free from duty. Retainers surround him, as constant guardians of his authority. He only issues abroad when the swarm is about to migrate, and when they start, each retainer wants to be next to him, delighting to be seen on duty; when he is tired, they support him by carrying him on their shoulders. Wherever the king alights is the camping place of the whole swarm. The capture of the leader holds up the whole swarm, and when they have lost him, they separate and migrate to other lords; in any case, they are unable to be without a king.

Other descriptions of Pliny are reminiscent of life in a military unit on the move: Beyond augury and its influence on bird classification, Pliny's outlook as a military commander sometimes trickles into his approach to bird behaviour. I do not have the details of how Roman infantry advanced in the field, but some things are classic in infantry practice to this day. An advancing military unit sends out scouts to high places to see the route. The leader marches in front while deputy commanders bring up the rear. When camping at night, the commanders station sentries to keep guard; sentries are bound to fall asleep, for which the commanders punish them; and a wise commander is alert during much of the night, to make rapid decisions in case the camp is suddenly attacked. In a similar manner, Aristotle, as mentioned above, wrote that the highly intelligent cranes fly up high to command an extensive view. They have a leader and patrols. When they settle down, the flock sleeps, while their leader keeps a sharp lookout; when he sees anything of importance, he signals it with a cry. Influenced by both Aristotle and, surely, much of his personal experience, Pliny wrote that migrating cranes fly high to see the route ahead. They choose a leader to fly in front and lead the others, who follow him; they station other birds of the flock at the rear to shout orders and keep the flock together with their cries. At night, while most of the flock is asleep, the cranes station sentries, each holding a stone in the claws of its lifted foot; if a sentry is overcome by drowsiness and falls asleep, the noise of the falling stone discloses his carelessness. Throughout the night, the leader of the group, with his neck kept erect, keeps a lookout to warn of encroaching danger.

Another case brought by Pliny: Quails *Coturnix,* fly with a corncrake *Crex crex* leading the way when migrating over the sea. As a hawk seizes the first bird approaching land, the quails are careful to secure a leader of a different "kind"—as a precaution during those first crucial moments. Upon being attacked, the remaining quails retreat and recruit an escort consisting of birds of another kind—they persuade the tongue-bird [unknown] and other birds to make the journey with them. At first, the charm of travelling lures the tongue-bird to sail on eagerly, but in the course of the flight fatigue sets in, it regrets its decision and deserts. This description again reminds us of the advance of infantry, with the corncrake representing the scouts, and with the concept that if infantry encounters a stronger force, it is wise to retreat and recruit additional forces. Pliny's description of the tongue bird is reminiscent of the recruitment of people to military service, some of whom start with great enthusiasm, but regret and desert after some time.

Today, the weak and the tired are placed in front of military marches. Pliny believed that geese marched on foot all the way to Rome from the northern coasts of Gaul (France or Belgium of today). He wrote that on these marches, the geese that tire are advanced to the front rank, and the rest drive them on by pressing forward from the rear. Another case: Today, a military unit when digging trenches roofs them over and leaves each trench with two openings. Pliny writes that white birds off the coast of Apulia act in a similar manner. When flying, they have two leaders, one leading the column and the other bringing up the rear. When nesting, they hollow out trenches with their beaks, roof them over with a lattice, and then cover them with the earth, which they have previously dug from the trenches. Each trench has two openings, an entrance and an exit; the birds lay their eggs in these trenches.

Some of Pliny's descriptions of battles among animals are reminiscent of combat among soldiers. In face-to-face combat, the soldier wears armour to defend his body; he wards off enemy blows; he waits for the enemy to expose a weak part, and then launches an attack on that weak spot. Similarly, Pliny wrote that the snake, when preparing itself for a fight against the mongoose, repeatedly plunges into the mud and then dries itself in the sun, thus equipping itself with a cuirass of several coats. During the encounter, it raises its tail and renders the blows it receives ineffective by turning away from them, all the while watching for its opportunity. When this comes, it attacks its adversary's throat with its head held sideways.

As another military case, Pliny noted that it is wise to wait patiently until the enemy slackens its guard, and then launch the attack. Similarly, when the mongoose attacks a crocodile, it waits patiently. When a plover comes to clean the crocodile's teeth, the crocodile opens its mouth wide for the plover to enter. When the mongoose notices that the crocodile is overcome by sleep in the midst of this dental treatment, "it darts like a javelin right through the open throat, and gnaws out the crocodile's belly". The Egyptian mongoose is intelligent in that, when it sees a serpent, it will not attack it before it summons other mongooses to help. Then, to protect itself from the bites of the snake, it plasters itself with mud, by first soaking in the river and then rolling on the ground.

Pliny's "military mind" also surfaces when, in describing terrestrial vertebrates, he mentions three cases of the use of the hide of wild terrestrial animals in human service. Two of these are for military purposes: Hides of the hippo supply an impenetrable material for shields and helmets, and the reindeer hide is so hard that armies use it for making cuirasses (the third animal is the hedgehog, whose hide the Romans used in cloth for garments). He further adds that elephants are the most important factor in eastern warfare, scattering the ranks before them and trampling armed soldiers underfoot.

Pliny's naval knowledge, in particular the need for ballast, may have influenced his biological thought regarding animal flight. He writes that, when cranes fly across a sea, they add bulk to themselves with pebbles and sand, which they discard when reaching land. He also wrote that bees flying during a storm steady themselves with the weight of a pebble held in their feet, and some authorities say that it is placed on their shoulders.

Thus, Pliny's military experience seems to have shaped his outlook on certain aspects of animal behaviour.

A major aspect in the *Historia Naturalis* that distinguishes it from the writings of Aristotle is Pliny's tendency to seek moral aspects in the behaviour of certain animals, which reflect the traditional, ancestral values of Roman society. The Roman "ancestral custom" (*mos maiorum*, Way of the Elders) was the unwritten code from which the ancient Romans derived their social norms. As the core concept of Roman traditionalism, the time-honoured principles, behavioural models and social practices affected private, political, and military life. The Roman patriarchal household was hierarchical, as was Roman society. The head of a household held absolute authority over his family, which was an autonomous unit within society, but he was expected to exercise this power with moderation and to act responsibly on behalf of his family. Roman conservatism finds expression in an edict of the censors from 92 BCE: "All new that is done contrary to the usage and the customs of our ancestors seems to be not right". The Way of the Elders was distinct from the written law but complemented it. Traditional Roman values essential to the Way of the Elders included trustworthiness, faithfulness and reliability; inner devotion and dutiful respect towards the gods, homeland, parents and family. It also included the observance and correct performance of rituals; discipline in education and training; dignified self-control with no admission of weakness, steadiness or perseverance; knowledge of what is good, evil, useless, shameful or dishonourable; worth, honour, and esteem.

I found two cases among the animals Pliny describes, which suggest that these Roman values and customs are part of the grand divine natural order of the world. One such an animal is the elephant.

First, for comparison, I present Aristotle's reference to the elephant. Aristotle wrote that the trunk of the elephant is long and strong, and the animal uses it like a hand to draw objects towards it, to take hold of them, and to introduce food into its mouth. It also uproots trees with its trunk, and when walking through water, it uses it to spout water. The elephant is the least hairy of all quadrupeds. Its front legs are much bigger than its hind ones, and it has five somewhat indistinctly defined toes. It has four grinding teeth on either side as well as two tusks, which in the male are large and curved upwards; in the female, they are small and point downwards. The elephant is born with teeth, except for the tusks. Its penis is disproportionately small, compared with the size of the animal; the testicles are internal. The female elephant has two breasts in the region of the axillae, insignificant in size and in no way proportionate to the bulk of the entire body; indeed, they are so insignificant as to be almost invisible. The structure of the gut is such that the animal appears to have four stomachs; its viscera resemble those of a large pig, with a small spleen and a huge liver, which lacks a gall bladder. It usually eats six or seven medimni of fodder in one meal, five medimni of wheat, and drinks 15 Macedonian metretae of water daily. The elephant does not sleep standing, as some suggest; it bends its legs and settles down, falls into a recumbent position on one side or the other, and in this position goes to sleep. When making sounds with its mouth alone (unaided by the trunk), the elephant sounds like the sound of a man who is panting or sighing; but, if it employs its trunk as well, the sound produced is like that of a trumpet. Male elephants become savage during the mating season (spring), fiercely fighting and stabbing opponents with their tusks; the defeated combatant becomes completely cowed and dreads the sound of the victor's voice. The female elephant becomes sexually receptive when 10–15 years old, and the male is sexually capable when 5 to 6 years old. Elephants copulate in isolated places, particularly by riversides; the female squats down and the male mounts her. The male allows an interval of 3 years to elapse after mating with a female: after having impregnated her, he has no intercourse with any female. The period of gestation is 2 years, only one young animal is produced at a time, and the newly born is the size of a 3-month-old calf. Elephants live for two to three hundred years. In short, Aristotle offers a rich spectrum of facts concerning elephant natural history.

Pliny, on the other hand, provides almost no proper zoological details about the elephant, only that their period of gestation is 10 years, and that they live up to three hundred years. Most of his text focuses on a comparison of the elephant and man. The elephant is nearest to man in intelligence; it understands the language of the people in its country, and it obeys their orders, it remembers duties it has been taught, and is pleased by affection and by marks of honour. It possesses virtues such as honesty, wisdom, justice, respect for the stars and reverence for the sun and moon. In the forests of Mauritania, herds of elephants pay respect to the moon when it shines by going down to the river and sprinkling water over themselves in a purification ritual. They do homage to their king by kneeling before him and offering him flowers, and when sent overseas, they refuse to board a ship before the captain promises that they will return. When fording a river, elephants put the smallest individual in front, so as not to wear the river bottom away by the tread of the larger ones, thus increasing the depth of the water. Owing to their modesty, elephants never mate except in private. Mating takes place only 5 days each year, and on the sixth day, they purify themselves by a showery bath in a river before returning to the herd; adultery is unknown. As people often hunt elephants for their ivory, the elephants place those with the smallest tusks in front when surrounded by hunters, to give the impression that the booty will be so poor that it would not be worth the effort to hunt them. When about to be caught, they break their tusks by dashing them against a tree, and they then ransom themselves for the desired booty. An elephant chancing upon a man wandering across its track is so good-tempered and peaceful that it will actually show him the way. But when it notices a man's foot-print before seeing the man himself, it stops in fear of an ambush, sniffs the scent, trumpets angrily, digs the footprint up and passes it on to the next elephant, and so on to the last of all, and then the column turns round and forms a battle line. Otherwise, the animal's natural gentleness towards those not as strong as itself is so great that if it finds itself among a flock of sheep, it will lift those that come in its way, so as not to trample them. The natural enemies of elephants are mainly serpents (who happen to be the natural enemies of man). In those distant countries where elephants dwell, the serpents are so large that they easily encircle the elephant in their coils, forming a twisted knot on top. As the strangled elephant falls, it crushes the snake coiled around it, so that in this duel, both combatants die together. Large snakes also poke their heads right up the elephant's nostrils, hindering its breath and lacerating its flesh; they submerge themselves in rivers and lie in wait for elephants, who come down to drink. They then rise, bite into the elephant's ear, and suck up all the elephant's blood. In this case, too, however, as the dying elephant collapses, it crushes the serpent, and in this manner, again, both die together. Tame elephants recognize their names when called, and Pliny relates the story of a leading elephant that was called by name and ordered to ford a river; he refused, whereupon the commander ordered another elephant to take the lead and gave this new leader a silver harness, "an elephant's greatest delight". The disgraced elephant preferred death by starvation to humiliation, for elephants have a remarkable sense of shame.

Throughout this text, Pliny thus describes elephants in human terms and attributes them a variety of (Roman) moral qualities. Indeed, their catalogue of virtues makes them almost appear good Roman citizens. According to Pliny, elephants are close to humans both with respect to their faculty of reason and with respect to their moral qualities. Like Man, they possess the ability to think rationally, to behave morally and to use language. They admittedly cannot speak as well as we can, but they can understand human language, and they can "write" words.

This point, of elephant communication without speaking, is carried a step further by a Roman who lived a century after Pliny, Claudius Aelianus (175-c. 235 CE; commonly known as Aelian). In his *On the Nature of Animals* (*De Natura Animalium*), he wrote that though elephants do not have the gift of active speech, they have their own ways of making themselves understood, in particular when they intend to articulate their ethical conscience. Thus, an elephant that was averse to any kind of evil drew the attention of the new wife of his trainer non-verbally to the fact that the latter had killed his wealthy previous wife, to gain access to her money. As soon as the new wife arrived, the elephant led her with its trunk to the grave of the dead woman and uncovered the hidden grave with its tusks (Fogel 2007).

A second case among the animals Pliny describes, which suggests that these Roman values and customs are part of the grand divine natural order of the world, is the lion. Lions also practice traditional Roman customs and values. Of all wild animals, the lion alone is so noble that it shows mercy to its supplicants; it spares persons prostrated in front of it, when enraged, it turns its fury on men rather than women, and it attacks children only when extremely hungry. Indeed, once a woman walking through a forest got away from a herd of lions by saying that she was only a woman, a fugitive, a supplicant to the most generous of all animals, the lord of all. Furthermore, once an assailant wounds a lion, the lion marks his assailant, knows him and finds him; however large the crowd; yet he merely seizes a person who had discharged a weapon at him but had failed to wound him, and flings him to the ground without wounding him. The lion's nobility of spirit can be seen in the way it protects himself by intimidation displays as far as possible; when it makes actual physical encounters, it broadcasts that it is doing so because it is enraged at the disrespect of those who dare to challenge him, and not because it is in any real danger. Its noble spirit can also be seen in cases in which numerous hounds and hunters beset it and it retreats contemptuously, constantly stopping; but once in the forest and out of their sight, it proceeds at top speed, as if aware that the bush conceals its disgrace. Similarly, when pursuing, it advances by leaps and bounds, but when retreating, it does so slowly.

Some of Pliny's descriptions concerning the natural history of birds seem, to our modern eyes, as ever so true, whereas others seem fictitious. He wrote (in the wake of Aristotle), regarding a mother hen saving her chicks, that if a fowler approaches the nest of a partridge, the mother bird runs to his feet. Pretending to be tired or lame, in the middle of a run or a short flight, she suddenly falls as if with a broken wing or damaged foot, and then runs forward again, continuously escaping him just as he is going to catch her, thereby leading him away from the nest—feigning injury. On this same subject, Pliny also wrote that when magpies notice a person observing their nest, they transfer the eggs elsewhere. As their claws are not adapted for grasping and carrying eggs, they place a sprig on top of two eggs at a time, and solder it with glue from their belly. Then, they place their neck under the middle of the sprig to make it balance equally on both sides, and carry it off somewhere else. Now Pliny's description of the partridge is true, while that of the magpie is, of course, not. However, how was Pliny to know?

Elsewhere, Pliny wrote that the male viper inserts its head into the female viper's mouth, and the female is so enraptured with pleasure that she gnaws it off. This description of decapitation is true, but for another animal, the praying mantis. Never having seen an elk, he wrote that its upper lip is exceptionally big and that it walks backwards when grazing, to avoid tripping on it when moving forwards. How was he to know?

Pliny described the remarkable friendship between a goose and a philosopher (Lacydes, head of the Academy at Athens in 241–215 BCE). The goose never left his side day and night, at home, in public and at the baths. Similarly, once a goose fell in love with the girl who played the harp for King Ptolemy Philadelphus, and yet another goose fell in love with a handsome youth in a town in the Gulf of Corinth. This text of Pliny's could well be the earliest written evidence of imprinting, a process in which a young animal (or person) acquires behavioural characteristics from its parent by rapid learning at a critically young age. The young animal learns the characteristics of some stimulus, which is then "imprinted" onto it. Imprinting is indeed obvious in precocial birds such as geese, which imprint on their parents and then follow them around. It is claimed to have been first reported in domestic chickens by the nineteenth-century biologist Douglas Spalding, and later studied extensively by Konrad Lorenz, who, working with geese, demonstrated that incubator-hatched geese would imprint on the first suitable moving stimulus they saw within a critical period of about 15 h after hatching. However, Pliny seems to have reported the phenomenon some 1900 years earlier. If we criticize Pliny for his writing that geckos have the nature of chameleons, in that they live only on dew, how should we consider Francis Bacon, who, living fifteen centuries later, still repeated Pliny's words, and wrote that chameleons feed on dew?

Pliny's writings also contain little treasures of ethnology. He wrote that in the Roman city of Cyrene (present day Libya), there is a law obliging war to be waged on locusts three times a year, the first time by crushing the eggs, then the grubs and finally the fully grown insects; a person who shirks this duty has the penalty of a deserter imposed on him. On the island of Lemnos (northern Aegean), there is a rule determining a specific quantity of locusts each person has to bring to the magistrates. The inhabitants also keep jays for this purpose, which meet and devour the locusts by flying in the opposite direction. People in Syria are commanded by military order to kill locusts, which are, however, an acceptable article of diet with the Parthians. He also wrote that in faraway India, there are locusts three feet long, with legs and thighs that one can use as saws.

Pliny believed that spontaneous generation *sponte natura*, is widespread in Nature. He erroneously believed, like Aristotle centuries before him, that organic matter, inorganic matter, even the wind, might spontaneously conceive a host of animals of different groups and habitats. Among partridges, a mere draft of air from males flying overhead, and often merely the sound of a crowing male, can make the female conceive. A snake can be born from the spinal marrow of a human being. Butterflies are born from dew, which condenses in the sun and shrinks to the size of a millet seed; a small maggot develops from this and becomes a caterpillar that grows, pupates and becomes a butterfly. Fly maggots are born in the flesh of dead bodies, and the clothes-moth breeds in sheep wool. The goat-moth is generated in wood (we today know that caterpillars of the goat moth *C. cossus* feed in trunks of trees, taking three or 4 years to mature). The wild fig tree breeds fig-gnats; some insects are born by fire in the copper foundries of Cyprus, flying in the fire and dying if they leave it. The tapeworm *Taenia* is born inside a human being. Aquatic animals reproducing by spontaneous generation include the frog, which, after 6 months of life, melts invisibly back into mud and is reborn again, in springtime waters. Mussels and scallops appear by spontaneous generation, and oysters are born from rotting mud, or from foam around ships that stay moored for some time, and especially around wooden stakes fixed in the ground. Marine snails are born from a sticky, sap-like saliva, whereas eels rub against rocks, and the resulting scrapings come to life.

Concerning sexual reproduction, Pliny described it rather superficially in dogs, cats, rabbits, elephants, horses and swine. He compared human sexual behaviour with that of animals and commented that whereas animals mate only in fixed seasons, humans mate at every hour of the day and night. In addition, whereas animals experience satiety in coupling, there is almost no satiety in man. As evidence, Pliny brought Claudius Caesar's consort, Messalina, who competed with the most notorious prostitute in Rome, who could have sex with more men; she beat the prostitute with a score of 25 men in 24 h.

1.3.4 Body, Soul and Generation

Galen of Pergamon, a prominent physician and philosopher (ca 162–ca 216), studied medicine in Smyrna, in Corinth and in Egypt (Alexandria), and aged 27 he returned to his hometown, as a surgeon to gladiators. Eventually the Roman emperors Marcus Aurelius and Lucius Verus recruited him to serve in the army, in their wars against the Germans. When the plague hit Rome, Marcus Aurelius appointed Galen as his personal physician; so also did Aurelius' son, Commodus.

Roman law prohibited the dissection of cadavers because the Romans believed that damage to the body of the dead enrages the gods, and this might bring a terrible disaster to their community. Galen had to do with dissecting animals and assuming that their bodies resembled those of humans. He dissected pigs, dogs, monkeys, goats, sheep, apes, horses, asses, mules, cows, lynxes, stags, bears, weasels, mice, serpents, fish, birds, and at least one elephant; modestly, he confessed that he had not dissected ants, gnats and fleas. In seeking to test the functions of organs and structures, he mastered the experimental approach of cutting nerves, cutting muscles, tying tubes, and observing the results. When a certain physician in Rome had asserted that blood excretes urine directly into the bladder, Galen set out to demonstrate that this is not so. The kidneys secrete the urine into the ureters, and from them, it drips into the bladder. To prove this, he carried out experiments in such detail and so wisely that I quote them:

> One must divide the peritoneum in front of the ureters; then secure these with ligatures; and next, having bandaged up the animal, let it go (for it will not continue to urinate). After this, one loosens the external bandages and shows the bladder empty, and the ureters quite full and distended– in fact almost to the point of rupturing. On removing the ligatures from them, one plainly sees the bladder filling with urine. Then, before the animal urinates, one has to tie a ligature around its penis, and then squeeze the bladder. Nothing goes back through the ureters to the kidney [...]. One now loosens the ligature from the animal's penis and allows it to urinate. Then one ligates one of the ureters and leaves the other to discharge into the bladder. Allowing some time to elapse, one now demonstrates that the ligated ureter is obviously full and distended, whilst the other, that from which we have taken the ligature, is itself flaccid, but has filled the bladder with urine. Then again, one must divide the full ureter, and demonstrate how the urine spurts out of it, like blood in the operation of venesection; after this one cuts through the other ureter also, and both being thus divided, one bandages up the animal externally. Then, when enough time seems to have elapsed, one takes off the bandages; the bladder will now be found to be empty, and the whole region between the intestines and the peritoneum full of urine, as if the animal was suffering from dropsy (Galen, *On the Natural Faculties;* in Keele 1961: 580).

I admire the beautiful design of these experiments.

Galen challenged Aristotle's views on many issues, the most central of which was his belief that the powers controlling animal life have one single soul, in the heart. For Aristotle, the heart was the chief organ, the source of all vessels and innate heat, the principal sensory and cognitive organ, the seat of the *pneuma* and, as such, of the soul itself. He believed that the body extracts the *pneuma* from the blood once it reaches the heart, and that the blood there vaporizes, by the inspired air and the heart's innate heat.

Galen, in rejecting this idea of Aristotle of a single, unified soul, claimed that the body operates through the activity of three souls, the natural, vital and animal ones. These souls form and reside in three distinct internal organs, with each part governing a distinct set of actions by means of a distinct set of vessels. In general, he claimed that as the incoming food passes first through the liver, then the heart, and finally reaches the brain, each of these organs purifies the food into a more subtle substance and simultaneously vitalizes it with a soul. The liver converts the food it receives from the stomach and intestine into venous blood and gives it life via the natural soul. In the heart, this venous blood receives a derivative of air and turns into arterial blood via the vital spirit, and carries the life-giving innate heat of the heart throughout the body. Finally, the arterial blood reaches the brain, which converts it into a sensitive animal soul, which passes out through the nerves and aids in sensation. The natural soul, in the liver, is for appetite and pleasure; the vital soul, in the heart, governs emotions; and the rational soul, in the brain, governs rational thinking, memory, and voluntary motion. These three souls are equipped with different vessels. Veins, for natural faculties, originate in the liver; arteries, for vital faculties, originate in the heart; nerves, for psychic faculties, originate in the brain.

As this (wrong) theory of Galen dominated Western thought for sixteen centuries, in the following three paragraphs, I present this general view of Galen in more detail, with the numbers referring to Fig. 1.38. All animals (including humans) must feed, and the food must somehow reach from the intestines to all parts of the body (Aird 2011). Galen thought that during this process of distribution, blood originates directly from elements in the food. Food traversing the alimentary tract is absorbed as "chyle" in the intestine (1), from where it is collected by a series of veins into the portal vein and conveyed to the liver (2), in which the portal vein splits up (3). Here, by body heat, the chyle is concocted into venous blood and is imbued with a "natural spirit" that provides the body with appetite and pleasure. Charged with natural spirit and with nutritive material from the food, venous blood flows through the veins and spreads, from the liver to all parts of the body (4). These parts continuously consume blood, which the ingested nutrients replenish, as an open-ended system. A major vein, the vena cava, gives off a large, blind branch, which is in effect the right ventricle plus the right atrium of the heart (5). Most of the blood reaching the right ventricle remains there a while, parting with impurities and waste vapours that it has brought from the organs. The pulmonary artery carries these impurities off to the lungs, from where the body exhales them (6). Having parted with its impurities, the venous blood ebbs back from the right ventricle into the vena cava and into the general venous system (7).

For a small fraction of the venous blood that enters the right ventricle, another fate is reserved. Drops of it, charged with the natural spirit derived from the liver, seep and trickle through minute, invisible channels in the wall separating the right from the left ventricle, thereby entering the left ventricle (8). Here it encounters concocted air, the *pneuma.* Air first enters the body through the windpipe, from where it passes to the lungs, and is concocted into a *pneuma*-like substance that goes down to the left ventricle of the heart (9). Here, in the heat of the heart, the *pneuma*-like substance mixes with blood and becomes *pneuma*. By the action of this *pneuma*,

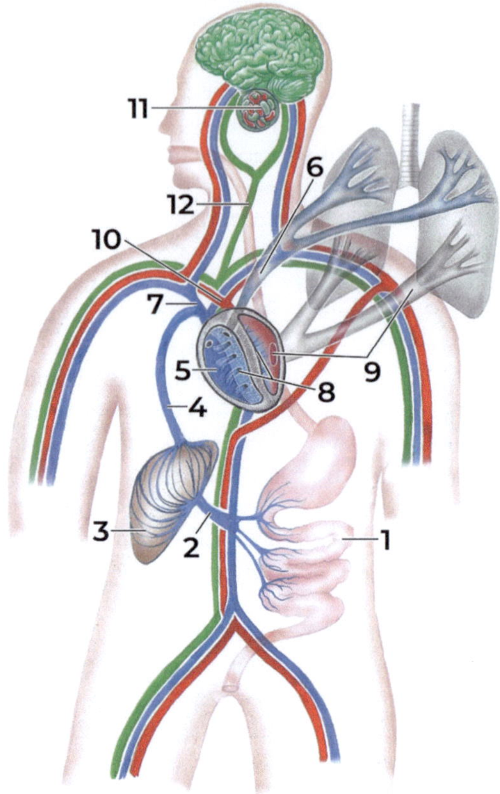

Fig. 1.38 Galen's three souls, the natural (blue), vital (red) and animal (green). Numbers refer to the accompanying text. (Tuvia Kurz)

the natural spirit in the blood in the left ventricle is mixed with and concocted into a higher and subtler form of spirit, the vital spirit *spiritus vitalis*, the arterial blood, to be distributed to the tissues via the arteries by means of their pulse (10). This vital spirit provides the body with life and motion. The heart is thus also the centre from which life (the vital spirit) disperses throughout the body. The lungs, surrounding the heart, function in cooling the blood.

Some of the arterial blood containing vital spirit ascends to the brain via the carotid arteries, each of which divides at the basis of the brain into minute tubes to form a *rete mirabilis*, a "marvellous network" (11). Here the blood again minutely divides, and here the vital spirit that it contains is concocted and refined into the third, highest and most subtle type of spirit, the animal spirit *spiritus animalis*. This substance enters the brain and is distributed to the various parts of the body via the minute channels that, Galen believed, the nerves contain (12).

Galen's theory, that veins carry the natural spirits and arteries the vital spirits, leads to the conclusion that the venous system is completely separate from the arterial system, except where they come in contact by tiny, invisible pores in the heart. The heart mediates the exchange between blood in the veins and air in the arteries.

Galen believed that the heart does not pump blood, it merely sucks blood in during diastole, and the blood then moves on by the pulsation of the arteries themselves (Aird 2011).

Hippocrates, in Greece of the fifth century BCE, had suggested that the human body contains four vital fluids, which he termed "humours" ("sap" or "juices" would have been better terms): blood, phlegm, yellow bile, and black bile (Kruk 1990). The four humours animating the body are each associated with the qualities of one of the four elements: Black bile with earth (cold and dry), Phlegm with water (wet and cold), Blood with air (hot and wet), and yellow bile with fire (hot and dry). Our body, for its good health, depends on the harmonious, balanced combination of these four humours, and an extreme excess, or deficiency, of any of these humours in a person would cause illness. Galen, coming six to seven centuries later, suggested that a moderate imbalance in the mixture of these fluids would produce different behavioural patterns. Thus, "sharpness and intelligence are caused by yellow bile in the soul, perseverance and consistency by the melancholic humour, and simplicity and naivety by blood. But the nature of phlegm has no effect on the character of the soul" (*On Hippocrates' The Nature of Man*). The four bodily humours theory was a central concept in medicine for almost 2500 years, up to 1858 (Kruk 1990).

Another matter on which Galen disagreed with Aristotle was on the mode of animal generation. Aristotle had believed that the female contributes the menstrual blood, which, however, has no active principle or virtue; and which provides merely the raw material, the Matter for the embryo. This passive menstrual blood is "informed" by the active male seed, which provides Form (*Generation of Animals*). Aristotle took this idea, of this male-female division of tasks, from the seed of a plant lying on damp soil. The soil, he thought, gives the seed matter, and the seed shapes this matter into the form of the growing plant, of one species rather than another.

Galen, coming some five centuries later, suggested that semen derives from the humours of the body as a whole, and that it consists of the most refined nature of the four body humours; nerves and blood vessels draw it from the whole body to the testicles, complete and with a soul. In addition, however, the female produces a true (though imperfect) sperm, which has the power of generation after mixing with the male seminal fluid. He reached this conclusion from finding, in the horns of the uterus of a barbary ape, a fluid that he mistakenly identified as the female semen; and he believed that such a fluid occurs also in other species. Consequently, both parents can contribute to Form, and offspring are, therefore, as likely to resemble their mothers as they are their fathers. During copulation, the womb receives the male semen, which here encounters the female sperm; and, aided by the heat of the womb, it develops into a fleshy embryo. When the embryo grows in the womb, like joins to like—finger to finger, artery to artery. If the sperm falls on the warmer, right side during intercourse, the baby will be a male; if it falls on the colder, left side, a female. If the male semen is more abundant than that of the female, the child will look like the father; if the semen of the mother is more abundant, then like the mother. It sometimes happens that in the sperm, the powers of one humour are more abundant than the other. The child will then follow that humour and be sanguine, or choleric, or phlegmatic or melancholy, and in no way like its parents. Women are colder than men and therefore smaller, with internal organs of generation (in men, the organs are forced out by heat). The female testes (the ovaries) are smaller and their seed is so imperfect as to need the influence of the male seed, for generation to take place and foetal development to start off (Smith 2006). Galen thus challenged Aristotle's views on the male and female roles in the sexual generation. His theory is remarkably similar to the Pangenesis theory of Darwin, some seventeen centuries later, in which he proposed that each part of the body continually emits its own type of small organic particles ("gemmules"). These migrate to the gonads, contribute heritable information to the gametes, and eventually, in the offspring, develop into their associated body parts as the offspring matures.

1.3.5 Plato, Aristotle and Epicurus in the Roman Empire

We recall that in Greece of the fourth century BCE, Plato suggested that beyond nature and beyond our deceptive sense impressions, there exists a realm of perfect, timeless, abstract Forms and Ideas. Aristotle's approach, however, had been that there are no dual realities of ideal versus real, no separations of spiritual versus material realms. There is one single world that runs by basic principles which we can understand by reasoning, from data presented by sense impressions.

Six centuries after Plato and Aristotle, in the Roman Empire of the third century CE and stretching into the sixth century, a series of philosophers revived these Plato-Aristotle disagreements. One of these philosophers, Plotinus, claimed that all reality derives from a single abstract principle, "the One", at the top of the Chain of Being. Plotinus (205–270 CE) came from Egypt. When 28, he began to study philosophy and, 11 years later, he sought to study Persian and Indian philosophy. Leaving Alexandria, he joined the army of the Roman emperor Gordian III as it marched on Persia. The campaign was a failure, and on Gordian's eventual death, Plotinus travelled, first to Antioch and then, at the age of 40, to Rome, where he attracted a number of disciples. He attempted to interest the Roman Emperor Gallenius in rebuilding an abandoned settlement in Italy, to be called Platonopolis, where inhabitants would live under a constitution that would follow Plato's Laws, but he did not get imperial permission. His many essays, which he used in his teachings and in debates, were eventually assembled under the heading *The Enneads* (from Greek *ennéa*, group of nine).

Considering himself an interpreter of Plato, Plotinus suggested that there is a supreme "One" that is neither any existing thing, nor the sum of all things (Rubenstein 2003). The One is prior to all existents, the One is sheer potentiality without which nothing can exist, the One fits our concepts of Good and of Beauty. The One is comparable to a Divine light and in addition, there is a Divine intellect that is comparable to the Sun, and there is also a Divine Soul comparable to the Moon (the Divine light exists, however, even without there being any celestial body; we recall the Old Testament, in which light was created on the first day, the sun and moon only on the fourth). The One, perfect and therefore unchangeable, is the source of the entire world, a world that emanates from it in succeeding stages of lesser and lesser perfection. These emanations occur constantly throughout time, yet like the Sun, which emanates light without thereby diminishing itself, the One is not diminished by these emanations. The first emanation is the *Nous*, the Divine Mind, of orderly thought and of Reason, which compares with the Demiurge in Plato's *Timaeus*. From the *Nous* proceeds the World Soul, from which individual human souls flow forth; and finally, matter flows forth, at the lowest level of being. Matter is thus the least perfect level of the world. Reading through Plotinus and his theory of the One we recall Anaximander of Miletus, who, some eight centuries earlier, had suggested the existence of an eternal, ageless, undifferentiated entity that has no beginning and no end, which is the source from which all existing things derive their existence, and to which they shall return upon their destruction (Rubenstein 2003). Plotinus' philosophy, which historians of the nineteenth century would name Neoplatonism, strongly influenced medieval Christian, Muslim and Jewish philosophy.

Half a century after Plotinus, Iamblichus of Syria (c. 245–c. 325) believed in a supreme One, from which everything emanates. Unlike Plotinus, he believed that matter is as divine as the rest of the universe. He placed the One at the head of a hierarchy of the world, beneath whom there is an assemblage of divine beings: 12 heavenly gods, 72 other gods proceeding from them, 21 chiefs and 42 nature-gods. These divine beings can influence natural events, they know and can tell the future, and humans can communicate with them by means of prayers and offerings.

Some two centuries after Plotinus, Proclus (412–485 CE) set forth an extremely elaborate system of Neoplatonism. Born in Constantinople to a family of high social status, he at first became a lawyer, but later in his life, realizing that philosophy was what interested him, he travelled to Alexandria to study philosophy and mathematics. Later, he travelled to Athens to study at the Academy that Plato had founded some eight centuries previously. Here, when aged 25, he became Head of the Academy—Plato's "successor".

Proclus, in his admiration for Plato, wrote that nothing in Plato's writings is unintended, or is there by chance; that Plato was divine and inspired by the gods; that the structure and content of Plato's texts imitate those of the universe. Accepting the philosophies of Plato and rejecting those of Aristotle, he commented that Plato, in explaining the world, had analysed broad, fundamental principles of all living organisms, never getting lost in over-detailed questions. Aristotle, however, had given his attention to ever so many detailed material components of ever so many animals that, in doing so in his over-extensive zoological studies, he had gone beyond what is needed. Proclus recognized three essential levels of spiritual reality: gods ("henads" in his terms), intellects and souls. These three correspond to the triad of Being, Life, and Intellect. By introducing this triad, he integrated the many gods of Greek traditional religion into the Neoplatonic concept of the One: The gods bridge the gap between the One and all the many things that come below it. In discussing the descendance of the human soul into a human body, Proclus argued, this time disagreeing with Plato, that human souls never enter animal bodies; only animal souls enter animal bodies. Some human souls, because of minor wrongdoings in this life, are doomed to live in animal bodies in their next life, but they are only "relationally" present in the animal's body. Proclus further maintained that the universe, with its celestial spheres, is eternal in that it never came to existence in time and will never cease to exist. It is also temporal, however, in that it possesses its being only through a continuous process of change through time. This applies also to the soul: All souls are immortal and indestructible, but they also continually undergo change. Proclus followed Plato in his belief that the heavenly bodies consist of the four elements with a preponderance of fire. However, as the celestial elements differ somehow from those on Earth, Aristotle was right in suggesting that the stars consist of an element that differs from the four earthly elements.

Having presented Neoplatonic views of people who seem to have been pagan philosophers, we now turn to the Christians. Tertullian (155–220), an early Christian of Carthage, rejected all Greek philosophy as supporting religious idolatry and heresy (Rubenstein 2003). A century or so later, however, Saint Ambrose (339–397), Bishop of Milan, claimed that Platonism converges on Christianity, in that both believe in other worlds: Plato had spoken of a realm of Ideas and Christ, likewise, had said that his kingdom is not of this world. Saint Augustine (354–430 CE), bishop of Hippo, argued that the Church has permission to convert the treasures of Greek philosophy to its own use. Plato's view of a Realm of Ideas in which perfect eternal originals dwell is close to what Christians call the Kingdom of Heaven. Plato's deity differs, however, in that it does not intervene actively in human deeds or in their salvation. Whatever information God intends us to have about the world, He has written; whatever is not obvious, He wants us not to know. The search for hidden laws of nature is irrelevant to people seeking salvation. Our knowledge is the result of divine illumination, not of reason, so Aristotle was wrong to suppose that rational thought alone could unlock the secrets of the universe (Rubenstein 2003).

We recall Epicurus and his belief that the world consists only of eternal atoms and of void, that all things are conglomerations of atoms, all events are random atom collisions, and that our world is a temporary agglomeration, which eventually will fall apart—nothing is eternal. Atomic collisions may form organs, and if an organ can fulfil a function, then the function sets in. There is no transmission of souls to other bodies, and no punishment of the soul after death. The teachings of Epicurus spread throughout the Greek world, and later they spread to Rome as well. One Roman convert to Epicurus' doctrine was Titus Lucretius Carus, better known as Lucretius (ca. 100–50 BCE). He wrote a poem, *On the Nature of the Universe* (*De rerum natura,* as translated by Latham 1951), glorifying Epicurus and presenting his philosophy. Totalling 7400 lines, it praises Epicurus for revealing the true nature of the universe. In transmitting the ideas of Epicurus (and of Democritus, to some extent), he wrote that there is no formation of something from nothing, no loss of matter. Matter consists of invisible, indestructible atoms that are constantly on the move, and beyond which there is only the void. The world has no boundaries, no centre, and chance alone guides it, not divine intervention. Indeed, Lucretius echoed the philosophy of Epicurus so closely that it is often impossible to separate the ideas of one from the other.

A creation story that Lucretius presented in his poem does, however, seem to be his own. It goes back to the faraway days in which the earth and stars were already in place, but there was no life. Mother Earth then created Life. According to Lucretius, Mother Earth first covered her hills and plains with green meadows, herbs and shrubs. She then enabled trees of every sort to sprout and to grow upwards (Singer 1941). Next, she created the various kinds of animals. To this very day, multitudes of small animals form spontaneously out of the earth, on warm days following rainy showers. In those faraway days when Mother Earth was young, not only small animals but also larger ones could take shape and develop inside her. At first, various kinds of birds spontaneously hatched out of eggs, in the same manner as winged cicadas today spontaneously crawl out "from their tubular integuments". Later, Mother Earth brought forth the first mammals. In those distant times, many places inside her were warm and moist. Wherever such a suitable spot occurred, a womb grew, which clung to Mother Earth by its roots. When their time came, embryos growing inside these wombs and struggling in need of air, split these wombs open. After they were born, Mother Earth's pores opened up and exuded a milky substance, which fed the young, who lay bedded in the "downy" leaves of Mother Earth's plants. The childhood days of the world experienced no hard frosts, no excessive heats, no stormy winds. Eventually, Mother Earth grew old and ceased to bear life, like an old woman worn out by age.

However, in addition to all forms of life which Mother Earth produced in those distant days and which we now see, she also produced a host of grotesque monsters. These included hermaphrodites, which were halfway between the sexes, creatures lacking feet or hands, dumb, mouthless brutes, blind-eyed creatures, and others. These and other monstrous creatures did not survive because they could not feed or could not reproduce. Which creatures survived? And by which means? Every living species that we now see has survived due to some specific adaptation: the lion by its prowess, the fox by its cunning, and the stag by its speed. In addition, many kinds of animals have survived under human protection because of their usefulness: the dogs, the beasts of burden, the sheep and cattle. All other kinds of animals, which were unable to live by their own resources or to contribute to human welfare, became easy prey for other animals and became extinct.

Did the list of grotesque monsters produced by Mother Earth also include the various fabulous creatures mentioned in mythology? On this matter, Lucretius is unequivocal: There never were and never can be creatures with a double nature, combining organs of different animals in a single body. Take, for example, the mythological Centaurs, half horse and half man. A horse reaches its vigorous prime in about 3 years, an age at which a boy still fumbles in sleep for his mother's breasts, and in later years, when a horse's muscles dwindle with the onset of old age, a boy has just reached his vigorous prime. Such a half-horse/half-man could never reproduce, and therefore, we need not believe that there ever was a centaur. Each species develops according to its own kind, and they all preserve their specific characteristics. Lucretius, in the first century BCE, thus clearly announced the ability of nature to preserve certain types and eliminate others.

In reflecting on Lucretius' story of creation, one noteworthy point is that the order of creation he presents closely resembles that of the Old Testament. Restricting ourselves to those elements specifically mentioned by Lucretius in his creation story, in *Genesis* (Chap. 1), the order of creation is: First heavens and the earth; then, plants, among which first the herbs were created and then the trees; then birds; then mammals, and then man. This is the same order given by Lucretius.

Another Roman follower of Epicurus was Marcus Aurelius, Emperor of Rome (121–180 CE; Fig. 1.39). Ruling in times of relative peace and stability of the Roman Empire, he would jot down various private, personal notes from time to time, including thoughts on the mysteries of life and death, ideas for self-improvement, quotations from texts that he happened to stumble on and a variety of other items, all written down as they occurred to him and which he probably never intended to be published. In this collection of notes, which have survived over the centuries and have received the title *Meditations*, he wrote, amongst other things, that everything, whether a horse or a vine, is created for some purpose. Nature always has a purpose in view, and this aim determines a thing's ending, its beginning and its duration. Everything has come from nature and everything shall return to it, in due time. Out of some universal substance, Nature, as out of wax, may first fashion a colt, then break it up and use the material to form a tree, and after that a man, and next some other thing; and none of these creatures endures for more than a brief span. Nature shuffles, transposes, interchanges, removes and transfers to another. Everywhere there is change, and yet we need not fear the unexpected, for all things are ruled by age-long wont, and even the manner of apportioning them does not vary. That which dies does not drop out of the world. It remains here and changes as it disintegrates, into several particles, into elements that go on to form

Fig. 1.39 Marcus Aurelius. (Tuvia Kurz)

the world. All things are in a process of change, indeed, each person is ceaselessly undergoing transformation and the decay of some of his parts, and so too is the whole universe. Everything in the universe decays in time or, to speak more accurately, suffers a change of form. In his days, already, people believed in the immortality of the soul, a belief that Marcus Aurelius strongly opposed. If souls survive after death, he argued, how has the air above us found room for them all, since the beginning of time? We might likely ask how the earth finds room for all the buried corpses, since immemorial ages; there, after a short time, change and decay make way for other dead bodies. Similarly, souls transferred to the air exist for only a while before undergoing change and diffusion, and they are then transmuted into fire and taken back to some creative principle of the universe, and room is made for the reception of others. "Such will be the answer to any believer in the survival of souls".

These musings of Marcus Aurelius largely reflect Stoicism, a philosophy in which he believed. (Here, in briefly presenting it, I am closely following Staniforth 1964, in his Introduction to the translation of *Marcus Aurelius: Meditations.*) This philosophy was founded some five hundred years earlier by Zeno (c. 334–262 BCE), a native of Cyprus living in Athens, and it had received its name from the "Stoa", a colonnade in Athens, where he had presented his ideas. Stoicism, by the time of Marcus Aurelius, held that everything in the universe can be referred to one single principle, the Mind as the Stoics termed it, and that everything is forever changing, into something different from what it was before. The Mind consists of the finest matter, of a Fire so ethereal that the term "heat" rather than fire would suit it better. The Mind-Fire possesses consciousness, purpose and will, and it is both the creator and the material of the universe. It gives all things their particular substances and forms, and thus, it produces all the diversity of all things in the world. Stoicism, in its faith in a Mind-God that has no separable existence outside all created things and, also, in its belief in a designed, ordered world, was mainly directed against Epicurus's philosophy, that chance is a dominant factor in the world, which consists only of atoms and empty space. There was thus a polarity between Epicureans and the Stoics (Mayr 1982).

The Stoics believed that the world was created when the Mind, in a state of intense heat and high pressure, began to expand. Some of it then cooled and became the element of earthly fire, and later, as expansion and cooling continued, fire partially condensed into air; and later still, parts of the air solidified into water and earth. The vital heat contained in these four elements now began to materialize into the diverse shapes and forms that compose the universe, the Mind realizing itself as various forms of plant life; then as the diverse forms of animals, those creatures having souls but without reason; and eventually as Man, who possesses a reasoning soul. The Mind finally took on the attributes of a World-Soul, holding the same relation to the universe as the individual soul does to man himself. This is how the many diverse forms of being were created. However, creation is cyclic, and a time will come when all the diverse forms and substances shall dissolve back into their original elements. Water shall evaporate into air, air into flames, and finally, the universe shall disappear, leaving no survival but for the primordial Mind. Thereupon, the entire succession of creation shall repeat itself once again and, under eternal and unchanging laws, every event that had happened in previous cycles shall repeat itself, down to the smallest detail. The world's life cycle is thus as finite as that of a human being. Just as a human foetus already contains the seed of its death, the beginnings of the world contain its end.

Man, in himself, reflects the vast organism of the universe, a microcosm that reflects the macrocosm. His body consists of the four elements, and that inside him which creates and controls is a particle of the Mind. The Mind is the soul of the world, and so here, residing in the human body, the human soul plays the same part for man, directing his life, his senses, his thoughts and his emotions. His soul is seated in the heart and nourished by blood. At an appointed time, Nature disperses the matter that has composed the human body to use it for other purposes, and this is what we know as death. The soul, after dissolution of the body, lives on in the upper regions of the air and is resolved into the World-Fire at the end of the cosmos' cycle. Some Stoics believed that it is only the souls of the good and wise that thus preserve their personal identity until the end of the world, the bad and unwise being allowed a brief period of survival before their re-absorption; still others held that in every case this re-absorption follows immediately on death; and others again believed in a purgatorial state in which the soul undergoes physical and moral purification as a prelude to its reunion with the world-substance.

References

Aird, W. C. 2011. Discovery of the cardiovascular system: from Galen to William Harvey. *Journal of Thrombosis and Haemostasis* 9 (Suppl 1): 118–129.

Aristotle. 2001. *De Partibus Animalium I and De Generatione Animalium I* (with passages from II, 1–5). Trans. D.M. Balme. Oxford: Oxford University Press.

Aristotle. 1974. *A selection of writings on biology*. Translated (to Hebrew) by N. Barri. Jerusalem: Magnes Press.

Aristotle, Physics. Translated by C. D. C. Reeve. 2018. Cambridge, MA: Hackett.

Aristotle, Parts of Animals. Translated by A.L. Peck. 1983. London: William Heinemann.

Aristotle. 1912. *The generation of animals*. Translated by A. Platt. Oxford: Clarendon Press

Aristotle, Metaphysics, Books VII-IX. Translated (to Hebrew) by H. Rosen, 1966. Jerusalem: Magnes Press/Hebrew University, .

Aristotle, The History of Animals. Translated by D. W. Thompson. 1910. Internet Classics Archive.

Aristotle, Meteorology. Translated by E.W. Webster, 1923. Oxford: Clarendon Press.

Balme, D. M. 1962. Development of Biology in Aristotle and Theophrastus: Theory of spontaneous generation. *Phronensis* 7:91–104.

Barnes, J. 1987. *Early Greek philosophy*. London: Penguin.

Barri, N. 1974. Translators Introduction. In Aristotle, a selection of writings in biology. Translated (to Hebrew) by N. Barri. Jerusalem: Magnes Press.

Bazaapoulou-Kyrkaniddou, E. 2001. Chimeric creatures in Greek mythology and reflections in science. *American Journal of Medical Genetics* 100:6–80.

Beagon, M. 2002. *Roman nature. The thought of Pliny the Elder*. Oxford: Clarendon.

Beagon, M. 2014. Wondrous animals in classical antiquity. In *The Oxford handbook of animals in classical thought and life*, ed. E. G. Campbell, 414–440. Oxford: Oxford University Press.

Bodson, L. 1998. Ancient Greek views on the exotic animal. *Arctos* 32:61–85.

Couprie, D. L., and H. J. Pott. 2005. Anaximander of miletus. In *Meet the philosophers of Ancient Greece*, ed. P. F. O'grady, 33–38. London: Routledge.

Diodorus Siculus. 1933. *Bibliotheca Historica*. Trans. C.H. Oldfather. London: Loeb Classical library edition.

Eom, K. S., and H.-J. Rim. 2001. Epidemiological understanding of Taenia Tapeworm infections with special reference to Taenia Asiatica in Korea. *Korean Journal of Parasitology* 39:267–283.

Franklin, K. J. 1961. *William Harvey, Englishman*. London: MacGibbon & Kee.

Fogel, T. 2007. Pliny the Elder's animals: Some remarks on the narrative structure of Nat. Hist. 8-11. *Hermes* 135:184–199.

Galen, On the Natural Faculties; in Keele K.D. 1961. Three early masters of experimental medicine – Erasistratus, Galen and Leonardo da Vinci. *Proceedings of the Royal Society of Medicine* 54:577–588.

Ganias, K., C. Mezarli, and E. Voultsiadou. 2017. Aristotle as an Ichthyologist: Exploring Aegean fish diversity 2,400 years ago. *Fish and Fisheries* 18:1038–1055.

Gaziel, A. 2012. Questions of methodology in Aristotle's zoology: A medieval perspective. *Journal of the History of Biology* 45:329–352.

Kingsley, K., S. Parry, and R. Parry. 2020. Empedocles. In *The Stanford encyclopedia of philosophy (Summer 2020 Edition)*, ed. Edward N. Zalta. https://plato.stanford.edu/archives/sum2020/entries/empedocles/.

Kocandrle, R., and Dl Couprie. 2017. *Apeiron: Anaximander on generation and destruction*. Springer.

Kruk, R. 1990. A frothy bubble: Spontaneous generation in the medieval Islamic tradition. *Journal of Semitic Studies* 35:265–281.

Lear, J. 1988. *Aristotle: The desire to understand*. Cambridge: Cambridge University Press.

Lennox, J. 2001. *Aristotle's philosophy of biology*. Cambridge: Cambridge University Press.

Lucretius. 1951. *On the nature of the universe*. Trans. R.E. Latham. London: Penguin.

Mayr, E. 1982. *The growth of biological thought*. Belknap: Cambridge, MA.

Mccrinde, J. W. 1882. *Ctesias: Ancient India as described by Ctesias the Cnidian. Being a translation of the abridgement of his 'Indika' by Photios, and fragments of that work preserved in other writers*. Calcutta: Thacker & Spink.

Marcus Aurelius. 1964. *Meditations*. Trans. M. Staniforth. Penguin.

Mojetta, A. 2004. *Sharks*. Vercelli: White Star Guides.

Pellegrin, P., 1986. *Aristotle's classification of animals*. Trans. A. Preus. Berkeley: University of California Press.

Plato. 1955–1979. *Plato's Writings*. Five volumes. Translated from the Greek into Hebrew by J.G. Liebes. Jerusalem: Schocken Publishing.

Pliny, Historia Naturalis. Eds. H. Rackham et al. 1938–1962, 10 volumes. Loeb Classical Library, Harvard University Press, Cambridge.

Prasad, B. 2012. Bioactive component, Cantharidin, from Mylabris Cichorii and its Antitumor activity against Ehrlich Ascites Carcinoma. *Cell Biology Toxicology* 28:133–147.

Randall, J. H. 1960. *Aristotle*. NY: Columbia University Press.

Romero, A. 2012. When whales became mammals: The scientific journey of cetaceans from fish to mammals. In *The history of science*, New approaches to the study of marine mammals, ed. A. Romero and E. O. Keith, 1–30. IntechOpen. https://doi.org/10.5772/5081. Ebook (PDF) ISBN 978-953-51-7039-6.

Ross, D. 1977. *Aristotle*. 5th ed. London: University Paperbacks.

Rubenstein, R. E. 2003. *Aristotle's children: How Christians, Muslims, and Jews Rediscovered Ancient Wisdom and illuminated the middle ages*. New York: Harcourt.

Scarborough, J. 1977. Some beetles in Pliny's "Natural History". *Coleopterist's Bulletin* 312:293–296.

Shaw, M. M. 2014. Aither and the four roots in Empedoceles. *Research in Phenomenolgy* 44:170–193.

Singer, C. 1941. *A short history of science to the nineteenth century*. Oxford University Press.

Sirks, M. J., and C. Zirkle. 1964. *The evolution of biology*. New York: Ronald.
Sloan, P. R. 1972. John Locke, John Ray, and the problem of the natural system. *Journal of the History of Biology* 5:1–53.
Smith, J. E. H. 2006. Introduction. In *The problem of animal generation in early modern philosophy*, ed. J. E. H. Smith, 1–18. Cambridge University Press.
Spiegel, N. 1981. *Epicurus, his life and thought*. Jerusalem: Magnes. (Hebrew).
Swynnerton, G. H. 1946. Vernacular names for some of the better-known Mammals in the Central Province, Tanganyika Territory. *Tanganyika Notes and Records* 21:21–38.
Taiz, L., and L. Taiz. 2017. *Flora unveiled. The discovery and denial of sex in plants*. Oxford University Press.
Voultsiadou, E. 2007. Sponges: An historical survey of their knowledge in Greek antiquity. *Journal of the Marine Association of the U.K.* 87:1757–1763.
Voultsiadou, E., and D. Vafidis. 2007. Marine invertebrate diversity in Aristotle's Zoology. *Contributions to Zoology* 76:103–120.
Wallace-Hadrill, A. 1990. Pliny the Elder and Man's unnatural history. *Greece & Rome* 37:80–96.
Wazana, N. 2009. Anzu and Ziz: Great mythical birds in Ancient Near East, Biblical and Rabbininc traditions. *Janes* 31:111–135.
Wightman, W. P. D. 1953. *The growth of scientific ideas*. Yale University Press.
Wilson, C. 1995. *The invisible world. Early modern philosophy and the invention of the microscope*. Princeton: Princeton University Press.

Chapter 2
The Dark Ages

Abstract To explain how animal diversity formed, some medieval scholars believed that Creation was the result of emanations of a divine substance, continuously flowing from Him. As the flow moves away from God, its divinity decreases; some animals are, therefore, less perfect, and this results in diversity. In man, nothing in the life of an individual is chance, but in plants and animals, Providence is responsible only for the survival of the diversity, whereas chance controls the fate of each individual. Some medieval scholars argued that additional worlds exist; others claimed that the universe consists of an Underworld of minerals, plants, animals and man, a Middle world of stars, and an Upper world of angels, beyond which is God. Angels move the stars and thereby influence creatures (and biodiversity) of the Underworld. There was not much active classification of biodiversity during medieval times. Some speculated that biodiversity was first formed by uncontrolled spontaneous generation; later, God complemented this early phase by miraculously creating additional animals that perpetuate by sexual generation. In each such animal, generation derives both from the male parent and from heavenly bodies, each of which is alive and has a soul. Astrology thus explains both life and diversity.

In the Western Roman Empire, Aristotle's views of biology waned from memory. Prevailing ideas were of a man in the centre of a world, which had been created to serve him. Early Catholic scholars classified animal diversity into eight major groups, thereby reflecting Roman classification, integrated with religious views. Also, a tendency to form natural groups mingled with a tendency to classify by habitat. Scholars labelled as vermin those animals created by spontaneous generation, and classified them by their source: whether generated in the air (spiders), water (leeches), earth (scorpions, slugs), leaves (silkworms) or flesh (tapeworms, lice). An Aquatic group included everything from crabs to whales; snakes received fantastic descriptions. The etymological meaning of Latin names was strongly emphasized, and it too was often fanciful. Bird names had become established as those of the present day by the ninth century, suggesting a good acquaintance with bird diversity. Plant classification consisted of eight major groups. In the Trees group, first came those representing heroic prestige, then trees with juicy fruits, trees with edible nuts, and finally trees in which wood was important. In the medieval approach to biodiversity, the order was very different from our order.

J. Heller, *Nature's Biodiversity*, https://doi.org/10.1007/978-3-031-89790-0_2

Medieval Christians saw in nature a book of moral lessons to enhance belief in Christianity. *Physiologus* was a treatise that applied animal lore to Christian indoctrination. In it, animal realities were not relevant beyond symbolizing religious values. Bees represent chastity, whales the devil and so on. In the thirteenth-fourteenth centuries, *Physiologus* developed into the *Bestiary,* in which accompanying illustrations, as a visual language for the illiterate, were not highly realistic, reflecting a poor knowledge of nature.

Hildegard von Bingen's medicinal text indirectly reveals the place of animal lore in medieval beliefs: In fishes, males partake in fertilization only sometimes, by spitting matter from their mouth over the roe; sometimes, females spit "seeds" from their mouth. Whales heave onto land where males discharge milt over eggs, which females then discharge. Some animal kinds are hybrids between fish and serpents, others are hybrids between mammals and birds, and adolescent girls attract unicorns and vermin bubbles from the soil. Hildegard may have described "helping": adolescents who remain in association with their parents, to help them raise subsequent broods.

Frederick II wrote on bird diversity, classification, food, flight, roosting, migration, mating and functional morphology. He classified birds into waterfowl, land birds and neutral birds. Being a falconer, he offered accurate descriptions of the raptors, dividing them by feeding mode. He generalized as to who migrates and when, in view of flight ability, site, climate and feeding habits; and which returning birds occupy the same nests for many years. Resident birds are polygamous and males do not assist females in nest building, egg incubating or feeding, whereas migratory males assist in brooding and raising young. Migrators nest early in the season, so their young would be strong enough to migrate, whereas resident birds nest later. Large birds lay few eggs, small birds many. Newly hatched chicks feed for themselves and follow their parent; fledglings have their parents feed them, and these include the raptors, who nest in lofty places. Hatched featherless high above ground, the young would fall to their death if they were to leave their nest. After the fledglings can fly and hunt, the parents drive them away to fare for themselves. Frederick's broad scope of natural history, orderly manner of presentation and reasonable generalizations mark the birth of ornithology.

During the High Medieval Europe began commenting on Aristotle's texts on biodiversity and reproduction. To put order into biodiversity, Albert the Great classified animals by a mixture of morphology, habitat, movement, senses, teaching and learning abilities. He also ranked groups on a scale of perfection: Man, Savage, Monkeys, Furry Quadrupeds, Birds, Aquatic Animals, Serpents and Relatives and Vermin. He emphasized the gap between man, who is perfect, and savages, who speak like man but not rationally, and who do not create art. Monkeys learn, but merely to imitate. Aquatic animals have limited learning ability, lack fine senses and are sluggish; the octopus is inferior to fish because it grows by radial symmetry, which is imperfect. Reptiles are imperfect, as many lack perfect organs for motion: their legs are often small or absent, and some creep on their bellies. Vermin (insects) cannot learn because they cannot distinguish words. Albert rejected beliefs in many mythical animals and in many strange ideas about animal behaviours. His treatise suggests that medieval naturalists were well aware of genus-level diversity in their surroundings. Names of European mammals have remained stable, at least since medieval times.

Albert believed that spontaneous generation occurs among terrestrial invertebrates, whereas sexual generation prevails in vertebrates. Semen contains one temporary substance responsible for individual change throughout a lifetime, and another substance that is eternal, unalterable and which is relevant to the individual's group. By receiving the form of a group, an individual participates in divine permanence and is part of the divine order. The spirit of the embryo forms inside the parents' testicles. The soul forms only when the embryonic body has been organized, as it requires a body distinguished in shape from that of other animals. It has (in humans) vegetative, sensitive and rational components, and a formative power generates the first two; the Prime Intellect directly inserts the rational soul into the embryo.

Matter is pure potentiality, with new embryonic parts developing by a gradual shift, from potentiality to actuality ("epigenesis"). Celestial bodies influence life—Albert believed in astrology, and he explained stillbirths and monstrous animal births by star movements. Throughout the Medieval, intellectuals, including Albert, believed that life on Earth is a microcosm within the macrocosm, and that humans follow patterns and cycles analogous to those of the world. Albert's text was, to some extent, also an applied zoology.

Early medieval law prohibited cadaver dissection. Knowledge of human anatomy narrowed to ancient texts and to dissected animals as human surrogates. Later in the medieval, cadaver dissection became essential for improving knowledge of human anatomy among medical students, increasing awareness of epidemics, improving forensic autopsy and enabling evisceration. Eventually, the church permitted cadaver dissection. Medieval anatomists viewed the human body as composed of three compartments. The abdomen contains the nourishing digestive viscera, which, under the control of the liver, transmits the natural soul through veins. The thorax contains the breathing lungs, supervised by the heart that transmits the spiritual soul through the arteries. The skull is controlled by the brain, which conveys the "animate" part of the soul through nerves.

Medieval physicians had difficulty distancing themselves from ancient texts, even when their observations were different. They described the human uterus as having two horns, as in farmyard mammals and that it divides into seven cavities: warmer cavities for male foetuses, colder ones for female foetuses and one for hermaphrodites. They thereby tended to describe texts based on animal anatomy, not from what they actually observed, in their human dissections.

Teaching presents comprehensive views of a discipline and its mainstream ideas. Fourteenth-century medieval lectures attempted to rationalize religious faith by integrating the Creation text with knowledge: the Biblical "Let there be light" lecture discusses light and perspective, and so on. Growth and generation are unique to life. Plants are generated by seeds that grow by attracting nutrients and converting them into tissues and another generation of seeds. Animals reproduce by eggs, which supply nutrition; they need an active male sperm to cooperate with a passive female one. Spontaneous generation also occurs, with different matter producing different types of creatures and thus increasing biodiversity. Environmental disturbances, combined with the distorted celestial influences, give rise to monsters. God's plan of Creation, as revealed by animal diversity, is a process from imperfect to perfect, from plant to animal and from animal to man. Within each of these links, an inner chain of perfection also occurs. Living creatures exist as an interdependent community that maintains a balance, with each member performing its role. God had created the world as a giant clock, and His presence ensures both that there is a clock and that it keeps running.

2.1 The View of Nature Through the Lens of Islam and Jewish Philosophers

Greek and Roman theories and knowledge cycled throughout the ancient world for centuries. In the year 380, Christianity became the official religion of the entire Roman Empire, and two centuries later, the Byzantine ruler Justinian closed (529) all Greek secular schools, which had thrived upon the spirit of contradiction and had cultivated a restless, wide-ranging curiosity. In biology, as in other subjects, the authority of the Holy Scriptures and of religion replaced research and reflection. The Biblical story of creation became the sacred and only criterion by which one could consider nature, as the Holy Scriptures dominate all powers of the human mind. In this manner, from then on and for a millennium to come, biology was closely entwined with religion. Within these limits, there was, however, some flexibility. Saint Augustine (354–430), for example, while recognizing the Holy Scriptures as the only authority, adopted a somewhat allegorical interpretation of Creation: At the beginning, God had created not only finished products, but much of His creation consisted in giving nature the *potential* to produce organisms. He had indeed created their essences in the beginning, but often He had activated them only much later. Accordingly, nature has the capacity to generate new entities well after the six days of Genesis, whether inanimate or alive. Spontaneous generation should therefore be no problem to the faithful, it can happen any day.

This overall attitude, that the sacred written words of the ancients were the single true basis for knowledge, had a far-reaching influence on the study of nature. The consideration of animals led to a consideration of how and why God had created diversity, how animals compared with humans, angels and demons, with stars and the four elements, and whether the stars determine animal generation. This was studied, however, not by direct observation of nature. Medieval philosophy was, largely, an endeavour into textual traditions and past opinions regarding truth. It sought to find solutions to problems stemming from the Bible. A deep connection therefore formed between biology and theology, with a tendency to focus on such aspects of biology as the creation of the world, how and why God had created animal diversity, how animals reproduce and the relationships between body, spirit and soul. Medieval scholars largely ignored the classification of animal diversity, beyond repeating the texts of Pliny.

In addition to religious changes, the Roman Empire also underwent political changes. By 285, the empire had grown so vast that it was no longer feasible to govern all its provinces from the central seat of Rome, and Emperor Diocletian divided it into an Eastern Empire, to be governed from Nicomedia (near Byzantium) and a Western Empire from Rome, each with its own leader. Originally, both parts were known equally as "The Roman Empire", but as time passed, the Eastern Empire adopted Greek as its language instead of Latin and lost much of the character of the traditional Roman Empire. The two parts of the empire continued to prosper together until they separated, upon the death of the Emperor Theodosius (395 CE). At the religious level, the communion between what is now the Catholic Church and the Eastern Orthodox Church broke down six centuries later (1054, the Great Schism), because of theological and political differences between the Christian East and West.

In the Greek-speaking lands of the Eastern Empire, many texts of Plato, Aristotle and Epicurus survived. Fortunately, some of these were translated into Persian and Syriac (a Semitic language spoken throughout the Fertile Crescent from the fourth to the eighth centuries). When Muslim invaders seized control of the Fertile Crescent and of Persia, in the mid-seventh century, they found many books of Greek philosophy in Persian and Syriac and had them translated into Arabic. In this manner, the Muslim Arabs acquired their subjects' cultural treasures along with their lands. Furthermore, many Jews lived in the lands conquered by the Muslim Arabs, and both Muslim and Jewish philosophers were almost constantly analysing, discussing, accepting or rejecting the thoughts of Aristotle, Plato and Epicurus. Seeking to assimilate the scientific knowledge of the civilizations they had conquered, the Abbasid caliphs of the ninth century established the "House of Wisdom" in Baghdad, a library and translation institute in which scholars translated, into Arabic, every piece of Greek, Roman and Persian classical knowledge they could lay their hands upon. It operated somewhat in the spirit of the early library of Alexandria. Among the books that they translated were Aristotle's zoological texts *Historia Animalium*, *De Partibus Animalium*, and *De Generatione Animalium*.

Many of the biology-oriented scholars of Islamic countries were strongly bound to Greco-Roman thought and merely wrote commentaries on these and other Greco-Roman philosophies, from one aspect or another; only occasionally were there original observations and reflections. However, Jewish and Muslim Medieval philosophers had one basic common outlook: they were deeply religious. While accepting many concepts of Greek philosophy, they believed that God, by His wish, had created the world with its diversity of plants and animals; hence, the universe has a limited timescale.

Medieval philosophers faced three classical opinions regarding the beginning of the world: Plato's opinion that the world was created from an eternal primordial matter, Aristotle's opinion that the world with its animals is eternal, and the Biblical opinion that the world was created from nothing, by the word of God. One major theological issue of the Christian, Muslim and Jewish religions was that if God is perfect, why and how did He create all those imperfect animals? This led to a fourth opinion: Emanation, as developed by Plotinus in the third century CE (from the Latin *emanare*, "to flow forth, out of"). Plotinus claimed that everything had always existed inside God and that the Creation of the world was the result of emanations of a divine substance continuously and eternally flowing from Him. Divinity decreases as the flow moves away from God (when a stone is dropped into water, the ever-widening circles from the point where the stone fell are emanations, becoming fainter and fainter); and accordingly, some animals are more perfect, others are less so. The emanation theory also solved the problem of how the unity of one God had created many different animals, the entire rich diversity of life.

Did God create the world? Do chance events govern it? Do laws of nature or God's will govern it? Lactantius (c. 250-c. 325; tutor to the son of Emperor Constantine) was not a biologist, but his theological criticism of Epicurus was on the battlefields of biology (*On the Workmanship of God*, Chaps. 2 and 6). He accused Epicurus, who had lived in a pagan world six centuries before him, of claiming that chance had created animals, rather than the workings of the Divine mind. He further accused him of claiming that when the world had formed, in addition to those animals that we see around us today, innumerable other animals of wondrous form and immense size had been created, but that they had all died out: their method of reproduction had failed them, and they ceased procreating. In attacking these opinions of Epicurus, Lactantius claimed that if Epicurus was right, then surely wondrous animals of this kind ought to appear daily, with diversity constantly expanding. That this is not the case proves that Epicurus' ideas were false and that chance had created nothing. Rather, God, who had created all creatures, had carefully planned and provided each animal with the manner of life that would suit it. He gave, for example, each animal a device that would best suit it for warding off predatory attacks, whether horns, sharp teeth, hooked talons, swiftness of flight, camouflage, or copious breeding. This is evidence of Providence, not chance. Yet another absurdity of Epicurus was his claim that eyes had not been created for seeing, nor ears for hearing, nor feet for walking; that all these organs of the body happened to be created by chance, and their function arose only later, as a by-product. First came Form, to which Function was applied, only later. This suggestion seemed very foolish to Lactantius: if eyes did not form for seeing, why do they do nothing but see? Indeed, if a chance collision of atoms had produced all things and not Providence, then why did it never happen by chance that an animal, of some kind or another, could hear with its nostrils, smell with its eyes and see with its ears?

This questioning, into the issue as to whether the world had been created by chance events or by a purposeful designer, surfaced again eight centuries later in Spain, where Yehudah Ha-Levi (1075–1141) explored the extent to which nature was formed by random chance events. Ha-Levi was a Spanish Jewish physician, poet and philosopher. He was educated in traditional Jewish scholarship, in Arabic literature and in the Greek sciences and philosophy, which were available in Arabic. Like most Jewish intellectuals of Muslim Spain, Ha-Levi wrote prose in Arabic and poetry in Hebrew. He was celebrated both for his religious poems, many of which appear in present-day liturgy, and for his secular ones (Halbertal 2009).

His greatest philosophical work was *The Kuzari* (five essays, written ca.1140). Written in Arabic, it has been translated many times into many languages, with many comments (Hirshfeld 1905, Schwartz 2017, Qafih 1997, Lasker 2008, Korobkin 2009, Shilat 2010; see also Gil and Fleisher 2001 and Goodman 2012). In it, Ha-Levi claimed, somewhat resembling Lactantius, that we can conclude that God had created all different kinds of animals (rather than by random creation) from the fact that wherever we turn, we see purpose. Domestic animals were created especially for Man, both for his use and for him to tend, as they cannot survive in nature on their own; and this surely is not random. Similarly, God had created water deliberately so that plants and animals could drink from it. He had covered the earth with grass deliberately so that domestic animals could feed. He had created high trees so that they would serve as nesting sites for birds, mountains as shelters for some mammals and rock crevices as shelters for others. He had created each creature to be adapted to its specific way of life, the lion with its claws and fierce bravery, the gazelle with its light feet and startled heart. Such differences, as between a vine and a palm tree, stem from their purposefully different design, not from some random combination of characters. However, within each kind of plant and animal, Halevi admitted to the existence of chance. The differences between one individual vine and another, or one individual palm and another, may be random. One individual vine may happen, by chance, to yield black grapes and the other white ones; or one vine may happen, by chance, to be sweet and the other not. This could be because, in some random event, some particular seed of some particular individual vine had happened to fall in one place and not in another, and hence, it grew as it did. In other words, the creation of each kind of animal is by divine will, but random processes may occur in nature at the level of what we today would term intraspecific variation. This is as far as random processes go in nature, according to Halevi.

Halevi took the mighty power of God a step further: He who has created all these great things is also the one who can change substances; He can even create new kinds, and thereby alter diversity. He does so only very seldom, and we use the term "miracle" to express such a change or new creation. To prove the existence of miracles, Halevi shifted from evidence from nature to evidence from history: God had altered the staff of Moses into a snake and had created the manna for the people of Israel to live on while wandering through the desert. We can recognize God and his divine omnipotence from the unpredictable, miraculous events that violate nature's customary order, better than from rule-bound natural phenomena. The evidence of miracles comes from our senses, not from reason. Reason merely explains knowledge as perceived through our senses; it offers theories to explain information obtained from our senses, but it cannot deny data, and it is not a reliable source of knowledge in itself. We might reason that everything on earth consists of the four elements, but in their pure form, they are not available to our senses; they are merely rational explanations for our feelings, and therefore, their existence is not certain. "Did anybody ever witness a body of fire, and a body of air, entering the substance of a plant or an animal, so we could confidently say that it is composed of all four elements?" Only information registered by the senses is trustworthy. The power of reason must not blind us, and when sensory experience points to a reality different from what is reasonable, we must believe our senses.

Halevi believed in the eternity of matter and that the Biblical Creation refers only to the beginning of humans; it thus does not contradict the eternal existence of matter (*Kuzari* 1). He accepted that the movement of the spheres created the four elements, whose fusions created all beings. These fusions varied with the environment, thereby giving matter the potential to receive from God a variety of forms, from the minerals, which are lowest in the scale of creation, up to man, on the highest scale because, in addition to the qualities of the minerals, vegetables and animals, he possesses an intellect. This intellect forms the rational soul and, being a spiritual substance and not a matter of chance, it is imperishable (*Kuzari* 5). In objecting to the emanation theory, he asked why the emanation should have stopped at the lunar sphere (*Kuzari* 5) (Halevi 1905, 1997, 2008, 2009, 2010, 2017).

There were also other opinions in the Judeo-Muslim world. Abraham Ibn Ezra (1089–1164) was born in Tudela, in the present-day Spanish province of Navarre. Later, he lived in Cordoba, and then he left Spain to escape persecution of the Jews by the new fanatical regime of the Almohads. He led a life of restless wandering that took him to North Africa, Egypt, the Land of Israel, Italy, France, England and back again to Navarre. He was one of the most distinguished Jewish philosophers of the Medieval. (Besides matters of philosophy, theology and the creation of the world, Ibn Ezra also wrote secular poems. My favourite one is that in which he laments his poverty and bad fortune: "if I were to sell candles, the sun would never set till the day I die; if I were to sell shrouds, nobody would ever die throughout my life".) Ibn Ezra favoured Plato's opinion: The world was created, but from some pre-existing matter which is eternal. He further suggested that the biblical story of *Genesis* deals only with the creation of our world, of our planet Earth, which was formed by carving it out of primordial matter. However, there are additional worlds beyond our world of Earth, worlds whose creation *Genesis* does not describe. As evidence for this claim, he argued that in *Genesis,* there is no indication of heavenly angels. It is therefore pointless to seek, in *Genesis,* a description of the manner of those additional worlds coming into existence. In Ibn Ezra's opinion, the whole universe is organized in an ascending order of three worlds. The *Genesis* world is the lowest; it is the Underworld, which consists of metals, plants, animals and man. The Middle world consists of the planets and fixed stars, and the Upper world consists of angels. In our earthly Underworld, each living body consists of the four elements, to which it will return upon death. In the Middle world, the star movement is the most perfect that one can think of—it is constant, circular and repetitive, without change, without beginning and without an end. These movements of the Middle world stars influence all creatures of the Underworld, including Man. The angels in the Upper world move the stars of the Middle world. There is no movement in the Upper world and no change whatsoever; everything stands in its perfect position. This is because (according to concepts of medieval times) a change indicates a fault, a position not complete or not perfect; a completely perfect thing does not require change. Things move and change because they seek perfection, or because they are inferior and subject to influences that move them. The uppermost world is dependent on God, it does not exist in its own right. However, as opposed to one God, there is diversity in the uppermost world—there are many forms of angels, even though we do not know their number. (We recall Iamblichus of Syria, who believed in a supreme One, from which everything emanates. He is at the head of a hierarchy of the world, and beneath him there is an assemblage of heavenly gods, nature-gods and other gods, divine beings that influence events on Earth and know the future.)

Maimonides (Moshe Ben Maimon, 1135–1204) was a commentator of Jewish law and a broad-minded philosopher, one or two generations after Ibn Ezra. Born in Cordoba, he migrated, first to Morocco in his early teens and then to Egypt as a young adult. A great admirer of Greek philosophy and of Aristotle in particular, he placed intellect at the centre of religion. He believed that the laws of nature were eternal and that they reflected divine wisdom; furthermore, even the Messiah would not alter the laws of nature, and wolves would continue to prey upon sheep (Halbertal 2009). He wrote *Guide for the Perplexed* ("*Morre Nevochim*", recently edited by Frank and Segal, 2021) for a select readership, knowledgeable in both Jewish religion and Greek philosophy, who were perplexed because of what seemed to be contradictory truths. Maimonides also probed into the question of whether everything in our world is determined by chance events or by a purposeful designer. He argued that nothing in the life of each individual man or woman is chance; everything happens through Divine Providence as either a reward or a punishment, and the outcome depends on each individual person. Even the smallest discomfort, such as a thorn in our hand, is punishment; even the smallest pleasure is a reward. However, only in humankind do rewards and punishments refer to each individual. In plants and animals, Providence refers not to the individual but to the entire kind, to prevent each kind of plant and animal from becoming extinct (and thus altering diversity). Beyond this point of survival of an entire species, chance reigns in the fate of its individuals. "I do not think that this specific leaf falls because of Divine Providence, and not that this specific spider predated this specific fly by God's verdict, and not that this specific fish, when it caught this specific worm, did so because of some Divine favour to this specific individual". Such events in nature are due to chance (Maimonides 2021). Thus, the Chance versus Providence controversy, kindled first by Epicurus in ancient Greece and later by Lactantius in ancient Rome, surfaces repeatedly in biological and theological thought (Lactantius n.d.).

Maimonides accepted most of Aristotle's theories without major reservations. These included, for example, the principle that movement causes change, that there is a transition from potentiality to reality, that each entity consists of two essences, matter and form, and that each material body needs movement to gain shape. The one and only Aristotelian principle that Maimonides did not accept was that Time and Movement are eternal, and hence, the world is eternal, having existed forever. Maimonides claimed that Aristotle's claims to the eternity of the universe had not been sufficiently proven, that they were open to doubt, and that it could well be that the universe was not eternal but had been created, as written in the Bible. Those who had argued for the eternity of the universe had relied only upon generalizations, principles and observations from the surrounding world, and they had assumed that these generalizations always existed and were always valid. However, it is possible that reality was once different. To illustrate this point, Maimonides offers one of the most beautiful allegories contradicting the Aristotelian eternal-universe outlook. Though somewhat lengthy, it is so powerful that I present it here in full.

Say, that shortly after a son was born, his mother dies, and his father brings him up on an island without women, indeed without any female animal. When the son grows up, he asks his father how he came to be. His father tells him that each one of us was formed within the womb of one of our kind, that is, a female, and she has such-and-such a shape. When inside its mother's womb, each child is at first very small, but already alive and moves about, and it slowly grows until, at a certain stage, an orifice opens at the lower part of its mother's body and the baby emerges from it. He continues to grow and develop until he reaches the size you are today.

Having heard that, the son would then ask whether a child inside a woman's womb eats, drinks, breathes through the mouth and nostrils and excretes faeces. The answer, of course, would be "no". Upon hearing this answer, the son would reject the whole story as untrue. Comparing himself to other males, he would claim that if our breathing stops for even a short time, we would die; how then is it possible that one of us would be within the body of someone else for months upon months, during which we live and move about? If we swallowed a live bird, it would choke and die immediately upon reaching our stomach. Furthermore, if we did not eat and drink, we would die within a few days—how could anyone remain alive month after month without food and drink? If we were to be prevented from excreting faeces, we would die in great agony—how can anyone exist without excreting faeces? If someone were to make a hole in our tummy, we would die within a few days—and here we are required to believe that the navel of this embryo would be open! To conclude, the child would claim that it would be impossible for any man to have generated, in this manner (Leibowitz 1991).

Maimonides then presented the interpretation of this allegory: The characters of fully and completely developed beings are not evidence of how they were before they became complete. Therefore, it is very possible that there was a Creation, and that heavens had formed before earth; that once there were skies without stars and that some kinds of animals were created before others. If Aristotle and his supporters were to ask how we know that the world was created, we would answer that we do not know, as there is no proof, but the possibility is there, and we believe in it. We believe that something, the world, had been formed from nothing. Aristotle had claimed that some primal matter is eternal, that everything had formed from it, and that everything returns to it. We believe that indeed, everything was created from such a matter, but that it too was formed by God. This is true for movement as well, which Aristotle claims is eternal, and we believe that it was created. From the reality of nature, as we know it, Aristotle inferred that the laws of nature were always the same, but it is possible that they were once different (Leibowitz 1991).

Maimonides thus emphasized the deep distinction between what exists on the one hand and the formation of what exists on the other hand, and he also argued that our modes of thinking about these two subjects must be very different. It is impossible to discuss the two with the same criteria. The hypothesis that the reality we know today was preceded by a time (or event) in which this reality was formed cannot be refuted by pointing to the nature of the present reality, since the hypothesis refers to a time when this nature did not yet exist. In other words, creation is possible and not, as Aristotle claimed, illogical and unrealistic. There can be no logical-scientific decision about the question of Creation. Fully and completely developed characters and organisms do not bear evidence as to how things were before they became complete. There is thus a fundamental incompatibility between Aristotle's claim that something cannot come from nothing and Maimonides' faith that God can produce everything from nothing (Leibowitz 1991).

Maimonides, in his earlier years, had an anthropocentric view of the world: nature's diversity exists only for Man. Animals were created to be eaten by man, to carry his burdens, or to facilitate his transportation from place to place; similarly, plants, too, were intended only as food and medication for Man. From elephant to worm, there is not a grass, fruit or animal that is not useful to man. In his later years, Maimonides changed his mind. He asked whether God could have created Man without creating all the creatures before him, and he believed that God could have done so. This means that all Creation before man occurred with no previous purpose. Therefore, we should not believe that the diversity of creatures exists to serve the existence of man. All creatures were created for themselves, and not for any other reason.

Gersonides (Levi ben Gershom, 1288–1344), an astronomer, astrologer, mathematician and philosopher coming from Provence (southeastern France of today), concluded that sense perception and experience-based inquiries are indispensable in any attempt to classify animal diversity and that no such classification can be accomplished by purely rational reasoning. Further, the dichotomous method must involve some sense perception in any case, since one must perceive the forms of an animal we wish to classify, which we cannot do unless we observe the animal by our senses (Gaziel 2012a). Beyond these sense-versus-reasons aspects, Judeo-Muslim medieval philosophers were not involved in any active classification of animals.

From animal classification, we now move on to spontaneous generation, to the issue of Life forming out of non-living matter. Scholars discussed this issue extensively throughout the medieval, often linking it with the issue of the biblical Creation. Medieval texts presented spontaneous generation as one mode of the creation of biodiversity; another mode being the creation of individual species by a divine force. Many scholars suggested that at first, uncontrolled spontaneous generation occurred in an eternal, chaotic world. Subsequently, God created a great order, which involved sexual generation (Saint Augustine of the fourth-fifth century, Avicenna, Samuel Ibn Tibbon 1165–1232, David Kimhi 1160–1235, Nahamanides 1194–1270). They claimed that in the very first origin of life on Earth, many forms of life emerged spontaneously, without any involvement of God, and that the biblical story of Creation does not mention these spontaneously generated organisms. Likewise, spontaneously generated creatures were not among the creatures in Noah's Ark because their recovery was ensured by their perpetual spontaneous generation, whereas in other creatures this was not so. In preserving the diversity of life on Earth during the Flood, Divine intervention therefore left them out (Kruk 1990;

Gaziel 2012c). Nahamanides (Moshe ben Nachman, 1194–1270 of Girona) believed that Man, besides possessing an animal soul as do all creatures, also possesses a special soul that is a direct emanation from God and which has existed since before the Creation. It enters material life through Man as a medium, and at the dissolution of its medium, it either returns to its original source or enters the body of another man. The resurrection, spoken of by the prophets and which will occur only after the coming of the Messiah, refers only to the body, which may, through the influence of the soul, transform itself into an essence so pure that it will become eternal. Nahamanides described God as having created all beings of heaven and earth from absolute non-existence. He brought forth a very thin substance, devoid of corporeality but having the potency to assume the form and thereby proceed from potentiality into reality. Having created this primary substance (termed *hyly* in Greek), He did not create anything else. He merely shaped and formed beings from it, thereby bringing all biodiversity of creatures into existence.

Medieval philosophers of Judeo-Islam believed that spontaneous generation was not merely a distant biblical event; it also occurred in their days. They believed that stagnant water or copious rain spontaneously generates frogs, that fermenting earth generates dung beetles and that rotting animal carcasses generate gnats and worms (Kruk 1990). Snakes come forth from decaying matter, or from a hair pulled from the scalp with its white root, field mice come from lumps of clay, bees from the flesh of oxen, wasps from the meat of horses and green algae from stones. Scorpions grow from figs, pearls from sea plants, and partridges are fertilized by the wind (Kruk 1990).

Avicenna (Ibn Sina, 980–1037) claimed that any kind of creature can be generated spontaneously, including humans, and furthermore, that all life on earth had originated in a spontaneous manner, and accordingly, its formation did not involve any supernatural entity. This deviates considerably from Aristotle's view that all animal species are eternal, but for a small number of low forms of life, which generate spontaneously (Belo 2004; Gaziel 2012c).

Maimonides claimed that animals divide according to whether their generation was an "intentional" or an "unintentional" by-product of natural processes such as putrefaction. This by-product occurs in the various kinds of worms, which are engendered from dung or in the intestines, and various kinds of insects engendered in putrefying fruit. The issue of spontaneous generation also appears in medieval Jewish law (*Halakha*). One of the many actions forbidden on the Sabbath is killing an animal, the technical term being the "taking of a soul". Do spontaneously generating organisms have a soul, and are they to be included in this prohibition? Maimonides ruled that whoever kills insects produced by the union of male and female, or which originate in the soil (as do fleas), is as liable as if he had killed a domestic animal or a wild beast. However, if one kills insects that originate in decaying matter, such as maggots in meat or weevils in peas, he is exempt. He also declared that the *Halakha* permits to killing of lice on the Sabbath because they originate in perspiration. In reference to dietary laws prohibiting the consumption of certain insects, he distinguished between organisms generated in garbage or carcasses, which are forbidden and those generated in fruit, fish, or water, which are permitted to be eaten. In this manner, biodiversity became entwined with religion (Gaziel 2012c).

Samuel ibn Tibbon of southern France (1165–1232, *Ma'amar Yeqqavoo ha-Mayim*) explained the Genesis narrative as alluding to a natural process by which dry land appeared and, subsequently, all life forms emerged spontaneously—the origin of life on earth was devoid of any involvement of a supernatural deity.

In ancient Greece, Aristotle had commented on human genesis that "man generates man—and the sun as well" (*Physics* 2). Each individual generates another individual without the intervention of any divine element, and the very modest role of the sun was to mix the four sublunary elements. Were it not for this mixture by the sun and the resultant four seasons, the world would have reached a stable state in which the four elements would have separated, and arranged themselves in four stable concentric spherical layers; there would then be absolutely no generation of life and also, no decay.

From Aristotle to medieval times, on the issue of how plants and animals are generated, Averroes at first believed that Form was carried by the seed or semen, which was derived from the male parent (Gaziel 2012a). To this, he added a mystic "external active intellect", a force that imposes Form directly onto Matter (we might view his external active intellect as close to the term God). Averroes later changed his mind to believe that Form derives from the male parent and, in addition, from the heavenly bodies, the sun, planets and stars, each of which has a soul. The soul-bearing celestial bodies thus became the equivalent of the active intellect of his early account. This, Averroes argued, is why Aristotle says that "man generates man and the sun as well". The sun and other celestial bodies are alive, they have a body and soul, and they can endow with life the entities down here, in our world. The heavenly bodies endow us, we sublunary beings, with form and soul. There is heat in semen, it has a formative power, and this soul-heat is derived both from the direct ancestor on earth and from the sun and from other heavenly bodies. The "heats" of the various planets and stars differ, and their effects vary as a function of their velocities and relative positions, and this affects the identity of the young of a species produced at any given moment. Just as the soul-heat in the semen is different for every animal species, so also the "heats" emitted by different stars are different according to their motions and different positions relative to other stars, at the moment of emission of the rays. Thus, the active factor in generation is not the soul heat in semen alone, but the semen's heat blended with the heat of the sun and of the other celestial bodies (Freudenthal 2002; Gaziel 2012a, c).

Closely following Averroes' opinion, Gersonides (Levi be Gershom, 1288–1344) claimed that different kinds of heating processes generate from different heavenly bodies and that these bring forth different kinds of spontaneously formed living creatures. Celestial influence governs the entire process, and hence, there is no need for arguments that divine creation has formed life. God does not create substances that can emerge naturally; He creates features that cannot come about through a natural process. Since one or two pre-existing individuals have produced most animal and plant species, at one time, miraculous events must have occurred to create the first ancestor of each biological dynasty. In this manner, Aristotle's philosophy of nature, written in ancient times, underwent "astrologization" in the Middle Ages (Freudenthal 2002).

The idea that the stars influence life and its diversity was widespread in medieval times, even centuries before Averroes. A Hebrew text from the fourth to sixth century says that there is not a single herb that does not have a star in heaven, which thrashes it, commanding it to grow (*Bereshit Rabbah* 10: 6). Gersonides adopts the notion of a Scale of Nature and recapitulates it in his account of the six days of creation: plants are lower than animals, imperfect animals are lower than flying animals, the latter to walking animals, and the latter to man, that walking animal that is also rational (Gaziel 2012b). The anatomical and physiological differences among animals are by virtue of their rank, higher-ranking animals having received, from nature, more superior body parts and functions for their survival (Gaziel 2012b, c).

We recall that Aristotle had urged his readers to consider all animal kinds, even the inferior and repulsive ones. Gersonides supports this appeal of Aristotle: investigators, were they to disregard these animals, would remain ignorant of the existence of these lowly animals and of the divine causes that supervise their preservation, both of the individual and of its species. We can recognize these causes only by observing the animal's organs and their uses, as these attest to the natural providence these animals enjoy. This idea turns up again in Gersonides' commentary on *Exodus* 34:6 ("The Lord God, **merciful** and **gracious**, long-suffering and abundant in goodness and truth"). He is merciful (*rakhum*) means he loves all beings and has mercy on them, to preserve them from injuries as much as possible, by endowing them with natural forces that preserve them. This is best seen in animals whom He has endowed them with many organs, to preserve them from harm and injuries to the greatest extent possible. He is gracious (*khanoon)*, meaning He acts graciously to creatures by granting them gradual perfection so that they progress from a deficient existence to a perfect one. To some of the creatures, He gives certain things by way of graciousness, that is, things that are not necessary for their existence but are present in them by way of betterment. This governance of mercy or grace is more perfect in the noblest of the natural beings, as explained by Aristotle: the more noble a creature is, the more providential care Nature invests in its formation. In another case of demonstrating providence in a biological context, Gersonides explains God's punishment of the woman in Eden (*Genesis* 3:16: "*Unto the woman he said*"). On account of her weak intellect, God had invested in her creation less than in that of her man. We can see this from the fact that many painful things befall her, such as menses, pregnancy and childbirth, and that her desire has been directed to serve her husband, who rules over her. This is so because the more noble a creature, the more God had invested in its creation. Man exceeds woman in intellectual excellence and hence, his bodily functions excel hers. The level of providential care corresponds to each creature's rank on the scale of nature, and well-adapted structures of plants and animals are evidence of God's attention and care for them (Gaziel 2012b).

Gersonides also believed that, as man is the most noble of sub-lunar creatures, nature exerts providential care over him and in his preservation also by the heavenly bodies, to the point that all his actions are ordered by them. If the activity of the heavenly bodies on the sublunar world were to cease even for a brief moment, all living creatures of the sublunar world would perish. Man is in particular need of this special care by the stars, because he consists of matter that is especially fine and susceptible to damage, and consequently requires special attention. Also, man's bad temper would lead to constant quarrels and fighting, which are prevented by celestial providence.

Features common to all members of a species derive from general providence. Humans, as they possess an intellect, enjoy the highest degree of providence as compared to other sublunar creatures, and they are at the top of the *scala naturae*.

We recall that Aristotle had urged his readers not to disregard low organisms since they, too, can provide information regarding the animal kingdom. In (by now Christian) Spain the Jewish philosopher Joseph Albo (1380–1444, *Book of Principles – Sefer ha-'Ikkarim*) interpreted Aristotle's call for investigating all animal kinds as referring to species that generate spontaneously because they represent the excellence of God, who infused the power of life into such inferior material so that from putrefaction and without the union of male and female, there arises an animal that is essentially similar to that which is born of the union of male and female. God's superiority as a craftsman is mostly apparent in the spontaneous generation since marvellous beings emerge from the vilest of all materials. Spontaneous generation is not merely a natural event, it manifests God's direct actions of providence (Gaziel 2012c). In general, the poor knowledge of generation and of heredity versus environment is reflected in a text by Ibn Khaldun (1332–1406, of northern Africa) in which the author claimed that negroes start producing white offspring when they settle in temperate or cold areas, whereas white people living in hot countries produce black offspring (Kruk 1990).

Aristotle had not suggested that higher animals (vertebrates) and even humans could come to be by spontaneous generation, but this idea crops up occasionally. Ibn il Hadid (1190–1258, of nowadays Iraq) wrote that when the elements had mixed and the composed bodies had come into being, so too did the human species, "just as maggots come into being in fruit and meat and lice in watermelons and putrid places" (Kruk 1990). Ibn Tufayil (1105–1185, of Andalusia) described how humans come into being by spontaneous generation ("without father or mother") on a certain equatorial island, off the coast of India, which has a most temperate climate; and a special supernatural light that streams down on it (Kruk 1990).

Abu Uthman Amr ibn Baḥr al-Kinani al-Basri (776–868), commonly known as al-Jāḥiẓ, was born in Basra. He later moved to Baghdad, where the caliph wanted him to teach his children, but then changed his mind when his boggled eyes frightened them. Eventually, he returned to Basra, where he lived with his concubine, her maid, a manservant and a donkey, and died when a pile of books, in his library, collapsed on him. His *Book of the Animals* (*Kitāb al-Ḥayawān*) is an encyclopedia of anecdotes, poetic descriptions and proverbs, with over 350 kinds of animals as points of departure. Aware of the struggle for existence, he wrote that every weak animal devours those weaker than itself. Rats eat vermin, small birds, their eggs and fledglings; snakes eat lizards; birds of prey, in turn, eat the rats; and foxes hunt snakes and lizards. Likewise, mosquitoes feed on blood, flies hunt mosquitoes, and lizards, spiders, and antlions hunt the flies. He thus introduced the concept of food chains. In a line that we, coming centuries later, would consider amusing, he wrote that since the cat resembles the king of beasts, one way of dealing with approaching war elephants is to release cats on them.

2.2 The View of Nature Through the Lens of Western Christian Philosophers

2.2.1 The Name of the Beast

The Roman Empire divided in the third century into an Eastern Empire governed from near Byzantium and a Western Empire governed from Rome, as we have already noted. For the Western Empire, as a result, Greek culture and language became somewhat remote. In 476, the last Roman emperor of the West was removed from power, and the new ruling eastern Goths ("Ostrogoths") sought to adopt conquered Roman culture in matters of government, architecture and art. They perceived much of Greek culture as a distant, vague legend from five or six centuries before their time. Catholic monks, between the sixth and eighth centuries, stopped copying most of the Greek classics, and Aristotle's Greek-written works waned from Western memory for almost a millennium; with them, faded his intrinsic and broad interest in nature. The only Aristotelian texts available in Latin were parts of *Organon* (Greek for "tool", "organ"), a collection of six works on logic that had been translated by Boethius (480–524), prime minister of the Gothic ruler Theodoric. He undoubtedly would have translated many more Greek texts into Latin, but he fell from royal favour, and the king executed him for treason.

The Church, during the first centuries of Christianity, welcomed Plato's philosophy that all earthly objects are merely fleeting, imperfect expressions of a perfect, abstract, eternal archetype that exists way beyond our senses and that we can grasp only by intellect. In Roman times, the Stoics had held that everything in the universe refers to one single creator but that man reflects the universe, a microcosm reflecting the macrocosm. We have further seen that Platonism converges with Christianity, in that both believe in other worlds. Pliny's *Historia Naturalis* indirectly preserved Aristotle's view of animals, to a certain extent. His thought, that man stands in the centre of the world with all the stellar elements, mountains, rivers, plants and animals created for the sole purpose of serving him, was broadly similar to the biblical concept (*Genesis*).

Christianity differed from these concepts of Pliny in its belief that God is very much involved in events happening in our world. Consequently, Christianity had to cope with the question of why loving God had created, alongside those animals beneficial to man, also many harmful creatures, such as serpents and thistles. The answer to this aspect of biodiversity was in the original sin: serpents and thistles exist because of that ancient sin committed by Adam and Eve; there were no harmful creatures before they had eaten from the Tree of Knowledge. This argument accompanies Christian theology throughout history. In the seventh century, Bede (673–735), an English monk, declared that all animals had been harmless until Adam's sin, and became harmful only afterwards, so that man would be aware of the final punishment awaiting him in hell because of Adam's sin. In the twelfth century, Peter Lombard (1096–1140), Bishop of Paris, stated that God had created all animals as harmless creatures. They became harmful to punish man for his sin of eating from the tree of knowledge. There was the belief that until God's curse, serpents stood erect, walked about and even talked.

Another theological issue was whether animals have a soul. Philo of Alexandria (20 BCE–40 CE) had already pondered why, during the Flood, all animals were doomed to die together with the sinful generation of man. If animals have no soul, they cannot sin, and so it would be pointless to punish them. Philo considered that perhaps animals do have a soul, though not as perfect as ours: the spider knows how to weave a web, the swallow how to build its nest, and we can teach a monkey to drive a carriage. These and other cases suggested to him that animals too possess wisdom, namely, reason, a soul. Over a millennium later, Thomas Aquinas (1225–1274) claimed that animals lack knowledge and have no ability to choose; their actions may seem to us purposeful, but that is because God decides for them, just as human decisions and actions cause movement in arrows and clocks. They, therefore, are not creatures worthy of charity. In general, in Catholic lands and as reflected in the literature of the time, the Medieval was thus an age in which many people viewed animals through a religious perspective. Aquinas further developed the concept of the great chain of being by placing God at the apex and an inanimate nature at the base. This concept closely mirrored the structure of the feudal society, with a king at the apex and peasants at the base, and that of the Church, with the Pope at the top and the laity at the base. It thus gave theological legitimacy to the feudal political and religious structures (Carrol 2000; Agutter and Wheatley 2008).

In the early Medieval, one trend of the Church was to focus on the opening sentence of the Gospel of Saint John (first century CE): *In the beginning was the Word, and the Word was with God, and the Word was God*. People of the Church sought to understand the meanings behind The Word, and behind words of the Latin language in general, and they presented and defined Latin terminology with lengthy etymological explanations. In line with this linguistic trend, Isidore of Seville (ca. 550–636) compiled his encyclopaedic treatise, *Etymologies*. Isidore, a Hispano-Roman, was born, brought up and educated in Seville, where, at about the age of 40, he became a bishop and, in practice, head of the Catholic Church in Spain. In his efforts to unite all groups in the kingdom around Catholicism and to wipe out all heresies of Christianity, his relations with the Visigoth court were particularly close (Barney et al. 2006; see also Albert 1990).

Isidore studied the histories of words in an attempt to trace their development back to their origins. God created the world by the use of words. By merely saying "Let the land produce vegetation: seed-bearing plants and seed-bearing fruit trees", vegetation was created. By saying "Let the water teem with living creatures, and let birds fly above the earth", the water and skies teamed with animals. Words were thus of cardinal importance in those early days of Creation. They were also important. Soon later, Isidore believed that Adam assigned the original names of animals in Hebrew. Adam had "called all the beasts by their names, and all the fowls of the air, and all the cattle of the field" (*Genesis* 2). God had endowed Adam with both reason and speech, and his insightful naming aligned him with God, unlike animals that lack reason. Each of Adam's Hebrew names encapsulated a compact piece of information about the nature of that particular animal. Indeed, the biblical subjugation of animals defines humankind, and Adam's naming was his first act of dominion over animals. Centuries passed, and when the Jews rejected the New Testament, it was given to other nations, and Latin became the (Western) Christian language. We must therefore understand the insightful meaning of the Latin names of animals. Isidore attempted to present the entire biodiversity of Nature from an etymological aspect, to explain the source of as many Latin terms of nature as possible. He argued that the term *Homo* (human being) derives from *humus* ("earth")—which happens to agree with the biblical narrative of Adam being created from earth. *Boa* (a kind of snake) comes from *bos* (cows) because this snake suckles on the udders of cattle. *Canis*, the dog, comes from the word *canor*, meaning sound, because dogs bark. Gastropods and bivalves (*Concha, Conchlea*) receive their name from the fact that during the waning moon, they are "hollowed out" *cavare*, that is, they are emptied out; their bodies swell with the waxing of the moon and shrink with its waning. Among plants, lily *Lilium* has a milk-white *lacteus* flower, which gives it its name, as if it were *liclia*, ivy *Hedera* is so called because it climbs and clings and *adheres* to trees; fennel *ferula* is from the verb "strike" *ferire*, for boys and girls would be flogged with it; and so on.

Having presented the etymology of each entry, Isidore took the purely descriptive, zoological part from earlier writers, mostly Pliny, and he sometimes added the animal's role in Christianity. He did not share Pliny's concept of Nature as a well-ordered and rational organizing system that had created everything for the benefit of Man. He replaced Nature with God, the divine power who is the creator of Nature. Accordingly, weasels chase, overcome and kill the ever-so-dangerous basilisk because the Creator of Nature has provided this solution, not because of Nature's initiative, as Pliny had written.

There are differences between Pliny and Isidore in animal classification as well. Whereas Pliny had classified all animals into four major groups according to their habitat (terrestrial, with a separate section of domestic animals, aquatic, aerial and insects), Isidore proposed an eight-group classification that reflected both a religious and a practical approach. He clustered all animals in his treatise into:

1. Grazing mammals and omnivores, both domestic and wild.
2. Carnivores, both domestic and wild and other large animals.
3. Small terrestrial animals.
4. Snakes and other reptiles.
5. Vermin.
6. Aquatic animals.

7. Birds.
8. Flying insects.

His religious approach seems to have relied on the Biblical distinction between pure animals, which we may eat, versus impure animals, which we may not. As presented in the Bible (*Leviticus* 11), all animals fall into one of seven groups:

1. Permitted mammals (only those with split hooves that also chew the cud), versus forbidden ones.
2. Permitted aquatic animals (those with fins and scales, namely fishes), versus forbidden ones.
3. Forbidden birds of various kinds (some 20 different kinds of birds).
4. Permitted flying insects versus forbidden ones.
5. Forbidden mammals.
6. Forbidden animals: small [quadruped] animals on the ground (mice, tortoises, various lizards).
7. Forbidden animals: those creeping on their belly, or crawling on four, or having many feet; or living close to the ground in any other way.

Isidore broadly attempted to keep this biblical list of groups and yet adapt it to Pliny's groups, which were more modern in his time. His first group included all grazing mammals and omnivores, both domestic and wild (Fig. 2.1). This group consisted of *lamb*, *goat*, deer, *hare*, *rabbit*, *pig* (and boar), *oxen*, *camel* and dromedary, ass, horse and onager (animals in italics are mentioned in *Leviticus*). Isidore translated the coney *shaffan* into a rabbit (the Vulgate had translated it to

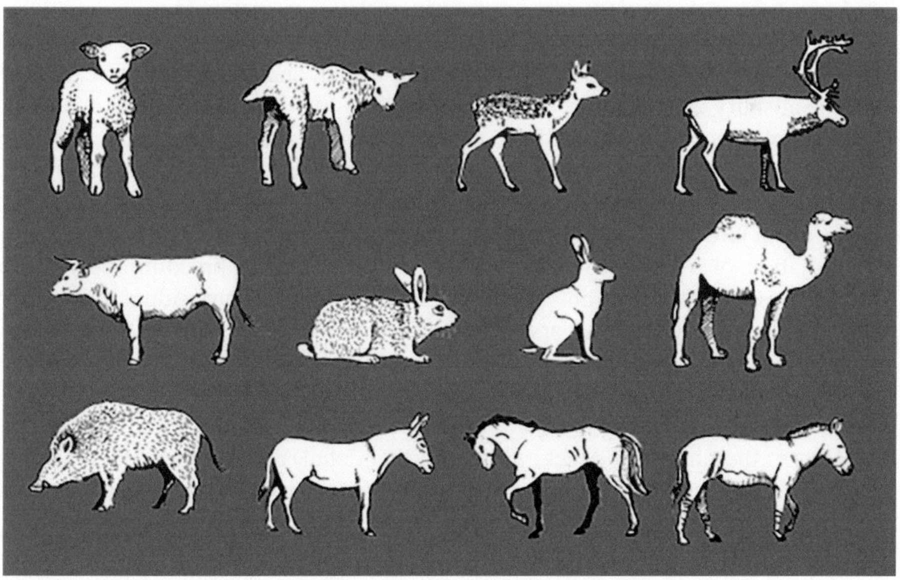

Fig. 2.1 Isidore: diversity of grazing animals. (Illustration by Tuvia Kurz)

chyrogrillus, which I cannot identify) and added deer, dromedary and equines to the list. The expanded group may perhaps reflect the European diet in those days (deer, equines).

Isidore might perhaps have arranged these permitted animals to present Christian thought. He placed the lamb as the very first animal, with goat and kid immediately following the lamb, and we recall the Lamb of God (Agnus Dei) as a title for Jesus ("Behold the Lamb of God who takes away the sin of the world"; John 1; see also the lamb, as mentioned in the New Testament's *Book of Revelations*). The third on Isidore's list of grazing animals is the stag (with young), which is referred to several times in early Christian lore. While hunting one day, Saint Eustace, a pagan Roman general under Emperor Trajan, saw a crucifix lodged between a stag's antlers; deeply impressed, he immediately had himself and his family baptized. He was martyred in 118. Four centuries later, the Merovingian King Clovis (466–511) prayed to Christ on one of his campaigns to find a place to cross a river, and as a divine sign, a deer appeared and showed him where he could pass with his army. The stag was a symbol of Christ, who tramples and destroys the Devil and who, the Catholic Church then believed, pursues snakes into their holes and devours them. This made the stag a symbol of the Catholic, in its battle against Evil. In this manner, theology perhaps might have influenced the order in which Isidore presented grazing animals, reflecting his opinion of their relative importance in descending order. Another reason for the prominent position of deer in Isidore's list may perhaps be its rising position in the medieval domestic household. There was intense deer farming or semi-domestication in the medieval ages, and many deer lived on the border between wild and domesticated. Laws, compiled by the Frankish King Clovis I around 500 CE, imposed strong penalties on anyone who stole a domesticated stag, and in the twelfth century, Gerald of Wales (translated by Thorpe 2004) described cheese made from the milk of tame deer. During the High Medieval, farmers fed hay or oats to deer in parks during winter, and cows would suckle deer calves (Salisbury 2011).

Deer never, however, reached the status of horses, which were considered the high-status domestic animals throughout the Middle Ages. In the thirteenth century, it was by horses that princes, magnates and knights were separated from lesser people, and a lord could not fittingly be seen among private citizens except through the mediation of a horse (Salisbury 2011). Selective horse breeding, during the Middle Ages, produced warhorses alongside breeds for ordinary riding, breeds for hunting and for farm use. Warhorses, especially, increased the status of their owners and were the most expensive. A good warhorse would cost as much as 800 times the cost of a peasant workhorse (Salisbury 2011).

After the grazing animals, Isidore listed the Biblical group of those walking on the palms of their hands and feet (without further detail) and he named them Beasts (Fig. 2.2). These were mainly carnivores and exotic animals, many of which he was probably not acquainted with: lion, tiger, pard, leopard, rhinoceros, unicorn, elephant, griffin, chameleon, giraffe, lynx, beaver, bear, wolf, dog, fox, apes, porcupine, ichneumon and cat. In beasts as in grazers, Isidore's order of presentation might have reflected their relative importance to Christian readers of his time, in that

Fig. 2.2 Isidore: diversity of beasts. (Tuvia Kurz)

first among the carnivores came the lion, that biblical symbol of the Tribe of Judah to which Jesus had belonged. The New Testament (Revelation 5) mentions that "one of the elders said to me, weep not: behold, the Lion of the tribe of Judah, the Root of David, hath prevailed to open the book, and to loosen the seven seals thereof". This Lion of the Revelation is widely regarded among Christians as a reference to Jesus. In line with these texts, Isidore wrote that the lion's nature is so majestic that it does not become angry unless it is hurt; being tender-hearted, it spares those who lie prone; it allows captives whom it meets to return home, and it never kills a human except in great hunger, and here we recall similar descriptions of the lion's noble, majestic nature, by Pliny and, before him, by Aristotle.

Isidore described two mythical creatures among his beasts, the griffin and the unicorn. The griffin is a creature with the body, tail and hind legs of a lion, the head and wings of an eagle, and an eagle's talons on its front feet. Pliny, almost six centuries earlier, had rejected the griffin as mythical, yet Isidore included it, probably because of its religious significance. The lion is the king of the beasts, and the eagle king of the birds, so the griffin, combining these two, is the king of all creatures, a symbol of divine power and a guardian of the divine. In Christendom, it was another symbol of the divine Jesus. To conclude this point, Isidore's major groups of animals and the order in which he presented them mirror a Biblical source, integrated as far as possible into Roman animal classification.

Another trend in Isidore's writing is his tendency to form natural groups. Among the grazers, the hare comes immediately after the rabbit. Likewise, he (correctly) grouped together bull, cow, buffalo and auroch; camel and dromedary; and the ass, onager, horse and mule. He tended to cluster animals into natural groups among carnivores, too. The tiger, pard and leopard immediately follow the lion (strangely, lynx and cat come only further on). He also clustered the bear, wolf, domestic dog and fox. Isidore correctly placed the badger, mongoose and cat among the carnivores. He erred in including the elephant, rhino, two rodents, the porcupine, apes, chameleon and beaver among them (he also upheld the fable that the beaver, when hunted, castrates itself, a fable that Pliny had already rejected).

Isidore then listed the Biblical group of small ground animals. The Vulgate specifies eight: weasel, mouse, crocodile, shrew, chameleon, stellio, lizard and mole. Isodore's list of small terrestrial animals included nine: *mouse*, *shrew*, *weasel*, solifuga, *mole*, dormouse, hedgehog, rabbit and ant (Fig. 2.3). Seven of these nine animals were mammals. Thus, Isidore formed a sequence of the three separate Biblical groups and conformed to Pliny's group of terrestrial animals.

Birds were another major group of animals Isidore described (Fig. 2.4), listing 34 of them. First among them was the eagle, a symbol of Christ and His Divine nature, of regeneration by Baptism and the emblem of Saint John the Evangelist. He clustered only the nocturnal raptors as a natural group (screech owl, horned owl, night heron and tawny owl).

In "Fishes", Isidore closely followed the groups recognized by Pliny, which included what we would today term aquatic mammals, fishes, sea snails, bivalves and crabs (Fig. 2.5). He erroneously failed to separate sharks from bony fish, as Pliny correctly had done. Among the aquatic fishes, his group of *amphibia* included

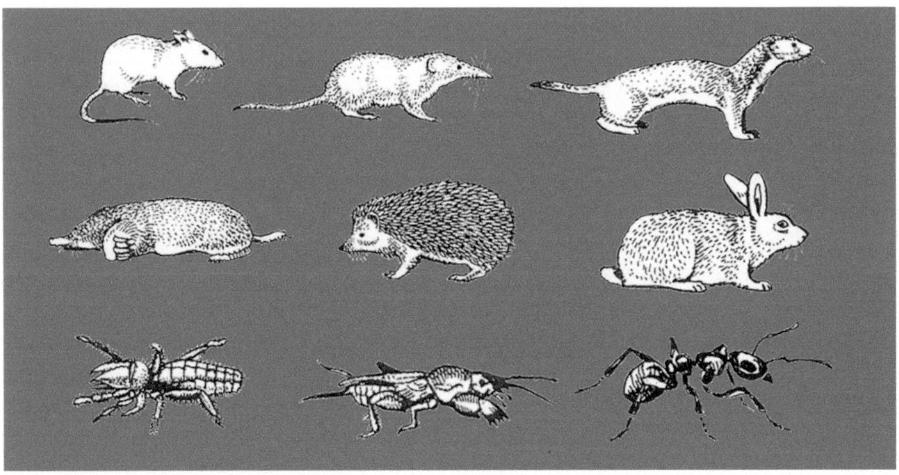

Fig. 2.3 Isidore: diversity of small animals. (Tuvia Kurz)

Fig. 2.4 Isidore: diversity of Birds. (Tuvia Kurz)

those living both on land and in the water, and here he included seals, crocodiles and hippopotamuses. Isidore, like Pliny, closed his list of aquatic animals with the sponges.

Fig. 2.5 Isidore: diversity of aquatic animals. (Tuvia Kurz)

Vermin (*De verminibus,* Fig. 2.6), defined by Isidore as those animals created by spontaneous generation, were classified by him according to their source, not by their similarities: vermin generated in the air (spiders), water (leech), earth (scorpion, cantharis, millipedes, slugs, fleas), leaves (silkworms, cankerworms) or flesh (tapeworms, lice, ticks, bedbugs, maw-worms, maggots).

Isidore's small flying animals included bees, wasps, butterflies, mosquitoes and others (Fig. 2.7).

Reptiles and amphibians (Fig. 2.8) differed from birds and mammals in that they were never valued in Europe as food or for sport. Aside from some limited medicinal concerns related to venomous species, there was little utilitarian value in studying them. Isidore suggested that snakes live for many years because when they slough their old skin, they also shed their old age. In a sentence that makes me contemplate the psychology of medieval bishops, he stated that a snake does not dare strike a naked human. Snakes had, however, considerable significance in folklore, and Isidore presented them very unfavourably, listing in considerable detail a great diversity of mythological snakes and their evil effects. He described dragons

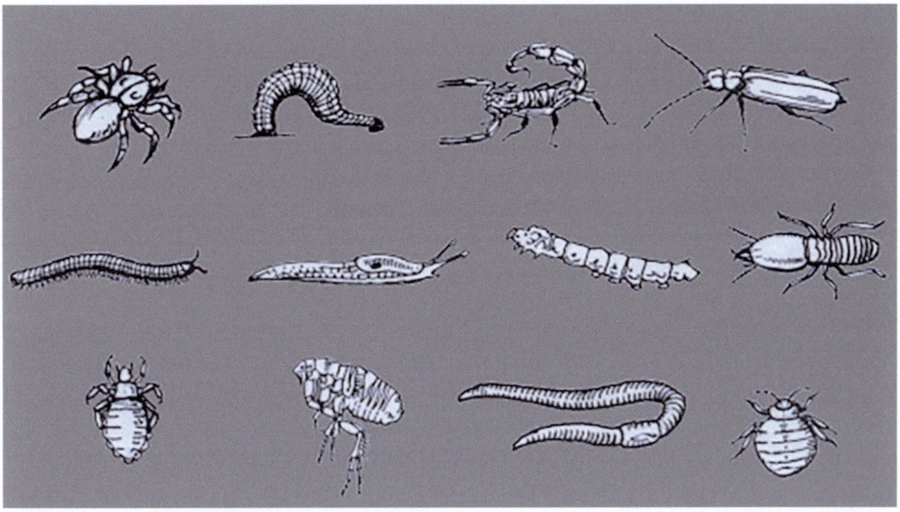

Fig. 2.6 Isidore: diversity of vermin. (Tuvia Kurz)

Fig. 2.7 Isidore: diversity of small flying animals. (Tuvia Kurz)

emerging from caves, soaring into the air and killing whatever they enveloped. *Sibilus* could kill by merely hissing at its victim, *Haemorrhois* caused the victim to lose all its blood, and thereby its life. The mouth of *Prester* was always open and steaming, and *Chelydros* caused the earth on which it crawled to smoke. The skin of *Scytale* gleamed intensely, attracting onlookers who would slow down to wonder at it, and the scytale would then capture them. *Boas* attacked herds of cattle and buffaloes, attached themselves to their udders and killed them by suckling on them. *Iaculus*, a flying snake, threw itself down from trees on animals happening by, thereby killing them; *Sirena*, another winged snake, could both fly and crawl faster

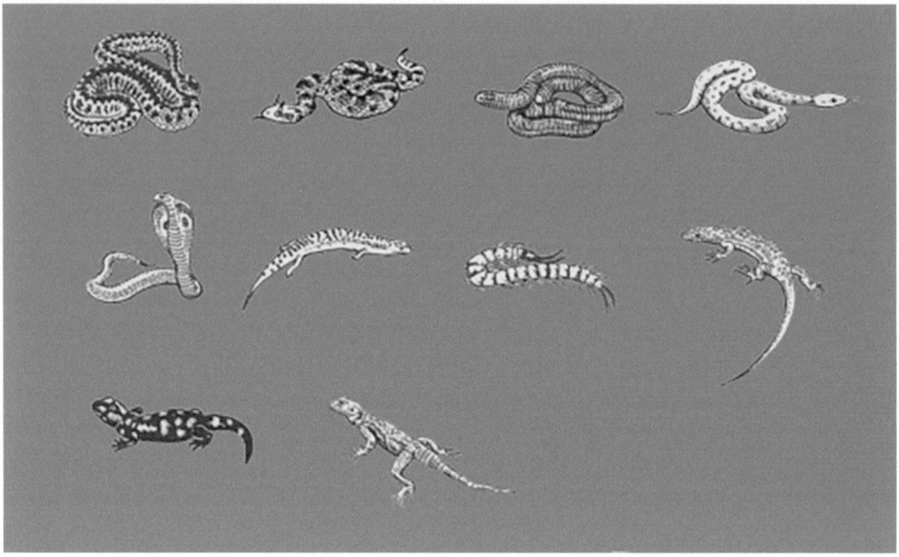

Fig. 2.8 Isidore: diversity of reptiles. (Tuvia Kurz)

than horses, and its venom was so powerful that when bitten, death ensued before the pain was felt. *Salpuga* was an invisible snake.

Salamandra was the most dreaded of all venomous creatures because it could slay many people at once, whereas other venomous animals kill people one at a time. It creeps into the crevice of a tree, injects its venom into all the fruit and so kills whoever eats the fruit; if it falls into a well, its venom kills whoever drinks from it (today we place the salamander among the amphibians, not among the reptiles, as Isidore did). To these vivid descriptions of fabulous serpents, Isidore added the strange behaviour of some snakes. An enchanter uttering magic words can draw *Aspis* out of its den, but it blocks one ear by pressing it against the ground, and plugs the other ear with its tail, so as not to fall under the spell of the magic words. *Amphisbaena* has two heads, one in the proper place and one on its tail, and it advances with both heads leading, its body trailing in a loop. *Parias* always travels upright on its tail, which ploughs a furrow in the ground. The king of all this diversity of serpents was the legendary *Basilisk* of Cyrene.

The basilisk kills other snakes with its odour and feeds upon birds flying above it by first burning and then devouring them. It can even kill by touching something that is in contact with the victim, such as a sword held in the hand, and not only by direct touch. Indeed, if a man on horseback killed one of these animals with a spear, the poison would run up the weapon and kill both the rider and the horse. The basilisk can also kill humans with its lethal stare. To understand this latter mode of killing, we must bear in mind that people of the ancient and medieval days assumed that vision requires physical contact between the viewer and the object he views. They believed that a visual ray streaming from the eyes has a material nature, and it extends in straight lines until it hits an object. Furthermore, this visual ray was also capable of sensation, of feeling the objects in its visual field, even at a great distance, somewhat like a hand: rays emitted by the eyes touch objects as hands do, and then return to the eye. The most powerful and effective eyebeams included the gaze of the gods, the gaze of love, of anger, envy, the evil eye, and eye-transmitted sicknesses. The ray could also broadcast a poisonous mixture of humour so strong that it may hit the man gazed upon, first destroying his eyesight and then leading to his death (Hub 2016). Snakes have no eyelids, thereby increasing both the eye's external appearance and the impression that they are continuously gazing, features that contribute to their "killing gaze".

Pliny noted that the basilisk, unlike other snakes, does not crawl; it "moves along upright and erect upon the middle". This sentence describes side-winding, a mode of movement typical of several desert-dwelling vipers of North Africa (*Cerastes vipera* and *C. cerastes*). Several venomous snakes (*Walterinnesia*, *Naja*) stand upright on their tails when threatened. These two (Fig. 2.9) could have played a role in the legend of the basilisk.

Pliny was not the first to believe in basilisk tales. Alexander the Great (fourth century BCE) ordered a mirror placed between his army and a basilisk that was defending a city. Upon seeing its reflection, the monster died instantly. Saint George (early fourth century) held his shield so that a basilisk saw its own image, causing its death. Many centuries after Isidore, Albert the Great (1200–1280) wrote of the lethal stare of the basilisk (*De animalibus*), and travellers in the Middle Ages would carry cockerels with them, because the basilisk was believed to be vulnerable to cockerels. The basilisk allegedly hatched by a cockerel from the egg of a serpent, part reptile and part bird, and was usually described as either a crested snake or a cockerel with a snake's tail. In 1474, the citizens of Basel (Switzerland) found a cockerel guilty of having laid an egg, which of course went against nature, and the citizens were afraid that this egg would hatch into a basilisk. They beheaded the cockerel following a proper trial and cast the unnatural egg into the fire. Today, the basilisk is the guardian creature and traditional symbol of the city of Basel.

Snakes also connect to ammonites. In Medieval Europe, people named them snakestones because they resemble petrified curled-up snakes. Saint Hilda, seventh-century Saxon abbess of Whitby (Yorkshire, England), rid the area of snakes by turning them into stone. The Jurassic ammonite commonly found around Whitby is *Hildoceras*, in honour of St Hilda, and The Arms of Whitby town council used to feature three ammonites; in 1935, the council changed them into three green snakes. With all these (few) real snakes, together with all these many mythological snakes and all these petrified snakes, snake diversity in the Dark Ages was thus quite considerable.

Moving now on from animals to plants, Isidore classified them into eight groups, as in his classification of animals (Book XVII Rural matters, *De rebus rusticis*):

Grain (about ten kinds).
Legumes (about ten).
Vine (one).

Fig. 2.9 Animals that may have contributed to the basilisk legend. From above: black desert cobra *Walterinnesia*; side-winding snake *Cerastes*. (Tuvia Kurz)

Trees (about 50).
Aromatic, exotic trees (about ten).
Herbs (aromatic or common; about 100).
Garden vegetables (about 20).
Aromatic garden vegetables (about ten kinds).

This classification has almost nothing in common with the previous classifications of Theophrastus and of Dioscorides. Isidore's Grain and Legumes are natural groups, and all the kinds he mentioned are field crops. In his Trees group, he first presented those representing heroic prestige: the palm adorns the hand of a conqueror and is the symbol of victory; the laurel crowns the heads of conquerors. Cultivated trees with juicy fruits come next: apple, the roundest of all fruits, quince, pomegranate, peach, lemon, plum, pear, cherry, fig, and others. Then come (mostly cultivated) trees with an edible hard nut inside the kernel: walnut, pine nut, almond, hazelnut, chestnut, holm oak, and others. Finally came not-farmed trees, in which the wood (not the fruit) was important, whether for timber, cork, resin, for carving statues, or making spears. Here included were cork, beech, pistachio, pine, fir, cedar, cypress, juniper, ebony, plane, oak, ash, yew, maple, alder, elm, poplar, linden, willow, tamarisk, myrtle, mastic, turpentine, box, oleander, reed (actually cane, not a tree), elder, briar, olive and others. Isidore's Aromatic, exotic trees included frankincense, myrrh, bdellium, mastic, pepper (actually a vine), aloe, cinnamon, and balsam. I have presented this list of trees in some detail to give an idea of the extent to which medieval people were familiar with tree diversity.

In his Herbs, Isidore included about 100 wild plants (other than trees), most of which had, in his days, value as medicine, poison, spice or perfume. These four categories actually merged into one another in those days, in that they served to ward off evil spirits of various maladies and misfortunes. For example, below are his first ten herbs:

Saffron helps breathing
Marjoram good for perfumes
Yew tree poison
Dittany drives iron from one's body and rids it of arrows
Mandrake a sedation for those about to undergo surgery
Poppy induces sleep
Dracontea feared by vipers
Henbane causes insanity or hallucinations
Orchid arouses lust
Birthwort purges the womb after birth

Beyond these ten, other examples would be Venus's hair which re-grows lost hair and discourages hair loss; hemlock, the poison from which Socrates drank and died; nightshade, which eases headache; knot-weed, which causes nose bleed; spleenwort, which takes away ill-temper, and bryony, which causes dry breasts to produce milk.

Among Isidore's Herbs, there were also plants for use in day-to-day crafts: Spanish broom to make into ropes, madder to dye wool red, mullein to make lampwicks, and valerian to repel scorpions. In one case, he described the mode of herb collecting: one collects scammony (*Convolvulus scammonia*) by digging a pit around its root and then by placing snail shells or leaves of a nut tree below, to collect the juice. Similarly, among his 20 garden vegetables, some had specific value beyond nutrition: radishes allow one to handle snakes with impunity, lettuce (*lactuca*) replenishes nursing women with milk and, in men, restrains sexual activity, and rocket (*eruca*) ignites the flame of lust. In his last group, that of aromatic garden plants, Isidore placed parsley first, as this was the vegetable that he believed represented heroic prestige. He explained that parsley received its name *apium* because in antiquity, the top (*apex*), that is, the head, of those triumphing would be garlanded with it. Hercules was the first to gird his head with this plant—sometimes he would wear poplar on his head, sometimes wild olive, and sometimes parsley. Parsley's roots are effective against poisons, coriander makes people prone to sexual activity, but in large portions it fosters madness; rue fends off poisons, as shown by weasels who protect themselves by eating it when they fight snakes.

Order in Isidore's approach to biodiversity was very different from our order, but there was order. His *Etymologies* stand out against the background of a millennium almost barren of any zoological and botanical research in the Western Roman Empire and Early Medieval Europe. He had to cope with such superstitions as the tale that a turtle's foot on board retardes the progress of a ship, or that bird flights predict the future ("it is a great sin to believe that God would entrust his councels to crows"; Barney et al. 2006). The *Etymologies* was very widespread, and it had an immense influence on Medieval European culture. Nearly a thousand manuscript copies survive today, a truly large number. In the fifteenth century alone, more than 60 manuscript copies of the entire work, as well as more than 70 copies of excerpts, were written, and it was among the earliest printed books (1472). During the High Middle Ages, between the eleventh and fifteenth centuries, people regarded *Entymologies* as a prime authority (Barney et al. 2006).

Etymologies was definitely an authority for the Benedictine monk Rabanus Maurus (ca. 780–856), abbot of Fulda and Archbishop of Mainz. He compiled an encyclopaedia of 22 books covering a large range of subjects, *On the Nature of Things* (*De rerum naturis*). Rabanus openly stated that his intent was to compile an encyclopaedic handbook for preachers; it was not for biologists. He drew much from the contents of Isidore's *Etymologies*, but the organization and presentation of the material were his own. Book 8, on animals, has seven chapters that refer to seven animal groups:

1. Beasts (mainly carnivores, with exotic animals such as rhinos, elephants, apes, chameleons, and two mythological ones, the unicorn and griffin);
2. Small terrestrial animals (mouse, weasel, mole, hedgehog, cricket, ant, frog);
3. Serpents (poisonous animals: Cerastes, asp, salamander, scorpion, and the mythological basilisk);
4. Worms (earthworm, tapeworm, slug, silkworm, millipede, spider, Spanish fly, *Teredonas* (shipworm?);
5. Fishes (crocodile, eel, bivalves, marine gastropods);
6. Birds (including one bat and two mythological birds, phoenix and cinnamon bird);
7. Small birds (bee, locust, fly).

These groupings are identical to those presented by Isidore two centuries earlier, but for his first chapter, on grazing mammals and omnivores, which is non-existent here. Rabanus also named some groups differently. His work clearly had no taxonomic aspirations: it lacks ungulates, it lumps the chameleon together with the lion, hedgehog with ant, spider with earthworm, crocodile with murex-shell. The only thing we can learn a bit about concerns taxonomic nomenclature. Almost every one of his 38 names for birds is used today, whether in scientific or vernacular nomenclature (Table 2.1). This suggests that by the ninth century, the Latin names of many European birds had become largely stabilized and familiar. This, in turn, suggests a widespread familiarity with bird diversity itself.

The animal inventory listed by Rabanus is similar to that of Isidore's *Etymologies*. Another book of the early Medieval was The *Physiologus*, to which we now come.

2.2.2 Gleaning Zoology from Moral and Medicinal Data

In the days of Rome, as we have seen, one feature of Pliny's *Natural History* was the tendency to seek moral aspects in the behaviour of certain animals, such as elephants, which reflect the traditional ancestral values of Roman society. In the intensely religious medieval days, this tendency led to the attaching of Christian moral, spiritual and even mystical features to animal lore, and to the claim that the "book of nature" has been arranged by God to provide a reservoir of parables, as a source of instructions to humanity. This approach was based, at least in part, on the biblical verse "ask the animals and they will teach you, or the birds of the air, and they will tell you; or speak to the earth, and it will teach you, or let the fish of the sea inform you. For which of all these does not know how the Lord has created the world?" (*Job* 12:7–9). Man is at the centre of the cosmos, and God created everything in the world he had created, just for man's instruction (as well as for his practical use and benefit). Animals have their characteristics because God created them with those specific characteristics to serve as guidelines for proper human conduct and to reinforce the teachings of the Bible. The young of the hoopoe cares for its elderly parents to show human children that they too should care for their parents; and the eagle, just as it rejects any of its young that cannot stare without flinching into the sun, shows men that God will reject sinners who cannot accept the divine light. Nature is a book of moral lessons to guide and improve our lives.

Because animals are neither wholly similar to nor entirely different from human beings, they provide people with a point of departure from which to explore what it means to be human. By the late Middle Ages, people tended to believe that the boundary separating humans and animals was narrow and not clearly defined, that similarities between humans and animals are more significant than the differences, and that under certain circumstances it is possible for one to transform into the other. Gerald of Wales tells the story of a man who was driven mad when a woman whom he had been kissing was suddenly transformed into a hairy monster (Salisbury 2011; Salter 2001).

The holiness of saints and the heroism of knights were revealed in medieval romance through miraculous encounters with wild beasts. One evening, in the fourth-fifth century, while Saint Jerome was sitting with his fellow monks in Bethlehem, a lion suddenly limped into his monastery, whereupon everyone fled

Table 2.1 Bird names used by Rabanus (in alphabetic order), modern vernacular names of these birds, and modern scientific or vernacular names (as presented by Peterson et al. 1964) preserve the name given by Rabanus

Bird name used by Rabanus	Modern vernacular name	Modern scientific or vernacular name preserving name of Rabanus	Remarks
Accipiter	Hawk	*Accipiter*	
Alcion	Kingfisher	*Halcyon*	
Anser	Goose	*Anser*	
Aquila	Eagle	*Aquila*	
Ardea	Heron	*Ardea*	
Bubo	Owl	*Bubo*	
Ciconia	Stork	*Ciconia*	
Cinomolgus	Cinnamon bird		Mythical
Columba	Dove	*Columba*	
Cornix	Crow	*Corvus cornix*	
Corvus	Raven	*Corvus*	
Coturnices	Quail	*Coturnix*	
Fenix	Pheonix	Phoenix	Mythical
Fulica	Coot	*Fulica*	
Gallus	Cock	*Gallus*	
Garrula	Jay	*Garrulus*	
Graculus	?	?	Not identified
Grues	Crane	*Grus grus*	
Hirundo	Swallow	*Hirundo*	
Lucina	Nightingale	*Luscinia*	
Mergis	Merganser	*Mergus*	
Milvus	Kite	*Milvus*	
Nocticorax	Night heron	*Nycticorax*	
Noctua	Owl	*Athene noctua*	
Olor	Swan	*Cygnus olor*	
Onacrotalus	Pelican	*Pelecanus onocrotalus*	
Passer	Sparrow	*Passer*	
Pavo	Peacock	*Pavo*	
Pellicanus	Pelican	*Pelecanus*	
Perdix	Partridge	*Perdix*	
Pica	Woodpecker	*Picus*	or *Pica*, magpie
Psittacus	Parrot	*Psittacus*	
Strutio	Ostrich	*Struthio*	
Turtur	Turtledove	*Streptopelia turtur*	
Ulala	Owl	*Surnia ulula*	
Uppupa	Hoopoe	*Upupa*	
Vespertilio	Bat	*Vespertilio*	A bat
Vultur	Vulture	Vulture	

except Jerome, to whom the lion showed his paw. Seeing that it was badly injured, the saint summoned his brothers and instructed them to wash and bind the wound. The monks restored the animal to full health and, from then onwards, it lost all traces of its former wildness and lived tamely alongside the monks. Power over savage animals was, thus, a sign of sanctity (Salter 2001); and the tale also suggests that any person, no matter how savage he is and how much wounded in body or soul, can find peace under the wings of the Church.

Eight centuries later, Saint Francis (1181–1226) was once in a town where a fierce wolf, who had killed both humans and animals, inhabited the countryside, so fierce that the town-folk feared to go beyond the city gate. St. Francis went out into the countryside and, when the wolf came charging at him with wide-open mouth, he made the sign of the cross and ordered the wolf, in the name of Jesus, to cease its assault. The wolf immediately closed its mouth, bowed its head, and lay down at the saint's feet. Francis then spoke to it, condemning it for its terrible crimes, and then marched back into town with the by-now tame wolf at his side. He told the town-folk to come back to the Lord and do penance, and God would then free them from the wolf in this world and from the devouring fire of hell, in the next world. Francis's supernatural power over savage animals was a sign of his sanctity, for he had miraculously restored Man's authority over the animal kingdom, as it was before the Fall (Salter 2001).

Other conversations, believed to have occurred between humans and animals, further narrowed the human-animal boundary. Saint Francis and his companions, while travelling through the countryside, came across a large flock of birds. Francis ran and greeted them, whereupon they all became alert, and turned toward him, and those perched on trees bent their head as he approached, directing their attention to him. Urging them to listen to the word of God, he said "Oh birds, my brothers, you have a great obligation to praise your creator, who has clothed you in feathers and given you wings to fly, who has provided you with pure air and who cares for you, without any worries on your part". The birds, while he was speaking to them, showed their joy in that they began stretching out their necks, extending their wings, opening their beaks and gazing attentively at him, none of them moving from the spot until he made the sign of the cross and gave them his blessing and permission to leave, upon which they all flew away together. We recall Christ's command to preach to animals (Mark 16: 15: "Go into the world and preach the gospel to every creature"; Salter 2001).

One of the first influential Christian texts to combine animal lore and religious allegories was *the Nature Inquirer* (better known as *Physiologus*). Written in Greek by an unknown author in Alexandria in the second century (or perhaps in Syria in the fourth), it was addressed to a Christian community as a work of religious instruction, using animal legends to inculcate basic tenets of the Christian faith. The zoological information in the texts was not original; we can trace much of it to earlier sources such as the Old Testament, Aristotle, Pliny, Aelian and others (Diekstra 1985). However, *Physiologus* was original in its deliberate application of animal lore to illustrate the Christian doctrine (Diekstra 1985). In the early Christian and medieval communities, symbols were believed to represent and express various aspects of a world that was perceived as widely and deeply meaningful. Animals are to be seen through, not to be seen, and they should be studied for the spiritual lessons they teach, not out of mere curiosity (Ogilvie 2006). Whereas Aristotle had aimed at a systematic investigation of nature, the *Physiologus* tried to explain and justify the ways of God to men (Houwen 1994). Nature is to be studied not for its own sake but for what it can tell us about God's purpose and about how we should conduct our lives. Nature is a book of metaphors for good Christians (Diekstra 1985; Houwen 1994). With this intention, it consisted of a series of brief descriptions of the natural history of some 30–40 existing and imaginary animals, together with a few gemstones. A symbolic interpretation and a Christian moral followed each description. From one animal to the next, the complete text was one long indoctrination of Christian symbolic aspects, and these were the essential texts, not the natural history aspects.

For example, the fox is sly and cunning. It wallows in red mud to appear as though it is bleeding, stops breathing and droops its tongue. Birds think that it is a carcass, come and sit upon it—whereupon it catches them and gobbles them up. Moral interpretation: Like the fox, the Devil too, when he finds a community of ardent believing Christians, behaves as though he is not dangerous; and like the flock of birds, so too the believers are tempted into sinful deeds and pleasures of the flesh—at which point the Devil seizes them and pulls them down to hell. In this animal lore, it is the devil and his ways that are the focus. Whether or not a fox ever wallows in red mud, stops breathing or droops its tongue is irrelevant; the animals themselves are mere vehicles for moral and religious teachings. From the moment that this fanciful hunting behaviour of the fox became fixated, as a behaviour with a religious message, cardinals, archbishops, writers, theologians, historians, and naturalists repeated it, time and again, deep into the Renaissance. (As to the pre-Christian origin of this mythical behaviour of the fox, the earliest text I could find is by Oppianus of Apamea, Syria, from the first century CE. He wrote that when the fox notices a flock of birds, it lies down on its side, sprawls its feet, and closes its eyes and mouth, a position suggesting that it is either deeply asleep or dead. Birds, upon noticing the fox, pounce on it and mockingly pull out its fur with their feet. However, when they approach their teeth, the cunning fox opens its mouth and seizes them. This absurd ancient description found its way into Christianity and into one of the editions of the *Physiologus*, with a significant addition.) A European Jewish version of the fox feigning dead associates this behaviour with a bird: Gershon ben Shlomo of France (second half of the thirteenth century, *Sha'ar HaShamayim* 22-1) wrote of a bird with a very long tongue that approaches ant nests, drops to the ground as though dead, and lets its tongue droop; the ants crawl over its tongue, whereupon it swallows them.

The animal cast of *Physiologus*, in an early surviving version from the ninth century, included about 30 real animals and half a dozen mythological ones, alongside animals that today we cannot identify. One of the real animals was the beaver of central and northern Europe, and obviously, no one in Alexandria or Syria would have seen a beaver. This is another indication that the author had compiled the text from earlier sources, which were also familiar with the fauna of central Europe. In somewhat later versions of *Physiologus,* the bee became a model for chastity and virginity because of the belief that it conceives asexually. The phoenix, which burns to ashes and then rises again, was nature's testimony to the resurrection; the beaver was a model of Christian resistance to sin, and so on. The chapters do not follow an orderly sequence of logical animal groups. Birds, mammals, sea creatures, reptiles and fabulous creatures follow one another with complete disregard for their animal group in modern taxonomy. Sometimes we can see a clear Christian reasoning behind the sequence. The cluster consisting of the pelican, night heron and eagle, deals with Christ's rejection by the Jews; the cluster of the partridge, weasel, aspis and ostrich deals with avoidance of the devil's snares by Christ and man; the sequence siren-centaur-hedgehog follows a series of creatures in the Septuagint translation of Chap. 13 in the book of Isaiah. The basis for the account of every animal included in the book was the Roman-Christian conviction that animals exist for human use. Romans (such as Pliny) believed they benefited man through material use and as moral symbols; medieval Christians believed that they consist also of a divine sign of religious significance.

We first hear of *Physiologus* in the fifth century, when Pope Gelas deemed its contents heretical and had it condemned. In the sixth century, after Christian authors added several more moralizations, the Church accepted it as a means of teaching its ways, and Pope Gregory lifted the sanction. The work's popularity was unprecedented, and it was one of the most well-known works of Christian Late Antiquity. By the early Medieval, translations into many European languages had appeared and, over the centuries, it was freely adapted and transformed.

In addition to the unrestricted addition of animals, the *Physiologus* underwent two major modifications between the tenth and twelfth centuries. One change was in the text, and occurred sometime before the year 1000 with the inclusion of excerpts from the *Etymologies* of Isidore into the text (Diekstra 1985). Etymologizing animals' names became a feature of *Physiologus* in the claim that the hare (*lepus*) derives its name from *levi-pes,* "light-foot". An unknown Englishman, in the twelfth century, introduced a second major change. He rearranged the animals of *Physiologus* into Pliny's four groups of beasts, birds, fishes and reptiles (Diekstra 1985). By this time, the number of animals had increased to over 100, and the *Physiologus* now went by the name "Bestiary", a book of beasts. The term derived from its opening Latin words "*Bestiarum vocabulum*..." Like the earlier *Physiologus*, the subsequent bestiaries also, were not books of natural history, the physical reality of the animals was of no interest in itself and animals were of value only insofar as they were signs for other things (Baxter 1998).

Sometimes the *Physiologus* and Bestiaries changed and created new traditions, through faulty transmission or misreading. Thus, the qualities of the stag in one bestiary were transferred to the elephant in another, because the Greek terms *elephos* and *elephas* are alike. Consequently, burnt stag antlers, which can drive away snakes in one bestiary, become burnt elephant bones and hairs with the same qualities in another. Similarly, in the bestiaries, in the (imaginary) *leucrota*, the mouth splits as far as either the ears or the nostrils, depending on whether reading *ad aures* or *ad nares*. The medical substance sought after in the caladrius to cure eye trouble is found, in some versions, in the marrow of its thighbone (*cuius femoris pars interior*), in others in its dung (*cuius interior fimus*). In Roman mythology, the Caladrius was an all-white bird that could take a person's sickness into itself and then fly away, thereby healing the sick person (and itself, later on). In medieval bestiaries, the Caladrius, if it looked into the face of a sick person, that person would live; if it looked away, that person would die. The bird symbolizes pure-white Christ who turns away from not-repenting sinners and they are doomed to hell, but he heals the souls of those to whom he turns his face and they repent.

Other *Physiologus* stories aimed to give credibility to legends in the Bible, in that they hinted that chains of events similar to those in the Bible also occur among animals. In one such story, the elephants, who dwell in faraway lands, are usually free of the wicked desire for sex. To produce offspring, they go eastwards, towards paradise, where a special plant grows, the mandrake. The female elephant tastes the mandrake and becomes aroused with lust; she persuades the male to eat it, and he too becomes aroused, and they mate. At her time to give birth, she enters a pond until the water reaches her teats, and there she gives birth. She does not give birth on land because she is unable to bend her knees and so cannot lift up the newborn elephant (Diekstra 1985). Now, what happens to elephants could well have happened also to humans, in days of old, and if we come across an ancient text describing such events, there is no reason not to believe it. Thus, in the book of *Genesis*, in the Garden of Eden, Eve ate the fruit of the forbidden Tree of Knowledge, persuaded Adam to eat as well, and their wicked lust was aroused, for which they were expelled from paradise (Houwen 1994).

In some cases, there were attempts to rationalize the zoological text. Thus, when a dog with a piece of meat in its mouth sees its reflection in the water, it opens its mouth to seize the other bit of meat and thereby loses the one it had. Alexander of Neckam (1157–1217) explained the dog's reasoning: A fish, like any other object, looks larger in water than out of water, and that is why the dog, swimming with a piece of meat in its mouth, is deluded by the reflection of the meat and drops it, vainly hoping it would be able to get a larger piece. Neckam's *De Naturis Regnum* (about 1190) contained many anecdotes, stories and popular beliefs about chameleons that live off air, swans that celebrate their own death in sweet song and unicorns that a maiden can trap. He further wrote that in winter the sparrow hawk holds a live bird beneath its feet for warmth; in the morning, mindful of the favour, it allows it to go away alive. In a somewhat unusual animal moralization, Neckam wrote that a goshawk in England one day unintentionally killed an eagle, the King of all the Birds; and that the king and his courtiers had actually witnessed this deed. The courtiers applauded the ferocity of the smaller and weaker bird, which had killed its adversary in self-defence. The king, however, interfered, reproved his followers for expressing sentiments which justified the employment of force by vassals against their sovereigns, and ordered the hawk to be hanged immediately, as guilty of treason.

Bestiaries were often lavishly illustrated, animal after animal. The pictures served as a visual language for the illiterate, who would learn the stories and remember the moral teaching when they saw the beast depicted. Indeed, a great part of the popularity of the bestiaries was as picture books. Medieval animal illustrations were usually not realistic, because often artists had but a vague notion of what the animal that they were supposed to illustrate looked like. The crocodile sometimes looked like some savage bull or a bear; occasionally it appeared with a headset the wrong way around, the whale in some bestiaries was a large-scaled fish, the ostrich had hooves and serpents had feet and/or wings. When it came to mythical animals like the unicorn, dragons or griffin, the illustrator had no choice but to follow earlier drawings.

Adeliza of Louvain, wife of Henry I and queen consort of England, owned one of the oldest bestiaries (Wright 1841). Dating from about 1125 and written in French, it described terrestrial animals from elephant to ant (including the flightless ostrich), then the aquatic whale, then birds, then gemstones—broadly the same sequence as presented by Isidore and previously by Pliny. We are already acquainted with some of the animals and their natural history through Isidore's *Etymologies* and Pliny's *Natural History*, but in Adeliza's bestiary, they received an additional religious significance. The phoenix signifies Christ, in that it too had the power to die of its own will and come back to life from death. Its two wings signify the two Testaments, Old and New. The unicorn is caught by placing a virgin in the forest with her breast uncovered, so that the unicorn comes to her, kisses her breast, falls asleep in her lap and can then be caught; the unicorn with its one horn signifies the one and only Christ, the virgin symbolizes Mary, her breast the Holy Church. The beaver bites its genitalia off, throws them to the hunter and flees, and thus, the hunter leaves the castrated beaver alone. The voluntarily castrated beaver signifies the pious man who abandons luxury and sin and then lives with God (Ogilvie 2006). This goes on and on, from one animal to the next (Wright 1841). In 1188, Gerald of Wales (*Journey through Wales*, translated by Thorpe 2004) confirmed that the beaver will "ransom his body by the sacrifice of a part", and added that once castrated, "he has the sagacity to run to an elevated spot, and there lifting up his leg, shows the hunter that the object of his pursuit is gone". Gerald also claimed that when constructing their dams, beavers "make use of the animals of their own species instead of carts". A few individuals receive on their bellies the logs of wood, cut off by their associates. Holding tight with their feet and having "transverse pieces placed in their mouths", other beavers, who fasten themselves with their teeth to the raft, draw these workers along with their cargo. Only in the seventeenth century did naturalists note that a beaver's testicles do not hang outside the body; they are internal. What people of the Medieval had confused for testicles were in fact small external bumps that connected to internal castor sacs, which produced castoreum, which was a prized oily matter. The beaver marks its territory with this oil's musky, vanilla-flavour scent (Gerald of Wales, 1188 2004).

The two examples below give an idea of the extent to which bestiary writers harnessed biology to theology, in Adeliza's bestiary. The whale *Cetus,* living in the sea, spreads sand over its back, rises to the sea surface and remains there, motionless. Seafarers, thinking it is an island, land on it and prepare their meals. Once the whale feels the fire and the people, it plunges into the deep, and they all drown. Furthermore, the hungry whale opens its mouth and sends forth a sweet and good smell, so that little fish, attracted by the smell, will enter its mouth, and then the whale gulps them down. The whale symbolizes the Devil, and the sea is the world; the sands are the riches of the world, and the fire is love for gold and silver. The Devil tempts people to enter his mouth with sweet attractions and then strangles them.

The pearl oyster (strangely indeed, described as a stone, not as an animal) receives the name *Unio* (as in Roman times, today a term for certain freshwater bivalves). *Unio* is the foundation, the most precious light. It grows in a little shell in the sea near a special island; its surfaces are as smooth as ice. The *Unio* stone opens up at its own will and then, as if it were a living creature, it receives the dew of heaven in the morning, which it encloses in itself, and becomes impregnated with it. Later on, it descends to the bottom of the sea. This dew remains inside the stone for so long that it petrifies into a pearl. Eventually, the *Unio* opens, emits the pearl and then closes again. Just as the stone opens itself when receiving dew, and when closing it joins its two parts without leaving a crack, so the Virgin was consecrated, and thus the Virgin who carried Jesus conceived and was with child. She conceived as a virgin, she brought forth a child as a virgin, and she remained, and shall remain, a virgin. *Unio* is Father and Son, *Unio* is the Holy Ghost; *Unio* is beginning, *Unio* is end; *Unio* is alpha and omega.

Queen Adeliza's bestiary concludes with the statement that in each of the three groups—beasts, birds and gems—there is a king: lion for the animals, eagle for the birds and diamond for the gems. This shows that God is king, in person He is Trinity and One only in divinity.

Around 1140, Hugh of Fouilloy, head of a monastery near Amiens in northern France, wrote another bestiary, the *Book of Birds* (translated by Clark, 1992). He wrote it for very specific monastic pupils, the lay-brothers (*conversi*). Typical lay brothers converted to religious life only as adults, so they lacked the education required of choir monks. In monasteries, they usually worked as manual labourers, thereby enabling the monks to give full attention to more important duties of the monastic community. The *Book of Birds* is a moralized treatise dealing exclusively with birds. It includes birds of the Bible that also occur in Europe, with which lay-brothers would be familiar, plus the fantastic phoenix, well-known from legend. It is in typical Isidore style; thus (Chap. 45) the kite *milvus* is weak, and it signifies those whom the weakness of desire tempts. It feeds on corpses because hedonists delight in the desires of the flesh. Flying around kitchens and markets to seize raw meat reminds us of those people concerned about their stomachs—people who pursue pleasures, frequent the markets and look longingly at kitchens. The kite ambushes chicks, seizing and killing the unwary, like the devil, who takes the weak, simple and careless people and drags them to evil practices (Hugh of Fouilloy, ca 1140 1992).

Bestiaries climaxed in the thirteenth and fourteenth centuries, mainly in England and France. Many of the illustrations were strongly anti-Jewish in that they illustrated, for example, Jews suckling from a sow and eating faeces directly from its rectum (Higgs Strickland 2008, Fig. 12). In 1290, in line with this trend, all Jews were expelled from England (and in 1306, also from France).

Books of a more secular nature appeared only gradually. Even then, they contained only knowledge based on those biological facts authorized by the church. They merely omitted the use of religious morals for each animal.

We can glean the place of animal lore in medieval beliefs also indirectly, as from medicinal texts. Hildegard von Bingen (1098–1179) was born a "tenth" child (a tithe) to a noble family in the Rhineland (Germany of today) and, as was customary with tenth children whom the family could not be sure of feeding, she was dedicated at birth to the church. She gradually climbed the church ranks and eventually founded two Benedictine convents on the Rhine, west of Mainz. Hildegard was closely in touch with the world outside the convent and corresponded with hundreds of people throughout Europe, including Henry II of England, Louis VII of France, the Holy Roman Emperor and the empress of Byzantium. She also embarked on a series of tours, preaching to crowds gathered in the cathedrals of Mainz, Cologne and Worms. Eight hundred years after her time (2012), Pope Benedict XVI named Hildegard a Doctor of the Church, one of only 35 people to have been awarded this distinction and one of only four women.

Hildegard wrote many books in which she developed a visionary, mystic version of the function of the four humours in the human body. At about the age of 60, she wrote a medicinal treatise, *Subtleties of the diverse Qualities of Created Things* (*Liber Subtilitarum Diversarum Naturatum Creaturarum*). The term "subtleties" refers to secret powers hidden in natural creatures for the use of humans. In a later version printed during the Renaissance (1533), this treatise was entitled *Physica*, as it is known today. I base this chapter mainly upon Hildegard's *Physica*, as translated by Throop (1998). While here gleaning biological data from her medicinal text, I emphasize that Hildegard had no intention to write a treatise in biology.

Physica consists of nine chapters, devoted to nine groups presented in an order that reflects the biblical Creation. As a brief reminder of the biblical narrative: on the third day grass and herbs and the fruit-yielding trees were created; on the fifth day, great whales and every other living creature in the waters, and every flying animal; on the sixth day beasts, and every other animal that creeps on the earth. Finally, on the sixth day, God also created Man to subjugate the earth and dominate everything else. Hildegard's nine chapters and the order in which she presented them reflect these biblical groups and the sequence in which they were created. Leaving out three irrelevant (to our book) non-living groups (elements, stones, metals), *Physica's* groups of Life are Herbs, Trees, Fishes, Birds, Animals and Reptiles.

Herbs and Trees Hildegard believed that some herbs grow from the air, others from the wind and yet others from the earth. In her belief that the earth consists (among other components) of sweat, moisture and juice, she thought that its sweat brings forth harmful plants, its moisture brings forth plants which are edible and otherwise useful, and its juice produces grapevines and fruit-bearing trees. Domesticated plants are (obviously) useful, wild ones are usually harmful, but some, when used in medicine, suppress man's harmful humours.

Hildegard presented her Herbs chapter as one long text of over 200 kinds, with no further divisions. First, clustered one after another and in full agreement with the biblical text, come the domesticated herbs which serve as food: the cereals (wheat, rye, oats, barley and others) and a cluster of three legumes (pea, bean and lentil). Her next group are wild herbs used as spices (ginger, pepper, liquorice, cinnamon, nutmeg and others). The four representatives of the carrot family, Umbelliferae, are clustered together (fennel, dill, parsley and celery), as are four species of mint (water mint, horsemint, field mint and spearmint) and four species of *Allium* (garlic, shallot, leek and onion). She also clustered edible plants with the roundish shape of the cabbage family Cruciferae (squashes, turnip, radish) and three exotic plants of sap and resin of the Burseraceae (frankincense, myrrh, balsam). It is noteworthy that in this chapter on herbs, Hildegard chose to present the healing qualities of honey, milk, butter, salt, eggs and sulphur as well. Among common people in medieval times, it was primarily the housewife who tended to beehives, farmyard mammals and domestic birds. This gives me the impression that Hildegard had mainly women in mind as the healing entity in describing her herb remedies.

In her Trees chapter, again in line with the biblical text, fruit trees come first (apple, pear, peach, walnut, cherry, almond, fig, olive and others). Then follow the conifers; cedar, cypress, fir and others, presented more or less one after another, as are two representatives of the Fagaceae (oak and beech). This chapter on trees also included an entry on smoke, as though intended for the person responsible for the household firewood. In summing up Hildegard's two plant-based medicinal chapters, though they have no purely botanical aspirations, we can trace some dim, hazy awareness of a systematic classification of herb and tree diversity.

Fishes Hildegard included all aquatic animals in Fishes: whale, seal, bony fishes, snails ("there is a kind of fish having a shell over it" *piscis conchas habente*) and crabs, altogether 36 aquatic animals. In the biblical text, God had blessed the fishes so that they would be fruitful, multiply and fill the waters in the seas. Accordingly, Hildegard found it important to explain how fishes fulfil this blessing and multiply. Her major problem was the whale, which, she wisely commented, resembles fishes on the one hand and beasts such as the lion and bear on the other. She described the whale as an omnivore feeding on anything from plants to fishes. Now, if the fishes are to multiply and fill the seas, is it not a contradiction that the whale had been created to eat them? Hildegard solved this paradox by claiming that this is so for the benefit of Man, not for the benefit of the fishes. After all, man is the ultimate goal of the Creation, and if the whale did not eat fishes and thus reduce their numbers, the multitude of fish resulting from unchecked reproduction would leave no passageway for his ships to cross the sea.

To become fertile, in some types of fish the male is involved in mating and fertilization (salmon, *copprea*, carp, sea bream, *Elsua*, cod, lamprey), but in others, he is not (turbot, *Grundula, punbelen)*. In some cases, the male spreads a matter coming forth from his mouth over the roe. The male seal swallows the female's "seed" and then spits it out, together with its milt. The male plaice fertilizes by means of foam it excretes from its mouth, which it then pours over the eggs. Both male and female tench, *Tinca,* emit foam from their mouths and breathe on it until the coagulated matter receives life-giving air. The female *rulheubt* (?) discharges its roe into the sea, and then the male pours out a little moisture (not milt) over it, which is so strong that her "seeds" are congealed. The female eel spits "seeds" the size of horse beans from her mouth, and then the male emits a milky substance from his mouth. Whales seek a special site on land in which "the sap from the earth" is stronger than elsewhere. They heave themselves onto this chosen spot, become saturated with the dew and sap of that earth, and their milt begins to increase. The female discharges her eggs, which she covers with earth, by using her fins. Then the male follows her and discharges his milt over the eggs. Only a single juvenile will eventually emerge from all these fertilized eggs. This description remotely resembles marine turtles coming to shore, digging a hole in the sand, laying their eggs and then covering them up with their flippers.

Hildegard also believed that some fishes form hybrids, particularly between fish and serpents. In her words, "Just as a human forsakes his nature by mixing himself with animals, so also animals bring forth a different kind by mixing together. Sometimes fish turn away from their own kind in pouring out their seed onto another kind, producing a variety foreign to their nature, as we can notice in eels and other fish". Hildegard thought that the eel *Anguilla* forms when a female aquatic snake rubs herself on stones or on sand in winter, thereby shedding her skin. In the following summer, she searches for a stone, breathes over it and spits "seeds" (from her mouth) the size of horse beans. Seeing her do so, a male fish hurries over to that stone and discharges a milky substance from his mouth over the seeds, covers them with his tail and writhes over them. Eventually, the young receive life-giving air and begin to live. Now aware that they have transgressed their normal nature, the eel and the snake flee in horror. Other kinds of fish that are born by unnatural hybridization between a male snake and a female fish include the lamprey ("*lampreda*"). Hildegard thought that lampreys come to be by a certain serpent that, when seeing an aquatic fish with eggs, chases it off, folds itself over the eggs and warms them. Lampreys (Fig. 2.10) are born from those eggs and, accordingly, they have the nature of both a fish and a serpent, since a lamprey's tail is (as she believed) poisonous. (Lampreys superficially resemble eels in that they have scale-less, elongated bodies. They live in oceans where they feed by sucking the blood of other fishes. To reproduce, they swim up rivers, where they spawn and then die. The larvae spend several years filter-feeding while burrowed in riverbeds. Eventually, they metamorphose, migrate down to the ocean where they feed by sucking blood and the cycle repeats itself. This life cycle of the lamprey was a mystery for naturalists of ancient and medieval times.) The unnatural hybrid creation goes on with the cod, which is formed from an eel, and further, both male and female cods emit a certain filthy substance from their mouth and from this, the fish named *punbelen* (?) is born. In her belief in strange hybrids, Hildegard was typical of her times. Gerald of Wales (1188, translated by Thorpe 2004), named a certain knight who gave birth to a calf, as witnessed by a great crowd of onlookers. He also described a mare that, after a stag had covered her, gave birth to a creature with front parts of a horse and haunches resembling a deer. A cow served by a stag gave birth to a deer-cow, its front bovine and its rump, tail, legs and hoofs like those of a deer. A bitch covered by a monkey once produced puppies that were ape-like in front, dog-like behind.

Fig. 2.10 The lamprey, *Petreomyzon.* (Tuvia Kurz)

Hildegard took great care to note whether each fish was active during the day or the night. The sturgeon "loves the nocturnal splendour of the moon and stars"; the salmon "likes the moon more than the sun"; the sea bream "loves the night and basks in the moon's splendour"; crabs are active both during the day and at night (the crab "walks forward with the sun before its face, and backwards following the moon").

Birds Hildegard's chapter on Birds actually contains all flying creatures and thus includes, besides 61 birds, one bat, nine insects (honeybee, housefly, cricket, grasshopper and others) and one legendary bird. The first three birds she presented were those of a mixed nature. The legendary griffin is both swift, after the nature of birds, and eats humans after the nature of beasts. Hildegard next presented the ostrich, which has the feathers of a bird but runs like a beast. Third on the list is the peacock, whose voice is, she thought, a mixture of that of birds and beasts. Hildegard did not look upon the peacock favourably, because its origin is disgraceful: Once upon a time, certain small animals capriciously mingled with certain birds in coitus, and thus created peacocks. Today's male peacock still has devious habits, and sometimes mixes in coitus with small animals and little beasts. Their offspring have the shape of their mother, not their father. When the peahen lays her eggs, she hides them from the peacock, who hates and breaks them if he finds them. She goes with her chicks to the male-only area when they have grown and can run. Seeing them, he recognizes them as his chicks, strikes them with his feathers and is joyful.

There were many other birds that, like the peacock, Hildegard did not like. She did not like ravens because they steal people's things; she disliked crows and jackdaws because they dare imitate human voices, carrion crows hurry to where they know there will be sadness, and magpies are mischievous and boastful. Screech owls and great horned owls are thieves who know of human death by the odour of the air and hurry to the cadaver even before lamentation has started. The hoopoe is unclean in that it dwells in filth and seeks the strongest filth in which to nest. As to the pelican, Hildegard had mixed feelings. She wrote that when the mother pelican first sees her chicks hatch, she kills them, but then she becomes so sad that she lacerates herself and resuscitates them with her blood. The pelican perceives and understands a person's sadness and happiness. Soaring high and above, it sees if the coming times will be joyful or sad. If they are happy, they rejoice by singing; if sad, they show themselves sad by being quiet. If it foresees that people will die, it indicates this with a few sounds and is then silent.

A very interesting and intriguing sentence in Hildegard's chapter on birds refers to the turtledove *turtur*: it always has a quasi-servant, *et quasi servum semper habet,* suggesting that for the turtle dove, there is another turtle dove who serves its needs. This calls to mind "helping", a social behaviour in which juveniles and sexually mature adolescents remain in association with their parents and help them raise subsequent broods, instead of dispersing and beginning to reproduce themselves. This simple form of cooperative breeding occurs in 3–8% per cent of bird species worldwide, I do not know if it occurs in turtledoves.

Terrestrial Animals Hildegard's chapter on Animals generally goes from the large, exotic and mighty mammals (elephant, camel, lion and bear) via the medium-sized and familiar ones (horse, deer, farm animals, wolf, cat) to the small mammals and to insects (mouse, shrew, flea, ant). It does not include reptiles, amphibians and venomous arthropods, which she classified as vermin (see below).

She described the mighty elephant in terms of a majestic ruler, strong and honest, yet sometimes of a fierce nature. When the elephant wishes to reproduce, it seeks that land which has, in Hildegard's words, the "water of paradise", and digs there with its foot until it smells paradise's water; the scent of this water arouses its sexual desire. Hildegard regarded the camel as very mighty and, mistakenly, described it as having three humps: one near its neck with the strength of a lion, the middle one with the strength of a panther and a hump near the tail with the strength of a horse. These three humps enable the camel to become both large and tall and to be so strong. The lion is very gentle when mating. When the lioness bears her cubs, she thinks they are dead and deserts them. Upon this, the lion immediately runs to them, roars loudly and wakens them. Once roused, the cubs emit loud roars. Upon hearing these roars, the lioness returns to the cubs, warms and takes care of them. The bear, according to Hildegard, has gentle habits, but he is angered if restrained in his lust. When a bear conceives, she is so impatient in childbirth that she aborts before the cubs within her have fully developed. The cub is born as a lump of flesh that does not move and does not have much shape. Mother bear licks it into shape until all its limbs are distinct, then covers it with her warm body, and the cub begins to grow.

Hildegard also mentioned the unicorn. In ancient Greece and Rome, the mythological unicorn had the body of a horse, the head of a stag, the feet of an elephant, the tail of a pig and a single horn, and it was impossible to catch one alive. People who drank from a goblet, made of this horn, became immune to stomach pains, epilepsy and even poison. In the first and second centuries, only kings could afford to possess a goblet made of the horn. The Septuagint translated the biblical Hebrew term *re'em* into *monokeras*—thereby fixing the unicorn as a biblical creature, created by God. Hildegard provided a rather early version of the widespread medieval belief about "How to Hunt a Unicorn". Once a year, the unicorn goes to the land that has the water of paradise, where it seeks the best plants, digs them up and eats, thus gaining great powers. Beneath the horn, it has an object as clear as glass, in which a person can see his own face as if in a mirror. The unicorn flees other animals and especially fears man, whom it shuns. However, just as the biblical serpent of the Garden of Eden shunned Adam and befriended Eve, so the unicorn avoids men but follows women. When girls walk by themselves to play among flowers, a unicorn may gradually draw near and, sitting on its hind legs, gaze at them from afar. Now, the girls must be noble, not country girls, and they should be adolescents, neither an adult nor a child. The unicorn loves them because he knows they are gentle and sweet.

We might perhaps believe that Hildegard's somewhat fantastic natural history consisted mainly of texts she had read about faraway, exotic and imaginary animals. This is not so. Inaccuracies in natural history are also present in Hildegard's descriptions of European animals, whether wild or domestic. She wrote that the wolf lies in wait for a human and will gladly tear him apart, even if not hungry. Certain "airy spirits" accompany the wolf, and when they see a person for the first time, they weaken his powers, so he does not know that the wolf sees him. However, if the person sees the wolf first, he holds God in his heart, and by that, both the wolf and his "airy spirits" flee. The dog, on the other hand, senses and understands man, loves him, willingly dwells with him and is faithful. The devil hates and abhors the dog because of its loyalty to humans. Recognizing hatred and wrath in a person, a dog will howl at him. If a person has treachery in him, the dog gnashes its teeth at him, even if that person loves the dog. If it knows there is hatred and wrath in a house, it quietly growls. If there is a thief in the house, it growls, gnashes its teeth and goes after him. Furthermore, the dog sometimes has a foreboding of future events, which it reveals in its behaviour. When future events are happy, it is happy and wags its tail; when sad, it is gloomy and howls. As for cats, when a cat becomes thirsty in summer, it licks toads or serpents. The moisture it takes from the toad or serpent is almost like poison, and its brain and flesh become harmful and poisonous to humans. Hildegard described the mouse as having insidious habits and devilish skills. When the female mouse is about to give birth, she seeks very small stones, takes as many of them as she can in her mouth, brings them to her burrow and spits them out. She then climbs on them, warms them up and immediately gives birth. One should tie such mouse stones over the umbilicus of a woman in labour, and she will then give birth. (This use of stones for birth remotely reminds me of the biblical delivery stools, where the Hebrew term for the delivery stool is *ovnaim*, literarily meaning two stones; *Exodus* 1.)

Reptiles For the last chapter in Hildegard's Animals, titled "Reptiles", a more suitable title would be Vermin. These correspond to the biblical creation of all creeping creatures of the earth, on the sixth day. Hildegard wrote that in the beginning, every creature was good, but then the devil used the serpent to deceive Man, whom God then expelled from paradise. After that, seeds of cruel and poisonous vermin rose up for revenge and punishment. Striking hellish terror in people, poisonous, deadly worms repeatedly bubbled out from the soil. These vermin drowned in Noah's flood, and their carcasses were scattered all over the earth. When the flood subsided, these poisonous cadavers decayed, and other worms of the same kind were born from this putrefaction and spread over the world. These are the vermin of today, some of which are deadly poisonous (spider, scorpion, toad), others of which cause disease. A certain serpent (*quoddam genus serpentis*), which lives in both land and water, breathes out a deadly poison. The basilisk is born from certain devilish toads. If the female toad sees the egg of a serpent or hen, she loves it and broods it, until the young in that egg hatch—as a basilisk that breaks through its shell and then sends forth a fiery breath resembling thunder and lightning, with which it splits the earth down to great depths. It then lies in this pit of moist earth until it grows to maturity. Then it climbs back up to earth and kills, with its breath, all creatures it finds. Hildegard's description of the basilisk being born from toads, remotely calls to mind the midwife toad *Alytes*, with its characteristic parental care: the male entwines and carries a string of fertilized eggs on his back (hence "midwife"). When they are ready to hatch, he wades into shallow water and the tadpoles leap out of their eggs. When threatened, the midwife toad inflates itself, filling itself with air, to make it appear as large as possible. Warts on the back of the midwife toad give off a poison so powerful that the toad has few enemies or predators.

Another animal on Hildegard's list of vermin was the weasel *Mustela nivalis*. The female weasel is very speedy, also she knows of a certain small tender herb that contains vital health. Upon seeing her young or another weasel in pain, she quickly seeks this herb, urinates all over it and thereby mixes her strength into that of the herb. When fully soaked with urine, she places this herb in the mouth of an ill weasel and thereby cures it. This herb has no life-giving properties in itself; it receives its power from the weasel's urine. The herb is unknown to humans and other animals, but even if they knew it, their urine would not be strong enough to saturate it with healing properties (Throop 1998). Hildegard's *Physica* gives some idea of whatever biological thought there was in the Dark Ages.

2.2.3 Ornithology Is Born

"*The Art of Hunting with Birds*" (*De Arte Venandi cum Avibus*; translated by Wood and Fyfe 2010) was written in the 1240s by Frederick II (1194–1250) Frederick II of Hohenstaufenh, 1241 (2010) Emperor of the Romans, King of Jerusalem and of Sicily and a lover of wisdom, a man with a philosophic and speculative mind (to quote his own words). Grandson of Frederick Barbarossa, Frederick II grew up in Sicily, where Italian, Norman, Greek and Arab civilizations intermingled. One of the towering figures in the Middle Ages, his brilliant court at Palermo was in the style of an oriental despot: he kept a harem guarded by black eunuchs, he had dancing girls, and he kept a zoo with elephants, lions, camels and many other animals. Frederick II led the Sixth Crusade to the Holy Land, during which he successfully negotiated with the ruler of Egypt for the peaceful surrender of Jerusalem, Bethlehem and Nazareth; then, in the Church of the Holy Sepulchre, he crowned himself King of Jerusalem. With a slight streak of megalomania, Frederick tended to think of himself in terms of the Roman Emperor Augustus (Fig. 2.11). At four different times, popes excommunicated him as a "heretic", a "blasphemous beast of the apocalypse", and even as an "Antichrist". He had scandalous sensual indulgences (in addition to his three wives in succession, he had nine mistresses).

Frederick was a man with an unlimited thirst for knowledge and learning, who spoke several languages and surrounded himself with the great intellectuals of his age. He was aware of Aristotle's studies in biology, probably through the translations of Michael Scot, a wandering medieval scholar who knew Latin, Greek, Arabic and Hebrew, and who stayed at the emperor's court as his physician and astrologer. Here, Frederick questioned him concerning the foundations of the earth, the geography and the governments of the heavens, what is beyond the last heaven in which God sits and the precise locations of hell, purgatory and heavenly paradise; he also asked him about the soul, and about volcanoes, rivers and seas. Encouraged by the emperor, Scot supervised a fresh translation of Aristotle from Arabic into Latin. An enlightened ruler (he was the first king to outlaw trials by ordeal), he encouraged scholarship and poetry and was an avid patron of science. He founded a university in Naples, today the bustling Università Federico II.

One of Frederick's favourite pastimes was falconry, a sport pursued in Europe since Roman times. He summoned many masters in the art of falconry from distant countries, and sought many opinions—indeed, he maintained up to 50 falconers at

Fig. 2.11 Frederick II. Attempting to renew the Roman Empire, he issued a gold coin called *Augustalis*, modelled on the gold coins of ancient Rome. The portrait and inscriptions link Frederick with Caesar Augustus. The front bears Frederick's bust and the inscription CESAR AVG and IMP ROM (Caesar Augustus Imperial Rome); the reverse bears the imperial symbol of an eagle and an inscription with Frederick's name FRIDE RICVS. (Tuvia Kurz)

a time in his court, and his letters requested arctic gyrfalcons from northern Germany and even from Greenland. His *The Art of Hunting with Birds* (also titled *The Art of Falconry*), which took 30 years to write and is largely a practical book, written by a falconer and for falconers, contains detailed instructions on the training of falcons and hawks and detailed descriptions of the various species. However, the book is far more than just a dissertation on hunting. When taken together, the first 57 chapters form a book of what we might term "everything on birds, and only on birds": their diversity and classification, food, flight, roosting sites, migration, mating and functional morphology. He drew heavily from classical sources (mainly Aristotle and Pliny) but added his own observations, and like Aristotle, Frederick sought broad, sweeping generalizations. Altogether, it is the earliest book devoted entirely to the classification and natural history of one single vertebrate group, the Birds.

Frederick used several classifications to divide birds into groups. First, he classified them into three large groups by habitat:

- Waterfowl, which habitually live in or near aquatic areas and are morphologically adapted to aquatic life (merganser, cormorant, swan, pelican, goose and duck).
- Land birds, which live on land and lack morphological adaptations to aquatic life (quail, partridge, pheasant, bustard, peacock, raptors).
- Neutral birds, which may shift from land to water (curlew, plover, lapwing, snipe, crane and stork).

Frederick very much believed in his triple classification, and it turns up time and again throughout his book. Thus, when describing birds that migrate locally, he mentioned both land birds that migrate from mountains to nearby valleys, and waterfowl that migrate to nearby thermal springs and warm rivers. When describing long-distance migrations, he noted that land birds fly in no special order, whereas aquatic birds preserve order, with one bird taking the lead. Evidently, Frederick's triple classification was important to him.

The raptors, which were of special interest to Frederick as a falconer, are clearly land birds. To single them out from other land birds, he suggested that another way to classify birds was by their food:

- Non-raptorial birds (whether aquatic, land, or neutral) live by eating grain and fruit, or on dead bodies and scraps (raven, crow, vulture, lammergeyer).
- Raptors, who prey on other animals (eagle, hawk, owl, falcon and others).

He very correctly noted that raptors differ from other land birds in seven clear morphological characteristics. Their beaks are more curved, stronger, harder and sharper, their talons are curved, hard and pointed, their eyes are deep in the head, they have a short neck, short legs, a very strong posterior toe and a definite number of large flight feathers on their wings and tail. Furthermore, they differ in their behaviour. They are strong in flight but walk badly. The raptor is also more keen-sighted and has more acute hearing than other birds. It flies alone, is long-lived and drives its young from the nest at an early age. The female is larger than the male. Raptors drink only a little and dislike being in the water because their feathers absorb it and they become heavy (with the exception of the osprey, which perches above water and feeds on fish).

He further divided diurnal raptors into three groups by their specific way of carnivorous feeding: vultures and lammergeyers feed only on carrion, never killing prey; kites and eagles prefer carrion, but sometimes kill for prey; falcons and hawks devour only what they themselves kill, never eating carrion. This multi-level classification is a very full and rich description of raptors as a group.

The Art of Hunting with Birds does not mention the mythological phoenix or griffin, but Frederick wrote of the mysterious barnacle goose, of whose nesting haunts there was no knowledge. He wrote of the popular belief that they fall off into the sea from nearby dead trees. The goose hangs from the dead wood by its beak until it is old and strong enough to fly. Accordingly, Gerald of Wales (1188) had described how Irish churchmen would eat the barnacle goose during fasts, arguing that "these birds are not flesh, because they are not born of flesh": they are born on logs of timber floating in the sea, and then "enclosed in shells to secure their free growth, they hang by their beaks, like seaweed attached to timber. In time, they become well covered with feathers, and either fall into the waters or fly away into free air [. . .] I myself have often seen many times, with my own eyes, more than one thousand, minute embryos of this kind of bird, hanging to one log on the seashore, enclosed in shells and already formed" (*Topographia Hiberniae)*. In a bestiary produced around 1250, the text repeats that these birds, found in Ireland, reproduce without copulation. Instead, trees from which they hang by their beaks generate them. They continue to hang there, protected by their shells, until they grow feathers and fly away. Because they are not born of flesh, the writer of this bestiary explains that barnacle geese are sinless and a sign of the reality of the virgin birth of Christ, placed in Nature by God.

Frederick sent special envoys to the North with orders to bring back specimens of those mythical timbers for inspection. His envoys brought back a driftwood plank with goose barnacles attached, but these bore no resemblance to any avian body. Rejecting the truth of this legend, Frederick wrote that this superstition arose from the fact that barnacle geese breed in such remote latitudes that men, in ignorance of their real nesting places, invented this explanation. Today we know that the barnacle goose *Branta leucopsis* breeds in the Arctic, in Iceland and Greenland and winters further south. Perhaps the fact that it was never seen breeding gave rise to a myth that it was, in fact, spontaneously generated from the goose barnacle or gooseneck barnacle *Lepas anatifera*, a marine barnacle that lives attached to driftwood and is occasionally washed ashore. This barnacle has a shell that somewhat resembles a tiny avian egg, from which a root, very remotely resembling a neck, emerges, and external gills, remotely resembling feathers, emerge from the opposite side (Fig. 2.12).

Frederick rejected the myth of barnacle geese reproducing spontaneously from drifting timber. Nevertheless, the barnacle goose myth was very much alive and remained so well into the sixteenth century, with Conrad Gesner and Ulysse Aldrovandi not hesitating in the least to give credit to the myth (Fig. 2.13). The English herbalist John Gerard (1545–1612) also confirmed the story, writing that, in

Fig. 2.12 Barnacle goose and goose barnacle. (Tuvia Kurz)

Fig. 2.13 Goose barnacles turning into barnacle geese. (From Aldrovandi 1602 *de insectibus*)

the Isle of Man, there is "a certaine spume or froth that, in time, breedeth unto certaine shels". These mussel-shaped shells would grow until they split open, revealing "the legs of the Birde hanging out...til at length it is all come foorth". The bird would hang by its bill until fully mature, then would drop into the sea "where it gathereth feathers, and groweth to a foule, bigger than a Mallard, and lesser than a Goose" (Woodward 1927).

Also, an impressive list of Jewish scholars dealt with the barnacle goose myth during the thirteenth century (Seide 1961). All scholars expressed no doubts whatsoever as to the authenticity of the barnacle goose story, but they had to decide whether the barnacle goose was a pure animal, which one could ritually slaughter for consumption, by observant Jews. They argued about what it should be considered—a bird that may be eaten like poultry after proper ritual slaughtering, a fish that one may not eat as it lacks fins and scales, a shellfish that is forbidden for consumption, or a fruit that may be eaten, as it grows on wood. The prohibition of the barnacle goose as food for observant Jews was repeated three centuries later in *Shulkhan Arookh*, the great code for all ritual and legal questions: "The kind of fowl which grows on trees and hang there by their beaks are forbidden by reason of being creeping things that crawl upon the earth" (Joseph Caro, 1488–1577). To stress further the uncleanness of the birds, the whole question was dealt with under the heading "worms". Some animal myths thus involved daily decisions of what to eat, for people of the Medieval (Seide 1961).

From barnacle geese to bird migration, Frederick generalized that:

- Slow birds (small birds, storks and herons) migrate as early as the end of August, better fliers (cranes) migrate during mid-September, and strong fliers (ducks and geese) leave only in mid-November.
- Within a kind, birds of northern latitudes migrate earlier than birds of southern latitudes, which have a shorter journey to make.
- Migrating land birds fly in no specific order, but aquatic birds preserve order: one bird forms the apex of advance and all others in the flock follow successively in arrowhead formation, one to the left and one to the right; occasionally, there is a single line or column. One member of the flock acts as a leader, always on the outlook for danger, of which it warns its companions; it also notifies them of any change that the flock should make in the direction of flight. The whole flock is thus entirely under the control of their leader or guide.
- Migrants that are incapable of withstanding severe cold fly to winter quarters in distant regions, in search of warmer climates (cranes, storks). Those who can withstand severe climates migrate to less distant latitudes (ducks and geese). Birds that feed on insects go as far south as where they can find such insects, from those living on worms to sites where the ground is not frozen.
- On their return journey, the stronger kinds of birds that were the last to migrate (goose and duck) are the first to begin the homeward journey. Following them come crane, stork, heron and quail; last of all are the smaller birds.
- The strongest herons, driven by the desire to nest, return earlier than the others do because they are able to withstand the fatigue of the encounter with extended winter weather; weaker birds linger on.
- Most returning birds seek their birthplace. Individuals of storks, crows, ravens and birds of prey occupy the same nests for many years in succession.
- Non-migratory birds are more numerous in temperate climatic zones.
- The more numerous the kind of any particular genus, or the individuals of a single kind that migrate from the north, the greater will be their numbers on

return to the north. This rule applies to aquatic and shore birds as well as to raptors and other kinds of land birds. Beyond the generalizations, themselves, the orderly manner in which Frederick presented them is also impressive.

Frederick was very happy to contradict Aristotle wherever possible. Contrary to Aristotle's generalization that bad fliers are good walkers, he noted that cormorants, grebes and divers are both bad fliers and bad walkers; he refuted Aristotle's generalization that good walkers are bad fliers, noting that partridges, quail, rail and pheasants are good walkers but only medium fliers. Against Aristotle's claim that owls see only at night, Frederick noted that owls see well both at day and at night. He correctly disclaimed Aristotle's assertion that the same leader heads the migrant column during their entire migration journey, stating that another experienced commander replaces the fatigued leader and assumes his duties, with the former leader retiring to a rear rank. He discredited Aristotle's statement that no one has ever seen, neither the nest nor the young of vultures, by reporting that he himself had inspected vulture nests and found an incubated egg in them. Whereas Aristotle had defined raptors as having hooked talons, Frederick argued that other birds also have hooked claws (jackdaw, swallow), and this definition is not sufficient. This brief list, which sums up all the anti-Aristotle points I could find in the 57 general chapters of *The Art of Hunting with Birds,* points to Frederick's critical approach to ancient texts. The fact that it is so brief and minor is, in my opinion, also a compliment to Aristotle, who lived and studied birds 1500 years before Frederick's time.

Frederick also conducted experiments: To settle the question of whether a vulture is attracted to carrion by its smell (as some writers maintained) or sight, he "experimented and observed that an assemblage of vultures who had their eyes covered but whose noses were not stopped up, did not smell the meat cast before them". To settle the question of whether vultures feed only on carrion or also on live prey, he offered them live chickens and noted that they "refuse to seize chickens when thrown to them alive and before their eyes, but which they ate after they have been killed". To discover whether the cuckoo deposits its eggs in the nests of other birds, he examined the nest of a small bird (mentioned by him as *praeneus*, perhaps the dunnock *Prunellus*) that he once inspected. He found that it contains the young of this bird together with a diminutive creature of dreadful, misshapen aspect that offered no resemblance to any avian species. This featherless chick had an immense mouth and long, thick hair-like down covered its whole head, obscuring its eyes and beak. "That we might eventually establish the identity of this strange nestling, we fed it along with the other young birds; and, behold, at maturity, we saw that it was a young cuckoo. By this experiment, we established the fact that these birds do not build nests of their own but make use of those of other birds in which to lay their eggs". Sadly, in his search for truth, Frederick also cruelly experimented on people. He shut a prisoner up in a cask to see if one could observe his soul escaping through a hole in the cask when the prisoner died. He fed two prisoners, sending one out to hunt and the other to bed, and then had them disembowelled to see who had digested their meal better. He imprisoned babies, forbidding them any contact with adults, to see what natural language they would develop.

As to Frederick's ability to generalize, we can see it in the following summary of his chapters on bird mating.

Mating Mating is preceded by the abandonment of the now full-grown young of the previous union. Raptors expel their young as soon as they can fly, and other birds might postpone driving their offspring off until the last moment before they breed again, but in every case, the parents eventually drive their fledglings away. Some birds are polygamous, and do not assist the female in building the nest, incubating eggs or feeding the young (quail, peacock, pheasant, bustard, grouse). This is true in general for most resident species, but the partridge, for example, which is a resident bird, is monogamous. On the other hand, in all migratory birds, the male assists the female in some of the many phases of brooding and bringing up the young. Mates court each other vocally until forming pairs. If a bird loses its mate, it will try to rob a new mate from an existing couple, and in this case, the males will battle.

Nests Many birds build nests, each with a shape typical of its species. Others, such as the bustard, dig a hole in the ground. The cuckoo deposits its eggs in the nests of other birds, such as blackbirds. Birds that migrate to distant countries build their nests early in the season, so that their young may become strong enough to combat the rigours of the autumn migration. Resident birds, or those that migrate short distances, usually build their nests later than long-distance migrants. Cranes and other large birds, such as swans, pelicans, bustards and vultures, lay few eggs; vultures in particular lay a small clutch, sometimes of one solitary egg. Smaller birds lay a larger number of eggs. Birds that normally nest once a year but lose their eggs or newly hatched young produce a second clutch. In some kinds of birds, only the female incubates (fowls, bustards). In others (pigeons), both males and females take turns incubating. The males of some species do not incubate but feed the incubating female. The ostrich does not incubate its eggs, lest it crushes them with its heavy weight; the sun warms the sand in which the female lays the eggs, thereby incubating them.

The Offspring Among land birds, many chicks run and follow their parent shortly after hatching. They can already feed for themselves, the parents merely clucking and scratching up seed for the youngsters (chickens, quail, partridge, starling, pheasants, peacocks and others). Similarly, many chicks of aquatic birds can swim about shortly after hatching, following their parents and feeding on their own.

Other newly hatched young must have their parents feed them. These include all raptors and other birds that build their nests in lofty places. Hatched featherless and in a nest high above ground, such young would fall and be killed if they were to leave their nest in search of food. Newly hatched starlings, woodlarks and other non-raptorial nestlings who, at hatching, are unable to defecate outside the nest, secrete a tissue-like glue that adheres to and coagulates about their excrement. The parents pick up these droppings and eject them from the nest without defiling their own beak. Eventually, when the young birds grow and are strong enough, they hold their rump over the edge of the nest and defecate outside the nest. The female raptor teaches her young to hunt for themselves. When the nestlings have reached a certain age and vigour, she brings them a dead bird and teaches them to deplume and feed upon it. Next, she fetches a live bird that she herself has partly plucked, so that it is unable to fly with ease. Showing this prey to her offspring, she permits it to fly off as best it can, instructing the fledglings to follow it. If one of the young birds succeeds in capturing it, she calls the others, allowing the successful hunter alone to feed upon the quarry. In this way, birds of prey teach their fledglings not only to fly but also to hunt. After the fledglings have learnt to fly and to hunt avian prey, the parent drives them away to fare for themselves.

This is a rich and varied description of bird reproduction, covering most of the important issues. Altogether, Frederick's remarkably well-organized *The Art of Hunting with Birds* presents a very broad scope of bird diversity, refers to the natural history of each species and formulates reasonable generalizations.

Christian scholars of the thirteenth century, who strongly believed in the theology of the resurrected body, wondered whether the flesh, in the risen body, would be that of humans, or that of oxen and of sheep which they ate. They decided that the risen flesh would be fully human, not animal. However, as people would no longer need food, Thomas Aquinas determined that there would be no animals in the resurrected paradise (Salisbury 2011).

2.3 Albert the Great

2.3.1 Classifying Animal Diversity

In the eighth century, Muslims crossed the Strait of Gibraltar, conquered the Visigoth Kingdom of Hispania and established a new, Umayyad caliphate in Spain. Their regime was enlightened and tolerant towards both Christians and Jews, and scholars of all three religions flourished side by side. Some 300 years later, during the eleventh century, Christian armies began moving southwards and conquering much of central Spain, and by 1080, Toledo was again under Christian control. The new Christian rulers of Toledo found themselves among highly cultured Muslim and Jewish communities, which centuries ago had translated Greek learning into Arabic and had commented extensively on Aristotle, Plato and other Greek philosophers, in light of their own commitments to monotheism. The large bulk of Aristotle's works translated from Greek into Arabic, in the Muslim libraries of Toledo and of other Spanish towns, was way beyond those few books on logic that Boethius had translated into Latin six centuries earlier (Rubenstein 2003). The Christians now set to translate many of these Aristotelian texts, from Arabic (or Hebrew) through Castilian into church Latin, and thus western Catholics became aware of these books for the first time. This immense Aristotelian corpus represented the most comprehensive, accurate and well-integrated account of the natural world that medieval readers had ever encountered, and Aristotle became, in their eyes, a true ancient master of knowledge, in zoology as in other fields. The dramatic reappearance of Aristotle's lost works was an unprecedented event for Europeans of the High Middle Ages—no intellectual discovery before had had anything like the impact of this remarkable find—and the recovered documents forever altered the European way of thinking about nature. Aristotle's newly discovered books included his three zoological tomes, *History of Animals, Generation of Animals* and *Parts of Animals,* translated into Latin in the 1220s. Also, now available in Latin were the comments of the medieval Muslims on Pliny's encyclopaedia (Rubenstein 2003). The time was now ripe for a new, updated overview of biological thought that would bring together all the accumulated information from the newly found Aristotle and the Arab commentators, and critically comment on it.

This overview was written, around 1260–1270, by Albert the Great (Albertus Magnus, ca. 1200–1280, Fig. 2.14), a German Dominican Friar who studied at Padua, where he became acquainted with Aristotle's writings. He taught theology at

Fig. 2.14 Albertus Magnus

Cologne and later at the University of Paris. At about 55, he became Prior Provincial of Teutonia, the German-speaking province of the Dominican order, which ranged from today's Holland through Germany as far as Latvia. This position mandated that Albert spend a great deal of his time travelling throughout the province visiting Dominican convents, priories and even a Dominican mission in Riga. Committed to humility, he refused to ride a horse and made all his journeys through his huge diocese on foot, observing, enquiring, examining and recording nature as he walked along, earning him the affectionate nickname Boots the Bishop. At about the age of 60, the church appointed him Bishop of Regensburg, but he resigned his see after a few years. Because of his lifetime scholarly legacy, contemporaries bestowed upon him the honourable surnames *Doctor Universalis* and *Doctor Expertus*, and late in his life, appended the term *magnus* to his name. Today, the Catholic Church honours Saint Albert as one of the 36 Doctors of the Church, and he is, as proclaimed by Pope Pius XII in 1941, "forever the patron before God of students of the natural sciences" (Albertus Magnus 1999).

Various leaders of the Catholic Church did not immediately receive, with open arms, the rediscovered and translated texts of Aristotle. Indeed, for decades, they censored his views because they discounted the notions of Creation. For Aristotle, since something must always come from something, there must always be something; and therefore, the universe must be eternal. How is one to reconcile Aristotle's claim that the world is eternal with the Christian belief in creation, meaning that the world has a temporal beginning of its existence? If there was no Creation, then there was no Adam and Eve, no primordial Sin, no Fall from Eden and no Last Judgement; and if nature is governed by laws of nature, then there is no room for divine providence or for miracles (Rubenstein 2003). The controversy was an encounter between those who claimed that we should find truth in faith and those who believed that we should base it on reason. If faith declares that the world has a beginning in time, can reason demonstrate that this must be true? We recall that in the Medieval Judeo-Muslim countries, Maimonides had already wrestled with this heritage of Greek science. In the thirteenth century, Christian thinkers confronted this question with the use of authority. In 1215, the Fourth Lateran Council solemnly proclaimed that God had created from nothing all that is [*de nihil condidit*] and that this creation occurred ab initio *temporis*. The University of Paris was barely ten years old when a council of bishops decreed (1210) that, because of the danger these texts might present to the Christian faith, neither Aristotle's books on natural philosophy nor commentaries on them are to be read publicly or privately in Paris and this under penalty of excommunication. Five years later (1215), they extended the ban and, in 1231, Pope Gregory IX renewed it once again. Only gradually did tolerance trickle in, and only in 1255 did an arts faculty statute of the Paris University stipulate that a master must lecture on some of Aristotle's works. In 1277, however, the Bishop of Paris, who was also chancellor of the University of Paris, issued a list of propositions condemned as heretical, among them the claim that the world is eternal.

It was in this atmosphere that Albert, then professor of Theology at the University of Paris, began his translations and commentaries, in order to present and interpret the newly expanded collection of Aristotelian writings. Beginning in the mid-1250s, he proceeded to comment on the entire Aristotelian corpus. This project, to make Aristotle intelligible to the Latin Church and to a Latin audience, remained Albert's preoccupation for the next 20 years. The result was his voluminous commentary *De Animalibus* (*On Animals*, translated and annotated by Kitchell and Resnik 1999) in which he interpreted, criticized and tested Aristotle's theories. His book thus reflects, besides Aristotle's thoughts, his own intellectual ambitions, curiosity and independence. Indeed, Albert has preserved and presented much of our modern knowledge of Aristotle.

Beyond Albert's ability to integrate and unify the vast amount of information on animals that had come down to him through previous centuries, he was a keen and devoted naturalist who sought to compare the written word to the real world and modify his beliefs according to his personal observations. Like Aristotle, he searched for concrete data, believing that only experience, one's own or that of others, holds the key to nature. Observing and reflecting on the habits of one animal after another, he gained prolific knowledge about a wide range of animals. People had claimed that the badger's left legs were shorter than its right ones, but he did not find this, even though he had examined this animal very often. Pliny had said that the eagle wraps its eggs in a fox pelt and hangs them in the sun on the limb of a tree until the young hatch, but Albert discovered none of this behaviour. People had said that barnacle geese are born in trees, hanging from branches and feeding on the sap in the bark—but this is entirely absurd since he himself had seen them lay eggs and nourish their young. Some people believe that in winter the sparrow-hawk holds a live bird beneath its feet for warmth; in the morning, mindful of the favour, it allows it to go away alive, but Albert himself had not experienced this. Similarly, some say that the ostrich eats and digests iron, but he had not experienced this to be so, since he had often spread iron out, for several ostriches, and they did not want to eat it, though they greedily ate rocks and dry bones. Likewise, he rejected the myth (mentioned also in the Talmud) that the ostrich lays her eggs in a glass box; and that when the enclosed chicks hatch, to extract them she uses the *samyr*, a creature that can split glass and rocks; she then warms the eggs by her eyesight alone, not by brooding. As to the claim that vultures do not copulate, this is false; many people have seen them copulate. People have claimed that the toad does not call outside of France, but throughout Germany, he heard them call.

Coming now to Albert's personal investigations, he wrote, concerning cicadas, that if he cut off their head, they go on singing for a long time. In studying the nesting and brooding habits of the golden eagle, he found that it produces two eggs, but has only one chick in its nest. But it is difficult to gain first-hand experience in such matters, because of the height of the mountains in which they build their nests. To test whether the mole has eyes, he dissected a mole. He placed one spider on a red-hot grill and another in the flame of a candle to test the claim that animals with very cold complexions, like spiders, are impervious to fire. He learned about bee metamorphosis by first-hand personal investigation of a bee nest. Albert's knowledge of natural history was thus nature-based wherever possible, and in this, it differed from the text-based investigations so typical of the scholarly medieval. Albert also exchanged views with other naturalists, hunters and farmers, thereby further enriching his knowledge of animals. Naturally, he could not see everything for himself and part of the difficulty for a naturalist, in Albert's eyes, was to examine and evaluate the reliability of other witnesses. In *De Animalibus,* he incorporated only observations reported by people whom he considered reliable. As for bees, he complemented his own investigations with what he had heard from people "who keep bees and busy themselves with the tending of their swarms".

Beyond these personal observations, one of Albert's major contributions to biology was his rejection of many beliefs about animal natural histories as false. These beliefs had accumulated since the days of Greece and Rome down to the High Medieval and had distorted biological knowledge. I have found some 60 false beliefs that Albert rejected. For example, he rejected the claim that the swamp hen *porfirion* (*Porphirio*) has one foot like that of a goose for swimming and the other like that of a land bird, with separate toes and for walking; and that it drinks by drawing water up its swimming foot, and putting food in its mouth with its terrestrial foot. As another example, he rejected the claim that among vipers, the male vomits up venom and calls the female with a pleasant hiss. When in coitus the female seizes his head, bites it off, and from this, she conceives her brood; the young eventually tear her viscera and eat their way out of her body, after which she dies. Furthermore, a viper exists that is like a human in the front of its body, and tapers off in the rear to a serpent. Albert's list of rejections included also mythical animals such as Pegasus, a horse with wings and horns; Harpy, a bird with a human face; Griffin, the half eagle and half lion; Siren, part woman, part bird and part fish; and *Scylla*, a young woman with the tail of a dolphin. Albert also channelled human thought away from the symbolic significance of bestiaries and into the world of real, flesh-and-blood animals. Nature is not (or not merely) a grand allegory that leads to knowledge of God through a series of symbols; it is of interest in itself. The bee, whether or not it is a model for chastity and virginity because it conceives asexually, also has a fascinating biology; so too does the beaver, whether or not it is a model of Christian resistance to sin.

To put order into biodiversity, he classified all animals by a mixture of morphology, habitat, movement, senses, teaching and learning abilities. Aristotle had written that some animals are capable of giving or receiving instruction, whether from one another or from humans. Such animals have such acute senses of hearing and seeing that they can understand the meanings of different words and gestures. This may well explain the emphasis that Albert put on learning in his classification.

In his classification of animals, he formed a top-to-bottom scale upon which he ranked all animal groups according to the extent of their perfection (Fig. 2.15):

Man
Savage.
Monkeys.
Furry quadrupeds.
Birds.
Aquatic animals.
Serpents and relatives.

Fig. 2.15 Albert's classification of animals. (Tuvia Kurz)

Vermin (terrestrial invertebrates).

Within each of these groups, he presented the various animals in alphabetical order. This order of presentation reflects the lack, during his time, of natural sub-groupings within each of the large groups. He did not recognize, say, carnivores versus rodents versus artiodactyls. In his times there was perhaps not much need for such sub-groupings: A local medieval naturalist would have seen only a limited number of quadruped genera, and it would not have been of much point for him to sub-group, say, local bison and local roe deer on one hand, versus local bear and local weasel on the other.

We shall now observe Albert's groups in further detail.

Man Man is uppermost on the scale. He is the perfect standard to which we should compare all other imperfect animals; he alone has all three souls (a vegetative soul for nourishment and generation, a sentient soul for awareness of his surroundings and a rational one for understanding). Man is also unique in his ability to learn and use language. Beyond this, Man's perfection is due to six unique characteristics. First, he has a perfect set of senses. Some animals have more acute senses than humans do, such as cats and dogs, which have a better sense of smell. However, man is best able to incorporate the information received from all his senses into memory and experience. Furthermore, Man's sense of touch is more developed. Now, the sense of touch is the most basic of all senses, and its intensities in man radiate into his other senses. In addition, Man alone has a hand that, being very flexible, serves many purposes. Moreover, man's body proportions are the perfect ones. Furthermore, he moves about in a flawless manner. Finally, "among all animals, the balance of his complexion belongs more to the nature of heaven".

Elsewhere in his book, Albert added five additional qualities that characterize man. He has the capacity to feel shame for an ill-done deed. Moreover, the power of discerning the difference between what is honourable and what is base belongs to humans alone, whereas all brutes merely pursue what is useful and what is pleasurable. Likewise, humans alone possess every conceivable type of friendship. Animals may appear to be friendly to one another, but this is only the appearance of friendship, which stems from the pleasure of those who have grown up together. Furthermore, only humans have the unequalled ability to speculate on intellectual theorems and delight in a pleasurable experience. Therefore, only humans are the perfect conjugal animal since only they make honourable marriages, which laws ordain. Simply put, Man is the only civil animal since he has sharing, separation, dwelling together, treaties and battles, all perfected and ordained by the laws of civilization. Lastly, only humans can learn by use of reason, be gentle by nature due to their civilized character and express their joy by laughter. In short, Albert made great efforts to separate (civilized) man from animal, as in Biblical *Genesis*, by showing that he possesses physiological, mental and psychological superiority.

Savage From one stage beneath man comes the savage, between man and the animal. Like Man, he can perceive much from audible signs upon hearing, and like him, he preserves much of his memories. Like Man, he can also speak—but not rationally. His speech is like the speech of idiots who are stupid by nature. He also lacks reason—he does not create art and has no philosophical contemplations. Though he speaks, he has only an obscure shadow of reason; he does not argue or speak of lofty philosophies, and he lacks the ability to extract the principles of art and science. He uses neither rhetorical nor poetic devices. Accordingly, the savage has neither laws nor a political system. He has no sense of shame; he dwells in the forests and merely follows his natural impulses, just like quadrupeds. Admittedly, he stands and walks erect, and therefore has better comprehension of his surroundings than other brute beasts, but he uses his hands only in the manner that animals use their paws, namely, to pick up food. He does not use his hand as an instrument of the intellect, for works of art, and therefore, he lacks the complete true sense of a human hand. The savage has a sense of instruction, which involves memory, a bit of experience and no reason, science or art. He is midway between the (civilized) human, who has divine intellect, and other mute animals, in which nothing of the divine light exists. Both Man and savage do not immediately imitate what they see, they make progress from instruction towards contemplative sciences, mechanical arts and moral virtues, all while using reason and intellect. Both Man and Savage have the ability to instruct and teach. (In referring to savages, Albert used the term “pigmy”. However, in no place did he refer to their short stature. I am not sure that “pigmy” meant to him what it means to us today.)

Monkeys Next down the rank come the monkeys, who can learn but cannot participate in the experience. Like Man and savage, the monkey has a sense of instruction, but it involves no experience, only imitation, without discerning as to what or whom it imitates.

Furry Quadrupeds are all those furry animals that bend down, due to both the weight of their head and the earthly nature of their body. In our modern classification, his quadrupeds would include most of our terrestrial mammals. Albert’s definition of this group as those with heads bent towards the ground neatly excludes Man and bats, and his reasoning regarding the earthiness of the body excludes marine mammals. His definition presented a conception of this group which prevented him from recognizing all mammals (rather than quadrupeds) as a natural group that also includes humans, monkeys, bats, seals and dolphins.

According to Albert, furry quadrupeds are not as teachable as monkeys. However, they are capable of remembering intentions. An indication of this is that animals can be instructed to do tricks in response to their master’s nods and commands. However, they respond only to his voice, not to his thoughts. (In this, Albert comments, they resemble those humans who perform a certain task all their lives without understanding its true nature.)

Albert presented 113 quadrupeds; many are from his native Europe, others from Asia and Africa. Some he described twice, under different names (such as the cat, once under *Feles* and once under *Cathus*). He believed many mythical mammal hybrids to be real; thus, *Cyrochrothes* is the offspring of a lioness and hyena. It never closes its eyes, has no gums in its mouth and has one continuous tooth, which fits into a recess in the lower jaw. *Leucrocotha* has the body of an ass, the hindquarters of a deer, the breast and legs of a lion, the head of a camel and a mouth that reaches all the way to its ears. *Maricon* has three rows of teeth, the feet of a lion, the face of a human and the tail of a scorpion. *Onocentaurus* is the ass-centaur, some of which have the head of an ass and the body of a human, whereas others have a human head and the body of an ass. *Tragelaphus* is a hybrid between a stag and a goat, *Unicornis* is the unicorn and *Zilio* is a hybrid between a hyena and a monkey. Albert was thus a twilight naturalist, with roots still deep in medieval superstitious zoology, but already with a selective rejection of many of these superstitions.

Birds come next. They resemble walking quadrupeds in having all five senses, in their ability to both walk and run and differ from them in that they can also fly. A bird has feathered wings and it flies by using them like oars, whereas other flying creatures have membranous wings and do not fly in this manner (butterflies, bees). Furthermore, birds never urinate; they lay eggs rather than giving birth and are, generally, more vocal than quadrupeds. After describing birds in general, Albert presented 114 different bird types. Some birds surpass other animals in their ability to imitate (parrot, starling, magpie, jackdaw, raven), others are very musical (calandra lark, skylark, nightingale), while others are mute (swan). Indeed, aquatic birds are less vocal than other birds, large birds are less vocal than small ones and the ability to sing also varies according to the structure of the tongue and palate.

He presented his Birds as one single monolithic group, with the various genera arranged in alphabetical order. One consequence of this was that he did not recognize bird subgroups, what we would call orders and families. For example, eagle *Aquila* and hawk *Accipiter* are separated from buzzard *Buteus* by 13 other birds that happen to have intermediate names in the alphabetic order (*Ardea, Anser, Anas, Alauda* and so on). Similarly, 25 names (*Ciconia, Cygnus, Corvus, Columba* and so on) separate the buzzard *Buteus* from the falcon *Falco*. With this order of presentation, he could not recognize the diurnal birds of prey as a natural group.

Occasionally, among the Birds, Albert followed Isidore in presenting the etymology of Latin bird names. The eagle (*Aquila*) derives its name from its sharpness (*acumen*), for it possesses sharpness of vision, wrath, talons and beak; the lark (*Alauda*) gets its name from praise [*alaude*] since its music celebrates quiet warm weather and so on.

Albert's work on raptors comprises roughly half of the text devoted to birds. Whereas most of his descriptions of birds are rather short (in many cases only a few lines), falcons, hawks and eagles are treated in unusual length and detail, often including comparisons between them (Oggins 1980). To gain an impression as to how accurate Albert's descriptions are, I present his description of the goshawk *Accipiter gentilis*, alongside a figure of the goshawk (Fig. 2.16). Albert wrote that the breast has yellow and black stripes in its first year, which become black on a white background with later moults. Black spots with a few white ones cover its back; it has a curved beak and yellow feet with large talons. Goshawks in northern countries are larger (according to what we, today, would call Bergman's Rule). Furthermore, Albert provided many comparisons of the goshawk with other raptors: its head is rounder than that of the eagle, and its beak is shorter but longer than that

Fig. 2.16 Goshawk. (Tuvia Kurz)

of a falcon. It has more pointed wings than eagles have but fewer than in falcons; generally, it resembles the sparrow hawk. Albert also included some aspects of natural history: it usually lays three to four eggs, and it usually flies alone. All these descriptions of Albert are borne out by modern authorities.

Regarding its hunting, Albert noted that the goshawk looks for the hearts of the birds it captures and hunts domestic fowl. He described in detail how it kills hares ("it fixes its left foot in the ground and holds its prey with its right; then, as quickly as it can, it plucks out its eyes and by this it kills it"). He listed the birds hunted by trained goshawks (crane, goose, heron, duck and coot), noting that it will also take birds larger than those taken by untrained ones (crows). Albert further noted that "since they are quite noble, they capture prey not for food but rather out of a sense of glory and because they delight in tyranny".

Elsewhere, he described falcons, and again he began with an overall description of the bird (compare his description with Fig. 2.17). He wrote that it has black spots on the jaws (the "moustache" so typical of falcons) and white ones around the sockets of its eyes and on both sides of its beak. It has black eyelids and a blackish ash-grey colour on its cranium, back, the upper part of its neck and the outside of its wings and tail. The falcon has a variety of descending stripes in other parts of its

Fig. 2.17 Peregrine falcon. (Tuvia Kurz)

body, broken off here and there. These stripes are usually black, but in its first year, there are also red and pink stripes, which become whiter as the falcon moults, from year to year. This is a very reasonable description of a generalized falcon.

In describing a falcon's hunting habits, he noted that first it rapidly ascends with its talons folded to its chest, and then descends (at an angle) on the bird flying beneath it with a rush so fast that it creates the sound of a raging wind. Striking the prey with its talons, it inflicts such a long wound that the victim sometimes falls split from head to tail. These descriptions of Albert are also all borne out by modern authorities. The peregrine falcon, which feeds on birds, dives towards its prey at an amazing speed of 320 km/hour or even 475 km/hour, tearing its victim with its talons.

From these general descriptions of a falcon, Albert proceeded to describe the various falcon species, some of which we recognize today.

In classifying animals into groups by habitat and movement, Albert placed the bat, a flying mammal, within the Birds, aerial vertebrates. He described the bat *Vespertilio* as a sort of flying mouse but with an elongated head containing many sharp teeth like that of a dog; it also has a voice resembling the barking and growling of a dog more than the sounds produced by a mouse. Sometimes the head bears four ears—Albert was obviously counting both each ear and its very long tragus (Fig. 2.18). The body is hairy, and it has membranous wings and a wide, membranous tail. When not flying, the bat clings to walls and hangs from them by means of

Fig. 2.18 *Plecotus* sp., a bat with a very large tragus. (Tuvia Kurz)

a sharp finger claw; it does not sit or stand, like other animals. In this description, Albert was clearly comparing the bat to other terrestrial quadrupeds, not to other birds; indeed, it is as if he were placing the bat among them rather than among birds. From his description, it is as if Albert had, in his mind, two classifications for the bat. In one, by its overall morphology, vocalization, and, we can presume, also the bearing of live young, he placed it among what we would call mammals. On the other hand, because of its overall aerial habitat and ability to fly, he placed it among birds. The bat's membranous rather than feathered wings, its sometimes four "ears" rather than two, and its inability to stand, walk and run, placed it as separate from both.

Remarkably, Albert only rarely referred to bird migration. He frequently explained the disappearance of birds during the harsh winter by dormancy in local hard-to-find places. Thus, the cuckoo lies dormant in the hollows of trees and rocks, which is when it loses and changes its feathers. Quails, too, lie dormant, as do other birds. The turtledove lies dormant in hollow fissures with its feathers gone, and the hoopoe sleeps (hibernates) in winter like a bat. Albert's argument that many birds disappear in winter because they lie dormant in hidden places might be an extrapolation of his observations that several other European terrestrial animals spend the winter dormant, such as bats, bees, wasps, ants and snails.

In conclusion, Albert's Birds were a monolithic group with no subdivisions and with bats as intermediate between birds and mammals.

Aquatic Animals I believe that Albert intended his descending scale beneath the birds to split into two separate scales according to habitat and not to morphology. He visualized an aquatic scale separate from a terrestrial one. In zoological terms, two scales definitely make more sense than a single one.

Albert found that aquatic animals are more sluggish compared to birds and furry quadrupeds, their senses are less refined, and they are generally dull-witted; hence, their ability to learn is limited. However, aquatic animals display such considerable differences among themselves that they form a scale on their own. The dolphin, whale and seal are on the uppermost rank on the aquatic scale, for they are cleverer than other aquatic animals. Beneath them come the bony fishes, and then the cephalopods (octopus and squid). Fishes are higher in rank than cephalopods because they have a finer set of senses, and because they grow mainly in length, whereas the cephalopods grow in a circular fashion, and this, Albert suggests, is an imperfect trait. Furthermore, fishes have flesh, a strong bony skeleton and distinct viscera and heart, whereas cephalopods have only a hard inner cartilage skeleton and lack a distinct head, intestine or heart—in short, a defective body with a defective soul. Albert wrote that admittedly cephalopods sometimes have foresight (thus, he claimed, cuttlefish tie themselves to rocks by their tentacles during a storm, so that they will not be cast away by the force of waves; and when afraid they emit ink so as not be seen), but fish are shrewder than cephalopods. Albert also admired the cunning of cephalopods. He wrote that the octopus hides itself not far from a closed bivalve, wraps a rock in one of its tentacles, and once the bivalve has parted its valves, it rapidly inserts the rock between them. The bivalve cannot close its valve because of the rock, and so the octopus sucks it out and devours it. The octopus thus has sharp senses, imagination, estimation, foresight and shrewdness of a certain inference. It rapidly changes its colour to imitate that of its environment, and it rarely lives beyond two years.

Lower still among the aquatic animals were the crustaceans (crab, lobster) that lack many of the sense organs and internal organs present in other aquatic animals. However, they are more perfect than sea snails and bivalves, in that they move about much faster.

Snails are very sluggish in their movements and beyond their sense of touch, have only a few sense organs. They move about very little and when they do, they move through expansion and contraction of their body, not by means of the proper organs of motion (feet, wings, fins). They are, therefore, even less perfect than crustaceans.

Below these come the sponges, which are immobile, putting them on the very lowest scale of animals, merely one stage above plants.

On this aquatic scale, starfish occupy a vague mid position, somewhere between octopuses and plants, whereas the octopus has eight long tentacles, the starfish has only five short ones. They are alike in the way each of them sucks up its food through its whole body. The octopus has suckers along the entire length of its tentacles, which (Albert believed) are “mouths” as it were, and it sucks its prey through them. As an indication of this, when the octopus clings to some animal, especially when it clings to a human, as soon as it wraps its tentacles around him, it draws out blood everywhere. The starfish does not have open round “mouths” of this sort, but it has a cleft along the entire length of all its five “tentacles”, in some of which there are small, hard little veins similar to small roots, and it sucks its prey through these.

Pliny, as we have already noted, had believed that “everything born in any department of nature also exists in the sea, as well as a number never found elsewhere”. According to him, there are creatures resembling sheep, and others with the heads of horses, asses and bulls in the sea. Beyond animals, the sea also contains various objects found on land, such as swords (swordfish), saws (sawfish) and cucumbers (cucumber-fish). Twelve centuries after Pliny, Albert produced a vast array of marine animals which, he thought, resembled terrestrial counterparts so closely that he named them sea ram, sea crow and so on. Table 2.2 presents Albert's names of over 20 animals that have terrestrial counterparts.

Serpents and Relatives Shifting back to Albert's terrestrial scale, beneath the birds come serpents and their relatives (our reptiles and amphibians): a mixture of crawling-creeping animals, some without legs and some with many: snakes, lizards, tortoises, salamanders, centipedes, crocodiles, toads and frogs. Of course, some of these are semi-aquatic. They are imperfect because, as a group, many of them do not have perfect organs for progressive motion: their legs are often small (skinks) or absent (serpents), and they creep on their bellies, sometimes assisted by their ribs.

Albert, surely influenced by Isidore, who lived six centuries earlier, believed in many fables about serpents. *Aspis* bears a precious stone on its head; feeling itself bewitched by the chants of a wizard, it presses one ear to the stone and plugs up the

Table 2.2 Names of marine animals with terrestrial counterparts: Latin name given by Albertus; translation into English, modern scientific name and modern vernacular name are based on Kitchell Jr. and Resnick (1999)

Latin name given by Albertus	Translation of Albert's name	Modern scientific name	Modern vernacular name
Anguilla	Sea snake	*Anguilla*	Eel
Aries marinus	Sea ram	*Orca gladiator*	Killer whale?
Corvi maris	Sea crow	**?**	?
Canes marini	Sea dog	*Phoca vitulina*	Seal
Draco maris	Sea dragon		Sea snake?
Equus maris	Sea horse		Fictitious
Equus nili	Nile horse	*Hippopotamus*	Hippopotamus
Equus fluminis	River horse	*Hippopotamus*	Hippopotamus
Foca = bos marinus	Sea ox	*Phoca vitulina*	Seal
Irundo maris	Sea swallow	*Exocoetus volitans*	Flying fish
Locusta maris	Sea locust		Lobster
Lepus marinus	Sea hare	*Lepus* sp.	Sea hare
Monoceras	Narwharl	*Monodon monoceros*	Narwhal
Monachus marinus	Sea monk	*Monachus monachus*	Monk seal
Porcus marinus	Sea pig	**?**	?
Pavo marinus	Sea peacock	*Scarus cretensis*	Parrot wrasse
Rana marina	Sea frog	*Lophinus piscatorius*	Fishing frog
Scolopendra	Sea centepede	Annelida: Nereidae	Polychaete
Scorpio maris	Sea scorpion	*Scorpaena porcus*	Sculpin
Tortucua maris	Sea tortoise	*Dermochelys coriacea***?**	Leatherback?
Vulpes marinae	Sea fox	*Alopias vulpinus*	Fox shark
Viperae marinae	Sea viper	*Echiichthys vipera*	Lesser weever

other ear with the end of its tail, so as not to hear the spell of the chants. *Armene* kills merely by its gaze and by the sound of its hiss. *Basilisk* has a deadly breath that burns every tree, herb and thicket, splits rocks and infects the air so badly that when a bird flies towards the serpent's location, it immediately falls down dead, as do approaching beasts. It also kills with its hiss and with its gaze. *Boa* stays close to the herds of cows. By suckling at one udder after the other, it grows to an immense size. *Celydrus* is a serpent whose strong poison causes the land where it passes through to smoke. *Dragon* has a long, hairy mane drooping from their neck, like that of a horse. Some dragons have wings, and some are marine. *Haren* is a serpent with a hairy head. *Hydra*, the most beautiful of serpents, lives along the banks of the Nile. When a crocodile sleeps on the bank with its mouth gaping, the hydra wraps itself with slippery mud to better slide into it, and enters the crocodile's jaws; when the crocodile awakens, it swallows the hydra, which then creeps about inside the crocodile and tears out its entrails. *Jaculus*, a winged serpent, lives in trees and infects the fruit so that a person who eats the fruit dies. *Phareas* moves about on its tail, almost upright. *A siren* is a monstrous serpent with a sweet hiss, like the sea sirens; some sirens fly. *Seps* bear precious stones in their heads or they have horns like those of rams, and they kill people by swinging at them with the horns. While Albert thus rejected many false stories about familiar animals, he continued to accept many fables about snakes he had never seen. As in mammals, and also in reptiles, Albert was thus a twilight naturalist, with roots still deep in medieval superstitious zoology, but already with a selective rejection of many of these superstitions.

Vermin Beneath the serpents and their relatives (reptiles and amphibians) and last and lowest on Albert's scale, come the vermin (what we would call terrestrial invertebrates). They are not subject to instruction by teaching, because they cannot distinguish between different words, though they can hear sounds. They, therefore, cannot learn, like other animals higher on the scale. They move about on many feet, which emerge from the chest, and their abdomen has no legs. Some vermin have double generations: they emerge from an egg as a caterpillar, which feeds on leaves; later, they emerge from what seems to be a second egg (we would term this the pupae) and become flyers that feed on nectar. Almost all vermin have all their senses. At the bottom of the vermin are the earthworms, born from subterranean vapours or from rotting roots, not by sexual reproduction. They have no sense organs other than the sense of touch, they do not see, do not hear, do not smell, do not taste. They move about, not by means of feet or wings but by contracting and then extending their body segments. They suck up their food and, like plants, they do not release excretions. Beneath the vermin, as a group, come the sponges, marine creatures that are completely immobile, and in this, they are on the lowest scale of animals, only one stage above plants.

To what extent was the medieval naturalist aware of animal diversity in his surroundings? In Table 2.3, I present all European furry quadruped genera (with body length over 12 cm) as listed by Albert, and compare them to a similar list available to a modern naturalist some 800 years later. For this purpose, I use Van den Brink's *Die Saugetiere Europas* (Van Den Brink 1955), from about 70 years ago, when

Table 2.3 European terrestrial mammal genera of over 12 cm: name given by Albert; Modern common name (from Van den Brink 1955); and Modern scientific name (from Van den Brink 1955) as relevant to the name given by Albert (whether genus or genus plus species)

Albert	Modern common name	Modern scientific name, as relevant to that of Albert	Remarks
ERICIUS	Hedgehog	***Erinaceus***	
	Pyrenean desman	*Desmana pyrenaica*	Endemic to Spain
TALPA	Mole	***Talpa***	
SIMIA	Monkey, barbary ape	*Macaca sylvana*	(= ***Simia*** *inuus*)
CUNICULUS	Rabbit	*Oryctolagus* ***cuniculus***	
LEPUS	Hare	***Lepus***	
PIROLUS	Squirrel	*Sciurus*	No relevance to modern name
VARIUS	Squirrel	*Sciurus*	No relevance to modern name
	Suslik	*Citellus*	
EMPTRA	Marmot	*Marmota*	No relevance to modern name
CASTOR	Beaver	***Castor***	
	Garden dormouse	*Eliomys*	
	Forest dormouse	*Dryomys*	Balkans, eastern Europe
GLIS	Fat dormouse	***Glis***	
	Dormouse	*Muscardinus*	
CRICETUS	Common hamster	***Cricetus***	
	Grey hamster	*Cricetulus*	Remote, in Greece
	Lemming	*Lemmus*	Scandinavia
	Vole	*Clethrionomys*	
	Snow vole	*Dolomys*	
	Water vole	*Arvicola*	
	Vole	*Microtus*	
	Muskrat	*Ondatra*	
	Field mouse	*Sylvaemus*	
	Rat	*Rattus*	
	Spiny mouse	*Acomys*	Crete
	Mole rat	*Spalax*	Eastern Europe
ISTRIX	Porcupine	***Hystrix***	
LUPUS	Wolf	*Canis* ***lupus***	
PAPIO	Jackal	*Canis aureus*	No relevance to modern name

(continued)

Table 2.3 (continued)

Albert	Modern common name	Modern scientific name, as relevant to that of Albert	Remarks
	Arctic fox	*Alopex*	Extreme northern Scandinavia
VULPES	Fox	***Vulpes***	
URSUS	Brown bear	***Ursus***	
URSUS	Polar bear	*Thalassarctos*	Presented by Albert together with brown bear
DAXUS	Badger	*Meles*	In German **"Dachs"**
ERMINIUM	**Ermine**, stoat	*Mustela* ***ereminea***	
MUSTELA	Weasel	***Mustela***	
GUESSELIS	**Weasel**	*Mustela*	
	Mink	*Lutreola*	
PUTORIUS	Polecat	***Putorius***	
FURO	**Ferret**	*Putorius putorius* ***furo***	Domesticated form of polecat
	Polecat	*Vormella*	In eastern Balkans
LUTER	Otter	***Lutra***	
MARTARUS	Marten	***Martes***	
	Glutton	*Gulo gulo*	
NEOMON	Mongoose	*Herpestes* ***ichneumon***	
GENETHA	**Genet**	***Genetta***	
CATHUS	**Cat**	*Felis* ***catus***	
FELES	Cat	***Felis*** *catus*	
LINX	**Lynx**	***Lynx***	
EQUI	Horse	***Equus***	
APER SYLVESTRIS	Boar	*Sus scrofa*	No relevance to modern name
DAMMA	Fallow deer	***Dama***	
CERVUS	Red deer	***Cervus***	
CAPREOLUS	Roe deer	***Capreolus***	
ALCHES	Elk	***Alces***	
RANGYFER	Reindeer	***Rangifer***	
BONACHUS	Bison	*Bison* ***bonasus***	
URNI	Auroch, **ur**	*Bos taurus primigenius*	Wild ox
OVIS	Sheep	***Ovis***	
IBEX	Alpine **ibex**	*Capra hircus* ***ibex***	
	Chamois	*Rupicapra*	
MUSQUELIBET	**Musk** ox	*Ovibos* ***moschatus***	
	Saiga	*Saiga*	Eastern Europe

In bold: Modern terms resembling those of Albert

classification still relied solely on external visual characters, and without the newer microscopic and molecular classification tools, which, of course, were not available in the thirteenth century either.

Albert's list includes some 42 European mammal genera compared to some 63 modern ones. About ten of the genera that he does not mention are rodents, which Albert would probably have lumped into "mice", a term for which he used *Mus* (this would include today's recognized genera *Cricetulus, Lemmus, Clethrionomys, Dolomys, Arvicola, Microtus, Ondatra, Sylvaemus, Rattus, Acomys*). Other genera are from remote fringes in Europe, which would not have been easily accessible to Latin naturalists of those days (our *Dryomys, Spalax, Vormella* and *Saiga,* of Eastern Europe, *Alopex* of extreme northern Europe). Bearing this in mind, Table 2.3 suggests that the medieval naturalist was very well aware of genus-level mammal diversity in his surroundings.

A second point emerging from Table 2.3 is that some 35 of Albert's 42 names resemble names used today, whether in scientific names (say for the bison, his Bonachus versus today's *Bison bonasus*) or in modern common names (say, for the badger, his Daxus versus today's German Dachs). Indeed, eight of his Medieval Latin names have been carried over to modern English. Altogether, most of the names we use today are identical, or at least similar, to those used eight centuries ago. This points to considerable stability in mammal names through the centuries, from Medieval times (at least) onwards, in line with the stabilization of European bird nomenclature by the ninth century. This, in turn, suggests a widespread familiarity with mammal diversity itself. When, five centuries after Albert, Carl Linnaeus presented a formal system of classification for the entire Animal Kingdom, he did not invent all the animal names: he used those Latin or common names used by European clergy and commons of his day, where possible.

Albert's concept of a scale of animals with man at the apex and vermin at the base was an excellent complement to the Scriptures in that it offered a complete model of animal life. Furthermore, it closely reflected the structure of the Catholic Church, with the Pope at the top and then the cardinals, archbishops, bishops, priests, deacons and the lay. In a way, his scale thus gave theological legitimacy to the hierarchical structure of the church.

2.3.2 Generation and Developmental Biology

Albert believed that spontaneous generation occurs in vertebrates, but rather infrequently. According to him, the only quadruped to do so is the mole, *Talpa*, which is generated from rotting earth soaked by rain. In addition, he mentioned a few cases among fishes: the eel *Anguilla*, he claimed, is born from earthworms rather than through sexual reproduction. Albert took this idea from Aristotle, who wrote that eels develop spontaneously from earthworms, which grow spontaneously in the mud and decaying matter of shallow waters along the seashore, in rivers and in coastal marshes. Indeed, Aristotle claimed, people have actually observed eels inside dissected earthworms. Albert suggested a scenario similar to that of the eel rising spontaneously as worms from decaying mud when wet. (In the ancient and medieval eras and up to the eighteenth century, the life cycle of the eel was a mystery to many naturalists because it lives in freshwater, but returns to the ocean to spawn and then dies. Consequently, anglers never caught young eels, and the eel life cycle was a mystery.)

Spontaneous generation was, so Albert believed, common among terrestrial invertebrates. Snails (under the Latin term *Testudo*, a term today reserved for tortoises) are generated from rotten vegetable matter and viscous dew; their terrestrial viscosity sometimes hardens in a spiral into the shell in which they dwell. Slugs arise from mud, on which they feed. Flies are born out of the putrescence of animals and their dung. Fleas are born from warm, moist dust. Lice generate from the putrescence at the openings of a man's pores, or from the warmth of his clothing, or the fur of an animal, or among feathers in birds of prey. Many ravenous boys have lice when they eat fruit, especially figs, because the coarseness of the pulpy acidic fluid, which passes from the stomach to the small intestine and consists of gastric juices and partly digested food, generates many lice. Salamanders are born from putrefaction. Horsehair snakes (*Seta* in Albert's terminology, today *Gordius aquaticus*, an aquatic worm) grow spontaneously from standing water that is not very foul. They are born from the hairs of horses, since "horse hairs that have been put in standing water take on life and spirit and move, as we have proven by test many times". *Tinea* (a rash of the skin caused by a fungus) arises from the vapours of rotten wool; it eats the wool itself, as well as hair. Shipworms (a wood-eating marine bivalve) generate inside wood and eat their way out of the wood through a hole. The earthworm is born from the exhalation of the earth. To conclude, Albert believed that spontaneous generation is something that occurs mainly among terrestrial invertebrates, is uncommon among fishes and lower quadrupeds and is present in only a single case among mammals. Spontaneous generation was a central issue in biological thought throughout the Medieval.

As to sexual reproduction, Albert devoted considerable attention to the male and female contributions to the formation and growth of the embryo. He presented Aristotle's ideas on embryology along with the views of Galen, Avicenna and Averroes, and he added comments of his own, which are not always clear. We focus on Albert's views on three major embryological questions: the origin, nature and role of the semen in the formation of the embryo in the uterus; the process and moment of animation; and the period of gestation.

Semen Albert realized that both male and female contributions are necessary to form "everything which is generated out of spermatic moisture" (what we would term the fertilized egg). This moisture, to have complete generative power, must have within itself the properties of both the male power, which creates actively, and the female power, which creates passively. Somewhat ambiguously, he mentioned Aristotle's example of the two parts involved in the substance of cheese, namely the rennet (corresponding to the male), which brings about the form of the cheese, and the milk (corresponding to the female), as material out of which the cheese is made.

This seems an acceptance of Aristotle's view, but on the other hand, he spoke, further on in the sentence, of two types of sperm involved in the basic substance of a fertilized egg, namely the male sperm, which affects the creation, and the female sperm, which undertakes the coagulation, the giving of shape and form. Albert thus thought, like Galen and unlike Aristotle, that both males and females produce "sperm" (what we would call a hereditary ability). He added that in some, or possibly in all animals, a third moisture is required, that which feeds the developing embryo. This third component is in the menstrual blood of mammals and the yolk of oviparous animals. Its presence where the two sperm interact is essential since otherwise, the fertilized egg would waste away, not having nourishment. Thus, Albert also adopted from Galen the division of the female contribution into two components, nourishing and material.

Focussing on vertebrates, Albert suggested that the male sperm is what is left after food has gone through a series of digestions, the last being in vessels of the testicles (he was not sure whether arthropods and molluscs had sperm). He adhered to Aristotle's view of matter as pure potentiality, with new parts developing by a gradual shift, from potentiality to actuality ("epigenesis"). This approach contrasted with the alternative idea, as though everything was latent but actually present in something precious, as though the seed actually contains tiny versions of all the organs within itself ("preformation"); Albert assigned this idea to Plato. These ideas, of epigenesis versus preformation, would occupy biological thought for centuries to come, and I shall present them in further parts of this book.

Albert further speculated that semen contains two distinct kinds of substance, of which one is temporary and variable, the other eternal and unalterable. All individual animals change throughout their lifetime through generation and destruction, but they are unchangeable in terms of the form of the permanently existing animal group to which they belong (say, a species). Animal groups maintain their eternity by perpetuating their unique form, and they do so via the constant generation of individuals, each group maintaining its soul and life via the soul and life of each of its individuals. By receiving the particular form of an animal group, the individual animal participates in a "divine permanence" and is, thereby, an integral part of the divine order, even though it is subject to individual change.

God has granted the stars a "formative power", *virtus formativa* that flows from them into the earthly elements, the semen (Takahashi 2008). Without this power, vegetative and sensitive souls cannot come into being. All creatures-to-be had existed already beforehand within Him, before the Creation. Eventually, He poured this plan, through the stars, into the particular elements on Earth, in a manner comparable to an architect who builds a house with a plan of that house already in mind. The formative power is comparable to the architect's hands and instruments, into which the plan pours from the mind, and through which the plan becomes a reality of a similar form. It is thus a kind of operative authority, conceived by God for both the one-time Creation and the current generation of creatures. This formative power, informed by the divine powers, produces the eternal species of animals, alongside the production of the corruptible individual. It resides in the semen and is composed of elemental heat, of the power of the stars, of the soul, and the power of God (which Albert terms the prime intellect or the prime intelligence). Animal generation is, therefore, a process under the supervision of God. For Albert, in conclusion, generation implies the intervention of a superior intellect, in the same way in which artificial productions require the artist's intellectual activity (Takahashi 2008).

The formative power generates the vegetative and sensitive souls, but not the rational soul. Three powers operate together in the formation of embryos, namely the powers of the elements, the soul, and the prime intellect of the universe. Every natural phenomenon is the work of this intellect, which sustains all operations in nature. This prime principle, the "universally active intellect", knows everything that comes into being. This knowledge is concerned both with the perception of beings and also with their production, as it is the forms or species known by and held in this principle that produce by themselves all entities, just as form conceived in the artisan's mind is the formative cause of artefacts. The knowledge contained in this prime principle is productive of all beings; it is the cause of being and of the order of all entities, including natural things. Since the agent principle cannot be plural, nature must be subordinate to the prime intellect. The diversity of nature is thus ultimately reduced to the prime principle. It is this formative power that relates the natural world to the prime principle, that is, to the universally active intellect.

Process How does the semen generate form? In elaborating on Averroes' ideas, Albert suggested that there exists a divine formative power, *virtus formativa,* in the male semen. Albert's concept of a formative power connected two complex embryological questions, about the origin of the male semen and about the induction of the soul. It also combined all universal forces into one grand system, which the one prime intellect, God, controls, from the movement of the cosmic stars *maior mundus,* down to the microcosm *minor mundus* of embryogenesis. This formative power pours down from the prime intellect of the universe, through the intermediary services of the celestial bodies; it introduces itself into an animal's semen and makes use of the semen's four elemental qualities (hot-cold, wet-dry) as its instrument to form bodily organs. The spirit forms inside the testicles: a certain moisture contained in the semen evaporates, stimulated by the heat of the genital organs. The formative power in the semen acts upon this vapour so that the spirit begins to "respire" inside the semen. Once inside an embryo's body, the spirit serves as a vehicle that carries the formative power. Thus, all forms of all animals take their origin from God; indeed, He has already preconceived the forms of all animals, just as an architect's mind preconceives an intellectual form, the structure of a house.

Regarding the soul, Albert followed the classical distinction between vegetative, sensitive and rational components. He argued that the formative power, "informed" by the prime intellect, generates the two inferior kinds of soul, the vegetative and the sensitive ones (*On the Nature and Origin of the Soul, De natuare et origine animae*). The rational soul differs from the other two in that matter cannot generate it; the prime intellect inserts it directly into the embryo. In this manner, the embryo gains life only gradually, by first attaining the vegetative soul, then the sentient one, and only then the rational soul. The soul forms only when the embryonic body has been organized because it needs a body clearly distinguished in shape from the bodies of other animals (Takahashi 2008).

Albert believed that to intercourse, in which the powers of both sexes mix, the Creator added the wondrous power of pleasure. If animals of a kind were to shun intercourse, they would surely die off, since reproduction involves being weighed down by pregnancy, pain in giving birth and getting busy with the care of nursing the young ones. Although these things are generally relevant to the reproduction of animals, Albert believed that they were particularly true of humans, whose nature is worthier and whose complexion is nobler. Indeed, Albert concerned himself mainly with human embryology.

Gestation Albert proposed a timetable for the development of successive embryonic stages in the uterus, taking human embryogenesis as an example (IX, 120–130):

1. Male semen and female humour joined	0 days
2. Embryo-to-be becomes frothy, with three large vesicles to provide space for the heart, liver and brain	6 days
3. Appearance of blood-like droplets and red threads	9 days
4. Coagulation	15 days
5. Formation of the heart, liver, brain and spinal cord	27 days
6. Embryo fully formed, with head, hands and feet	36–40 days

It thus takes some six weeks for the organs of the human embryo to develop fully, and from then on, the embryo merely grows (we today consider this stage, of a foetus in which all major body organs are present though not yet fully developed, to begin from the ninth week after fertilization). In its eleventh week, the foetus begins moving inside the womb, but this may vary according to the extent of happiness between the couple. If the wife is on good terms with her husband at the time of impregnation, and the husband presents himself most happily to his wife, the foetus may start moving eight days earlier. Albert knew this because "in matters such as these, a great deal of credence must be placed in women worthy of trust, who have given birth to a number of children".

Not all pregnancies end in the happy birth of a baby after nine months. In medieval days, there was a commonly held view that premature deliveries were most successful in the seventh month but were likely to be stillbirths in the eighth. This view arose from popular beliefs in the perfection of the number seven on the one hand, and in the malignity of Saturn's control of the eighth month, on the other. Albert, in discussing the influence of celestial bodies on pregnancy (in *De fato*, as presented by Demaitre and Travil 1980), identified several distinct physiological changes influenced by lunar revolutions. The moon modulates fluid motion such as menses and sperm production; in conjunction with Mercury, it promotes the mingling of the spasms; it has the power, derived from the sun, to instil life into moisture; and under the influence of Venus, it determines the development of the embryo. As the moon controls all movement and change of the embryo, and as the number seven is of special significance, Albert concluded that embryogenesis must be completed by seven lunar revolutions, a stage at which the embryo is complete and has all things necessary for its being. Why, then, do most human births occur only during the ninth month? This is so because at this stage, there might not yet be a sufficient balance between its formative power, on the one hand, and its material power, on the other. An abundant formative power could compress undersized material, and then a small but agile premature baby would be born in the seventh month. Normally, the material is adequate, and the formative power gains strength by resting, through one revolution of the moon, so that birth comes in the ninth month. If the material is uncooperative, it is unable either to complete the movement started under lunar impetus in the seventh month or to recover in the eighth month, and then the infant is born and, in most cases, dies.

Albert's embryology thus revealed his deep interest in astrology, a central component of his thought. Throughout the Medieval, intellectuals believed that life on Earth is a microcosm within the macrocosm (the cosmos itself) and that the celestial bodies follow patterns and cycles analogous to those on Earth. With this belief, it seemed reasonable to assert that astrology could also predict and determine the probable future of a creature. Albert gave much attention to the production of monsters in nature as a visible manifestation of the forces of accident and chance. He explained monstrous animal births, such as Siamese twins, dwarfism or surplus fingers, by the movement of stars (Takahashi 2008).

Beyond being merely academic, *De Animalibus* also had practical aspects. Albert repeatedly referred to culinary aspects of animals and whether their flesh was tasty (sturgeon, carp, salmon, buzzard, bittern, hazel hen, domestic fowl, pheasant, thrush) or not (rays, wild goose, swan). In addition, he repeatedly referred to medicinal prescriptions based on animal ingredients. Of his 140-odd remedies, about 60 were from domestic animals. Domestic animals had many uses in medieval times. Accordingly, Albert placed heavy emphasis on farm animals: he devoted 28 pages to the horse, only three to the red deer; seven pages to the dog and only one to the fox. Albert's *De Animalibus* was thus not only for nature lovers, it was also an applied zoology for farmers, meat-eaters and pharmacists.

2.4 Generation: Anatomy

The roots of the medieval attitude to human anatomy go back to ancient Rome, where the law prohibited the dissection of cadavers; thus, Galen, the greatest authority on the form and function of the body, based his conclusions on the dissection of animals. After the fall of Rome, Christianity (and subsequently, Islam) absorbed the negative Roman attitude to cadaver dissection. As the dissection of humans was not possible, the study of functional human anatomy involved mainly farmyard animals, as human surrogates; I am not aware of any dissection of wild animals, such as Aristotle's dissection of a snake.

Medieval physicians were generally content to rely upon ancient Greek and Roman texts, which were passed down over the generations. Galen's legacy lived on in Arab and Persian cities, where physicians translated his work, pored over it, and embellished it with interpretations and commentaries. However, Galen's spirit, with its emphasis on observing rather than relying on authority, was lost. Christian thought emphasized the soul, not the body in which it was contained, and which it considered a mere temporary cask for the soul. Under such circumstances, there was no incentive for personal observation. Indeed, for a physician, detailed knowledge of functional anatomy was unnecessary because medicine relied mainly on seeking balances between the body's four humours. A physician who knew the general positions of individual organs, and their very broad presumed functions, and who mastered Galen's theories of functional anatomy, had all he required. He could obtain all this knowledge from a translation of an Arabic summary of Galen's anatomical treatises, such as that of Avicenna. The study of anatomy was thus narrowed down to the reading of a Latin translation of an imperfect Arabic summary of Galen's teaching. This complete submission to the dictates of antiquity characterized the medieval period and gave the classics an authority and finality that suppressed all stimulus for further inquiry. Everything worth learning was in Galen's books, so there was no point in directly observing the morphology of the body.

In line with this general negative approach to the practical study of human anatomy, in the twelfth century, various monastic orders that practised medicine prohibited members from dissecting human bodies, under the threat of excommunication. The strange claim was that they, as holy men, must not spill human blood ("The church abhors blood", *Ecclesia abhoret a sanguine*; see Ghosh 2015). In 1299 a Papal Bull of the Catholic Church prohibited the dismemberment of corpses and the boiling of bodies to reduce them to bones: "all forms of mutilation and tampering with cadavers, boiling, and embalming bones from human bodies, are prohibited, apart from those judicial interventions to ascertain the cause of death". This bull was aimed to prohibit trade in the bones of crusaders as holy relics, who required that in case they lost their lives in distant lands, their remains would be sent back to their homeland for burial in their native lands (Renghachary 2007, Ghosh 2015; for example, the cases of Emperor Frederick Barbarossa, who drowned when crossing a river in Turkey in 1190, and of King Louis IX, who died on a failed crusade to Egypt in 1270). The papal bull was not opposed to dissection for educational purposes. The prohibition against human dissection in Islamic culture meant that Arabic knowledge of anatomy, too, came predominantly from Greek and Roman sources. In a reality in which human dissection was illegal, and in which medieval physicians could rely only on the morphological texts of eminent figures from the past, medieval physicians turned to the dissection of animals.

Anatomia Porci (*Anatomy of the Pig*) was written in Salerno around 1100–1150 by an unknown person, with the statement, in the Introduction, that "Since the internal organs of the body of Man were not known, the ancient physicians, and especially Galen, undertook the presentation of the internal organs by the dissection of animals. Pigs resemble us in their internal organs, and therefore, we are going to dissect a pig". The book mostly describes the position of the various organs, but in the description of the female genitalia, we get some idea about the horizons of a medieval physician. In reference to the dissection of the sow's uterus, the text guided the reader to cut the uterus along the midline and find inside it the two "testes" (ovaries), which transfer the female semen to the uterus, where it joins the semen of the male. It then informs the reader that "there is also a long tube, the umbilical cord, which breaks when the embryo is born. When the embryo is born, the midwives break the umbilical cord and tie it, at a distance of four fingers from the newborn". This mixed pig-human text exemplifies the use of the pig to explain the anatomy and physiology of Man, in a reality in which dissections of the human body were not favoured and not common.

Later, during the High Medieval, the negative approach towards dissecting corpses began to fade. The newly founded universities had a growing demand for first-hand knowledge of anatomy for training students in medicine, and this meant the dissection of corpses. Accordingly, Frederick II, the well-educated and wise Holy Roman Emperor, issued a decree (1231) that mandated that the medical school of Salerno must dissect a cadaver at least once every five years for anatomical studies; and attendance would be compulsory for all those wishing to practice medicine or surgery (Ghosh 2015). This was the first royal decree that required all medical students to study anatomy first-hand, as part of their curriculum. Spain passed a similar law a short while later (1238), and Venice (1308), Germany (1347) and Naples (1365) all followed, within a century. In Bologna, the authorities permitted cadaver dissection, on condition that the body belonged to a person who had lived beyond a 50 km radius of the town; and a short time later, cadaver dissection became permitted in Padua and Pisa. During the fourteenth century, whatever the religious restraints imposed on dissection and autopsy that had then existed had thus relaxed considerably, and the church was broadly in favour of putting cadaver dissections into the curriculum, as long as the corpse would then receive a Christian burial. In those days, 20 students would witness the dissection of each male body and 30 of each female, because of the difficulty of procuring female cadavers; however, female cadavers were more readily dissected because of a prevailing belief that women do not have souls.

Another cause that facilitated cadaver dissection was the need to become alert to the potential presence of epidemics. In 1286, there was, in several cities in Italy, a great mortality among both humans and hens; a doctor had some of the hens opened and found that there was a vesicular aposteme on top of each hen's heart. "He also had a dead man opened and found the same thing", a coincidence so suggestive that it moved a Venetian physician to issue a bulletin warning of the dangers of eating chickens and eggs. From this public health context, the practice of autopsy to determine the cause of death shifted rapidly to a forensic one, of suspected murder, such as by poison (Park 1995). Then there was evisceration, to preserve the bodies of people who had died some distance from their chosen place of burial, and the bodies of notables whose funerals might take some time to organize. These various causes led to it, and eventually, Pope Clement VII formally accepted the teaching of anatomy by dissection (1537). In many European universities, however, no dissections took place for very many years, whether because of a lack of cadavers, a lack of motivation or for some other reason.

Mondino de Luzzi (ca 1270–1326) was born in Bologna, where he also obtained his medical degree, and it was here that he was a professor of surgery. His father and grandfather had both been pharmacists in Bologna, and his uncle was a professor of medicine, so Mondino was continuing a family trade (Mavdrovi and Paraskevas 2014). He was one of the first to incorporate a systematic study of anatomy and dissection into a medical curriculum, and to do so with full permission of the authorities: in 1315, he performed an officially sanctioned dissection of a human cadaver in full public display, in the presence of medical students and of other spectators. His treatise *Anathomia corporis humani* (1316) is one of the first medieval texts devoted entirely to human anatomy, and it soon became the standard anatomical textbook in most European universities for over two centuries. It was first printed in 1478 and ran through 40 editions, and Mondino was regarded as the grand old master of anatomy (in his days another atlas of human anatomy, containing 24 plates, was composed by Guido da Vigevano (1280–1349), who dedicated it to Philip the Fortunate, King of France; Rengachary et al. 2009). Beyond anatomical descriptions, the book elaborately explained the dissection procedure, as well as techniques used to visualize specific anatomical structures. For his classes, Mondino would boil parts of a corpse to separate the bones, and sometimes he would dissect the body only after soaking it in water to achieve better visualization of the muscles and nerves (Crivellato and Ribatti 2006).

Mondino's *Anathomia* opens with the assertion, similar to the views of Albert the Great, that human beings are superior to all other creatures because of their intellect, reasoning, tool-making abilities and upright stature. Man is worth studying because he possesses these noble qualities. Broadly following in the footsteps of Galen, Mondino theorized that the human body was composed of three distinct compartments. The most inferior compartment was the abdominal cavity, which housed the digestive viscera that nourished the body; it was under the control of the liver, which contained the "natural" part of the soul, which transmitted through the veins. The second compartment is the thorax, with the organ of respiration, and is supervised by the heart, which is transmitted through the arteries, those vital "spiritual" parts of the soul that keep the body alive. The third compartment was the skull, controlled by the brain and responsible for mental activity, motion and sensation, and the organ which conveys the "animate" part of the soul through the nerves (Infusino et al. 1995; Mavdrovi and Paraskevas 2014; Mavdrovi and Paraskevas 2014).

Mondino described the compartments of the body in the order in which they presented themselves during the dissection process, beginning with the viscera of the abdominal cavity. To give a taste of Mondino's high standards of observation, I present his very detailed description of the dissection of the muscles of the abdominal wall (as given by Crivellato and Ribatti 2006):

> There are eight of these muscles [...] Two muscles are longitudinal; indeed [in the *rectus abdominis*], their fibres run longitudinally from the xiphoid appendix of the sternum to the pubis [...] In addition there are two oblique muscles that are superficial [the external oblique muscles], one on the right side and one on the left. Both originate from the ribs and terminate with their tendons on the pubis; the right tendon runs downward and to the left, whereas the left one runs downward and to the right. The deep muscles [the internal oblique muscles]

> proceed upward in opposite directions, their origin is on the pubic and iliac bones, and they insert with tendons into the distal parts of the ribs, so their tendons intersect each other, making an X picture. Beyond these muscles come the transverse ones [the *transversus abdominis*], one to the right and the other to the left, whose fibres proceed sideways. In the upper part they are more dorsally situated and intersect the longitudinal muscles, forming a right angle.

Considering that this is one of the earliest texts in descriptive anatomy, its detailed accuracy is impressive.

A close examination of *Anathomia* reveals that while relying on his dissections of the human body to a considerable extent, Mondino found it difficult to distance himself from Galen's descriptions, even when his own personal findings were different (Infusino et al. 1995). This is obvious in his description of the uterus: he wrote that the human uterus has two horns, as in sows, cows and other farmyard mammals. (Today we know that the two uteri in the most primitive mammalian situation are quite separate; in most mammals, the distal parts of the two uteri fuse, leaving two non-fused, hornlike uteri; but in higher primates, including humans, there is complete fusion to a single uterus, which completely lacks hornlike structures, Fig. 2.19). Furthermore, Mondino maintained that the human uterus divides into seven cavities, that the three warmer cavities are for male foetuses, the three colder ones are for female foetuses and the seventh is intended for a hermaphrodite. Mondino also followed Galen in (erroneously) maintaining that a *rete mirabile* or wonderful network, which is found only in the brains of ungulates, is present also in the human cerebrum. Such texts are clearly not based upon what one observes in dissections (Infusino et al. 1995; Mavdrovi and Paraskevas 2014). (A *rete mirabile,* occurring in some vertebrates, is a network of small arteries and veins lying very close and parallel to each other. With their blood flowing in opposite directions, it acts as a counter-current exchanger of heat, gas concentrations, or blood pressures.)

Early in his career, Mondino probably performed all the dissections by himself. Later on, he most likely removed himself from the hands-on work and did so only occasionally. Local public authorities provided bodies of executed criminals for dissections at medical schools. These dissections were public, and the professor of anatomy (*Lector,* lecturer) would sit on a high chair, a distinct physical entity placed well above the cadaver, and from this lofty position would read out an authoritative anatomical text and comment on it (say, Mondino would read aloud from Galen;

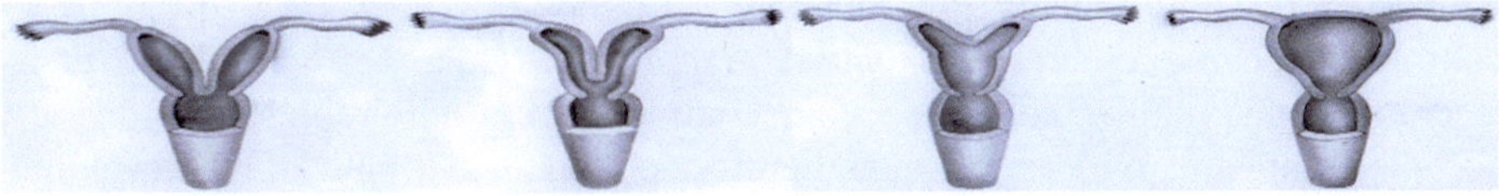

Fig. 2.19 Four main forms of the mammalian uterus, representing progressive fusion of the ends of the oviducts. From left to right: The two uteri are wholly separate (rodents, bats). The two uteri are separate for most of their length but share a single cervix (ruminants, cats, horses). The upper part of the uterus remains separate, the lower parts are fused into a single structure (dog, pig, elephant, whale, dolphin). The entire uterus is completely fused into a single organ (higher primates, including man). (Tuvia Kurz)

later, other lecturers would read from Mondino's *Anathomia*). On the floor, a surgeon (*Sector*), who was a layperson of the commons, often a barber-shaver with his sharp yet delicate knives, would be doing the dissection itself (and from here, the German family name *Schneider*). The professor would be lecturing, say, on the heart, and the surgeon would rummage through the body, dissecting and then lifting out the heart for the audience to see (Fig. 2.20). Sometimes an *Ostensor* would be present as well, a person who would point out with a pointer, those specific parts of the body that the professor was describing. In these public dissections, the surgeon would quickly mutilate the interior of the corpse, and it would have been difficult for students to see much when looking at mangled and decomposing organs. No wonder, therefore, that Galen's text was held to be indispensable and more important than anything that might be observed in the mutilated corpse. Thus, the tradition of relying on the written word remained dominant, and there was little encouragement to conclude from dissection, even when the lecturer's own personal observations were in conflict with Galen's (Rengachary et al. 2009).

Due to natural decomposition, cadavers were suitable for dissection only during the first three to four days following death; after this, the stench became too much for the dissector to bear. As a cadaver would decompose even faster in warm weather, medical schools preferred to dissect it in the winter months. Dissection was carried out in rapid sequence using crude dissection tools such as cleaver knives, and they were conducted throughout the night so that the study of the entire body could be completed in four days. Food and wine were provided for students and instructors during this unpleasant ordeal. The practice of cadaver dissection led to cadaver robbing, for dissection in private houses, which involved informal teaching between a lecturer and his students. In 1319, four students of a certain Master Alberto, who was a lecturer at the University of Bologna, were persecuted for robbing a grave and bringing the corpse to the house where Alberto lectured (Rengachary et al. 2009; Ghosh 2015).

Mondino's *Anathomia* became the standard anatomical textbook for over two centuries.

2.5 Teaching: Generation and Diversity

University teaching usually presents a broad panoramic view of a discipline, in contrast to specific research that focuses more narrowly. Furthermore, as university teaching usually aims at non-specialist students, it often represents generally accepted mainstream ideas. We may therefore obtain an overview of medieval biology from the way faculty taught it at universities of the time. By 1300, almost every Western European country had at least one university. Education included Roman law, ecclesiastical law, medicine, works of the Church Fathers and the Scriptures. European universities were where people came to train for occupations in theology, law and medicine.

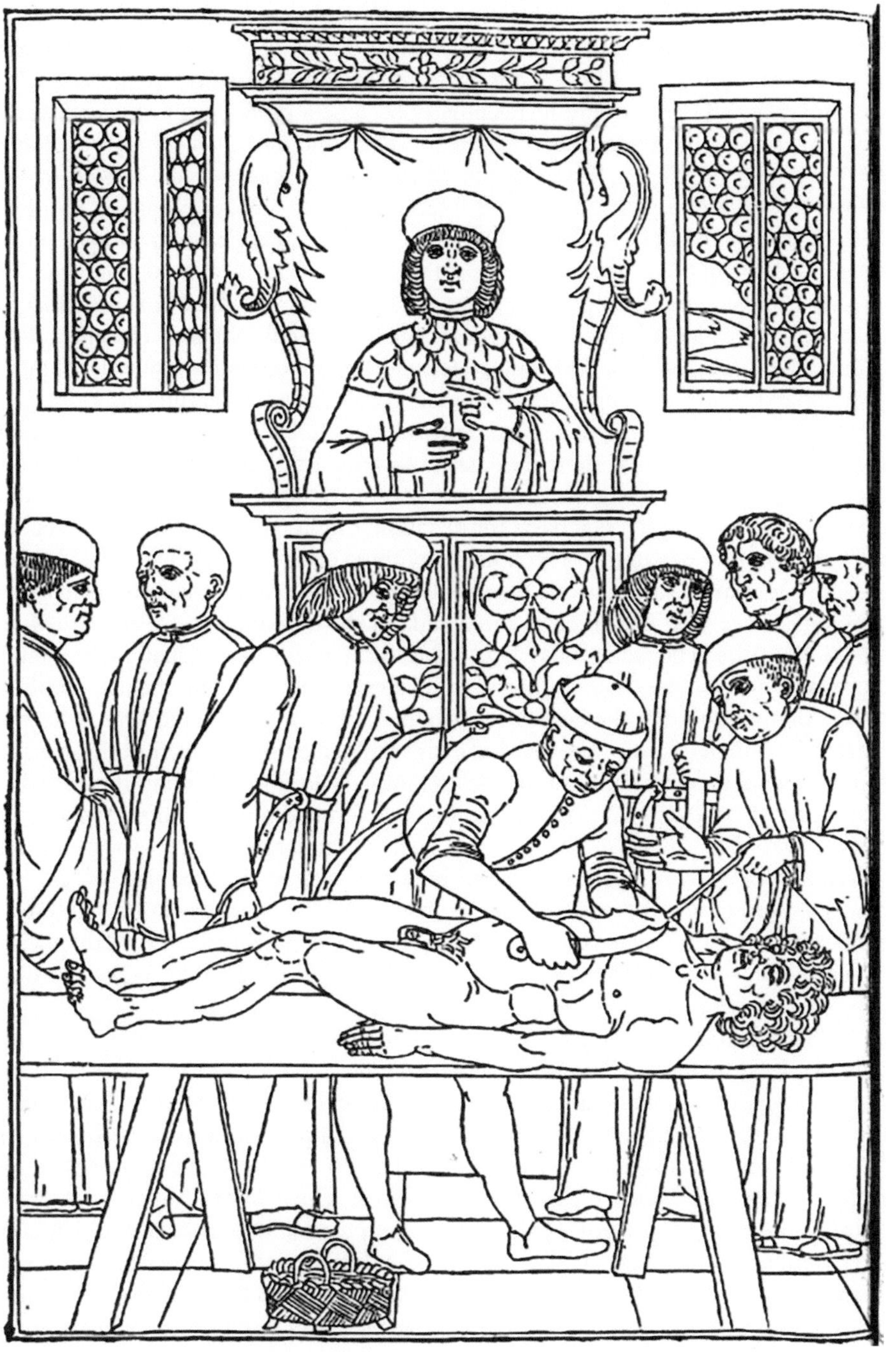

Fig. 2.20 Cadaver dissection during the High Middle Ages

We are fortunate to have the lecture notes of a course titled *Lectures on the Creation* (*Lecturae super Genesum*) as taught at the University of Vienna by Henry of Langenstein (c.1325–1397). Coming from Hesse in Germany, Henry studied at the University of Paris, where he eventually became a professor of philosophy and theology. During the Western Schism, he left Paris and took up a faculty position at the University of Vienna. There he taught theology, exegesis and Canon law. His course served him as a framework in which he attempted to rationalize religious faith by integrating the Creation text with knowledge. Henry's *Lectures* were popular in his days. In the following pages, at the price of somewhat repeating what has already been said throughout this Part of the book, I briefly present some excerpts of these lectures as described by Nicholas Steneck (1976), whom I closely follow. (The *Lecturae* spread over 850 folio pages and contain one million words; Steneck summarized Henry of Langenstein's ideas and condensed his arguments.)

Henry set the *Lectures* in the framework of the Genesis creation story, which proceeds from higher and more ethereal to lower, earthly matters. Within the context of the six days of Creation, he did not ignore any aspect of knowledge. The first day, "Let there be light", is a basis for discussing light and perspective, the second day (heavens) for discussing astronomy, day three (meteorology) is a basis for discussing geographic variation and the fourth day (creation of the stars) for discussing astrology. Once God had filled in the details of the celestial region, He turned his attention to Life. The waters (day five) bring forth all forms of aquatic life, including reptiles (crocodiles) and monsters, while air becomes the natural place for birds. Lastly, on the sixth day, the earth gave rise to mammals, and then to Man, who was created as though he were "the end and cause of the creatures of the inferior world". Man is God's greatest creation; he is the king and lord of all living things in the world and of the elements. Even more, he is the embodiment of all that there is in the universe. Man is the microcosm that mirrors the larger macrocosm of all that exists. Like angels, he has intellect and freedom; like animals, he has senses; like plants, he grows and reproduces; like the heavens, he is composed of a mixture of elements. Even if his state is not as spiritual as that of the angels, it is Man, not the angels nor any other creature, who was created "according to God's image and likeness" (*Genesis* 1).

In shaping the human body, God made it perfect in every way. Legible figures are written as the features of the head. The eyebrow and bridge of the nose form the capital letter "M"; the forepart of the ear is in the shape of "d" (as written in Gothic Latin), and the nostrils and their intervening part appear in the shape of a capital "E", while the eyes make an ""i" when closed and two "o's" when open. Again, we recall the opening sentence of Saint John's gospel, that *in the beginning was the Word, and the Word was with God, and the Word was God*; and written words consist of letters. Moreover, each of the limbs extends the same distance from the centre of the body. To prove this, Henry of Langenstein repeated the common suggestion that a circle drawn with a compass that is at the navel would exactly enclose the ends of the outstretched hands and feet. Finally, the proportions of the human body correspond precisely to another object inspired by God's command, Noah's ark. Its proportions—300 cubits by 50 cubits by 30 cubits (*Genesis* 6: 15)—correspond proportionally to the human body, which is six times as tall as it is wide and ten times as tall as it is from front to back. With evidence such as this, there can be little doubt as to the nobility of the human body. God had commanded all living creatures to thrive and multiply, and accordingly, generation and growth are responsible for the uniqueness of life and for distinguishing the animate from the inanimate.

Much of Henry's biological interest in the *Lectures* focused on generation. Plants generate by means of seeds, which depend on the availability of the four elements for their nutrition. When seed and soil come together, and with the added help of celestial influences, the seed germinates; the germinated plant then grows by attracting nutrients and converting them, through a "digestive power", into foliage, flowers, fruits and another generation of seeds. Animals differ from plants in that they generate by means of a surrounding egg, which supplies the elements for embryo nutrition (rather than direct contact with the elements, as in plants). As to the mode of generation from this egg, three types of semen are involved. One type, the most active in fertilization, passes from the male to the female. This active type cooperates with the second type, the passive semen that resides inside the matter, to produce progeny of a specific form; together, these two types produce progeny. In special, very rare instances, there is a third type, *seminales obedentialis*, which resides inside matter and cooperates directly with God, to generate very exceptional individuals. In addition to such generation by means of fertilizing male semen, spontaneous generation occurs. Worms of decay propagate themselves by arising from occult or hidden semen (*seminales materiae occultae*) which resides in the elements and germinates under special celestial or supernatural influence. Since the process of becoming an animal involves mutual interactions, differences in the matter in which the generative process takes place may lead to the generation of different types of creatures: the rotting flesh of a cow gives rise to bees; that of a horse to beetles, mule flesh to locusts, rotting crabs give rise to scorpions. In this manner, animal diversity builds up.

Moreover, since these two processes take place at the mercy of the elements, changes in the disposition of the elements can drastically alter the nature of the progeny generated in various animals. Indeed, this may well be the origin of sea monsters. Since fishes disperse their semen in water, a slight turbulence in the water, combined with the distortion of celestial influences resulting from their passing through water, can easily distort the original course of generation and give rise to monsters instead of fishes. However, such chance generation happens only among lesser, imperfect creatures in which the species is not fixed. They do not happen in higher, more perfect animal creatures, in which the species are fixed. Seed and semen thus provide the mechanism, as Henry of Langenstein said, through which.

> God, the Creator, wishing there to be as many different types of things as were needed for the proper ordering of the universe, established in living creatures a marvellous array and diversity of parts and members, both internal and external, as well as a varied and wonderful placement of organs.

This "marvellous array" included, in the microcosm of the body, the one organ that mirrors God's role in the macrocosm: the heart. God rules the worldly machine, *machina mundi*, and so too is the body ruled and enlivened by the heat of the heart; without its action, the body would die.

How should we classify animal diversity? Henry spent some time discussing classification in his lectures because he felt it added to the understanding of God's plan of Creation. The general outline and course of Creation is clear: God proceeded from the imperfect to the perfect, from plant to animal and from animal to man. In accordance with this general order, there is an inner chain of perfection also within each of these three links, each creature being better than the one created before it. Thus, there would be a full-length great chain of being, from imperfect-to-perfect plants, from there onwards from imperfect-to-perfect animals, and from there onwards again, to perfect men. Taking his students on a tour through the world of living creatures (Fig. 2.21) and starting with aquatic animals, his first group, the fishes, reproduces by spreading their many "seeds" through the water. Noting their prolific generation, Henry remarked that just one single fish, when spawning even just once, can produce up to one thousand progeny. Predators, such as sea monsters, keep fish population explosions in check. These huge predators, however, fail to devour all fish and to cause entire species to become extinct, because their large size prevents them from reaching shallow waters; small fish instinctively flee to shallow waters, where they are safe. This second group of aquatic animals, the sea monsters (whales, which he termed *Cetus, balaena,* or *leviathan*), are the largest of all animals to exist: a single sea monster occupies as much as four acres. This immense size has led some people to suggest that once they multiply, whales would devour all fish and make the sea impassable for shipping. Henry calms us: When God had created the whales, He made them infertile, reserving their terror for a coming age (*in futuro saeculu*).

A third group of animals living in water, and in this case on land too, is the reptiles. Focussing on their manner of motion, Henry noted that no reptile, including those with feet, moves in the manner of men and of other quadrupeds. Newts and lizards use their feet for only brief periods; serpents have no feet and move by using their ribs (*vi costarum*); worms move by drawing themselves along. Fish, which Henry in this case also included in the reptiles, use their fins and tails to propel themselves through water. The turtle and crocodile are also reptiles, as when they walk, they keep their hips close to the ground, unlike other quadrupeds.

A fourth group of animals, living equally well in water and on land, are the amphibians. Henry mentioned two "amphibians", the crocodile (which he had already classified as a reptile) and the hippopotamus. The crocodile has immense jaws, of which only the top one moves (good old Aristotle, once again). Henry of Langenstein classified all animals with wings which move through the air as birds. The type of wings they have is irrelevant, and he considered birds both those with membranous wings (flies and bats) and those with feathered wings. Since they eat constantly and must move about to get food, they are oviparous, as a pregnant bird would be too heavy to fly. Their light nature makes birds talkative, as proved by the fact that lighter, smaller kinds of birds chatter more than heavier ones. Cranes provide an example of an efficient military organization, as their night guards voluntarily patrol their grounds and wake the next shift of guards to replace them before

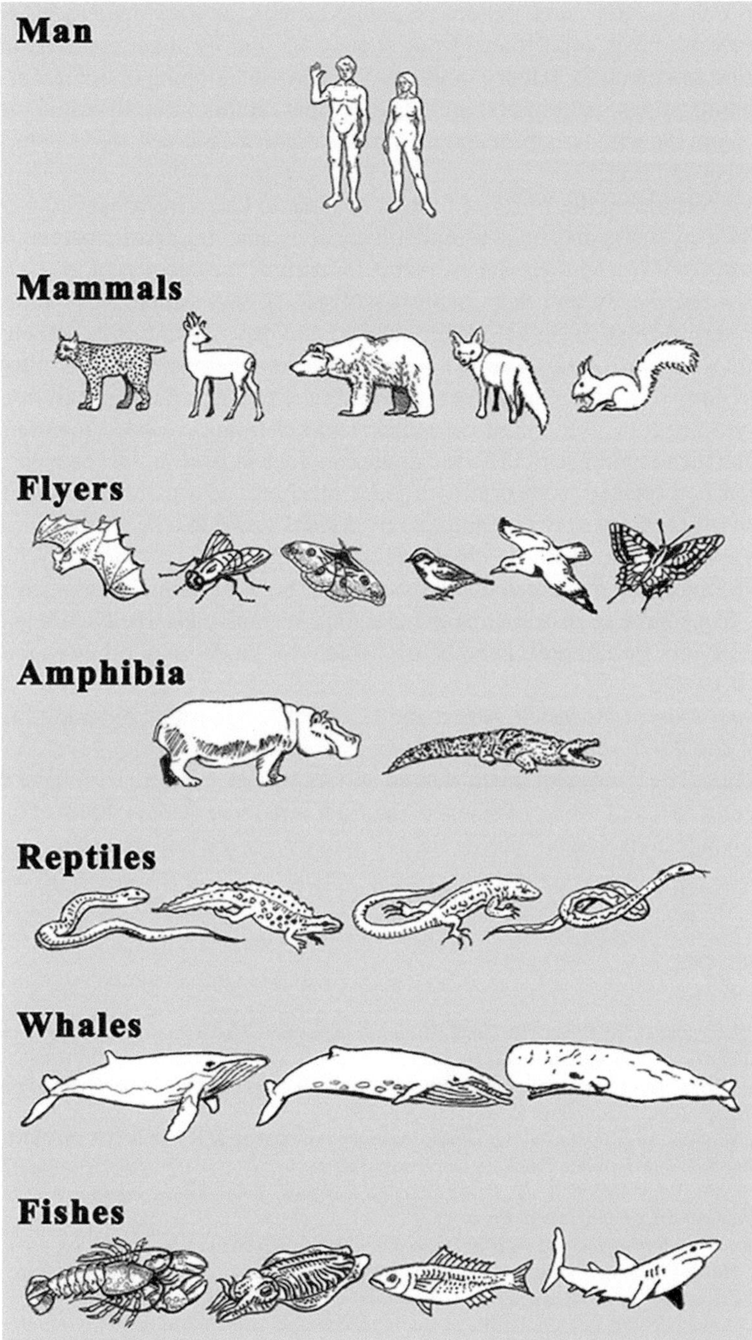

Fig. 2.21 Henry of Langenstein: the classification of animal diversity. (Tuvia Kurz)

retiring. Falcons and other raptors resemble tyrants, as they oppress other birds. Consequently, birds expel them from society to live by themselves in the high mountains and on cliffs. (Henry might have taken the mobbing of nocturnal raptors and generalized it to all raptors, including diurnal raptors.) Finally come the quadrupeds, from the wild lion, bear and wolf to the domestic animals, here including the elephant, as in India.

Henry considered the world of living creatures to be an interdependent community. While creatures may exist as individuals, they also must participate in preserving nature. That is why there are predators in nature, "the cat toward the mouse, the spider toward the fly and dogs against wolves" (Henry called these animals the *correctores*). Animals also help keep balance and order within the elements; thus, frogs and toads purify the air of venomous spirits and help keep it clean. Just as the different parts of each organism function and fit together, with each part depending on every other part, so too does the larger world of living creatures maintain a balance, with an assemblage of different creatures, each performing its own designated role. God had created the world as a giant machine, as a giant clock, and God's presence is necessary to ensure that there is a clock and that it keeps running (Henry of Langenstein alluded to the clock metaphor on several occasions). Within this great machine, God had created and placed the "book of nature", arranged by Him to provide a source of instructions to humanity. We must observe Nature and study it carefully to find moral lessons by which to guide and improve our lives (Steneck 1976).

In the studies of Frederick, Albert and Henry of Langenstein, old traditions in the spirit of the *Physiologus* and *Bestiary* coexisted with more high-level, advanced approaches. The biology of the thirteenth and fourteenth centuries offered a delightful blend of fact and fancy, of science and traditional lore (Crane 2008). This blend would change considerably during the following Age, the Renaissance.

References

Agutter, P. S., and D. N. Wheatley. 2008. *Thinking about life. The history and philosophy of biology and other sciences*. Springer.

Albert, B. 1990. Isidore of Seville: His attitude towards Judaism and his impact on early medieval canon law. *Jewish Quarterly Review* 80:207–220.

Albertus Magnus. 1999. *On animals*. Translated and annotated by K.F. Kitchell Jr. and I.M. Resnik. Baltimore/London: Johns Hopkins University Press.

Barney, S. A., W. J. Lewis, J. A. Beach, and O. Berghof. 2006. *The etymologies of Isidore of Seville*. Cambridge University Press.

Baxter, R. 1998. *Bestiaries and their users in the middle ages*. Stroud: Sutton.

Belo, C. 2004. Ibn Sina on chance in the physics of as-Sifa. In *Interpreting Avicenna: Science and philosophy in medieval Islam. Proceedings of the Second Avicenna Study Group*, ed. J. McGinnis, 25–41. Leiden: Brill.

Carrol, W. E. 2000. Creation, evolution, and Thomas Aquinas. *Revue des Questions Scientifiques* 171:319–347.

Crane, S. 2008. A taxonomy of creatures in the second-family bestiary. *New Medieval Literatures* 10:1–48.

Crivellato, E., and D. Ribatti. 2006. Mondini de Liuzzi and his *Anathomina*: A milestone in the development of modern anatomy. *Clinical Anatomy* 19:581–587.

Demaitre, L., and A. A. Travil. 1980. Human embryology and development in the works of Albertus Magnus. In *Albertus Magnus and the sciences: Commemorative essays*, ed. J. A. Weisheipl, 405–440. Toronto: Pontifical Institute of Medieval Studies.

Diekstra, F. N. M. 1985. The *Physiologus*, the bestiaries and animal lore. *Neophilologus* 69:142–155.

Frederick II of Hohenstaufenh, 1241. 2010. *The art of falconry by Frederick II of Hohenstaufen*. Translated by Wood C.A., and F.M. Fyfe. Ishi NY.

Freudenthal, G. 2002. The medieval astrologisation of Aristotle's biology: Averroes on the role of the celestial bodies in the generation of animate beings. *Arabic Sciences and Philosophy* 12:111–137.

Gaziel, A. 2012a. Questions of methodology in Aristotle's zoology: A medieval perspective. *Journal of the History of Biology* 45:329–352.

Gaziel, A. 2012b. Gersonides' naturalistic account of providence in light of the *book of animals*. *Aleph* 12:243–271.

Gaziel, A. 2012c. Spontaneous generation in medieval Jewish philosophy and theology. *History and Philosophy of the Life Sciences* 34:461–480.

Gerald of Wales, 1188. 2004. *The Journey through Wales, and the Description of Wales*. Translated with an Introduction by L. Thorpe. Penguin.

Gil, M., and E. Fleisher. 2001. *Yehuda Halevi and his circle*. Geniza Documents. Sources for the Study of Jewish Culture 6.

Goodman, M. 2012. *Halomo shel HaKuzari (the king's dream)*. Kinneret, Zmora-Bitan, Dvir; Or Yehuda (Hebrew).

Ghosh, S. 2015. *Writing the Barbarian Past: Studies in Early Medieval Historical Narrative*. Brill.

Halbertal, M. 2009. *Maimonides*. Jerusalem: Zalman Shazar.

Halevi, Y. 1905. *Kitab el Khasari (Kuzari)*. Translated by H. Hirshfeld. Wikisource.

Halevi, Y. 1997. *Sefer Ha-Kuzari: Maqor ve-targum*. Translated by J. Qafih. Makhon Mishnat Ha-Rambam, Kiryat Ono (Hebrew).

Halevi, Y. 2008. *Sefer Ha-Kuzari*. Translated by J. Ibn Tibbon updated with Introduction by D. Lasker. Am Ha-Sefer. Yediot Aharonot ve-Sifrei Hemed (Hebrew).

Halevi, Y. 2009. *The Kuzari: In defense of the despised faith*. Translated by N. D. Korobkin. Northvale: Jason Aronson, 1998. 2nd revised edition, 2009 Feldheim, Jerusalem (English).

Halevi, Y. 2010. *Sefer Ha-Kuzari: Precise Hebrew translation in the style of the period of its composition*. Translated by I. Shilat. Shilat, Ma'aleh Adumim-Jerusalem (Hebrew).

Halevi, Y. 2017. *Kuzari. The Book of the Rejoinder and Proof of the Despised Religion*. Translated by M. Schwartz. Beer Sheva: Beer Sheva University Press.

Higgs Strickland, D. 2008. The Jews, Leviticus and the unclean in medieval English bestiaries. In *Beyond the yellow badge: Antijudaism and antisemitism in medieval and early modern visual culture*, ed. M. B. Merback, 203–232. Leiden: Brill.

Houwen, L. 1994. Animal parallelism in medieval literature and the bestiaries: A preliminary investigation. *Neophilologus* 78:483–496.

Hub, B. 2016. Aristotle's 'bloody mirror' and natural science in medieval and early modern Europe. In *The Mirror in medieval and early modern culture*, ed. B. Hub, 31–71. Turnhout: Brepols.

Hugh of Fouilloy, ca 1140. 1992. *The Medieval Book of Birds, Hugh of Fouilloy's De Avibus*. Translated by W.B. Clark. Binghamton: Medieval & Renaissance Texts and Studies.

Infusino, M., D. Win, and Y. O'Neill. 1995. Mondino's book and the human body. *Vesalius* 1:71–76.

Kitchell Jr, K. F., and Resnick, I. M. 1999. *Introduction: The life and works of Albert the Great*. The writings of Albertus Magnus","The De animalibus","The Historical Context of De animalibus", y "De animalibus: Method, Goals, Importance", secciones en Albertus Magnus, On animals: a Medieval Summa Zoologica, 1.

Kruk, R. 1990. A frothy bubble: Spontaneous generation in the medieval Islamic tradition. *Journal of Semitic Studies* 35:265–281.

Lactantius. *On the Workmanship of God.* Translated into English, Online version, Bible Hub.

Leibowitz, Y. 1991. Bri'at Ha'Olam Be'emunato shel Ha-Rambam (the creation of the world in the belief of Maimonides. pp. 71–78 in *Bri'at Ha'Olam: BaMada, BaMitus BaEmuna (creation of the world in science, myth and faith)* (L. Mazor et al., ed.). Magnes, Jerusalem (Hebrew).

Maimonides. 2021. *Guide of the perplexed.* See Frank D. and A. Segal (eds): *Maimonides' guide of the perplexed: A critical study*. Cambridge.

Mavdrovi, A., and D. Paraskevas. 2014. Mondino de Luzzi: A luminous figure in the darkness of the middle ages. *Croatian Medical Journal* 55:50–53.

Neckam, A. ca. 1190. *De Naturis Rerum libri duou.* Ed T. Wright. London: Longman Roberts and Green, 1863.

Oggins, R. S. 1980. Albertus Magnus on falcons and hawks. In *Albertus Magnus and the sciences*, ed. J. A. Weisheipl, 441–462. Toronto: Pontifical Institute of Medieval Studies.

Ogilvie, B. W. 2006. *The science of describing: natural history in renaissance Europe*. University of Chicago Press.

Peterson, R. T., G. Mountfort, and P. A. D. Hollom. 1964. *A field guide to the birds of Britain and Europe*. London: Collins.

Rabanus Maurus. *De rerum naturis*. Online version copyrighted by W. Schipper 1996.

Rengachary et al. 2009. Development of Anatomic Science in the Late Middle Ages: The Roles Played by Mondino de Liuzzi and Guido da Vigevano. *Neurosurgery* 65:787–794.

Rubenstein, R. E. 2003. *Aristotle's children: How Christians, Muslims and Jews Rediscovered Ancient Wisdom and Illuminated the Middle Ages*. Orlando: Harcourt.

Salisbury, J. E. 2011. *The beast within: Animals in the middle ages*. New York: Routledge.

Salter, D. 2001. *Holy and Noble beasts. Encounters with animals in medieval literature.* Woodbridge, Suffolk: Boydell & Brewer.

Seide, J. 1961. The barnacle goose myth in the Hebrew literature of the middle ages. *Centaurus* 7:207–212.

Steneck, N. H. 1976. *Science and creation in the Middle Ages. Henry Langenstein (d. 1397) on Genesis*. Notre Dam, Indiana: University of Notre Dam Press.

Takahashi, A. 2008. Nature, formative power and intellect in the natural philosophy of Albert the Great. *Early Science and Medicine* 13:451–481.

Throop, P. 1998. *Hildegard von Bingen's Physica; the complete English translation*. Rochester: Healing Arts.

Van Den Brink, F. H. 1955. *Die Saugetiere Europas. Ein Taschenbuch fur Zoologen und Nature.* Berlin: Paul Parey.

Chapter 3
Renaissance in the Study of Nature

Abstract In the fifteenth century and in new lands, Europeans discovered a new diversity of plants and animals. This did not immediately influence overall classification, as they erroneously placed completely new animal types into old, existing categories. There were vague beginnings of a modern classification hierarchy, with a clear recognition of Species, Genus and Order, and dim recognition of Family and Class. As cheap printed texts began to be produced and dispersed widely, knowledge based on ancient classics blended with new knowledge and more people could gain a view of biodiversity in faraway countries. The outlook on diversity blended the discovery of new animals with the belief in fantastic creatures and with the belief that terrestrial animals have marine counterparts. A major Renaissance innovation was the collection of natural history objects that one could inspect and study in the future. Such collections eventually became museums, which were to exist far beyond the collector's life and enable improved future study of animal diversity. Renaissance naturalists did not conduct intensive studies of animal behaviour, but they were aware of satellite copulation, whereby a non-dominant male on the periphery of a dominant male's territory sneaks up to copulate with a female in the latter's harem.

During the early and mid-Renaissance, people followed ancient texts that viewed plant diversity in medical contexts. Universities began teaching medicinal plants, undertook plant-collecting excursions, planted botanical gardens, maintained herbaria and wrote herbals. Much of the herbal knowledge was eventually channelled into one book in which, in frequently updated editions, first came uncritical acceptance of ancient texts; then critical appraisal and finally additional, original observations. Botany, as a discipline independent of pharmacology, may have first emerged when, in the late Renaissance, Cesalpino established a classification of plant diversity based on essential features of the plant, not on medicinal value. His essential, defining functions of plants were dictated by the vegetative soul, which is typified by nutrition (= growth) and reproduction. The medulla performs nutrition, so he divided all plants into those with a hard, woody medulla (trees and shrubs; they live many years and bear fruit) and those with a soft, fleshy one (herbs; they die after seed formation). These two groups are divided further by their fruit, which has three essential features: number, location and shape. He subdivided his two groups by the

J. Heller, *Nature's Biodiversity*, https://doi.org/10.1007/978-3-031-89790-0_3

floral organs' position on the seed case, number of seeds in the fruit and number of compartments per fruit.

During the Renaissance, a "fossil" was any hard object dug up from the ground. Intellectuals believed that fossil shells inside rocks had never been living beings. They formed by an incomplete process of spontaneous generation, inside stones. Naturalists classified fossil diversity into absurd groups, of sea urchins resembling serpents, sea lilies resembling stars and belemnites resembling human artefacts. Plato's philosophy, that objects in the world are imperfect expressions of eternal, perfect ideas, supported the view that animal-resembling fossils are not animals. Some believed that fossil 'seashells' form according to male (sulphur) and female (mercury) principles, sulphur because of its power, heat and activity; and mercury because it is cold, passive and flows, taking the form of active powers. The sulphur principle dictates the shape, and the mercury principle supplies matter. Intellectuals classified mineral diversity within this ideological frame. In metals, sulphur and mercury combine in varying proportions and levels of purity. In imperfect minerals, the contribution of sulphur or mercury weakens, because a salt principle mixes in. In sedimentary, fossils containing rocks, sulphur, mercury and salt principles mix with "fluores". Fossils are, therefore, inorganic minerals, whatever their resemblance to animals. They grow within enclosing rocks, as during pregnancy.

During the Renaissance, some people viewed the universe as a series of spheres ("worlds"), one enveloping the other. The upper world, of pure eternal spirituality, consists of angels. The middle world, of eternal spirituality, mixed with eternal matter, consists of stars. The Sun is at the centre of the world of stars, at the mid-point between spirituality and materialism, between God and Man. It is the residence of the Holy Spirit, and the spirit of life emanates from it. The underworld consists of temporary matter with little spiritual essence. Entities of these materials may change, from one species to another, and they decay over time. Spheres of animals, plants and minerals surround the Earth, which consists of the earth and water elements. Man-the-Creator, in the sphere of animals, imitates God's creation by creating things himself, such as agriculture and machines. The male takes the right side of the sun. His animals include a lion, snake and dolphin; his plants are mainly trees, and his minerals are gold and lead. The female takes the left side of the moon. Her animals are eagle, snail and fish, her plants are wheat, flowers and roots, and her metals are silver and copper. The male-female division is thus an essential feature of the underworld.

Anatomy, more than other branches of biology, at first sought to glorify and restore Greek and Roman observations; later to break away from them; and finally, to develop original views. Early in the Renaissance, Leonardo da Vinci illustrated conventional viewpoints, gleaned from classic and medieval texts. Assuming that the human body and that of many animals are identical, he formed a distorted illustration of male and female human genitalia, "planted" structures of animal organs also elsewhere in the human body and believed that the male organ possessed some individuality, independent of the body. Vesalius, in the mid-Renaissance, based human anatomy on his own dissections and observations, not on texts, and he accurately illustrated the human genitalia. Though noting many errors in classic literature, he too erred in assigning animal structures to human anatomy; consequently, his observations were not always accurate. By insisting on detailed, accurate nomenclature for every structure, he advanced anatomy as a discipline. Late in the Renaissance, Casserio compared vocal and hearing organs in humans, dogs, cows, birds, amphibians, fish and insects and produced one of the first deliberate studies in comparative anatomy. During the late Renaissance, the study of human and animal anatomy began drifting apart.

3.1 Coping with Nature's Diversity

3.1.1 All Animals of the World

In Ancient and Early Medieval times, awareness of diversity was limited to what people saw in their immediate environment, and there was not much need for a detailed systematic hierarchy to master Nature's variety and diversity. During the High Medieval, European people gradually began emerging from their close surroundings and travelling to distant lands. This trend, which began with the Crusades to the Middle East (twelfth century) and continued with longer journeys of Italian merchants to China (Marco Polo, thirteenth century), peaked with Columbus' discovery of America (1492). Soon later Vasco De Gama sailed to India around the Cape of Good Hope (1498), and Balboa discovered the Pacific Ocean (1513). These all-important geographical discoveries led to Cortez conquering Mexico (1520) and Pissarro conquering Peru (1533). The discovery of a hitherto completely unknown diversity of plants, animals and cultures followed this discovery of unknown seas and lands, and a flood of specimens poured into Europe, where people began collecting them and attempting to classify this rich, newly discovered biodiversity. A new awareness of biodiversity thus developed in the Renaissance.

In the twilight between the Medieval and Renaissance, the earliest illustrated travel book (in print) I am aware of appeared, *Wanderings through the Holy Land* (*Peregrinatio in Terram Sanctem* (1486). Its author (a certain Bernhard von Breydenbach from Mainz) had journeyed to the Holy Land accompanied by an artist. His illustrations (Fig. 3.1) present a giraffe (misspelled *scraffa*; none have existed in the Holy Land over the past 100,000 years); a crocodile; and a unicorn ("we saw a large animal, larger than a camel, and our guide told us that it was truly a unicorn" p. 35). It further presents a camel driven by a humanoid form with a strange mane, a tail and a walking stick. The caption on this assemblage of animals reminded the reader that the pilgrims had depicted these illustrations as they had actually seen them, in the Holy Land, and that the figures are on-site transcriptions of what they saw. This vagueness of knowledge, as to what was real, what was wishful thinking, what was simply untrue and what was mythic, went on and on. Some 150 years after Breydenbach's report of his observations of giraffes and unicorns in the Holy Land, a certain Eugene Roger would write that he had seen, in the Holy

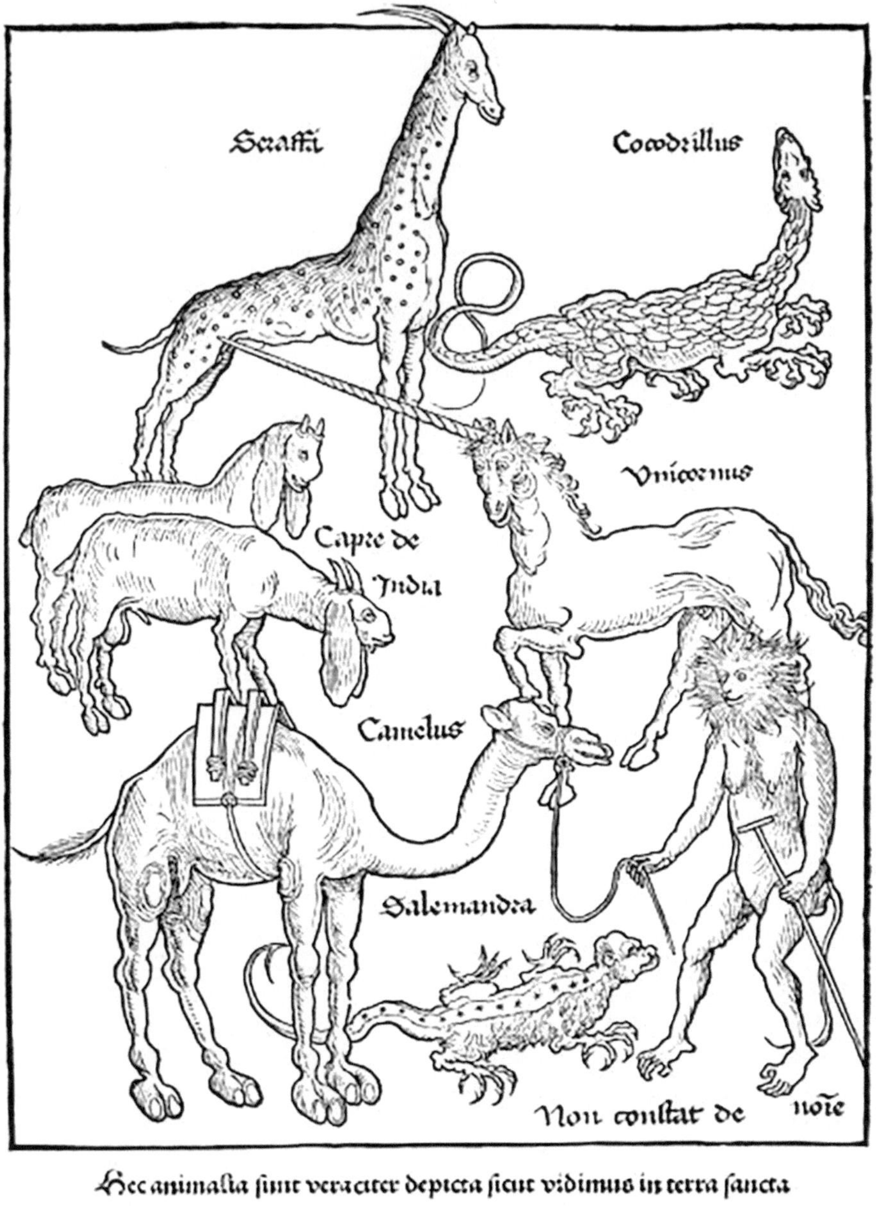

Fig. 3.1 Animals of the Holy Land, Bernhard von Breydenbach (1486), *Peregrinatio in Terram Sanctem*

Land, a basilisk (*Travels in Palestine*, 1631), its shape that of a crocodile, and it had a killing gaze. He himself had seen one specimen, which fortunately was dead, but a live specimen had killed many people by just gazing at them. In Rome, the Pope

(Leo IV) had destroyed this living basilisk by his prayers and by his use of the Sign of the Cross.

Also in this century, as early as 1440, both paper manufacture and typesetting had advanced and become so widely applied that many books were printed. Knowledge began spreading to larger readerships. As cheap printed texts became widely dispersed, classical biological texts flooded Europe and kindled a broad interest in natural sciences. Indeed, by 1500, there were already 11 printed editions of Aristotle and 39 of Pliny. Knowledge based on these old, ancient classics eventually blended with new knowledge. More people could read, and more easily, about the diversity of animals and plants and gain a better view of nature in faraway countries. The print also enabled people to publish their own discoveries and ideas, which, if not printed, had little chance of influencing others. Furthermore, the multiplicity of print stimulated comparison, comparison stimulated criticism and revision, revision stimulated progress, and knowledge thus advanced (Wilson 1992).

Printing also resulted in considerable advances in the illustrations of plants and animals as well. Previously, the medieval scholastic tradition of illustrations had referred to authorities and to texts, rather than to nature itself. Instead of drawing plants and animals from nature, scholars would copy from manuscripts generation after generation, and this resulted in a rapid deterioration of realism, the illustrations becoming stylized rather than resembling the actual plant or animal. In Schools of Medicine, where teaching was often done without any subject before the students, illustrations of human anatomy were diagrams copied from previous texts, not from independent observation (Locy 2005).

By the end of the Medieval and early Renaissance, however, the printing of illustrations by use of woodcuts, a technique originating in China that reached Europe by the fifteenth century, enabled the easy spread of identical illustrations, some of which had astonishing beauty. Furthermore, there was a newfound impetus to work from direct observation, and illustrations of nature became more realistic. Better drawings called for keener, more accurate observations and indeed, the Renaissance artist Albrecht Dürer (1471–1528), in his figure of a young hare and again in his figure of two squirrels munching on nuts, produced masterpieces of observational art with almost photographic accuracy (Fig. 3.2). He also drew the first ever picture of a Rhinoceros, in his engraving of Adam and Eve he inserted a parrot, an elk, ox, cat, rabbit, mouse, chamois (and snake) and he sketched dancing apes, leading us from reality to fantasy (he also pictured plants in detail). Dürer thus brought animals into art, thereby increasing awareness of their diversity. Another person fond of animals was Maximilian I, Emperor of the Holy Roman Empire. An ardent hunter, he introduced severe laws to protect ibexes from extinction, and he loved to have his room full of birds (Waetzoldt 1950). In Dürer's marginal drawings of Maximillian's prayer book, the emperor could feast his eyes on species of game to be found in his forests, with birds on every side—cocks, hens and pheasants; swans, ducks and geese; eagles, goshawks and falcons; screech owls, barn owls and great horn owls; storks, king-fishers, spoonbills and cranes.

Drawings and paintings were important for depicting the organism as it appeared in life, especially in the case of animals that did not keep their form or colour when

Fig. 3.2 Dürer: a young hare; two squirrels eating nuts

preserved, such as worms, slugs, fishes and amphibians. During the later fifteenth and early sixteenth centuries, several illustrated natural history books appeared in which the artists made new drawings from life, rather than copying from old texts. These illustrations, beyond delighting the eye, could convey what words, still insufficiently subordinated to technical needs, could not (Hall 1962). In those days, there did not yet exist a technical language, ever-so-accurate in meaning, universally known, and suited to explain in detail the necessary description of animals. Pictures were often the only possible means of identifying a plant or an animal whom an author had loosely, or inaccurately, described. Whereas the quality of the Renaissance illustrations of European animals was often very good, those of animals from distant countries varied, due to the poor quality of preservation techniques and to shipping constraints of the specimens. Usually, only dried skins, shells and skeletons survived the journey. Naturalists in the field would preserve the specimens they collected by coating them in lacquer, and consequently, they were shrunken, distorted and dark by the time they reached a collection (the preservation of animals in alcohol began only around 1600).

At this point, the European artist's cultural background often influenced the illustration of new kinds of animals, and this combined with the fact that the naturalists of those times lacked a classification system that could place new animals into new groups. In at least some cases, known European animals served as a basis for the illustrations of unknown animals, and the original descriptions of the traveller who had observed, collected and described the exotic animal were merely "sprinkled" over them. For example, some of the earliest illustrations of sloths from America presented them as standing on all four legs or sitting upright, since the Europeans thought that sloths were primates. Nobody at the time imagined that sloths form a separate yet unknown group of mammals and that they live by hanging upside-down, grasping branches with their long, strong nails. This error would go on for another three centuries, deep into the nineteenth century (Fig. 3.3).

Similarly, artists of the Renaissance sometimes illustrated the armadillo as a horse that has osseous armour (Beustern 2017). The Spaniards, upon first encountering it, and not imagining that this strange animal belonged to a separate, unknown group of animals, thought that because of its scales, it resembled a little armoured horse. Next, they convinced themselves that its feet and head also resembled those of a horse; and furthermore, they believed that it even grazed like a horse. They

Fig. 3.3 The sloth, illustrated as climbing above branches and not below them; from Lenz H.O., 1834–1839, *Gemeinutzige Naturgesichte*, Hebrew translation

called it a horse that is *armado*, a word used to describe the Spanish armoured horse guard; eventually, in 1574, they coined the term *armadillo*. This name created its own dynamics: Naturalists of the sixteenth century soon began describing armadillos as resembling a fully armoured horse with scales hanging down its sides, even including the articulated hinges of its armour harness. Deep into the mid-eighteenth century, an illustration appeared of an animal that in every respect looks like a horse, with a tail, hooves, a horse's mane and an equine shape. However, in place of a saddle, it has scales, intended to show that it is a naturally armoured horse—the illustrator's version of an armadillo (Fig. 3.4).

Later in the sixteenth century, the Flemish painter Maarten Vos depicted a woman, as an emblem of America, riding a huge (for its natural size) armadillo. She is carrying a bow and arrow and there are three battle scenes and one hunting scene in the background—indeed, the scenery in which one would expect to find a little armoured horse (Fig. 3.5). People in the Old World thus perceived the armadillo as an odd, made-in-America horse (Beustern 2017). Such descriptions and illustrations, which should be judged in the context of their times, illustrate how the knowledge of known animals was extrapolated to less-known ones.

These new discoveries also raised doubts as to the way by which this newly discovered diversity fits in with biblical narratives. In his encompassing *Historia natural y moral*, José de Acosta (1540–1600), a Spanish Jesuit missionary and naturalist in Latin America, assumed that all animals had come from Noah's ark, but he wondered how, from it, they had reached America. He rejected the possibility of

Fig. 3.4 Armadillo from the Turkish translation of Francisco López de Gómara, *Historia general de las Indias* (1552): Tarih-I Hindi-I garbi (Constantinople: Ibrahim Butafarrika, at the Imperial Press, c.1730). (Reproduced under Creative Commons License, courtesy of the John Carter Brown Library, Brown University, Providence, Rhode Island. From Beustern 2017)

Fig. 3.5 America riding an armadillo side-saddle (by Adriaen Collaert, 1560–1618, from his series *The Four Continents*). The Met

swimming as an explanation and suggested a former northern terrestrial continuum around the world. However, he wondered: if this was the case, why did certain animals migrate in one direction while others were abundant in another?

Another development of the Renaissance was that naturalists of different countries would frequently correspond (mainly in Latin) with colleagues from other countries. This international network of correspondence resulted in a trans-European community of naturalists, an international "Republic of Letters" (Ogilvie 2006).

During the early Renaissance, Pliny fell out of grace (Ogilvie 2006). A professor at the University of Padua claimed (1493) to have corrected 5000 errors in Pliny's text concerning plants, inaccuracies that he blamed on the medieval scribes who had copied the manuscripts over time. Another professor, at the University of Ferrara, attacked Pliny himself (1492, 1509). Referring to the Roman encyclopaedist's blunder that the moon is larger than the earth, he argued that if Pliny had erred on such a fundamental fact, then he could not always be trusted, and we should examine his work for further errors. Furthermore, Pliny had claimed that ivy (an evergreen climbing or ground-creeping woody plant) had both a male and female version. The professor of Padua argued that Pliny was confusing two very different plants with similar Greek names: ivy (*kissos* in Greek) and cithon (*kisthos* in Greek; today *Cistus*, rockrose, a perennial shrub with flowers ranging from white to purple and dark pink; Fig. 3.6). He suggested that this confusion reflected the fact that Pliny did

Fig. 3.6 An ivy (left) and rockrose. (Tuvia Kurz)

not know what he was writing and that one should replace Pliny's flawed treatise, regarding medicinal botany, by the original Greek texts (Ogilvie 2006).

The trend, during the Renaissance, to encourage the study of nature rather than ancient texts, is well expressed by the Renaissance satirist François Rabelais (1483–1553), who wrote of two grotesque gigants, Gargantua and his son Pantagruel. When Pantagruel was to leave for Paris, Gargantua advised him to devote himself to the natural world. "Let there be no sea, river or book, whose fish you do not know. Nothing should be unknown to you–all the birds of the air, each and every tree and bush and shrub in the forests, every plant that grows from the earth, all the metals hidden deep in the abyss, all the gems of the Orient and the Middle East – nothing." In line with this trend, Renaissance artists began accurately picturing plants, animals and landscapes.

Men in the fifteenth and sixteenth centuries believed that they were on the verge of founding a new stage in history that would rival that of classical antiquity in brilliance, learning and glory (Hall 1962; Ogilvie 2006). They regarded themselves as in rebellion against scholasticism, the intellectual discipline of the medieval schools, which they saw as concerned with theology rather than with secular studies. It seemed to them that the work of the immediate past was inferior to that of natural philosophers of Graeco-Roman antiquity and that the last few centuries were merely a "middle age", an unfortunate break between the glorious achievements of the distant past and the glorious potentialities of the present. New maps and exploration made geography an even more popular subject. The Renaissance was an age in which travellers discovered strange and previously unknown animals in newly discovered lands.

Naturalists in Europe had only fragmentary evidence of these animals, most of which they had never seen alive. These newly discovered animals—are they real or imaginary? If real, then what about all the fantastic animals recorded by medieval scholars—are they real or imaginary? Two of the first to cope with these questions were Conrad Gesner and Ulisse Aldrovandi, each of whom attempted to catalogue and describe the entire world of animal diversity (Ogilvie 2006).

Conrad Gesner (1516–1565, Fig. 3.7; in modern studies sometimes spelt Gessner) was the son of a furrier in Zürich, from whom he may have gained his keen observation of animals. His uncle, who brought him up, had cultivated and collected medicinal herbs, and so Conrad was familiar with many plants and their medicinal values, as well. With a broad interest in plants and animals, Gesner collected, throughout his life, both specimens and descriptions of wildlife, which he gained through extensive correspondence with other scholars. He also had a keen interest in classic languages, and at the age of 19, he travelled to Strasbourg to study 'the three languages' (the contemporary term for Latin, Greek and Hebrew). Eventually, he travelled to Paris, where he read Latin and Greek literature, rhetoric and natural and moral philosophy. Religious unrest drove him back to Switzerland, and he became a professor of Greek at the newly founded Academy of Lausanne. From there he travelled to Montpellier, where he met the French naturalists Guillaume

Fig. 3.7 Gesner

Rondelet and Pierre Belon (we shall meet them further on, in this chapter). At the age of 25, having received his medical degree at Basel, he settled in Zürich and became both the town physician and the lecturer of Aristotelean physics at the "Carolinum", which was the precursor institute of the University of Zurich. He had a remarkably broad knowledge of zoology, botany, medicine and linguistics, and indeed Emperor Ferdinand himself, upon granting him a rank of honour, remarked that "Gesner is scholarship itself". Gesner died in Zurich during the Bubonic plague, when only 49 years old. By the time of his death, he had published 72 books and had written an additional 18 unpublished manuscripts. His first treatise was *Bibliotheca Universalis* (1545), comprising an alphabetic guide of all authors who had written in any of the three languages, Latin, Greek and Hebrew. For each author, he presented a biographical sketch and a list of works, including both those already printed and those existing only in the manuscript. Soon, libraries throughout Europe were organizing themselves according to Gesner's book.

Another of his published treatises was an immense encyclopaedia of animals *Historia Animalium,* with 1500 animals, a comprehensive description documenting virtually every animal species known at the time—a study of animal diversity. Gesner's encyclopaedia marks, perhaps, the beginning of zoology as a separate discipline. In general, men of the Renaissance regarded their period as a time of rebirth and recovery of the values and achievements of classical antiquity. Gesnera Renaissance scholar of ancient languages more than he was a zoologist, intended his *Historia Animalium* to be a worthy successor of Aristotle's great work with the same name, but he set Aristotle's observations alongside those of his own contemporaries (Kusukawa 2010). Following Aristotle's classification, the first volume of his *Historia Animalium* dealt with viviparous quadrupeds, the second with oviparous quadrupeds, the third with birds and the fourth with aquatic animals. The full set of volumes, consisting of over 3500 folio pages, was sold for almost 7 florins; a coloured set went for 19 florins, the worth of two copies of the best coloured Bible in those days (and with a purchasing power of about 19,000 modern US dollars). It was advertised as a book of grammar and rhetoric, to be used like a dictionary that one dips into, from time to time, not as a book to be read from front to back. Though containing much on animal behaviour and *materia medica,* it is predominantly on animal philology and etymology. Animal natural history thus developed, in the Renaissance, out of philological investigations (Ogilvie 2006; Kusukawa 2010).

Gesner aimed to collect everything ever written about any animal and, in this sense, he was building on his earlier *Bibliotheca Universalis*. He did not reduce or eliminate similar, contradictory or false descriptions, and did not omit descriptions of animals whose existence was uncertain. *Historia animalium* included familiar domestic species such as horses, cats, dogs and mice, alongside mythical beasts like unicorns and griffins and exotic animals such as crocodiles and sloths. The text was longest for domestic animals: 197 folio pages on dogs, 178 pages on horses and 80 pages on cows (Kusukawa 2010). A collection of the woodcuts alone was issued as *Icones animalium* (1553 and 1560) and *Icones avium* (1550, 1560). A fifth volume, on serpents, was published posthumously (1587). Gessner wrote the book with a coloured copy in mind, and accordingly, woodcuts were repeated for sub-species of martens and weasels, to indicate varieties differing only by colour. He took pictures from books and manuscripts, usually stating their source (but not for domestic or well-known wild animals). His illustration of a porcupine was of an animal shown around in Zurich by a beggar; illustrations of fishes were often based on dried specimens.

While copying northern beasts such as reindeer, gulo and whales from Olaus's map of the Northern Lands, Gesner wrote that he does not guarantee the accuracy of these illustrations, many of which seem to him to have been depicted following seafarers' tales, not from life—some of the heads given by Olaus are too similar to those of terrestrial animals and the feet are armed with claws (Ogilvie 2006). Olaus, in his *Description of the Northern Peoples* (1555) had also pictured reindeer as having three antlers, not two; and he had also described men who had been transformed into werewolves, who raid remote human settlements and devour their occupants, differing from natural wolves in that they drink heavily; and in that they gather by their thousands at a ruined castle, where they whip those who fail to jump over a wall (Ogilvie 2006). And then, from Olaus's illustrated map of 1539, there was the myth of the wolverine *G. gulo* (named also Glutton) that eats to gluttony and then forces out its faeces so as to eat even more, a myth that stems from its German name *Vielfrab,* meaning "devours much". This German name may have come, however, from the Old Swedish *fjellfras*, meaning "mountain cat".

Gesner also took from recently published works of Rondelet and Belon; a giraffe from Breydenbach, a hamster from a German book on animals and his woodcut of a toucan was based on a beak that someone had sent him, the rest of the body being reconstructed from a book (Kusukawa 2010). When a full reconstruction from a part of an animal was impossible, he showed just the part that he had, for example, the pelt of a genet. A "very learned" person had sent him a picture of the footless bird of paradise, whose authenticity this learned person and many other trustworthy witnesses had confirmed (on this occasion, he reported that a bird of paradise was on sale in Nuremberg, for 800 talers). His figure of a rhinoceros was that of Albrecht Dürer. A unicorn's horn was out of his reach because of the extraordinary price charged by apothecaries made Gesner state that its medical efficacy should be tested by those richer than himself (Kusukawa 2010).

Given Gesner's urge to be comprehensive, it was inevitable that the quality of his illustrations would vary. The sources of his illustrations included live, dried or partial specimens, images from other printed books, manuscripts, maps and prints, drawings made by artists at his request and drawings sent in by his friends. Their qualities also varied, from accurate to tolerable, false or uncertain. Inaccuracies arose from using distorted dried specimens, from his limited knowledge of some animals, and from relegating the supervision of pictures to his printer, which led to mishaps such as a changing of places between woodcuts of a hen and a cock. He had live specimens of the guinea pig *porcellus indicus* (today *Cavia porcellus*) but provided an illustration sent to him by a nobleman in Paris. In the *Icones* (1560), "more accurate" images were added of the camel, elk and zibeth cat, but in each case, he retained the older woodcuts. Likewise, concerning textual descriptions, he did not discard older or less accurate ones when finding newer or more accurate ones of an animal. In the case of the fire salamander, Gesner included the erroneous image of a hairy creature with stars on its back, copied from Breydenbach's *Travels to the Holy Land*. He described it as depicted recently, but unreliable. In the Appendix to the first and second volumes (1554), he again included the Breydenbach salamander, now labelled as "false". This "false" woodcut nevertheless reappeared, alongside the "true" image of the salamander in the *Icones* (1560). The use of a wide range of sources for illustrations with a varying degree of quality becomes understandable within Gessner's enterprise of gathering everything that was ever written or known about every animal (Kusukawa 2010).

Gesner attempted to cover all knowledge of the entire animal world. His encyclopaedia was, of course, mainly a review of the literature and compiled the knowledge of others, combining data from such ancient sources as the Old Testament, Aristotle and Pliny, together with medieval folklore and bestiaries and with (then) modern observations by German, French, and Italian authors. He organized all animals broadly according to Aristotle's groups, namely viviparous quadrupeds (mammals), oviparous quadrupeds (our reptiles), birds, fishes (including marine mammals and aquatic invertebrates) and, finally, insects. Where possible and relevant he gave, for each animal, the name in all languages; a description; diseases, habits, instincts, behaviour; general use in the service of man; use as food; use as medicine; and presence in human languages, customs, religion, poetry, proverbs, names of places, sculpture, etc. He was critical of Albertus Magnus, whom he considered a man who knew a lot about many things, but who clumsily abused their name.

The encyclopaedia appeared (1551–1558) during a time of religious tension, and it was included in the Roman Catholic Church's list of prohibited books: Gesner was a Protestant, and Pope Paul IV felt that the religious convictions of an author contaminated all his writings. Nevertheless, Gesner maintained friendships with people on both sides of the Catholic-Protestant divide.

Many scholars, during the Renaissance, considered the study of languages as a key to unlocking the past. Gesner, as a linguist aware of the many European languages and the many vernacular names given across Europe to each specific animal, presented each animal by its Latin name (which would, two centuries ahead, become the scientific name) and, in addition, as many of vernacular names as he could trace (Fig. 3.8). As an example, in describing the crested ibis (which has since become extinct) he wrote that this bird is usually named the forest crow but also the rock crow, in Bavaria the crevice crow, in Luthringia the marine crow and in Italy the forest crow. In his entry for the mackerel (*Scomber*), the entire text consists of different vernacular names. Thus, each animal gets a list of vernacular synonyms. One reason for this was that Albert the Great had sometimes discussed the same animal in different places under different names, and this, Gesner tried to avoid (Ogilvie 2006). He was so concerned with animal names that he named his encyclopaedia a *"Nomenclature"*, and the front page of his book on aquatic animals loosely translates:

> *Nomenclature: Pictures of aquatic creatures in the sea and freshwater, some 700 names for different genera. For each genus, the name given is in Latin, Greek, Italian, Spanish, French, German and English.*

112 **Auium**

LATINE Gallinago uel Scolopax minor, ſiue Ruſticula minor. Bellonius Aſcolopacion etiam appellat, ficto nomine diminutiuo ab Aſcolopace.

ITALICE Piczardella.

GALLICE Becaſſon, Becaſſe petite, Becaſſine, Becaſſeau. Vocantur & aliæ duæ auium aquaticarum ſpecies à Gallis eodem nomine Becaſſine. quarum prior ea uidetur eſſe, quæ à nobis Sametbünle nominata eſt ſupra pagina 101. ipſe Cinclum Ariſtotelis eſſe ſuſpicatur. poſterior Gallicis tantùm nominibus ab eo uocatur Becaſſine la plus petite, id eſt, Gallinago minima: et Deux pour un, quòd cum dimidio minor ſit Gallinagine maiore, duæ huius generis precio unius maioris uæneant: & obſcœniùs Fouton, nomine Oceani accolis uſitato, quoniã caudã præ cæteris ferè auibus ſubinde agitet: unde Cinclum minorem appellare licebit: & Germanico nomine ein Halbſchnepflin. Vtraq; à Bellonio deſcribitur, ſed non pingitur. Lege etiam infra in fine paginæ 123. in Motacillæ genere, &c.

GERM. Herrſchnepf/ Härſneff/ Harſchnepf/ Grasſchnepf/ Schnepflin.

Fig. 3.8 Animal names in Latin and various modern European languages: The woodcock. (From Gesner's *Historia Animalium*)

Gesner's encyclopaedia also presented the first vague beginning of a modern classification hierarchy: in his book on aquatic animals, he assigned each kind of animal to a genus or species, and, furthermore, he established the rank "Order". For example, in his classification of fishes, for the group consisting of eels, other elongate fishes and sea snakes, the title was *Ordo VI. De Piscibus longis, spinosis*; then came the Order of Marine Fishes that are flat and have spines, *Ordo VII, de Piscibus marinis spinosis planis* (Fig. 3.9); for another group, consisting of lizardfishes, the title was *Ordo VIII. De Lacertis similibus que piscibus, et aliis quibus*; and so on (see Fig. 3.10, for birds).

As in genera and species, some of his names for the orders serve us to this very day. Gesner also divided some of his Orders into chapters (what we might call Families, in modern classification). Though he did not use the term, his overall groups correspond to what we would term a Class, for example, the overall group of Aquatic Animals would be a Class as compared to another Class, say the Birds. Every Genus gets a noun, and every Species an adjective. He thus used a binomial nomenclature, with one name for the genus and another for a species. For example, he named all tits *Parus* (as today) and the different tit species as *Parus major*, *Parus*

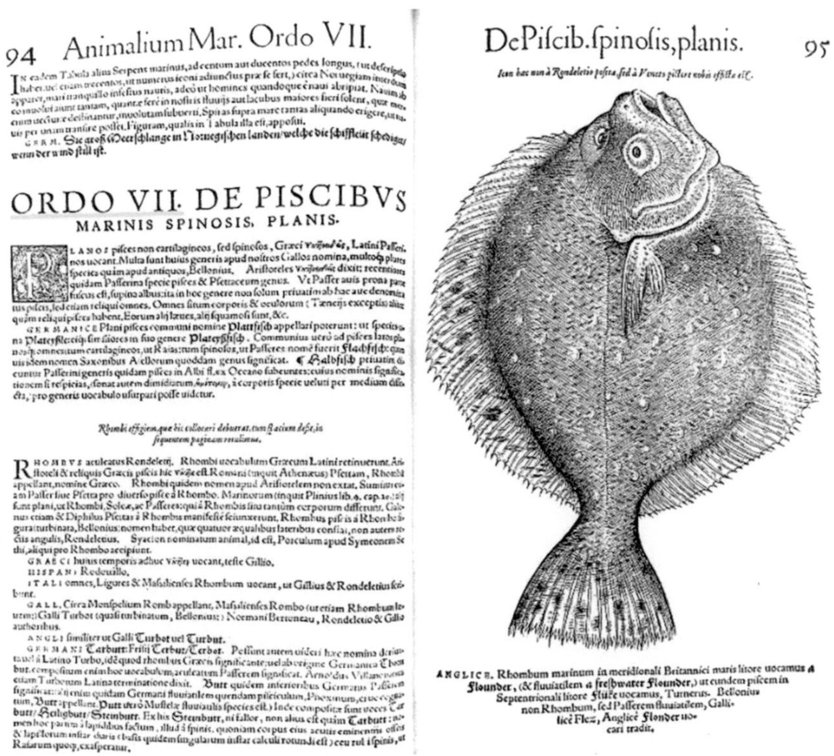

Fig. 3.9 Gesner's use of the term "Order", to describe a hierarchical systematic group (here, marked by me in yellow): Fishes, seventh Order. (From Gesner's *Historia Animalium*)

Fig. 3.10 Gesner's use of the term "Order", to describe a hierarchical systematic group (here, marked by me in yellow): Birds, eighth Order. (From Gesner's *Historia Animalium*)

coeruleus and *Parus cristatus* (as today). The modern systematic hierarchy was actually born in Gesner's encyclopaedia.

Gesner provided beautiful descriptions of domestic mammals, and a few excellent ones of wild mammals, such as the deer. We are grateful to him for writing also of the now-extinct wild bull *Bos primigenus*, which still lived in Poland until 1627. "The *Uro* is very strong and also very fast, and it has no mercy on any man or animal which crosses its path". His accompanying illustration depicted a wild bull charging a tree, behind which a hunter is hiding (Fig. 3.11). His descriptions are reasonable also when extending a bit beyond Europe, thanks to the classic ancients. For example, he described the jerboa *dipodides* (today *Jaculus dipodides*) and the spiny mouse *echines* of northern Africa (today *Acomys*).

In his main treatise on Birds (*Historia Animalum* III, 1555), Gesner did not give any classification; there is no hierarchy above the genus level, and he merely arranged all birds in alphabetical order. However, in that same year, he also published a treatise on Birds and their illustrations (*Icones avium omnium quae in historia avium Conradi Gesner: descripuntur*), and here the Birds are arranged in eight orders:

1. Diurnal raptors with hooked talons.
2. Nocturnal raptors (including bats).
3. Large non-raptorial strong fliers (raven, woodpecker).
4. Small non-raptorial strong fliers, which feed on fruit or worms.
5. Terrestrial weak fliers that bathe by wallowing in dust, and are domestic (doves).
6. Terrestrial weak fliers that bathe by wallowing in dust, and are wild.

Fig. 3.11 Gesner's wild bull, *Bos primigenus*. (From Gesner's *Historia Animalium*)

7. Web-toed aquatic birds.
8. Shore-dwelling birds that are not web-toed and usually have long feet.

Gesner took Orders 1, 2, 5, 6, 7 and 8 from Aristotle, who, we recall, had mentioned birds wallowing in dust. Gesner surely would not have referred to this marginal, unimportant behavioural pattern of wallowing in the dust as significant for bird classification, had it not been clearly stated by Aristotle.

In his volume on aquatic animals, Gesner relied heavily on Rondelet (1507–1566, Fig. 3.12). Guillaume Rondelet was born in Montpellier. When 18, he went to Paris to study at the Sorbonne, returned to Montpellier to study medicine, went back to Paris to learn Greek and anatomy and then returned to Montpellier. His medical practice was not a success. He managed his finances badly, and he outraged the citizens of Montpellier when he publicly dissected his infant son, in an attempt to determine the cause of his death. Eventually, he became a teacher at the medical

Fig. 3.12 Rondelet

faculty, but the plague reached Montpellier a few years later, and he found himself with only three students to teach. His fortunes turned when he gained a powerful patron, a cardinal whom he attended as a personal physician. He left Montpellier to travel in the Cardinal's entourage, and journeyed widely around France, in Belgium (of today), and in Italy, staying in Rome for three months. While in Italy, he was able to indulge his interest in natural history by visiting the coast. When 38, he received an appointment at Montpellier as Regius Professor of Medicine (a university professor with royal patronage or appointment). He constructed aquaria and fishponds in the gardens of his country house, with fresh and salt water, with flowing water supplied by an aqueduct, and here he kept and observed fishes, crustaceans and other animals. Rondelet's treatise on aquatic animals, *The Complete Story of Fish* (*Libri de piscibus marinis in quibus verae piscium effigies expressae sunt,* 1554/5), covered all aquatic diversity, from seals and whales to crustaceans, molluscs, worms and sponges, with freshwater fishes in a section of their own.

Gesner, in his encyclopaedia, closely followed Rondelet. Many of the genera described and illustrated by Rondelet and presented by Gesner are so good that we use their scientific names to this day (*Anguilla, Atherina, Blennius, Chromis, Colias, Gobi, Raia, Sardina, Solea, Sparus, Torpedo*, to mention but a few). This leads us (again) to conclude that people of the Medieval and Renaissance were excellent naturalists who knew well how to distinguish among different species and genera, in the surrounding European fauna. Sometimes Gesner's genera divide into species, each with a different name, but sometimes not: in the genus *Anthia* Gesner mentioned four species as the first, second, third and fourth species described by Rondelet, but without naming them.

Gesner classified all aquatic animals into 17 orders:

Order 1. Small marine fishes, the classification of which is unclear;
Order 2. Marine rock-dwelling fishes;
Order 3. Harp-shaped fishes;
Order 4. Flat fishes with golden scales;
Order 5. Round-mouthed fishes;
Order 6. Long, spiny fishes; marine snakes;
Order 7. Flat fishes;
Order 8. Lizardfishes;
Order 9. Flat cartilaginous fishes [rays];
Order 10. Roundish cartilaginous fishes [sharks];
Order 11. Roundish fishes;
Order 12. Whales and monsters;
Order 13. Molluscs;
Order 14. Crabs;
Order 15. Testaceans;
Order 16. Marine insects [worms];
Order 17. Zoophytes [sponges].

Many of these orders seem natural to our modern eyes (flatfishes, cartilaginous rays, cartilaginous sharks, whales, crabs, sponges), some represent only a habitat (marine rock-dwelling fishes). Gesner further subdivided many of these orders into 'chapters' (such as the "testaceans", which he split into gastropods and bivalves). His classification of aquatic invertebrates followed that previously described by Albert the Great. He named and described properly marine bivalves (*Mytilus, Chama, Clycimeris, Pecten, Pinna*) and then, after them, marine gastropods (*Purpura, Murex, Aporrhais, Nerites*). He placed these two natural groups into one large group, the Testacea, together with other shelled invertebrates such as (and here he erred) sea urchins and barnacles. In another group, the Molluscs, he placed cuttlefish, squid, octopus, argonauts and sea hares, together (again, an error) with sea anemones and medusas. In another order, he placed freshwater bivalves and land snails.

Gesner also classified the entire spectrum of aquatic animals by their various habitat, as those either of the littoral or of the sea. Littoral aquatic animals he further classified as dwelling either on the sand, in marshes, on vegetation or in rivers. The sea dwellers he further classified as those dwelling on mucous algae, on the sea floor, on vegetation, inside shells, on rocks, on both rocks and sand, pelagic animals and migrating ones. Several orders inhabited the freshwaters: fishes, crabs, testaceans, insects and amphibians, the latter group including the otter, beaver, hippo, crocodile, lizard, aquatic turtle and frog—all the freshwater air-breathing vertebrates he could think of. Frogs he further classified as born either of a female and then they are terrestrial, such as toads and fossils, or born spontaneously in rain puddles, appearing abundantly after rains and are aquatic. Gesner's combination of Aristotle as an authority on bird diversity on the one hand, and of Rondelet as an authority on aquatic diversity on the other hand, somewhat symbolizes the Renaissance approach of combining rebirth and recovery of classical antiquity, with contemporary knowledge. In certain parts of his book, Gesner was, however, carried away by names. Thus, immediately after describing the "sea horse", he described the "river horse" (*Hippopotamus; On the Nature of Fishes and Marine Animals*, volume IV).

It is fine to present knowledge about well-known European animals, about whom the literature offers sound knowledge of natural history. But what about well-known European animals, of whom the literature contains fantastic knowledge? How shall we refer to everything that every previous naturalist has ever written? We recall the fox that, in medieval *Physiologus*, feigns death and thereby attracts birds, which he then gobbles up; likewise, the devil attracts innocent believers and pulls them down to hell. Cardinals, archbishops, writers, theologians and historians had repeated this fancy hunting behaviour of the fox—and so did Gesner (Fig. 3.13), though without the religious aspect.

Gesner's encyclopedia also included fantastic animals. During the Renaissance and even concerning well-known European animals, naturalists developed a belief in the jumar—a cross between a mare and a bull. The earliest description I know of the jumar is from Gesner himself (*Historiae Animalium* (Book I, *De Quadrupedibus viviparis,* 1551, pp. 19, 106). He wrote that he had heard of a peculiar sort of mule in France, near Grenoble, which was born of a she-ass and a bull, and which the French call *jumarts*. He further claimed to have learned, from credible witnesses, of

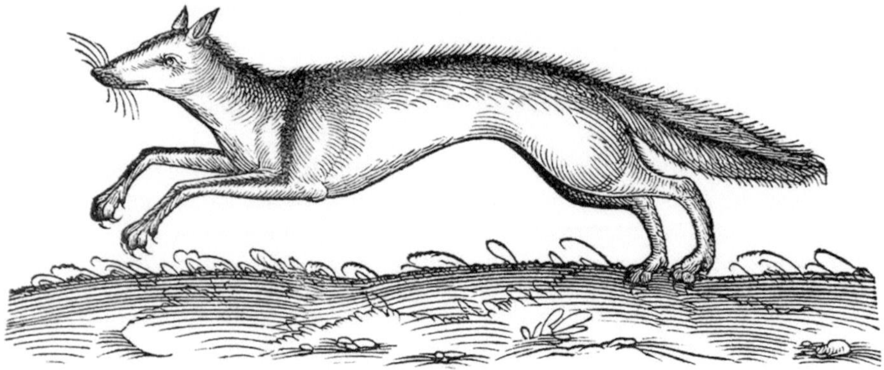

Fig. 3.13 Gesner's fox. (From Gesner's *Historia Animalium*)

a foal that was once born from a mare sired by a bull in the Swiss Alps. Later authors followed Gesner in claiming, in considerable morphological detail, that in Switzerland a mare sired by a bull gave birth to a foal with the hooves of a horse, but with the shape, hair and tail of a cow. It had traces of its father's horns in the shape of two bony lumps, each the size of half a walnut, on its forehead; also, the upper incisors are missing, and hence the lower jaw protrudes. Other authors claimed that such a hybrid lives in the palace stables of King Louis of France, that they had seen it, or ridden it, or that the masses had seen it in Rome, or that it lives in Turkey. Furthermore, there were detailed descriptions of anatomical dissections of the jumar, claimed to have taken place in veterinary schools. Two centuries after Gesner, even the twelfth edition of *Systemae Naturae* (1767, after Linnaeus' death) described the jumar. The Jumar myth, started by Gesner and his generation, endured for 250 years, until the beginning of the nineteenth century. *Jumar* is a distortion of the Greek *chimaera*, a fire-spitting monster that was a mix between a lion, a goat and a dragon: Chimaera-Cuimera-jumar.

The distance from such myths, concerning the natural history of well-known animals and their fantastic hybrids, to fantastic animals such as the unicorn is not great. I have already described the unicorn myth once, as it began in ancient days, and then again, as it developed in medieval times. Here I present it again, from its early beginnings, through the Medieval and, this time, into the Renaissance. The earliest unicorn I am aware of is from a ninth-century BCE archaeological site in Syria, at the springs of the Khabur River (Tell Halaf), which was the capital of an Aramaean principality. A basalt relief from the palace of an Aramaean prince presents what may well be a unicorn (Fig. 3.14). The unicorn myth surely would have sunk into oblivion, were it not for the Septuagint translation of the biblical Hebrew term *re'em*, which appears seven times in the bible. The Septuagint translated this term into *monokeras*. In the Vulgate, the Hebrew plural *re'emim* (Isaiah 34:7) was translated as "**And the unicorns shall go down with them** (*et descendent unicornes cum eis*)"—thereby establishing the unicorn as a biblical creature, created by God. Whoever denies the existence of unicorns defies the existence of God. The

Fig. 3.14 Unicorn. Khabur River (Tell Halaf), Pergamon Museum, Berlin. (Illustration by Tuvia Kurz)

Babylonian Talmud, compiled between the third and fifth centuries CE, suggested that the biblical *takhash*, whose hides Moses used to cover the Tent of the Tabernacle, was an animal with a single horn on its forehead. It had existed in the days of Moses and has since then been hidden (Mishna: Shabbath 28b).

During Medieval times, people ascribed medicinal value to the unicorn's horn. They described the unicorn as a creature so very aggressive (or so very shy) that, to capture it alive, one needs a special trick. Men lead a virgin to places of which the unicorn is fond, and there they leave her, and wait in ambush, nearby. The unicorn, upon noticing the virgin all alone, runs toward her and places his horn in her lap. She fondles it until he falls asleep, and the hunters then capture it. In centuries to come, people further elaborated this myth, which closely resembles the Biblical story of Samson and Delilah. A *Physiologus* from the fourteenth century mentions a large lake, at which many animals gather to drink. However, before they gather, a snake comes and throws its poison into the water. The animals, upon noticing the poison, do not dare drink and instead, they wait for the unicorn. The unicorn, upon coming, makes the sign of the cross with its horn over the water, thereby neutralizing the poison, whereupon all other animals drink from it (Fig. 3.15, from Bosch, the *Garden of Earthly Delights*). Symbolically, the snake poisoning the water is the Devil, and the unicorn represents the Saviour.

During the fourth crusade (1204), the Doge placed unicorn horns in Venice to prevent poisoning of the water. In 1389, a certain Joken von Hesse reported that he had seen with his own eyes a unicorn rising from the sea to purify impure water, to enable animals to drink. There were unicorn horns in the cathedrals of Saint Dennis in Paris, in the Basilisk of Saint Martin in Venice and in Westminster Abbey in London. The unicorn was a remedy for measles, fever, stomach pains, leprosy,

Fig. 3.15 Bosch, The Garden of Earthly Delights (in the Prado Museum). Detail

scorpion poison, viper poison and against the plague. Unicorn horn powder, when mixed with egg white, was used against leprosy, a belt from a unicorn hide defended against the plague and fever and shoes made of unicorn hide kept away leg diseases. Indeed, the horn of a unicorn was one of the most expensive and prestigious medications, to be found only in kings' palaces. In Denmark the horn served to build the throne, in the Austrian empire it served as the royal staff and was also set into the crown, in Burgundy it was made into the sword sheath of Charles the Bold (1433–1477), and Queen Elizabeth I (1533–1603) of England acquired a unicorn horn for 10,000 pounds, the price of a castle. German merchants sold the Pope a unicorn horn for 18,000 English pounds in 1560; in France, the Pope gave a horn to Francis I (1533), and the King would go nowhere without his little bag of unicorn powder. The great Inquisitor Torquemada (1420–1498) carried a unicorn horn everywhere, lest he be poisoned or murdered. Emperor Ferdinand I (1503–1564) possessed a unicorn horn (actually, the tusk of a narwhal *Monodon monoceras*, in which the left upper canine tooth grows into a huge "horn" that is used in the male's combats). In Scotland, in Rothesay Castle (from before the fifteenth century), a unicorn was carved out of stone as a symbol of the noble house, and King James III (1466–1488) minted golden coins with a unicorn on them. At the time King James VI of Scotland succeeded Elizabeth I of England, effectively uniting the two countries, the Scottish Royal Arms featured two unicorns as shield supporters. In a gesture of unity, King James replaced the one on the left with the English lion. (Today, the Royal Coat of Arms of the United Kingdom of Great Britain still has the English lion on the left and the Scottish unicorn on the right, and the Royal Coat of Arms in

Scotland has them the other way around. The heraldic unicorn is chained, because according to folklore, a free unicorn is a dangerous beast.) In Brussels, several huge tapestries presented (1495–1505) the hunting of the unicorn. There were also Jewish versions, though a shade different (Gershon Ben Shlomo 1288–1344, of France; *Shaar Hashamaim* 28: 2). Could it be that the unicorn, a creature of enchanted forests, which the Holy Scriptures had described, and whose horn is deposited in cathedrals and in kings' palaces—does not exist? Gesner, though he never saw one, believed that there were several kinds of unicorns, which he described and illustrated (Fig. 3.16).

Whereas the capture of a unicorn requires the recruitment of just one single virgin, the capture of an elephant requires two (naked) ones, according to the *Gesta Romanorum* (a collection of tales compiled in the late thirteenth century in England, Germany or France, by an unknown author). One virgin carries a basin, the other a sword. The elephant, attracted by their song, licks the breasts of one of them and then falls asleep in her lap; the other takes her sword and kills him, after which the first maiden fills her bowl with the blood shed by the elephant. The elephant symbolizes Christ, the two virgins are Mary and Eve, and the breasts are the Old and New Testaments (Houwen 1994).

Other European animals about whom the literature contained fantastic knowledge were the scorpions. Pliny had written that some scorpions possess wings, so Gesner presented a figure of winged scorpions. That scorpions could have wings may well have stemmed from scorpion flies, with their scorpion-resembling

Fig. 3.16 One of Gesner's unicorns. (*Historia Animalium*)

Fig. 3.17 Scorpions. Left: Gesner's illustration including a winged specimen. (*Historia Animalium*). Right: *Panorpa communis*, a scorpion fly (by Tuvia Kurz)

genitalia (Fig. 3.17). During the sixteenth-seventeenth centuries, knowledge of the natural history of scorpions was very imperfect, and Gesner wrote that they climb up to the roofs of houses, and if they can find any broken tile they remove it, and "one of the strongest of their captains" hangs down by his claws, his tail hanging down. Then another scorpion climbs down his back, and takes hold by the "captain's" tail; then a third takes hold of the second's tail and so does all the rest, until the scorpion ladder reaches the bed; they come down and sting the person lying asleep in his bed, and then climb up again, on the scorpion ladder. Why the scorpions do all this is a question Gesner did not address.

He did, however, address the question of scorpion generation, which, he stated, may be either spontaneous, in that scorpions arise from rotting carcasses of crocodiles or snakes, or by copulation. Having copulated, they produce little white eggs, which they brood. The hatched young, when they mature, drive their parents away and destroy them. Some naturalists believe in the opposite: The parents devour all their young, but for one that, being more cunning than the rest, hides itself among its mother's legs and thus escapes her sting; it eventually avenges the death of all his brothers by (climbing on her back and) killing her. Today, we know that scorpions reproduce either sexually or through parthenogenesis, and all are viviparous. The young resemble their parents, and their mother carries them about on her back until they have undergone at least one moult. They grow by splitting the old exoskeleton and emerging from this split.

If about well-known European animals, the literature contains fantastic knowledge, then what are we to believe about those vaguely known marine animals of northern Europe? The Scandinavian lands were very poorly known during the Medieval and also during the Renaissance; most knowledge of their fauna was based on travellers' tales. A Scandinavian (Catholic) archbishop, Olaus, had visited Scandinavia around 1518, and he had described the walrus *Odobenus rosmarus* from the arctic and subarctic north as a mammal with four feet, with fins and upward-pointing teeth, by which they hang from rocks. A French maritime explorer had, during his voyage through the St. Lawrence Strait (1534), encountered "several great beasts like great cattle. They have two teeth in their mouths, like elephants, and they go into the water". Out of these two descriptions and from these two illustrations, Gesner had to illustrate a walrus He did not accept that a walrus could hang from rocks with upward-pointing teeth, so he illustrated it with teeth pointing downward, and with wing-like fins separated from its feet as a distinct organ, instead of between its fingers (Fig. 3.18; Ogilvie 2006).

Gesner valued Olaus as "a man who is worthy of faith on account of his erudition and his experience of the region" (Ogilvie 2006), but Olaus described, from Norway, a huge, black, 90-metre-long sea serpent with a long mane, sharp scales and fiery eyes. On clear summer nights, it devours calves, lambs, octopi, lobsters and crabs, and it also coils around a ship's mast and lunges at sailors on the deck. Gesner seems to have doubted the existence of this sea serpent because while illustrating it in his *Historian Animalium* (1558), he noted that Olaus was responsible for the accuracy of this picture (Fig. 3.19).

We have mentioned, above, Rondelet's book on aquatic animals, with its excellent descriptions and illustrations of so many species. This book also had a handful of entries on creatures of dubious authenticity of which he had seen pictures and heard accounts, but which, he emphasized, he had never seen (Fig. 3.20). Rondelet included in his book a "leonine monster", an anthropomorphic-looking "bishop-fish" and a "monk-fish", in a category of declaredly dubious entries. He made it clear that he had never seen anything like these creatures. They were included based on no evidence other than reliable hearsay. Of the leonine creature captured not long before the death of Pope Paul III (1549), he had a portrait made (from a painted picture that circulated). Rondelet noted that "the description I was given came from men of learning, worthy of trust, but I think that the painter added features of his own invention, which are not found in nature". The leonine monster reminded him, he wrote, of the fanciful whales and other sea creatures that one finds painted in the southern corners of maps and in Munster's *Cosmography*.

The "marine monster dressed like a monk" was washed up by a storm on a beach in Norway, horrifying the onlookers with its rough and repellent humanoid face,

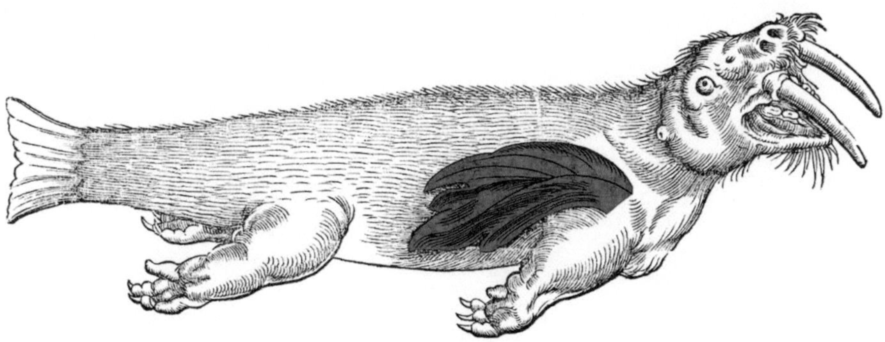

Fig. 3.18 Gesner's walrus, with fins above the arms. (*Historia Animalium*)

Fig. 3.19 Gesner: Olaus's sea serpent. (*Historia Animalium*)

Fig. 3.20 Gesner's sea bishops. (*Historia Animalium*)

bald pate and long fins like a monk's habit. A man who claimed to have seen the creature presented one portrait of it to Emperor Charles V, and another to Marguerite de Valois, Queen of Navarre, who gave it to Rondelet, who speculated that this was possibly a merman, or Triton, of the kind mentioned by Pliny. The "marine monster dressed like a bishop" was said to have been fished up in Poland in 1531, pictured and presented to the Polish King. These strange, fantastic creatures later found their way and trickled from Rondelet into Gesner's encyclopaedia (Fig. 3.20).

In Medieval times, Albertus Magnus in his *De Animalibus* had described a sea monk *Monachus marinus*, occasionally seen in the British sea, as a fish with white skin surrounded by a dark circle on the top of its head, like the head of a tonsured monk; however, with the mouth and jaws of a fish. The sea monk entices those travelling on the sea until it lures them in, takes them to the sea bottom and eats their flesh. The sea monk surfaced again sometime between 1545 and 1550 when the Danish king Christian III sent the Holy Roman Emperor Charles V (then in Spain) drawings of a strange animal that had been caught in the straits between Denmark and Sweden. Described as a sea monk, this strange creature had "a human head and face, resembling in appearance the men with shorn heads, whom we call monks because of their solitary life; but the appearance of its lower parts, bearing a coating of scales, barely indicated the torn and severed limbs and joints of the human body". This find was commented on and illustrated by two of the best ichthyologists of the day, Pierre Belon (1553) and Guillaume Rondelet (1554), and eventually, it too appeared in Gesner's (1558) *Historiae Animalium* based largely on Rondelet's and Belon's descriptions.

Gesner also described another marine monster, a combination of Pan and a marine satyr, an *Ichthyo-centaurus*, generated from a marine demon. The figure Gesner presented was, so he commented, of an individual from Antwerp, but also people in Norway and Germany who had seen similar creatures. Gesner described this creature as having the arms and horns of Pan and a marine satyr, who resembles a monkey and is similar to a whale in that it has no scales. Pope Eugene the Fourth himself had seen this strange marine man give birth to descendants. Sometimes the sea waves sweep it to the rocks, where it cuddles. Its face is like that of humans, the skin of its face resembles that of an eel, on its head there are two straight horns, its hands bear organs resembling fingers, its arms bear wings like those of a bat, and its legs have fins and its tail ends in strings (Fig. 3.21).

There was also a giant ray, beneficial to man. If someone fell into the sea, and many fishes pounced upon him to gobble him up, the giant ray would come and save him. Then, there was a whale so large that it resembled a small island rather than a fish; sailors, believing it was indeed a small island, climbed upon its back and cooked their meals—but the whale would then ram their boat (Fig. 3.22). This is reminiscent of Adeliza's bestiary, in which the whale spreads sand over its back, rises to the sea surface and remains there, motionless; seafarers, thinking it is an island, land on it and prepare their meal. Once the whale feels the fire and the people, it plunges into the deep, and they all drown.

We recall Pliny's opinion that everything born in any department of nature exists also in the sea—marine creatures resembling sheep, horses, asses and bulls. This

Fig. 3.21 Gesner's *Ichthyo-centaurus*. (*Historia Animalium*)

Fig. 3.22 Gesner's whale. The crew, thinking it is an island board it and cook a meal; the whale meanwhile rams the ship. (*Historia Animalium*)

opinion of Pliny would continue for centuries upon centuries, throughout the Medieval, with scholars claiming that a vast array of marine animals closely resemble terrestrial counterparts, at least from the navel upwards. There are sea pigs, sea dogs, sea monks, king-fishes wearing crowns and so on—marine animals that have terrestrial counterparts. This belief of ancient Roman and Medieval times continued during the Renaissance, with people believing that every terrestrial animal had its parallel in the sea, usually with the tail of a fish. There was a goat-fish, a horse-fish, a cat-fish and a dog-fish. Accordingly, Gesner depicted a bullfish (Fig. 3.23) and a horsefish.

There was also a pigfish, a sea hog of which, Gesner claimed, one specimen was observed in England in 1532 and another in Germany in 1531. It had the head of a

Fig. 3.23 Gesner's sea bull. (*Historia Animalium*)

Fig. 3.24 Gesner's sea hog. (*Historia Animalium*)

boar, a body covered with scales, the forked tail of a fish, the feet of a dragon with webs between the fingers, horns of a cow on its shoulder and strangest of all: in addition to the two eyes on its head, it had three eyes on each side of its body (Fig. 3.24). Gesner clearly reported the existence of almost all the imaginary sea animals of which he had read.

Gesner was not the last to believe in monsters. People during the Medieval and the Renaissance considered monsters to result from some sinful sexual behaviour by the mother, or her state of mind. They believed that the cravings and experiences of an expectant mother, especially those arousing strong passions such as lust or fear, inscribe themselves directly on the body of the foetus, thereby producing its deformities. People also considered monsters as signs of God's wrath, an omen of divine displeasure and future misfortune: war, death of great men, the fall of empires and religious upheavals.

During the German Protestant Reformation, a misshapen calf was born in Saxony (1522) with oddly shaped legs (its hind legs were straight, as a human's) and with a fold of skin over its head shaped like a hood, hence its comparison to a monk (Fig. 3.25; Park and Daston 1981).

An illustration of this monk-calf made its way to a Prague astrologer, who declared that the monster signified someone terrible—Martin Luther. Luther quickly responded that the "monk-calf" stood, in its monstrosity, for the Catholic Church, a sign of a false and idolatrous religion that will soon be thrown aside, and the cloistered men and women shall be revealed as soul-killers (Walton et al. 1993). Luther's anti-papist pamphlet (1523) illustrated another monster, the Pope-Ass: a hybrid between a man and a donkey supposedly found near Rome after the 1496 flood, its various bestial parts corresponding to the bestial vices and errors of the Church (Fig. 3.26). This pamphlet, frequently reprinted in the sixteenth and seventeenth centuries and translated into French, Dutch and English, was received with high

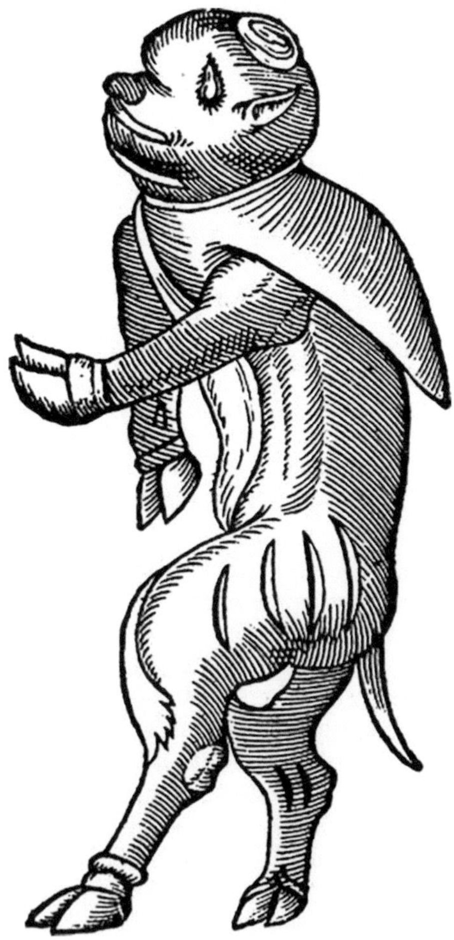

Fig. 3.25 Paré: the monk calf. (From Park and Daston 1981)

Fig. 3.26 The Pope-Ass. (From Park and Daston 1981)

excitement by learned and barely literate alike. It also served in Lutheranian wars against Calvinism in France during the Wars of Religion, and against separatism in England, during the Civil War. Monstrous births also marked the accession of Elizabeth I of England (1558), as a warning to "Catholics and other sinners" (Park and Daston 1981).

Ambroise Paré (c. 1510–1590) was a French barber-surgeon whose major contributions were surgical techniques of battlefield medicine. In spite of being a Protestant, he served four French Catholic Kings (Henry II, Francis II, Charles IX and Henry III). In addition, he compiled much of what was known in the sixteenth century about birth defects and their causes (*Des monstres et prodiges*, 1573; translated by Pallister 1982 as *On Monsters and Marvels*). He thought that an overabundance of semen would result in the birth of Siamese twins, as it would cause superfluous parts of the body to develop and thereby produce conjoined twins. In Verona, indeed, he wrote, in 1475, two girls were born who were joined at the rear. Their two heads had the same desires to drink, eat and sleep, had identical speech and the same emotions. Their parents toured them around the towns of Italy, exhibiting them for money, but in Bavaria, the Duchess drove them out, lest they distort the bodies of future babies, by women who, by seeing them, might imagine all sorts of things and give birth to similar monsters. Had not a black queen who ruled Ethiopia alongside her black husband, given birth to a white child (Chariclea), who resembled Andromeda, whose portrait was before the queen's eyes during conception? (Bitbol-Hesperies 2006). And then, near Fontainebleau in France, in 1517, a boy was born with the face of a frog, because his mother had slept with her husband while holding a frog in her hand, to cure her fever. This belief, in the influence of a mother's imagination on the appearance of the future newborn, went on for over a century at least. In England, the *Philosophical Transactions* of the Royal Society (1687) published that a pregnant mother, when milking a cow, was hit by its teat on her left temple. In this very same place, her newborn daughter ("a girl called Elizabeth Dooly") had a piece of flesh growing exactly like the cow's teat, in size and shape. Further, in the seventeenth century, René Descartes thought that there was nothing impossible in the claim of the fable that the urine of a man who has been bitten by a rabid dog contains crude, nasty figures of little dogs. He believed that it is the mother who, by her imagination, determines the external appearance of the foetus developing inside her; it scorches onto the surface of her developing foetus a likeness either to herself and other humans or, if her imagination is harmed by some extremely unusual impression, a likeness to some other species, and the child will be monstrous (Smith 2006). There were also tales (1694) of such inherited abnormalities as that of a woman in France, stabbed to death; an embryo taken out of her body was marked with as many spots as she had received stabs and in the very same parts of the body. And later still, Malebranche (whom we shall meet in Part 4 of this book) described a child born with bones broken in exactly the same place where a publicly executed criminal—whose execution the pregnant woman had witnessed—had had his own bones broken (Bitbol-Hesperies 2006).

Other birth defects were due to deficiencies in semen, and bestiality between women and dogs would generate some monsters (Fig. 3.27). He further noted that a mare had once foaled a colt with the face of a man and the body of a horse, and soon after that, war began between the Florentines and Pisans. Also, just one month before the Ravenna battle (1512) between Pope Julius II and Louis XII of France (which the Pope's forces lost) an androgynous monster, which symbolized Louis, was born in Ravenna with a horn on its head, two wings, a single clawed foot resembling that of a raptor and an eye in the knee (Fig. 3.27; recently, Walton et al. 1993 have suggested that the monster of Ravenna was an actual birth). Then, there was a child conceived by the union of a woman and a dog. It had a human upper body and a canine lower body.

As for a most peculiar monster reportedly found in Africa (Fig. 3.27), Paré acknowledged his inability to give any functional explanation for the multiplication of its parts. "The only thing I can say is that Nature was playing, to make us admire the greatness of her works". In this terminology, emphasis falls on the works of Nature rather than the works of God, and monsters are treated as jokes, or "sports" of a personified Nature, rather than as divine prodigies. They signify her abundance of creative invention, and through her, God's own fertility and creativity rather than

Fig. 3.27 Paré: From left: Siamese twins; woman-dog bestiality; the Ravenna monster; colt with human face; African monster. (From Ambroise *Paré Monsters and Marvels (1585)*, translated by J. Pallister, 1982; Author's library)

his wrath. Natural monsters were nature's artefacts, in a certain sense, and nature became the artisan par excellence. Paré improved the credibility of his monster descriptions in two ways. One manner was by providing very precise details of the time and place where the monster was born. "In the year 1564, in Brussels, at the house of a man named Joese Dickpert, who lived on Warmoesbroeck Street, a sow farrowed six pigs. The first of these was a monster having a man's face as well as arms and hands, representing humanity in a general way from the shoulders up; but the hind legs and hindquarters of a swine, and having the nature [genitalia] of a sow" (Fig. 3.28). Ever so conveniently, monsters always turned up elsewhere.

In addition, if possible, a credible person was mentioned as a witness: "In this present year 1577, a lamb was born in the village named Blandy, a league and a half from Melun, having three heads in one, and when one of the heads bleated, the others did the same. Master Jean Bellanger, a surgeon living in the city of Melun, affirmed that he had seen it" (Fig. 3.28). If there were no credible persons as direct

Fig. 3.28 Paré: man-pig; three-headed sheep. (From Ambroise Paré *Monsters and Marvels*, translated by J. Pallister 1982; Authors library)

Fig. 3.29 Paré: Monster Snail. (From Ambroise Paré *Monsters and Marvels*, translated by J. Pallister 1982; Authors library)

witnesses, then high-ranking people were indirectly tied to the event: "in the time of Pope Martin V, there was found enclosed in a large solid stone a live snake". The other method used by Paré to increase the reliability of his monster descriptions was to mix them with descriptions of "marvels", of real, albeit strange animals from exotic countries such as the ostrich, giraffe, elephant and chameleon. As these animals undoubtedly exist, no matter how strange they are, so too did the monsters.

Paré also bridged the gap between monster and marvel by describing and illustrating creatures resembling standard animals but of strange appearance (snails, Fig. 3.29). Some 25 years later, Aldrovandi (whom we shall meet presently) joined Paré in the monster parade by describing and illustrating chickens with three legs, others as Siamese twins, and still others with additional deformations and monstrous horses.

Paré was not the only Man of the Renaissance to believe in monsters (Bitbol-Hesperies 2006). Jakob Rueff (1500–1558), a physician and surgeon in Zurich who wrote a practical handbook on midwifery (1554), also thought that humans can conceive with animals, due to an "attractive virtue" that is common to both humans and animals. He wrote of a hybrid in Switzerland, between a mare and a stag. Then there was Conrad Lycosthenes of Alsace (1518–1561), who wrote of comets, earthquakes and mythical animals as omens for catastrophic events throughout history (*Prodigiorum ac Osentorum Chronicon,* 1557). He noted that in the year 1557, he saw both a triple parhelion and the birth of a boy with neither a forehead nor a neck, two closely linked events that were related to God's infinite might and human impotence. Caspar Bauhin (1560–1624), professor of anatomy, botany and medicine in Basel, today remembered mainly for his descriptions and classification of plants, studied also monsters (*De hermaphroditorum monstrosorumque partum natura et theologia*…1614). He classified monstrosities as resulting either from superior causes, which come from God or under the influence of the stars, or the winds, or inferior causes, some of which are internal and some external. The internal causes are divided into those concerning matter (of the father or the mother), those concerning place (the size of the womb) and those arising from the weakness of a "formative faculty". The external causes are divided into those which we can ascribe to both parents and those which we can ascribe to the mother alone. As to compound monsters, those consisting of parts of two animals, Bauhin considered them impossible: no animal can be born after a length of time that is different from its natural gestation. In going back to Aristotle's concepts of material and shape, Bauhin thought that in monsters, matter is conserved, whereas the natural form is changed: monsters with four eyes, four arms and four legs consist of the same material as those with only two sets of each. Fortuno Liceti (1577–1657) was, throughout his career, a professor of logic, philosophy theoretical medicine, in several universities in northern (Italy Pisa, Padua, Bologna). He thought (*De monstrorum causis, natura et differentis*) that the duration of gestation in women may vary from seven to twelve months, which could allow them to be made pregnant by any of the animals belonging to a species whose gestation occurs within these limits. His treatise offers a number of examples arising from the "unfortunate coupling of a human and a beast". Besides the famous child with the canine lower body (already shown by Paré) and the elephant-headed boy, Liceti showed in his book, among other examples, a pig with a human head, hand and feet and a monster that is half-pig, already in Paré (Bitbol-Hesperies 2006).

With this knowledge about monster diversity in Europe, what did European naturalists know about strange animals from faraway continents? Back to Gesner and his encyclopedia, we may consider Gesner's treatment of the birds of paradise (Paradisaeidae, a family of 40 species native to New Guinea) in which the male has a splendid plumage (Fig. 3.30). Gesner wrote of a bird that he named *Paradisea avis*, from islands off far eastern parts of Asia (Moluccas), that this bird has transparent wings and two long feathers, which are as hard as a horn. It has no legs because it flies perpetually, and rests on trees by means of these two feathers, by which it hangs from trees. Over the sea, its flight is so swift that no ship can overtake

Fig. 3.30 Bird of paradise *Paradisea apodus*. (Tuvia Kurz)

it. Aristotle did not believe that there are legless birds, but Gesner commented that he himself had seen this bird already three times and indeed, it has no legs. Gesner further wrote that it spends its entire life on air. The female lays eggs into a depression on the male's back. She herself has a depression on her abdomen, and both male and female together brood the egg on the male's back. The male has, in his tail, a string resembling a shoelace, and during brooding, he ties the female to him by means of this shoelace feather, lest she fall. The moment he spreads his wings the air immediately lifts and carries it so that his flight does not cause him any fatigue. Paradise birds eat and drink only the dew of the sky. The kings of the Moluccas believe that these birds represent the souls of people now in paradise, they name it the bird of the gods and it is sacred. In Europe, King Charles V has five specimens of this bird, sent to him by the king of the Moluccas. Gesner added that the Turkish janissaries decorate themselves with the feather of a bird named *Rhyntaces*, sold to them by the Arabs. It too has beautiful feathers. People sometimes name the *Rhyntaces* as *Apodem* [= that without legs] but Gesner, following Belon, believed that this is the phoenix of ancient days (Ogilvie 2006).

How did Gesner come to this fantastic description? The earliest similar description I know of, for this bird, precedes Gesner by three centuries and is that of Gershom Ben Shlomo (second half of the thirteenth century, *Sha'ar HaShamaim* 22). He described a bird species that has no legs. This species is born in the air, and as it does not rest on the ground, it does not need legs, which it does not, since nature does nothing without a purpose. When the female lays her eggs, she warms them beneath her wings, and after the chicks have hatched, she lifts them onto her back until they grow and can fly by themselves. Three centuries after Gershom Ben Shlomo and throughout the sixteenth century, spice merchants returning from the Far East described preserved birds that natives had brought to them from the inner parts of the islands. These natives would stuff spices into a dead bird's skin, and then chop off its feet, wings and even the head (Ogilvie 2006). As a result, descriptions and illustrations of paradise birds based on their skin did not resemble the live birds. Gesner described *Paradisea avis* (3.31) as an individual without wings, whereas others illustrated the bird with wings but without legs and without the two long tail feathers. Furthermore, the natives claimed that these reddish-golden birds, upon reaching old age, fly very high and straight into the sun, and hence they named them the birds of the gods.

When the stuffed birds of paradise reached Europe, the red-golden plumage combined with the myth of their flight into the sun suggested to the French naturalist Belon that these birds were actually the mythical phoenix. As a youth, Pierre Belon du Mans (1517–1564) had travelled widely through Flanders, England and Germany. When about 25, he studied medicine at Paris and then undertook extensive scientific voyages through Greece, Crete, Asia Minor, Egypt, Arabia, northern Italy, Savoy, the Dauphiné and Auvergne. Cutthroats murdered him one evening while he was walking through a park, not too far from the gates of Paris. On his travel to Egypt, Belon had found that the headdress of the Turkish janissaries (the Sultan's palace guards) consisted of ostrich feathers, together with "feathers of a bird named *Rhinactes*. These are beautiful feathers which [...] grow from a small creature from which only the skin is left [...] I believe this is the phoenix". Two years later, in a chapter on the phoenix, in a treatise on birds (1554), Belon repeated this opinion: "This bird has no legs, but in nature, in its attempt to cover this weakness, it develops two feathers on the two sides of its tail. These feathers are hooked at their end and very stiff, and it is by these means that the bird hangs itself from the trees [...] Nature has granted the phoenix this special character to save it from the hostility of predators. As to the mode by which the female broods her eggs, some believe that she broods them in a depression on the back of the male while sitting upon his back. Others think that she gathers twigs, which are ignited by the sun's heat, and from these emerges a grub, from which the phoenix develops". These descriptions by Belon are the basis of Gesner's description and illustration (Fig. 3.31). Live birds of paradise reached Europe only towards the end of the seventeenth century. Two centuries after Gesner, Linnaeus would name the largest of the birds of paradise *Paradisea apoda*, "the paradise bird without legs", and we now understand this term of Linnaeus (Ogilvie 2006).

Belon was typical of the Renaissance scholar in that he took an interest in many disciplines, including zoology, botany and classical antiquity. He forwarded the idea of all vertebrates sharing a basic plan, by pointing out the similarities between the skeleton of a bird and that of a human. Posing the two skeletons side by side, he used the same lettering to indicate the corresponding bones in the two skeletons

Fig. 3.31 Gesner's bird of paradise. (*Historia Animalium*)

(1555; Fig. 3.32). This is one of the earliest ventures into pan-vertebrate comparative anatomy.

The Asian rhinoceros reached Lisbon in 1515, as a gift to King Manuel I of Portugal from the Sultan of Gujarat (a kingdom in north-western India). A pen and ink sketch of this rhino, made in Lisbon by an unknown artist, reached the German artist Albrecht Dürer in Nuremberg (Bavaria), who made a pen and ink drawing of his own, and then printed a mirror image of it (1515) as a woodcut. Unfortunately, Dürer, otherwise so meticulously accurate in painting every hair in the fur of a hare and of a squirrel, every plant on a piece of turf and every vein in the leaf of a violet, was not accurate in representing the Asian rhinoceros. He depicted the rhino as having hard plates covering its body like sheets of armour, with a gorget at the throat, a breast-plate, and rivets along the seams; he placed a small twisted horn on its back, gave it scaly legs and saw-like rear quarters. None of these features is present in a

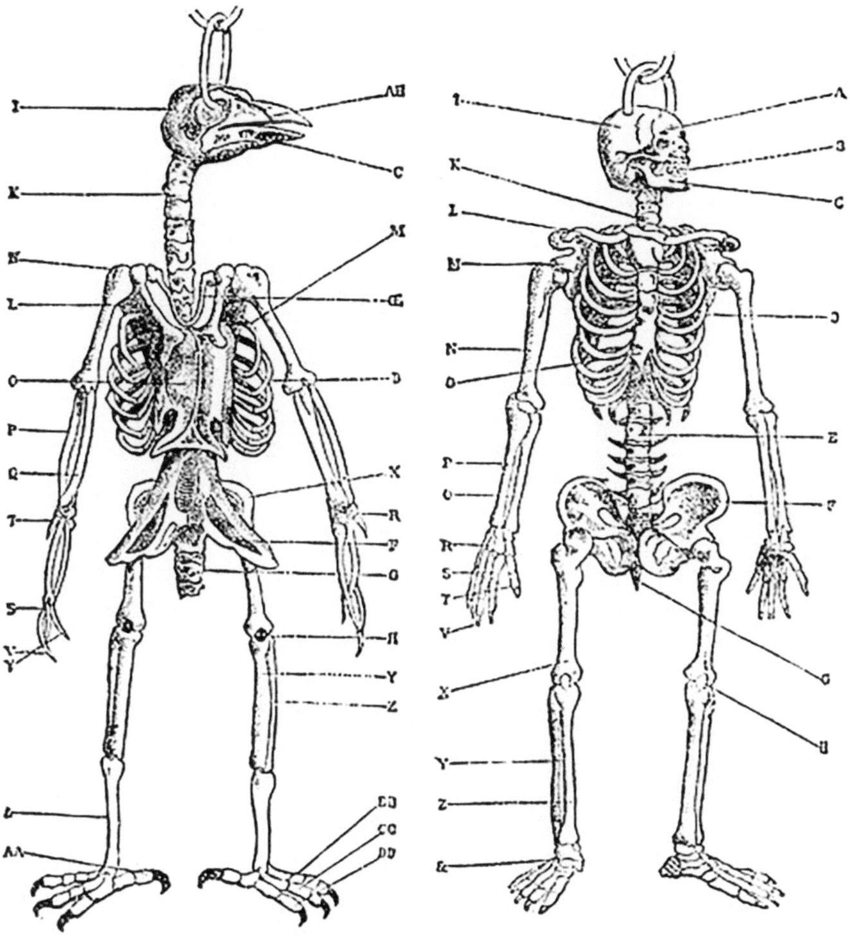

Fig. 3.32 Belon's comparison of the skeleton of a bird to that of a man

real rhinoceros. Dürer also changed the angle of the horn and added a beard and large ears. This was the wood print that Gesner copied and printed (Fig. 3.33).

If we accept fantastic animals from the Old World, creatures we are not sure whether they exist or not, what about animals from the recently discovered New World? Are they real or imaginary? The opossum, *Didelphis,* was first discovered in Brazil (of today) in 1502 and was described as a tree-dwelling animal living on fruits, the female having, beneath her abdomen, a large sac in which she carries her descendants, lest they fall and get lost. A female opossum with two young was shipped to Portugal in 1511. The young died during the journey, the female survived somewhat longer and was displayed in Granada. An opossum was first illustrated, as a motif in a map in 1516 (Waldseemueller's *Carta marina*), and this illustration became the model for many later illustrations of *simivulpe*—the vulpine monkey, or

Fig. 3.33 Dürer: Rhinoceros

Fig. 3.34 Gesner's opossum. (*Historia Animalium*)

monkey fox. When Gesner illustrated the opossum in 1551, he drew from that illustrated map but added the lactating young and also a row of imaginary teats (Fig. 3.34). I wonder how many of us would recognize the opossum in Gesner's figure.

Gesner also presented the first illustration of the raccoon *Procyon*, though only in a later edition. He named it *Mus indicus* and classified it into a group he named *Mus*, which included mammals with a long-pointed muzzle, such as the sable. Today, the term *Mus* refers to only one group of long-muzzled mammals, the mouse

Fig. 3.35 Gesner's turkey. (*Historia Animalium*)

Mus. Gesner named the raccoon *Mus indica* because Columbus thought he had reached India, and Gesner was perhaps not yet sure that this was not so. There were several other cases during the Renaissance in which naturalists referred to American animals as Indian. They had observed the turkey *Meleagris gallopavo* already during the conquest of Mexico (1520; Fig. 3.35). To this day, people still call the turkey *Indichke*, the Indian bird, in Yiddish (Medieval German mixed with Slavic and Hebrew); in Euskara (Basque language), it is *Indioilar arrunt.* (The English name for this species results from an early misidentification of the bird with an unrelated species imported to Europe through the country of Turkey.) In this case, of the turkey, Gesner's illustration was reasonably accurate.

Turning now from Gesner to Ulisse Aldrovandi (1522–1605, Fig. 3.36), he was an Italian naturalist from Bologna. His father, a notary and secretary of the Senate of Bologna, died when he was six, and his mother brought him up. He enjoyed an adventurous youth: At the age of 12, he left for Rome, without money and without telling his mother, but returned home after a few months. When 16, he again travelled to Rome, and from there he journeyed all the way to Spain. Bandits in the Pyrenees robbed him, in the mountains of Galicia he starved for two days, and pirates attacked the ship in which he sailed from Marseilles; eventually, after a year, he returned home. Originally versed in humanities, law, philosophy, logic and medicine, at 27 he was sent to Rome for questioning, under accusations of heresy. It was

Fig. 3.36 Aldrovandi. (From Findlen P., 1996: *Possessing Nature*. Author's library)

during this third visit to Rome that he met Rondelet, who was then planning his book on fishes, and the two spent some time together watching Roman fishermen at their nets and inspecting their catches. At the age of 32 (1554), he was appointed lecturer on logic and philosophy at the University of Bologna. Then, for a two-year period (1560–1561), he was appointed *lectura philosophiae naturalis ordinaria de fossilibus, plantis et animalibus*, the first-ever professor of *natural sciences*. He created and directed a public botanical garden, which today is the *Orto Botanico dell' Universita di Bologna*, and organized several expeditions to the Italian mountains, countryside, islands and coasts to collect and catalogue specimens He wrote over 400 volumes which, he estimated, would take over a century to print; his home contained a huge library and a staff, of scribes, painters, designers and engravers. As an ardent Aristotelian, in describing when to propose a new name, he explained that "A philosopher is allowed to invent names where there are none, by the testimony of our philosopher Aristotle". Indeed, he felt that he was extending, if not completing, the scientific work of Aristotle. In the frontispiece of the first volume of his treatise on Ornithology (1599), he placed his own portrait, and beneath it, in Latin, he modestly wrote: "This is not you, Aristotle, but an image of Ulysses: the faces are dissimilar, the genius is the same" (Fig. 3.36).

Aldrovandi's book on aquatic animals came some 60 years after Gesner's book on this subject (*De piscibus libri V*, published in 1613, ten years after his death). It differed from that of Gesner in that it included only aquatic vertebrates and had fewer fabulous monsters. He followed Gesner in using the taxonomic rank of Order, and he grouped fish diversity according to five aquatic habitats: those of marine rocky environments, of the marine littoral, of the marine pelagic environment, fishes of (what seems to me as) both marine and freshwater habitats and of freshwaters; he presented marine mammals as a sixth group. Many genus-level names of fishes in Aldrovandi are also those of today. (*Acipenser, Anguilla, Anthias, Atherina, Barbus, Carpio, Chromis, Clupea, Congrus, Cottus, Cyprinus, Gladius, Gobius, Iulis, Lampetra, Mola, Monoceros, Mullus, Muraena, Ostracion, Pagrus, Perca, Phoxinus, Raia, Remora, Sarda, Sargus, Scarus, Scomber, Scolopax, Silurus, Solea, Sparus, Thunuus, Tinca, Torpedo, Trutta, Uranoscopus*). This suggests that by the seventeenth century at the latest, the genus-level taxonomic nomenclature of many European fishes had become stable. This further suggests that naturalists of those days were quite familiar with fish diversity. Some of the illustrations are so good that we can easily recognize the genus. The high accuracy of the illustrations of his well-known fishes surely lent credibility to his imaginary ones. Aldrovandi also presented fossil fishes together with what he believed were their modern genera. We shall discuss the Renaissance views on fossils in some detail, further ahead.

Aldrovandi further covered the entire marine invertebrate landscape of his days (1602, 1618). He closely followed Aristotle's four large groups: Molluscs, Crustaceans, Testaceans and Zoophytes. Within these large groups, his classification was poor in that he lumped all invertebrates with shells into one group ('Testaceans') that included, besides the shelled gastropods, also shelled cephalopods (*Nautilus* and *Argonauta*), sea urchins, barnacles, goose barnacles (with an illustration of the barnacles becoming geese) and tube-dwelling worms. He classified squid and octopus together with sea hares; and *Nautilus* and *Argonauta*, which are cephalopods, together with snails; he grouped limpets, abalones and cowries as testaceans with uncoiled shells, separated from other marine snails with coiled shells, by the bivalves.

Coming now to terrestrial invertebrates (*De animalibus insectis Libri VII*, 1602), Aldrovandi classified their diversity into six groups, by the morphology of their legs and wings, by their readiness to take to the air and by their "nobility". Bees (six legs; four membranous wings) come first as they are the most noble group of insects, just as eagles come first as they are the most noble of birds. They are the first among insects because they construct hives, produce honey and have a complex social system governed by a king. This group includes, besides bees, wasps and hornets. Dragonflies, cicadas and butterflies (six legs; and four wings which are sometimes membranous) do not construct hives, do not produce honey and do not live in a monarchical society. Within this group, dragonflies and cicadas have membranous wings like bees, but butterflies have scaly wings. Then come flies and mosquitoes (six legs; only two membranous wings). Grasshoppers, including mantids and beetles, have six legs and four wings, of which only the two hind wings are membranous, whereas the two forewings cover them (Fig. 3.37). Though capable of flying, they do so less readily than bees, dragonflies and mosquitoes. Ants, bugs, earwigs, scorpions, spiders, millipedes and centipedes are groups in which something is faulty in either the wings, the number of legs, or both. In ants, wings are present for

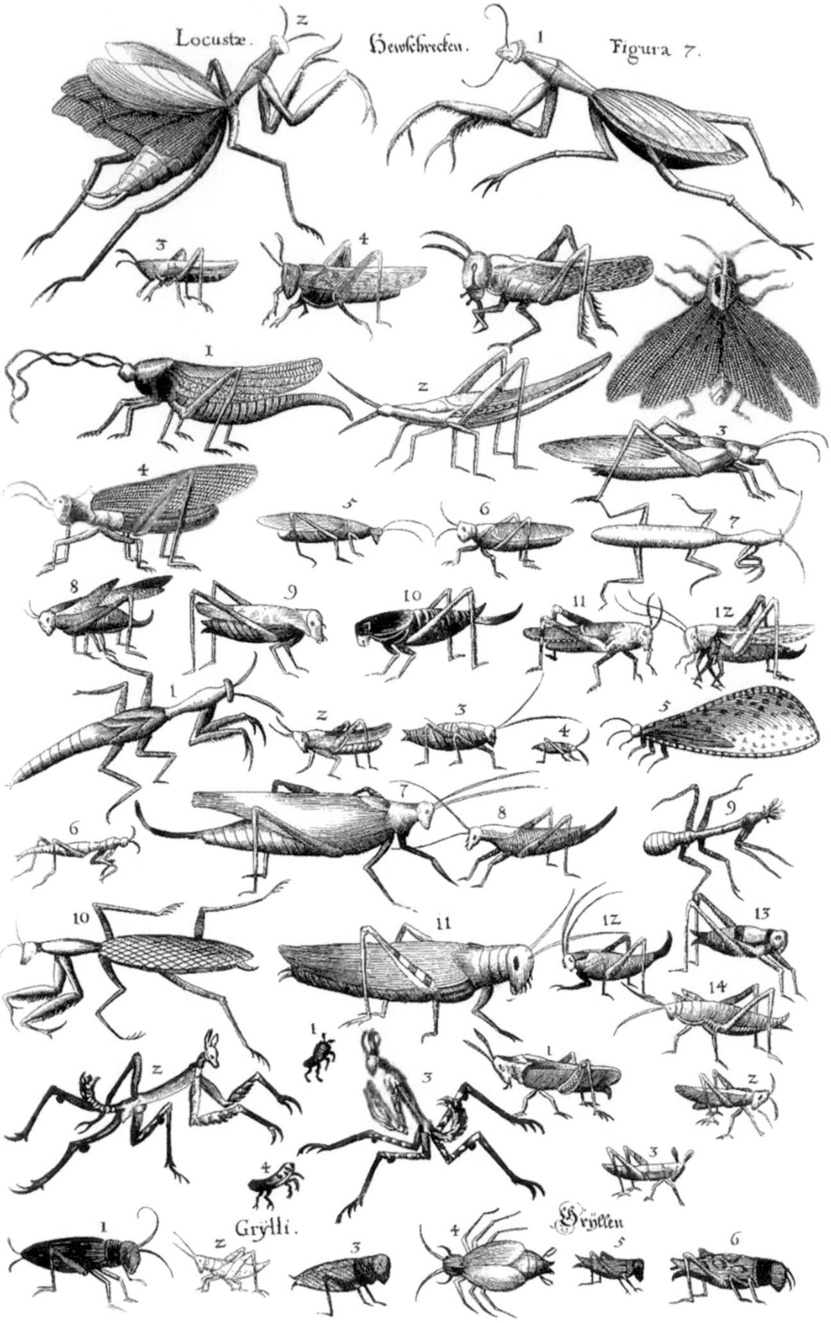

Fig. 3.37 Aldrovandi, insects: grasshoppers and mantids

only a very short while, during their nuptial flight. In bed bugs, the forewings are vestigial and lice and fleas are wingless. Scorpions and spiders have eight legs and mostly do not fly (Aldrovandi believed that some scorpions have wings). Millipedes and centipedes have no wings and ever so many legs. These groups are predominantly terrestrial. Worms and slugs (no legs, no wings) are subterranean; various worms may occur not merely inside the soil but also inside man, animals, plants and even rocks and metals, as parasites. One consequence of this classification was that Aldrovandi presented the butterflies in one chapter and the caterpillars in another, three chapters later. He sometimes described which butterflies develop from which caterpillars. This separating adults from their larvae is meaningful, in that caterpillars and butterflies are very different; and also, in that they appear in different seasons, so in nature, one would confront caterpillars with other caterpillars, not with butterflies. Not every caterpillar becomes a butterfly. Some die, and some others form a cocoon that, Aldrovandi wrote, produces not a winged insect but several worms. Furthermore, not every butterfly comes from a caterpillar; some come from worms (1602). Beyond matters of classification, Aldrovandi's natural history reflects how little was then known about the life of insects. For example, he claimed that the praying mantis can deliver a painful bite, that it feeds on vegetation, that it has large front legs because it flies more than it jumps and that its scissor-like forelegs allow it to grasp and hang onto its perch as it alights.

Aldrovandi, like Gesner, believed that there is a sea version of every animal occurring on land. His belief surfaces in his fish names, which referred to a marine peacock, marine thrush, marine swallow, scorpion, sparrow, spider, lizard, weasel, dog, cat, fox, monkey, marine frog, wolf and marine parrot. In a similar manner, he proceeded to bestow names on marine and fluviatile invertebrates that correspond to the terrestrial fauna. In the parallel panorama of aquatic invertebrates, there was a sea flea, an aquatic Spanish fly, a sea louse, a sea botfly, a sea centipede, a sea leach, a sea earthworm and a sea slug; and, of course, a sea star.

Aldrovandi included, in his illustrations, Paré's monster snail. His most interesting and fantastic animal was, however, his dragon (Senter et al. 2013). On 13 May 1572, when Aldrovandi was about 50 and on the very day that Ugo Buoncompagni happened to have chosen to return to his hometown Bologna to be invested as Pope Gregory XIII (1572–1585; his coat-of-arms happened to depict a dragon), a fearsome two-footed dragon appeared in the countryside near Bologna. A party set out to overtake and capture the dragon, and carried it inside the walls of the city, where they entrusted it to Aldrovandi, who, in addition to being a collector of strange and wonderful things, happened to be a second cousin of the newly elected pope. Aldrovandi promptly displayed it in his cabinet, where many came to see it. He almost immediately set about writing a treatise on dragons, including examples of all other reptiles he had ever seen or heard of, and he even had an artist illustrate "his" dragon (Fig. 3.38). The specimen was a long-necked, long-tailed, scale-covered biped with a thickened torso and a forked tongue, unlike any bipedal animal known at the time, and Aldrovandi used this opportunity to reopen the debate as to whether or not dragons generated spontaneously, from an alleged rooster's egg. Quick to write, he was, very wisely, slow to publish. His *Serpentum et Draconum*

Fig. 3.38 Aldrovandi, the Bologna dragon. (From Senter et al. 2013)

Fig. 3.39 Aldrovandi: Hen and rooster, Persian breed

Historiae Libri Duo was not published until 1640, 35 years after his death. (Recent studies show that the specimen was a taxidermic hoax, made by attaching the forelimbs of a common toad to a European grass snake, the midsection of which someone had replaced with that of a fish; Senter et al. 2013).

Not all of Aldrovandi's books were diverse. One book published during his lifetime was a monograph on the chicken in which he drew upon all the texts available to him from Greek times onwards, as well as on his very few personal observations (*Ornithology Vol. II Book XIV*; Fig. 3.39). As a typical Renaissance scholar, his approach was to present his subject in a historical and cultural context as well, and consequently, he overburdened the text with such non-biological details as the history of cockfighting, the use of chicken in sacred rites, mystical and moral interpretations, dreams, riddles, fables, as images on coins. Beyond this, however, the book is rich in well-presented biological information and etymology (whether the biblical and Talmudic terms *Sechvi, Tarnegol, Barbur, Gever, Doochifat* are synonyms of the

Greek *Alektor* and Latin *Gallus*). It presents a detailed definition of the chicken via the rooster's external features (erect reddish crest, wattles beneath the chin, a tail with two large, soft feathers bent like a curved bow, spurs) and the hen's internal ones, especially her genitalia. This is a reasonably full description of a species, in line with the best Aristotelian tradition. Beyond defining the chicken, Aldrovandi mentioned its keen sight, sharp taste and a voice that, he claimed, might be very expressive (he claimed that a hen, when unwillingly mounted, utters a weak cry). As for reproduction, according to Aldrovandi, the rooster mounts his many "wives" over 50 times a day, whereas other bird species are content with a single female and with less sex. The rooster also fights other males over the control of his females, not allowing any of his hens to copulate with another male. When fighting, the roosters raise their neck feathers, vibrate their tail feathers and then dash at each other, beak against beak and spur against spur. The jubilant victor announces his victory by crowing and stamping over his conquered and now groaning, cowering enemy. "Not for almost any other cause than for his wife does he enter battle, fearing lest another rooster rob him secretly of her embraces". In a similar and remarkable sentence, Aldrovandi commented that one rooster fiercely chastises any other rooster, lest he in secret tries to copulate with one of his wives. This reference, to a non-dominant rooster secretly mating with one of the dominant male's females, might perhaps refer to what we today term satellite copulation: a reproductive tactic whereby a non-dominant male, on the periphery of the territory of a dominant male, sneaks up to copulate with one of the females, in the latter's harem.

The rooster will mount even males, indeed even females of other genera such as pheasants and partridges. However, such interspecific copulations are not restricted to roosters. Basing himself on one of his very few personal observations, Aldrovandi writes that once "there were in my house an unmated male dove and a widowed hen. The dove was quite ample of body and lustful; the hen was small. The two copulated and the hen laid eggs and incubated them; the chick that hatched was mixed: its size, head and beak were those of a dove, and its feet and plumage were those of the hen. The chick did not sleep anywhere except in the bedroom or on the bosom of a woman". Back to normal reproduction, Aldrovandi was aware that hens laid eggs either after mating with a rooster, or without copulation; and that only eggs resulting from copulation with a rooster produce chicks. Eggs produced without a rooster are infertile; they are termed wind eggs, as a specific southern wind, Zephyr, conceives them in autumn. Infertile eggs are also affected by contact with dust, in which the hen bathes. Closely walking in Aristotle's footsteps, Aldrovandi carefully studied the daily development of the foetus in a hen's egg (Lind 1963).

Fully aware of intra-specific variation in size, song, feathers, reproduction and temperament, Aldrovandi described and illustrated several breeds of poultry. These included Italian, Persian, Turkish and Indian races, even a wool-bearing hen (of the existence of which he had learned only from an illustration in a map) and distant relatives of the chicken, such as the guinea hen. He also presented freak individuals with three legs, two bodies and one head, or a quadruped tail (Fig. 3.40).

One remarkable Renaissance innovation was the cabinet of curiosities. The term *cabinet* originally described a room rather than an item of furniture, and these rooms

Fig. 3.40 Aldrovandi, monstrous chickens: with two bodies; and with three legs

contained encyclopaedic collections of objects related to natural history, geology, ethnography, archaeology, religious or historical relics, works of art and antiquities. It is sometimes claimed that those who collected items for their cabinet regarded it as a microcosm of the world, that it was symbolic of the collector's control of the world through its indoor, microscopic reproduction, and that the cabinet is for the collector what a dollhouse is for a little girl. Being myself an ardent amateur collector (antique maps, fountain pens, stamps, kitchen chairs, porcelain, shells), to me collecting is a desire to possess and to master the diverse. In a way, my little collections are part of my self-image and self-identity. From a more scientific perspective and in general, cabinets enabled researchers to inspect specimens with their own eyes and analyse them with their own intellect, rather than blindly accept scholastic stories as true. Gesner had, we recall, described and sometimes illustrated various fantastic animals on hearsay, from anyone over the centuries. With collection-containing cabinets, one could look hands-on, at all evidence concerning a specific animal. Evidence for the existence of unicorns should rely upon real specimens, in cabinets. Collections were thus a stored, accumulated potential, for future research.

During the Renaissance, cabinets of curiosities were in the possession of rulers, aristocrats and members of the middle class (Findlen 1994). The cabinet of Rudolph II, Holy Roman Emperor (ruled 1576–1612) in Prague, was unrivalled north of the Alps. Rudolf's uncle, the Archduke of Austria Ferdinand II, also had a collection, with a special emphasis on paintings of people with deformed bodies. The Habsburg Imperial collection included such important Aztec artefacts as the feather headdress of Montezuma, now in the Museum of Ethnology, Vienna. The earliest pictorial representation of a natural history cabinet is the engraving in Ferrante Imperato's

Dell' Historia Naturale (Naples 1599; Fig. 3.38; Biblioteca Universitaria Bologna). An apothecary in Naples, he maintained his collection not only as a place to conserve raw ingredients for medicines, but also as a place of study that many scholars of the day could visit. Imperato published a book describing the collection, and his son Francesco assisted in writing up his observations, and in maintaining the collection; in Fig. 3.41 he is seen pointing out details of the specimens to two visitors with his father Ferrante looking on. Preserved fishes, stuffed mammals and curious shells occupied every surface of the vaulted ceiling, with a stuffed crocodile suspended in the centre, stuffed birds standing against panels inlaid with square polished stone samples and various corals on the bookcases. The room was fitted out with a range of built-in cabinets filled with small mineral specimens, and below them, a range of cupboards contained specimen boxes and covered jars. A large library of voluminous books covered an entire wall (Findlen 1994).

Aldrovandi was obsessed, both with assembling what had ever been said before him about each and every living creature, and with collecting specimens of nature. He compiled one of the better-known Cabinets of Curiosities of the Renaissance, with thousands of specimens of natural history, housed in several rooms, in his family palace. He commissioned professional artists to create detailed sketches of specimens, thus leaving a great legacy of pictures of various species, after his death. Indeed, he assembled a body of drawings of the natural world so extensive that he

Fig. 3.41 The curiosity cabinet of Ferrante Imperato. (From Findlen 1994: *Possessing Nature*; Author's library)

only succeeded in having a fraction of it published during his lifetime. Ever so modest, he described his cabinet as the eighth wonder of the world.

Aldrovandi's great contribution to the study of biodiversity was his foresight, which led to his private collection becoming a public museum. He formalized an agreement with the Senate of Bologna to have all his vast animal and plant collections transferred to the seat of government in the centre of town, after his death, on condition that the Senate persevere in publishing his collections-based works. Indeed, for the next 50 years, custodians of the *Studio Aldrovandi*, as it was now called, edited and printed several impressive volumes based on his manuscripts. Bologna kept the collections for three centuries, and it was only during the nineteenth century that the city distributed its items, among various other libraries and institutions. Aldrovandi's agreement with the Senate of Bologna marked the birth of the public natural history museum, as a cultural institute that was to exist permanently, beyond the life of the collector, as a cultural treasure destined to be a source of future research. Over the following centuries many collections, that began as private cabinets of curiosities, evolved (or fused) into university, municipal and national museums. In these institutes, which gradually accumulated and housed extensive comparative collections, future studies of animal diversity flourished.

The community of naturalits did not immediately accept everything that any naturalist or observant traveller had claimed to have seen in their travels. Their works went through a careful process of evaluation before being either assimilated into the body of knowledge, or rejected as untrustworthy. One (erroneous) such case would be Aldrovani's account of the circumpolar reindeer *Rangifer tarandus*. He wrote that Olaus had described it as having three antlers whereas others had claimed that it has three *pairs* of antlers. Following Olaus "on account of his erudition, his way of living and his experience of the region", Aldrovandi decided that reindeer have three horns (Ogilvie 2006).

3.1.2 All Plants of the World

Plants are much easier to observe, sample, preserve and illustrate in intricate detail than animals, and their study relies less on fantastic legends. However, because the study of plants was largely associated with medicine, botany was slow to develop as a field of enquiry in its own right. The texts of only two classic botanists, Theophrastus and Dioscorides, were at the disposal of naturalists of the Renaissance. Theophrastus, in the fourth century BCE, had discussed such general subjects as the influence of climate on plant growth, and differences between wild and cultivated plants; he classified the vegetable kingdom into trees, shrubs, under-shrubs and herbs. Dioscorides, in the first century CE, had written an encyclopaedia on medical material (*De Materia Medica*) that included 600 kinds of plants, and about 1000 medications that could be extracted from them. His book is divided into five volumes that do not correspond to any modern classification. He grouped plants by those producing aromatic oils; cereals producing medications; plants in which the medication is in the roots and seeds; and grapevines.

Throughout the Medieval and deep into the Renaissance, for more than 1500 years, *De Materia Medica* never left circulation and it attained canonical status, with both Arabs, Jews and Christians regarding Dioscorides as the ultimate authority on plants and medicine, and they copied and recopied his treatise by hand incessantly in Latin, Greek and Arabic. It was printed first in 1478, and soon after, it was translated into Italian, French, German, Greek, Hebrew and Spanish as well. Dioscorides' influence only gradually weakened, as the sixteenth century herbalists learned to add and substitute their own observations.

Since the late Medieval Ages, any book describing the use of herbs and other plants in medicine was termed a herbal. A herbal (such as *De Materica Medica*) was a treatise that would contain the names and descriptions of plants, with information on their medicinal, toxic, hallucinatory, aromatic, or magic powers. Herbals would also classify the plants they described; they would sometimes also include mineral and animal medicaments. Throughout the medieval, there were quite a few herbals. Many of them were richly illustrated, but these illustrations completely depended on the artistic interpretation of the illustrator; or, sometimes, on interpretations and imaginations of the botanist himself. Some herbals were in vernacular language. In an age when learned physicians' fees were unaffordable to most people and plants compounded most medicines, vernacular herbals provided a relatively cheap source of up-to-date medical information, and the illustrations would help those with difficulties in reading. In medical faculties and for academic teaching, on the other hand, herbals were in Latin.

It was in this context that another development took place in the Renaissance: The Herbarium (sometimes termed the "dry garden" *hortus siccus*; Ogilvie 2006). This was a collection of plants whose form and colour were preserved by pressing them between sheets of absorbent paper; these dried specimens were then mounted on sheets of stiff paper, and labelled with such essential data as date, location found and special habitat conditions. The university herbarium was invented in the 1530s and 1540s, to preserve specimens collected on field trips of the academy. Dried flowers were known earlier, commonly for domestic use, or as souvenirs of pilgrimages—there were dried flowers in the Shroud of Turin, a linen cloth bearing an image of a crucified man and first recorded in France in 1350s; but the first herbarium for scientific use was created during the first half of the sixteenth century, by an Italian professor at the University of Bologna (Luca Ghini, d. 1556). It consisted of some 300 specimens. By the late-sixteenth century, there were herbaria already all over Europe and they allowed work with dried plants, in the absence of anything better. One could also send individual dried plants of an herbarium on loan to other naturalists, and by this means, local botanists could integrate into an all-European network. One disadvantage of herbarium-based botany was its strong dependence on plant morphology, with little regard to colour or scent, and with little attention to where and when they grew (Ogilvie 2006).

During the Renaissance, leading medical faculties came to require courses in medical botany, *materia medica*, for their students. The first such courses began at

Padua (1533) and Bologna (1534), and by the 1550s, they were routine in many European medical faculties. However, by then people could no longer identify many of Dioscorides' plants and they misidentified many of them, so they often concocted wrong plants and even poisonous ones, as medications. Furthermore, Dioscorides had described plants he had collected in the Middle East, whereas by the sixteenth century, his book was in use all over Europe and across the Islamic world. This resulted in people attempting to force a match between the plants they knew and those described by him, leading to what could be catastrophic medicinal results. For these reasons, people of the Renaissance considered it their mission to rediscover precisely to which medicinal plant Dioscorides was referring. This mission inspired plant-collecting expeditions to the Middle East, where Dioscorides had been active. This was the reason for establishing pharmacology university chairs in which the lectures revolved around Dioscorides, and this was the reason for planting botanical gardens, in Italian universities (Padua 1546, Pisa 1547). Fortunately, many kinds of plants missing in the literature of the 1540s (cinnamon, parsley, myrrh, frankincense and others) were re-discovered thanks to ambassadors, physicians in Venetian ships and consulate staff in various cities of the Orient, who volunteered and harnessed themselves to the mission of searching for those long-lost plants of the Classic Age.

The first university botanical gardens were planted in the 1530s (Padua, Pisa), and the newly appointed professors undertook field excursions throughout Italy, to collect herbs and trees for these new gardens, in which they could investigate them as live plants. Some of them even sailed abroad, thus the Keeper of the Garden of Padua spent three years in the Levant (1558–1561) and planned to continue his travels even further, to south-eastern Asia; unfortunately, pirates kidnapped him along the shores of Africa, and he was released only after a heavy ransom. Another professor from Padua travelled to remote mountains in Wales, to interview Welsh herb-healers in Welsh that, he believed, related to ancient Greek. These professors sent home the booty of all these collecting expeditions, to the botanical gardens and to the herbaria—seeds, leaves, twigs and complete plants. In Leiden, the University Botanical Garden opened in 1594, for students to come and learn about medicinal plants and also for citizens of the town, to leisurely stroll among the flowers (Sirks and Zirkle 1964; Fig. 3.42a); the garden also displayed animal curiosities such as the jaw of a bear, the armour of a sea turtle, stuffed crocodiles, sawfish, blowfish and huge bats.

The development of a growing international community of botanists accompanied these developments, and this community bridged religious and national conflicts: botanists united in a frequent exchange of letters regarding new findings, as well as the exchange of seeds, bulbs and live or dried plants. The collecting and study of plants thus became a lively medical discipline that interested a broad lay audience, not just physicians and pharmacists. Educated nobles and townspeople indulged in a refined pastime that combined aesthetic pleasure, the passion to collect and lofty ideals of improving the health of humanity.

During the century that followed the first printed edition of Dioscorides (1483) many herbals appeared in many countries, in many languages. Notable among these many books was that of Leonhart Fuchs (1501–1566), a Bavarian physician and

Fig. 3.42 The Botanical Garden of the University of Leiden. (**a**) The Botanical Garden of the University of Leiden, detail

botanist. At the age of 32, he was called by the Duke of Wurttenburg to help in reforming the University of Tubingen. He created its first medicinal garden in 1535 and served as chancellor seven times, spending the last 31 years of his life as professor of medicine. Fuchs' *Notable commentaries on the history of plants* (*De Historia Stirpium Commentarii Insigness*, 1542) described about 400 wild and 100 domesticated plant species and their medical uses, in alphabetical order: he made no attempt at presenting them in a natural system of classification. The text, based mainly on Dioscorides, contains very beautiful, accurate and detailed pictures of plants, largely growing locally (Fig. 3.43). Fuchs emphasized high-quality drawings as the most effective way to specify what a plant name stands for. Its appeal to gardeners, botanists, bibliophiles.

Most of this botanical surge channelled into a Commentary on Dioscorides, written by Pietro Andrea Mattioli (1501–1577, Fig. 3.44). The son of a physician, Mattioli studied medicine at Padua, and when in his forties and practicing as a physician in north-eastern Italy (of today), he published the first version of his *Commentaries on Dioscorides* (1544: *Di Pedacio Dioscoride Anazarbeo Libri cinque Della historia, et materia medicinale tradotti in lingua volgare italiana*). In doing so, he was one of numerous humanists who, since the late fifteenth century, had aspired to translate and comment on the ancients. This earliest version of

Fig. 3.42 (continued)

Mattioli's commentary was a relatively modest work, devoid of illustrations and lengthy criticisms, written in Italian for the local readership. Mattioli presented himself in the book as an herbalist, who had entered a vast and long-neglected garden filled with broken, tangled and unappreciated plants, which he had begun to weed with Dioscorides as his guide. In later editions, in addition to identifying the plants originally described by Dioscorides, Mattioli added descriptions of plants not mentioned by Dioscorides and not of any known medical use. He thus advanced from the study of plants as an auxiliary field of medicine to the study of plants as a field of interest in its own right. All in all, he described about 100 new kinds of plants. The third edition (1554) was in Latin and contained beautiful illustrations, so that (as he wrote in the Introduction) "now, whoever cannot go out to Nature—here he has a garden in his home". The Latin, combined with the magnificent illustrations, elevated the standards of the book well beyond local readership, to international significance. In later years, he had the book translated into German, French, Spanish, Czech and Hebrew. As one edition soon followed the next, the text gradually diminished the status of Dioscorides and elevated that of Mattioli. In the 1552 edition, he described himself as the restorer of ancient learning, who "with measured reasons and faithful authority" corrected errors of the past and the present. By 1565, he stated that he had outstripped Dioscorides in his knowledge of nature, and by 1568, he accompanied the Italian edition by the modest statement that "if the mind could be portrayed, a portrait of Dioscorides will be that of Mattioli". His growing fame made him both the major interpreter of Dioscorides of his day, and the undisputed authority on the natural history of plants. His book had become the talk of papal and imperial courts, the one book that any aspiring naturalist and physician had to possess. As the most published natural history treatise alongside Pliny's *Natural History*, and the only ancient natural history that constantly changed to incorporate new information, it became a book of record, of scholarly knowledge of plants. Eventually, as he published many subsequent editions, Mattioli became the grand coordinator of medical botany of his time. The book was such a great

Fig. 3.43 Fuchs: Pepper

success that the first ten editions alone sold 32,000 copies, and the later, grand editions sold even more. It brought Mattioli such fame that Ferdinand II, Archduke of Austria and later his brother, the Holy Roman Emperor Maximilian II, appointed him as imperial physician. He now moved to Prague, where he employed two artists and five engravers, and his *Commentaries* peaked in a grand new edition of 1500 pages, packed with huge illustrations. A clause in the will of a British diplomat (dated 1637) reveals the high estimation in which even later generations held his *Commentaries*. "I leave, to our most gracious and Virtuous Queen Mary [wife of Charles I] *Dioscorides*, with the Plants naturally coloured, and the text translated by

Fig. 3.44 Mattioli

Matthiolo, in the best Language of *Tuscany*, whence her said Majesty is lineally descended; for a poor token of my thankful devotion, for the honour she was once pleased to do my private study with her presence".

Mattioli himself did no fieldwork beyond a small study when he was young. His main talent was in recruiting others to carry out research. Comfortably settled in his home, he would grumble to his readers about the many petty things in which he was currently involved and which prevented him from travelling to Cyprus, to the Holy Land and to other countries in the footsteps of those ancient giants of botany. At the age of 54, he grumbled to his readers that his advanced age did not enable him to travel abroad. "But", he wrote to Aldrovandi, whose new findings he wished to incorporate into his new, 1554 edition, "as long as I am alive, [...] I shall lift my eyes to you and to your friends to send me the fruits of your work. If you have anything to add to my new *Dioscorides* which shall soon be printed, I shall give you credit, as I give to all those who send me material". From his armchair, Mattioli became the author of a book that summarized all the deeds and aspirations of a whole generation of botanists.

Now at the peak of his fame and power, Mattioli tolerated neither rivals nor corrections, and naturalists and physicians who dared disagree with or correct him, did so at their own peril. As "Emperor of Botany" of his day, he tyrannized the botanists of the Renaissance. Not many botanists could afford to be excluded from his book, and not many could afford to quarrel with him. He either ignored or attacked publically on a personal level, in the next edition, whoever had not supplied new data, or had criticized his book. Mattioli's tyranny enabled the long-term dominance of his version of *De Materia Medica* throughout Europe. For example, Amatus Lusitanus (1511–1568) was a Portuguese-Jewish physician living in Italy, who delivered lectures on medicinal plants, attended the family of Julius III and published a Latin translation of Dioscorides some eight years before Mattioli's first edition. He found and published (1556) some 20 mistakes of Mattioli, noting that "Mattioli contradicts himself", "Mattioli errs" and "Mattioli ineptly proves Theophrastes". Mattioli was furious. In his *Defence against Amatus Lusitanus* (1558), he claimed to have uncovered ten items to correct, and 100 mistakes to censure, in Amatus' work. He also levelled the full weight of his authority in the medical circles of Catholic Italy against this Portuguese immigrant of dubious Christianity, who attempted to usurp Mattioli in defining the state of botanical knowledge. Accusing his Portuguese rival of everything from crypto-Judaism to heresy and apostasy, Mattioli likened any criticism of himself to an abuse of the Catholic faith: "Just as there is no faith and religion within you, so in truth you are completely blind as to the medical art which you unworthily possess. [...] Beware that your envy and calumnies do not further increase and elevate my glory among learned readers and critics". These were the dark days of Pope Paulus IV, who sentenced and burned converted Jews at the stake, erected the Jewish Ghetto in Rome, forced Jews to wear distinctive hats and forbade them to trade in everything but food and secondhand clothes.

Another example was Melchior Wieland of Prussia (1520–1589), a botanist who travelled to Greece, Syria and Egypt between 1558 and 1560 and returned to Padua laden with specimens for the university's botanical garden. In his *Defence against Pier Andrea Mattioli* (1558), Wieland criticized Mattioli's 1544 Italian translation of *Dioscorides*. Again, Mattioli was furious. He stated that Wieland was a hermaphrodite, a sodomite to be banned from the botanical Eden of Mattioli's *Dioscorides*. When he heard that Wieland was off to Constantinople in search of plants from the East, he imagined his trip as a form of penitence that Wieland had undertaken for his sins against his botanical superior. Of course, when Wieland returned only to publish further rebuttals of Mattioli's views, redemption became out of question. Then there was Cesare Odone (d. 1571), whom Mattioli labelled as "piggish, disgraceful, vice-ridden, inhuman, envious and pugnacious", for refusing to introduce Mattioli's *Dioscorides* into the Bologna University curriculum.

However, not everyone adored Mattioli. The Bavarian herbalist Leonhart Fuchs (1501–1566) wrote to a friend, that "that trifling Italian will learn that there are men in Germany who recognize his trifles and paint him in his true colours" (Ogilvie 2006). Nationalism was on the rise during the Renaissance and many Italians disliked any non-Italian, feeling perhaps that other nations were stealing from them the wand of Renaissance progress (Ogilvie 2006).

Konrad Gesner of Basel criticized Mattioli, on a far more central issue: at least some of the illustrations in his 1554 edition were fictitious illustrations based on the verbal descriptions by Dioscorides, not on living "flesh-and-blood" plants. As a specific example, he presented the leopard bane, a member of the daisy family, *Aconitum pardalianches* (today *Doronicum pardalianches,* Fig. 3.45). Dioscorides had described it as a plant in which the stem is as high as the palm of a hand, with three to four slightly hairy leaves somewhat resembling those of a cyclamen but smaller, the glistening root resembling the tail of a scorpion. It overpowers scorpion

Fig. 3.45 Leopard bane, Doronicum pardalianches

stings and reduces pain in the eyes. Fifteen centuries later, Mattioli identified *Aconitum pardalianches* as a plant, which he himself had found in mountains north of Venice and of which he presented a detailed illustration. This identification came under heavy criticism by Gesner, who claimed that Mattioli's illustration was a fake, his too-beautiful illustration deriving from the verbal description of Discorides, not from a real plant. Gesner was thus accusing Mattioli of dishonesty. Mattioli responded (1558) that Gesner himself had presented fictitious illustrations in his books, and he calls both the artists who drew the plant and the mountains in which he found it, to evidence the existence of his *A. pardalianches*. Gesner responded in a call to Mattioli's conscience (Introduction to his Fishes book), and the whole issue soon became an issue if it is at all permissible to disagree with Mattioli. Gesner was deeply involved in the study of plants, with two texts (*Enchiridion historiae plantarum* 1541 and *Catalogus plantarum* 1542) published in four languages. He also wrote a very comprehensive manuscript about plant life, with about 1500 wood engravings of plants and their important flowers and seeds, but it was not published until two centuries after his death (in Nuremberg, 1751–1771). The Renaissance was an age in which nationalism began to rise, and Mattioli seems to have had feelings of contempt for all non-Italian naturalists, among them the herbalists of German-speaking countries. Seemingly in response to these feelings, the herbalist Leonhardt Fuchs of Bavaria, in a letter to a colleague, wrote of Mattioli that "that trifling Italian will learn that there are men in Germany who recognize his trifles and paint him in his true colours". It was in this manner that the medical use of plants advanced, more as an over-authoritative sub-discipline of medicine and less as a discipline in its own right, and amid bitter quarrels. It was during the sixteenth century that botany, as a discipline independent of pharmacology, was born.

Andrea Cesalpino (1519–1603, Fig. 3.46), who studied at Pisa, took charge of the garden at Pisa seven years after it was established. In addition to being Director of the Botanic Garden and professor of Medicine at the University of Pisa, he was one of the first to establish an herbarium (it still exists, in Florence), and was one of the first generation of educators to include the use and preparation of medicinal drugs (*materia medica*) in the medical curriculum. After 30 years of teaching, he wrote up his lectures in 16 books published as a single volume (*De Plantis libri XVI*, 1583). At the age of 73, he moved to Rome, as a physician to the Pope and lecturer at the Sapienza University. Today, the plant subfamily Caesalpinioideae, with its genus *Cesalpina*, is named for him. Cesalpino's most important medical contributions concern the anatomy and physiology of blood movement. He described the cardiac valves and pulmonary vessels connected to the heart, recognized that the heart is the centre of blood circulation and accepted the existence of anastomoses between arteries and veins. His principal contribution to biology was in the classification of plant diversity.

Cesalpino argued that all living matter, both plants and animals, contains a single, undivided "living principle", located at some definite spot in each organism. In higher animals (our vertebrates), this living principle is in the heart, which is at the centre of their body; without a heart, they die. Lower animals and plants, which can

Fig. 3.46 Cesalpino

live after we slice them up, have a less centralized structure. Cesalpino thought that the living principle of a plant is at the point at which its stem and root join (the "root collar"). Propagation by means of cuttings shows, however, that this central point of the plant is not as absolute as that of the animal. This is the centre only in actu, that is, at any given moment; *in potentia* the centre is anywhere within the body since a cutting will form a new individual, whose centre of life will once again be where stem and root join. Cesalpino thus considered the root collar, which he called the "heart" (*cor*), to be the seat of the plant's "soul" (*anima*), though the soul is present throughout the plant.

Taking further this point, of a single living principle for both animals and plants, Cesalpino next argued that all living matter must be organized according to the same basic pattern, and consequently that the organs of plants correspond to those of animals. He suggested that just as the plant's root collar corresponds to the animal's heart and soul, so also various vessels extending from the root-collar correspond to the animal's veins, the root corresponds to the animal's digestive system, the fruits to the animal embryo and stalk and stem, as (he believed) they produce the fruit, belong to the reproductive system. The leaves serve merely to shade the shoots and fruit from strong sunlight, and the flower petals are no more than modified foils. Not aware of the existence of sex in plants, he thought that the fruit is formed from buds produced from (our modern terms of) the medulla and the phloem; the medulla is the most vital part of the plant, and it provides the actual ovule; the phloem gives rise to the flower-leaf. The stamens serve only to protect the young fruit.

Having established a set of comparisons between plants and animals, Cesalpino next sought to establish a classification of plant diversity by some natural order, a classification which would be based on essential features of the plant, rather than on extraneous characters such as habitat or medicinal value (or alphabetical sequence), as was customary in so many herbals. (As a reminder, Aristotle had believed that for any given natural group of living creatures, some features are essential: they appear in all its members, and in no other group, and they define the group.) To do so, he turned to the two defining functions of the vegetative soul as given by Aristotle, nutrition (namely growth) and reproduction. Indeed, most (if not all) organs of a plant correspond to these two functions. Cesalpino therefore based his classification on the organs of nutrition and of reproduction: root, flower and fruit. By focusing on these operations of the vegetative soul, he sought to eliminate non-essential properties such as medicinal virtues, so central to the *materia medica* of Dioscorides and his Renaissance followers (Sloan 1972; Ogilvie 2006). In Cesalpino's classification, the organs of nutrition determine the first subdivision of plants. As the medulla of plants performs nutrition, he classified all plants into those with a hard, woody medulla which he named Trees *Arbores*, and those with a soft one, the Herbs *Herbae*, groups that Hildegard had established, four centuries earlier. Next, he subdivided these two groups by three features of the fruit that he viewed as essential: number, location and shape. By the number of seeds in a fruit, and the number of compartments per fruit, he classified plants as those with fruits as single (maple, oak, chestnut), and those having several seeds in a single container (rose, fig, pine; Sloan 1972). Next, the latter could be further split into bipartite, tripartite, quadripartite, or multipartite. Colour, flavour and odour are non-essential features of the fruit and are not to be used in classification. Some plants do not form fruit (ferns, algae, lichen) and Cesalpino believed that they generate spontaneously, from rotting matter. By first classifying all plants by one set of essential features, involved in nutrition, and then dividing the resulting sub-groups by another set, involved in reproduction, he formed a subordination of characters, one set below another. By this concept of character subordination, the reproductive parts of the plant are "essential", in the Aristotelian meaning of the term. A classification could now yield knowledge of the essences, thereby warranting the claim to the title of a "natural" classification (Sloan 1972, Ogilvie 2006).

Moving on in the classification of plants, about one generation after Cesalpino we come to Caspar Bauhin (1560–1624), a Swiss botanist, the son of a French physician who, being a protestant, fled France. Aged 22, Caspar became a Professor of Greek at the University of Basel and, six years later, he took the chair of anatomy and botany too, became the city physician, professor of medicine, rector of the university and dean of his faculty. His *Pinax theatri botanici* (1623) is an index to the plants in the works of Theophrastus, Dioscorides, Pliny and later authors, up to the (then) present century (Cain 1994). In it, he described and classified some 600 genera and 6000 species of plants. He divided his *Pinax* into 12 books, each with six sections, a division so symmetrical as to suggest that it has little to do with taxonomy. Indeed, only very occasionally do the books and sections coincide with natural groups. Far more often, they are merely numbered, and the section titles are simple lists of items contained in each book or section. Moreover, the divisions into books and sections do not usually coincide with marked discontinuities in the sequence of items, and they are irrelevant to questions of taxonomic rank. Each section consists of a series of what vaguely corresponds to our genera, many further divide plants into groups vaguely corresponding to our subgenera. Bauhin's genera and subgenera consist of groups vaguely corresponding to our species, each with a name that may be up to 13 words long. Bauhin had, however, no conception of the Species and the Genus as ranks, in the modern sense. He handled these terms interchangeably, in that he divided species into species. Thus, under the species *Lamium purpureum vel album non foetens folio oblong*, he says that as to the flower, there are three species. The whole of Bauhin's usage of genus and species is (loosely) consonant with the terms of Aristotelian logic: the genus is the general kind at any level under investigation; the species are particular sorts within it. The shifting use of genus and species sufficiently indicates his Aristotelian use of these terms. To Bauhin anything except an individual could be a species or genus, according to the point of view (Cain 1994). At a broader level, he retained the sequence of major groups of Theophrastus (trees shrubs and herbs) but revered it to start with the best-known plants, grasses. Not aware of the importance of details of the flower and fruit, he used mainly characters of leaf shape.

Naturalists of the Renaissance and early modern Europe knew very little about the basic physiology of plants. Jan Baptiste van Helmont (1580–1644) was a Flemish chemist. To determine how plants gain growth, he grew a willow tree in a pot, adding nothing but water for five years, and measuring the amount of soil, the water he added and the final weight of the tree. After five years, the plant had gained 74 kg, and the amount of soil was closely the same, as it had been when he started his experiment. Helmont deduced that the tree's weight gain had come from water. From this, he argued that no solid could be generated "but by getting the water with childe" (Sirks and Zirkle 1964).

3.1.3 The Study of Fossils

In 1551–1558 Conrad Gesner, we recall, had published his immense encyclopaedia of animals, *Historia Animalium.* Some years later, in 1565, he published *A Book on Fossil Objects, chiefly Stones and Gems, their Shapes and Appearances* (in brief, *On Fossil Objects*), The word "fossil" originally had, in those days, a much broader meaning than today. It meant any interesting hard object dug up from the ground or found on the surface. This included fossil remains of organisms in the modern sense (bones, teeth, shells), but it also embraced a broad spectrum of mineral ores, crystals, gems, rocks and archaeological artefacts such as pots, pans, hammers and nails. Some of these objects closely resembled organisms but others did not, and between these two extremes was a wide range of objects having only some degree of resemblance to organisms, but not more. Which of this broad range of "fossil objects" were organic and which were not?

Living entities grow, and some stones—we can almost see them growing. Travertine is a form of sedimentary limestone that forms by a process of rapid precipitation of calcium carbonate in mineral springs, especially hot ones. In limestone caves, it forms stalactites that grow from the perpetually dripping water, and in some springs, it coats a layer of stone around objects such as plant roots. Then, miners of those days believed that ores that they were mining steadily replenished themselves, deep in the earth. Furthermore, corals showed the power of a petrifying agency in the sea, pearls and gallstones proved that stones could form and grow within the bodies of animals and even in man himself; and meteorites suggested that stone materials could also form above the Earth's surface, somewhat like hailstones. The decay of some minerals likewise suggested an analogy with disease, old age and death. A petrifying agency thus seemed to pervade the entire sublunary sphere, with fossils originating above, on, or in the Earth. To the extent that life was characterized by growth, decay and death, the distinction between living plants and animals on the one hand and non-living stones on the other hand, was thus vague.

Furthermore, for us today, a sharp line separates living from non-living entities, in that living entities can generate. During the Renaissance, however, naturalists believed that generation could occur also in stony matter. They believed in a magical stone *aetites*, often containing a smaller stone within its central cavity, thereby suggesting it is in the act of generation. Such great authorities of the past as Theophrastus, Pliny and Dioscorides, all had mentioned it, so there was no doubt of its existence. It promoted healthy childbirth, prevented abortion and shortened labour. Pliny noted that it occurs in the nests of eagles, who cannot propagate without it, and Dioscorides recommended it also for the treatment of epilepsy and, when mixed with meat, for betraying a thief. The belief that it often contains a smaller stone within its central cavity, fed into the belief that some minerals have male and female forms. If shell-producing animals can be born and grow, why not stones? So-called "fossil shells" grow right inside the rocks in which we find them, by mysterious vital forces combined with invisible emanations from the stars. The so-called "seashells" inside rocks and on mountaintops are not real seashells at all, despite their close resemblance to animal shells, in that no clams had ever lived inside them and no sea had ever covered them. One supporter of this approach was Avicenna (980–1037), who believed that a "shaping force" within the Earth had formed stone into the shapes of bones. Another supporter was the medieval nun Hildegard von Bingen (1098–1179) who wrote of stones as being born, of growing and of doing so in very recent times. Green emerald grows during the morning hours, when the natural vigour of the earth and grasses is particularly vital and the herbs vigorously imbibe vitality. The magnet stone is born from the spume of certain poisonous worms that inhabit sand and water. Another "stone", mother of pearl, comes from shell-dwelling animals in the sea and rivers. They draw into themselves some poisonous matter from the river bottom, and when they spew it out, mother-of-pearl congeals from the poison. Thus Hildegard's "fossils", whether of stone or shell, were objects which were born and which grew on a day-to-day basis.

A further support for the idea that generation could occur also in stony matter was in the belief that simple organisms form by spontaneous generation from non-living material. If so, then certain specific forms might develop spontaneously not only on the Earth's surface or in the sea, but also within the Earth, inside stony materials; or they may form by an uncompleted process of spontaneous generation and remain inside stones. Complex organisms might also be able to grow within the Earth, if their specific "seed", that held their potential specific form, happened to penetrate there, perhaps in percolating ground water. For example, if the specific "seed" of a certain fish was washed into the ground, it might grow from the stones and form a fish-like fossil inside the rock. Such a fossil could have grown by a process directly comparable to the growth of a living fish, and it would owe its specific (say) carp-like form to an identical "seed", but its matter would be different. It would not be the remains of a fish once alive. Many naturalists believed that there is a direct subterranean circulation from the oceans, by which "seeds" of marine organisms could have lodged in the earth.

Thus, while it stood to reason that among the whole range of "fossils" there should be some with forms resembling living organisms, it was not necessary to invoke a hypothesis that they were actually the remains of once-living animals and plants.

The distinguishing feature of all "fossils" of any kind was stony matter, but the stones were very variable. Accordingly, Gesner divided all "fossils" into 15 categories:

1. Stones whose forms are based upon, have some relation to, or suggest the geometrical conception of points, lines or angles.
2. Stones that resemble or derive their name from some heavenly body or the elements.
3. Stones that take their name from something in the sky.
4. Stones that are named after inanimate terrestrial objects.
5. Stones that resemble certain artificial things.
6. Stones that resemble things made artificially out of metals, stones, or gems.
7. Stones that resemble plants or herbs.
8. Stones that have the form of shrubs.
9. Stones that resemble trees or portions of trees.
10. Corals.
11. Stones that resemble sea plants of a stony nature.
12. Stones that resemble men or four-footed animals, or are found within these.
13. Stones that derive their names from birds.
14. Stones that resemble things which live in the sea.
15. Stones that resemble insects or serpents.

Gesner arranged these categories in descending order, broadly from the heavens downwards, by the positions of their analogues in the cosmic scheme. He began with stones having shapes related to geometrical figures or to elements, the most fundamental entities in the universe (and we recall Plato, who, in ancient Greece, had argued that each of the basic elements consists of tiny, geometrically unique atoms). The atom of fire is a four-sided pyramid, the Earth is a six-sided cube, air is eight-sided, water is twenty-sided, and heaven is twelve-sided. Gesner then descended through those resembling the heavenly bodies and those related in some way to the realm of meteorites, next coming to those resembling terrestrial objects. First among these came those relevant to man, in that they resembled artefacts that he creates, whether for daily use, such as wheels, or for special occasions, as cut gems. Then come those resembling terrestrial plants, and then marine plants (with a stony nature). Then come the animals, first quadrupeds, then birds, fishes, and finally come the lowest of fossils, those resembling serpents and insects.

Into these categories, Gesner set his fossils (Fig. 3.47 is an assemblage of some of his figures). In many cases, this was not a difficult task, especially when the fossils were of geologically recent origin. He placed fossil wood, bones and teeth into chapters devoted, respectively, to objects resembling trees or parts of trees and to objects resembling parts of quadrupeds. Likewise, he placed Devonian fishes into the category of Stones Resembling Aquatic Animals, which also included shark teeth, molluscs and crabs. Many of his specimens were, however, more difficult to interpret. He correctly recognized the flint cast of a fossil sea urchin as resembling a living sea urchin, stripped of its spines. However, he was familiar only with sea urchins with small, delicate spines, so he failed to recognize the fossil remains of large spines. He considered these spines, when preserved detached from the shell, as acorn-like in shape, and placed them in the category of those resembling wood.

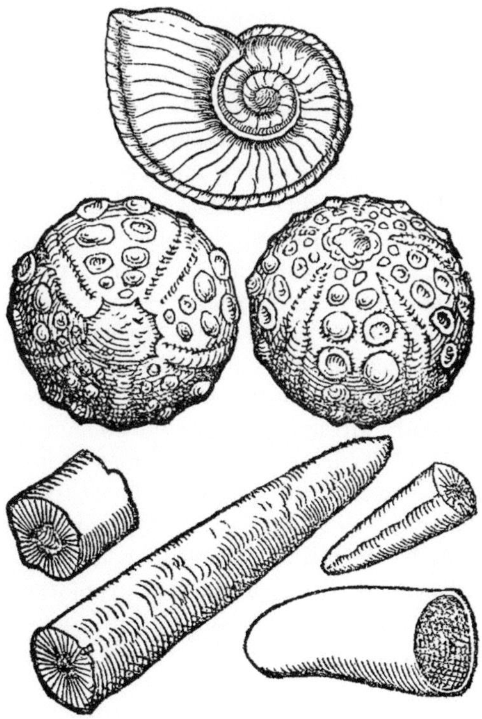

Fig. 3.47 An assemblage of Gesner's figures, of fossil invertebrates

Not acquainted with the entire range of sea urchins as represented by their skeletons, he thought that some were serpents' eggs and assigned them to the category of Stones that Resemble Serpents and Insects. He placed in this category also a loosely spiralled ammonite, as it resembled a coiled snake; whereas he considered a more tightly spiralled ammonite as resembling the shell of a gastropod. Belemnite fossils (especially the guard) resemble darts or arrowheads, so he placed them in the category of Stones Resembling Human Artefacts ("Belemnites" comes from the Greek *bélemnon*, "dart, arrow", and was termed some decades before Gesner, by Georg Agricola). Some circular ossicles of sea lilies, his *trochite*, resembled wheels so he placed them in the category of Stones Resembling Human Artefacts, whereas some star-shaped ossicles, *astroite,* resembled stars, so he placed them in the class of Stones Resembling Heavenly Entities. Gesner's classification of fossils thus ranged from the obvious to the obscure. One source of difficulty in perceiving organic resemblances in fossils arose from the fact that some fossils were of extinct groups (belemnites, ammonites). Similarly, not all extant groups were well known, thus living stalked crinoids were discovered only in the middle of the eighteenth century. The fact that many fossils were preserved as different materials, further hindered Gesner's assigning similar natural groups into the same category. Ammonites may reach us as casts in hard rock, or with the shell replaced by other materials (pyrite), or as paper-thin impressions flattened on the surface of a shale. This fact distorted the view that all ammonites are of the same group.

Plato's philosophy supported the outlook of animal-resembling fossils as not being animals. We recall that in ancient Greece, Plato believed that beyond all real objects and creatures in our ever-changing world, there is an unchangeable reality, an eternal, perfect, abstract model of each group of objects, an idea (*eide*). Every object in the world is merely a fleeting, imperfect expression of an eternal, perfect idea. There are thus two kinds of worlds: the earthly, apparent, ever-changing, always-becoming world, which we observe with our senses; and an eternal, unseen, unchanging world of timeless perfect forms and ideas, which we grasp by our intellect. Each *eides* is constant and discontinuous with other *eides*. As applied to animals, every mortal animal of a species is only the external expression of an ideal, eternal idea of that species.

In Gesner's world of the Renaissance, a *trochite* could resemble a manmade wheel because both embodied, each in its different sphere, the same abstract idea of a circle and an *astroite* was star-shaped because it owed its form to stellar influence. Similarly, all stones that resembled animals could owe those resemblances to their bonds of affinity with the same idea. Platonic thought thus provided an explanation for the resemblances between some fossils and living organisms. As abstract ideas determine all entities in the world, it would be natural that many fossils should resemble the forms of living animals, without necessarily being predecessors. Thus, it was not necessary to invoke a hypothesis that "fossils" were actually the remains of once-living animals and plants. A dug-up hammer was not the predecessor of a recent one.

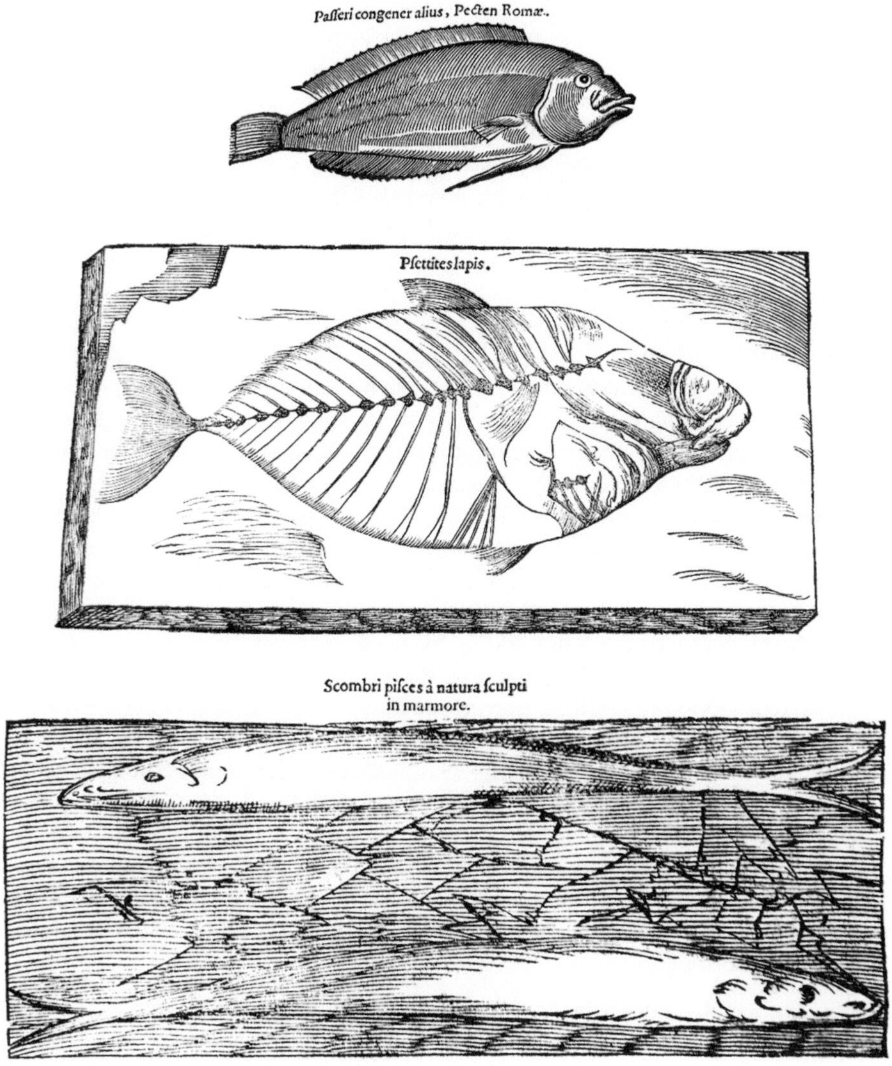

Fig. 3.48 Aldrovandi: a sea sparrow (above), and two fossil fish

Aldrovandi presented fossil fishes together with what he believed were their modern genera (*Scomber*, *Pastinaca;* Fig. 3.48). However, he regarded fossils as, either figures somehow sculpted into stones by nature; or as incomplete animals in which spontaneous generation had failed to reach full completion. He did not consider them as the remains of normal living forms.

Another basic question arose from the places in which fossils were found, far from the sea, far up at the tops of mountains. Two opposing outlooks on the history of life dominated thought on this question, ever since ancient times. One was the biblical outlook: God had created the world and its life several thousand years ago. The rising waters of the Great Flood brought bivalves and marine snails to the mountaintops, where their empty shells have remained to this day. In line with this outlook, an early Christian writer (Tertullian, c. 155-c. 222 C.E.) wrote that water had once covered the Earth. Similarly, but much later, a thirteenth-century Italian monk (Ristoro d'Arezzo), having picked up quantities of cockle and mussel shells near the top of a very high mountain, noted that they were mixed with sand and cobbles, as if they came from rivers; "and this is a sign that the mountain was made by the Deluge". In a secular variation of this approach, Aristotle believed that the world and its animal and plant groups are eternal and unchanging. The waxings and waning of the seas are part of the world's "vital process", and these natural inundations have occurred repeatedly over the ages. A past rise of the sea would have brought seashells to mountaintops, like the much more moderate rising of the tide, which carries shells up the beach. The opposite outlook to the Great Flood was, as we have presented in the above paragraphs, that fossils were not and never had been living organisms; they were rocks of unusual shape.

It was in the context of this intellectual debate between the two opposing outlooks on fossils that Encelius, a German mineralogist of the early Renaissance, one day found a fossil scallop (1551, a few years before Gesner's treatise on fossils). Like so many intellectuals of the Renaissance, he too sought to reconstruct the wisdom of the ancient world (Gould 2002a, b). In seeking among classical texts for a name for his fossil, the only major source for terms on types of rocks and minerals in the ancient world was Pliny's encyclopaedia of natural history, Chap. 37. Pliny had presented a list of rocks and minerals arranged in alphabetical order, accompanied by very brief descriptions. This was not much for a mineralogist of the Renaissance, but if the ancients had indeed learned all that is worth knowing, then Pliny's telegraphic, incomplete descriptions must at least hint at some essential character of each of the minerals on (and in) the Earth. It was therefore Encelius's duty, as a Renaissance mineralogist, to match his object with Pliny's names. Pliny had listed three minerals with closely similar names, for each of which he gives a rather poor description:

Chelidonia, which he described as having two types.
Chelonia, which he described as having "the eye of an Indian tortoise".
Chelonitis, which he described as resembling a tortoise (*Testudinium similis*).

This classic description does not offer a very broad basis for further advance, but the sages of the classical world added nothing further, and from this point onward, the Renaissance scientist had to continue. Encelius thought he had correctly solved the identity of Pliny's *Chelonitis* and sought to express his solution in an illustration to accompany his verbal description. We have no idea why he thought his fossil represented something that Pliny had described as resembling a tortoise (perhaps he confused the Latin *Testudinium* with *Testa*, a shell). Encelius' illustration enabled future discussions and criticisms, unlike verbal descriptions, the significance of which one can twist this way or that. In the following paragraphs, I am closely following Gould (2002a, b).

Encelius believed that his fossil scallop belonged to the kingdom of minerals and that it formed inorganically. Naturalists of the Renaissance considered the Earth as an entity without history or, at most, as preserving such prominent biblical events as the flood of Noah. They also thought that the Earth was created according to a very rational plan, thought out by a divine architect, with all its components wisely interlocked by common principles. Moreover, they considered the entire world as the "big world" macrocosm, whose composition, structure and function correspond to that of the human body, the "little world" microcosm. Man, the little world, is the essence of the entire universe; and the structure and organization of his body parallels the structure and organization of processes on the big world, the universe. For some believers in the microcosm/macrocosm idea, the human organs corresponded to the celestial bodies; thus, for example, the heart corresponds to the sun. Furthermore, even diseases correspond to natural occurrences. For example, dropsy (the swelling of soft tissues due to the accumulation of excess water) corresponds to rain, and phthisis (a disease similar to tuberculosis, characterized by the wasting away of the body) to drought. This big-world, little-world idea was one of the ways in which people sought to define a unity between themselves and the world in which they lived. According to this idea, the mountains (representing the element earth) build a supporting frame for the Earth, which parallels the human skeleton; the volcanoes (fire) parallel human heat; the rivers (water), which ensure the flow of water, so vital for all growing creatures, parallel human blood. Consequently, knowledge about ourselves enables us to approach knowledge about the universe, about how it is structured and how it progresses. Furthermore, as man and the universe are comparable and essentially rather similar, man can control the forces of nature in the universe. The idea, developed by ancient Greek philosophers to describe human beings and their place in the world, was one of the expressions of the unity of the universe. It enjoyed a long life, from Rome during antiquity, throughout Europe during the Medieval (including the kabbalah, a Jewish movement of mysticism), and from the Renaissance to early modern times.

Encelius began his theory on the origin of metals and of other minerals (including fossils and anything dug up from the earth) in the very opening sentence of his essay. There he declared that the principles of generation of all animals are based upon male and female, and indeed the embryo forms by the sex products of man and woman. Following the microcosm/macrocosm doctrine, he next expanded this outlook to the mineral kingdom by identifying sulphur as the male principle and mercury (quicksilver) as the female principle. Encelius was referring to ideal, abstract values, not only to the elements we now refer to by these names. Sulphur represented the male because of its power, heat and activity; mercury represented the female because it is cold, passive and capable of flowing, thereby taking the form of more active powers. The basis for this distinction goes back to Aristotle, who had claimed that male semen imposes shape while the female gives only matter. Encelius applied this principle to the mineral kingdom, in which the sulphur principle, as male, dictates shape and the mercury principle, as female, supplies only matter. As the male semen is heavier and warmer, it gives the foetus shape and movement, as against the female contribution, which is lighter and colder, and which can contribute only matter. In this manner, God had created metals along the same reproductive principle of male and female, with sulphur as male and mercury as female.

Encelius now organized his classification of the mineral kingdom within this general ideological frame, and this classification, eventually, led him to decipher the fossils as inorganic. True metals are the most noble groups in the mineral kingdom, since both sulphur and mercury are combined in their creation, though in varying proportions and levels of purity, from the most perfect (gold, of course) to the least so. Encelius developed an idea of inherited mineralogy, in analogy to human lineages: Pure and impure metals form according to the purity and impurity of the sulphur and mercury, just as parents with leprosy will procreate leprous offspring. Encelius next treated imperfect minerals, in which the contribution of sulphur or mercury is weak or even absent, because a third component, salt, is mixed into them. He finally reached the least noble of minerals, in which the principles of sulphur, mercury and salt are mixed with an even more inferior component, "fluores", which dissolves both in fire and in water. Sedimentary rocks belong here, including those that contain fossils.

Encelius' classification of the mineral kingdom and his theory on their human-like creation explain why he considered fossil shells as completely inorganic and formed within the mineral kingdom. Even though minerals are macrocosmic versions of the same principles that form microcosmic life, Encelius still considered fossil shells as entirely inorganic minerals, whatever their resemblance to animals. As mineralogical entities that grow within rock, Encelius saw the rock as enclosing these fossil shells, "as during pregnancy". "By breaking the rock open, we find it to be pregnant with these chelonites; three or four of them in full and perfect form" (Gould 2002a, b).

Other intellectuals of the Renaissance had different theories regarding the meaning of fossils. Georgius Agricola (1494–1555), a German miner, and the botanist Mattioli, embraced the notion that a certain "fatty matter" gives birth to fossil organic shapes after heat sets it into fermentation. Gesner, for his part, criticized Encelius' identification ("Encelius is wrong in presenting this specimen as Chelonites"). He compared Encelius's figure with the figure of a bivalve described by Rondelet as *Striate*, noted the almost perfect resemblance in shape, and concluded that the striated shell presented by Rondelet is the correct genus (*Conchae striatae a Rondeleto dictae genos*). For the first time, a direct comparison of figures created key information that led to a correct taxonomic conclusion regarding a synonym. The debate over the origin of fossils continued throughout the Renaissance and well afterwards.

3.2 Animals, Plants and Sex: Their Place in the Universe

In ancient Greece, as we recall from the first part of this book, Plato had claimed that the universe is an eternal living creature with a body and a soul, which contains reason (=logic). Man is also composed of reason, soul and body; but only divine reason in him is imperishable. There is thus both a contrast and an analogy between macrocosm and microcosm. Man differs from the universe only because he dies, and his body then dissolves back into the elements. We cannot say that the universe dies and its elements dissolve into something else. Plato's arguments somewhat resemble the biblical text in which "God created Man in his own image".

People, over the centuries, elaborated on these ideas of Plato and the Bible. The entire world is to be considered as the "big world" *macro-cosmos*, whose composition, structure and functioning correspond to that of the human body, the "little world" *micro-cosmos*. Man, the little world, is the essence of the entire universe; and the structure and organization of his body parallels the structure and organization of processes in the big world, the universe. The Big World surrounding us, and the Little World inside Man, are closely linked and, as God had created both, He influences Man both directly and via other creatures, such as the stars. Throughout the Medieval Jewish, Christian and Muslim thinkers elaborated on the idea of the human as a microcosm, at the centre of the Creation. (In the kabbalah, this approach takes a step further: Man can influence the stars since all is the same system. In this manner, the relation between man and stars is mutual, not one-way directed from the stars downward, as in Greek philosophy.)

Renaissance philosophers accepted these ideas, but they also believed that medieval scholars had distorted the wisdom of the ancient philosophers, and a mission began to restore those lost original Greek and Roman sources available to the ancients, and to revive what those golden eras had enlightened, and which were stifled by the later, darker ages.

Later during the Renaissance, however, cracks appeared in the belief in ancient facts and theories. One crack appeared in the adherence to Aristotle's cosmic theories. A basic assumption of Aristotle was that the Earth is at the centre of the universe, and another was that the stars never change, as they consist of a fifth element, *aether*, which never changes and never moves. Accordingly, he divided the universe into the "sub-lunar world" (Earth), which is constantly changing, and the world beyond the moon, which is eternal and unchangeable. Comets admittedly appear occasionally and unpredictably, but Aristotle believed that they were atmospheric events of the sub-lunar world. Many centuries after Aristotle, Christian theologians welcomed his cosmic theories and, in addition, viewed the eternal heavens as pure and the earth as corrupt and doomed.

During the Renaissance, however, Copernicus (1473–1543), a Polish-Prussian diplomat who was also an astronomer, struck a hard blow to Aristotle's cosmic theory. He placed the sun at the centre of the universe, not the Earth, and he offered evidence that the planets, including Earth, encircle the sun. Earth, which is only the centre of gravity and the centre of the cycle of the moon, rotates on its axis once a day, whereas the skies do not move at all. Copernicus' new cosmic theory was revolutionary in that it contradicted Aristotle and, what was far more radical, it contradicted Christian theology. Furthermore, it contradicted our intuition, as based upon our senses, by which Earth is at the centre of the world, and that the skies revolve around us. Accordingly, not everyone welcomed it. Martin Luther, for one, claimed that "this fool seeks to turn on its head all the science of astronomy, but the holy scripts teach us that Joshua had commanded the Sun to stand still, not the Earth". However, the new cosmic theory gained more and more supporters, among them Galileo.

In line with Copernicus, one generation later and perhaps under the influence of Lucretius, Giordano Bruno (1548–1600), a Dominican friar, argued that Earth is no more than yet another stellar body, and so too is the Sun. Throughout the universe, there is an infinite number of worlds populated not by Man but by other intelligent creatures. God is infinite, and the universe reflects this fact in that it too is endless, without boundaries or movement. The stars we see are actually distant suns like our own sun, each encircled by planets in which creatures other than us might live. Indeed, the basic unit in the universe consists of a sun with its planets. Time and space are endless, so there is no place for claims of a Divine Creation, nor of a doomsday that will come at the end of all days. Accused of heresy, the church burned Bruno at the stake.

Another blow befell Aristotle's cosmic theory when two supernovae (exploding stars) suddenly appeared in the skies beyond the moon and even beyond the sun, the first one in 1572 and the second in 1604. To the many people who observed them in those days, they seemed like new stars. In addition, a comet appeared in 1577, and it too was calculated and found to be far beyond the moon and the sun. These cosmic events bore evidence that the skies, too, undergo change. Following these blows, people began doubting Aristotle's cosmic theory about an ever-changing central Earth versus never-changing eternal stars.

Theology underwent further entanglements. In 1417, in the twilight between the Medieval and the Renaissance, Lucretius' book *On the Nature of Things* was rediscovered in a monastery in Germany. We recall that Lucretius had suggested, following Epicurus, that everything in the world is made of minute, invisible particles constantly in motion, clashing with one another, coming together to form new shapes, coming apart and recombining again. These elementary particles of matter ("the seeds of the things", to use his terminology) are indestructible and eternal, they move about in an infinite void with no fixed points and no limits, they collide, combine and then separate, in endless combinations and at unpredictable times. There is no purpose to this ceaseless creation and destruction, which chance governs. The world, which thus consists only of matter and space, has no creator, and consequently, it was not created for humans to govern. All living beings, from plants to animals and to man, have formed through this random process, and they are temporary. The human soul consists of the same matter as the human body, and upon death, when the matter of the body disperses, the soul, which is part of the body, dies as well. These ideas of Lucretius had faded away into history and were forgotten for 14 centuries. In 1417, they were rediscovered, printed and distributed throughout much of Europe, cracking the blind belief in Creation, in an end of the World, in a Soul that is eternal and in an afterlife. A century later, the Protestant Reformation, which Martin Luther began with his 95 Theses (1517), rapidly spread throughout much of Europe and created an additional atmosphere of liberation from religious dogmas, which had governed the minds of people for over 1000 years. Towards the close of the Renaissance, at the beginning of the seventeenth century, intellectuals sought a general view of the universe and of Man's place within it.

Paracelsus (1493–1541, Fig. 3.49), a physician, was born Theophrastus von Hohenheim in Switzerland. In Basel, where he lectured at the university, he was prone to outbursts of abusive language and, in a display of contempt for conventional medicine, he ceremoniously burnt Galen's texts in the market square. He also heavily criticized the town's physicians and apothecaries. Threatened with an unwinnable lawsuit, he left Basel, first for Alsace and then to Nuremberg, to Beratzhausen, then St. Gall, Appenzell, Innsbruck, and on to Meran, but because of his bad reputation, the medical professionals everywhere barred him from practising medicine. As a pseudonym for his publications, he began using the name *Paracelsus*. Aged 47, he moved to Salzburg, where he died.

In searching to understand what life is, Paracelsus explained that fire, to burn, must have wood, otherwise, there would be no fire. Fire is thus something vital, yet it cannot live without wood. Likewise, among animals, the body is wood, and life within it is fire, but here comes the difference: Life derives its substance from the body, but the body is not consumed; it continues to exist. Consequently, the body

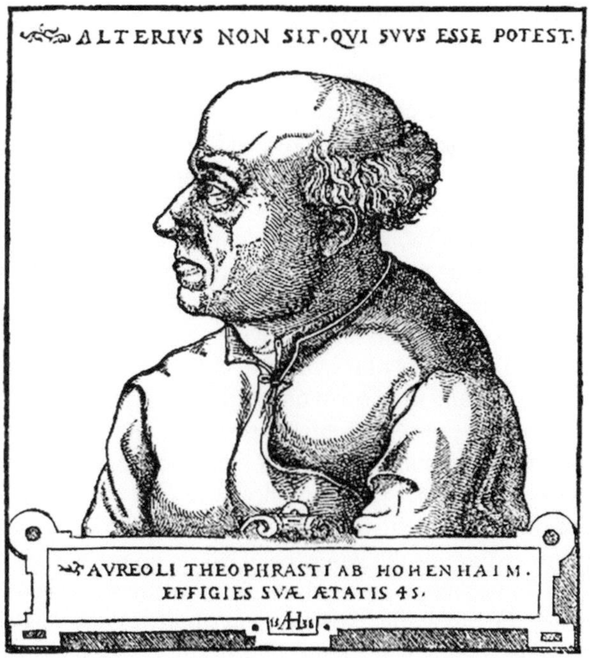

Fig. 3.49 Paracelsus. (Wikipedia)

must possess something that prevents it from being consumed by life. Paracelsus called this something the *mysterium*, and we today may view his idea of it as an invisible, metaphysical atmosphere that occurs throughout the universe (King 1971). This *mysterium*, which supports all creatures in heaven and on earth and through which all creatures live, may, however, be polluted and thereby change and may thus influence living creatures (including man). Many influences derive from the stars, which form a stage in the transmission of divine influence. There exists an *Ens Astrale*, an invisible power deriving from the heavenly bodies that sustains life, in us and in everything alive and sentient. Now stars have their own natures and moods, just like people on earth, and they too undergo changes, for better or for worse. If they are well disposed, nothing evil emanates from them. But when they are evil-disposed, their wickedness comes to the fore. The stars surround the world like the shell of an egg. Air penetrates this shell and passes through it to the centre of the world. Poisonous stars pollute the air with their poison, and wherever the poison penetrates, on that same spot, diseases will crop out, in conformity to the character of the star in question. One single star does not poison the entire air in the world, but merely a part of it, depending on its strength. *Ens Astrale* thus signifies the vapour of the stars as mixed with air, and it is in this manner that we get cold, heat, dryness and moisture. The polluted *mysterium* may pollute and weaken living creatures. In this manner, the *Ens Astrale* influences life, for good or for ill (King 1971). Furthermore, the liver corresponds to Jupiter, the gall bladder to Mars; the heart is the sun, the brain is the moon, the spleen is Saturn, the lungs to Mercury and the kidneys to Venus. All these organs perform planetary movements of the body and, if they come into an unfavourable position, disease arises (Nordenskiold 1928).

Paracelsus further claimed that every creature is perfectly adapted to its specific way of life. However, it is also a link in a food chain, and as such, its specific adaptations may not be relevant to other creatures in the chain; they may be damaging to the animal that consumes it. The ox, for example, was created perfectly adapted both to its specific way of life and to serve as food for man's nourishment. Had it been created only on man's account and not also for its own sake, it would not have possession of horns or hooves, as these do not constitute food for humans (as its flesh does). Now, when a man enjoys an ox as food, he also eats organs that are contrary to his own nature, its "horns and hooves" (to use the term as a paraphrase). Though the ox is not poisonous in the least, specific parts of its body that man does not have, its "horns and hooves", may be poisonous for man. Paracelsus thus believed that parts of the internal chemistry of one animal species differ from those of another, and may harm other species (King 1971).

God had created all things in heaven and on earth—day and night, the four elements and all animals. After all these were created, God made man. All things were made of nothing, by a word only, save man alone. God made man out of *something*, out of dust, which was *something* (Hall 1951). He created man's body as a most subtle extract of the essences of all creatures and all elements on earth and of all stars in heaven. As an extract from all these, man is a microcosm, a little world inside the big world. Now, as man's body is from the earth and from its elements, it must be nourished by the world from which he had sprung. Accordingly, his nourishment and drink come from the ground. The body of man takes its food and drink from the earth, and animals do the same. If man were structured not from the earth but from some heavenly substance, he would be taking celestial bread from heaven, along with the angels. Man takes from the fire his heat and from the air his breath. Man is not, however, only flesh, blood, heat and air; he also has intellect—his features of wisdom, of intelligence and of arts, and these, alongside his senses, come from the stars. Man is thus divided into an elemental body, which must be nourished, and into spirit, whence he sustains from the spirit of the stars (Hall 1951).

A man and a woman alone cannot beget a child; along with these two, the elements are required, and so too is the spirit of the stars. These four make up a man. The semen is not in man, save insofar as it enters him elementarily. When, in the act of conception, the elements do not operate, a body is not begotten. When the star does not operate, no spirit is produced. Both body and spirit must be there. Whatever is begotten, without the elements or without the spirit of the stars, is contrary to nature. Semen is something that man attracts from without as if it were a potion or a principle of food or drink. Thus, man becomes a microcosm. The father and mother are made from the universe, and the universe is constantly contributing to the generation of man. Eventually, the body returns to its "primal body", and the senses to the "primal sense". They die, pass away and depart, never to return. The ashes cannot again be made wood, nor can man come back to be a man again, from that state in which he is ashes (Hall 1951).

At a theological level, Paracelsus believed that the virtues that make up natural objects had existed inside God before the Creation. When Earth and Heavens shall eventually disappear, they shall, therefore, continue to exist and shall simply return to God. He described four beings, each corresponding to one of the four ancient elements: salamanders correspond to fire; gnomes correspond to earth; Undines to water (these were aquatic nymphs); and Sylphs, corresponding to air (imaginary spirits of the air; King 1971).

Paracelsus noted that though the virtues of different herbs are invisible, man can detect them. It is God's will that all things exist for man's sake. Like the configuration of stars that spell out a message, the external forms of plants are signs of their inner powers, their physiognomy that bears evidence of their essential features. Nature provides texts in the patterns of stars and in the mysterious shapes of plants (and the flight of birds, and swarming of bees), and we must read these disguised texts to grasp the connection between external traits and internal virtue (Wilson 1992).

Didacus Valades, of the viceroyalty of New Spain (Mexico), the son of a Spanish conquistador and an Indigenous mother, joined the Franciscan Order and travelled to Rome on their behalf. In 1579, he published *Rhetorica Christiana*, with an illustration that seems to me a mishmash of the biblical text of Creation, of the Fall of the Angels and of the great Chain of Being (Figs. 3.50 and 3.51). In the book of *Genesis*, we recall that God had created light and separated it from the darkness, gathered all the water into one place and let dry ground appear. Next, He let the land produce diverse plants and trees and created the sun, the moon and the stars. God then let the water teem with animals, let birds fly above and let the land produce diverse animals that move along the ground, both livestock and the wild ones. God

Fig. 3.50 Valades: The Great Chain of Being?

Fig. 3.51 Valades: The Great Chain of Being? (detail)

created Man and then, on the seventh day, he rested. Later, while the man was sleeping, God took one of his ribs and made a woman from that rib. Valades illustrated all these. He illustrated God surrounded by light at the top of the illustration and the dark underworld of hell at the bottom, he presented the rivers as separate from the dry mountains, with various kinds of plants on the mountaintops, he illustrated the sun, the moon and the stars, he pictured diverse creatures of the sea, birds, wild animals and livestock; and the woman arising from one of Man's ribs. In addition, however, between God and Man, he illustrated three rings of angels. The upper ring consists of seven arch-angels, each representing, perhaps, one of the seven days of the week. He also pictured angels falling down to hell, where they are tortured by the devil and his servants. In the centre, one of the high angels is handing down a long chain—could this be Valades' presentation of the Great Chain of Being? Another possible source for this picture could be the *Book of Revelation* of the New Testament, written toward the end of the first century. Valades pictures seven arch-angels, and *Revelation* repeatedly mentions the number seven: seven churches, seven spirits before God's throne, seven stars, seven seals, seven angels who stand before God with seven trumpets, the prostitute of Babylon sitting on a beast with seven heads, and seven angels with seven bowls full of seven final plagues. Furthermore, Valades illustrates God sitting beyond the clouds, and *Revelation* describes someone seated on a throne in heaven. Valades illustrates the fallen angels, and *Revelation* describes a war in heaven, between Michael and the dragon; the dragon lost and was hurled out of heaven, and his angels with him. There is a little square within a circle, I do not know its significance.

Robert Fludd (1574–1637) was the son of a high-ranking English government official and Member of Parliament. He studied medicine and chemistry in Europe and at Oxford and entered the College of Physicians (after six failures); from then on, he became a prosperous London doctor. Fludd developed a medical practice almost devoid of chemical remedies, which depended almost solely on prayer and on the use of the name of Jesus. He supported his devotional medicine with a theology derived from the secret mystical teaching of Judaism, the Kabbalah, which Fludd employed in a Christianized form. He equated Jesus Christ with the Kabbalistic angel Metatron, the heavenly form of the Jewish Messiah, the soul of the world. Fludd stated that 'Hochmah' (wisdom of the Kabbalistic view of life) is the same as "verbum" (the "word" in the Christian Gospel of John, who is identified with Jesus Christ). The "verbum", or Metatron-Christ-Messiah, is the form of God, residing in the sun. Whereas Bacon's views of nature were based on experiments, Fludd's views were based on mystical thought.

Fludd presented his philosophy in a grand treatise, *The metaphysical, physical and technical history of the two worlds, namely the greater and the lesser* (*Utriusque Cosmi, Maioris scilicet et Minoris, metaphysica, physica, atque technica Historia*, 1617), in which he discussed the two worlds, the microcosm of human life on earth and the macrocosm of the universe. He saw the source of all life in the universe as a trinity consisting of God, the Word and Light. In referring to the four elements of earth, water, air and fire, he claimed that earth is no more than highly condensed water; water is no more than very condensed air; air is very condensed fire; and fire and light are the same. Therefore, the Divine Light is fire, and its source is fire, and it represents the active element in the creation of the world. The macrocosmic Big World is organized as a series of spheres contained within each other, falling into three distinct groups: the underworld of the four elements (earth, water, air, fire); above it, the star world; and above it, the upper world of the heavens beyond the stars. All three groups of spheres were created out of the Divine Light. There are creatures in all three groups of worlds: In the high heavens, there are angels; in the star world, there are "demons"; and in the underworld, there are plants, animals and men. These three groups of worlds receive the Divine Light according to their place in the hierarchy, and as the Light lessens, illnesses and death increase. The three divisions represent three levels of existence: the physical level (of nature) in which we live, the heavenly one (of the stars) and the spiritual one, that of the angels. Beyond all these is God. This is the structure of the Big World. Whatever occurs in it influences the Little World, and whatever happens in the Little World reflects changes in the Big World.

The Little World of Man is also divided into three categories: Man's head and intellect correspond to the spiritual field. His torso corresponds to the stellar field, and more specifically, his heart corresponds to the sun, the physical representation of God in the visible world. The heart is the site of the spirit or soul. Man's genitalia and viscera correspond to the physical field: they are the crudest parts of human anatomy, and they are devoted only to such physical needs and demands as feeding, excretion, reproduction and desire. Genitalia and viscera have no divine function; they have no connection to spirituality or logic. They are part of our beastly entity, and those people incapable of seeing beyond their physical desires are at a level that is only slightly above animals. Fludd's triple division of the Little World of Man echoes the medieval triple division of the human body.

Triangles cutting one another are among Fludd's common symbols. One triangle represents the descent of the Divine Light into matter, and it is wide above and pointed downward. The upper worlds contain more Divine Light than the lower ones, which contain very little light. The second triangle represents the ascent of matter into light, and it is wide below and pointed upwards. All worlds contain at least a trace of matter, except for God himself, who is beyond the system of worlds. Matter exists mainly in the lower world and becomes rarer as we ascend. Together, these two triangles present the balance between matter and spirit. More materialistic entities are less spiritual, and vice versa. The sun, at the centre of the system of the Big World spheres, has an equal measure of these two components. It is at the mid-point of the series of spheres, in the mid-point between Man and God, at a point where soul and matter are in complete balance. The Sun is the tabernacle tent of God; it is his residence when He descends into the world. The Sun gives light, without which plants and animals cannot exist, as it gives them something vital, the spirit of God. We recall Aristotle, who named Earth's *mother* as being female, and the Sun's *father* as being a male and initiating generation; the influence of the Sun triggers the generation of plants on Earth, and thus, the Sun generates another, whereas Earth generates in itself.

According to Fludd, everything in the world had already existed before being created. During Creation, everything flowed out of God as a series of spheres, one outside the other, one enveloping the other; consequently, the world is a series of spheres, each surrounding the previous one with God surrounding everything. (We recall Plotinus' theory of emanation.) During Creation, the lighter elements rose upwards towards the purer upper world, whereas the denser ones remained below. Accordingly, more light occurs in the upper worlds and more muddy sediment in the lower world. The Big World has a soul, *Anima mundi*. Man, also, has a soul, *Anima humana*, a complete microscopic reflection of the soul of the world. In Man, the harmony between body and soul unites them into one entity.

Many of Fludd's ideas about the organization of the world appear in a large illustration that he designed (Figs. 3.52 and 3.53). The illustration presents the universe as clearly divided into three parts: God, the Big World and the Little World. God, here in his Hebrew name *Jehovah*, dwells in a divine cloud, beyond the world. He is the only entity in the universe consisting of pure spirit, without even a drop of matter. The Big World consists of three great groups. The outer one is that of the angels, who consist mainly of spirituality, with almost no matter; they exist forever, they never decay, and they are beyond Man's senses. The sphere of angels has three sub-spheres. The outermost sub-sphere is adorned only with winged faces. These angels only sing songs of glory to God and are not involved in the fate and destiny of Man. The human figures in the inner sub-sphere of angels signify that these are angels involved in Man's deeds. There are nine types of angels, and they are organized in three groups. We recall Iamblichus of Syria, who, in the second century CE, believed in the One at the head of a hierarchy of the world, beneath whom there is an assemblage of heavenly gods, other gods proceeding from them, chiefs and nature-gods.

The World of Stars *Caelum Stellatum* (the fixed stars, planets, sun and moon) is inside the sphere of angels. The stars contain much more matter than the angels do; they have a material existence to such an extent that we can see them, but this matter never decays and never transforms from one shape to another. The fixed stars are all at the same distance from us, and they all move as one unit, as a bridging sphere between the world of the stars and the world of the angels. Star movements are predictable, and they never move from their track. Inside the sphere of fixed stars come the planets that influence Earth. To illustrate the influence of stars on the lower, physical world in which we dwell, the symbol of Saturn (on the left) radiates its influence on objects in the more materialistic underworld. The world of the stars also bridges the world of spirit and the world of matter in which humanity dwells, and it shapes the lower world according to the dictates of the upper world.

Earth is at the centre of this Figure, and the planets are arranged correctly regarding their place in the solar system, with Saturn as the most distant and Mercury the closest to the sun. However, the sun is not in place. The Sun is exactly at the centre of the world of stars. It consists of half spirit and half matter. The sun glows with a light that is both spiritual and material, as it is the midpoint between the two contrasts, spirituality and materialism. The Sun had special significance in Fludd's cosmology: it is the residence of the Holy Spirit, and light and the spirit of life emanated from it. For man, life on Earth could be possible only through the inspiration of the

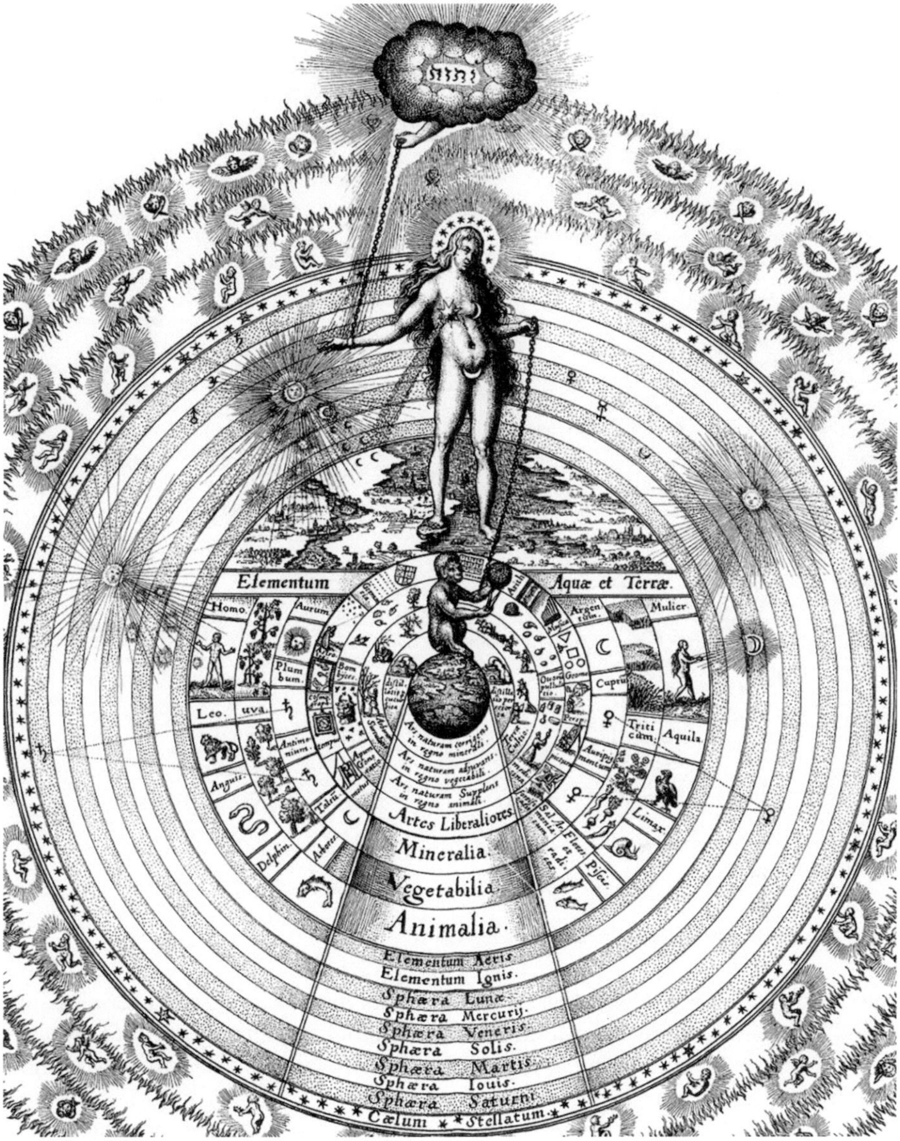

Fig. 3.52 Fludd: the universe

Holy Spirit, from the atmosphere. The circular movement of the sun imposes a circular motion also on the Holy Spirit. This Holy Spirit influences man: our blood, which carries the spirit inside us, must also have a circular action. (This is a mystic description of the blood circulation. Fludd was aware of Harvey's studies on blood circulation at the Royal College of Medicine in London. We shall meet Harvey in the next part of this book). He believed that the moon is the most "terrestrial" of

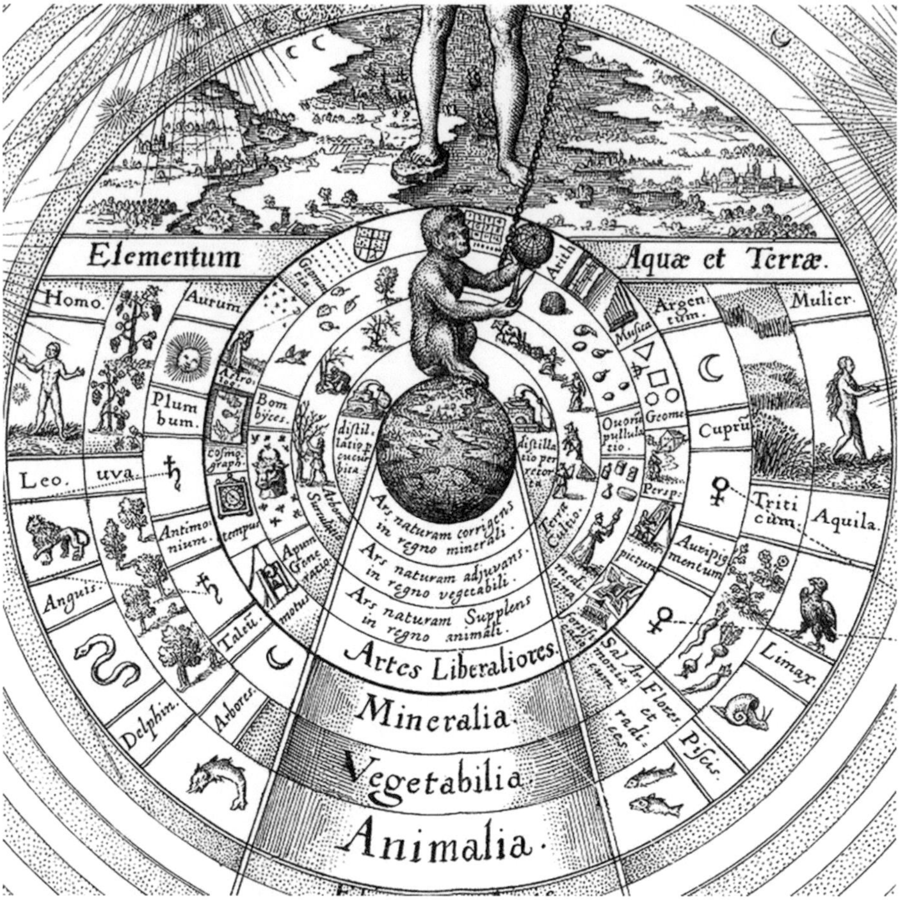

Fig. 3.53 Fludd: the universe, detail, showing Earth

stars, and that, in the tidal cycles, it expresses its influence on the material reality. Fludd's emphasis on the sun is reminiscent of Plotonius' philosophy (third century), which likened the One to the Sun, which emanates light indiscriminately without thereby diminishing itself,

The Little World, our material world, is the lowest and innermost of worlds, consisting mainly of matter with very little spiritual essence. The material in our world is transitory: its entities are subject to decay and putrefaction, they do not survive forever, and they change by transmutation, from one material species to another. The outer spheres of our material world consist of its two lighter elements, fire and air; water and earth are further in. Earth and water are surrounded by the three spheres of nature, a sphere of animals, a sphere of plants and finally one of minerals. *Homo* is in the sphere of animals, the male on the right side facing the sun. Beneath him is a lion as the king of animals, a snake representing powerful, hot, fast-crawling animals (or perhaps the psychological, sexual significance of a snake?)

and a dolphin representing hot aquatic animals. Bees are on his side, as a symbol of a monarchy ruled by a king, and he is capable of understanding the significance of the star movement. Plants at his side are mainly trees; the minerals on his side are gold, lead, antimony and talc. Among his machines, a mortar symbolically expresses the male, the active attacker. The female is on the left side and faces the moon. Beneath her is an eagle as king of the birds, a snail representing cold, damp, slow-crawling animals and a fish representing cold aquatic animals. Plants on her side include wheat, flowers and roots; metals on her side are silver, copper, auripigmentum and ammoniacum, and she has healing (medicinal) powers. A fort symbolically expresses the female as the passive defender. The male-female division is thus an essential feature of the underworld.

Spheres further inside show Man-the-Creator represented as a monkey, who imitates God's creation by creating things himself: machines, buildings, agriculture and so on (the expression of man imitating God "as a monkey after Man" comes directly from the Kabbalah). The upper centre shows a female image, her head among the angels and her feet planted in the material world. Her right hand is in a chain, which descends from God—she does things only at his will. This might represent the *Book of Revelations* Chap. 12, in which "a great sign appeared in heaven: a woman clothed with the sun, with the moon under her feet and a crown of 12 stars on her head". (The *Book of Revelations*, written towards the close of the first century, describes a series of prophetic visions, culminating in the Second Coming of Jesus.) Alternatively, it might represent the *Book of Wisdom* (or the *Wisdom of Solomon*), a Jewish text of the first century BCE written (Greek) in Alexandria. In this text, Lady Wisdom, who has existed since the Creation, is guided by God and comes to the aid of the righteous. It might also represent the *shekhinna*—the face of revelation of God to humans, which is also a feminine and maternal face. According to the Kabbalah, the *shekhinna* gives life and vitality to the entire universe; it is the spirit and soul of the world. Through this face of divinity, God presents Himself and communicates with men. (In the kabbalah, the *shekhinna* is termed "kingdom", because it is the noble, spiritual queen of the world, through which God commands reality.)

Reading through Fludd's opinion of the structure of the world, we recall Abraham Ibn Ezra (1089–1164), who argued that the whole universe is organized in an ascending order of three worlds. The *Genesis* world is the lowest; it is the Underworld, which consists of metals, plants, animals and man. The Middle world consists of the planets and fixed stars, and the Upper world consists of angels. In our earthly Underworld, each living body consists of the four elements, to which it will return upon death. In the Middle world, the movement of the stars is constant, circular and repetitive, without change, without beginning and without an end. The angels in the Upper world move the stars of the Middle world. The uppermost world is dependent on God.

Towards the close of the Renaissance, the big-world little-world analogy was a central element in theories concerning the significance of Man and his position in the universe. Walter Raleigh expressed this idea very well. He wrote that "my blood, spreading throughout my body, parallels the waters carried by the springs and rivers throughout the Earth. The breath in my nostrils resembles air. The heat of my body resembles the heat deep inside the earth, my hair to the grasses which cover the face of the Earth and the beauty of my youth to the flowers of spring, which wither away so rapidly" (*The History of the World*, 1614). From a slightly different angle, William Harvey wrote that the heart is "the sun of the microcosm", and King Charles I is "the sun of his microcosm, the heart of the commonwealth" (*de Motu Cordis* 1628). The microcosm-macrocosm idea thus assumed correspondences between two uneven layers of reality. In later centuries, it gradually lost its metaphysical importance and became marginal.

3.3 Man Differs from Animal: Anatomy

Anatomy, of all biological branches, typifies the Renaissance approach in that it first came glorifying acceptance of ancient texts, then a critical appraisal and eventually an original outlook. The tendency to uncover the ancients' wisdom manifests itself in a drawing made by Leonardo in 1490, of a man in two superimposed positions with his arms and legs apart, inscribed in a circle and square (Fig. 3.54). In the accompanying text, he noted that this drawing is a study of the proportions of the (male) human body as described by Vitruvius (a Roman architect who lived during the reign of Augustus). Vitruvius had advocated ideal proportions in architectural designs and, in support of this idea, had maintained that ideal proportions also occur in natural objects, such as the human body. For example, the length of the body is eight times the height of the head, and six times the length of the foot. Leonardo's drawing illustrates two of Vitruvius' ideal proportions. First, in a man standing erect with arms stretched horizontally, the distance between the fingertips will equal the height of the body (and the figure is, therefore, inside a square). Second, in a man lying with arms and legs spread, a circle is drawn with the umbilicus as the centre will touch the tips of the fingers and the toes. Leonardo typifies the attempt to understand conventional viewpoints in classic texts, his "Vitruvian Man" reflecting the search for absolute laws during the Renaissance, in the hope of finding them in scripts of the Classical world.

Leonardo da Vinci (1452–1519) was the illegitimate son of a peasant girl and a Florentine notary. He was born in Vinci, near Florence, and his father was quickly married off to a girl from a respectable family, his mother to a cowherd. At 14, Leonardo was apprenticed to a leading Florentine painter and sculptor, Verrocchio, who required him to study human anatomy in depth as an essential part of his training as a future painter. After becoming a painter at the age of 20, he spent much of his life in the service of the rulers of Milan, Rome, Florence, Venice and France. Beyond painting (Mona Lisa, The Last Supper), his areas of interest included architecture, science, music, mathematics, engineering, literature, cartography, geology and astronomy. He designed flying machines, armoured fighting vehicles, concentrated solar power, adding machines, moveable barricades, a scheme for diverting the flow of the Arno River and a plan for a 220-metre bridge to span an inlet at the mouth of the Bosporus. He recorded many interests in sketches, which he made and

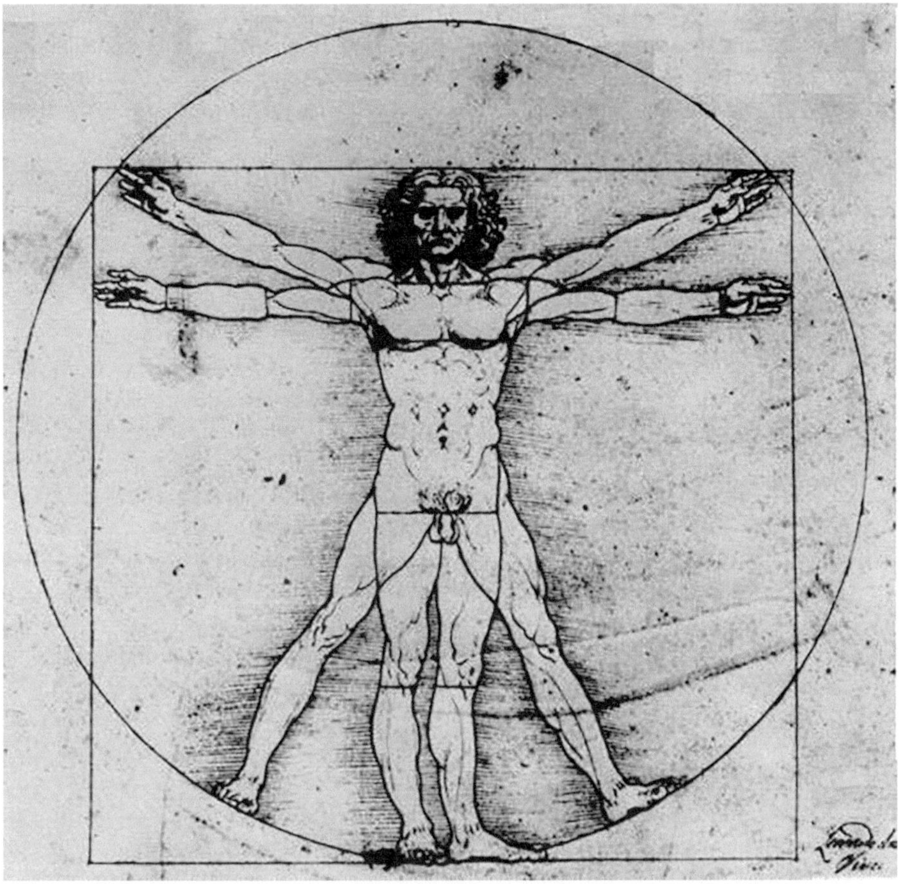

Fig. 3.54 Leonardo: Proportions of the human body, after Vitruvius. (*Leonardo da Vinci*. Author's library)

maintained in notebooks or on separate pages. These included designs for wings, shoes with which to walk on water, studies of faces and emotions, of animals, babies, plants and of whirlpools. Leonardo tried to understand phenomena by depicting them in utmost detail, indeed, he was more a man of pictures than of words. Being a left-handed person, writing in mirror-image cursive was perhaps easier for him than writing from left to right. He never published his findings, so we must be aware that we are probing into his intellect, not into any influence that he intentionally sought to exert, on his own generation or on following ones.

Leonardo sought to understand the nature of life, and as such, he was interested in anatomy. He observed the anatomy of cows, horses, birds, monkeys, bears and frogs, sometimes in comparison with the anatomy of humans, which he dissected at hospitals in Florence, Milan and Rome. At the age of about 60, he created over 240 detailed anatomical drawings of the human body. His underlying theory of life was the microcosm-macrocosm theory, in which he compared the blood to the ocean and its rivers. Man has within him a pool of blood wherein the lungs, as he breathes, expand and contract; likewise, Earth has its ocean, which tidally rises and falls, with the breathing of the world. Blood vessels emerge from the blood pool of the heart and branch throughout the human body, and the ocean fills the Earth with an infinite number of veins of water (McMurrich 1930; Del Maestro 1998; Jose 2001; Dolezal et al. 2012).

Leonardo's observations were not always as accurate as might be desired. In his thirties and before his many detailed dissections, when much of his information on the human body was gleaned from textbooks, he spent much of his time producing drawings in which he apparently tried to explain, to himself, the conventional viewpoints presented in these texts. Chief among the textbooks at his disposal was the 1493 edition of a compendium of medical information, *Fascicolo di Medicina*, which included a complete presentation of Mondino's *Anathomia*, from 1316. This edition of the compendium was the first anatomical textbook in Italian, and Leonardo's poor command of Latin had made its earlier versions unavailable to him. This text enabled Leonardo to gain access to the ideas of Hippocrates, Aristotle, Plato and Galen, as reflected in the later views of Avicenna and Mondino. (Direct translations of the original Greek text into Latin and European vernacular languages were not available until 1525, six years after Leonardo's death.)

We focus, first, on Leonardo's sketches of the human genitalia, in which he sought to explain (or understand) the viewpoints presented in these medieval texts. Fig. 3.55 reveals the amazing tubing he designed to describe "textbook" copulation. He pictured *two* ducts in the penis, the lower one connected to the urogenital tract via the urethra, the upper passing to the spinal cord by means of three vessels (Morris 1986; Noble et al. 2014). We should view this strange tubing in light of medieval opinions, which sought to settle the controversies between Aristotle and Galen. In Greece in the fourth century BCE, Aristotle believed that the testes played no role in procreation other than to provide saliva-like humour that would lubricate the vagina during intercourse and that the animal spirit provided the essence of the offspring, which spread from the base of the brain to all parts of the body via the nerves. Accordingly, Leonardo presented a spinal connection to the penis. However, in ancient second-century Rome, Galen believed that the testes concoct sperm from the blood. Accordingly, to include both views, Leonardo added a large blood vessel from the heart to the testicle. By combining the two arguments as they had reached medieval thought, Leonardo's sketches provided an appropriate anatomy for both theories.

It was a common belief in the Medieval that the male organ, because of its involuntary behaviour, possessed a certain amount of individuality independent of the body, and both Plato and Aristotle spoke of it as being to some extent an independent animal. Leonardo was sufficiently medieval to accept this idea: "This animal often has a soul and intellect apart from the man". The fallacious idea of the testis filtering off the sperm from the blood brought down to it by the testicular artery lasted deep into the Renaissance. He made a later drawing (after 1500) in which the description of the anatomy is remarkably accurate and which must have been based on careful dissection (Noble et al. 2014).

Medieval thought is also present in Leonardo's depiction of female anatomy. Figure 3.55 clearly shows that the uterus is made up of several cavities, and this strange uterine morphology is once again drawn from the description by Mondino, who maintained the common medieval misconception of a womb with seven chambers. The addition of a blood vessel connecting the womb with the breast, in Fig. 3.55, is even more extraordinary. This, too, seems to stem from Mondino, who wrote that "nature arranged it so that tortuous veins lead from the uterus to the breasts". As Galen has said, they are torturous for the blood to be constantly refined and well-dispatched.... In medieval thought, this hypothetical vessel carried the suppressed menstrual blood to the mammary glands and caused the enlargement of the breasts and the production of milk. The noteworthy absence of ovaries in Leonardo's drawings reminds us of Aristotle, who had considered the female as merely the soil into which the male seed was planted, with the ovaries having no reproductive role (Galen's approach had been that the ovaries produced their own "sperm", which intermingled with the masculine sperm in the uterus). Finally, in Fig. 3.48 Leonardo erred in believing that during copulation, the penis penetrates far beyond the vagina and deep into the uterus.

Leonardo, like Galen many centuries before him, assumed that where the functional demands on animal organs are identical to those of humans, their structure (as seen in dissection) would also be identical—a structure that works in one animal would work in another animal as well. Accordingly, he drew the hyoid bone of a dog in a human throat (Fig. 3.56).

To conclude, Leonardo's drawings were mainly attempts to illustrate and clarify for himself anatomy and physiology as presented in textbooks of his time, which relied upon medieval scholasticism. Beyond the mere anatomical data, he also described, in his diaries, the difficulties of studying and illustrating the anatomy of the human body. He wrote of the aversion from staying near a stinking corpse, and of the repulsion of spending many long hours of the night in the horrifying presence of a cut-up cadaver with a peeled-off skin. This persistence, in seeking to find and illustrate the truth of anatomy in spite of these unpleasant experiences, was even more characteristic of Vesalius, a quarter of a century later, to whom we now turn.

Andreas Vesalius (1514–1564, Fig. 3.57), who came a generation after Leonardo, based his anatomy more strictly on his own dissections and observations of the human body, and less on scholarly texts. He was born in Brussels into a well-to-do Flemish family of men of medicine. His great-great-grandfather was a physician of reputation who had gathered a large and costly collection of medical treatises and had written a commentary on Avicenna. His great-grandfather had been physician to the city of Brussels, his grandfather had been a prominent court physician to Mary of Burgundy, and his father was imperial apothecary to Margaret of Austria and then to Emperor Charles V. It therefore came as no surprise that young Andreas had medical ambitions. At the age of 14, he studied art, rhetoric, logic and philosophy at the University of Louvain, studies which included classical Latin, Greek and Hebrew. In line with the new spirit of the Renaissance, the professors at Louvain

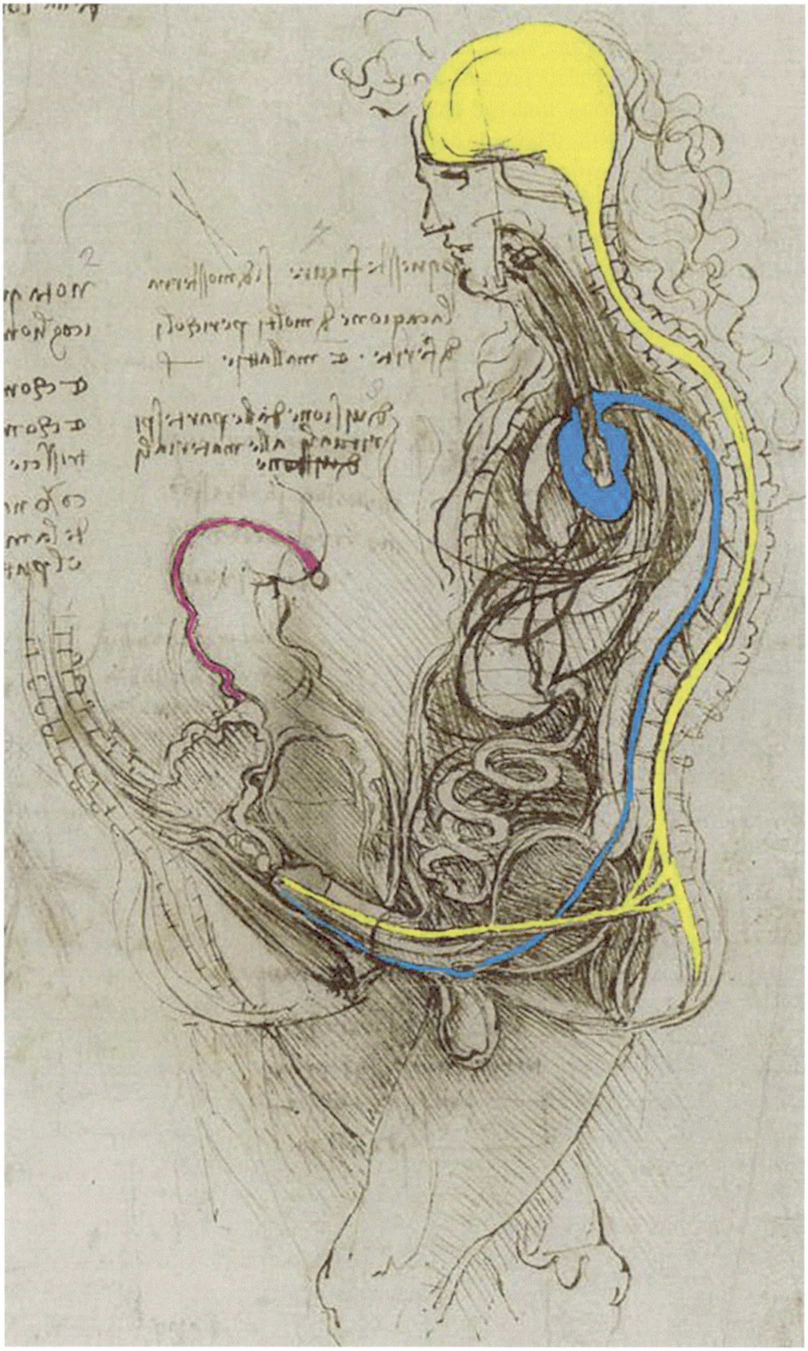

Fig. 3.55 Leonardo: a sketch of a couple copulating. Male: The penis contains two tubes, one from the heart (here emphasized in blue), and another from the brain (yellow). Female: A blood vessel (pink) runs from the uterus to the breasts. (From Noble et al. 2014, with colour added)

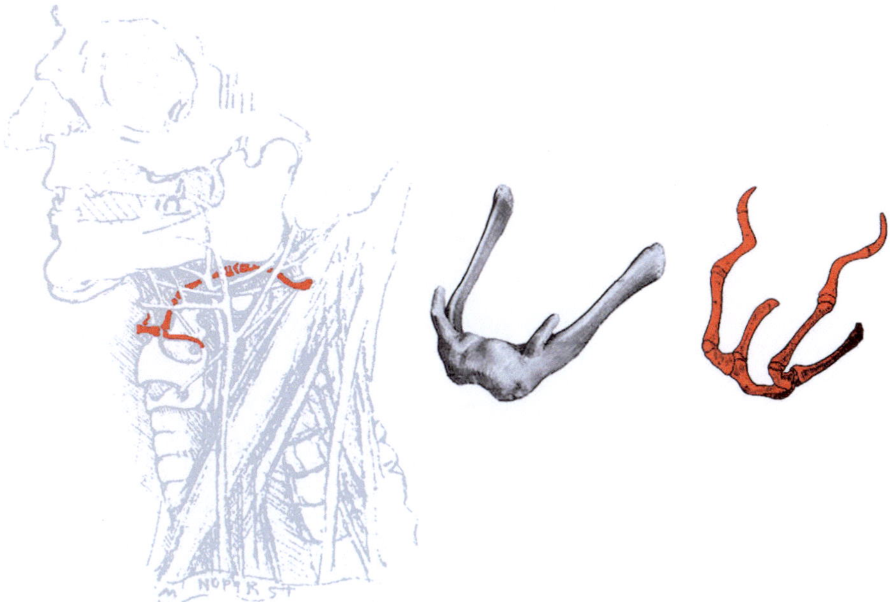

Fig. 3.56 Left: Leonardo, human neck (the hyoid bone in red). Middle: the hyoid bone of a human. Right: hyoid bone of a dog. The hyoid that Leonardo placed in the human neck is that of a dog, not a human

sought to educate their students in understanding the ancient texts. They also sought to revive the original Latin language of 1500 years before their days, hoping that it would displace the vulgar and distorted Latin that had developed during the medieval "dark" age. Furthermore, they emphasized the need for original Latin that would be flowery, elaborate and elegant. This was the Latin which Vesalius absorbed and which, eventually, he would use in his texts. At 18, he enrolled at the medical school of the University of Paris, where his studies included lectures on the classical texts of Hippocrates and Galen, and of medieval physicians such as Ibn Sina. In Paris, as in Louvain, the professors sanctified the classical texts, which they considered as The Original Truth. They believed that the Medieval Latin versions handed down to European physicians contained many mistakes, stemming from errors in the translations and copying of Galen. They wished to return to those pure classical sources and understand what Galen really had written, to discover and add to their libraries more and more undiscovered Greek texts rolling about among the ruins of fallen Byzantium, which would give them a clearer understanding of the principles of anatomy and of medicine in general. They would correct errors in medieval translations by the linguistic accuracy of a new translation, and thereby they would restore the medicine of ancient days. They believed that Galen might have erred in minor details (such as failing to describe the thymus gland), but he did not err in the major things. Although he had, in fact, dissected mainly animals, nobody in the fifteenth and early sixteenth centuries doubted that he had also dissected human

Fig. 3.57 Vesalius

bodies and had described human anatomy from first-hand observation. Printed versions of Galen's texts in Greek (1525) and in Latin (1529), of Hippocrates (1526) and other Greek authors flooded the shelves, and revived the teaching of classical texts, with an emphasis on Galen.

This general approach charmed young Vesalius, who had mastered the classical languages. Naturally, like any other young student of his time, he accepted Galen's anatomy. He could hardly do otherwise, since there was no other anatomy, and his anatomy teachers (among them a certain Jacobus Sylvius) considered the writings of Galen as the highest and undisputable authority on anatomy. Once, during the dissection of a cadaver, a student drew Sylvius' attention to the fact that the leg muscles of man differ from Galen's description. The teacher responded, "well, the muscles of the leg have changed since the days of Galen, presumably due to the wearing of very tight trousers" (Sirks and Zirkle 1964, Tubbs et al. 2007). So great was Galen's influence. Students did not participate actively in the dissecting of human bodies and Vesalius, displeased with the way in which human anatomy was studied and frustrated by this limited access to real human material to observe, would make unauthorized trips to charnel houses outside the walls of Paris, to obtain skeletal material and bodies of executed criminals left hanging to disintegrate.

The outbreak of the Franco-German War (1536) forced Vesalius, now an enemy alien, to leave Paris before finishing his medical degree and return to the University of Louvain. Here, he continued his nocturnal countryside forays. On one such night, Vesalius found the rotting body of an executed criminal hanging on a pole beyond the city gates, one arm and one leg having already fallen off. He overcame the stench of the body and the threatening growls of packs of hungry dogs, took down the stinking cadaver and prepared a skeleton from it. Beneath another pole, he found the arm and leg of another executed criminal, prepared a skeleton from them as well and added them to the former, to obtain a single almost complete skeleton. Only then did he apply to the authorities and request permission for his actions, claiming that the study of human anatomy helped him cure by bloodletting; they granted him retroactive permission. Vesalius's growing reputation in dissection led to an invitation to conduct the first public anatomy dissection in Louvain in 18 years, and he personally performed both the dissection and delivered the accompanying lecture.

Vesalius moved to Padua to complete his studies, and within a year, he received his Doctor of Medicine with honours. At the age of 23, the University appointed him to the chair of surgery and anatomy. Being empirically oriented, he was constantly on the search for facts. He performed detailed public human and animal dissections daily, over several weeks, events that demonstrated anatomy to a large (and paying) audience. Vesalius broke with tradition by offering a new anatomy based on his own observations of the human body. Rather than sitting on a high chair well above the stench of the corpse and reading out an authoritative anatomical text with a lay surgeon doing the dissection itself, he personally carried out the dissections for his students, lecturing as he did so. Vesalius also enlivened these dissections through his novel use of accurate drawings. Based on sketches he had made during dissections and with artistic assistance, he prepared, printed and handed out to his students a set of six plates of anatomy that accompanied his lectures (*Tabulae Anatomicae Sex*). These artistic plates set a new standard for biological illustration. Vesalius's career in Padua flourished, and when just 27, he began a grand book on anatomy, *Seven Books on the Structure of the Human Body* (*De humani corporis fabrica libri septem*) (Vesalius 1453). Commonly known as the *Fabrica*, it was a commentary on Galen, based on Vesalius's own dissections and observations. He now left Padua and journeyed to southern Holland, where the Emperor Charles V recruited him into his imperial team of physicians, thus continuing the strong traditional ties of Vesalius' family with the Habsburg court. Vesalius thereby left the academic world of dissecting and lecturing to accompany the emperor on his many battles and to delve into military medicine, the practical aspect of human anatomy. In 1556, he left the imperial court and returned to Brussels. Eventually, he embarked on a pilgrimage to the Holy Land. On his return, he became ill and died.

Written in the flowery Latin he had learned in his youth, Vesalius accompanied his *Fabrica* text of 659 pages with 227 magnificent illustrations printed from woodcut engravings. He sought to illustrate every organ in perfect precision, and he labelled the engravings accordingly and explained them with great accuracy. The anatomical nomenclature was comprehensive, orderly and all in Latin. He rejected any non-Latin term, reviving old Latin names and not hesitating to create new ones when he did not find suitable terms in the ancient writings. Indeed, his attention to distinct nomenclature is a key aspect of *Fabrica*, in which he combined the text and illustrations very closely. Nomenclature is of utmost importance in any profession, in that it creates a commitment to accuracy in communication (Singer 1925, 1941, 1946; Saunders and O'Malley 1973; Silverman 1991).

Vesalius did not intend his splendid anatomy book for the standard physician. He intended it for the wealthy intellectual, for the physicians of the elite who could value the content and splendour of the book and could pay, accordingly, five golden coins, each of 3.5 gr. Through his remarkable control of the dissecting scalpel, Vesalius created a detailed and thorough body of knowledge at a level that did not previously exist. *Fabrica* also reflected a deep knowledge of Galen's texts. This was the first time that anatomy was anchored in a very accurate, detailed presentation of the human body, based (mainly) on observation of the human body, not on the teaching of the ancients. *Fabrica* was not a book of ideas; Vesalius was not presenting new interpretations; he was offering thorough observations at a level that had not existed before.

Vesalius's search to understand the relation between a woman's genitalia and her embryo required a continuous hunt for cadavers. Occasions on which he was able to legally acquire and dissect the body of a pregnant woman were so rare that he resorted to grave robbing. This was no new malpractice, as already two centuries before Vesalius, in 1319, four students of a certain Master Alberto, who lectured at the University of Bologna, were persecuted for robbing a grave and bringing the corpse to the house where Alberto lectured (Ghosh 2015). Back to Vesalius, it once so happened that a pregnant woman, the mistress of a monk, had died during childbirth. Hearing of her death, Vesalius and his pupils snatched the body from the tomb, but the monk and the girl's parents complained of this outrage to the city magistrates. Vesalius and his students had to dismember and free the body from all its skin as rapidly as possible, to prevent it from being recognized. Since they had stolen the body specifically to examine the woman's reproductive organs, they hurriedly encircled the external genitalia with a knife, split the pelvis and severed the urethra, then excised the vagina and uterus in one piece. Illustrating the genitalia of a woman "of very tall stature", he could demonstrate that the uterus has only one cavity, not seven. (Fig. 3.58, which assembles his figure of the genitalia of a woman as compared to a female cow and dog.

Vesalius and his students developed a frenzy for grave robbing. The *Fabrica* features many historiated initials that describe the preparation of human cadavers for dissection. In some of the scenes, "putti" (chubby naked boys) play the roles of doctors, students, soldiers and priests (Fig. 3.59). The historiated initial **I**, shows a group of putti robbing a grave at night to obtain a body for the anatomy table (probably from a Jewish cemetery, to judge by the "O" on the banner held by the flying putto; a special badge in form of a circle was imposed on Jews by the Senate of Venice in 1374. Paduan Jews possessed the privilege that the bodies of their dead were not to be touched; a privilege for which they paid large sums. "**L**" shows people taking a female corpse down from the gallows, with a priest with his crucifix present to legalize the act. "**C**" shows men carrying a corpse in a perforated coffin;

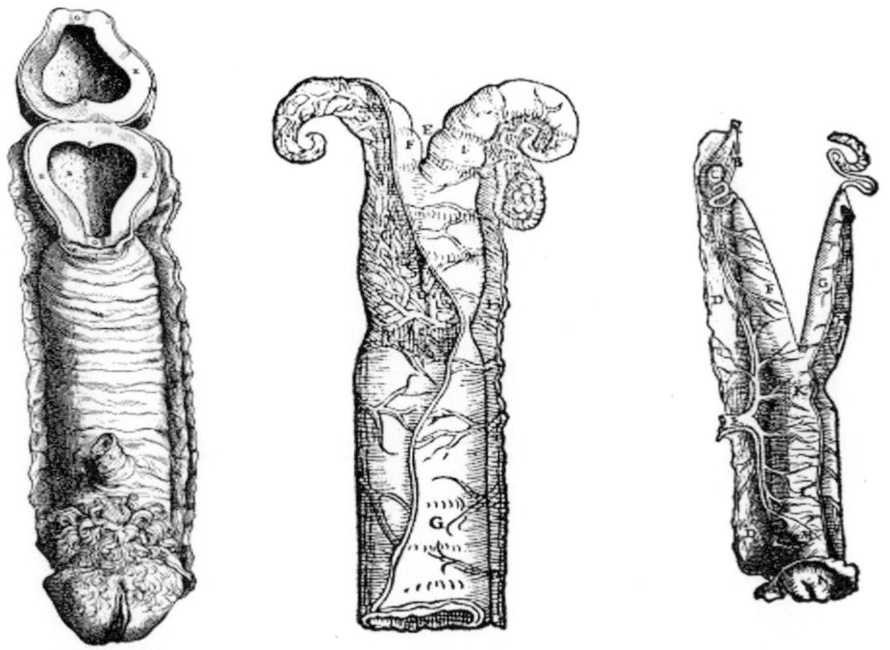

Fig. 3.58 *Fabrica*: uterus and vagina, an assemblage. Left: of a woman 'of very tall stature' who had often given birth. Middle: of a cow. Right: of a dog

they would place it in a stream to remove soft tissues from the skeleton "**O**" shows putti receiving a head from the executioner, with a soldier to safeguard the deed.

In Paris, *Galenus dixit* (Galen has spoken) was the sole permissible dictum and also Vesalius, when a student in Paris, was an ardent follower of Galen (Sirks and Zirkle 1964). However, as he became more familiar with the human body and began to notice that Galen had made major errors, he made it his main goal to overthrow Galen's authoritative anatomy and relieve anatomy of its blind and total faith in Galen's ancient books (Sirks and Zirkle 1964). Carefully noting down 200 of Galen's errors, he launched a full assault on him. Galen had said that the human breastbone consists of seven segments, but Vesalius noted that there were only three. Galen had claimed that minute pores connect the right and left ventricles of the heart, but Vesalius showed that no such pores exist (Aird 2011). Galen had claimed that the upper arm bone (humerus) is the longest bone in the human body, but for the femur, Vesalius noted that the shinbones (tibia and fibula) were longer. Galen had described the pre-maxillary bone and suture of the dog as though present in man, Vesalius illustrated both the skull of a dog and that of a man (Fig. 3.60) to show this was not the case and so on.

In this twilight age, Vesalius did take from Galen, however, large parts of his book. As an example, Galen had used piglets when conducting experiments on nerves that controlled breathing. During one of these experiments, on a piglet who

Fig. 3.59 *Fabrica*, historiated initials: **I**. Putti digging up a corpse at night. **L**. A female corpse taken down from the gallows. The priest with his crucifix is there to legalize the act. **C**. Carrying a corpse in a perforated coffin, to be placed in a stream to remove soft tissues from the skeleton. **O**. putti receiving a head from the executioner, with a soldier to safeguard the deed

was struggling and squealing while being strapped down, Galen accidentally cut nerves that innervate the larynx, whereupon the piglet immediately stopped squealing, though it continued to struggle. Galen next experimentally eliminated vocalization by cutting the laryngeal nerves in a wide array of animals (long-necked birds, dogs, goats, bears, lions, cows, monkeys), and he described cases in which humans had lost their voice, because of accidental injuries during surgeries to remove thyroid enlargements. This was the first publicly repeatable evidence that the brain controls animal behaviour. Vesalius, coming 14 centuries after Galen, repeated his pig squealing experiment (Fig. 3.61). Indeed, Vesalius accepted most of Galen's account of how the human body functions.

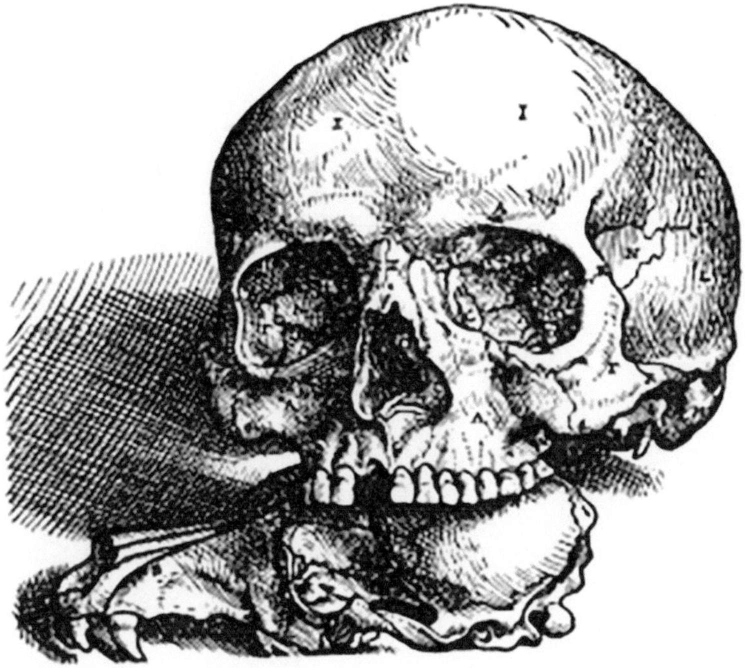

Fig. 3.60 *Fabrica*: "We have placed the skull of a dog beneath that of a man so that anyone may understand Galen's description of the bones of the outer jaw without difficulty"

Fig. 3.61 Vesalius' illustration of the dissection of a squealing piglet strapped down, before cutting his laryngeal nerves

Vesalius fell, however, into the same mistake as Galen had, of assigning details of animal structure to humans. For example, in describing the deep muscles of the human abdomen, he presented the straight muscles of the abdomen (*abdominis rectis*) as reaching up to the clavicles (Fig. 3.62). Now this is so in dogs and monkeys, but not in man.

Another example is Vesalius' error concerning the "marvellous network", the *rete mirabilis*. In ancient and medieval times, the method by which a mental process

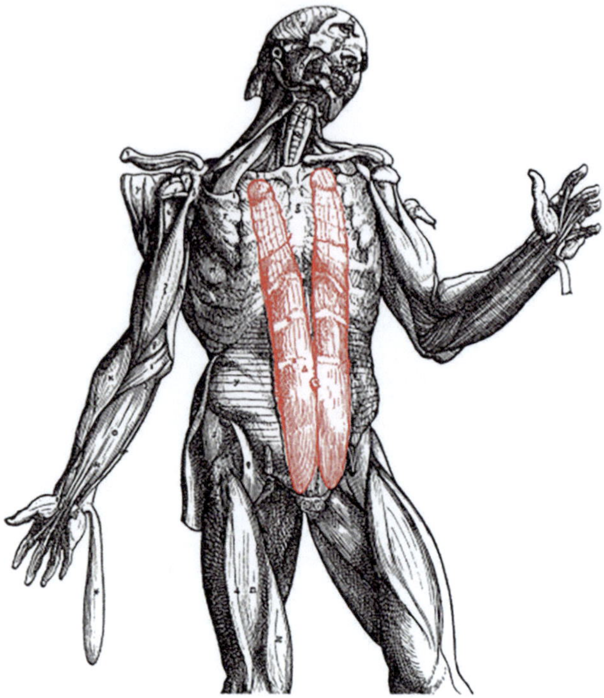

Fig. 3.62 Vesalius: deep muscles of the human abdomen. The abdominis rectus (red) reaches up to the clavicles, as found in monkeys and dogs, but not in humans

initiates and controls action was a mystery. In Galen's opinion, the *pneuma* of the outer world was modified into a vital spirit, which was supposed to come to the *rete mirabile* for transformation into the animal spirit, which nourished mental powers that produced voluntary motion. The *rete* was essential to this view, as it contained the central mystery of conscious life**.** Now the *rete* is a conspicuous feature in oxen, sheep and pigs, in which Galen had observed it; there is no trace of it in man. From his observations on cattle and pigs, Galen had erroneously generalized that it is present in man as well; and the existence of a *rete* in man was not doubted during the medieval. Vesalius, too, coming in the Renaissance, expressed no doubt regarding the existence and form of the *rete* in man. He even portrayed a reticular plexus at the base of the brain, as a circular organ from which two pear-shaped processes arise and pass into the brain (Fig. 3.63). When coming to treat the *rete* in the *Fabrica* five years later, he acknowledged that he had been describing what was not. "I cannot sufficiently marvel at my own stupidity... in no way do the carotids form a *plexus reticularis* [in man], as Gallen alleges" (*Fabrica* 1543: 642).

Another error concerns his illustrations of the human foetus, in which he illustrated the annular placenta of the dog as part of the human membranes. Some three years later, he excused himself, on the grounds that he had had no opportunity to examine the human foetus and that one of his teachers in Paris had informed him that the arrangement of the dog was also true for man; he corrected this and other errors in the second edition. Vesalius' sweeping generalizations from animals to

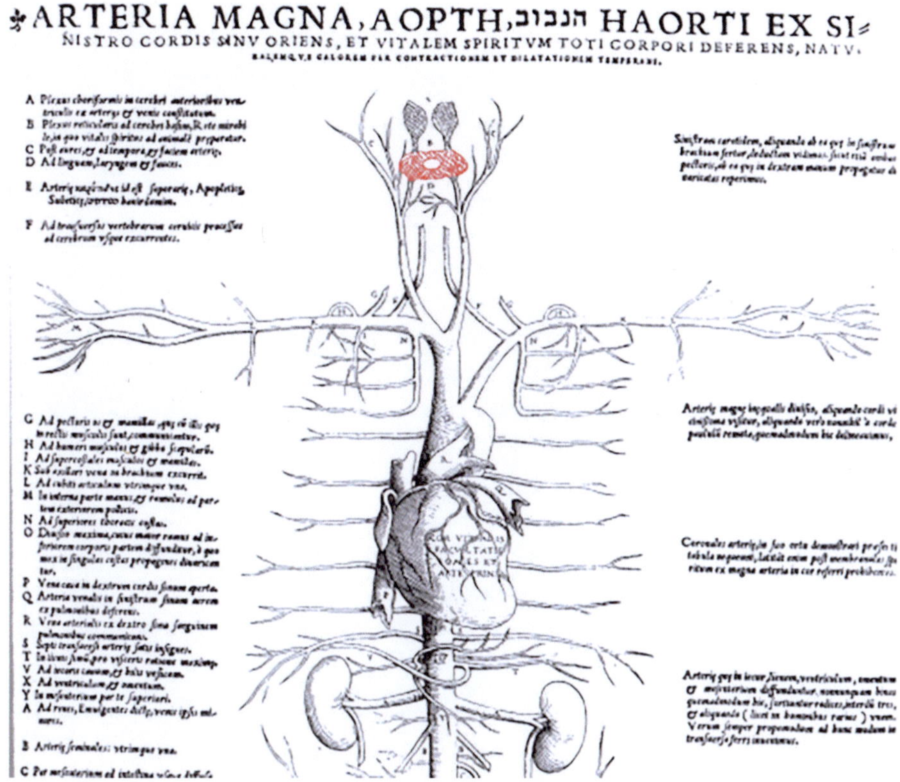

Fig. 3.63 Vesalius, erroneous illustration of a human rete mirabilis (red)

man are irritating, especially in view of the self-righteousness with which he accused Galen of this very fault.

Vesalius typified both a man of the Renaissance, seeking to glorify and enliven standards of the classical Greek and Roman cultures, and of the post-Renaissance, seeking to break away from the ancients and discover new horizons. He revolutionized descriptive anatomy in his great desire for thorough accuracy in detail, in accurately and precisely describing and illustrating his own findings and in his insistence on naming every structure.

It was perhaps due to the confusion between human and animal anatomy that comparative anatomy was born, a generation or so after Vesalius. Giulio Cesare Casserio (1552–1616, Latinized into Iulius Casserius; Fig. 3.64), from a poor family in Piacenza, moved to Padua as a servant in the house of Gerolamo Fabrici d'Acquapendente (Fabricius, 1533–1619), professor of Anatomy and Surgery, who encouraged him to pursue medical studies. Upon Fabrici's retirement, Casserio succeeded him in the chair of anatomy at the University of Padua. Casserio attempted to explain human anatomy by reference to other animals, and his *De Vocis* (1601; Figs. 3.65, 3.66, and 3.67), containing comparative studies of the vocal and auditory

Fig. 3.64 Casserius

organs, represents one of the sixteenth century's earliest investigations in comparative anatomy. In comparing the human vocal and hearing apparatus to those of other mammals, and of birds, amphibians, fish and insects, he described the complex sound-producing organs of the cicada, gave a detailed comparative account of the auditory ossicles and the mammalian osseous labyrinth, and noted that fishes lack an external ear, which is present in cetaceans. In humans, he described the auditory structure of the foetus, newly born, the child and the adult. In accuracy and artistry, Casserio's anatomical illustrations rank with those of Vesalius and, like the Vesalian illustrations, they provided a model and a standard for subsequent drafters. This was one of the first deliberate studies in comparative anatomy. The late Renaissance was a period in which the study of human anatomy, on the one hand, and animal anatomy, on the other, began drifting apart.

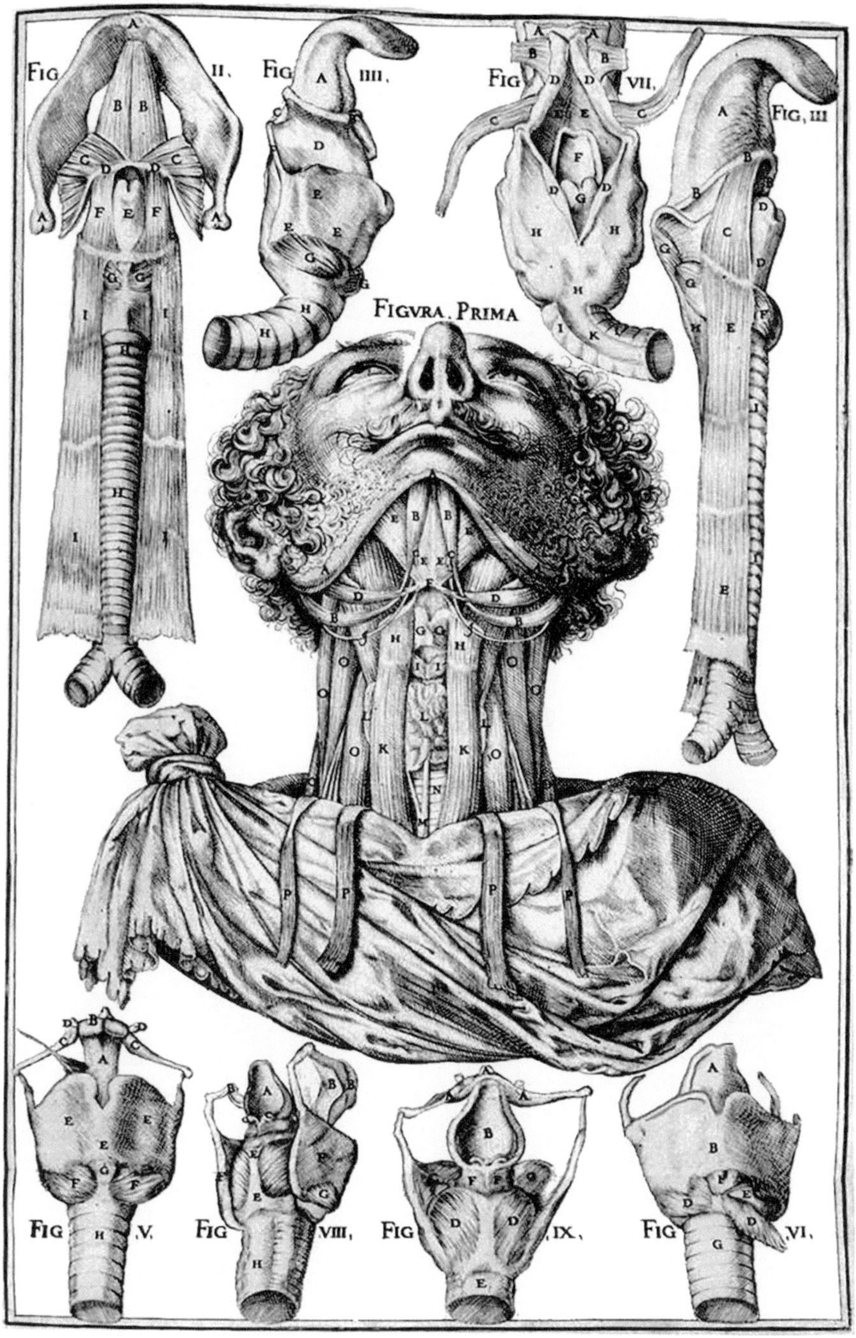

Fig. 3.65 Casserius, vocal organs in man

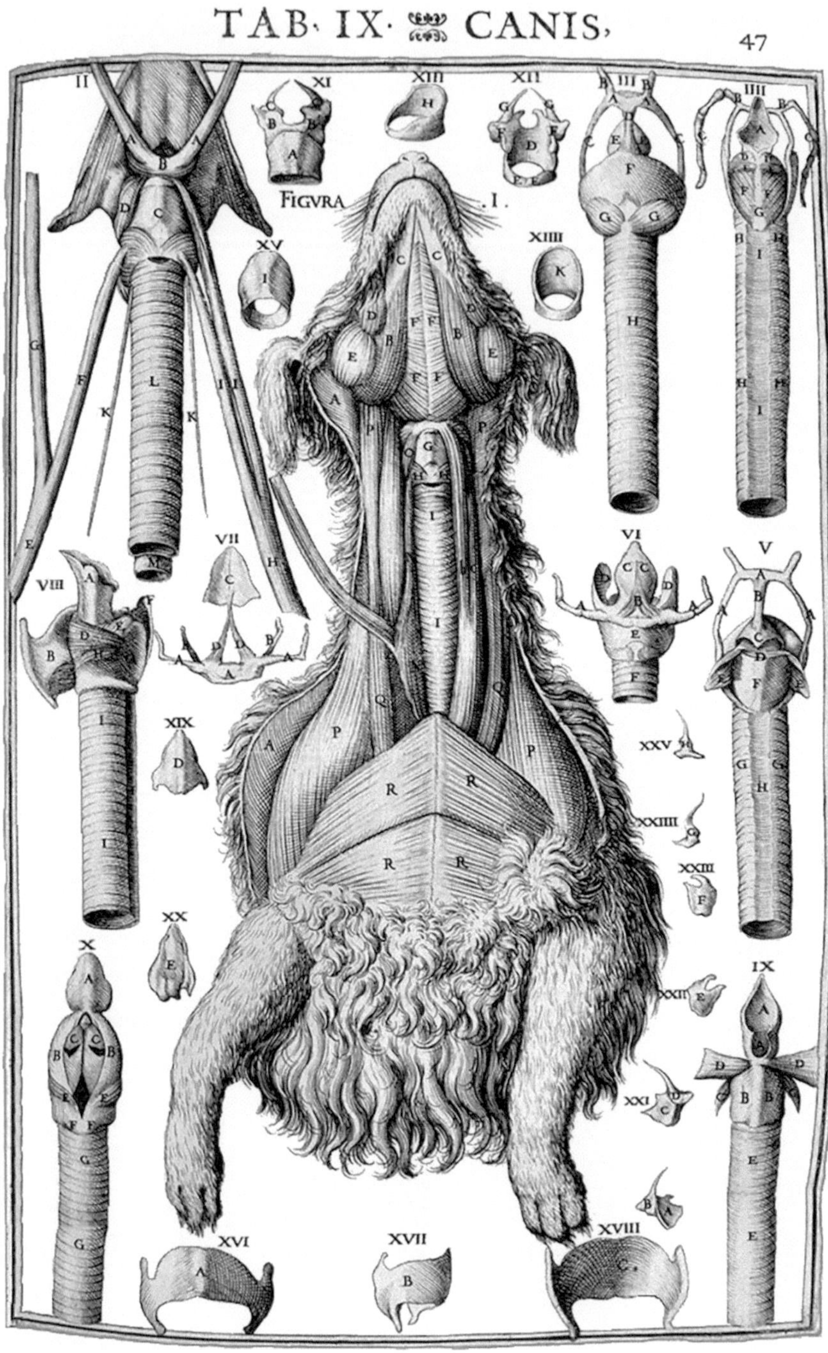

Fig. 3.66 Casserius, vocal organs in a dog

Fig. 3.67 Casserius, vocal organs in cow

References

Aird, W. C. 2011. Discovery of the cardiovascular system: From Galen to William Harvey. *Journal of Thrombosis and Haemostasis* 9 (Suppl.1): 118–129.

Beustern, J. 2017. The armadillo: Spain creates a curious horse to belittle America. *Bulletin of Spanish Visual Studies* 1:27–52.

Bitbol-Hesperies, A. 2006. Monsters, nature and generation from the renaissance to the early modern period: The emergence of medical thought. In *The problem of animal generation in early modern philosophy*, ed. J. E. H. Smith, 47–62. Cambridge University Press.

Breydenbach, B. 1486. Peregrinato inn Terram Sanctam.

Cain, A. J. 1994. Rank and sequence in Caspar Bauhin's Pinax. *Botanical Journal of the Linnean Society* 114:311–356.

Del Maestro, R. F. 1998. Leonardo da Vinci: The search for the soul. *Journal of Neurosurgery* 89:874–887.

Dolezal, A., Z. Skorepova-Honzlova, and K. Jelen. 2012. Leonardo da Vinci and Kethem – Kiveris vena. *Neuro-Endocrinology Letters* 33:732–735.

Findlen, P. 1994. *Possessing nature: Museums, collecting, and scientific culture in early modern Europe*. London: University of California Press.

Fludd, R. 1617. *The metaphysical, physical, and technical history of the two worlds, namely the greater and the lesser (Utriusque Cosmi, Maioris scilicet et Minoris, metaphysica, physica, atque technica Historia)*. London.

Ghosh, S. K. 2015. Human cadaveric dissection: A historical account from ancient Greece to the modern era. *Anatomy and Cell Biology* 48:153–169.

Gould, S. J. 2002a. Both neonate and elder: The first fossil of 1557. *Paleobiology* 28:1–8.

Gould, S. J. 2002b. Drawing a gloriously false inference. *Paleobiology* 28:179–183.

Hall, T. S. 1951. *A source book in animal biology*. New York: McGraw-Hill.

Hall, B. M. 1962. *The scientific renaissance: 1450–1630*. New York: Dover.

Houwen, L. A. R. J. 1994. Animal parallelism in medieval literature and the bestiaries: A preliminary investigation. *Neophilologus*, 78:483.

Jose, A. M. 2001. Anatomy and Leonardo da Vinci. *Yale Journal of Biology and Medicine* 74:185–195.

King, L. S. 1971. *A history of medicine: Selected readings*. Penguin.

Kusukawa, S. 2010. The sources of Gessner's pictures for the Historia animalium. Published online.

Lind, L. R. 1963. *Aldrovandi: On chickens, the ornithology of Ulisse Aldrovandi (1600) Vol. II, Book XIV*. Norman: University of Oklahoma Press.

Locy, W. A. 2005. Anatomical illustration before Vesalius. *Journal of Morphology* 22:945–987.

Mcmurrich, J. P. 1930. *Leonardo da Vinci the anatomist (1452-15-9)*. Baltimore: Williams and Wilkins.

Morris, A. G. 1986. On the sexual intercourse drawings of Leonardo da Vinci. *South African Medical Journal* 69:510–513.

Noble, D., D. Difrancesco, and D. Zancani. 2014. Leonardo da Vinci and the origin of semen. *Notes and Records of the Royal Society Journal of the History of Science* 68:1–12.

Nordenskiold, E. 1928. *The history of biology: A survey*. Alfred E. Knopf.

Ogilvie, B. W. 2006. *The science of describing, natural history in renaissance Europe*. Chicago: University of Chicago Press.

Pare, A. 1585. *On monsters and marvels*. Translated with an Itroduction and Notes by J.L. Pallister, 1982, University of Chicago Press, Chicago.

Park, K., and Daston, L. J. 1981. *Unnatural conceptions: The study of monsters in sixteenth- and seventeenth-century France and England*. Oxford University Press.

Saunders, J.B.deC.M., and C.D. O'Malley. 1973. *The illustrations from the works of Andreas Vesalius of Brussels*. New York: Dover.

Senter, P., L. C. Hill, and B. J. Moton. 2013. Solution to a 440-year-old zoological mystery: The case of Aldrovandi's dragon. *Annals of Science* 70:531–537.

Silverman, M. E. 1991. Andreas Vesalius and *De Humani Corporis Fabrica. Clinical Cardiology* 14:276–279.

Sirks, M. J., and Zirkle, C. 1964. The evolution of biology. https://www.cabidigitallibrary.org/

Smith, J. E. H. 2006. *The problem of animal generation in early modern philosophy*. Cambridge University Press.

Singer, C. 1925. The evolution of anatomy: A short history of anatomical and physiological discovery to Harvey. In *Being the substance of the Fitzpatrick lectures delivered at the Royal College of Physicians of London*. London: Paul, Trench and Trubner.

Singer, C. 1941. *A short history of science to the nineteenth century*. Oxford University Press.

Singer, C. 1946. Some Galenic and animal sources of Vesalius. *Journal of the History of Medicine* 1:6–24.

Sloan, P. R. 1972. John Locke, John Ray, and the problem of the natural system. *Journal of the History of Biology* 5:1–53.

Tubbs, R. S., S. Linganna, and M. Loukas. 2007. Jacobus Sylvius (1478–1555): Physician, teacher, and anatomist. *Clinical Anatomy* 20:868–870.

Vesalius, A. 1453. *De Humani Corporis Fabrica*. Basel.

Waetzoldt, W. 1950. *Durer and his times*. Translated from German by R.H. Boothroyd. *Phaidon*, London.

Walton, M. T., Fineman, R. M., Walton, P. J. 1993. Historical essay: Of monsters and prodigies: The interpretation of birth defects in the sixteenth century. *American Journal of Medical Genetics* 47:7–13.

Wilson, E. O. 1992. *The diversity of Life*. Penguin Group

Chapter 4
Early Modern Biology

Abstract Bacon saw matter as an entity that creates potential worlds by drawing atoms together, thereby forming compounds. Constantly changing matter, with inter-atomic connections that at first were brief, formed our world; gradually, the connections lasted longer, the world stabilized and we now randomly have certain compounds and not others. Not all things that could have formed did indeed form, and consequently, our world with its biodiversity represents only one of many abilities of matter. Furthermore, even today, many species are transient, unstable units that can become other species (maggots to flies). We should apply this secular theory to the creation of a new biodiversity for our benefit. Man should deliberately transform species into other ones by building new combinations and changing forms. Monsters and hybrids hint at the hidden potential of nature, and their study is, therefore, important. We should create new species through parentless generation and then let them develop sexual generation to reproduce. Bacon thus envisioned man as leading drastic diversions of nature's biodiversity. He demonstrated this vision by describing a utopian, government-supported research institute, which improves diversity by developing formulas by which materials would be mixed to obtain specific creatures.

Harvey considered embryonic development as potential realized into form. The egg of the farmyard hen forms and grows by an innate principle that flows from the female's entire body into her ovaries as innate heat. Yolk and egg white are matter that, after fertilization, gain form by nourishing the foetus, first the yolk and later the egg white. The cock contributes to fertility by transferring an invisible, immaterial entity to the egg, a divine spirit that conveys fertility. Semen also transfers the male's hereditary traits, as an idea that the uterus "sees" and transforms into matter. The fertilized egg then forms a worm-like embryo, and accordingly, in the embryonic development of complex animals, the early stages represent simple animals. Beyond matter, which eventually gains form, the fertilized egg also has a soul, defined as the ability to produce offspring of the same species. The Sun, as an instrument of God, also influences fertilization. The male receives generative power from the approaching sun and therefore from God. Blood begins life, and since it is the first to appear, it contains the soul. Form development continues well after hatching, in forming crests, combs, wattles and spurs.

J. Heller, *Nature's Biodiversity*, https://doi.org/10.1007/978-3-031-89790-0_4

Harvey studied mammal generation and embryological development in the king's deer herd. He observed satellite copulation, namely, non-dominant young males secretly mating with the dominant male's females. In search of semen and some beginnings of life, he dissected the uteri of mated females daily. Finding nothing, he concluded that no conception occurs immediately after coitus. Seven weeks onwards, however, he found pregnant hinds with embryos in the uteri. He generalized that, in both viviparous and oviparous animals, the foetus does not arise from sperm emitted during coitus. He rejected the involvement of ovaries in fertilization.

Harvey presented a new definition of the Egg, as the very first stage of embryonic development (the zygote, in our terms). His Egg was an inanimate lump lacking a defined structure, in which the foetus was already present. Life develops from a lifeless, inanimate lump lacking a definite shape. Any primordium from which a living creature generates is thus an egg, whether in the form of a seed or a worm. Mammals produce something resembling an egg, just as plants produce seeds and trees produce fruits. At first, nothing materializes in the uterus; eventually, the "egg" appears. Scientists discovered mammalian egg cells only two centuries after Harvey.

Epigenesis, according to Harvey, refers to developing embryos that attract and use material as it comes in. Formation and growth occur simultaneously, and successional embryonic parts differ in matter and form. An embryo progresses from an unorganized, non-functioning body to an organized, functioning one, gradually growing and forming itself, each stage developing into the next. Epigenesis occurs in (the more perfect) vertebrates, which require well-determined matter for growth. The heart, as an instrument of the soul, directs the successive development of the embryo. Metamorphosis, for Harvey, includes both sudden, profound changes in form (insects) and sudden birth directly from matter by spontaneous generation. All matter for a developing embryo is present from the beginning. Consequently, embryonic matter is homogenous, and not much growth occurs. In these lesser animals, created by chance, species are less constant. An egg is a potentially living primordium, whether perfect in the womb, where it reaches final size when laid (birds), or imperfect as it is released immaturely and continues growing after having been laid (insects). From imperfect eggs, imperfect animals ensue (worm, caterpillar) that change into a chrysalis (pupa), which is the perfect, true egg; from it, the prematurely hatched embryo is reborn as an adult.

Descartes envisioned the universe as an enormous clockwork, moved by fixed laws, and animals as complex machines that automatically respond to various stimuli, mindlessly following mechanistic rules that move them towards objects to obtain food and generate a way to avoid danger. They have no souls, and accordingly, they are lifeless, self-growing, sometimes self-moving machines devoid of mind, consciousness, sensation, passion, judgement, reason or intelligence. Though acting in ways resembling man, they cannot think. Descartes thus eliminated Life as a category of nature, drawing the line not between living and non-living but between those with a soul (man) versus those without. Man differs from animals in possessing a soul (= mind) that thinks logically and consciously, using senses, images, memories and language to express ideas. The soul, in the pineal gland, is a non-material entity that does not follow the mechanical laws of nature. Descartes' followers added that animals, having no soul, cannot suffer and therefore do not suffer pain. God created the world along eternal mathematical and mechanical laws that now restrict Him—He cannot break them. Today, God behaves only according to mechanical principles; he has no power to change the course of nature. Consequently, the cosmos is a meaningless entity.

During early modern Europe, the detailed study of animal diversity began extending beyond the vertebrates into invertebrate groups. The study of mollusc diversity began with shell classification, the "shell-animals" group including also goose barnacles and tube-dwelling worms. Octopus, squid and terrestrial slugs were, however, included, though without a shell. Shell animals did not include sea slugs, chitons and ammonites, which naturalists regarded as inorganic rock ingredients. Shell classification was global from the start, with each shell represented by a figure, a describing sentence and the locality. Several groups were presented in increasing order of spire height.

Terrestrial arthropod and worm diversity is classified by leg and wing morphology, readiness to fly and "nobility". Socio-political yearnings for a constitutional monarchy trickled into ideas of bee natural history: the hive ruler is an elected monarch; his subjects honour him but keep ancient privileges. He prescribes laws, imposes missions, determines the hive's timetable and leads the army in battle. He is merciful but intolerant of rebellious bees. Kings form a royal race in which princes are born winged, while commons hatch as worms. Bees generate spontaneously from various carcass parts: kings and princes from brains, common bees from flesh. Drones are weak and impotent.

Microscopes first came into biological use in the 1620s, and they served to study the microanatomy of animal organs (capillaries, renal glomeruli, skin microstructure, insect tracheae and leaf pores). They also served in the study of the external morphology of tiny creatures, displaying the complexity of "simple" animals compared to man-made objects. A major contribution was in discovering the vast diversity of unicellular creatures and comprehending their tiny size and huge numbers. Microscopic animals occur in water, soil and inside big animals; some survive years of desiccation in a dried state, "returning to life" when returned to water. By this, temporary pools are colonized, and not by spontaneous generation. Microscopes also enabled the discovery of sperm, which some naturalists then believed to be minute (male and female) organisms. Sperm occur in the uterus after mating, and they contain the potential for forming all the limbs an animal will have at birth.

An association of British intellectuals sought to establish a society devoted to the pursuit of experimental knowledge, and Charles II granted royal charters to establish "The Royal Society of London for Improving Natural Knowledge" (1662–1663). These charters espoused many of Bacon's precepts and objectives, his vision of future science thus becoming a reality. Beyond lectures, discussions and attempts to base knowledge on observations and experiments, the Society also provided a continuing international forum for the practice and development of science through sponsorship of the *Philosophical Transactions of the Royal Society* (1665). By these means, scientific communication became more dynamic. The *Transactions* developed a new scientific outlook: articles were to contain no high speculations and no moral lessons; discussions were to be pragmatic interpretations of the results. Through the new science emerging from the *Transactions,* the study of nature gradually became secular. *Transactions* also established principles of scientific priority and peer review. The editor became an arbiter, with authority to discredit observations. The Royal Society of London also actively approached scientists to study and publish on specific matters.

Several early modern naturalists believed in spontaneous generation. Inferior beings arise from rotting matter by seminal entities that drop from dying plants or animals. Traces of a Force of Life remain in carcasses and generate new living creatures, similar but inferior to the original animal. Animal faeces generate insects, which resemble the animal in whose excrement they were born. Bees are born from cattle manure, and their head resembles that of a cow; stag beetles are born from stag manure, and they resemble a stag's horns. Animal diversity also consists of many strange hybrids, which develop into dragons.

Eventually, careful observations of mosquito life cycles led to doubts about spontaneous generation, and careful experiments found that rotting meat does not generate maggots. There is, thus, no spontaneous generation. Consequently, creatures of one species do not transform into another species—the species is a stable, unchanging entity. Paradoxically, naturalists also believed that insects are born from plant galls and that the production of worms and winged animals may be the primary purpose of a plant, fruit production being only secondary. In all spontaneous generations, however, life arises only from matter that was once alive; it never arises from inorganic matter.

Swammerdam opposed Harvey's theory of the pupa being a second egg. He believed that insects develop gradually, like vertebrates; being as complex as higher organisms, they reflect nature's unity. The "nymph" is a life stage in which insects have reached adult dimensions. Insects vary in the timing at which they reach it, and this is the basis of their diversity. Four "orders" correspond to different developmental life histories:

1. A miniature version of the wingless adult emerges from the egg and increases in size throughout succeeding immature (nymph) stages.
2. A miniature version of the winged adult emerges but lacks wings, which develop throughout successive nymph stages. Functioning wings and genitalia appear in the final moult.
3. Adult structures (wings, legs, antennae) appear gradually within the nymph, before pupation. The pupa, a transitional stage between larva and adult, has internally developing wings and modified larval eyes. Nymph state is reached during pupal development.
4. Metamorphosis involves the use of larval cuticle to form a case, the puparium.

Spontaneous generation is blasphemy, as it implies that part of nature exempts itself from God's law and order. His laws of nature are immutable, apply throughout the Creation, and there are no chance events.

During the sixteenth century, naturalists debated the essence of fossils. Some claimed that so-called fossil shells are not real seashells but inorganic minerals, growing inside the Earth under the inspiration of mystic powers. Steno found that modern shark teeth resemble fossil ones, and he claimed that rock, once liquid, had solidified around these solid ancient shark teeth. Furthermore, heavy particles sink before fine silt, and similarly, in fossil-bearing deposits in nature, big particles occur below fine ones, suggesting that water had laid them. This explains why fossils in rocks are of aquatic, not terrestrial, creatures. Steno further argued that lower layers are older than higher ones, and so are their fossils. Seashells reached mountaintops by water rising to cover land or by land sinking beneath water. The Earth's crust, an archive of its ancient history, presents evidence that the world had undergone drastic changes. Steno's theories were not well received, both for personal animosity and because they were not relevant to ammonites, which do not resemble shells of any living creature. On religious grounds, naturalists rejected the possibility that ammonites represent extinct animals.

Ray formulated a reproductive definition of a species, as opposed to definitions of essence. Classifying bird diversity by habitat, feeding mode and morphology, his resulting groups resemble many avian families that we accept today. Aware of natural history, he described anti-predator distraction displays and injury feigning. Ray rejected Epicurus' theory of randomly colliding atoms as unable to explain the perfect structure of plants, animals and man. He also rejected Descartes' theory of the mechanics of matter in motion as explaining everything in nature—it does not explain plant and animal diversity, nor why one species cannot form from another. Descartes' claim that animals lack a soul is also wrong; animals possess rational thinking. God's will to sustain each species is what rules the world, not mechanical laws. God created Earth as a planet intended just for Man, with the endless diversity of plant and animal species created specifically for his benefit, whether today or in the future. The perfect design of man and animal is evidence of God's wisdom. This argument, of God's intended design for a specific purpose, stems from the claim that all phenomena fulfil specific purposes and that an all-knowing entity, God, creates each phenomenon.

4.1 Improving Nature

4.1.1 The Grand Vision: Man-Made Biodiversity

By the seventeenth century, the Renaissance had pretty much run its course. The upheavals of the Protestant Reformation had mostly subsided, but the Age of Enlightenment was barely on the horizon. During the seventeenth century, biology (as we term the discipline today) advanced on a very broad front by seeking to develop empiric rather than scholastic knowledge and to expand knowledge into applied fields. It also began developing organized institutes for carrying out research.

Within the discipline, morphology progressed from a rather descriptive to a more functional level, focusing mainly on the blood system and the heart. Scientists used an experimental approach to explore the possibility of spontaneous generation, and some philosophers argued that animals are merely mindless, complex, organic machines, obeying fixed mechanistic laws. Naturalists sought to understand why, in some animals, embryonic development is gradual, whereas in others it involves an abrupt transition from one growth phase to another. The development of the microscope enabled both the exploration of animal organs in minute detail and the unsuspected discovery of unicellular creatures, previously invisible.

Francis Bacon (1565–1626, Fig. 4.1; "knowledge is power") was a British statesman who served as a member of both houses of Parliament for over 30 years. He was also an advisor to Elizabeth I and Attorney General and Lord Chancellor of England to her successor, James I. His life at the court also involved duties beyond medial practice, such as the investigation of witches in Lancashire (11 went on trial, ten were hanged), the autopsy of a man who had lived to the age of 152 years (his testes were sound and he was sexually active until beyond the age of 100) and participation in diplomatic delegations (Franklin 1961). At the age of 55, he was impeached by Parliament and stripped of all public office for taking a bribe, following a political conspiracy in which he was used as a scapegoat to protect Lord Buckingham (of *The Three Musketeers* fame). An erudite man, Bacon wrote on issues of law, state, religion and policy, as well as on the possible structure of an ideal society and issues of ethics (*The Advancement of Learning* 1605, *Novum Organum* 1620, *The New Atlantis* 1626, *Sylva Sylvarum* posthumously 1629 and others). He argued that the only knowledge of importance is that which one could discover by observation—"empirical" knowledge rooted in the natural world, rather than scholastic rephrasing of ancient scripts.

The impressive title page of his *Sylva Sylvarum* depicts his ideas (Fig. 4.2). It presents two Corinthian pillars representing the Pillars of Hercules—the cliffs flanking the Straits of Gibraltar, the limits of the world in ancient times. The seascape beyond them represents the infinite sea of knowledge that awaits us to discover and understand it. Between the pillars is a globe inscribed with "the world of knowledge". A ray of light, originating in God's name, illuminates the upper part of the globe. Below it are the biblical words "And God saw the light, that it was good". The image of sailing and discovery is clear: in sailing beyond the boundaries set by the ancients and into the world of light, Bacon and his followers would map the world of knowledge. The cherubs present proximity to divine knowledge: the philosophy of nature is to progress, together with the belief in Creation.

His *Sylva Sylvarum* (*Trees of the Woods,* 1629) is a sort of encyclopaedia. It consists of a thousand paragraphs, each stating some fact, a titbit of knowledge condensed from books. Some facts offer experiments; others are observations. Yet others are musings on a variety of topics ranging from acoustics to botany to the essence of fire and the like, thoughts of ancient philosophers (Aristotle, Pliny), of contemporaries (Galileo) and experiments of Bacon himself, combining his thoughts with the sayings of the ancients. Only a few of Bacon's experiments focused on

Fig. 4.1 Bacon

biology. He repeated the old claim that the chameleon lives off air and that the salamander lives in fire. There is no evidence that he himself had experimented. *Sylva Sylvarum* was widely read in the seventeenth century; it went through 16 editions in just 2 years (Gaukroger 2001).

Fig. 4.2 *Sylva sylvarum*, title page

We recall (this book, parts 2–3) that in seeking to understand the creation and structure of the world, philosophers of mediaeval times believed that during Creation, everything flowed out of God as a series of spheres, one enveloping the other and thereby forming the series of spheres of which the world consists. In the late Renaissance, as we have seen, Fludd believed that spheres of animals, plants and minerals surround the Earth and that Man-the-Creator imitates God's creation by creating things by himself. Bacon was of the same generation as Fludd, and he published his (different) views on the structure of the world a decade or so later.

To explain the creation and structure of the world, he proposed the existence of one single principle: matter. He also undertook to explain the myth of Eros, the ancient Greek god of passion (Cupid in Latin), who, he believed, embodies the atom, and he thus symbolizes both matter itself, as a substrate, and the atom's potential force of combining with other atoms. Eros/Cupid participated in the creation of the world, and to this day, he still participates in its maintenance. Bacon believed that matter has the power to create potential worlds. The beginning of all movements is in the movements of the atoms. Creation is a process involving a realization of the force of Cupid, who draws all atoms together by adding, joining, building and fusing at certain points and not others, resulting in new bodies that are completely different from the atoms that compose them. The result is (what we would term) elements and compounds that attain a stability that is only relative, in that they involve potential, conflicting movements. No element has absolute stability. The different atom-consisting compounds are endlessly battling each other in what seems to be a constant effort to win. The stronger entities bind the weaker ones into apparent inactivity, into a momentary state of tension. We can, Bacon wrote, compare this tension to a man pinned to the ground in a wrestling match: although his opponent holds his hands and feet in place, he tries with all his might to get up, and his resistance is not less because he does not succeed. Change occurs when movements do not bind in such dynamic tensions: growth, decay and death. Bacon thus saw matter as an essentially active, creating entity. We recall Lucretius of ancient Rome, who had suggested, following Epicurus of ancient Greece, that everything in the world is made of minute, invisible particles constantly in motion, clashing with one another, coming together to form new shapes, coming apart and recombining again; and that all living beings, from plants to animals to man, have formed through this random process.

Regarding the creation of the world, Bacon suggested that in the beginning, there was an uneven, constantly changing lump of matter, a dark, mysterious disorder—and there was chaos, namely a situation of constant general change, in which matter lacked stable form (he may have had in mind a semi-fluid volcanic mass). Cupid, the god of passion who attracts opposites, was born on the very day on which this initial chaos had formed. When he first emerged from the darkness, he was very weak, and so at first, no stable compounds formed, and whatever connections formed immediately disintegrated. Cupid gradually grew stronger; his uniting force increased, connections became longer, and so, eventually, Earth was formed. However, there remained in matter a constant desire to move and to change, a tendency symbolized by the god Saturn. Saturn's destructive tendency was suppressed because of Cupid's power of attraction. The stability of the world is therefore dependent on Cupid's ability to vanquish Saturn in the constant battle raging between them. From the moment the present world attained a stable general form, only local changes occurred, and matter remained mostly unchanged. This transition, from general to local change, is the transition from chaos to order.

In this theory of the creation of the world, Bacon was echoing the ancient Greek sages and their theories. We recall Empedocles, who regarded love-hate and the attraction-rejection of these forces; Democritus and his theory of atoms, which are eternal, but whose cosmic systems, the worlds, are transient; and Epicurus, with his theories regarding worlds as always forming and disintegrating. Actually, the roots of Bacon's creation myth go even deeper, to *Theogony*, a poem by Hesiod composed about 700 BCE, describing how the Greek gods came to be and how they established permanent control over the cosmos. In Hesiod's mythical cosmogony, as in Bacon's, the initial state of the universe was chaos, a dark, undefined void. Then came Gaia (Earth), a more orderly and safe foundation that would serve as a home for the gods and mortals, and Eros, an archaic god who represented sexual desire and the urge to reproduce sexually. Only later did Gaia give birth to Uranus (Sky) and Pontus (Sea). Later still, Kronos (Time) castrated his father Uranus, and eventually, Kronos' son Zeus overthrew him and gained control of the world. As for Saturn, in ancient Roman mythology, he was the god of Time (identified with the Greek Kronos, and Jupiter was called the son of Saturn). He was both a god of destruction and dissolution and a god of generation, plenty, wealth, agriculture, periodic renewal and liberation. Bacon's creation theory was thus deeply rooted in ancient mythology.

A key point in Bacon's theory of Creation was the possible existence of an alternative world: in the original creative process, of harnessing the power of matter, beyond things that did indeed form, there were things that could have formed. There are endless possibilities for alternative associations and combinations of matter. Our world, with its natural biodiversity, represents only one facet of the potential biodiversities, and this opens the door to man-made, applied diversity (Wilson 1995). Nature, with its current species, represents only one of many abilities of each species. Man should exert force on these natural species to change and rebuild their form. Freaks and monsters hint at the hidden potential of nature, and therefore, they are of major importance in building up a man-made biodiversity. In England of those days, monsters were of considerable public interest, with discussions as to whether hermaphrodites could marry, whether both heads of Siamese twins deserved baptism and in what form they would be resurrected. Furthermore, books of travel described such monstrous tribes in faraway lands as men with a single giant foot, men with huge ears, or without heads, or possessing horns, or giants or dwarfs (Park and Daston 1981). We must make a collection of all monsters in nature.

According to Bacon, there are hints of the hidden aspects of nature in natural aberrations, in what we might call monsters, namely animals or humans born with some severe birth defect: a two-headed cow, a chicken with two sets of wings and the like. These aberrations "leads the mind from what exists to what could exist"; they represent the dormant force of matter that can burst forth even in the existing natural system, testifying to the fact that the fertility of nature extends far beyond its regular course. We recall that people, during the Mediaeval and the Renaissance, had considered monsters to result from some sinful sexual behaviour by the mother or her state of mind. The cravings and experiences of an expectant mother, especially those arousing strong passions of lust or fear, inscribe themselves directly on the body of the foetus, thereby producing its deformities. Monsters were also signs of God's wrath, an omen of future misfortune. We recall also the detailed monster descriptions of such Renaissance people as Gesner, Paré and, to a lesser extent, Aldrovandi.

Hybridization is another source of creating novelty for developing rare and unusual creatures. There are many intermediate, "hybrid" creatures in nature, such as moss, which is between putrefaction and plant; comets, between stars and blazing meteors; flying fish, between birds and fish; bats, between birds and quadrupeds; and apes, between monkeys and Man. These intermediates hint at the possibility of one creature changing into another one, even a very distant one (*The New Organon* 1626).

The issue of species transmutation fascinated Bacon. He believed that insects are imperfect animals in that they lack blood and bones and in that they generate spontaneously from putrefying matter. Furthermore, he believed that some insects change their species: in summer, they change from maggot to fly and in winter from fly to maggot (obviously, an error). This issue of species transmutation has a long history. In ancient Greece, we recall, Theophrastus had claimed that plants change their species when transported to different soil and climates and that seeds of cultured species germinate and revert to their wild state. In the Mediaeval period, with the rise of Christianity, the belief in conversion from one species into another took spiritual directions: the species of a given organism is its exterior appearance and not its essence or inner spirit; it is the body of the organism, not its soul. (*Specere*, from which comes our term "species", means to see, as in "spectacles"). The species of man is his transient physical appearance, unlike his immortal spirit, and the werewolf is a man by day and wolf by night. At a different level, in the bread and wine ceremony, the species of the bread and wine do not change—chemically they are still bread and wine, yet their inner essence transforms into flesh and blood. Some species were thus considered to be transient and to change in time and space, as circumstances required.

Bacon, in line with this, believed that throughout nature, many species are transient, unstable units that, if forced to do so, can become other species. In plants, the seed, by enclosing the embryo, restricts it and prevents it from changing. However, plants also grow from soil that (he thought) lacks seeds, and this suggested to him that plants could convert from one species to another. Any person can transform a plant from one species to another by adhering to some basic rules. We should create unfavourable living conditions so that the forces of the food in the environment would overcome the (hereditary) forces in the seed: we should plant marsh plants on hilltops and hill plants in marshes. Next, we should plant the seeds together with other species of plants. Then, we should mix plant leaves or roots of other species into the soil. We should force the plant to develop contrary to its nature; thus, we should force a creeping plant to climb. We should keep the experiment away from light, in a barrel placed at the bottom of a lake, or inside a hollow tree trunk or at the bottom of a cave or hung down a well. By adhering to these rules, any person would be bound to see with his own eyes how a plant of one species converts into another one. True innovations are beyond the routine course of nature, and man should deliberately create constraints to produce them. He should use his own force on matter to torture it and thereby force it to change its form, to take it apart and then rebuild it. Bacon thus envisioned an entirely alternative universe in which human sovereignty over nature would be restored to that biblical power that Adam had had when, by giving animals their names, he had power over them (Gaukroger 2001).

Elsewhere (*The New Organum* II), Bacon called for the study of not only form and function but also of processes that create forms. Studies of plant growth should involve continuous observations from the moment we sow a seed to learn how and when the seed begins to bulge and swell, when it begins to break the husk and put out shoots, some downward for the roots and some upwards as stems and so on. He also suggested studying the hatching of eggs to view the very beginning of life and the processes of shape formation: which organs come from the yolk, which from the egg white and so on. He furthermore encouraged the study of animals arising from putrefaction. Indeed, he suggested keeping a kind of round-the-clock watch on everything in nature.

For Bacon, monsters represented what could occur if we forced nature to divert from its course. Bacon would have said that if chickens can occasionally have three legs, then man could deliberately force nature to produce a breed of three-legged chickens. Three legs—is this the limit, or can we produce chickens with 20 legs? The possibilities are endless. It would be an easy transition, from observing the wonders of nature as expressed by extreme variation within a species to the intentional cultivation of such wonders by forming new breeds and, thereby, a new diversity. He accordingly adopted the study of monsters as part of his vision for a better world. He advised that the first step in advancing knowledge should be to compile a list of all monstrous births, indeed of every abnormality in nature. Such a list must, of course, be credible, so it must contain, as already suggested by Paré, only reliable reports (*Novum Organum* II: 29). Reports of monsters were to identify the authority or witness from whom the description was originally obtained, to assess the reliability of the source, to state how the source had come by the information (eyewitness, oral or written) and to judge whether additional corroboration was required (*Preparation for a Natural and Experimental History*). Furthermore, lists of monsters should not include cases known only from religious stories, as these are not true and not natural. In line with his call to study monsters, Bacon also urged the study of those rare and extraordinary creatures that share two worlds in that they have the morphology of one group but the habits of another. Such would be flying fish, between bird and fish; bats, between bird and quadruped; apes, between man and animal; moss, between decaying matter and plants (and comets, between stars and blazing meteors; *Novum Organum* II). Bacon, to conclude this point, puts emphasis on the unusual and the anomalous: we can learn more from exceptions than from general rules.

Bacon outlined his vision for the future of human discovery and knowledge in a utopian novel (*New Atlantis,* 1624) that describes a crew of a European ship that, lost in the Pacific Ocean somewhere west of Peru, happens to discover the utopian island of Bensalem. The inhabitants of this island possess qualities of generosity, enlightenment, dignity, splendour, modesty, piety and public spirit, and they exceed Europeans in terms of material wealth, medicine, technology and general education. The natives of the island have a government-supported research institute, Solomon's House, that observes nature to understand it theoretically and then to improve it. To achieve this, each member of the institute does not act individually but is part of a system of teams:

- A team that collects literature and background materials.
- A team that collects experiments from background literature.
- A team that conducts new experiments.
- A team that organizes the results in tables.
- A team that drew conclusions from the experiments.
- A team that integrates these conclusions into a general theory.

To achieve its goals, Solomon's House has instruments to create heat, sound, movement and the effects of light. There are also parks and enclosures for observing and dissecting animals, kept so as to test various poisons and medicines. In these parks, the animals are cultivated to become taller or smaller, more fertile or more barren and of different colours, shapes and patterns of activity. There are also orchards and gardens for studying the effects of different soils on plant growth, for grafting and developing new plant strains in which flowers appear earlier in the season than in nature and for producing larger and tastier fruit. The island also has kitchens for developing new foods and drinks and shops of medicine that offer a large variety of plants and living creatures. The scientists of the House of Solomon have the means to make diverse plants germinate from mixtures of Earth without seeds and to make one kind of plant turn into another. They also create meteors, snow, hail and thunder and lightning and can generate frogs and flies. They also have the means to hybridize different animal species to produce new kinds. Furthermore, they generate various kinds of serpents, worms, flies and fishes out of putrefaction. Once created, some of these parentless animals advance to a higher level of their animal being by developing sexual generation. They copulate and propagate as do "perfect" creatures, such as beasts or birds. There are formulas for which materials to mix to obtain each of these parentless, a-biogenetically created animals. *New Atlantis* was thus a vision of biology and nature for the benefit of man.

Bacon was keenly aware of the vast options that the recently invented microscope opened. He wrote (*The New Organum* II) that by increasing the size of a specimen, the microscope could reveal the structural complexity of such tiny creatures as the flea and the fly, tiny creatures whose subtle structures are bound to be far more complex than those of any artificial texture of man-made arts, such as linen cloth; it could also discover irregularities of gems, liquids, urine, blood, wounds and many other things (Bacon 1626, *The New Organon*). Four decades after his death (as we shall see further on), Bacon's followers closely followed this list, from cloth via gems to urine.

Despite his closeness to James I and his persistence in trying to obtain a hearing for his projects, Bacon's ideas were never taken seriously by the king or by anyone in a position to realize them (Gaukroger 2001).

4.1.2 The Royal Society of London

The Royal Society of London grew out of an informal association of a dozen intellectuals keenly interested in science, who sought to establish a society devoted to the pursuit of experimental knowledge. In 1662–1663, Charles II (Fig. 4.3) granted two royal charters to establish "The Royal Society of London for Improving Natural Knowledge". The purpose of the society, as stated in these charters, was to promote and carry out experiments and encourage argument, discussion and the dissemination of scientific knowledge; its motto, *Nullius in Verba* ("take nobody's word for it", Fig. 4.4), symbolized the determination of the society to base knowledge only on observations and experiments, not on authority. Bacon had preached that one should admit nothing but on the faith of eyes. The Society's charters explicitly espoused many of Bacon's precepts and objectives, such as gathering collaborative information on "what Nature does willingly, what constrained; what within its power, what by the succours of Art; what in a constant rode, and with some kind of sport and extravagance...". The "productions and rarities of Nature and Art" were to be studied with the aim of censoring errors and discovering useful information. The founders of the Royal Society advocated a communal approach to learning, citing Bacon's view that if the company heightens the emotions, it must do the same for the intellect. Bacon had demanded that reports of the monstrous and unusual should

Fig. 4.3 Charles II. (By Tuvia Kurz)

Fig. 4.4 Emblem of the Royal Society of London

list names of witnesses and particulars of time, place and circumstances, with descriptions and illustrations drawn from life. Bacon's vision of future science thus became a reality. The king himself took a lively interest in science and enjoyed experimenting in his own laboratory. Bacon's symbolic role as a philosophical founding father, based on the groundbreaking inductive method, is aptly summed up in the celebratory ode with which a certain Abraham Cowley prefaced an early *History of the Royal Society* (1667). He described Bacon as the saviour of mankind from the long, erroneous paths in which its predecessors went; it is Bacon who, like Moses, led science across the barren desert; it is Bacon who, like Moses, reached the border of the blessed Promised Land (of knowledge), saw it and showed it to us. For the founding members of the Royal Society of London, Francis Bacon was the "Father of Modern Science", a patron saint.

The Royal Society of London sought to advance its goals through formal meetings and through informal networks that it fostered. At its formal meetings, there would be a lecture followed by a discussion on the significance of the subject for the advancement of humankind, as befitted the spirit of Bacon's vision. Such was, for example, one focusing on the flight of birds (February 11, 1674, the protocol).

A lively debate followed the lecture on the possibility of man himself flying. Members of the society discussed whether one could slow down the movement of things shot by gunpowder, just as a shell, which, when flying through the air, is slower than a bullet shot from a musket. Some members noted that it would be very beneficial to humankind if one could shoot a person so high that he could fly over a wall so that people in a besieged city could collect and pass on information to other forces.

The Royal Society of London also provided the first stable and continuing international forum for the practice and development of science through its sponsorship of the *Philosophical Transactions of the Royal Society*, a scientific journal to inform fellows of the Society and other interested readers of the latest scientific discoveries. Started in 1665, today the *Philosophical Transactions* is the oldest continuously published scientific journal in existence. It published observations and experiments that could be relatively specific and small (in terms of printed volume), matters that would be of the academic scale of an article (of today) rather than of a thick multi-volume book. By these means, scientific communication became much more dynamic and fluid. Furthermore, the *Transactions* developed a new scientific outlook: articles were to contain no high flights of speculation, no great moral lessons; discussions were to be pragmatic interpretations of the results. Robert Hooke's letter to the Royal Society (*Micrographia* 1665) emphasized this new trend: "The rules you have prescribed yourselves in your philosophical progress do seem the best that have ever been practiced. And particularly that of avoiding dogmatizing, and the espousal of any hypothesis [that is] not sufficiently grounded and confirmed, by experiments".

The new science emerging from the issues of the *Transactions* was thus very different from that of just a century before. Nature gradually lost its pretensions to divinity and became secular. *Transactions* also established the important principles of scientific priority and peer review. The new, open science that was the creation of the scientific journal enforced the new idea of accountability. The editor positioned himself as an arbiter, with authority to discredit observations whose reports he considered not credible or not interesting for his readers. Beyond publishing submitted articles, the Royal Society of London also actively approached scientists to study and publish on specific matters. For example, just a few years after being founded, it approached the Italian microanatomist Marcello Malpighi to write on the anatomy of the silkworm. The *Philosophical Transactions* and other scientific journals were important in the rapid dissemination of science. Since then, almost every important advance in science has been in an article, rather than a book.

The prevalence of reports in the early volumes of the *Philosophical Transactions* is hardly surprising. Fellows and correspondents regularly sent in accounts, which scrupulously followed Bacon's format, listing names of witnesses and particulars of time, place and circumstances, with painstaking descriptions and illustrations clearly drawn from life. Robert Boyle's reports on a monstrous colt and calf are typical: a certain Mr. David Thomas and a certain Dr Haughteyn of Salisbury were cited as witnesses to the calf, whose "hinder leggs had no Joints, and whose Tongue was, Cerberus-like triple" and the reader was referred to the doctor for further information (Park and Daston 1981). (Robert Boyle, 1627–1691, was one of the founding members of the society; he discovered a basic law of chemistry, that in gas, pressure increases as volume decreases.)

Despite the energetic efforts of the Baconians of the early Royal Society to realize this programme, the study of marvels—and the emphasis on nature's irregularities—proved fruitless in the fields of both invention and natural history. By 1700, science had integrated the study of monsters into broader theories. Abandoning Bacon's plan for a distinct history of the "new, rare, and unusual in nature", they based their methodology on nature's uniformity and order.

4.2 Generation

4.2.1 The Farmyard Hen

William Harvey (1578–1657, Fig. 4.5), the son of a merchant, was born in south-eastern England. Having studied the classics, rhetoric and philosophy at Cambridge, he travelled to Padua to study medicine. Padua was home to the greatest medical school of those days, characterized by religious tolerance, an international spirit and freedom of thought and there Harvey became familiar with comparative anatomy and embryology. He dissected many animals and learnt how their various organs operate and the connections between form and function. He was in Padua when Fabricius, at the height of his career, was conducting detailed studies on the blood system and on the embryonic development of the chick. Upon completing his studies and earning his medical degree, Harvey returned to England, became a Fellow and lecturer at the Royal College of Physicians and, by the age of 40, physician to the kings (first James I, then Charles I). He first gave public notice of his theories on the circulation of blood in lectures begun when he was 38 but published his treatise only when he was nearing 50: *Anatomical Exercises on the Motion of the Heart and Blood in Animals* (1628). In this treatise, he proved that blood circulates constantly in a closed system. The heart pumps and forces it through the lungs, where it cools, and then returns into the heart; from there it distributes, through the arteries, to all parts of the body. Here, he suggested, it makes its way into tiny (unseen) vessels of the flesh. From them, it then flows into veins, from the lesser to the larger ones, and is finally discharged into the heart. The heart pumps blood into the lungs, and the circulation repeats itself. Harvey stated that blood circulation was a necessity because death is a corruption that takes place through a deficiency of heat. Since all living things are warm and all dying things cold, there must be a particular seat and fountain, a kind of home and hearth where the heat is stored and preserved and from where it is dispersed to all parts, as water from a fountainhead, and upon which concoction and nutrition depend. This place is the heart. Blood movement generates and maintains heat and "spirits" (what we would term oxygen). Having reached the extremities of the body far from its fountain, blood must move to return to the heart;

Fig. 4.5 Harvey

otherwise, it would congeal and be robbed of its spirits in the cold of the body's extremities, just as in a dead body. It is therefore of vital importance that it should return to its fount and origin to once again receive heat and spirits. Harvey's scheme of the blood's circular movement contrasted with Galen's view that envisioned an ebb and flow, with blood passing from the right to the left side of the heart through invisible pores in the septum.

Harvey's wide zoological horizons are impressive. Beyond man, he also studied other vertebrates, beyond which he studied insects (bees, flies, hornets, lice) and snails. His research on the circulation of blood is a peak of the empiric approach, of observing and believing what we can see with our own eyes only, not of believing in scholarly texts. It is noteworthy, however, that he based his claim, of blood making its way into tiny vessels of the flesh from where it then flows into veins, on an interpretation of facts, not upon something that he saw.

Having completed his research on the functioning of the circulatory system, Harvey turned to another, perhaps even more central question in biology:

generation, the mystery of the creation of life. How does the life of an individual organism begin? To answer this question, he conducted observations mainly on two model animals, the farmyard hen and the deer, with complementary observations on humans and their embryos gained from his experience as a physician, and, though to a lesser extent, on ostriches, cassowaries, horses, hares, rabbits, moles and mice. At some point in time, during the 15 years between 1628 and 1642, he assembled his notes into a coherent, complete text, which took many years to write. Unfortunately, it was then that in England the nine-year-long series of civil wars broke out (1642–1651).

Harvey, the king's physician, a member of his inner circle and his staunch supporter, was heavily fined, and his house was looted by Parliament's command. Many of his notes were destroyed, mainly those summarizing years of toil on the generation of insects. Harvey retired from all scientific endeavours. Only one or two decades after he had written his book did friends convince him to publish it. *Disputations Touching the Generation of Animals, The Organs, Membranes and Humours of the Uterus and Conception* (*Exercitationes de Generatione Animalium (Quibus accedunt quaedam de partu; de membranis ac humoribus uteri; & de conception)* was eventually published in 1651 (usually cited as *de Generatione*; an English translation appeared 2 years later). *De Generatione* contains many careful observations, together with many arguments on the generation theories of Aristotle, whom he greatly respected ("this most outstanding philosopher"); and those of his former teacher in Padua, Hieronymus Fabricius of Aquapendente, whom he greatly disrespected ("Fabricius has stepped aside from the truth" and so on). Fabricius had written a book on the development of the chick not too many years before Harvey's treatise (posthumously, 1621), and it was at this book that Harvey, in his own book, aimed much of his fire. He observed a point of pulsating blood in the yolk already on the third day (Wilson 1995), evidence that Fabricius had been wrong in claiming that the developing foetus resembles a house or a ship, whose frame or skeleton is laid down first and the soft parts later. It is the heart and blood, where the soul resides, that form first, and only then do the ribs and chest bone form and wall the embryo all around (Wilson 1995). Harvey also referred to Aldrovandi's observations on the chick, but only occasionally.

For Aristotle, embryonic development meant the actualization of the potential form of a plant or an animal. This actualization would proceed stepwise: during development, the human embryo would successively actualize the functions of a hierarchy of souls (vegetative, animal or sensitive and rational), reflecting the hierarchy and order of plants, animals and man. It would first pass through the state of a plant, rooted in its mother's womb by the umbilical cord. Later during development, the animal soul would become active, and the embryo would acquire sensitivity and movement ability. The higher, rational soul would go into action only after postembryonic development. Thus, all organisms vegetate through their lifetime, but animals add sensitivity and motion capability to this vegetation, and man adds rationality to animal life. Embryonic development mirrors the ascending order of natural things. This idea of progress was essential to Aristotle's views.

It was essential also to those of Harvey, who considered the hen's egg as a microcosm, mirroring the macrocosm. This correspondence of microcosm and macrocosm stemmed from Plato's theory of the existence of two worlds, one eternal, forever being, the other temporal, always becoming (Rieppel 1985). The becoming of an organism, as gained from the development of the chick, provided a visualization of the actualization of the eternal form, potentially contained in the semen of the male, within the female substance. Ancient and Mediaeval philosophy had aimed at a parallelism between the developing microcosm on the one hand and the macrocosm on the other hand, the universe being born and developing much as the chick. It further claimed that in the embryonic development of complex animals (mammals, birds), the early stages represent low, simple animals. Accordingly, Harvey also suggested that the first stage in the development of the chick is that of a maggot. In doing so, he was going back to the teachings of Aristotle, that a worm-like stage produces all creatures, even the most perfected ones (Rieppel 1985). Harvey also claimed (*De motu cordis*) that in the early chick embryo, the heart appears as a simple vesicle and thus corresponds to the adult structure of the heart in animals of lesser perfection (Harvey 1651).

Harvey believed that the driving force behind the creation and growth of the chick egg is some innate Natural Principle that flows from the entire hen into the very first rudiments of the yolk. It remains there as an innate heat, enabling yolk nourishment and growth. He carefully described egg formation, from the very first appearance of tiny "pimples" of yolk in (what we nowadays term) the hen's ovary, and then the yolk's descent in the uterus, where it is enveloped, first by egg white, then by membranes and finally covered by a hard shell. The hen alone makes and forms the egg, he stated, and its yolk and white are merely matter, which only achieves form after fertilization by the male. When eventually the foetus appears, the yolk will nourish it right from the start and later on the white, which is the egg's purest part. As (he claimed that) it is also the warmer part, it flows around the entire circumference of the egg while still in the womb, while the yolk, the earthlier part, withdraws to the centre. Indeed, the outer part of the white might perhaps be even purer than the inner.

In the process of the creation of life, according to Harvey, the cock contributes only fertility to the egg. Harvey's belief stemmed from the fact that he (erroneously) found no sperm in the hen's uterus immediately after copulation, nor did he find in it any element of the female's egg. Lacking any observed traces of semen in the uterus, his explanation for the creation of life called for the passage of some non-visible, immaterial entity, which passed from the ejaculate to the egg without any mechanical action, without even touching it. Accordingly, he suggested that the male delivers his semen imbued with some Divine Spirit, and it is this spirit that conveys fertility: something not corporeal passes from the cock to the hen in this momentary event of fertilization. This moment is comparable to a spark struck from a flint, which suddenly causes an object to burst into flame, or to a lightning flash from a cloud, which sets something on fire or to an iron touched by a magnet, immediately becoming magnetic. By merely touching the female, the sperm passes on to the hen that virtue of fertility by which, Harvey argued, not only the egg but also the uterus, indeed the entire hen, all become fertile at once (Harvey 1651). Something divine, spirituous and "analogous to the essence of the stars" resides in the sperm and sets the process in motion (Wilson 1995).

Fertility and the transfer of sperm also involve the transfer of the hereditary traits of the male. To explain hereditary similarities between father and son, Harvey proposed that the womb is subject to immaterial conceptions, resembling those of the brain (*Of Conception*). Our imagination, for example, springs from some conceptions in the brain, which in turn spring from some external objects; similarly, we create a piece of art according to an idea in our brain, which some external object has inspired. Likewise, in a woman's uterus (and resembling these processes in the brain), something invisible and immaterial arises from the male, "as being the more perfect animal and, as it were, the most natural object of desire", and it is from this immaterial conception that she produces offspring resembling the male. The mere idea, or appearance, of the genitor that remains in the uterus generates a foetus resembling him by imposing his immaterial appearance. The semen transfers an idea, which is "seen" by the uterus and transferred to the matter. In this manner, the male ejaculate transfers its idea via the female reproductive system, without any actual passage of matter. The egg then forms the embryo step-by-step, according to a preformed plan, like an artisan creating his work according to a concept in his mind. Since the offspring also resembles the mother, the fertilized egg must contain something of her spirit as well. However, her maternal contribution is not complete without the male, and the evidence for this is that an unfertilized egg does not produce offspring. Thus, the contributions of both sexes are necessary for the development of the offspring.

Harvey found no difference whatsoever between the morphology of a non-fertilized egg and a fertilized one. How, then, should one define a fertilized egg? His definition: "The [fertilised] egg is a certain conception proceeding from the male and from the female, endowed equally with the virtue of both, out of which, being one, one animal is made. It is the beginning of a new offspring about to be generated, but it is the fruit of both parents". An egg is the joining and combination of male and female; it is the earliest stage of conception for creating the foetus of the offspring. Because it contains both the matter and hereditary power of both sexes, it has the virtue of both, and accordingly, it produces a foetus resembling one or the other. Furthermore, the egg is at the bridging point between animate and inanimate: not absolutely endowed with life, yet not absolutely deprived of it.

Beyond matter, which will eventually gain form, Harvey also suggested that the fertilized egg has a soul of its own. The soul is "the form and species of an animal", namely the ability to produce offspring of the same species. By this definition, the fertilized egg with its soul, as determined by Harvey, is what we would call the genome of the zygote. Even when within its mother's body, however, the soul of the fertilized egg is not part of the mother or her soul, any more than fungi growing on trees are part of the tree, even though they adhere to the tree and receive nourishment from it. An unfertilized egg, whatever kind of soul it might or might not have, differs from a fertilized egg in that it does not produce offspring. Harvey thus saw the soul through Aristotle's eyes, as the realization of the living body (Harvey, 1651).

He dismissed the idea, popular in his days, that during copulation both male and female deliver liquids into the womb, and that the predominance of one liquid over the other will determine whether the offspring will be male or female. The copulating hen emits no "seed" whatsoever because, Harvey claimed, she lacks organs necessary for its generation, namely female "testicles and spermatic vessels". Harvey also dismissed the idea suggested by Fabricius that the chalazae are, in some way, the semen of the cock when inside the egg (chalazae—two spiral bands in the white of a bird's egg that suspend the yolk and hold it in place in the centre of the white). He correctly noted that the same chalazae occur in both fertilized and unfertilized eggs, and hence this idea is not valid. Another common opinion of his times was that, immediately after coitus and as the sperm enters the uterus, something begins in the uterus. Here Harvey noted his (alas false) observation that the sperm of the cock does not enter the cavity of the uterus, nor is anything of the egg to be seen in the uterus immediately after coitus. He also rejected the idea that the fertile egg comes from (what we today name) the ovary. How can one believe that an element of fertility so elaborate, so concocted and so vital, which sometimes exceeds that of the male in strength, in spirit and in generative potency, could possibly come from such inconspicuous parts?

Fully aware that he did not know everything, Harvey put forth a series of open questions. One of these was the issue of hybrid eggs, such as those formed between mated cocks and pheasants—do they possess the soul of one of their parents? Of both? Another question concerned those animals that, he believed, grew out of the bodies of other animals and yet had their own souls: tapeworms, roundworms, lice, mites and the like. Furthermore, there were many insect larvae, which, again, as he believed, arose from plants, such as we find in oak acorns, kermes berries, dog roses and many others. Animals born in oak galls do not "live by the soul of the oak", although they live in the oak and feed on it. This also applies to plants and fungi growing on trees. Furthermore, Harvey (erroneously) wrote that this also applied to cancers and tumours, which feed and grow inside our own bodies, as if they possessed their own vegetative soul (Harvey 1651).

In addition to sperm, Harvey believed that an external factor influences fertility: the Sun. Spring and autumn follow the Sun as it approaches or recedes, and it is mainly during these seasons that animals generate. The Sun is, therefore, he emphasized, the common father and begetter of all life; or, at least, it is the instrument, in this world, of the supreme Creator, in his generation of plants and of animals, males and females, indeed even of those creatures created spontaneously.

> As if at the approach of this noble planet, there descended from the sky sweet Venus and her retinue of Cupid and the Graces, and incited all living things with the hot vehemence of love to seek their own eternity [...]. Also, the cock and hen become fruitful mainly in springtime, as the Sun, or Heavens, or Nature, or the Soul of the Universe, or Almighty God (for all these words present but the same thing) are a superior and more divine cause in generation than they.

In this manner, Harvey argued, both the Sun and man beget a man, or, in other words, the Sun uses man as his instrument. Both the Creator and the cock beget a fertilized egg, and from the egg, a chick, by the constant advance and recession of the Sun. Harvey concluded that the male, though he was an earlier and better instrument than the female, received his generative power from the approaching Sun and therefore from God. His emphasis on the importance of the Sun is closely reminiscent of Fludd's cosmology (1617), as we have already seen, in which the Sun had special significance as the residence of the Holy Spirit, with light and the spirit of life emanating from it.

Harvey incubated fertilized eggs and opened them on consecutive days. This allowed him to compose a detailed picture of the dynamics of embryonic development, stage after stage, beginning with a tiny drop of blood from which other organs form in a regular, consistent process. He noted that the heart, or at least its first embryonic rudiment, is one of the very first organs to develop. Having appeared, it sets up the rest of the body as a future home for itself, surrounding itself with a wall by adding ribs and the sternum, like fortifications for a building of its own. From there, it gives and governs the life of the embryo and eventually of the chick and adult. Like some household god, it is the first dwelling of the soul, the first receptacle of innate heat, the fount and origin of all faculties and the only solace in difficulties and unpleasant situations. Harvey's careful observations revealed that actually it is the blood, not the heart, which begins life since it is the first to appear in embryonic development. Consequently, it is blood that contains life, and it is blood that also contains the soul, as indeed we read in the Scriptures (*Leviticus* 17, "For the soul of a creature is in its blood"). The heart develops shortly after as the blood's instrument for circulation. Harvey saw the formation of the chick foetus as proceeding in four stages.

1. Formation of the blood and of the heart with its veins. The blood and heart make use of all other parts as their instruments, and they are the origin of all other parts. The blood, veins and heart appear together on the fourth day; of these, Harvey believed blood appears first.
2. Formation of a mass of jelly-like, homogeneous material by the accumulation of matter (after the fourth day). Within this mass the embryo develops, which at first resembles a worm or a maggot ("for it seems to enjoy the life and obscure movements of a maggot"). This worm soon develops a larger part in front, from which the brain, beak and eyes eventually form, and a smaller part in the rear, resembling the keel of a ship, above the vena cava and extending along its length (the backbone). The limbs, with their bones, cartilages, muscles and sinews, develop at this stage as well.
3. Formation of the viscera (sixth-seventh day). As soon as the sternum is set up, the heart enters the chest, from where it undertakes the governance of the surrounding mansion. The beak and claws form at this stage.
4. Down, quill feathers and other structures of minor rank form (at a later stage).

The development of the embryo thus occurs step after step, in an orderly, regular manner. Furthermore, while these four stages of formation occur while the foetus is inside the egg, development does not cease upon hatching. After hatching, crests, combs, wattles and spurs form in the male. With these, Harvey writes, the male both entices his hens to the service of Venus and fights other males to the death. He who wins in battle has the prerogative to propagate the species before all others. Indeed, all males are more armed and more aggressive than other males and eagerly enter battle. They do so either to protect their unborn young, their territories or their food, but chiefly—to conquer females for posterity's sake. The victor in battle, flaunting his victory, exults in the enjoyment of his captive females. These lay fertilized eggs, and the cycle goes on. The common product of both sexes, an animal's fertilized egg, is the centre and very hinge about which the generation of every species revolves (Harvey 1651).

Harvey's view of the egg was rather abstract: the egg is not a concrete object; every beginning capable of life is an egg. He thought that there was a stage in the development of all animals during which they are undifferentiated living masses and that this stage constitutes the egg. In birds, these eggs are most clearly perceptible; in mammals, they constitute the first stage of the embryonic development; in insects, the larvae change from an "incomplete egg" to a "complete egg", the pupa.

It is perhaps in this light that we should view Harvey's title-page statement, *ex ovo omnia*. Harvey considered the hen's egg a model for the generation and development of all oviparous epigenetic creatures, including other birds, snakes, tortoises, lizards, frogs and even fish to a certain extent. *De generatione*'s title page (Fig. 4.6) shows God sitting on a column base with the book's name inscribed. In his left hand, he is holding an egg, its upper part removed, from which many forms of life emerge (a tiny man, bird, deer, fish, lizard, butterfly, snake, grasshopper,

Fig. 4.6 *De generatione,* left: title page; right: detail. God is holding an egg, from which many forms of life emerge (a tiny man, bird, deer, fish, butterfly, snake, grasshopper, crocodile and spider); *ex ovo omnia* (all life originates from the egg) is inscribed on the egg

crocodile and spider; interestingly, no farmyard hen). The phrase *ex ovo omnia* (all life originates from the egg) inscribed on the egg reflects Harvey's assertion that the egg is the origin of the entire animal world and that only one kind of generation occurs throughout the animal world, whether the animal is oviparous or viviparous. A trained eagle is gazing up at its divine master, symbolizing Christianity—great power subjugated to the will of God.

Harvey next turned to the study of viviparous animals. We now come to his study of generation in the deer.

4.2.2 The Deer

Mammals are the dominant viviparous group, yet work on generation and embryological development in mammals presented Harvey with greater difficulties than the study of the chick. Eggs were cheap and easy to come by, whereas producing a sufficient number of pregnant females of any mammal species to conduct a systematic study of foetal development was more difficult. Fortunately, Harvey was the king's physician, and this enabled his investigations into mammals. Deer hunted by the king and his entourage were put at Harvey's disposal daily to dissect and examine their genitalia in search of the beginning of life. Besides being large mammals and, thus, easier to observe than small ones, the study of the origin of life in royal deer undoubtedly added considerable prestige to Harvey's research.

Believing that conception occurs either in the uterus, as in women, or in the uterus horns, as in many other mammals, Harvey described the uterus of the hind, closely comparing it to that of a woman and, to a lesser extent, to other mammals (cows, goats, apes, horses, moles, mice). In this context, he also described the female "testicles". He noted that they resemble glands, but they correspond more to the prostate in males than to their testicles. He (erroneously) rejected the idea that these organs have anything to do with fertilization because he observed that during coition, whereas the stag's testicles swelled and were filled with semen, the hind's ovaries and uterus horns remained unchanged, as though completely uninvolved in generation. They did not swell, nor did they vary in any way in structure, either before or after coitus, and there was no indication of their having any function for copulation or for generation. In moles and in mice too, the male testicles were greatly distended during coition, and there were vast changes in the genitalia, while "these small glands which men consider to be the testicles of females keep their original appearance without any alteration whatsoever". Clearly, he did not recognize the mammalian ovary as such, and his opposition to two-semen theories prevented him from attaching significance to the organ itself.

In ancient Greece and in Rome, the ovary, because its position in some mammals resembled that of the testes, was termed the "female testis", but there was only a vague awareness of its function. During the Renaissance, Vesalius had described the ovary, and Fallopius (1534–1562), his pupil, had discovered the uterine tubes. Fabricius (1533–1619), Fallopius' pupil and successor to the chair of anatomy at

Padua, had claimed that in a hen's egg the ovary deposits the yolk, while the white, egg membranes and shell form when the egg passes through the oviduct. Thus, by the mid-seventeenth century, naturalists had recognized the ovaries as distinct structures, but their function was not clear. Harvey was Fabricius' student, so, from this aspect, there was a Vesalius-Fallopius-Fabricius-Harvey lineage.

Progressing from anatomy to reproductive behaviour, Harvey noted that in England, the stags were kept apart from the hinds during most of the year and were very shy and timid. However, during the brief 1-month-long rutting season (mid-September to mid-October), they congregated with the females, jealously keeping them in harems. If a strange stag drew near and attempted to mate with them, the hinds fled to their own mate and took refuge with the herd "to guard their chastity and their affrighted modesty". However, if a young bachelor stag found a female straying alone, he swiftly ran in pursuit of her, and when she wearied of the chase and could flee no further, he mounted and enjoyed her by force. This description, of a non-dominant young male secretly mating with one of the dominant male's females, is what we now term satellite copulation: a reproductive tactic in which a non-dominant male, on the periphery of the territory of a dominant male, sneaks up to copulate with one of the females in the latter's harem. Some 50 years before Harvey referred to this tactic in mammals, Aldrovandi, as we have already mentioned, had described satellite copulation in birds.

Harvey sought to tackle the function of the "male principle" in the conception process. During the rutting season, he daily dissected the uteri of many females that had just mated in search of semen and some beginning of life; he found none. The King (Charles I, Fig. 4.7), who was very interested in the research and to whom Harvey reported this unexpected observation, told this to some of his court. Far

Fig. 4.7 Charles 1. (Tuvia Kurz)

from being a passive patron, the king was thus closely involved in Harvey's research. A very lively conversation soon ensued, with some accepting Harvey's position (including the king) and others claiming it impossible for any embryo to form without a trace of the male's seed surviving in the uterus, and hence Harvey's observations were false. To settle this question, the court decided to conduct an experiment, and the king ordered that around the beginning of October, some hinds that had just mated should be fenced up in an enclosure.

Harvey immediately dissected some of the hinds to examine whether their uterus retained semen, but he found nothing. He dissected other hinds as the experiment proceeded over the following 7 weeks but did not discern any traces of semen nor any sign of a foetus; there were, however, changes in the uterus that hinted at gestation. From that time onwards, the hinds were found to be pregnant "as if by some contagion" and eventually gave birth to their fawns. This comparison of fertilization to a contagious disease, in which it takes some time from the time of infection to the moment of outbreak, may reflect the medical physician inside Harvey's personality and his sensitivity to disease. From mid-November onwards, an elongated, sausage-like structure appeared in the hind, adhering to the sides of the uterus that contained it but easily separated from it. This was Harvey's first observation of conception (in our modern terms, this would be the blastocyst). As he found nothing in the hind's womb until mid-November, and likewise, he found nothing in the uterus for several days after copulation in dogs, rabbits and other mammals as well, Harvey concluded that no conception occurs immediately after coitus. He generalized accordingly that in all viviparous animals, as in oviparous ones, the foetus does not arise from the male sperm emitted during coitus, nor from any female sperm, nor from a combination of the two, not even from menstrual blood, as Aristotle had thought. In addition, there is no matter waiting ready in the uterus, which the male sperm could concoct, coagulate or shape into an actual foetus during copulation (Harvey 1651). He now defined an egg—"an egg is that out of part of which an animal is begotten, and the remainder is the food for that which is begotten. Because it is the rudiment of the future foetus, it is called the egg of these animals". Harvey's egg would be the zygote, in our terms. He presented a new concept of an egg as the very first stage of embryonic development.

Harvey's definition of an ovum is therefore different from our modern definition. His egg was an inanimate lump lacking any defined structure, in which the foetus was already present. Life develops from a lifeless, inanimate lump lacking definite shape, thus "animal from which is not animal, entity from non-entity". Thus, any primordium (an organ or tissue in its earliest recognizable stage of development) from which a living creature is generated is an egg, whether in the form of a seed, an egg or even a worm. The hind produced something resembling an egg, the rudiments of a conception, just as plants produce seeds and trees produce fruits. During a long latent phase, lasting a whole month after copulation, nothing materializes in the uterus; only then does the "egg" appear.

Unfortunately, Harvey erred in his observations. We should bear in mind that he carried out his observations on the deer in 1630–1635, several decades before the microscope became widely used in scientific research. This lack of a microscope prevented him from seeing the male sperm as Leeuwenhoek would, 20 years after Harvey had died. He did not have any insight into gametes, in the form of egg cells that combine with sperm cells at fertilization.

So far as fertilization, we now turn to embryonic development.

4.2.3 Epigenesis Versus Metamorphosis

Moving on from fertilization to ontogeny, Harvey compared two modes of embryonic development, epigenesis and metamorphosis. He coined the term epigenesis as referring to a developing embryo that attracts, prepares, elaborates and makes use of material as it comes in as a homogenous substance. Formation and growth occur simultaneously: material is, at the same time, both made and shaped. He presented this process (already established by Aristotle) as comparable to a potter who shapes a form out of clay by adding parts, by giving it a shape while providing material, part after part, while increasing its size. In such a developing embryo, parts formed in succession differ from one another both in matter and form, and they distribute differently into various organs as these develop. As the material is at the same time both made and shaped, the development of the embryo occurs step after step, in an orderly manner, and the developing embryo is composed of different parts. Before beginning to develop, the egg does not contain any preformed organs, only homogenous matter (in the chick: yolk and white). It then gradually progresses from the state of an unorganized, non-functioning body to an organized, functioning body by a Moulding Force that allocates material from the homogeneous matter to the various organs, transforming it into a heterogeneous embryo with all its organs. In this manner, the embryo gradually grows and forms itself, stage by stage, each stage developing into the next in sequential growth and development. The heart is the first to appear in the chick embryo; it is the embryo's starting point that, as an instrument of the soul, directs the successive development of the whole animal from an initial homogeneous and soft jelly-like substance. Epigenesis, Harvey declared, occurs in vertebrates ("animals with red blood") that require for their propagation "an identity of species, rather than any supply of fitting material". Lions or cocks, he wrote, are typical cases of creatures formed by epigenesis.

Metamorphosis was, for Harvey, a term that included both the sudden, profound change in form from one stage to the next in the life history of an insect (as from caterpillar to pupa and from pupa to adult butterfly) as well as the sudden transformation of inanimate, undifferentiated precursors into organized beings by parentless, spontaneous generation. Creatures may spring directly from matter rather than by coition. Rotting matter, mire, dew, excrement and parts of plants or animals—all these may be the origin of many insect "worms". Harvey referred, in metamorphosis and spontaneous generation, to a developing embryo in which all matter is present from the beginning, as a stone from which a sculptor will eventually make a statue. This material receives form all at once, as though cast in a mould. Consequently, this matter is homogenous, and as such, it is distributed to all organs, no matter how dissimilar they are. The animal that was born as a worm, whether from an egg or from pre-existing rotting matter by chance, grows to full size, and then, by metamorphosis, it becomes a bee, wasp, butterfly or fly, which does not grow any further. Harvey considered these animals as "less perfect, less capable of preserving their own kind and less enduring than the red-blooded terrestrial or aquatic animals, which have an unambiguous source [of generation], that of the same species, [the generations] following one another to all eternity". He believed that animals developing by metamorphosis did not preserve their own race, as they formed by chance from any suitable material happening to be in the environment (Harvey 1651).

Harvey considered any primordium living only in potentiality as an egg, whether perfect in the womb, in that it reaches its final size when laid, or imperfect, in that it is released while immature, not yet having reached its full size, and continues growing outside the womb after having been laid. From an imperfect egg, an imperfect animal generally ensues—a worm or a caterpillar. This imperfect animal is an intermediate between a perfect and an imperfect egg because, in comparison with its egg or primordium, it is an animal endowed with senses, motion and the ability to nourish itself. Compared to a butterfly, which possesses a potential primordium, it is no more than a crawling egg, providing for its own growth. Such a caterpillar, having acquired its proper size, changes into a chrysalis—a perfected egg in that, like any normal egg, it ceases to move and is thus a potential animal. The insect chrysalis, the "second" egg, is the true, completed ovum. The egg is an interim, vegetative individual, which precedes epigenesis or metamorphosis. New life throughout nature develops in a similar manner. The cocoon, or chrysalis, which a caterpillar enters during its pupa stage, is a "second egg" in which the prematurely hatched embryo is born again. Thus, an imperfect egg hatches as an imperfect animal, a worm, larva or caterpillar, which turns into a perfect egg, the chrysalis, which hatches as a perfect animal, the winged adult (Harvey 1651).

4.3 A Grand New Theory of Life

In the decades between Bacon and his vision of a new science on one hand and the foundation of the Royal Society on the other, Descartes published his *Discourse on the Method of Rightly Conducting the Reason* (1637). He regarded the universe as a vast mechanism in which the only truly significant factors are matter and motion. Nature is stable, orderly and uniform, governed by laws that apply at all times and in all places. The universe is essentially an enormous clockwork, a meaningless mechanism obeying fixed laws. Planets, clouds and animals consist purely of moving particles, and animals are merely complex machines whose actions mindlessly follow mechanistic rules. The one major difference between animal and man is that man has a thinking soul.

René Descartes (1596–1650, Fig. 4.8), the son of a lawyer, was born in mid-west France. The Jesuit College, where he studied until the age of 18, exposed him to

Fig. 4.8 Descartes

Aristotle's theories. He then studied law at the demand of his family, but he never practised. Instead, at the age of 22 and seeking a professional military career as a mercenary, he first joined the Protestant Dutch States Army, which was fighting the Catholic parts of the Netherlands (today in Belgium), and later entered the service of the Catholic Duke Maximilian of Bavaria. Two years later, he left the military profession, sold his inherited property and invested in bonds, which provided him with a comfortable income for the rest of his life. At the age of 32 he moved to the Dutch Republic, where over the next 20 years he wrote his major works: *L'Homme* and *Le Monde* (1630–1633, published posthumously 1662 and 1664), *Discours de la méthode* (1637), *Principia philosophiae* (1644), *La description du corps humain* (1648, posthumously 1667), *Les passions de l'âme* (1649) and others. He was 53 when Queen Christina of Sweden invited him to her court to organize a new scientific academy and to tutor her in his ideas about love. Shortly after moving to Sweden, he contracted pneumonia and died. In the years following his death, many universities throughout Europe widely taught and discussed Descartes' new, comprehensive theory of nature. However, at the University of Utrecht, it was condemned (1643) in Rome; the Pope placed his works on the Index of Prohibited Books (1663), and in France, Louis XIV forbade lectures on his theories (1671). Today, most of us know him from his phrase "I think, therefore I am".

Descartes sought to develop a general and comprehensive philosophy that would be an alternative to that of Aristotle, a completely new, mechanistic approach to nature. Aristotle, we recall, had claimed that an active principle (form), governing a passive principle (matter), controls all of nature, physical and biological. One distinction between living creatures and non-living objects is life's ability for self-movement (including growth), due to an intrinsic power, the presence of which distinguishes living from non-living things. Non-living objects are moved only when they interact with other objects and forces, whereas living creatures can move by themselves, apart from their interaction with other objects. The principle, or mechanism, responsible for self-movement is the Soul—the Principle of Life, which takes charge of all phenomena of life in all organisms, from the lowliest plants, via animals, to man. Different organisms have different levels of souls: a vegetative soul (feeding, growth, reproduction) found in plants, animals and man; a sensitive soul (appetite, movement) only in animals and man; and a rational soul, in man alone.

Descartes suggested that there is no such thing as an active principle. There is only matter, which possesses four traits: size, shape, position and movement (Des Chene 2004). Bodies fall to the Earth because the contact of particles results in movement, and the motion of one (sometimes invisible) particle pushes the other visible particles, which we can see moving. Magnetic movement is also the result of invisible particles that flow from the north southwards or in the opposite direction, causing the visible magnetic needles to straighten mechanically. In the living world, there are, he argued, mechanistic explanations for moving towards certain objects to obtain food and away from other objects to avoid danger, even for feeding, growth, reproduction and sensory functions. He rejected the very existence of vegetative and sensitive souls: plants and animals have no souls, and nothing sets them apart from the inanimate, physical world. Indifferent to the distinction between living and non-living, he drew the line not between living and non-living but between those beings with a soul versus those without a soul. Plants and animals are merely self-growing and sometimes self-moving machines, devoid of mind and of consciousness. Admittedly, some animals resemble us in so many ways that we tend to believe that they too have a soul like ours, but in fact, there is nothing in them that resembles the soul—they have no sensations, no passions, no judgements, no reason or intelligence. These animal-machines are powerful enough to act in many ways that resemble man but not powerful enough to think. Whatever sensations and perceptions animals might seem to have are actually automatic responses to internal and environmental stimuli, as they are no more than complex machines. In animals, purely mechanical processes explain nutrition, growth, generation, functioning of the senses and even the ability to respond via sense organs to specific situations, whether to approach beneficial things (food) or avoid dangerous ones (predators). To a certain extent, Descartes' "mechanical processes" merged into what we would call instinctive reflexes. As examples of automatic instinct rather than reason in determining animal behaviour, he enlisted the return of swallows in spring and the scraping of the floor by dogs and cats to bury their excrement, even though they never bury them. Animals carry out all these activities only by instinct, not by thought. Descartes thus eliminated Life as a distinct category of nature, consisting of those entities that live. The science of Life should henceforward be merely an extension of physics, moving uninterruptedly from explaining the behaviour of magnets to explaining that of butterflies. The Principle of Life is no longer a form peculiar to living things. Descartes insisted repeatedly that nature's laws are indifferent to the distinction between a living, intact animal and a mutilated or dead one (Des Chene 2004).

Furthermore, humans also respond mechanically to signals coming from their senses as part of their behavioural repertoire. When a person stumbles, he extends his arm to protect his head. This movement of the arm is a mechanistic response. No logic guides him to do so, but merely the sight of the fall reaching his brain, which sends signals to the nerves that operate muscles to produce the protective movement. Indeed, Descartes declared, most human behaviour is subject to a simple mechanistic explanation like that of animals and occurs with no interference from logic. Skilfully designed automats could successfully mimic nearly all of what we do. As these actions, which we attribute to instinct, are indeed thoughtless, matter and its properties will suffice to explain them even in humans. Machines, following mechanical laws, can perform actions once attributed to the somewhat mystic vegetative and sensitive powers of the soul. Descartes thus attempted to explain in mechanistic terms all those involuntary functions of the body that occur in us without thought, and which Aristotle had assigned to the vegetative and sensitive parts of the soul: nutrition, growth and generation on the one hand; sensation, passion and locomotion on the other. The man-machines and animal-machines with their pores, nerves and valves were systems that controlled their own instinctive operations. Operating in fixed patterns, they created the conditions under which the pattern can begin anew (Des Chene 2004).

To prove that the living body is essentially a machine of complex engineering, Descartes analysed the process of digestion. In the stomach of a man-machine (or animal-machine), there are liquids that, coming among the food particles, separate and ferment them. The fermenting particles descend toward the conduit, through which the large ones leave; whereas the more subtle ones encounter an infinity of little holes, through which they flow into the branches of a large vein that carries them toward the liver (what we today would call the hepatic portal vein). Here, the incoming liquid evaporates, and the remaining matter takes on a red colour, thereby becoming blood. This newly formed blood passes from the liver to the heart, into which it drips. The heart contains one of those fires without light, so hot and ardent that the blood, as it enters it, automatically heats and expands. The heart is entirely passive, its only power being that of resisting expansion, like a bladder filled with air; and yet, Descartes argued, the heat of the heart is the driving motor of the entire process of digestion and blood circulation. It is broadly the same as the heat found in fire and in steaming lava, but it is a fire without light, like the heat in rotting hay. In this manner, Descartes transformed the entire concept of Life into something strictly mechanical: fermentation and particle size are the only qualities distinguishing food particles entering the liver from those eventually excreted, and blood flows due to the physical properties of heat. Nothing in the animal-machine resembles a soul. This was the keystone of the argument, which allowed Descartes to do away with the vegetative and sensitive souls (Des Chene 2004).

Beyond the machines of nature, it is the human soul (= mind) that thinks logically, and that is what differentiates Man from animal. Only humans have a soul in the context of logical, intelligent thinking, with a conscious use of senses, images and memories. Man has reason, and he has language too, to record his thoughts for the benefit of others. To base his assertion on the argument that animals cannot use language or communication signals the way we do, Descartes argued that, in anatomical terms, there is a need for only a little morphological "investment" to develop the ability to talk. Thus, it would have been easy for animals to overcome these minor morphological deficiencies—if they only had a soul to develop language. Taking this argument from its other side, the lack of language in animals is evidence that, though they have cravings and instincts, they do not possess a soul and therefore are not capable of rational thinking. Hence, animals are, in effect, automata.

Nicolas Malebranche (1638–1715) later added that, under the rule of a just God, no creature suffers unless it deserves it. Since animals have no soul, they cannot sin, and therefore they cannot suffer. Therefore, animals do not suffer pain. Malebranche was concerned lest God might seem unjust if animals were to suffer without having sinned. According to him, there are only minor anatomical differences between man and animals, the only major distinction being that man has a thinking soul.

This basic indifference to animal suffering had deep ideological roots in Christian thought. Four centuries before Descartes and Malebranche, Thomas Aquinas (1225–1274, *Summa Contra Gentiles*) had claimed that God supports all creatures not for their own sake but for Man's benefit. Man has, therefore, no obligations to them. Any remarks in the Bible regarding avoiding cruelty to animals are only allegories for avoiding cruelty to humans. Animals lack knowledge and have no ability to choose; their actions may seem to us purposeful, but that is only because God decides for them, just as human decisions and actions cause movement in arrows and clocks. They, therefore, are not creatures worthy of charity. Aquinas added that animals are without intellect, and whatever animal behaviour occurs is merely instinct, as animals (unlike humans) are incapable of acting beyond instinctive behaviour. An animal's instinctive reaction is, as stated by Aquinas, "as inevitable as the upward motion of fire". Hence, whatever the explanation for certain patterns of animal behaviour, it must not involve them having rational thought. For example, why does a sheep run from a wolf? It is not feasible that a sheep would have enough intellect to interpret danger and flee from an anticipated danger. Aquinas solved this dilemma by positing a sixth sense, *estimativa*, which could perceive the intentions of surrounding objects. This sixth sense of the sheep could detect the wolf's malice, and the sheep responds to this sense by fleeing, in the same way that it might instinctively recoil from the heat of a fire. It was, thus, easier to assume the existence of a sixth sense than to assume logical thinking in animals. Similarly, according to Aquinas, animals lack the property to command because command is an act of reason. Aquinas' arguments remind us of Philo of Alexandria of the Hellenic period, who, in seeking the theological border between man and animal, had suggested that perhaps animals do have reason, though not as perfect as ours.

Descartes suggested that the soul, which controls the body, is non-material and does not follow the mechanical laws of nature. The site at which the soul interacts with the body, the material “seat of the soul”, is primarily in the pineal gland. Near the centre of the brain and near its ventricles, the gland protrudes near the entrance of the canal from the third to the fourth ventricle. Today, we know that it secretes melatonin in different amounts over the day and over the seasons, thereby regulating circadian and annual rhythms. Descartes believed that sensations delivered by the nerves to the pineal cause it to move in some manner, which in turn gives rise to the emotions and causes the body to act. Why did Descartes consider the pineal gland to be the place where the soul resides? By way of a brief historical summary, in ancient Rome, Galen had explained that the pineal gland owes its name to its resemblance in shape and size to the nuts in the cones of the stone pine *(Pinus pinea)*. He believed that the surrounding ventricles contained “airy thoughts” because he thought they were filled with a fine, volatile, airy or vaporous substance, a “spirit” responsible for the ability of self-movement. In late Roman times, Nemesius of Emesa (about 400 CE) had assigned various functions to the brain’s ventricles: the anterior ventricle was the organ of imagination, the middle one was the organ of reason and the posterior ventricle was the organ of memory. Anatomists of the Renaissance discovered that the ventricles were filled with fluid, not with some airy spirit, and rejected the theory as though the pineal gland could regulate the flow of spirits in the ventricles of the brain. A century later, Descartes repeatedly discussed the pineal gland’s function (*Dioptrics* 1637, *Treatise of man* posthumously 1662, various letters 1640–41, *The passions of the soul* 1649). Noting that it is small, light and (as he erroneously thought) easily movable, he suggested that it was involved in sensation, imagination, memory and bodily movements. Movements of the pineal gland (eventually resulting in bodily movements), he suggested, may bring it close to some valves in the walls of the ventricles. Spirits continuously flowing from it may then push these valves open, enabling some of the various animal spirits in the ventricles to escape through these valves. These spirits would flow to the muscles by means of the hollow, spirit-filled nerves; here they would open or close certain valves in the muscles and thus bring about muscle contraction or relaxation. A second argument in favour of the pineal as the seat of the soul was that other parts of the brain are not suitable candidates for a unifying principle since they are themselves duplicated (Lokhorst 2018). The pineal gland is the only logical candidate for the unifying principle of sensation to perception. Since it is the only solid part in the whole brain that is single, it must necessarily be the seat of common sense, namely of thought and consequently of the soul (1640).

Unfortunately, some of Descartes’ basic anatomical and physiological assumptions were totally mistaken by our standards and in light of what physicians and naturalists already knew in his time. He thought that the pineal gland was in the middle of the ventricles, but it is not, as Galen had already pointed out. He thought that the pineal gland was full of animal spirits, brought to it by many small arteries, but Galen had already pointed out that veins surround it, not arteries. He described these animal spirits as a very fine air, or a very lively and pure flame or a wind inflating the ventricles like the sails of a ship inflated by the wind, but naturalists had already discovered a century earlier that the ventricles contain liquid, not anything like air. Elsewhere, he claimed that the pineal gland is unique to humans, whereas it occurs in all vertebrates. Descartes was not a very careful observer.

Descartes, coming a generation after Bacon, differed from him in that God was central in his philosophy. Accepting the biblical story of Creation, he believed that the universe began as a chaotic soup of particles in motion and that everything else formed within this moving matter: many suns, around each of which many planets coalesced, in each of which mountains, seas, metals and magnets came into being, as well as atmospheric phenomena such as clouds and rain. God acted in complete freedom and great wisdom in creating the world along eternal laws of mathematics and mechanics. However, the Creation, Descartes claimed, had consequences for God himself. His wisdom in formulating the mathematical and mechanical laws of creation meant that these laws now restricted God himself—he cannot break them. Today, God behaves only according to mechanical principles. He is unable to perform miracles, and he has no spiritual powers to change the course of nature (*The World*, ca. 1650). Strict laws, applied at all times and in all places, govern nature, and there is no place for variation: the cosmos is essentially a meaningless entity. This perspective of Descartes threatened the Christian point of view. Mechanical laws limit the freedom of God in that they leave no place for miracles, for faith or for divine inspiration.

Having accounted for nutrition and growth, two of the three basic operations ascribed to the powers of the vegetative soul, Descartes discussed the third operation, generation. He refrained from delving into fundamental problems of generation, beyond the question of parentless, spontaneous generation. He argued that so little is required to make an animal that it is not surprising to find so many worms and insects spontaneously forming in rotting matter. However, he did devote considerable thought to embryonic development. Des Chene (2004) has presented Descartes' views on generation in considerable detail, so here I present only his ideas concerning the embryonic development of sexual differences. These, again, reflect his ideas regarding simple, mechanistic explanations as governing the living world.

According to Descartes, the foetus develops a penis, with its base towards the mother's back and a tip growing toward her navel. If the embryo's head faces the mother's navel while its buttocks face her spine, the foetus will become a male and the penis will grow outside its body. If, on the other hand, the embryo's head faces her spine and the buttocks away from it, it will be female, since the penis curves back toward the navel of the mother and toward the interior parts of the embryo (Fig. 4.9). The vagina is thus an inverted penis. A few paragraphs later, he offered a different explanation, involving the strength or weakness of the foetus. The strong foetus excretes liquid, retains solid and becomes male; the weak foetus excretes solid, retains liquid and becomes female. Interestingly, both explanations for sexual differences have nothing to do with the soul. There are no male or female souls; there are only bodies that happen to become one or another as they develop in the womb.

Descartes' theories, which emphasized the similarity of biological functions and machines, influenced many biologists. Among them was Malpighi. In his

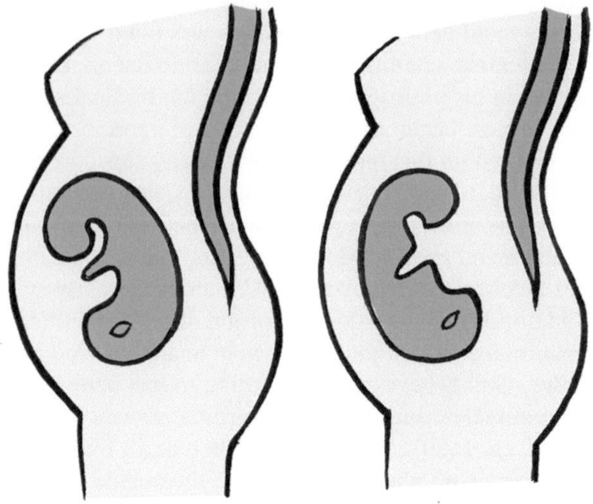

Fig. 4.9 Descartes: the position of the embryo determines its sex

description of the development of the chick embryo, he drew an analogy to an artisan who "in building machines must first manufacture the individual parts so that the pieces are first seen separately, which must then be fitted together" (*Discourse letter on the formation of the chicken in the egg*; *Dissertatio epistolica de formatione pulli in ovo*, 1673). Nature, in order to carry out the various functions of animals and plants, has "constructed their bodies with a very large number of machines, made up of extremely minute parts, so shaped and situated as to form a marvellous organ, the structure of which is usually invisible to the naked eye, without the aid of the microscope". He noticed that within 30 hours of incubation, the heart of a chick begins to beat, even before it becomes red with blood. In an observation that became controversial, he stated that he had seen the complete form of the embryo in an unincubated egg. Later, as we shall see further on, Bonnet took up this erroneous notion of "pre-formation" to support the theory of development called preformationism. According to this theory, embryos are preformed entities that occur in either the maternal egg or the paternal sperm and need only to grow.

Descartes was also interested in reproduction (Pyle 2006). He believed in the occurrence of spontaneous generation: as only a few things are necessary in the making of an animal, it is not surprising that we see so many different animals, whether worms or insects, forming spontaneously in one kind or another of putrefying matter. The generation of higher animals, however, occurs by a simple, straightforward mixture of male and female semen, two liquids that, when warmed, some of their particles become dilated and press on the others, thus gradually pressing them into the required organs. It is not necessary to assume any strange, unknown powers.

The formation of the different organs is a true epigenetic process: the heart is the first organ to be formed. It, in turn, produces a distillation of the blood, from which other organs condense out in a fixed order. Blood currents to and from the heart become veins and arteries, respectively; blood streaming through pores in the walls of blood vessels deposits muscle fibres, and "spirits" in the brain produce the nervous system. Descartes' programme of embryology thus explains epigenesis (the successive appearance of organs out of a homogenous mixture of fluids) in purely mechanistic terms (Pyle 2006).

A far more modern approach than that of Descartes to animal biomechanics was that of Giovanni Borelli (1608–1679) of Naples, the son of a Spanish infantryman and an Italian mother. He studied physics and mathematics in Rome, became Professor of Mathematics in Messina and then shifted to the University of Pisa. Tracking the path of a comet, he concluded that it was moving in an elliptical curved orbit around the Sun. This was against the scientific theory imposed by the Catholic Church, that the Earth is the centre of the universe. In those days, it was dangerous to oppose theories of the church, and Borelli therefore published his findings under a pseudonym (Pier Maria Mutoli). Aged 60, he returned to Messina, in the midst of an uprising against Spanish presence in Italy and, joining the anti-Spanish forces, he changed his Spanish surname (from his father) to a variation of his mother's name. Under suspicion of involvement in a conspiracy to free Sicily from Spanish rule, he was exiled to Rome and here he became acquainted with ex-Queen Christina of Sweden, who had been forced to give up her crown and who was exiled to Rome two decades prior for converting to Catholicism, and he was her personal physician and scientific consultant. He lived his last years in poverty, teaching mathematics at the school of the convent where he had been allowed to live. During these last years, he worked on his book *De Motu Animalium* (*On the Movement of Animals*), which was published in 1679, shortly after his death. It was dedicated to Christina of Sweden, who financed its publication.

De Motu Animalium, an investigation into biomechanics, discusses how muscles move while animals walk, run, swim, jump and fly (Fig. 4.10). Borelli recognized that forward motion entailed the movement of a body's centre of gravity forward, which was then followed by the swinging of its limbs in order to maintain balance. Moving on to bird flight, he presented how this ability is supported by a combination of the immense power of the pectoral muscles and the light weight of the skeleton (Hall 1951). He then described how a bird can fly perpendicular to the horizon with a transverse movement parallel to the horizon; how birds, without wing flapping, can rise into air both horizontally and obliquely upward; and why it is impossible that men should be able to fly, by their own strength (and hence the Icarian invention is entirely mythical; for it is impossible to increase a man's pectoral muscles or to diminish the weight of the human body; whatever apparatus is used, although it is possible to increase the momentum, the velocity can never equal the resistance; and therefore wing flapping by the contraction of muscles cannot give enough power to carry up the heavy body of a man; Hall 1951).

In coming to deal with the causes of the contraction of muscle fibres, Borelli suggested the existence of "animal spirits" characterized as a "very subtle self-moving fluid" that we cannot see and which is transmitted from the brain along the hollow nerve fibres. This transmission works on the hydrodynamic analogy of a tube filled with liquid, from whose further end a quantity may be squeezed out by a light

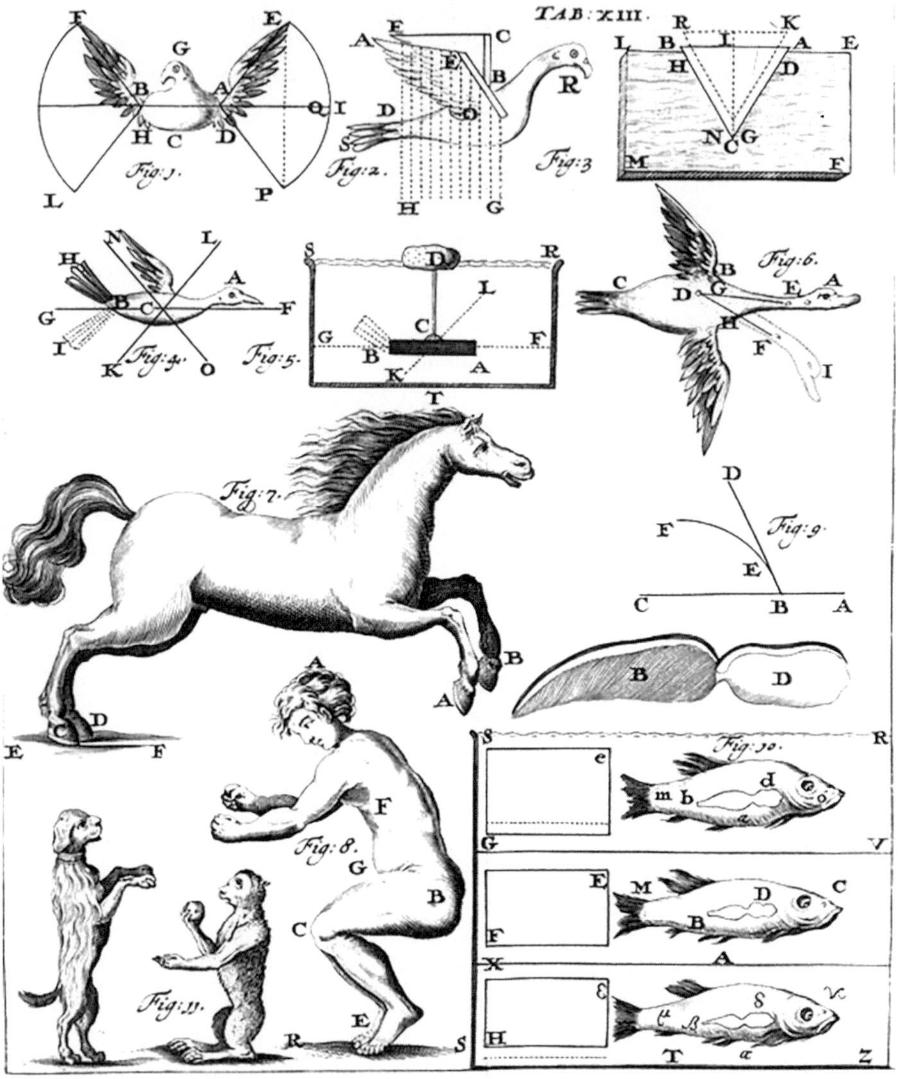

Fig. 4.10 Borelli's *De Motu Animalium*: biomechanics of birds in flight, in different positions of the head and of the tail; biomechanics of terrestrial mammals, their hind legs serving to push forward while running and to push upwards when rising; and biomechanics of fishes, their swimming bladder in different volumes

pressure at the near end. The brain, through repeated overall action on the near ends of the tubes, "learns" by trial and error to squeeze only the appropriate tubes, as may be inferred by watching the blundering efforts of small children to walk.

4.4 Diversity of the Small

During the late seventeenth century, the detailed study of animal diversity began extending beyond the well-studied vertebrates into invertebrate groups. One group of invertebrates in which such a study was then advanced was the molluscs, with their protective calcareous external shell, both beautiful and hardy enough to survive long voyages. With the discovery of new lands during the sixteenth-seventeenth centuries, travellers began importing shells to Europe for the curiosities market. A hobby began of building up a collection devoted specifically to shells, rather than a general collection of all kinds of curious natural objects. The resulting shell cabinet created the need to organize and present the shells by one system or another, and this called for a classification. Fortunately, a shell often gives some insight into mollusc taxonomy.

One of the earliest books on shells was Lister's *Historiae Conchyliorum* (1685–1692; "conchology" from the Greek *Konkhos*, a cockle). Martin Lister (1639–1712) of Buckinghamshire was the son of an MP in the Long Parliament and nephew of a royal physician under whose direction he was educated. To study medicine, he travelled to Montpellier, famous for its university, where he spent 2 years but did not graduate. It was there that he first met another Englishman, John Ray (whom we shall meet further on), and this became the beginning of a long and fruitful intellectual association between the two; 20 years later and in London, they would cooperate in the production of the lavishly illustrated *Historiae Piscium*. At Montpellier, Lister also met Nicolas Steno, then an expert at dissection, who eventually turned to the study of fossils, and the disagreements between the two I shall also discuss further on. Aged 31, Lister returned to England and settled at York to practice medicine, and a year later, he became a Fellow of the Royal Society. Eventually, he moved to London, became a Fellow of the Royal College of Physicians and at age 70, he received an appointment as physician to Queen Anne.

Beyond being a busy physician, Lister was devoted to the study of shells, and he was one of the first, in early modern times, to publish on their systematic classification. In 1685, he distributed among his friends a modest volume of copperplate engravings of land shells, and he published further volumes over the following 7 years. Eventually, he combined them all into one large treatise, the *Historiae Conchyliorum*, published between 1685 and 1692. This treatise is an illustrated guide for shell collectors, a collection of nearly 500 folio leaves containing over a thousand engraved plates of shells. His two daughters carried out the drawings and engravings, and their quality is such that many of them are today recognizable down to genus and often even to species level. Lister accompanied each engraving, not by the binomial nomenclature as of today, but by descriptive sentences engraved on the plates of the figures. The book preserves the old name for cowries in European culture, *Conchae Veneris*, which expresses the shell's resemblance to a woman's genitalia (Fig. 4.11).

Lister's book is indirectly associated with the Dutch painter Rembrandt, who in 1650 engraved a print of the marine cone shell *Conus marmoreus*. Rembrandt was

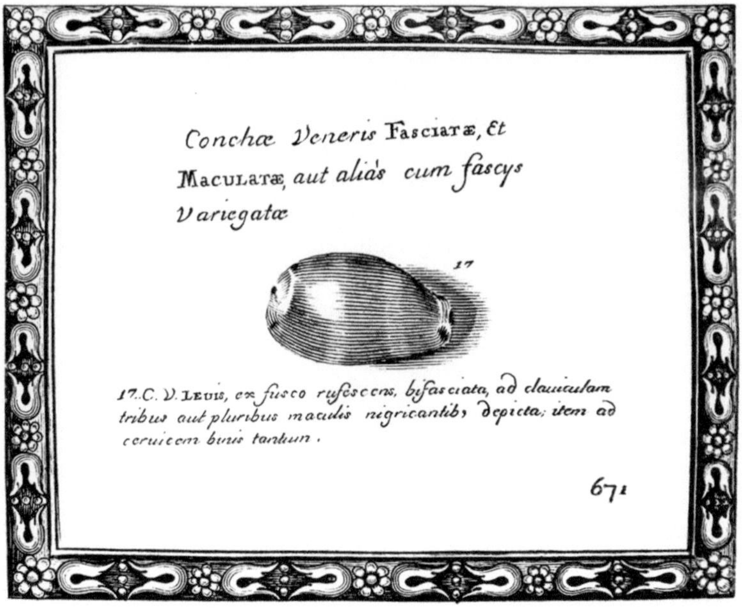

Fig. 4.11 *Historiae Conchylium*: a cowry, here in its older name *Conchae Veneris*. (Courtesy Mollusc Collection, Hebrew University of Jerusalem)

careful to engrave his name on the copper plate in mirror writing but forgot to engrave the shell itself in mirror image, so the print showed a left-whorled shell. Lister's daughters noticed this detail while preparing the illustration for their father's book. They used Rembrandt's drawing but turned it over in the right direction. In the lower margin, they noted that their figure was adapted from "Rambr. V. Ryn" (Fig. 4.12).

The general picture emerging from leafing through the book is that Lister's *Historia Conchylorum* broadly parallels our current phylum "Mollusca". It rightly included snails, bivalves, cephalopods and tusk shells. As *Historiae Conchyliorum* was concerned with shell classification with little awareness of the animal within, it encompassed, alongside gastropods, bivalves and relatives, such other animals with calcareous shells as goose barnacles (Fig. 4.13) and tubeworms. Among terrestrial snails, Lister, very wisely, did not regard the shell as a single, group-defining character. He also considered internal characters, mainly of the genitalia. Consequently, beyond snails with external shells, he also included slugs. Likewise, among (recent) cephalopods, he included both those with an external shell (*Nautilus*) and those with their inner remnants (squid). He did not include sea slugs, however, which also lack shells, probably because shell collections of those days did not include them. He also did not include the chitons, today included in the molluscs, again, perhaps because shell collections of those days did not include them.

In Lister's treatises, a dim, emerging classification of shell diversity began, and it was global from the start. He represented each shell by a figure, a describing

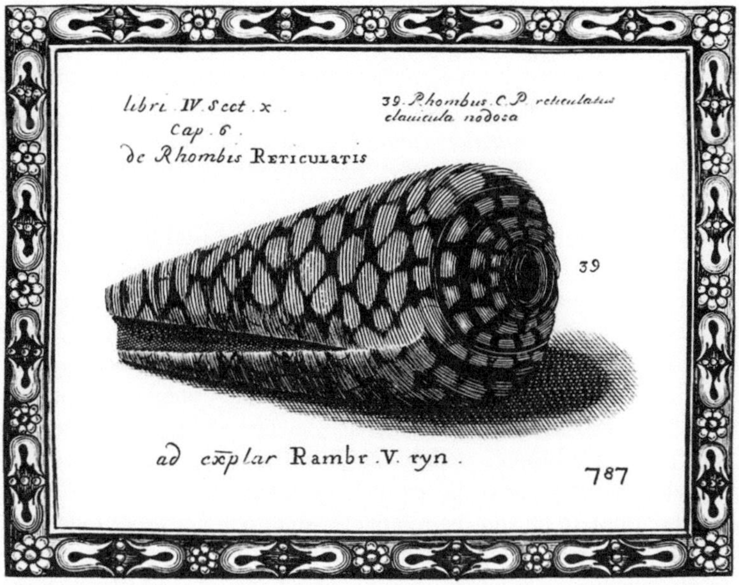

Fig. 4.12 *Historiae Conchylium*: Rembrandt's engraving of *Conus marmoreus,* flipped to its correct view. (Courtesy Mollusc Collection, Hebrew University of Jerusalem)

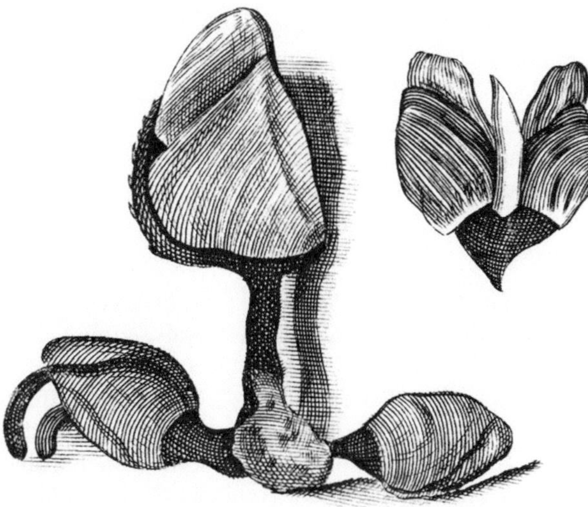

Fig. 4.13 *Historiae Conchylium*: a goose barnacle. (Courtesy Mollusc Collection, Hebrew University of Jerusalem)

sentence and, where known, the crude locality: England, Mediterranean Sea, Africa, Jamaica, Carolina and so on. He classified the shells by their major habitat (terrestrial, freshwater, marine) and then, in each of these habitats, by classes (of today: gastropods and bivalves). In each of these classes, he seems to have recognized genera, which he distinguished by a name taken from the Latin. As there were only a few such Latin names available in the ancient Roman texts, this resulted in many shells being described without being named. The various shells were described by a term (the genus) followed by a specific name composed of several words that described the species; thus, *Buccinum septem orbium fascys fere bicoloribus interdum tricoloribus iridus instats depictum* is for a terrestrial snail from Barbados. In vertebrates, naturalists in Lister's days had already displaced such polynomial names with binomial ones.

Historiae Conchyliorum leaves the impression that Lister had started off by taking a very broad outlook of the entire shell collection of all habitats at his disposal; he then clustered all these shells into groups (genera), and only then did he settle down to illustrate and briefly describe them. Accordingly, he was aware of the resemblance between, say, freshwater nerites, which he described in one section, and marine nerites, which he approached in a much later one; and so, he termed the freshwater ones *Nerites fluviatilis*. Likewise, aware of the resemblance between marine and freshwater limpets (today, incidentally, recognized as two completely different phylogenetic groups), he named the latter *Patella fluviatilis*. Indeed, it takes a broad outlook to cluster the relatively small pelican's foot together with the huge and very different conchs, as he (correctly) did. From Lister's clusters, it appears to me that among bivalves, he recognized at least the following groups:

Unionidea	Freshwater mussels
Pectinidae	Scallops
Ostreidae	Oysters
Spondylidae	Thorny oysters
Anomoidea	Saddle-shells
Pterioidea	Pearl oysters, hammer oysters
Arcoidea	Arc shells
Solenidae	Razor shells
Telinidae	Telins
Pinnidae	Pen shells
Pholadidae	Piddocks, angel-wings

For one of the very first books on bivalves, this is a very good classification. Further, within each of these major bivalve groups, Lister sought to form subgroups by shell details. He thus divided the scallops into groups based on the equality or inequality of the valves and "ears" and further differentiated the species by reference to the number of ribs. Among gastropods, he distinguished between limpets, keyhole limpets, hoof snails, cowries, olive shells, abalones, nerites, cones, conchs and many others. Here and there along the book, he admittedly presented also figures of live snails and bivalves, but his main interest was in classifying shells, not animals.

In attempting to follow Lister's logic in organizing shell diversity within each of these major groups, I focus on the order in which he presented gastropod shells in some sections in the fourth part of his book. He began by clustering all those with virtually no spire: limpets, keyhole limpets and hoof snails (section III). He placed here also sea tusks because of a morphological continuum from limpets, in which the non-spiral shell has only a single aperture; via the keyhole limpets, in which there is, in addition, an opening at the apex; to the tusks, in which there is one aperture at the base and another at the apex. He placed in this section also *Tenagodus* (Siliquariidae) by slightly expanding the definition of this section to also include those in which an originally spiral shell secondarily loses its spiral during growth and becomes non-spiral, long and pipe-like. Next in his book (section IV, and again we must remind ourselves that Lister is interested mainly in classifying empty shells, not living molluscs) came the shell-possessing cephalopods (*"Polypis testaceis, sive Nautilis*"), as they too lack a spiral shell. In this section he placed several *Nautilus* shells, the egg brooding case of the pelagic paper nautilus *Argonauta* (an octopus in today's classification) and also a tiny white, barely coiled shell of a tube-worm (Polychaeta, perhaps *Spirorbis* of the lower littoral and sub-littoral; it usually lives attached to seaweeds; Fig. 4.14).

Then (section V) come several gastropod shells presented in an increasing order of spire height relative to the aperture: moon shells, violet shells (he here, exceptionally, included also a live snail), some top shells and turbans, wentletraps and, finally, auger shells. This group is very weak in that, among its trochoideans, there are what seem to me to be a few misplaced terrestrial gastropods. Section VI is devoted to a global view of the nerites, section VII to a global view of the abalones, then (section IX) comes a global view of cowries, section X to global views of the olives and cones and so on via the conchs to ceriths. The only case I could trace of the splitting up of a major natural group was that of the Trochoidea (presented once in section V and then again in section VIII).

Lister studied both recent and fossil shells. In addition to presenting the shells of live molluscs, he included in his treatise two appendixes (*de Conchitis: ysve lapidubus qui quandam similitudinem* cum *conchis marinis habeant*; and *de Buccinitis: ysue lapidibuus qui buccina omnigena vale referant*) in which he presented beautiful illustrations of fossil gastropods, bivalves and ammonites (Fig. 4.15). While he recognized the similarity of fossil molluscs to living forms, he could not bring himself to admit that they were a group of fossils; he regarded them as inorganic imitations produced by the rocks. We shall come back to Lister's ammonites further on.

Beyond molluscs, another major group of invertebrates into which the detailed study of animal diversity began extending during the late seventeenth century was the terrestrial arthropods. We recall Aldrovandi, who, during the late Renaissance, in 1602, had classified them into groups by legs and wing morphology, by their readiness to take to the air and by their "nobility". First come bees, wasps and hornets; then flies and dragonflies, mosquitoes, butterflies, locusts, grasshoppers, cockroaches, beetles and crickets; then the ground-dwellers with legs (spiders, earwigs, scorpions, ants, lice, caterpillars and ants) and various vermin (clothes moths, fleas, lice and ticks); then worms (terrestrial and parasitic).

Some decades after Aldrovandi, another book on terrestrial arthropods and worms came out, *Theater of Insects* (*Insectorum sive Minimorum Animalium Theatrum*; the Latin version came out before the English version). The complex history of this book goes back four decades before Aldrovandi's book. Thomas Penny (1532–1589), an English physician and naturalist, collected, studied and

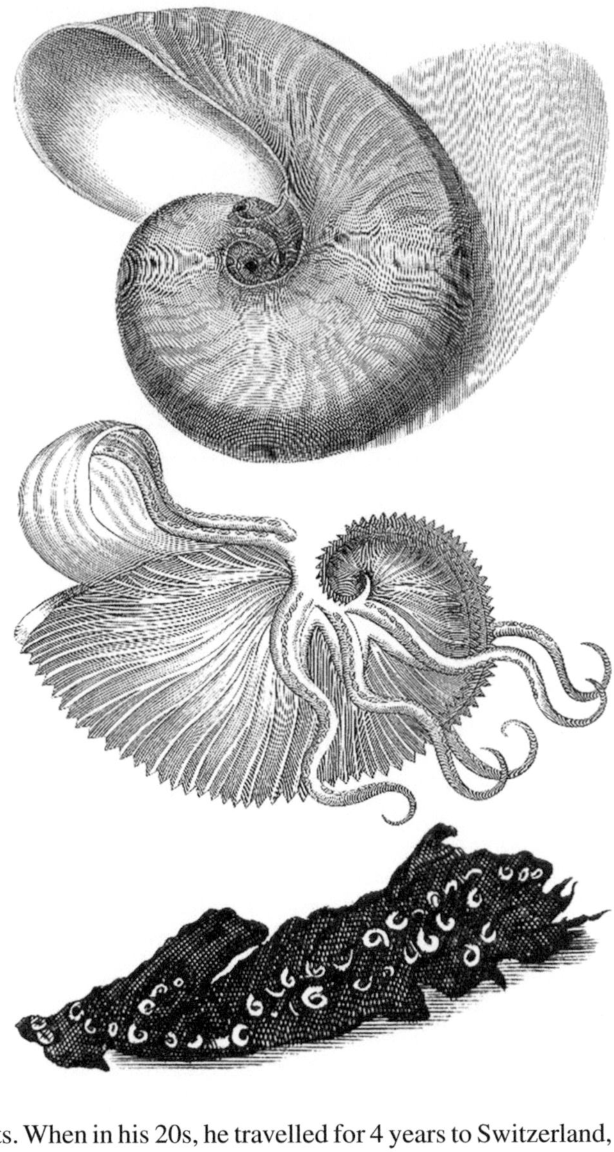

Fig. 4.14 *Historiae Conchylium*: above, *Nautilus*; middle, *Argonauta* (paper nautilus, an octopus) in its egg brooding case; below: *Spirorbis*, a tubeworm, on an alga. (Courtesy Mollusc Collection, Hebrew University of Jerusalem)

drew both plants and insects. When in his 20s, he travelled for 4 years to Switzerland, Germany and France, and, as one of the many collectors who had sent plant specimens to the great Swiss naturalist Conrad Gesner, when in Zurich (1565), he presented Gesner with further drawings and specimens of English plants. Gesner, we recall, had attempted to cover all knowledge of the entire animal world and had already published four volumes (1551–1558, on mammals, lizards with amphibians, birds and aquatic animals). Papers left after his early death contained enough material on "serpents" and "dragons" to be published posthumously (1567). Gesner also wrote drafts on terrestrial invertebrates. In exchange for Penny's botanical materials, Gesner made a gift to Penny of pictures and notes on terrestrial

Fig. 4.15 *Historiae Conchylium*: an ammonite. (Courtesy Mollusc Collection, Hebrew University of Jerusalem)

invertebrates he had been gathering over the years. Gesner's notes and illustrations would later form the nucleus of Penny's "insects" book. Upon returning to England, Penny began exchanging specimens, images and notes with a circle of natural history correspondents. Sometimes he would accept just the image of an insect standing in for a specimen. He died early, and upon his death, his friend and collaborator Muffet edited and completed the manuscript.

Thomas Muffet (1553–1604), like Penny, was an English physician and a naturalist; in his later years, he was a Member of Parliament. Upon inheriting Penny's collection of insect materials, he continued Penny's practice of obtaining information through correspondence with a wide range of people of widely varying backgrounds—physicians, apothecaries, naturalists, noblemen, merchants and travellers. Editing and expanding the book, he did not live to see its publication, which occurred only 30 years after his death. The manuscript, after Muffet's death, passed first to his son, then to his apothecary; and eventually a certain Huguenot physician who had settled in London, Theodore de Mayerne, came to own it. In 1634, he finally had the manuscript published as *Insectorum sive Minimorum Animalium Theatrum*, a book (in Latin) with three authors: Edward Wotton (of whom I have no information), Gesner and Penny; Muffet was the editor (1658).

Theatrum insectorum was thus a product of Gesner's notes as a dominant basis, upon which Penny and Muffet, in their efforts to gain specimens, images and information, placed a wide range of information culled from numerous classical and contemporary sources, created by the collaborative effort of a wide range of naturalists. Preparation of the book spanned several generations, ranged widely across Europe and encompassed a variety of scholarly backgrounds. The title page is representative in that "Insects" include such terrestrial invertebrates as scorpions, millipedes and spiders and in that the book places heavy emphasis on bees (Fig. 4.16).

INSECTORVM
SIVE
Minimorum Animalium
THEATRVM:
Olim ab
EDOARDO WOTTONO.
CONRADO GESNERO.
THOMAQVE PENNIO
inchoatum:
Tandem
THO. MOVFETI Londinatis opera sumptibusq; maximis concinnatum, auctum, perfectum:
Et ad vivum expressis Iconibus suprà quingentis illustratum.

Londini ex Officinâ typographicâ *Thom. Cotes.* Et venales extant apud *Benjam. Allen,* verticulo, quod Anglicè dicitur *Popes-head Alley.* 1634.

Fig. 4.16 *Insectorum Theatrum*, title page

The story goes on. Coming decades after Muffet's death, an English cleric, by the name of Edward Topsell, had published *The History of Four-footed Beasts* (1607), which was a translation into English of Gesner's treatise on mammals and *The History of Serpents* (1608), an English translation of Gesner's reptiles together with Gesner's (unpublished) manuscript on scorpions. There were also three sections (on bees and related insects, spiders and earthworms). Five decades later (1658) and a century after Gesner's death, he reprinted these two treatises together (*The History of Four-Footed Beasts and Serpents*, 1658) and with them an English translation of the Wotton-Gesner-Penny-Muffet treatise, *Theater of Insects*. Topsell's book is thus in part a translation of Gesner's published works, in part a translation of unpublished manuscripts written by Gesner and in part an assemblage of several subsequent writers.

Topsell was not a naturalist, and he crammed into his 1100-page treatise many bizarre ancient and fantastic myths. The weasel gives birth through its ears, elephants become pregnant by chewing on mandrake and the adder, when thirsty, casts up its venom at the waterside, lest, while drinking, it would swallow and kill itself; having quenched its thirst, it licks up its venom again, like "a Souldier re-armed after he was disarmed". Topsell further presented tales of elephants falling in love with women, of a dragon avenging the death of his human friend and of a lion persuading a woodsman to cut down the tree in which a bear that had killed the lion's cubs had taken refuge. Besides many actual animals, he included such fabulous creatures as gorgon, sphinx, manticore, winged dragon and the unicorn. Coming now to his treatment of animal diversity, Topsell listed his animals in alphabetical order by their English names, in mammals from "Antalope" to "Zibeth" and in serpents from "Adder" to "Viper". He grouped, however, monkeys and baboons under the A's with apes (in which he included also satyrs and sphinxes), camel with dromedary and hyaena with crocuta. He did not group dog with fox or wolf, ass with horse or mule or cat with lion or tiger. His book on serpents included our reptiles (snakes, lizards, crocodiles and tortoises) and amphibians (frogs, toads and salamanders), together with several terrestrial invertebrates (bees, wasps, hornets, Spanish flies, caterpillars, scorpions, spiders and earthworms). After these two volumes came a third volume, an English translation of the book on terrestrial invertebrates: Penny and Muffet's *Theater of Insects*. Though it did not offer a formal classification of terrestrial arthropod diversity, the text broadly followed Aldrovandi's order. Taking the section of wingless invertebrates, the book classified them first by habitat (terrestrial or aquatic) and then by feet: all insects without wings either belong to the

Earth,

With feet,

Many feet: Caterpillars, beetles.
Eight feet: Scorpion, spider.
Six feet: Wasps, glow-worms, the female meloe; worms in wood, trees, roots, fruits, meats, garments, chambers.

Without feet: Parasitic-worms, earthworms.

Water,

With feet,

Six feet: Shrimp, lake-scorpion, *Notonectus*.
Many feet: Sea scolopendra, many-footed shrimp.

Without feet: Horse leech, hair worm.

We feel Gesner's dominant influence in that for each animal, where possible and relevant, comes the name in both classic and several modern languages; then a detailed description of the body, habits and behaviour; and then use in the service of man, mainly in medicine. The book, with its excellent mastering of languages and of classic literature, leaves the impression that it is mostly that of Gesner; Topsell only added notes here and there.

The text echoes, however, also the flow into biological thought of socio-political yearnings of Topsell (and others) for an idealistic constitutional monarchy. This flow is to such an extreme extent that in the following paragraphs, I present it in some detail.

Bees live in a monarchy, in which they receive their king by election, not by succession or by casting lots. The elected king is deeply bound to his bees by an oath while they still keep their ancient liberties and privileges. He is merciful, and though possessing a sting, he seldom uses it against his bees. However, he will not tolerate rebellious, obstinate and unruly bees, which he first wounds and torments with his sting and then expels from the hive. As captain general of the whole swarm, he is twice as tall as the rest; his wings are shorter; his legs are strong; his gait, pace and manner of walking are lofty, stately and upright; on his forehead there is a white spot like a shiny diadem; and he differs from the commons in his comeliness, beauty and honour. All bees in the hive lovingly accept and honour him as their monarch. The kings of each hive form a distinct royal race in which the young are born winged and immediately able to fly, in contrast to the commons that are born as worm-like larvae. Each hive also has dukes and princes, who are larger and stronger than the commoners are. Education of the royal princes is very strict, and the king, if he finds any of his sons to be a fool, or unhandsome, or rugged, rough, ill-tempered, ill-shaped or ugly—destroys that son by common consent and by parliamentary authority.

To govern the hive, the king prescribes laws, and he imposes missions on each bee: who shall be his bodyguard, who shall bear water to the king and to those that are at work, who shall build and maintain honeycombs and who shall clean them. Some are to gather honey, others to concoct it or bring it to specific cells; some must seek distant pastures. Some care for the young, lest they suffer cold or hunger. Other bees undertake to cure those that are bedridden, feeble and aged by collecting medicinal nectar (saffron, violet) and mixing it into a medicinal honey. Some must drive away spiders that might seek to cover the entrance of the hive with cobwebs. When a bee dies, special body-bearers meet and carry the corpse on their shoulders outside the hive, lest it pollute or defile their clean and neat hives. So great is the workers' care for the well-being of their king that whenever he goes abroad, they gather close and guard him as he flies along. If he tires of flying, the guard bears him home. He appoints the young and strong members of the hive to labour. If they stumble during these labours, the king steps in: though exempt from physical work, when needed, he applies himself to the task. Young or middle-aged bees go abroad and bring home those things that the king demands. In sum, the bee commonwealth is a wonderful and well-ordered world. The hive reflects an ideal society run by kings (Prete 1991).

In battle, the king marches in front of his winged army, being the first to expose himself to perils. Bees wage wars that are either civil or against foreign enemies. Civil wars result from diverse causes, one being the nobility being treacherous to king and state, another being narrowness of space. If the nobility multiply and become too many, the king puts to death all the superfluous, lest in their increasing numbers they join forces and overpower him. Bees cannot endure more than one king, and the swarm will destroy any usurpers. The swarm also kills drones, both in want of room and because they steal honey. When their honey supply fails, bees fight amongst themselves, as it were, for life and blood, and the winning side puts all losers to the sword, yielding no quarter or truce. In war with foreign enemies, bees boldly attack man, beast, bird or wasp that molests them, wounding them with their stings. They do not endure unclean persons, heavily perfumed people, those wearing curled or ruffled locks, or red clothes or base and vile people. If any of them have lost his sting in a skirmish, as a soldier having his arms taken from him, he is quite disheartened and, living not long after, dies with grief. Bees, when they go forth to battle, surround their beloved king. Occasionally a king becomes negligent of the commonwealth or becomes a tyrant. In such cases, his subjects do not immediately kill him; they merely crop his wings, and if he amends his manners, they love and respect him once again. The workers never desert a good king. When he dies, both the nobles and common people lament him. Afterwards, they carry him out of the hive, and eventually, by grief and hunger, they all die. Take the king, and you take the entire swarm.

In building their hives, bees fence themselves against weather, vermin and enemies. The royal quarters are of a very fine wax, and a defensible wall fences them in, in the highest and largest part of the hive; those that are aged and stricken in years (and serving as counsellors of state) have their lodgings near the king's court, and the king's children sleep nearby. The whole honeycomb contains four types of cells: for the bees, the drones, the gentles and honey. When night approaches, a honey pipe gives a sign, a watch is appointed and the bees go to bed. In the mornings, they are all still and silent until the king gives three hums and wakes them up and then everyone hastens out to his duties. In the evening, when they return home, they at first make a great noise, but after a while, they quieten down, till at length the captain of the watch flies about and makes a buzzing, commanding them to their rest. After this signal is given, they are all so hushed and still that if you were to put your ear to the hive's entrance, you would hear nothing.

Bees generate spontaneously from rotting carcasses. The best and most noble bees generate from the carcass of a lion; they have lion-like courage and fear nothing (we recall the biblical story of Samson, who killed a lion and, after some time, found a swarm of bees in its carcass). Next come bees generating from the carcass of a bull, another strong and valiant beast. Then come cow bees and ox bees, which are very industrious in work and are not inclined to anger. Calves' carcasses generate the softest and most tender bees, who are excellent makers of honey but do not endure hard work. In each carcass, the kings and princes, who excel the others in comeliness, size, prudence and strength, generate from the brain, whereas the common, ordinary bees generate from the flesh. Some of the individuals in the hive are males. These are the chief leaders, and they are stronger, greater and more able than the others, and they stay at home with their many females. Wasps and hornets differ from bees in not producing honey, in being flesh-eaters (including bees) and in that they generate spontaneously in the dead bodies of horses, not bulls.

The drone cheats the bees since, under the pretext of keeping the hive warm (which is its duty), it ransacks the combs. It is without a sting, and it is idle. Drones are weak and impotent; they are generated from old, exhausted bees and are not fit to propagate their species. This is why the other bees rule over them, putting them to work and punishing them without mercy. Two or three (especially of the younger fry) may drag out a drone, buffet him with their wings, gore him with their stings, cast him down upon the ground and finally kill him. There are reasons for this expelling and killing of drones. If their number increases too much, there will not be enough room left for the bees and if the honey supply fails, the bees will be in want of provision. The drone hides all day among the honey cells and preys upon their honey at night when the bees are asleep. It may prey upon the honey also during the day, when the bees go to pasture. If the bees catch a stealing drone, they crop its wings, thrust it out and banish it (Prete 1991).

The Almighty, just as He had created so many other things for the use and service of man, had created bees for us as examples of political and economic wisdom, instructing us in divine knowledge to deepen our understanding of our own socio-economic systems. He created bees, honey and wax for use in medicine (Gesner, Penny, Muffet and de Mayerne were all physicians). Bees are also foretellers of events to come. When Leopold of Austria marched on Sempach (1385), a swarm of bees flew into the town, thereby foretelling the people of the invading force. The Swiss, in the battle that followed, defeated the Austrians and Duke Leopold lost his life. Flies are another group of insects that intervene in human fates. When Charles, Count of Valois of France, led an army into Spain against King Peter of Aragon with 16,000 cavalry, 17,000 crossbowmen and 100,000 infantry, suddenly an army of flies set upon the French and slew them (1285); this is just so, as written in the *Theater of Insects*.

Whatever the *Theater of Insects*' contribution to the broad, general view of entomology, in the specific case of bees, it was far outdated. An English parish priest, Charles Butler, thought that bees live under a monarchy, which he considered the "most natural and absolut' form of government". Relying on his observations and dissections, he recognized that the queen and the workers are female and the drones are males; he believed, however, that worker bees lay eggs. He further observed that bees produce wax from their bodies and not, as previously thought, that they gather it from plants (*Feminine Monarchy*, 1609, revised 1623 and 1634). Butler was not the first to describe the largest honeybee as a queen rather than a king—the Spaniard Luis Mendez de Torres (1586) had stated clearly that the leader of the hive lays the eggs, from which all workers, drones and future queens develop. Coming some 30 years after Butler, Richard Remnant (*A Discourse or Historie of Bees,* 1637) already understood the hive to be a "female Monarchie", ruled and run by females. The females produce the drones, or males, only when they need them to fertilize the eggs and keep the hive warm. Beyond these two tasks, the males do no work, living on the labour of the rest and in the autumn, they are killed by the females because they are the great "eaters and wasters of the winter provision". Remnant was uncertain about the generation of the queen bee. He reasoned that if the hive represented a monarchy, the monarch certainly could not be born of a commoner. However, he reasoned, if we argue that the queen bee gives rise to new queens, we face the impossible task of finding her a royal mate. After all, the males of the hive are all lazy, common drones; there is not a regal one among them. Remnant wondered whether the existence of a feminine monarchy among the bees is an indication that "it is lawful for women to govern in a commonwealth [or] in private houses and families". He answered that a woman may govern the home, if God has made her able, only in her husband's absence or weakness (Prete 1991).

4.5 Diversity of the Tiny

4.5.1 Looking Deep into the Invisible

One event in the mid-seventeenth century that considerably influenced biological thought was the development of the microscope. Magnifying lenses made of polished beryl were used to magnify manuscripts as early as the eleventh century in the Mediaeval (hence the German *Brillen*, spectacles; rock crystal was also used as a magnifier). To increase the power of a magnifying lens, one had to reduce the focal length, but this required a reduction of the lens diameter and after a point, the lens became difficult to see through. The invention of the compound microscope system in the seventeenth century solved this problem. It incorporated more than one lens so that one lens further magnified the image magnified by another lens. The first compound microscope, with two glass lenses positioned one in front of the other, was built in the 1590s in Holland; its magnification could be adjusted between ×3 and ×9. These first microscopes were more of a novelty than a scientific tool, as maximal magnification was only small and the images were somewhat blurry. Eventually, in 1609, Galileo Galilei made a microscope by inverting one of his telescopes and by 1624 he was already in possession of an instrument "of no greater height than a dining room table" that could enlarge by ×36, showing "a fly as big as a hen". These and other early microscopes were cumbersome because they required the user to push or twist an inner tube into or out of an outer one, which was not easy to do without jarring the specimen. In spite of these difficulties, Francis Bacon (*New Atlantis* 1624) presented a vision of the importance of the microscope in future service of biology. In his utopian island, he wrote, "We have also glasses, and means to see small and minute bodies perfectly and distinctly; as the shapes and colours of small flies and worms, grains and flaws in gems, which cannot otherwise be seen". This very list of Bacon would turn up a generation later, as we shall presently see.

The first actual use of a microscope in a biological context was by Federico Cesi (1585–1630), Duke of Acquasparta. Cesi grew up with a life ambition to establish a group whose members would devote themselves to secular learning (Freedberg 2002). When only 18, he founded a scientific society devoted to the understanding of nature through observation, experiment and the inductive method. He named it the *Accademia dei Lincei*, the Academy of the Lynxes, because of this animal's reputed keen vision (in a classic poem by Horace). The academy's motto was: "take care of small things if you want to obtain the greatest results" (*minima cura si maxima vis*). Cesi intended that communities of Lynxes should establish themselves in every city, on the model of an active monastic order, to create a widespread network of scholars devoted to new ideals in natural science. In reality, he founded only one such colony, in Naples. The handful of academy members remained scattered singly, each in his town, and the unifying force of the group was Cesi himself, who corresponded copiously with them, supporting the society not only with his interest but also with his money. There were four founding members (one was his friend Francesco Stelluti, to be mentioned later); a fifth joined in 1610, and in 1611, Cesi took on his sixth member, Galileo Galilei. Cesi's Academy was responsible for publishing two of Galileo's books (*Letters on Sunspots*, 1613, and *The Assayer*, 1623); the lynx icon, together with the term "Linceo", appeared on their title pages (Fig. 4.17). Furthermore, Cesi helped Galileo in overcoming the ecclesiastical authorities. Galileo had obtained telescopic evidence that Copernicus was right in claiming that the Sun, not the Earth, was the centre of the world, but in attempting to publish his conclusions, he ran into censorship problems. Cesi then advised him to avoid the censor by omitting all scriptural references and referring to Horace instead ("But Worth opens wide the gates of heaven to those who have done immortal things"; Horace was a Roman poet of the first century BCE). Cesi died suddenly at the age of 45, and the *Accademia dei Lincei* dissolved soon after his death. It was recreated by Pope Pius IX in 1847 and given the name *Accademia Pontificia dei Nuovi Lincei*, the Pontifical Academy of the New Lynxes. Today this Academy is an independent body within the Holy See.

Cesi, like Aldrovandi before him, set out to document in drawing everything in the world around him, and this included the world of nature as seen through the microscope. He was a lifelong friend of Galileo, who sent him (1624) an instrument that he termed an *occhialino*, for observing small objects. It had a tube whose length could be adjusted and a movable plate on which the object to be viewed could be positioned. Galileo's accompanying letter described how, looking through the *occhialino,* he had observed very many small animals, among which the flea was the

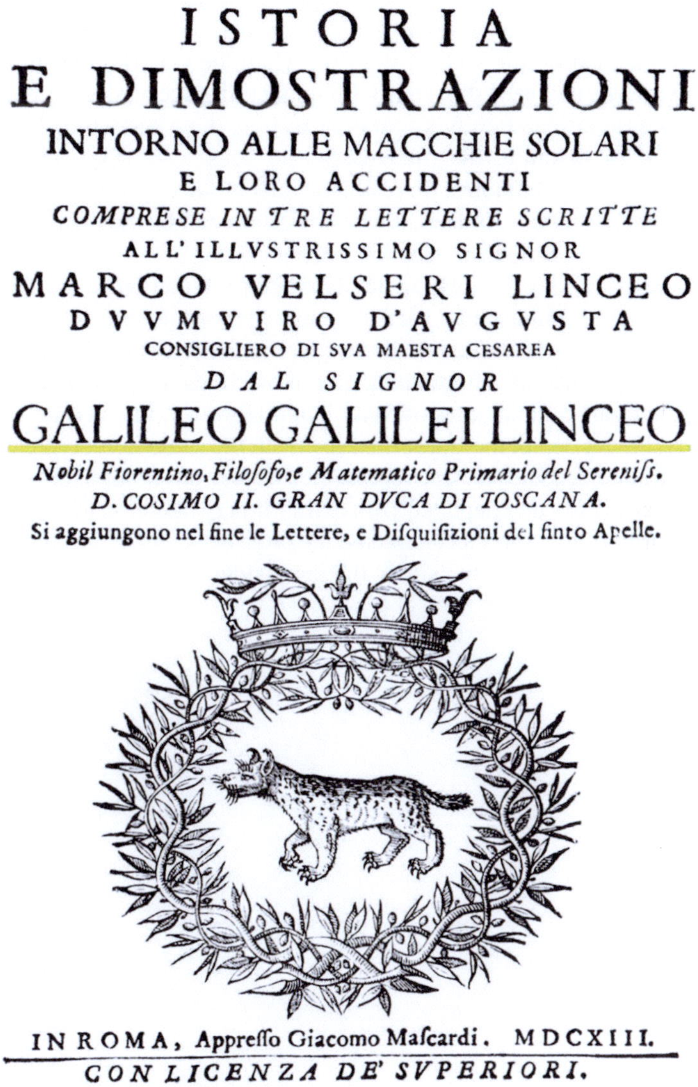

ISTORIA
E DIMOSTRAZIONI
INTORNO ALLE MACCHIE SOLARI
E LORO ACCIDENTI
COMPRESE IN TRE LETTERE SCRITTE
ALL' ILLVSTRISSIMO SIGNOR
MARCO VELSERI LINCEO
DVVMVIRO D'AVGVSTA
CONSIGLIERO DI SVA MAESTA CESAREA
DAL SIGNOR
GALILEO GALILEI LINCEO
Nobil Fiorentino, Filosofo, e Matematico Primario del Serenifs.
D. COSIMO II. GRAN DVCA DI TOSCANA.
Si aggiungono nel fine le Lettere, e Difquifizioni del finto Apelle.

IN ROMA, Appreffo Giacomo Mafcardi. MDCXIII.
CON LICENZA DE' SVPERIORI.

Fig. 4.17 *Letters on Sunspots* by Galileo Galilei "Linceo"

most horrid and the mosquito and the moth were the most beautiful, and how flies and other little animals walk on mirrors and thus can also be seen from below. A few months later, one of the Linceans replaced the term *occhialino* with a new term, *microscope.*

Cesi took great interest in this new instrument, as it occurred to him that members of his society might examine, through the microscope, yet another tiny creature, in addition to fleas, mosquitoes, moths and flies: the bee. In 1623, just before

Maffeo Barberini of Tuscany was elected to the papacy (as Urban VIII), the story goes that a swarm of bees entered the Vatican palace from the meadows facing Tuscany and settled on the wall of the cell, where he was waiting. It seemed that divine providence had sent this sign to announce the succession to the papacy of Barberini, whose Tuscan family's coat of arms bore three bees (Fig. 4.18). Within a few years of his election, the three Barberini bees were displayed everywhere throughout Rome: on fountains, on restored walls of the city, on tombs, on ceilings, in churches and on statues. Indeed, it was impossible to go anywhere in Rome without encountering the Barberini bees (Freedberg 2002).

The Linceans too set to work on an elaborate presentation of the Barberini bee emblem, and they published (1625) a large engraved sheet that they named *Melissographia* and which they dedicated to Urban VIII (Fig. 4.19). At its centre were three bees framed by a flourishing pair of bay branches, with two putti holding aloft the papal tiara and the keys of Saint Peter. However, this was no standard bee triad, and it did not depict the Barberini crest with the same stereotypical bees as on the coat of arms. The Lincean bees represented a close observation of bees from life, magnified many times and depicted with extraordinary attention to precise anatomical detail. Furthermore, the sheet depicted the three bees: one as viewed from above, one from below and one on the side—the three sides from which the bee could be

Fig. 4.18 The coat of arms of Urban VIII. (Tuvia Kurz)

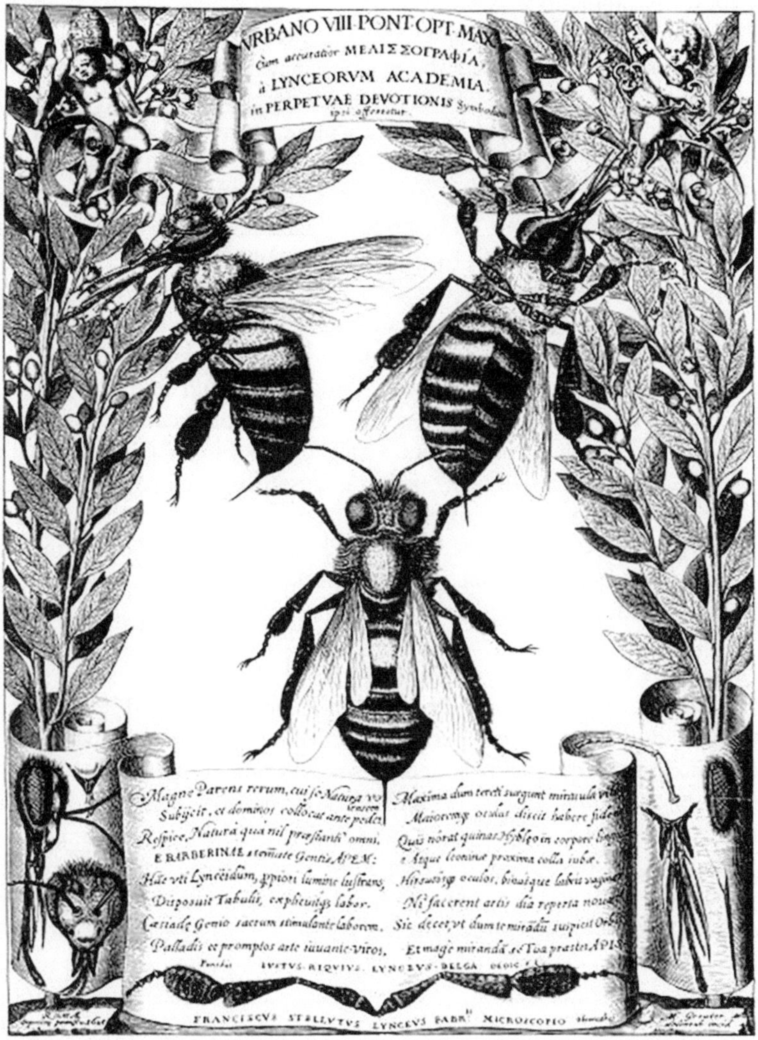

Fig. 4.19 *Melissographia*

viewed in life (or under the microscope), rather than presenting all three as viewed from above, as in the coat of arms. In addition to these three whole-figure bees, the scroll that unfolded across the bottom of the page displayed specific organs of the bee. On the left was a profile and a frontal view of the head with its eyes, antennae and the proboscis; on the right, an antenna, a single compound eye, the sting and the mouthparts forming the proboscis; and in the centre a pair of hairy posterior legs. This figure of the three bees, in the *Melissographia* sheet, was the first-ever printed illustration of a creature that anyone had made with the aid of a microscope. The inscription at the top of the sheet reads "To Urban VIII, Supreme Pontiff, an accurate description of a bee as offered to him by the Academy of the Linceans as a symbol of their perpetual devotion". On the scroll below, the text reads that nature has nothing more remarkable than the bee and that the Linceans, led by the genius of Cesi, have keenly appraised it with polished glasses and observed how the neck resembles a lion's mane, the eyes are hirsute, there are two sheaths in each mandible and the tongue is of complex structure. The text also noted that while the world looks up to the Pope in wonder, his bee shows itself even more worthy of wonder. At the very bottom of the page, an inscription stated that these figures of bees were "observed by Francesco Stelluti, Lincean of Fabriano, by means of a microscope" (*Franciscus Stellutus lynceus Fabr. microscopio observabat*). *Melissographia* was thus a sheet of carefully planned Lincean propaganda, aimed at flattering the Pope as well as looking into the future, that of microscopes (Freedberg 2002).

While the *Melissographia* was merely what we would term a poster, another tribute to Urban VIII was the *Apiarium*—the first-ever publication, which describes observations made with a microscope (Ball 1966). Published in the same year (1625), Cesi had intended it to be a chapter in a Latin encyclopaedia on natural history that he planned to write. This project was never realized, and the chapter on the bee was the only part ever published. Cesi dedicated the *Apiarium* to Pliny, who had written, back in Roman times, that the big secrets of nature lie in its smallest forms. If only Pliny would have had a microscope, what would he have said about the very tiny structures of the bee?

In *Apiarium*, Cesi presented a very broad range of references to bees from antiquity onwards and, with it, his new knowledge on bee anatomy, described with infinite care and attention to detail, bee habits, hive building, government and family life, various kinds of bees as reported by ancient and contemporary authors, and wax and its production. A separate section described bee anatomy: head, mouthparts, legs and divisions of the body. In this section, the details of miniature organs are remarkable. Cesi described the compound eyes as beautiful golden dice boxes, arranged in a network and separated by hairy columns, which form little lines. He described such hairy columns also on the wing fringes, where they are small and thinner. He described the parts of the mouth as instruments for honey making and noted that curved sheaths in four cavities enclose the very large and strong jaws. The four outer parts of the mouth form a sheath around the longer tongue in the middle, just as a flower bud is enclosed by its outer leaves; two of these parts have rigid tips and can move apart in a scissor-like motion. The bee uses hairy ridges on the four outer parts of the mouth to scrape nectar from the tongue. Then came a detailed description of the jointed legs, their three digits and the nails of these digits. This first-ever descriptive microanatomy is of a very high standard.

In strong contrast to the accurately detailed description of the microanatomy, the (non-microscopic) section on bee generation and reproduction was highly speculative. Cesi believed that the king [=queen] bee gave rise to all its workers and subjects, who protect the king and are protected by him. The process of generation starts with the honey itself, the basic formative material—from which the bee emerges. Freshly collected from flowers, the still liquid honey is in a constantly active "bubbling" state, containing a mass of disorderly particles, rudimentary but still unformed "figures" (Cesi's term). Gradually, by means of "struggles and embraces", these agitated and conflicting particles come together and interconnect; the material is progressively stabilized, order is established within it and the "figures" of the future parts of the bee slowly begin to take shape. The parts and "figures" of its body are formed by a process in which the basic particles both move and are moved, dissolve and are dissolved, until necessary connections for the formation of the animal are coordinated, ordered and established. The bubbling activity of the honey is now sedated and calmed but still contains some gentle, vital heat. This description of Cesi brings to mind the theories of Empedocles from 2000 years earlier, who, as we recall, had suggested that when animals form, at first only individual limbs and organs form and these wander around aimlessly; then, under the combining power of love, the limbs come together and join to form individual animals. It also closely resembles Francis Bacon's theory on the creation of the world (Ball 1966).

In Cesi's bees, the substance from which the bee was eventually to emerge had to undergo a kind of fermenting process, from which the living form would appear. At this point, the work of the father, the king bee, is crucial. He provides additional warmth for the production of what Cesi called the "seminal spirits", which he then introduces into the primordial substance. By this means, the substance is set into still faster motion, and the various figures now combine to complete the parts and organs of the bee. The father, king bee, thus creates offspring without engaging in any sexual act—the actual actions of the father in producing his manifold progeny are pure and free from any lust. The father further procreates by introducing his own vital spirits, which give the progeny their life. Having received the vital spirit of the father, the egg begins to form its round shape, and the embryo it contains, taking strength from the matter of the egg, puts forth its legs, arms and wings. It is at this moment, Cesi says, that "little mothers" stand by. They incubate the royal foetuses until they are complete.

Cesi intended the *Apiarium* to please the Pope and to solicit his support for the Lincei, which Cesi was rapidly finding himself unable to continue patronizing. Upon the death of his father, Cesi took over the management of his family's affairs at Acquasparta. The estate had been squandered by the elder duke, and the financial burden of publications was probably too great a drain on his resources (Freedberg 2002).

Five years after the *Apiarium*, in 1630, Francesco Stelluti published *Persio tradotto in verso schiolto e dichiarato*, a translation into Tuscan dialect of the works of a Latin poet, Aulus Persius Flaccus. He dedicated the work, which also displayed the Lincean emblem on the title page, to Cardinal Franscesco Barberini, the Pope's nephew and a member of the group, in an attempt to gain his patronage for the Academy. An obscure reference in one of Persius's poems to what may have been the ancient town of Eretum gave Stelluti his pretext for including the bee images since the former Eretum was then presumably Monterotondo, seat of the Barberini country estate and the Barberini family had adopted the bee as its emblem. Stelluti's bee image (Fig. 4.20) was similar, but not identical, to that in the *Melissographia*. It was smaller and unadorned and showed far more attention to detail of the antennae,

Fig. 4.20 Stelluti, *Persio*

legs, sting and head. Moreover, having described each organ clearly, he noted its colours with great precision (this attention to colour is almost absent in the *Apiarium*). In this account of the microscopic examination of the bee, he did not mention even one ancient writer. Furthermore, and most importantly, it was written in vernacular Tuscan, not in Latin. It was as though microscopic biology was, for the first time, addressing a different audience.

So far for bees. With Cesi, Stelluti had collected hundreds of specimens of (what we would consider as) fossil woods from the hills near Aquasparta. Upon inspecting them, he claimed (*Trattato del Legno Fossile Minerale* 1637; see Fig. 4.21) that these structures were not fossil plants; they were a type of clay, which slowly

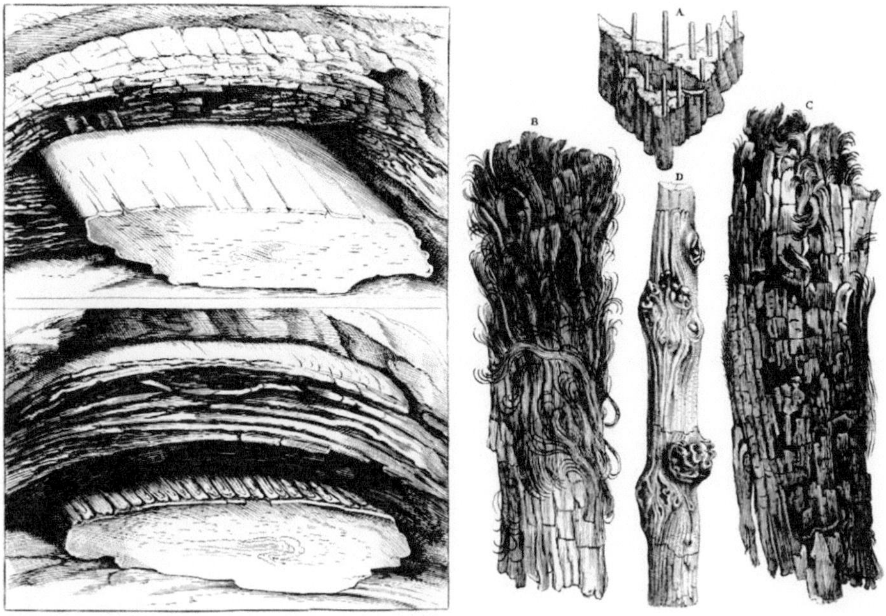

Fig. 4.21 Stelluti, fossil wood

became transformed into what seemed to be wood by underground heat and sulphurous waters. This would agree with the *Genesis* text, that man's life came from clay. Other naturalists coming just two decades later would question this opinion of Stelluti, as we shall see further on (Freedberg 2002).

From the first decades after its invention and onwards, the microscope assisted in two different directions of animal study. One direction was the study of the micro-anatomy of animal organs, and the dominant person in this direction was Marcello Malpighi (1628–1694). The son of an estate owner near Bologna, he was born in the same year that Harvey published *De motu cordis*, graduated as a physician from the University of Bologna and soon after joined the university medical staff, first at Bologna, then Pisa, Messina and finally in Bologna again. At the age of 63, he became chief physician to Pope Innocent XII in Rome.

Malpighi was the first to look into the microscopic structure of internal organs. He dissected and observed beneath the microscope the lung of the frog (*De Pulmonibus* 1661), and here he discovered the missing link in Harvey's theory of blood circulation—a network of tiny, thin-walled microtubules forming artery-vein connections, allowing the blood to flow back to the heart (Fig. 4.22; West 2013). He named them capillaries and concluded that blood always passes through tubules and vessels, thus confirming Harvey's theoretical prediction that arteries and veins are continuous. He found the lung to be an aggregate of very thin membranes forming ever so many cavities, the alveoli, the surface of which was covered by a rich network of blood (in image II at the bottom of Fig. 4.22). In explaining the differences

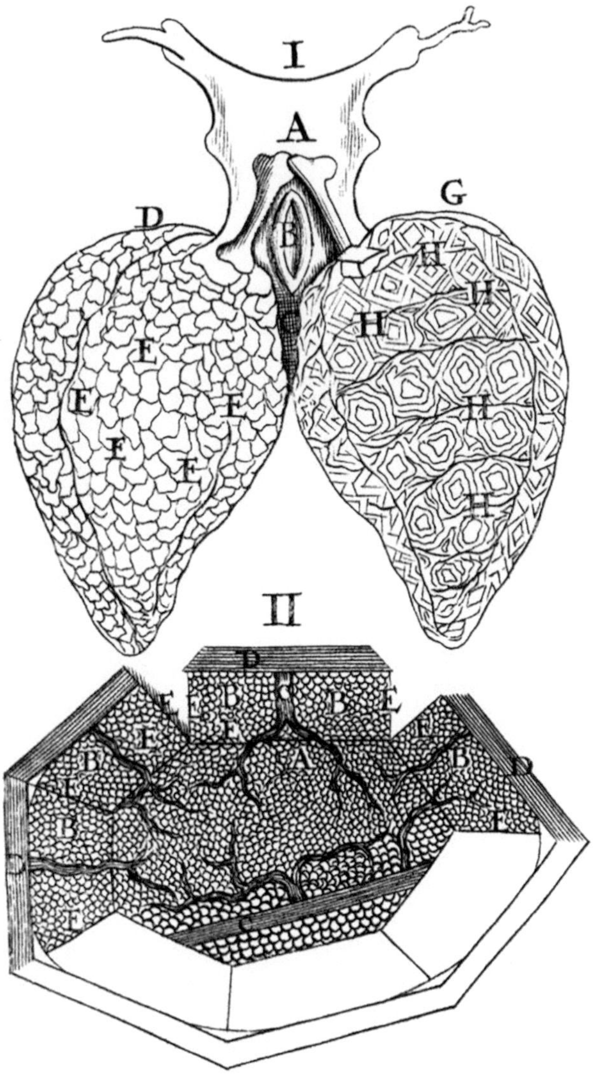

Fig. 4.22 Malpighi, the two lungs of a frog. Above: general view (left: the alveoli, right: the capillaries). Below: one of the alveoli with its base and sides opened up to show the dense network of capillaries in the walls

between the lungs of frogs and mammals, he erred in suggesting that Nature first tries out the basic form and function of its organs on low animals, and only then does she apply advanced models to more advanced animal groups. He thus used the term Nature, where religious people would use the term God (West 2013).

The Royal Society of London, founded 2 or 3 years after Malpighi's discovery of blood circulation in the lung, invited him to send them his scientific correspondence (1667). Malpighi maintained strong links with the Royal Society, publishing most of his studies through it. With him becoming a Fellow and publishing through the Society's journal, the Society became an international institute, which could boast top-level scientists among its ranks. The first of the Malpighian studies in the Royal Society's journal concerned the anatomy of the silk caterpillar (Cobb 2002). He discovered that the distribution of air within the insect occurs through a system of tubules (the tracheae), which, from their openings along the sides of the body (the spiracles), go on dividing and diminishing in size until they can hardly be seen by microscope. He compared the insect tracheae with the lungs of vertebrates and found that when he occluded the spiracles with oil, the caterpillar immediately died "during the time it takes to recite the Lord's Prayer *intra Dominicae orationis spatium*" (1669).

Malpighi also had a strong interest in the anatomy of plants. He discovered the pores of leaves, that the layers of tissue in leaves are continuous with those of the main stem, and that the ovule develops into a seed. His studies on plants led him to believe that tiny tubes found in many plants perform the same function as tracheae in insects. Based on this conclusion, Malpighi erroneously hypothesized that plants and animals had similar breathing mechanisms (1675). He believed that in plants, the seed is (what he named) a "compendium" of the entire future plant, comparable to the foetus among animals in that both seed and foetus need a surrounding matrix, whether of Earth or of a womb. The seed of the plant is an ovum containing a foetus made up of its more important parts, and likewise, the cicatrix of the hen is a "compendium" of the animal, in that it has the first outlines of its major organs (Wilson 1995).

He went on to conduct extensive studies on the morphogenesis of the chick embryo. The Dutch physician Regnier de Graaf (1641–1673) used the term *ovarium* (site where the egg lies) in his treatise on the female reproductive system (1672). Due to the similarity in size of the large, yolk-rich ova of the chicken ovaries to the follicles in the mammalian ovary, he erroneously thought that the entire follicle was an egg. He even provided biochemical evidence to support the likeness: when cooked, the liquid in the female testes acquires the same taste and texture as the yolk in the chicken egg. He believed that every follicle was an egg, that fertilization occurred within the ovary and that fertilized eggs were released from the ovary, passed through the uterine tubes and reached the uterus. Malpighi (posthumously 1697) correctly observed that the follicle never leaves the ovary; it merely serves to protect the real egg embedded in it (and which he, of course, never saw since the diameter of a mammalian egg cell is 0.1 mm). Only well over a century later, in 1827, was the egg cell discovered. In the unfertilized egg, Malpighi found no peculiar structure, suggesting that the matter of the egg is not pre-differentiated (Wilson 1995). In fertilized eggs, however, at first, he saw the spinal cord and primordia of the brain and only a little later the head, neck, thorax, beginnings of the eyes and motion of the heart. It was now no longer possible to maintain that heart and blood were the architects of the body, as claimed by Harvey, who saw the pulsating heart as first.

As Malpighi was a very productive scientist, many structures that he was the first to document now bear his name, such as the Malpighian corpuscles (=renal glomeruli) of mammals, the Malpighian corpuscles in the spleen, the Malpighian layer in the skin and the Malpighian tubules of arthropods.

Another direction in which the microscope assisted in the study of animals during the first decades after its invention was in exploring the external morphology of tiny creatures. This took place in England. Back in Bacon's visionary *New Atlantis,* there was a Custodian of Experiments. Four decades later, we find, accordingly, that the Royal Society established a paid position of a Curator of Experiments. Robert Hooke (1635–1703) had previously been Boyle's assistant and had constructed the vacuum pumps used in Boyle's experiments, which had led to Boyle's renowned gas laws. He was the first to fill this position, which he did for 40 years. His job was to present an "experiment" or observation every week, whether of his own or one suggested by others. Following the presentation, there would be a discussion on the significance of the experiment for the advancement of humankind, as befitted the spirit of Bacon's vision.

Hooke began experimenting on a broad range of subjects related to the essence of air (implosion of glass balls full of hot air at increasing atmospheric pressure), the force of gravity, measurements of stellar movements and the force of gunpowder. He also conducted observations with a microscope (Fig. 4.23). The Royal Society had purchased a microscope for the then astronomical sum of £2000 (the entire

Fig. 4.23 *Micrographia*: Hooke's microscope

annual income of 200 labourers), and Hooke was required to present, each week, at least one microscopic observation. Below is an example, taken from the notes of the Royal Society:

April 8 1663	Moss under the microscope
April 13 1663	Cross- and longitudinal sections of cork
April 22 1663	Leeches preserved in vinegar; mould on skin
April 29 1663	Diamonds in flints; a spider with six eyes
May 6 1663	A flea, male and female
May 20 1663	The head of an ant; a flea-like fly; the point of a needle
May 27 1663	Petrified wood; male flea
June 10 1663	Leaves
June 17 1663	Petrified wood with holes

These observations so impressed the society members that the Society's Council instructed Hooke (1665) to show them to the King in a book to be written specifically for this purpose. Charles II loved science, and he would invite researchers to his special laboratory to show him their findings. Hooke added other weekly observations, wrote captivating descriptions, created beautiful copper engravings and wrote *Micrographia*—the book of illustrations of the microscopic world, the world stretching far beyond the vision of the eye. To strengthen and emphasize the great enlarging power of this modern instrument, the microscope, he presented huge illustrations of tiny creatures with which everyone was familiar (the flea, in a super-large engraving presented on a foldout four times the size of the book, is 45.5 cm long—as big as a cat; Fig. 4.24). To give his readers a hint of the experience of observing through the lens of a microscope, he often enclosed the illustrations in round frames. Each illustration was the combined result of many observations and sketches from different perspectives, in different lighting conditions and with different lenses. Many observations also included texts such as "being looked upon with a microscope". The new term *microscope* was always in italics, thus emphasizing the presence of the instrument at the time of the observation.

Fig. 4.24 *Micrographia*: a flea

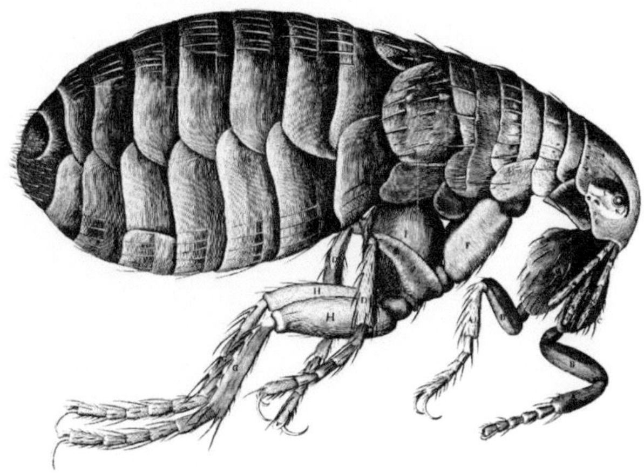

Hooke also stressed the difficulties of interpreting what he saw through the microscope: the difficulty, in some objects, of distinguishing between a prominence and a depression, between a shadow and a black stain, with the transparency of most objects rendering them yet much more difficult than if they were opacous. Depending on the light, the eyes of a fly appear almost like a lattice, drilled through with an abundance of small holes, or like a surface covered with golden nails or like a surface covered with pyramids or with cones. He declared, in his Preface to *Micrographia,* that mankind can behold the works of nature and consider, compare, alter, assist and improve them for various uses. We again recall Bacon and his applied approach that we can study nature and improve it for our use. The title page of the book sports a quote from Horace's First Epistle: "While your eyes cannot match the vision of a lynx, you will salve them with medicine when swollen"—a tribute to the Lynceans who first began scientific use of the microscope.

Micrographia showed what the microscope could mean for naturalists. Its superb illustrations, revealing structures of minute animals that are way beyond the resolution of the human eye, are remarkably accurate (Fig. 4.25, of a tiny pseudoscorpion; Fig. 4.26 of a fly, with enlargements of its eyes and a wing). Further, captivating verbal descriptions accompanied these illustrations. Thus, he described a flea as wearing a curiously polished suit of armour neatly jointed and beset with multitudes of sharp pins, shaped almost like porcupine quills.

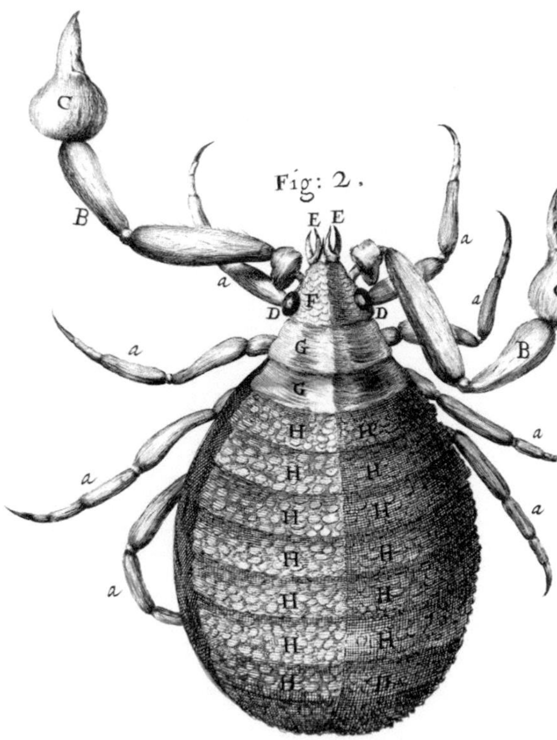

Fig. 4.25 *Micrographia*: a pseudoscorpion

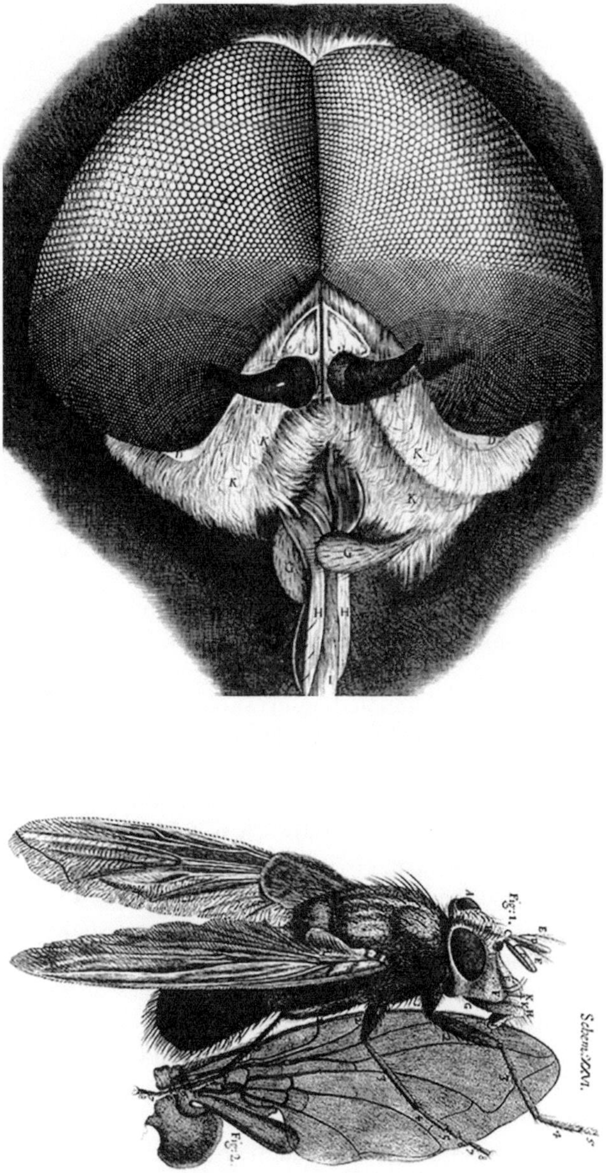

Fig. 4.26 *Micrographia:* a fly and its eyes

Observation 18 is his famous observation on the microscopic units making up a slice of cork (Fig. 4.27). Seeing "little boxes" without passageways between them, he coined the term "cell". *Cella* is Latin for a small room, and Latin-speaking people applied the word *cellulae* also to the six-sided cells of the honeycomb. By analogy, Hooke applied the term "cell" to the thickened walls of the dead cells of the cork. Although Hooke used the word differently than later cytologists (he thought of

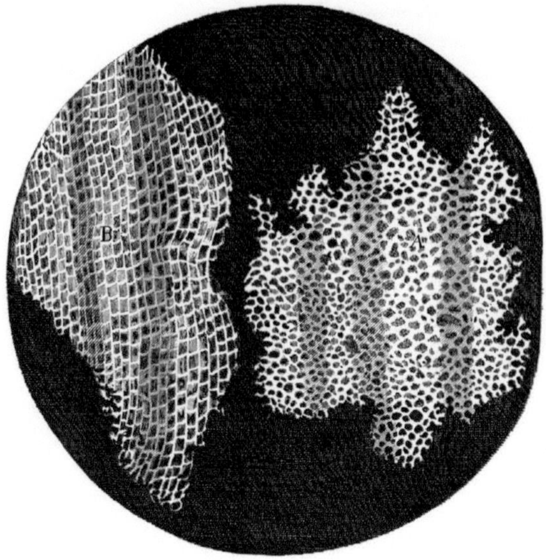

Fig. 4.27 *Micrographia*: cork cells

the cork cells as passages for fluids involved in plant growth), the modern term "cell" comes directly from his book.

Hooke wrote *Micrographia* in the overall style of Bacon's *Sylva Sylvarum*: a series of short, unrelated observations. Unlike *Sylva*, and better than it, its observations were all made by Hooke himself and therefore far more trustworthy. First came observations related to inanimate objects: the point of a needle, the edge of a razor, linen cloth, silk, glass pipes, colours in general, sand, kidney stones, frozen water, coal and finally fossils. Then, followed observations on plants: cork, mould, sponges (a mistake, obviously), seaweeds, leaves, barley seeds, thyme and poppy. Then came the animals: hair, fish scales, a bee's sting, the structure of the feather, peacock feathers, various insects, molluscs, spiders, mites and silverfish. Finally, came telescope observations on the heavenly level: how light passes through air, the planets and the moon. Hooke thereby showed that viewing objects through the microscope displayed the amazing complexity of objects in nature, compared to which man-made objects are wretched and "the more we see of their shape, the less appearance will there be of their beauty". We recall Bacon's vision that in the future, the microscope would reveal the structural complexity of tiny creatures such as the flea and the fly; illustrate that their subtle structure is far more complex than man-made arts and artificial textures, such as linen cloth; and serve in the study of gems, liquids, urine, blood and wounds. Four decades later Hooke, closely following the path envisioned by Bacon, observed and illustrated, one by one, most of the items in this text by Bacon: the flea, the fly, linen cloth, gems and urine.

Yet, the appeal of *Micrographia* also lies in the fact that many of its observations were random and not planned. Hooke described how he was sitting in his room reading a book when suddenly a silverfish appeared from between its pages, and he took it to the microscope (Fig. 4.28). He also described how he saw an unknown

Fig. 4.28 *Micrographia*: a silverfish

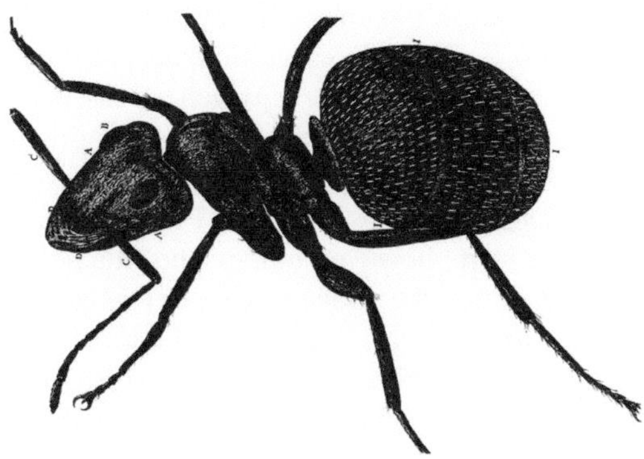

Fig. 4.29 *Micrographia*: an ant

creature on a vine, which he also took to the microscope. And the problems in trying to observe an ant, which does not rest still and quiet in its natural posture even for a moment, under a microscope (the solution: sedate it in a drop of wine; Fig. 4.29). His style created an atmosphere of here and now, of unique enthusiasm, of an informal portrayal of things based on oral presentation of evidence. No theories, only the joy of discovering a new world that can really be anywhere, popping out of books or creeping over plants and rushing "to tell the guys". We see this joy of discovering a new world once again 300 years later, with the invention of the scanning electron microscope.

Micrographia made the world aware that, even among its most minute creatures, nature is complex. By collecting tiny creatures and observing them through a compound microscope, Hooke revealed structural details previously unknown in these (broadly) well-known creatures. With much of the invisible becoming visible, this called for a recalibration of human knowledge with respect to the very small animals in nature (Wilson 1995).

Hooke also studied fossils under the microscope. Naturalists of the Renaissance still believed that God had created all animals during the 6 days of creation, with no new species created since. They considered fossils as draft models, created by the Creator and then either accepted or rejected by him, for the future Creation. We recall that Stelluti had claimed that fossil wood is generated from a type of clay, which slowly transforms into wood: life comes from clay, not from plants that do not generate fossils. Hooke, coming some two decades after Stelluti, noticed, however, the great structural similarity between fossilized wood on one hand and modern rotting oak trees on the other. He concluded that this fossil tree had been lying in some place where there was water from stones, that is, water soaked with rock particles, which was gradually conveyed into the microscopic pores of a fallen tree. This, he believed, was true also for fossilized shells found on mountaintops. Moreover, he suggested (in a posthumous paper) that most of the land was once underwater and that it emerged above water by a series of earthquakes. Most importantly, he suggested that many fossils are of animals that no longer occur today and that many species of the past no longer exist; furthermore, he raised the possibility that many species of our day did not exist from the beginning of time. These beliefs of Hooke were not immediately accepted. Other members of the Royal Society wrote that there was no such thing as petrified shells. There are "stones unto themselves", *lapides sui generis,* which are not part of any animal group.

Micrographia captured the public imagination, and print runs sold out rapidly. A high-level officer in the British Navy, who kept a personal diary at the time, called it "the most ingenious book that I ever read in my life".

There were, however, others who ridiculed Hooke for wasting £2000 on microscopes to study the nature of eels in vinegar, fleas in cheese and blue in plums. The greatest disparager of the involvement of the Royal Society in basic research was Jonathan Swift, the author of *Gulliver's Travels.* In 1726, six decades after the foundation of the society and publication of the *Micrographia*, he wrote his satirical work on Gulliver's Travels. Gulliver's first and most well-known journey was to Lilliput; his third and far less known journey was to Laputa, an island that floated in the sky without direct ties to the Earth, with a kingdom dedicated to music, astronomy, technology and mathematics, as its people were incapable of anything practical. They carried out pointless research at their academy, such as producing sunbeams from cucumbers, softening marble to fill cushions with it, mixing colours according to their odour and exposing political plots by examining the inhabitants' excrement.

At the time of the publication of *Micrographia*, insects, along with mites, water fleas and nematodes, were the smallest known organisms. A decade later, it occurred to an inquisitive Dutch draper to observe a drop of lake water through a simple, homemade microscope. He discovered the entirely new world of tiny unicellular creatures and of bacteria. It was in 1673, 8 years after the publication of *Micrographia*, that the Royal Society received its first communication from a Delft shopkeeper (Wilson 1995).

Antonie van Leeuwenhoek (1632–1723; Fig. 4.30) was born in Delft, the son of a basket maker. He received only a primary education and did not learn Latin, which in those days was still dominant as a scientific language. At 16, he was an apprentice in a linen-draper's shop in Amsterdam, where salesmen would use magnifying glasses to count the threads in cloth. Six years later, he returned to Delft and opened a draper's shop of his own. When 28, he received a job as chamberlain for the assembly chamber in the city hall; 9 years later, he was also appointed as a land surveyor and at the age of 47, he combined these with yet another municipal job, that of the official wine gauger of Delft. With his income secured, he could devote

Fig. 4.30 Leeuwenhoek

time to his great passion of manufacturing lenses and looking at tiny objects through them. He had seen Hooke's *Micrographia* (1665), and although he could not read the book's English text, he became intrigued with the illustrations of microscopic investigations. Leeuwenhoek incorporated the powerful lenses he began making into small hand-held single-lens microscopes, which he constructed, each about 5 cm in size (they magnified 270 times, whereas most microscopes of his day reached only ×50). He made his observations with these magnifying glasses ("vergroot glasen"), rather than with compound microscopes. He was secretive about his process of forming them, but he was more forthcoming with his illustrations, and he showed some illustrations of his observations (on mould, bees and lice) to a friend, the Dutch physician Reinier de Graaf.

The recently founded British Royal Society was eager to communicate with anyone working for the promotion of natural knowledge, regardless of his nationality. It succeeded in this through the efforts of its energetic secretary, Henry Oldenburg, who also edited the *Philosophical Transactions of the Royal Society of London*. De Graaf wrote to Oldenburg endorsing Leeuwenhoek's microscopic observations, and shortly after, Oldenburg published these observations (1673). Furthermore, he invited Leeuwenhoek to submit more observations, which Leeuwenhoek did indeed do. For the rest of his life, almost all his observations appeared as letters to the Royal Society. Leeuwenhoek wrote only in his own colloquial Dutch, and the editor translated his letters into Latin or English and often condensed them before publication. An inquisitive person with a somewhat adventurous approach, Leeuwenhoek, like Hooke, studied the microscopic structure of whatever he could lay his hands on:

bark, salt crystals, chalk, kaolin, domestic animals, humans (brain, spinal cord, eye, fingernails, bone, blood, milk, tears and sweat) and other objects (Mayr 1982; Wilson 1995).

It occurred to Leeuwenhoek to observe a drop of lake water through a microscope. What he discovered was an entirely new world—that of unicellular creatures and of bacteria, all of which he named "animalcules", tiny animals. These first-ever observations of unicellular creatures were very naive, as we can see from his description of *Vorticella* (a bell-shaped ciliate with a ring of beating cilia and a contractile stalk that attaches itself to substrates). Leeuwenhoek carefully and correctly described the body as containing several vacuoles, sometimes with two horns sticking out and continuously moving "like the ears of a horse" (these would be the wreath of cilia around the peristome). At its hind end, each animal had a "tail", four times as long as the whole body, as thick as a spider's cobweb. However, he could never see the creature using this tail for movement, as in other tail-possessing animals. This seemed strange to him because, at the time, a common definition of animals was that they were creatures endowed with the ability to move. Noting that these little animals were the most wretched creatures he had ever seen, he observed that the moment the base of their tail happened to hit a particle or filament, it got entangled in them. To get their tail loose, these animalcules would extend their body into an oval shape and stretch themselves, after which their tails would coil up serpent-like, like "a copper or iron wire that, having been wound close about a round stick and then taken off, kept all its windings". It did not occur to Leeuwenhoek that the contraction and extension of the stalk ("tail") of *Vorticella* could have any significance other than that of a tail serving for movement. The idea of a stalked sessile animal never entered his mind, and consequently, he jumped to the incorrect conclusion that the animals were "trying to get their tails loose". Some 30 years later (1702), Leeuwenhoek again published illustrations of *Vorticella*, and by now he had arrived at a more correct interpretation of the tail's function (1712). He concluded that they were animals in spite of the fact that they did not chase their food. Sessile and fixed by their tail (= stalk), they stirred water to obtain food from it by coiling and then discharging their tail or by a circular motion of the anterior part of their body (Wilson 1995) (Fig. 4.31).

In 1675–1676, he carried out an impressive series of observations in which he searched for tiny animals in every type of water he could find: rivers, wells, melted snow, stone troughs, gutters, rainwater dripping from roofs into tubs, rainwater dripping from vines, water from within the mantle cavity of freshwater mussels, seawater and water to which he added cloves or nutmeg. He would put aside these types of water for a few days and then scan them for tiny animals. In addition to describing his animalcules, Leeuwenhoek devoted considerable efforts to convincing his readers of their tiny size (Egerton 1967). He did so by presenting objects for comparison. He described one animalcule as being 10,000 times smaller than a Daphnia (which one can see with the naked eye). Of another, he wrote that 10,000 individuals were not as big as a grain of sand. One of his animalcules was described as being to a mite as a honeybee is to a horse; some animalcules are 50 times smaller than the eye of a louse; others are as thick as a single small thread of a silkworm; and so on.

Fig. 4.31 Leeuwenhoek's rich microscopic fauna. (A modern painting by Tuvia Kurz)

He also devoted considerable efforts to convince his readers of the vast numbers of tiny animals, calculating that in a drop of water the size of a pea there were 2,730,000 individual animalcules (Egerton 1967). In terms of numbers, as in terms of fauna, his microscopic life was an entirely new world (Wilson 1995).

Leeuwenhoek's descriptions and calculations were not immediately accepted. When he sent the Royal Society his observations of teeming microscopic life (1676), some of its members met them with severe incredulity. After all, nobody at that time had ever seen any microorganism. Leeuwenhoek's observations were fully acknowledged and published by the Royal Society of London in 1677 only when he enclosed letters from eight respectable gentlemen in Delft, all confirming his claim of a microscopic world of life (Egerton 1967; Wilson 1995). Three years later, the Society elected him as one of its fellows, although he did not attend the induction ceremony in London, nor did he ever attend a Royal Society meeting.

Leeuwenhoek persisted in his research on microscopic life in water for 40 years. He also described what we today call rotifers (Ford 1982; Tunnacliffe and Lapinski 2003). Rotifera is a phylum of tiny multicellular animals with a complete digestive tract that includes both a mouth and anus and a crown of cilia around their mouth that beats in synchronization, appearing to whirl like a wheel; hence "Rotifera", wheel-bearers). Mostly only 0.2–0.5 mm long, these multicellular creatures are the size of unicellular ones. They dwell in water, in soil crevices and among mosses. Leeuwenhoek, having discovered rotifers, was the first (1702) to describe a remarkable phenomenon: when he rehydrated dry and apparently lifeless dust from a roof gutter, many small rotifers became active within an hour. They would adhere to the glass vial, creep along or swim about. He replicated this observation repeatedly and experimented with leaving dried individuals on a sheet of paper for 5 months. Time and again, when placed in a glass vial and having boiled cool water poured over them, the rotifers would come to life within an hour. This was astonishing because, without water, there can be no life. "I never thought that there could be any living creature in a substance so dried as it was", and yet here was Leeuwenhoek observing creatures that had survived such severe desiccation and "returned to life" (Ford 1982; Tunnacliffe and Lapinski 2003). Today, we know that rotifers can maintain viability in a dried state for extended periods of at least 9 years; Leeuwenhoek had observed the rotifer *Philodina roseola*. He concluded that in pools and marshes that dry up in summer, many kinds of animalcules survive in the bare soil and repopulate the water when the winter comes. Waterfowl bring other kinds into the pools in the mud, sticking to their feet and feathers. This, he argued, is how temporary pools were colonized, and not by spontaneous generation, which Leeuwenhoek strictly opposed (Wilson 1995; Tunnacliffe and Lapinski 2003).

Eager to impress his readers with the very rapid population build-up of microscopic creatures, in 1716 Leeuwenhoek described the phyto-flagellate *Polytoma* (a genus of green algae lacking chlorophyll, with two flagella, that obtains its nutrients from decaying organic matter). He found that they had a peculiar form of reproduction: after living no longer than 36 hours, they "fixed themselves upon the glass, stopped moving; and soon later their body burst asunder, and they divided into eight portions; and these were actually young animalcules, because in five-six seconds some of them swam off". He calculated that in just 9 days, one parent *Polytoma* could thus produce 262,144 animalcules (Egerton 1967).

Leeuwenhoek had no names for his animalcules, so he just described them. Two and a half centuries later, a Fellow of the Royal Society, Clifford Dobbel (1932), succeeded in identifying many of Leeuwenhoek's animalcules down to genus level. This is a mark of honour to Leeuwenhoek's ability to describe accurately what he saw. Table 4.1 presents most of Dobbel's eukaryote genera and their modern classification to give some impression of the taxonomic range of the microscopic animals Leeuwenhoek had studied. His research extended over at least nine phyla, 16 classes and 21 orders.

In addition to life in water, from 1674 onwards, Leeuwenhoek also searched for tiny animals inside bigger animals. He found unicellular organisms and bacteria in the guts of cows, sheep, rabbits, fowl, turkeys and horseflies; in the excrement of

Table 4.1 Leeuwenhoek's aquatic animalcules as identified by Dobbel down to genus level (in bold), their modern classification and common names

Common name	Phylum	Class	Order	Genus
Green algae	Chlorophyta	Chlorophyceae	Volvocales	*Volvox*
			Chlamydomonadales	*Haematococcus*
				Chlamydomonas
				Polytoma
Ciliates	Ciliophora	Spirotrichea		*Kerona*
		Prostomatea	Prorodontida	*Coleps*
		Oligohymenophorea	Peritrichia	*Carchesium*
			Mobilida	*Trichodina*
			Pleuronematida	*Cyclidium*
			Sessilida	*Vorticella*
			Hymenostomatida	*Colpidium*
		Spirotrichea	Sporadotrichida	*Stylonchia*
				Oxytricha
		Hypotrichea	Euplotida	*Euplotes*
Amoebas	Amoebozoa	Tubulinea	Euamoebida	*Amoeba*
	Euglenozoa	Euglenoida	Euglenales	*Euglena*
Diatoms	Ochrophyta	Bacillariophyceae	Pennales	*Asterionella*
				Fragilaria
		Bacillariophyceae	Fragilariales	*Meridion*
		Chrysophyceae	Chromulinales	*Anthophysa*
Forams	Foraminifera	Globothalamea	Rotaliida	*Polystomella*
		Eurotatoria	Flosculariaceae	*Filinia*
		?	**Monogononta**	*Melicerta*
Rotifers	Rotifera	Bdelloidea	Bdelloida	*Philodina*
Hydra	Cnidaria	Hydrozoa	Anthomedusae	*Hydra*
Daphnia	Arthropoda	Crustacea	Cladocera	*Daphnia*

cows, hens and pigeons; in the urine of horses; and in the tartar on human teeth. Until then, naturalists did not know of animals other than worms that live inside the bodies of humans, and the existence of hordes of microorganisms within the bodies of healthy animals was wholly unsuspected. Nowhere in his writings did Leeuwenhoek associate protozoa or bacteria inside a body with disease. He found them in normal hosts, and it probably never occurred to him that any of them might have pathogenic properties. He also found yeast in fermenting beer.

4.5.2 *Tiny Animals Are Living Inside Sperm!*

In 1677, Leeuwenhoek discovered and described human sperm (Wilson 1995; Karamanou et al. 2010). A professor of medicine had sent a young student to Leeuwenhoek with a flask containing the sperm of a man suffering from gonorrhoea. The student claimed to have observed tiny living creatures in this semen, animated due to their tails. Could Leeuwenhoek inspect the matter? Leeuwenhoek inspected the sick man's sperm, then he inspected his own sperm and then—he confirmed the student's observation. In a letter to the Royal Society (1677), he wrote that he had seen a multitude of such live animalcules; "a million of them would not equal in size a large grain of sand, and moving. Their bodies were rounded, blunt in front and running to a point behind, with a long thin tail, about five or six times as long as the body, and very transparent, and with the thickness of about one twenty-fifth that of the body [...]. The animalcules moved forward with a snakelike motion of the tail, as do eels when swimming [...]. They lashed their tails some eight or ten times in advancing a hair's breadth". Somewhat embarrassed by looking into such a private issue as sperm, Leeuwenhoek wrote that he would receive with understanding any decision of the Society not to publish his observation, if disgusting or offensive to the learned members of the Society.

Very wisely, the secretary of the Society (by then Nehemiah Grew) encouraged Leeuwenhoek to expand his observations to other quadrupeds. Leeuwenhoek followed this advice, and finding animalcules in the semen of dogs and rabbits as well, he wrote (1678) that he believed that they appeared in all male animals. Later on, during his lifetime, he would describe the sperm of many kinds of animals—mammals, birds, amphibians, fishes, arthropods and molluscs (Fig. 4.32). In cod, he

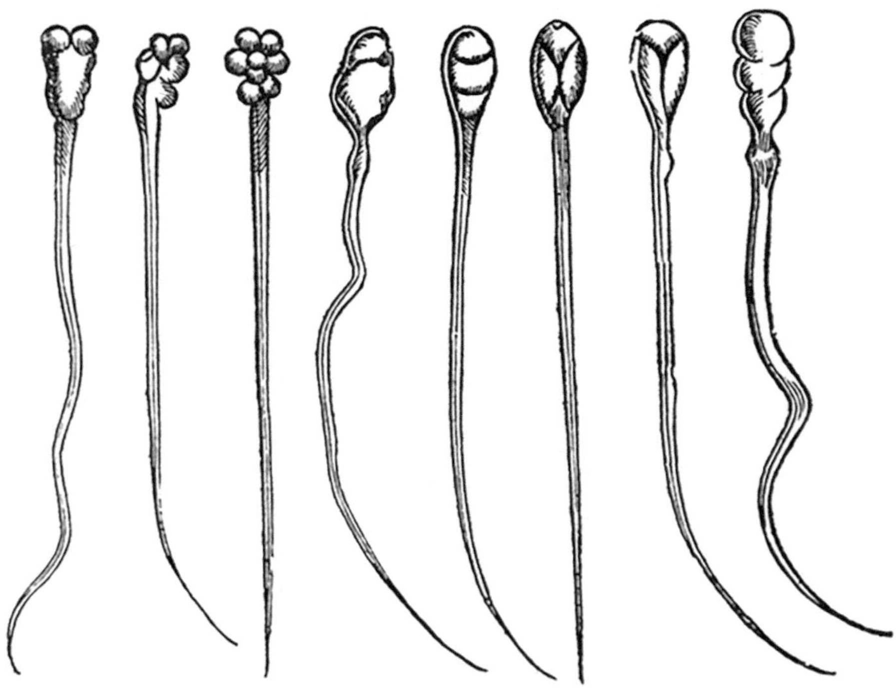

Fig. 4.32 Leeuwenhoek: the sperm of various animals

calculated there were 150,000,000,000 sperm in the milt of a single fish. Estimating the number of humans on Earth at this time at 13,385,000,000, he suggested that in the milt of a single cod there are more than ten animalcules versus each human being on Earth.

Why so many animalcules? Leeuwenhoek answered by referring to an apple tree in full blossom: thousands of blossoms, each of which could develop into an apple, each of which will contain six to eight seeds. However, not all these numerous seeds develop into apple trees—some land on places without enough light, others on some other unsuitable sites and so on. Furthermore, he said, if you place six seeds into a hole in the ground, only one of them will sprout—the rest will be crushed by this single dominant individual. This justification for the numerous blossoms on an apple tree also applies to the male of these animalcules (Karamanou et al. 2010).

A usually excellent observer, at this point Leeuwenhoek erred: in the liquid of the semen, he believed that he saw "all manner of great and small vessels, so various and so numerous that I doubt not that they are nerves, arteries and veins (Fig. 4.33). I have indeed observed these vessels in such great numbers, that I believe I have seen more in a single drop of semen than an anatomist would meet with in a whole day's dissection [...] In no human body, are there any vessels which may not be found likewise in sound semen". He further erred in that he thought he was able to identify male and female spermatozoa. The idea that spermatozoa were organisms was entrenched to such an extent that, over 150 years later, in 1837, a certain

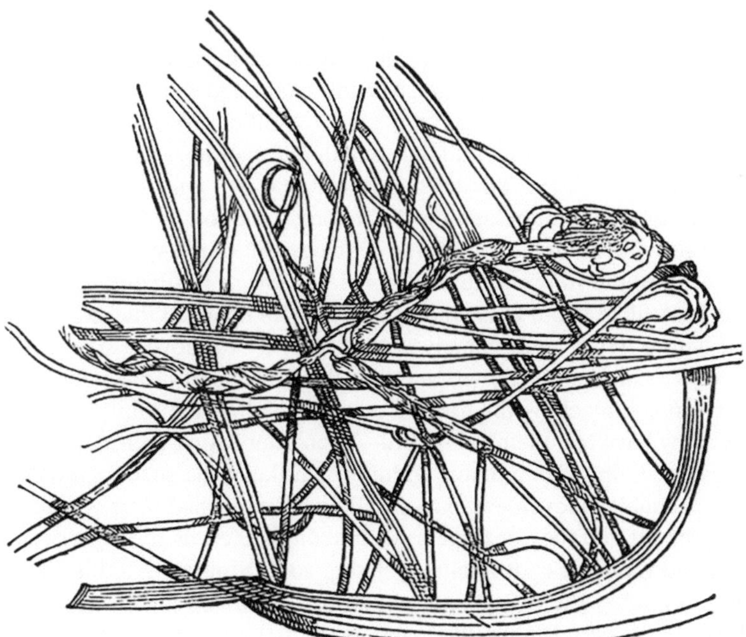

Fig. 4.33 Leeuwenhoek: structures he thought he saw in semen

physiologist by the name of Gustav Valentin, a professor of physiology at the University of Bern, described the mouth, anus, stomach and early stages of evolutionary development in bear sperm (Valentin 1837).

By then, investigators had generally accepted the preformation theory of development (as we shall discuss further on) and began turning their attention to the question of whether the individual was preformed in the father's sperm or the mother's ovum. The ovum theory stated that only a vapour from the male seed penetrates the womb and effects fertilization. Supporters of this theory based themselves on the fact that Harvey had never found sperm in the uterus after copulation and on de Graaf's claim that he had found the mammalian ovum in the ovary, which suggested that from there it descends and enters the uterus through the fallopian tubes. Leeuwenhoek disputed this theory (1685). Using powerful magnification (which Harvey did not have), he found an abundance of sperm all along the two horns of the uterus of a dog and argued that this agrees with the basic fact that throughout nature the male genitalia are long and thin, adapted to transporting the sperm right up into the uterus, where it is discharged. As for de Graaf's suggestion, Leeuwenhoek correctly claimed that de Graaf's "egg" was far too big to pass through the tubes, which, as he proved by pressing quicksilver through them, were extremely narrow. De Graaf had described the eggs of rabbits as reaching the size of a rapeseed, of sheep as being as big as a pea and of cows as larger than a cherry. We now know that, in fact, he had found not the tiny mammalian egg but the much larger ovarian follicle, an aggregate of cells surrounding the egg.

Leeuwenhoek thought that, though "learned people adhere to the opinion that the fallopian tube sucks up or pulls off the egg from the ovary and transmits it to such a narrow channel, the human [embryo] originates not from an egg but from an animalcule formed in the male sperm". Having examined the ovary of a sheep and finding how difficult it is to detach an "egg" (= follicle) from the ovary, he called upon his readers to note "how little time a bull, rabbit or boar require for copulation. [Are we to believe that] in this brief time the one-two eggs or six-eight eggs are sucked off through the fallopian tube?" The only function of the ovary, in his opinion, was to provide nourishment for the development of the spermatozoa.

In advocating the importance of the male sperm in fertilization, Leeuwenhoek sought to generalize about tails among animals. He believed that every animal was endowed with a living soul, namely with the ability to move. He argued that tadpoles emerge from their eggs with a tail as their means of movement. When sufficiently grown and reaching the form of a frog, they lose their tail, but this does not mean that they have lost their living soul—even after losing their tail and becoming motionless, the living soul remains within it. Similarly, the sperm animalcules have a tail, which enables movement, but once they have reached their destination in the uterus, it falls off and they become oval. In animalcules, as in frogs, even after the tail is lost and motionless, it retains its living soul. Similarly, and in line with this argument, in poultry the living soul within the sperm animalcule of the cock does not transfer to the egg. Rather, the contents of the egg pass over to the cock's animalcule. The hen's egg merely nourishes and develops the animalcule, which is possessed with a living soul, which has come from the cock. Leeuwenhoek is echoing the Aristotelian doctrine that the female merely supplies food to the young organism. The "principle of movement", the soul, resides in the male seeds. It is the food that goes to the spermatozoa (the living soul), and not the other way around.

When Leeuwenhoek dissected a 17-day-old sheep embryo, he found it to possess all major organs. Hence, he concluded that a similar creature, even if a 1000 million times smaller, might already be equipped with all its organs. Thus, a sperm animalcule, of whatever animal it may be, contains the *potential* for forming all the limbs that an animal will have when it is born. Lest anyone say that he claimed that the human sperm is replete with small babies, at this point Leeuwenhoek was very explicit: the sperm is full of animalcules with tails and nothing else. It would be wrong, he wrote, to assert that the animalcules in the human sperm are babies, even though a child eventually forms from them. Likewise, it would be wrong to say that some worms swimming in the water are flying creatures, even though winged creatures eventually emerge from them, or that the pip of an apple is a tree, though a tree will eventually grow from it. This echoes Aristotle's theories. As to parentless "spontaneous" generation, Leeuwenhoek claimed that no animal, however small it might be, originates in putrefaction, but exclusively by procreation. This is so because all animals have animalcules in their sperm, even tiny animals the size of a flea.

Leeuwenhoek expanded his research far beyond the study of microscopic animalcules and studied a broad range of biological phenomena, including parthenogenesis in aphids, flukes in sheep, insects in galls and the banded pattern of muscle fibres (1682). Seeking to know not only the size and quantity of organisms but also to determine their age, he proposed using annual rings to determine the age of trees and growth annuli in fish scales to determine the age of a fish. He also calculated the potential rate of increase for several insect species and speculated on factors that limit this increase, usually food and climate. In a study of the grain weevil *Sitophilus granarius* (Fig. 4.34), he found that they laid very few eggs. When he compared this with the silkworm, which lays many eggs in 1 or 2 days, after which it quickly dies, he concluded that grain weevils must live longer as adults and lay eggs several times (Egerton 1967). We today describe modern aspects of this approach as two alternative reproductive strategies of living organisms under the terms iteroparity and semelparity. A species is iteroparous if it carries out multiple reproductive cycles over the course of its lifetime and semelparous if it carries out a single reproductive episode before death.

Over the years Leeuwenhoek's reputation as an expert in microscope observations grew to such an extent that, by the end of the seventeenth century, Hooke lamented that the field of microscopic study and discovery had come to rest entirely on one man's shoulders. Leeuwenhoek was visited by such notables as the Russian Tsar Peter the Great, James II of England, William III of Orange and his wife Mary II of England, the Duke of York, the ambassador of France, the mayor of Amsterdam and the philosopher Gottfried Leibniz. All came to gaze at his tiny creatures. A devout Christian, he considered his discoveries as proof of the great wonders of God, who so carefully designed and created creatures both so great and so small.

Fig. 4.34 The grain weevil (Tuvia Kurz)

Leeuwehoek's research is more adventurous than that of Hooke. Hooke took tiny creatures he happened to find and examined how they look when magnified many times. Leeuwenhoek took things in which one could observe nothing with the naked eye and inspected them under magnification. Looking back over the years at Leeuwenhoek, his lack of academic education is remarkable. He is, I believe, the first key personality in biology who lacked higher education. Indeed, we are lucky that he was not a scholar: not knowing what the texts of earlier experts would have trained him to expect and observe, he observed and thought "outside the box". Furthermore, he showed that fascinating studies could progress without hypotheses, without research questions that one must answer. Only curiosity per se led him: let us examine lake water and see what we observe; let us look at the plaque of teeth.

Another point regards the Royal Society. It was one of the first scientific establishments, and it is difficult not to admire the enlightened, liberal manner in which it operated, encouraging scientific creation in those who did not speak its language and did not think in the way that it did, without religious, national or conservative fanaticism. This spirit of liberal open-mindedness in the early members of the first scientific-biological establishment still wafts through the corridors of biology. For centuries and down to our generation, the Royal Society shaped the norms of the biological establishment.

4.6 Spontaneous Generation: Theories and Experiments

The years after the publication of *Micrographia* witnessed important observations on parentless, "spontaneous" generation. We recall that ever since the time of Aristotle, naturalists had believed that many organisms could come to be through spontaneous generation. Aristotle had described the generation of insects from animal dung, flesh, mud, dew and other organic and inorganic substances and of eels from riverbeds and sand. He thought that mosquitoes came from worms in the mud of wells. The mud first becomes white by decomposition, then black and finally blood red. At this stage, small red worms grow out of it. At first, they cling to the ground; later they detach themselves and swim about in the water. After a few days, they come to the surface of the water, become stationary and harden and a while later the mosquito leaves its split shell and flies off.

Belief in spontaneous generation persisted over the centuries, with claims deep into the seventeenth century that mice may be produced from piles of grain in a glass plugged with dirty linen and that scorpions are formed from a herb (basil) when crushed in a cavity between two bricks (Jean Baptista van Helmont, 1577–1644).

Athanasius Kircher (1602–1680, Fig. 4.35), a German Jesuit scholar, was born near Fulda, where he studied philosophy and theology. He fled to Cologne during the Thirty Years War to escape advancing Protestant forces, narrowly escaped death after falling through ice when crossing the frozen Rhine and Protestant soldiers caught and nearly hanged him later on. At the age of 26, he became a professor at the University of Wurzburg; at the age of 31, he moved to Rome, where for several years he taught mathematics, physics and Oriental languages at the Collegio Romano (now the Pontifical Gregorian University) before being released from teaching duties to devote himself to research. His investigations culminated in his major treatise, *The Underground World* (*Mundus Subterraneus*, 1665), in which he presented his main theory: Earth is hollow. It consists of huge empty spaces and hidden burrows, in which it holds many great oceans and fires, interconnected by passageways that reach down all the way to its core. Volcanoes are Earth's vent holes, her breath pipes: wherever there is a volcano, there is a storehouse of fire under it. The tides, caused by the moon, push an immense bulk of water through hidden passages at the bottom of the ocean, and the resulting winds stir up and feed the subterraneous fire, like huge bellows. The seas, which would freeze without the fires, keep the fires from getting out of hand. Earth's mountains are hollow and serve as reservoirs, and hot springs form where underground water passageways come near the fire channels.

To investigate whether Vesuvius had some subterrene connections with Stromboli and Etna, Kircher climbed Vesuvius and entered its crater, where he claimed to have found a steep hollow chamber, its boiling bottom spewing streams of smoke and fire, emerging from 11 points, amidst horrible roars and a terrible stink (Fig. 4.36).

Kircher was an ardent believer in spontaneous generation, a belief closely linked to his beliefs in Creation (Hiarai 2007). To prove spontaneous generation, he

Fig. 4.35 Kircher

suggested taking dead flies, sprinkling and soaking them in honey water and then placing them on a warm copper plate. Minute, microscopic worms will soon appear, which little by little shall grow wings and assume the shape of very small flies that shall slowly attain adult size. (To this, Redi, whom we shall meet further on, responded by noting that the honey water must have attracted living flies to drop their eggs in the corpses of the dead ones; he also noted that large flies do not grow from small flies).

To explain the spontaneous emergence of life, Kircher argued that God had inserted into matter a Universal Seed of all Nature, *panspermia*. This seed possesses a "plastic power" (*virtus plastica*) to give all living things their form and colour, according to the nature of each species. Insects and other inferior beings arise

Fig. 4.36 Mount Vesuvius. (Mundus Subterraneus)

spontaneously from putrefied matter thanks to an additional seminal entity, a Separated Seed *semen decisum*, which drops from the body when a plant or inferior animal dies. The plastic power is present in vegetable seeds and in animal semen, and it is protected by envelopes that maintain its original heat. Separated Seeds have no enevelopes; their unprotected power is weak, and, therefore, they give rise only to imperfect living beings of degenerate form. The weaker their force, the more inferior the being they will produce: a zoophyte, a rudimentary plant, a mushroom, a fossil resembling a plant or even an inanimate body, if this force is totally lacking. Thus, the difference between sexual and spontaneous reproduction is in the strength of the seminal heat. The Separated Seeds are minute corpuscles, easily dispersed by wind and rain. Upon finding a proper substrate, they can form, with their plastic powers, what he named a Web of Life, *tela vitae*. Furthermore, traces of the Force of Life remain in the carcasses of creatures that were once alive. These traces too can generate living beings, similar but inferior to the original animal. Something in the body of living beings persists in the dead body and retains a portion of the seminal plastic power even after death. Living beings therefore arise, however, only from matter that was once alive.

Kircher next considered the emergence of life from these seminal corpuscles, which come from the bodies of former living beings and some of which are dispersed everywhere in the world. When fermented by ambient heat, these corpuscles, coming from previously living beings and retaining only a small, weakened part of the material soul, acquire a degree of heat that enables the material soul hidden in them to emerge in the form of life. In the mass of matter concocted by ambient heat, this soul takes on a concrete form, which gives life to the matter and creates a new living being. To prove his point, Kircher claimed that one could take a vial of water, mix it with Earth, add some pieces of a chopped serpent and expose it to the air, after which little spheres will appear, which will turn into worms [snakes?] and eventually develop wings. Thus, Kircher suggested, the difference between sexual and spontaneous reproduction is in the strength of the seminal heat (Hiarai 2007).

Mundus Subterraneus also included a chapter on dragons, which I here briefly present as part of the diversity of the animal world, in which people of early modern Europe believed (Fig. 4.37). Kircher began by counting dragons through a survey of literature, from the Book of Daniel through Aristotle, then the Mediaeval period and into the Renaissance. He then presented dragon observations from his own seventeenth-century times. In 1602, people found a dragon skeleton in a cave in Switzerland; the dragon died when the cave collapsed during an earthquake. In 1609, a noted Swiss prefect reported that one night he saw a brightly glowing dragon flying from a certain large mountain cliff (Mount Pilate) to a cave on the opposite cliff (the Flue Cave). Swiftly flapping its wings, "in the midst of its flight it spewed out sparks from its body, like embers flying when smiths beat glowing iron". In 1660, an inhabitant of Rome ran into a dragon the size of a very large vulture when he was in coastal marshes trapping birds. He killed it and brought the dragon's head to the city, where it is now on display for all to see, in Kircher's own museum. Furthermore, to this very day, winged snakes fly from Arabia to Egypt after the flooding of the Nile, where their offspring are born as insects in decaying matter in the muddy slime left behind by the flood. Kircher thus presented evidence that dragons occur everywhere, in his days as in the good old days. How do dragons come into being? Kirscher described a dragon slain in 1345, in the Isle of Rhodes. It had a head like a large horse but as wide as that of a cow and with scales like those of a

Fig. 4.37 A dragon. (Mundus Subterraneus)

snake. It had elongated ears like those of a mule, a horrific mouth with massive teeth, oversized eyes and a fiery breath. It had four feet with claws like those of a bear and a tail similar to that of a crocodile; a tough hide of overlapping scales protected the entire body; and it had two membranous wings. Its sides were coloured like dolphin fins, with blue on its back and a golden-yellow underside. The remainder of the body had a mixture of these two colours. The Rhodes dragon thus possessed a mixture of the characters of many animals: horse, cow, snake, mule, bear, crocodile and dolphin. From this mixture, Kircher formulated a theory as to how dragons come into being.

To begin with, he argued that in animal species of hybrid origin, a mixture of characters occurs: mules from a combination of horse and donkey, giraffes from a combination of camels and lions, leopards from a combination of lions and other cats, armadillos from a combination of turtles and porcupines and even humans (who commit bestiality). Next, he discussed form in animals produced by spontaneous generation: the bee, born from cattle manure, has a head similar to that of a cow; the scarab beetle, born from horse manure, has a head similar to that of the horse; and the stag beetle, engendered from stag manure, resembles the head and horns of a stag. Indeed, the faeces of many animals generate some kind of insect, which invariably resembles the very animal in whose excrement they were born. Moving on from this point of spontaneous generation, Kircher noted that remote cliffs and mountains, in which dragons commonly dwell, are also the haunt of eagles, vultures and other birds of prey. These large raptors customarily take all kinds of vertebrate prey, such as snakes, birds, rabbits and lambs; they even seize children and take them to their nests up in the high mountains as food for their young. Consequently, a pile of rotting prey remnants accumulates in their nest. This heap inevitably becomes a mass of decaying matter, which is suitable for generating other forms of life. Because some portion of its semen stays in every corpse after death, an animal can come into being in this mass of fermenting matter: a worm generates, which resembles the raptor's prey. If the prey was a rabbit, the worm acquires elongated ears from the force of the rabbit's semen. If the semen of a bird is also present in this mass of decaying matter, it joins with the rabbit semen to produce a winged worm. If a snake corpse is present, its semen also imparts the head, tail and neck of a snake to the nascent worm. The resulting hybrid embryo, once formed from the variety of semen, grows into a dragon. If many dragons of both sexes come into being by generation in a mass of decaying matter, they will also be able to reproduce sexually, as do insects that come into being in this way. Lest this fecundity result in too much damage to the environment, nature has provided an excellent law by which only one dragon at a time can generate by the mixture of semen in an underground cavern. Among the hybrid-formed dragons, Kircher mentioned the basilisk, born from a chicken eating a snake's egg, who is a combination of both chicken and snake. Dragons may also come from a marine source: ducks and geese that lay eggs on the ice of the freezing northern sea. In spring these eggs are broken open by the melting ice, and their contents washed up by the tides, on the shores of the Orkneys and the Hebrides. Here the content of duck and goose eggs are brought together, and in this manner the remoras (suckerfish), which cling to the undersides of ships, are formed and brought to life. In 1609 a certain Claude Duret, a French judge, botanist, historiographer and linguist, published his *Histoire Admirable des Plantes* (1609). Among other remarkable tales, he described and figured a tree “not, it is true, common in France, but frequently observed in Scotland”. From this tree, leaves are falling. On one side, they strike the water and slowly transform into fishes; on the other side, they strike land and turn into birds (Fig. 4.38). Seven decades later, Kircher in 1678, similarly, described orchids giving birth to birds and even to very small men; this occurs when they touch the ground, where a sort of fecundation takes place.

Kircher believed that God’s words present the structure of the world because for Him, to speak means to create (see *John* 1:1: “in the beginning was the Word, and the Word was with God, and the Word was God”). Creation was the expression of syllables of which the terms are formed, and thus, words—in God’s language—are the primary realities (Breidbach and Ghiselin 2006). Natural objects are therefore to be ordered according to the grammar that determines how the words are used, and the etymologies of words are indicators of the nature of the things to which they

Fig. 4.38 Duret’s tree. Leaves when falling on one side become fishes; those falling on the other side become birds. (From *Histoire Admirable des Plantes*, 1609)

refer. Kircher further believed that God's language was that in which He had spoken to Adam and Eve (and the snake) in the Garden of Eden. Adam adapted his own language to that of God, to such an extent that the rules, grammar and vocabulary of the two languages were one and the same. The by-now lost languages of God and of Adam would have given the fundamental laws of the universe and would have provided access to reality. A reconstruction of God's language, through comparative linguistics, might gain a formula for knowing everything. Thus, the rules that permit identification of how the terms are combined are essential for understanding the order of nature (Breidbach and Ghiselin 2006). Kircher's emphasis on the mystic levels of words calls to mind the Kabbalah, a current in Mediaeval Judaism that sought, by attaching mystic significance to words, to explain relationships between the unchanging, eternal God and the ever-changing, finite universe.

By the mid-seventeenth century, the rich diversity of life in the New World, when added to that of the Old World, was casting doubts on the biblical story that after the Flood, all life on Earth originated from that one single point of dispersal, Mount Ararat (Breidbach and Ghiselin 2006). Kircher (*Arca Noë* 1675, Fig. 4.39), in addressing doubts as to how so many different kinds of animals could have fitted into the Ark, began by describing the Ark as intended to serve mainly Old World species. Thus, on its lower deck, one side of the Ark housed beavers, otters, crocodiles, hippopotami, goat-stags, gazelles, elks, bison, goats, sheep, cattle, deer, reindeer, wild goats, gazelles, chamois, fallow deer, dogs, aquatic hare-hounds, molossian hounds, Maltese dogs, Indian dogs, seals, turtles, hedgehogs, porcupines, badgers, dormice, martens and weasels. The other side housed small Indian pigs, rabbits, hares, squirrels, monkeys, apes, cats, asses, donkeys, horses, dromedaries, camels, elephants, rhinoceroses, lions, bears, tigers, panthers, leopards, unicorns, other horned animals from Africa, lynxes, gluttons, wolves, foxes, wild boars and domestic pigs. On the upper deck, he housed birds.

Next, Kircher argued that only pure species boarded the Ark, not hybrids. There were no mules and no giraffes (since their Latin name *Camelopardis* suggested that they are hybrids between camel *camel* and leopard *pardalis*). There were also no armadillos, as they seemed to him to be hybrids between turtle and porcupine; and no alpine marmots, as he thought that they were a cross between squirrel and badger. Not included on the Ark were also those "lower animals" such as voles, mice, lizards, scorpions and insects, animals that did not need the Ark since they reproduce themselves from mud or from dung by spontaneous generation. (This is a slight leaning on evidence, since the Bible specifically refers to "every creeping thing" as having been on the Ark). Kircher did, however, include serpents among the passengers of the Ark, both to remind mankind of the sin in Paradise and, by their medicinal properties, to cure illnesses during and after the voyage. He also included the unicorns, gryphons and sirens—but not the phoenix, because it had violated the command to enter in pairs. By excluding most animals of the New World, hybrids and lower animals, Kircher thus reduced the burden of animals entering the Ark. Kircher considered Noah's Ark as mankind's first attempt to collect nature, perhaps the first-ever museum of natural history. He also presented the exact year of the Flood (2396 BCE), the Ark's precise dimensions (198 metres long,

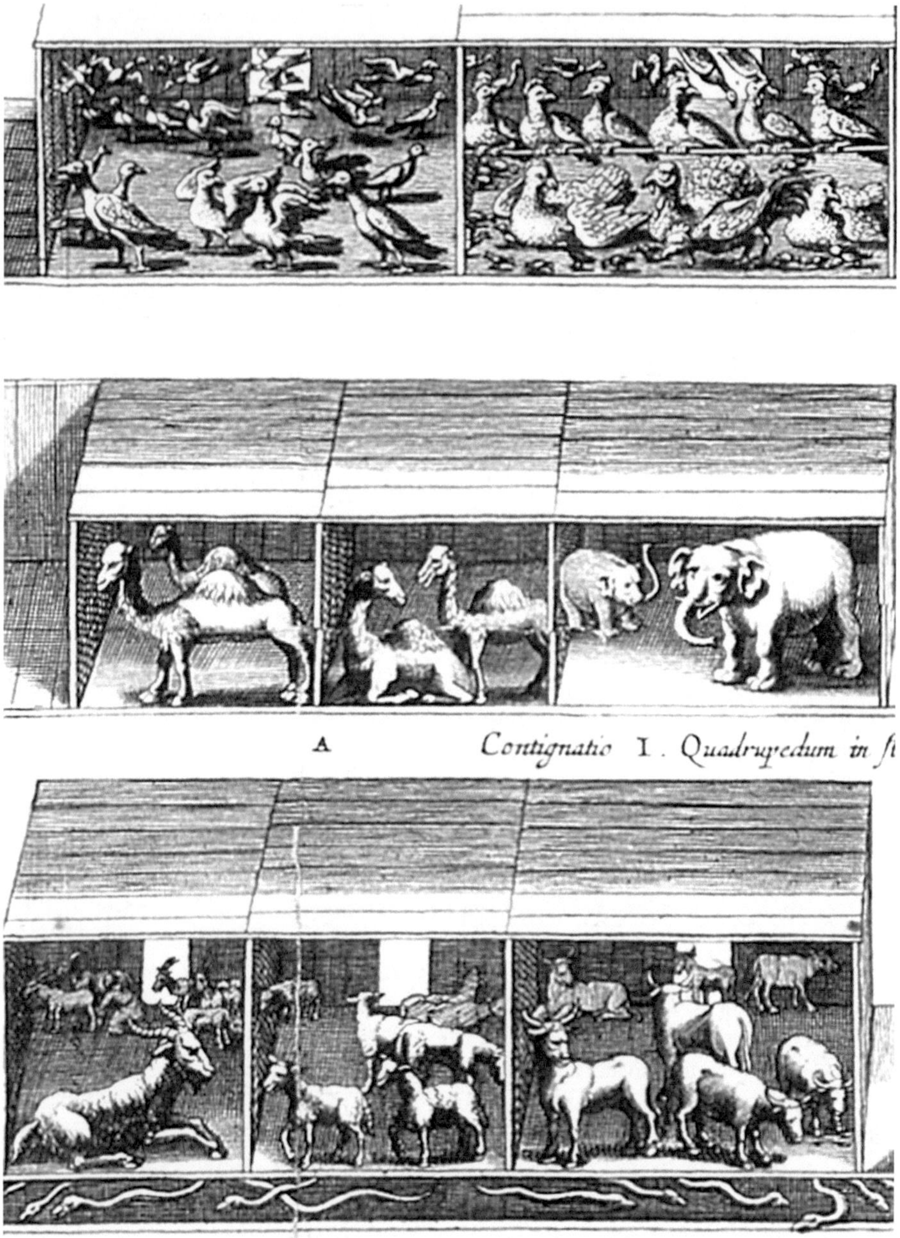

Fig. 4.39 Kircher, Noah's Ark: detail

33 metres wide, 19.8 metres high) and the period, from the day of the first rain to the day when Noah emerged from the Ark and set his foot on dry land (365 days; Breidbach and Ghiselin 2006).

By the mid-seventeenth century, however, doubts began about the occurrence of spontaneous generation. Hooke had observed mosquito larvae in his microscope, and although he did not see the tiny mosquito egg, he did observe how a first-stage larva develops into a more advanced larva and how eventually the adult hatches and flies off (*Micrographia* 190; Fig. 4.40). This observation led him to doubt the entire matter of spontaneous generation in amphibious creatures and to propose certain speculations as queries that he had not yet had sufficient opportunity and leisure to answer from his own experiments or observations. Was it possible that mosquitoes, as well as other aquatic creatures believed to breed spontaneously, actually breed by flying above the water and dropping eggs into it? Are the eggs of such aquatic creatures so very small that we only see the young hatching from eggs, but not the eggs themselves, under the microscope? Are the eggs dropped into the water by the mosquitoes themselves, or are they brought into the pond by falling rain? Could the mosquitoes be ejecting their tiny eggs into the air, where the wind carries them over vast distances, until eventually the rain washes them down? Could all those multitudes of other tiny aquatic creatures also eventually take wing and fly into the air? Or perhaps dig themselves into the Earth?

Reservations regarding spontaneous generation appear again in Hooke's description of mites (Fig. 4.41). He had observed mites on all kinds of mouldy or putrefying substances, sometimes enmeshed in the thickets of mould, sometimes lying

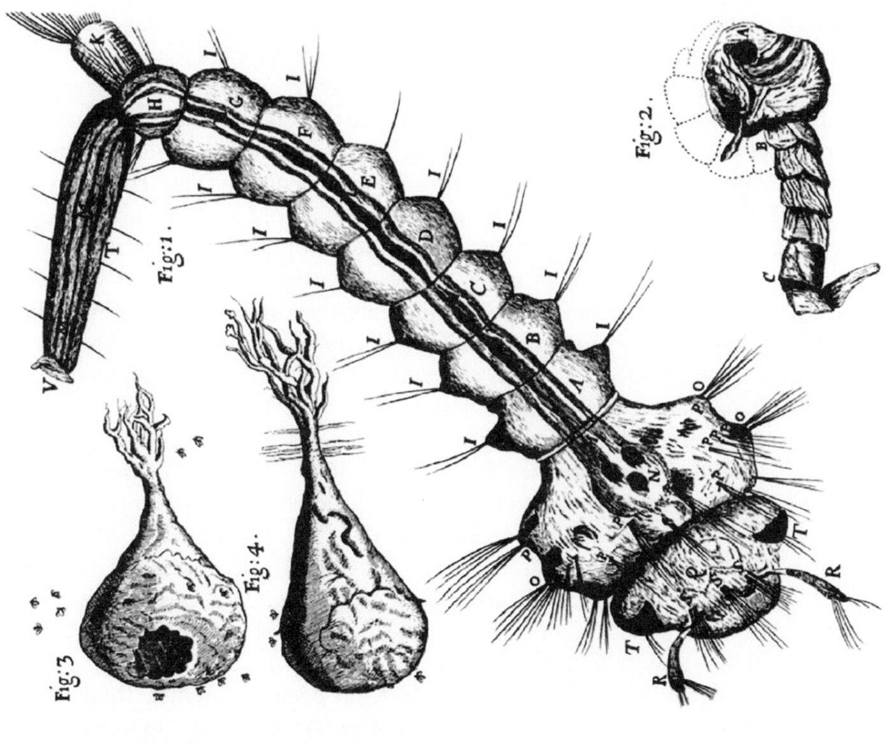

Fig. 4.40 Larva and pupa of the mosquito *Culex*. (*Micrographia*)

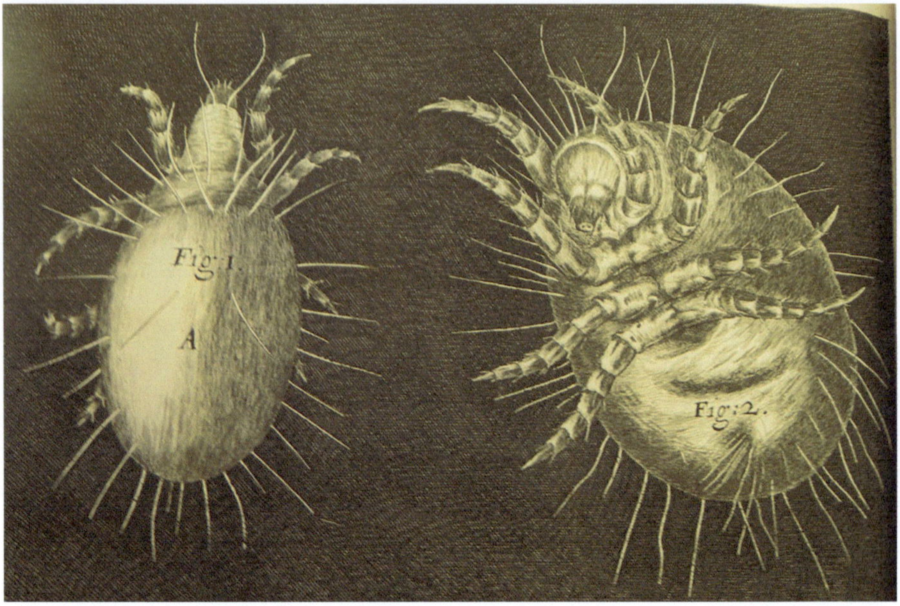

Fig. 4.41 Mites. (*Micrographia*)

dormant beneath them. It was clear to him that they feed on that vegetating substance. Now it was commonly claimed that mould is appropriate food for spontaneously generated animals, but do mites or any other creatures indeed generate spontaneously? It seemed quite probable to Hooke that some wandering mite might have reached a place suited for laying its eggs. The resulting offspring might have had a different shape than the parents at first, but eventually they could grow to resemble them. Alternatively, perhaps, one of these altered progeny, upon wandering from its native soil, could by chance reach the birthplace of its parent and settle there, producing a generation of mites of the same shape as that first wandering mite. Hooke concluded from these speculations that mites might perhaps have sprung from "one and the same sort of mites at the first", and that beyond mites, this would apply to other sorts of animals generally believed to generate spontaneously in putrefying substances (Schneider 2018).

These observations, doubts, arguments and thoughts of Hooke could well have stimulated the work of Francesco Redi (1626–1697, Fig. 4.42). The son of a renowned physician from Florence, he served as head physician and superintendent of the ducal apothecary at the Medici Court, to the Grand Duke of Tuscany Ferdinando II de' Medici and to his successor, Cosimo III. He was one of the founders of the Florentine *Academy of Experiments* (*Accademia del Cimento*), a scientific institution promoting the study of natural phenomena, and he devoted himself also to poetry and the study of European languages.

To study whether animal life can emerge from decaying matter, Redi investigated whether maggots generate from rotten meat. He did so by dividing six jars

Fig. 4.42 Redi

into two groups of three. In the first jar in both groups, he put an unknown object; in the second, a piece of dead fish; and in the third, a piece of veal. He then covered the first group of jars with gauze so only air could pass through and left the second group uncovered. After a few days, he observed maggots in the open jars, which flies had been able to enter, but none in the gauze-covered jars.

In a second experiment, he kept meat in three jars: one open, the second closed with a cork and the third covered with gauze. Flies could enter only the uncovered jar, and indeed this was the only jar in which maggots appeared. In a third experiment, he captured maggots and observed how they metamorphosed into flies. Finally, when he placed dead flies and maggots in a closed jar with meat, no maggots appeared; yet when he repeated the process with live flies, maggots appeared. Redi repeated this experiment over the course of many years. He used hundreds of different kinds of plant and animal matter, including flowers, grasses and an impressive array of animal and fish corpses, from dog and eel to lion and water buffalo (as a member of the Tuscan court, he had access to a vast collection of exotic animals). He concluded, from his experiments, that rotting organic matter does not generate insects; it serves only as a potential nest for the development of eggs into flies. He summarized his findings in a statement that *omne vivum* ex vivo, all life comes from life (*Experiments on the Generation of Insects*, *Esperienze intorno alla generazione degl'insetti*, 1668). Living things, be they animals or plants, never generate from non-living matter (Fig. 4.43). As a side effect, Redi's experiments seriously challenged the belief that creatures of one species could transform into another species.

There is no spontaneous generation of species, nor is it possible for a species to develop from the seed of a different species (Parke 2014). This paved the road to a stable, systematic classification of plants and animals. The species is a stable classification unit, constant and unchanging, created in the 6 days of Creation.

Redi, by proving that the dogma of spontaneous generation of small animals from decaying matter was groundless, placed a milestone in the history of reproductive biology. Strangely, however, while rejecting generation from decaying matter regarding flies, Redi also believed that insects are spontaneously born from galls—namely, from abnormal growths of plant tissue formed when insect and mite parasites lay eggs in the plant. We, today, know that there are 1500 gall-producing species, mostly gall midges (Cecidomyiidae) and gall wasps (Cynipidae), different species forming different galls, which vary in size, shape and position on the plant. People in classical times had known oak tree galls quite well, perhaps because of their importance as a source of tannin. Theophrastus described the external morphology of different types of oak tree galls, and we, today, can identify with certainty nine of the ten kinds of oak galls he had mentioned, including the well-known

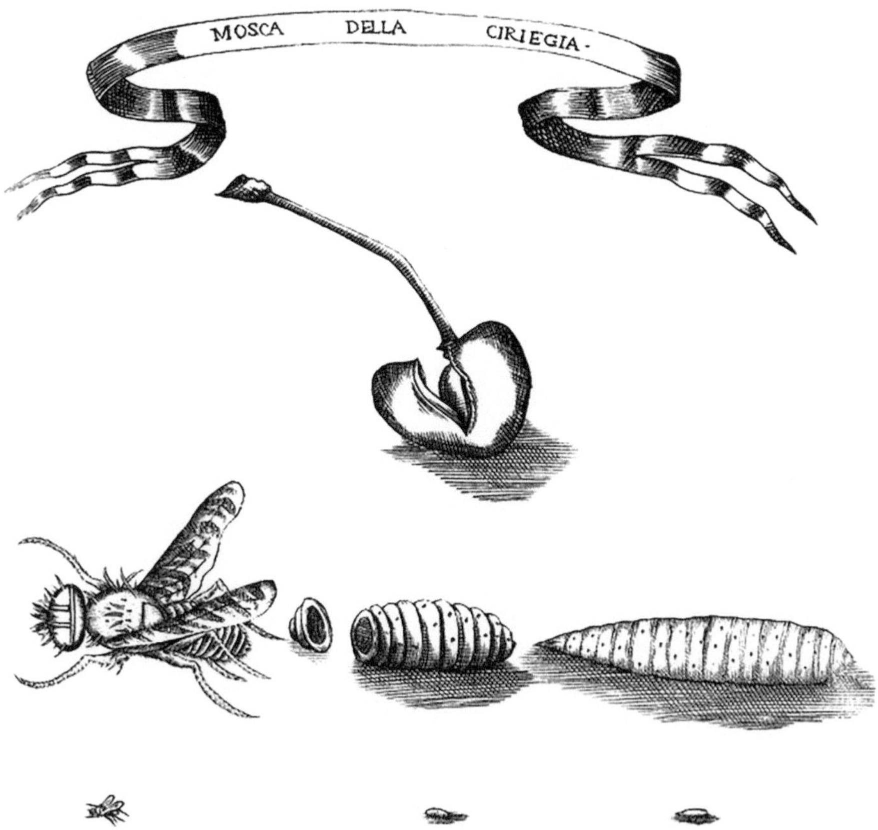

Fig. 4.43 Flies: maggot, pupa and adult. (Redi, *Esperienze Intorno alla Generazione degl'Insetti*)

scarlet "berry" gall (Fig. 4.44). The ophrastus was well aware of the difference between fruits and galls, and he noticed the presence of insects within them, resembling mosquitoes and flies. In the seventeenth century, people commonly observed insects emerging from galls and commonly believed that the gall itself was responsible for their generation (Thanos 2005). Hooke (1665) described the bark-like character of the gall tissue. All hole-less galls that he dissected contained maggots, whereas those with holes did not. From this, he inferred that the maggots had eaten their way to the outside (it is, actually, the adult that does this), taken wings and flown away. In line with this common belief, Redi too claimed that plants can either become infested by maggots coming from the outside, eating their way into the plant and forming a gall, or they can form gall tissues in which they generate insect eggs. At first, Redi accepted only the first alternative, which argued that gall insects come only from eggs laid by parents in the gall. Later, however, he observed that galls brought to his room hatched flies, that the many different types of galls corresponded with different types of insects and that he had never found a single gall not containing an egg or larva. Consequently, he changed his mind and accepted the second alternative as well, that plants make gall tissues to generate eggs inside them (Parke 2014).

To understand Redi's contradicting conclusions regarding houseflies versus gall-flies, we must bear in mind that one of his opponents was William Harvey, who had accepted that animals may be generated from decaying matter. His second opponent was Fortunio Liceti (1577–1657), an Italian physician who taught at the universities of Pisa, Padua and Bologna and who was concerned with matters of generation and development. Regarding spontaneous generation, Liceti argued that life could generate from decomposing plant or animal material in which part of the vegetative or sensitive soul remained. However, a vegetative soul could not produce an animal soul, for lower forms cannot produce higher ones. The generation of animal organisms must therefore proceed, Liceti argued, either from the remnants of the animal soul in decaying matter or animal excrement or from living atoms flying about

Fig. 4.44 A scarlet berry gall and a gall midge (Cecidomyiidae). (Tuvia Kurz)

through the air (*De spontaneo viventium ortu libri quatuor* 1628). Redi, by criticizing Liceti, embraced the idea of the production of animal life from a vegetative soul. Indeed, he wrote, the production of worms and winged animals may be the primary purpose of the plant, with the fruit only secondary:

> The same soul or power in living plants which generates their flowers and fruits, generates also their maggots. Who knows? Perhaps many tree fruits are produced not as primary ends in themselves, but rather for the secondary, subordinate end of generating maggots, serving as a womb in which they reside for a fixed amount of time, after which they emerge to enjoy the sun.

Furthermore, he extended his view of spontaneous generation from living matter to parasitic-worms as well:

> ... if a thing is living, it can produce within itself a little maggot, as in the case of cherry, pear, and plum trees [...] In this same manner it could perhaps be true, and I am inclined to believe, that in the intestines and other parts of human bodies, worms are born (*Generazione degl'insetti* 1668: 63).

Thus, a thing can generate new life if, and only if, it itself is living.

Redi may possibly have gotten the picture of gall insect generation wrong because he studied them on his bench in the Medici Court and not in the field in their natural environment. By neglecting nature, he gained knowledge only of the larvae inside galls and their emergence as flying adults. He did not see where the eggs and larvae came from. He never saw the gall midges and gall wasps buzzing around plants and depositing their eggs in places where galls would later form. His contemporary Malpighi, who worked on the same questions at the same time and in the same country, got the gall insect story right. Another person who strongly rejected spontaneous generation of any kind was Swammerdam.

4.7 Development Determines Diversity

Aldrovandi, we recall, viewed insects in light of their diversity. Naturalists of the seventeenth century began viewing them in light of their life cycles. Johannes Goedaert (1617–1668) was a painter and engraver in Zeeland, the Netherlands, who loved to raise insects. He would collect larvae in the field, keep them in glass containers, feed them on different kinds of leaves and observe them as they underwent metamorphosis. He drew them as larvae, noted when they changed, drew their pupae, noted how long they remained in pupal form and when they emerged and drew what emerged, whether an adult butterfly or a parasite. His book, *Metamorphosis naturalis* (1662–1669), consisted of engraved plates presenting the insect itself in its different phases, a complete life history that involved the caterpillar, the cocoon and the emerging adult. This was one of the first studies of insects that single-mindedly emphasized their metamorphosis. Goedaert drew the caterpillar and waited, then he drew the pupa and waited, then he drew the emerging animal—a butterfly, a fly or a worm. Having no training in natural history and being unaware of the parasitic behaviour of ichneumon flies and blowflies, he drew what happened—the first naturalist to devote himself to the detailed study of insect metamorphosis and to see his results through to the end. He did not arrange his material in any methodological order, and we do not know whether he had any broader interest in natural history. Goedaert was an accurate observer who all too often did not understand what he observed. One of Goedaert's critics was Swammerdam, who used metamorphoses as the basis for a new classification of insect diversity.

Jan Swammerdam (1637–1680), the son of a Dutch apothecary in Amsterdam, loved animals from his early years. He graduated from the University of Leiden as a physician, with a dissertation on respiration, but devoted his time to research instead of establishing a medical practice, as his father had expected. This caused a rift between Swammerdam and his father, who cut off his financial support. In his early thirties, inspired by Hooke's *Micrographia* (1665), he started combining his dissection skills with microscope observations to study insects and reveal their complex internal organs (Cobb 2000). Swammerdam would use single-lens microscopes, practically a very small magnifying glass in which the object almost touches the lens, and the observer, to see the object, must place his eye very close to the lens. He observed only under direct natural light, only during summers and only bareheaded, to allow maximum light to reach the lens. In autumn and in winter he stopped his observations for want of light (Cobb 2000). In his late thirties he fell under the sway of a bizarre mystic cult headed by a French woman, Antoinette Bourignon, and under her influence he renounced science, destroying some of his manuscripts (she had described his work as amusements of the Satan). She did allow him to publish his study of the mayfly (*Ephemeri vita*, 1674), but it was undoubtedly her influence that led him, in this text, to smother his findings with pages upon pages of religious reflections (Cobb 2000). After 9 months of living on her island off the coast of Schleswig, a disillusioned Swammerdam returned to Amsterdam; he died 2 years later of malaria, aged 43.

Much of Swammerdam's research focused on William Harvey's few paragraphs on insect metamorphosis, a term including both the sudden profound change in form from one stage to the next (as from caterpillar to pupa and from pupa to adult) and the sudden transformation of inanimate precursors into living beings by spontaneous generation. Harvey had claimed that any primordium living only in potentiality is an egg. Insects produce an "imperfect" egg in that an imperfect animal ensues—a "worm" (= grub or caterpillar). In comparison with its egg, the worm is an animal; compared to the adult, it is a crawling, feeding egg. This worm, having reached its proper size, changes into a pupa, cocoon or chrysalis—which Harvey considered the perfected egg because it is a potential animal. This "second" egg is the true, completed ovum. In Harvey's metamorphosis (and spontaneous generation), a developing embryo, in which all matter is present from the beginning, forms all at once. Consequently, this matter is homogenous, and, as such, it reaches all organs, no matter how dissimilar they are. An animal born as a worm, whether from an egg or from pre-existing rotting matter, is not as perfect as the red-blooded animals. Harvey thus distinguished between two groups of animals, according to their mode of embryonic development: the perfect, epigenetic ones (vertebrates) and the lower, less perfect ones, which develop by metamorphosis (insects).

Swammerdam (1737) strongly opposed this theory of the pupa being a second, perfect egg, as it relied only on the shape of the pupa, ignoring its contents (Sleigh 2012). He suggested considering the pupa a "nymph", a term that had been coined by Aristotle: that stage in the life of insects in which they "have received the outlines of shape they are afterwards to wear", namely their adult shape. (In today's terms, a nymph is the immature form of those insects undergoing gradual metamorphosis before reaching the adult stage). He considered the timing at which the nymph state is reached as the basis of all insect diversity. Insects, he argued, differ from one another in the stage of development at which they attain the nymph state. Accordingly, he arranged insect diversity into four "orders", corresponding to four different developmental pathways, based on the extent of metamorphosis (Cobb 2000):

1. Those in which a miniature version of the wingless adult emerges from the egg and simply increases in size throughout succeeding immature (nymph) stages. These wingless insects often continue to moult after reaching sexual maturity (silverfish, springtails; Swammerdam included spiders, scorpions, snails and lice; today we term such insects the Ametabola; Fig. 4.45).
2. Insects born in the shape of their winged parents as miniature versions of the adult but lacking wings. A young nymph hatches, and its wings gradually develop and grow throughout successive nymph stages. The fully functioning wings and the genitalia emerge in the final moult (dragonflies, locusts, mayflies; we term them Hemimetabola; Fig. 4.46).
3. Those with a pupal stage in which adult structures, such as wings, legs and antennae, appear gradually within the nymph prior to pupation (butterflies; the Holometabola). The immature stages, called larvae in this case, pass from the larval to the pupal stage and then from pupa to adult. Larvae of holometabolous insects differ from the nymphs of hemimetabolous insects by the presence of internally developing wings, modified larval eyes and reduced or absent larval legs (Fig. 4.47). Swammerdam, in a dissection study, showed that the pupa is not egg-like; it is a transitional stage between larva and adult, "a butterfly enclosed and hidden in a caterpillar and perfectly contained within its skin". It is thus during pupal development that the nymph state is reached.
4. An extreme, derived form of Holometabola: insects in which metamorphosis involved the use of the larval cuticle to form a case, the puparium (flies; we term them Holometabola too; Fig. 4.48).

Swammerdam (1737) claimed that insects develop in the same gradual manner as other animals, in a growth process resembling that of a chick developing epigenetically. Inside the pupa, adult development is continuous rather than abrupt: wings, antennae and other parts typical of the adults gradually grow and accumulate under the skin, just like organs gradually developing in the chick foetus. (He described the pupa in poetic human terms: "the worms [of insects], after the manner

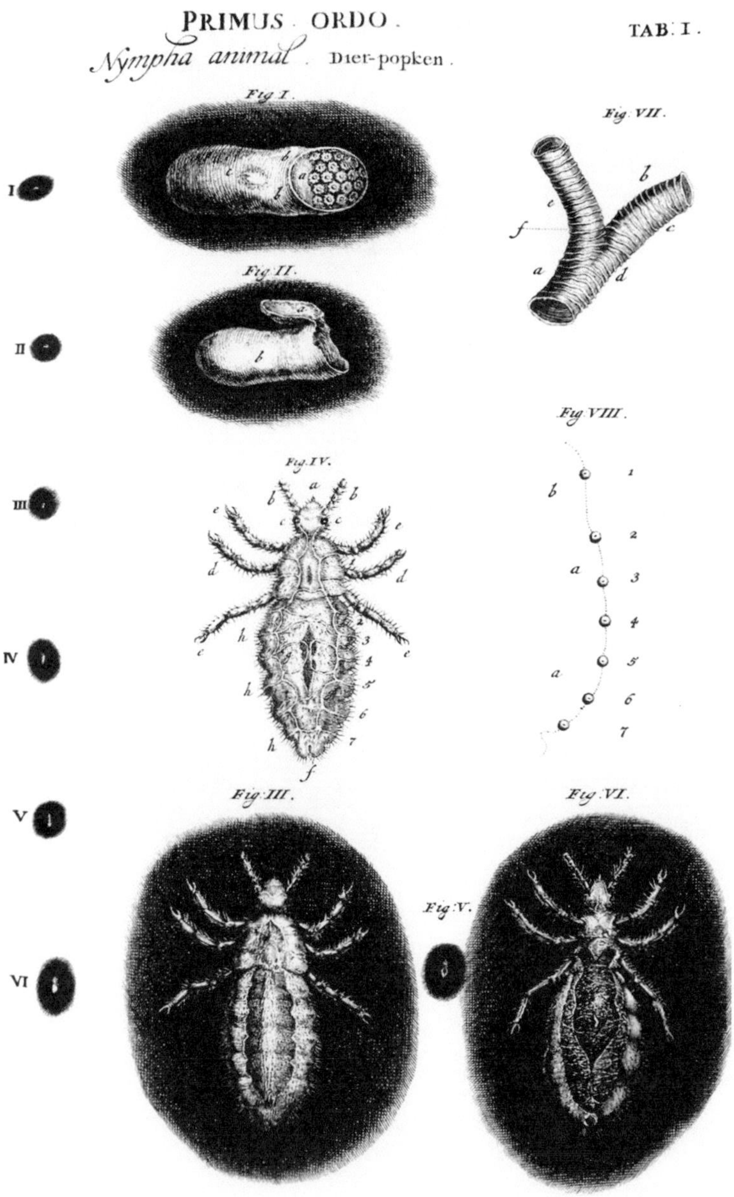

Fig. 4.45 Swammaerdam's first Order (Ametabola): lice

of the brides in Holland, shut themselves up for a time, as if to prepare and render themselves more amiable, when they are to meet the other sex in the field of Hymen"; Cobb 2000; Sleigh 2012). Therefore, there is no room for theories about insects differing from vertebrates in that they abruptly metamorphose into adult

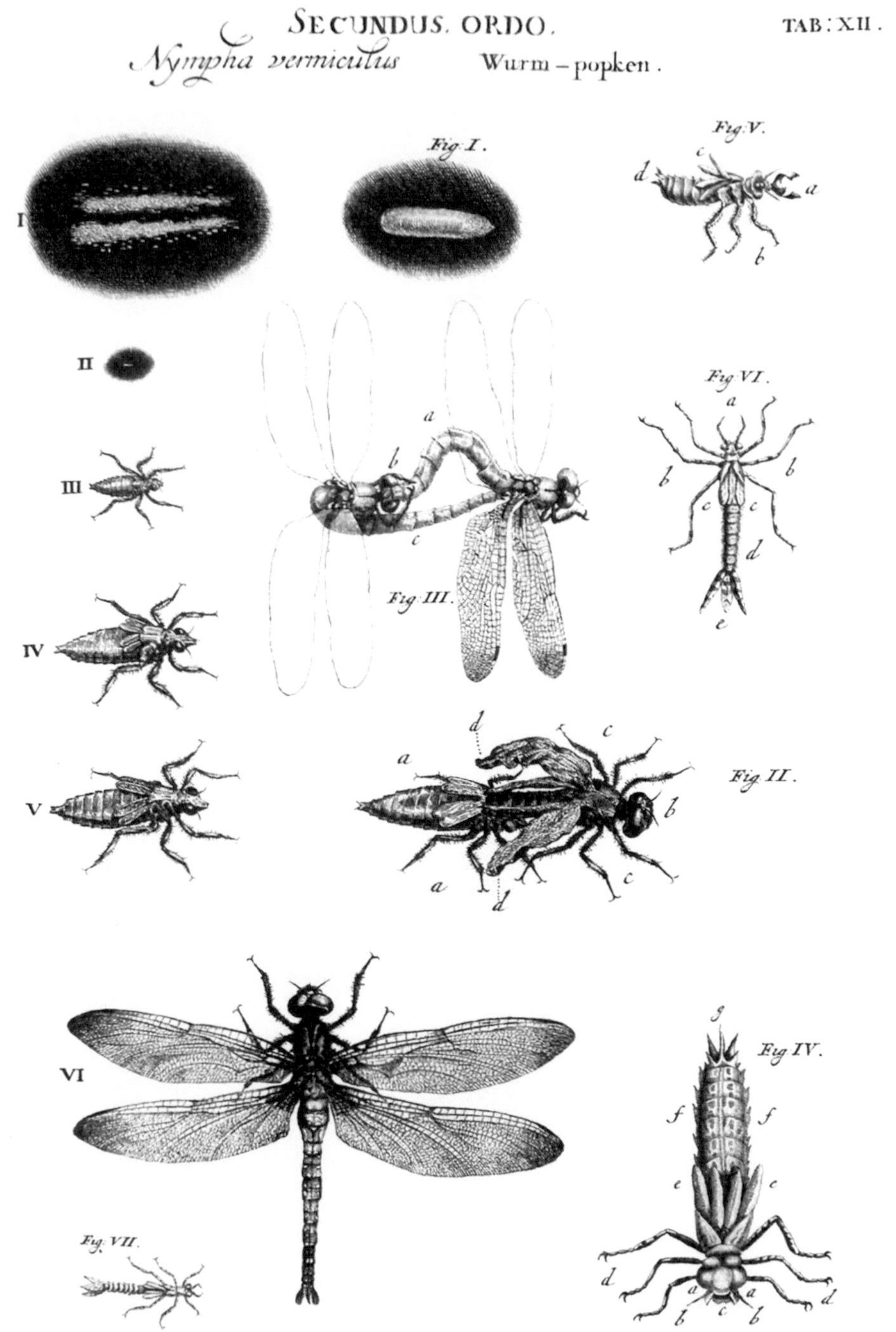

Fig. 4.46 Swammaerdam's second Order (Hemimetabola): dragonflies

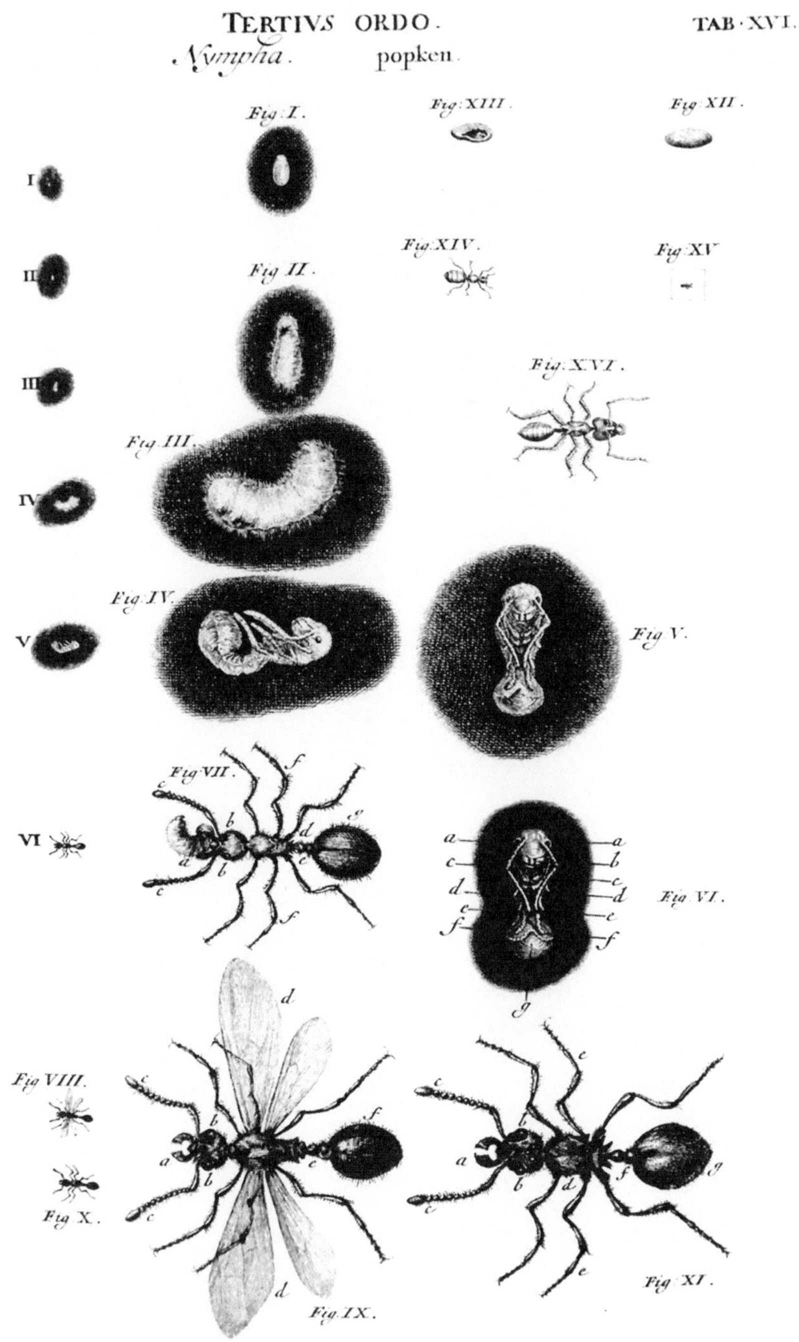

Fig. 4.47 Swammaerdam's third Order (Holometabola): ants

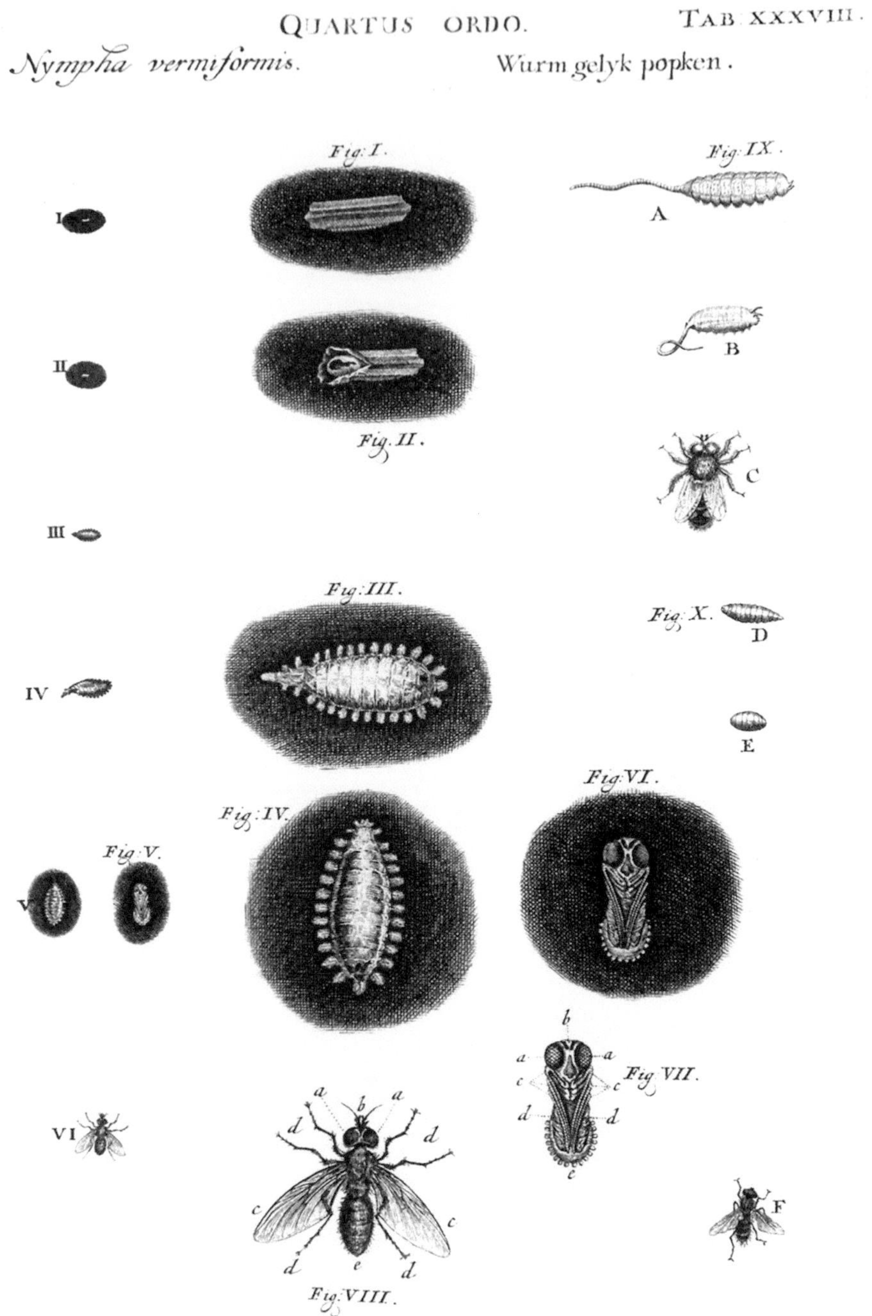

Fig. 4.48 Swammaerdam's fourth Order (Holometabola): flies

form. Some of the illustrations in his book show, step by step, how various insect species change their shape and size, starting from an egg. The life-sized depiction of the entire process enhances the realistic effect, and the reader sees not just the egg from which the caterpillar emerges but also those parts of the butterfly already present in the caterpillar, along with the pupa and the butterfly itself (Cobb 2000).

To emphasize the uniformity of nature, Swammerdam also included an illustration showing the stage-by-stage development of a frog, pointing out the similarities in the growth processes and using it as an intermediate animal, midway between Harvey's ignoble insects and his larger, red-blooded, noble animals (Wilson 1995; Sleigh 2012). The development of frogs is epigenetic, as is that of all four insect types, a gradual growth and development of the insect's parts, all of which were present in earlier stages. This study of frog development took Swammerdam into theological spheres, in which he claimed that the encumbrance created by the tadpole's long tail reflected the difficulties humans seek to shake off in their journey through life. "His appendage of misfortunes and trouble, like the tail of the frog, yet adheres for a long time to him, for he is full of misery, and is born in tears; and it is very long before he comes to maturity of understanding, and full growth of body", and the male frog clasping a female in the act of coition resembles a peasant's hands, clasped in prayer (Sleigh 2012). Swammerdam sought to reorder all animals, from insects to humans, within one glorious, God-ordained natural history and philosophy. He therefore linked insects to frogs through a demonstration that both underwent epigenesis and frogs to humans through a demonstration of their identical muscular activity (Sleigh 2012). Interestingly, the figure containing the epigenetic frog also depicts a carnation, the traditional symbol for the incarnation or passion of Christ. The bloodless insects progressed on to the red-blooded frog and up to the fully red Christ flower (Fig. 4.49).

Swammerdam further argued that Harvey's doctrine of metamorphosis being inferior to epigenesis was dangerously close to the doctrine of spontaneous generation. "We may easily see how rash and precipitate their opinion is, who esteem the larger creatures only as perfect, and the less as scarce worthy to be classed with animals; but, as they say, produced by chance, or generated by putrefaction; rendering, by such reasoning, the constant order of nature subject to chance". Swammerdam claimed that belief in spontaneous generation is blasphemous, as this would imply that part of nature exempts itself from the order and laws laid down by God. If the generation of life is subject to spontaneous chance, then what would prevent Man from also being easily produced in the same manner? (Cobb 2000). In asserting that insects are as complex as higher organisms, he felt that he was asserting the unity of the natural world. Nature is orderly, and God's laws of nature are fixed, immutable and apply throughout the entire Creation. The same rules apply everywhere at any time, and there can be no such thing as chance. Just as humans give birth to humans and sheep to sheep, so do lice give birth to lice. God might create some creatures according to one law and others by another, but He creates all creatures with equal care, and they all function in the same divine way. Swammerdam would repeatedly stress that the small size of frogs, reptiles, fungi and, above all, insects makes their anatomy all the more admirable. In his wish to disprove spontaneous generation, he

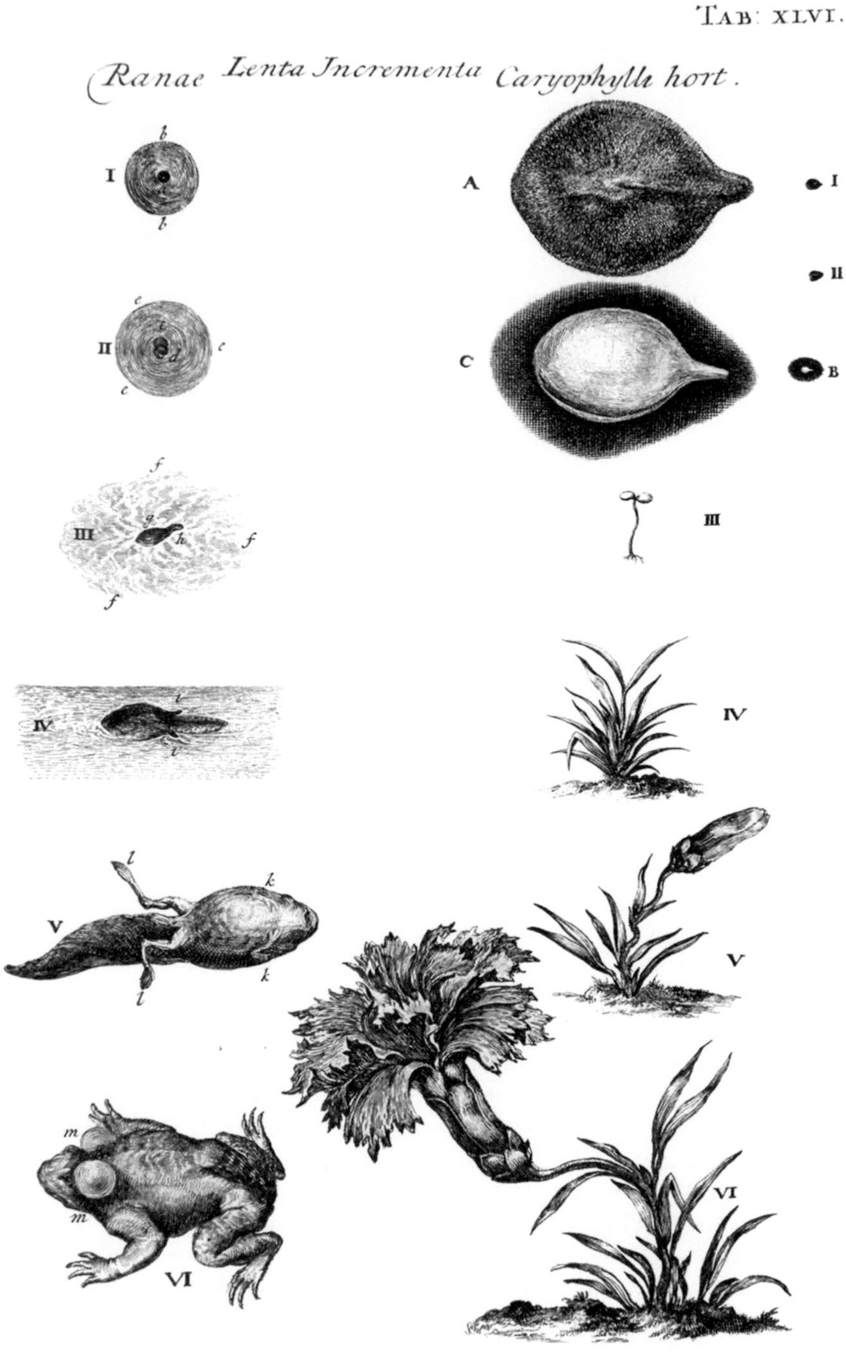

Fig. 4.49 Swammerdam: frog metamorphosis and a carnation

sought to prove the unity of God's creation: all animals had been created with equal care, and they work in the same divine, natural way. He disliked the idea that God might have created some creatures by one law and others by another (Sleigh 2012).

Swammerdam was, like so many of his generation, highly critical of those who blindly followed the views of the ancient Greeks and Romans. However, he himself accepted stories of the Old Testament. In his *The Book of Nature* (posthumously 1676–1679), he discussed at length the story of Samson and the lion, the bees that nested in the dead lion's carcass and how they could have produced honey so rapidly (Cobb 2000). Having noted that metamorphosis in the butterfly leads to a change in the digestive system, he wondered whether King Nebuchadnezzar, who in the Old Testament had turned into a grass-eating beast with a hairy hide and long talons (*Daniel* 4), "did not suffer a change in the internal parts correspondent to that which appeared in his external form". Swammerdam thus used the wonders revealed by the microscope and dissecting instruments as proof of the glory of the Creator (Cobb 2000). In *The Book of Nature,* Swammerdam also confirmed, by microscopic dissections, that drones are male and the workers and queen are female (Cobb 2000). Unfortunately, the mode of honeybee reproduction escaped him ("I do not believe the male bees actually copulate with the females").

We today know that metamorphosis in insects arose some 285 million years ago. By decoupling larval from adult development, larvae and adults are able to occupy separate niches. In addition, internalization of the developing wing primordia allows the worm-shaped larvae to burrow, and these juveniles are therefore able to live inside ephemeral food sources such as fruits, detritus and those animals that they parasitize.

4.8 Fossils: What Are They?

Throughout the Renaissance, we recall, debates over the essence of fossils raged wide and far, with intellectuals of the sixteenth century asking why fossil remains of seashells, if indeed they are seashells, are far from the sea, sometimes in solid rocks on high mountaintops. Some, in line with Aristotle, thought that the seas wax and wane cyclically and that a past rise of the sea would have brought seashells to mountaintops. Then, there were those who thought that so-called fossil shells are not real seashells at all; they are inorganic minerals of unusual shape that grow inside the Earth, under the inspiration of some mystic power.

In the seventeenth century, one man, Nicolas Steno (1638–1686, Fig. 4.50), pioneered the huge leap forward in the understanding of fossils. The son of a goldsmith, he was born in Copenhagen (under the Danish name Niels Steensen), and it was here that he began studying medicine. During the wars between Denmark and Sweden (1655–1660), besieged and starvation-ridden Copenhagen was not a good place for academic education, and Steno, to continue his medical studies, left for Leiden. It was here that he developed his skill at dissecting, discovering the ducts of the parotid salivary gland and then those of the tear-producing glands (people had

Fig. 4.50 Steno

previously assumed that the brain, by the force of grief, would squeeze the tears out). Gaining a reputation as a keen anatomist, he moved from Leiden to France, where he put to the test Descartes' theory as to the seat of the soul. Descartes, we recall, had claimed that humans differ from animals in that they have souls and that the pineal gland, in the centre of the brain, is the link between soul and body. Twisting and turning in response to the soul's demand, it pulls strings that control the body's movements. Steno, coming some 15 years after Descartes' death, dissected and demonstrated that in an undamaged human brain the pineal gland is immobile, as it is held fast by surrounding tissues, and therefore there is no way it could perform the gyrations that Descartes' theory required. After Paris and a brief stay in Montpellier, he set off to Florence, where, some 8 years earlier, the Grand Duke Ferdinando II de Medici and his younger brother Leopoldo had established the Experimental Academy *Accademia del Cimento* (1657). Ferdinando nominated Steno to the Academy and granted him lodgings in the palace. Here, Steno tested another of Descartes' theories, that the heart was a heat-generating furnace and not a pump (since it lay deep inside the chest, isolated from all muscles and moving parts, with no mechanical way for motion to be transmitted to it). Steno tested this theory of Descartes by dissecting the heart of an ox. He found that its fibres were arranged in a way that, when shortened, it would function like a pump that squeezes blood through the arteries. Descartes was thus wrong on yet another aspect of his mechanical philosophy.

In 1666, and soon after Steno's arrival in Florence, it so happened that anglers caught a huge shark of about 1600 kg off the coast of Tuscany. The Grand Duke, who was a generous patron of the sciences, ordered the shark's head to be brought to Florence for his court scientists to examine, and it fell upon Steno to do the dissection. The shark's head was badly desiccated, but Steno succeeded in dissecting the jaws, eye and ear. Then, gazing into the shark's huge mouth, Steno saw teeth—hundreds of teeth in 13 rows, each about 10 cm high (Fig. 4.51). They immediately reminded him of *glossopetrae*, "tongue stones", which he had seen in the pharmacy cabinets of his professors in Leiden and which people believed to be the petrified tongues of dragons and snakes. The Leiden professors were interested in the

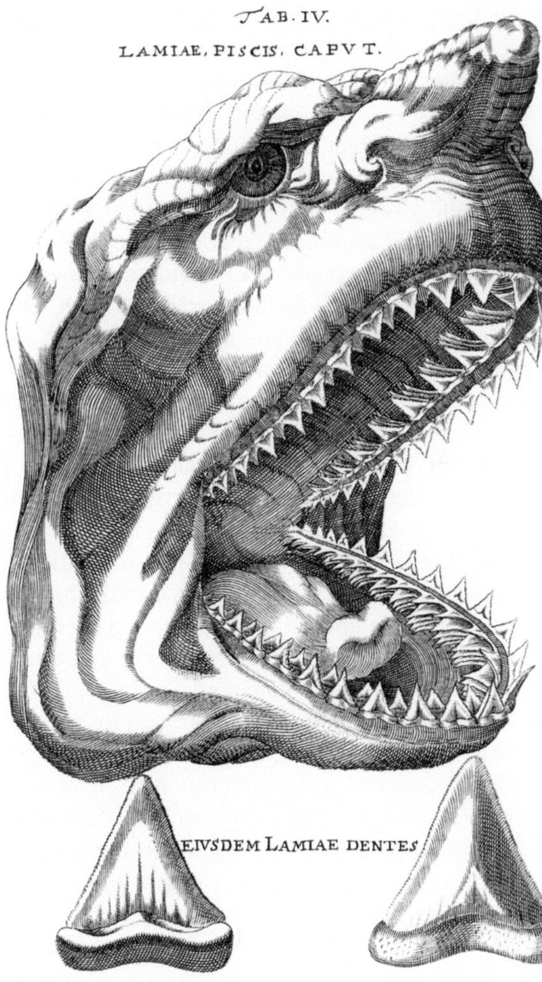

Fig. 4.51 The head and teeth of a shark (Steno 1667)

curative powers of *glossopetrae*, not in their origins and they had never seen shark teeth. They, and other physicians throughout Europe, believed that tongue stones cure snake venom, can ward off sorcerer's spells and help "those that court faire women"; others credited tongue stones with the power to restore speech. People mounted them in silver and retained them as amulets worn about the neck as a pendant or kept them sewn into special pockets as good-luck charms. When powdered, they sold them as remedies for plagues, fevers, poxes, labour pains, epilepsy and bad breath. The best and most famous tongue stones came from the island of Malta, where they were found scattered on the ground or just beneath the surface.

Steno reasoned that these tongue stones, found on Malta and flooding the apothecaries of Europe, must be shark teeth. (We today know that sharks continually shed their teeth; some shed 35,000 teeth in a lifetime, new ones replacing those that fall out.) How, then, could they come to be on dry land? People assumed that tongue stones fell from the sky on dark, moonless nights or that they were broken pieces of lightning bolts. There was also the classical opinion that fossils grew naturally inside rocks, wherever (the as-yet-unknown) conditions were ripe. Steno's contemporary, Kircher, had attributed fossils to a mysterious power diffused through the whole world (1664). In Kircher's opinion, the Earth was alive with many mysterious spirits that enabled plants, animals and, indeed, every imaginable shape to grow from the Earth, even within stones. Fossils take their shapes from what Kircher called a plastic virtue, which gives form to all things, whether mineral, vegetable or animal. It is active to this day, generating flies in dung and moulding shells in stones. Seashells inside rocks could not possibly be real seashells; no snails had ever lived inside them because God had created Earth with its rocks on the third day, but He created seashells, alongside all other forms of life, only on the fifth day.

Steno, however, argued that *glossopetrae* looked like shark teeth because they were shark teeth. They came from the mouths of once-living sharks and had come to be buried in mud or sand that was once undersea and is now dry land. He was not the first to link tongue stones with sharks' teeth; a century before him, Rondelet (1554) had noticed that the teeth of large sharks closely resembled tongue stones. However, Rondelet did not explain how the teeth had turned into stone, nor what they were doing on land. Steno, in opposing the theory that fossils grew inside rocks, noted that in none of the places where he had found them did they seem today to be growing in the ground. Furthermore, a shell growing inside solid rock would necessarily crack the rock, but he never found cracks around fossil shells. Then, shells occurring in softer ground not only showed no signs of growing; they appeared instead to be slowly disintegrating due to erosion. Furthermore, tree roots growing in soil, when in hard ground, twist in countless ways and assume shapes quite different from those growing in soft ground. Tongue stones and fossil shells are always the same, however, whether dug up in soft ground or hewn from hard rocks. Furthermore, some fossil seashells lie broken with the missing part found nearby, like modern broken seashells on a beach. From these and other arguments, Steno concluded that *glossopetrae* actually were sharks' teeth. Rather than growing inside rocks, the rock had solidified around the teeth. The undistorted shapes of the fossil teeth (and shells) implied that when becoming entombed, the surrounding rock was not yet solid; namely, the rock did not yet exist. In his student days at Copenhagen, Steno had learnt how easy it was to dissolve various powders in warm water. When the water had cooled or partially evaporated, solids would precipitate and settle to the bottom, forming a soft ooze that would eventually harden into a solid mass. Steno thought that a similar process could have happened in nature, though on a much larger scale. A rock containing fossils could have originally been a soft, semi-solid aquatic ooze in which some creature might have lived. Only later on did the ooze harden into rock. Therefore, fossils must have solidified before the formation of the host rock containing them. Steno formulated a general principle: a solid object will cause any solids that later form around it to conform to its own shape.

Fossil-bearing rocks gave yet another clue that water had deposited them. In sediment-laden water, the particles sink to the bottom in order of their weight—larger and heavier particles drop out first, and fine silt remains in suspension much longer. Steno saw the same arrangement of sediment grains in fossil-bearing deposits, big particles below, finer ones above and this suggested to him that water had indeed laid down these deposits. The theory that rocks formed from water laden with sediments also explained another important observation; that the fossils entombed in rocks were of marine, not terrestrial, creatures. In addition, the fact that fossils consisted of the durable parts of animals—shells, bones and teeth rather than soft, rapidly disintegrating tissues—supported his theory that the sediments accumulated gradually around animal remains, gradually burying them.

Why do seashells occur far from the sea, sometimes embedded in solid rock at mountaintops? Steno reasoned that they could have reached there either by the water once rising to cover the land or by the land once sinking beneath water. Both possibilities seemed reasonable; indeed, the Bible confirms that water had covered the land at least twice, once at Creation and then again during Noah's Flood. The island of Malta, from which so many tongue stones came, was a good example. It had sprung up from the depth of the sea, and "Perhaps at one time when this land lay under the sea, a place where sharks lurked, their teeth were buried freely in the muddy sea-bed, but now they are found in the middle of the island owing to a change in the position of the sea-bed" (Steno, as quoted by Cutler 2003). Steno thus reasoned that our seemingly perfectly designed, unchangeable world might actually have undergone very drastic changes.

Steno's work on shark teeth led him to the more general question of how nature produces solids. He argued that a solid grows by the addition of new particles, which a fluid brings to it. Because solid particles cannot penetrate one another, these new particles accumulate at the surface of the existing solid, attaching to it one layer at a time, one atop the other, with the lowermost layer being the oldest. We today refer to this generalization as Steno's Law of Superposition: layers of rock form a time sequence, with the oldest on the bottom and the youngest on the top, unless later processes disturb this arrangement. Accordingly, in mountains and hills that consist of several layers of solids one on top of the other, the lowermost layer would be the oldest—and so too would be its fossils. Steno thus grasped that mountain layers represent a sequence and that the Earth's crust contains an archive of its ancient history. The world had undergone colossal changes over its long past. A dynamic, evolutionary concept of the world had to replace the old, static concept. Steno's law is a statement of relative, not absolute, time: two rock layers, in principle, could have formed millions of years apart or a few days apart. Steno himself saw no difficulty in attributing the formation of many rocks to the biblical Flood. Noticing that of the two major rock types in the mountains near Florence, the lower layers had no fossils, whereas the upper ones were rich in fossils, he suggested that the upper layers had formed in the Flood, after the creation of life, while the lower ones had formed before life had existed.

Steno presented his new scientific principles in two treatises, as an addendum to his *A Shark Head Dissected* (*Canis carchariae dissectum acaput*, 1667) and in *A Preliminary discourse to a Dissertation on Solids Naturally Enclosed in Solids* (*De solido intra solidum naturaliter contento dissertationis prodromus*, 1669). At about that time, he converted to Roman Catholicism, abandoned science, became a bishop and spent the rest of his life ministering to the minority Roman Catholic populations in northern Germany, Denmark and Norway. Two years after its publication, the first copies of his *De solido* arrived in England, and the Royal Society's secretary had the entire book translated into English (1671). Fossil shells were topics of considerable interest among the Society's members, who regularly brought petrified shells to meetings for examinations and discussions. The investigation of all petrified objects submitted to the society fell to its curator of experiments, Robert Hooke, who, having discovered the microstructure of plants, had viewed under a microscope a piece of petrified wood and found that it had the same structure as living wood. Now in 1637, Francesco Stelluti had observed, by microscope, petrified wood from clay hills near Rome and had found a continuum of transformation, from solid rock to fresh wood. He had concluded that petrified wood grows inside clay and then changes into wood.

Hooke, having seen the cells in his petrified wood, thought that Stelluti had it all backward. Stone did not turn into wood; wood turned to stone. Buried in the clay, the wood had soaked in "mineral juices" that hardened it to stone. Hooke also criticized the idea of petrified wood and of fossils in general as being nature's jokes. Why, he asked, is it specifically those hard, decay-resistant parts of plants and animals that nature chooses to sport with? Why does nature never amuse itself by creating, say, subterranean roses? Perhaps the species in ancient seas were simply different from modern ones. With such a rich diversity of species living in such a rich variety of soils and climates, God could easily have done the same with ammonites, albeit on a larger scale. Unfortunately, Hooke was notorious for his feuds with other members of the Society, accusing them of stealing his ideas. He saw Steno's entry into the debate on fossils less as support for his position than as an encroachment onto his turf. He even accused the secretary of sending word of Hooke's ideas to Steno "by stealth" to undercut his own research.

Another response to the translated *De Solido* was a letter from Martin Lister, whom we have already met, a newly admitted member of the Society who had met Steno in Montpellier and who, upon returning to England, had begun building up his reputation as a top authority on shells. Lister did not doubt that in some Mediterranean countries one might find seashells of modern species inside rocks, especially along the coasts. But as for English inland quarries, their "cockle-like stones" were stones, had always been stones and had never been part of any animal, whatever their superficial resemblance to shells. Lister's criticism is understandable. In Italy and Malta, fossil-bearing strata are only 2–5 million years old (Pliocene). Clams, snails and shark teeth in these young strata resemble living creatures very closely. In England, however, the fossil-bearing rocks in which Lister found fossils are mostly from the Jurassic and Carboniferous, as much as 300 million years old. These rocks bear fossils, some of which are very different from those of Italy: ammonites (Fig. 4.52). Distantly related to squid, their spiral, chambered shells, sometimes as large as a waggon wheel, do not resemble the shell of any living creature. They were never alive.

Fig. 4.52 Lister: Ammonites

Another possibility was that ammonites represented animals now extinct. Hooke was perfectly comfortable with this idea, but to others, such as John Ray, the loss of even a single species was unacceptable on religious grounds. On his way back to England, fresh from a fossil-collecting trip to Malta, he too, like Lister, had met Steno in Montpellier. Ray read and praised *De Solido*. From what he himself had seen in Malta, he agreed completely with Steno's arguments; the tongue stones were shark remnants, and the shell-like fossils found together with them were real shells. However, ammonites could not be real shells because if they were, they must be extinct today. Ray believed that "divine providence is especially concerned to preserve and secure all the works of creation". Extinction was simply impossible.

Starting from a shark that happened to be caught off the coasts of Tuscany, Steno, who then happened to be in Tuscany at first, explained tongue stones and then shifted on to explain the broader problem of how fossils got inside rocks. From here, he went on to explain the even more general problem of rock formation in layers and of how solids grow in nature. I often wonder how palaeontology and stratigraphy would have developed as disciplines had two anglers not happened to catch a shark off the coasts of Tuscany just when Steno happened to be there.

4.9 Biodiversity Exists Just to Serve Man

John Wray (1627–1705, eventually John Ray; Fig. 4.53), the first to emphasize that the study of nature must be based on the ability to identify species, was the greatest naturalist of his time. His father was a village blacksmith, and his mother was a herbalist from whom he absorbed his love of plants. He studied and taught at Cambridge (Greek and mathematics), where he devoted most of his time to a study of the local flora (*Catalogue of Cambridge plants,* 1660). Later on, he was the first to formulate a biological concept of a species and to offer a clear-cut definition:

> ... no surer criterion for determining species has occurred to me than the distinguishing features that perpetuate themselves in propagation from seed. Thus, no matter what variations occur in the individuals or the species, if they spring from the seed of one and the same plant, they are accidental variations and not such as to distinguish a species... Animals likewise that differ specifically preserve their distinct species permanently; one species never springs from the seed of another nor vice versa. (*History of Plants* 1686).

This reproductive concept of a species is radically different from the essentialist concept in that it is applicable only to organisms and does not apply to inanimate objects. It is thus a *biological* concept since only living organisms reproduce, whether or not two individuals of a (sexual) species are the same.

Elsewhere, in line with this biological definition of a species, he wrote that he considered all dogs to be the same species because they interbreed. Indeed, Ray was the first to put strong emphasis on the species rather than on higher systematic levels. As to the higher levels of classification, in plants his major distinction was between herbs and trees, and in each of these, putting aside those imperfect herbs

Fig. 4.53 Ray

that lack seeds, he next distinguished between monocotyledons and dicotyledons (*Historia Plantarum* 1685–1703):

Herbae (Herbaceous plants)

- Imperfectae (Cryptogams)
- Perfectae (Seed plants)
  - Monocotyledons
  - Dicotyledons

Arborae (Trees)

- Monocotyledons
- Dicotyledons

(Monocotyledons are flowering plants whose seeds contain only one embryonic leaf, or cotyledon. The remaining flowering plants have two cotyledons, and therefore, scientists have traditionally classified them as dicotyledons. Molecular research shows that while the monocotyledons are monophyletic, dicotyledons consist of a number of lineages and are thus paraphyletic.)

Regarding animals and their classification, one person who would become Ray's close friend was Francis Willughby, a wealthy nobleman very interested in nature and one of the founding members of the Royal Society. He met Ray when he arrived at Cambridge as a student, and the two set out together on collecting expeditions to northern England, Scotland and Wales; Ray would be responsible for the study of the plant specimens collected, and Willughby for insects and birds. Later, they both set out again with another two pupils, this time on a 3-year collecting expedition to continental Europe (Belgium, Holland, Germany, Switzerland and Italy). Upon returning to England, Ray joined Willughby's household staff as the teacher of his children, with an annual income that enabled him to devote himself to research, and the two friends began working on their collections. However, Willughby died unexpectedly, which immediately changed Ray's status. Willughby's widow harassed him, forced him to marry one of the household staff (he was by then 46), and eventually threw him out of the house. At the age of 52 he returned to his home village, where his four daughters were born, and it was here that he wrote most of his books. One of these, *The Ornithology of Francis Willughby* (Willughby and Ray 1678), was in memory of his pupil, patron and close friend.

Ornithologia focused on birds but differed from previous treatises on birds in that it centred mainly on material collected by the authors themselves. The classification system was based first on habitat, then on feeding and then on morphology. This scheme broadly resembles that of Gesner, and through him, that of Aristotle. (HA VIII3):

- **Terrestrial**; living chiefly on dry land whether:
  - Carnivorous; feeding chiefly on flesh
  - Phytivorous; feeding on vegetables, whether:
    - Of short round wings, less fit for flight
    - Of long wings and swift flight, having their bill either
      - Long and slender - pigeon and thrush kind
      - Short and thick - bunting and sparrow
  - Insectivorous; feeding chiefly on insects, whether:
    - Big
    - Small
- **Aquatic**; living either
  - about and near the water
  - in the water:
    - Fissipedes, having the toes of their feet separate
    - Palmipedes, having the toes united by a membrane

Further down the classification, the species and genus-level classifications were based strictly on morphology—the shape of the bill, the legs, body size and plumage—and the resulting species groups broadly resemble the avian families we know today. Although the book focused on taxonomy, it included wonderful observations in natural history. The lapwing *Vanellus vanellus* builds its nest on the ground, exposed to view. However, "The eyes being so like in colour to the ground on which they lie, it is not easy to find them. The young, upon hatching, instantly forsake the nest and run away, sometimes with the shells still upon their heads, for they are covered with a thick down, and follow the old ones like chickens. If one approaches the nest then the mother lapwing, the nearer she is to her nest the quieter she is, the further away the more clamorous, that she may draw you away from the true place, and induce you to think it is where it is not" (1678:308). Today we know that among birds, Distraction Displays are anti-predator behaviours of parents to attract the attention of an enemy away from the nest or young. These displays frequently take the form of injury feigning.

Ray also wrote that among puffins, *Puffinus puffinus,* then nesting on the Isle of Man, the parent birds leave their nests and young early in the morning and spend the whole day fishing in the sea (1678:333). Today we know that a puffin's diet consists of various kinds of fish. They have the unique ability to hold several fish in their bills at a time: spiky spines on the roof of their mouths help grip each fish while still allowing the bird to open its mouth and catch more, about ten fish per dive. They take about 6 hours to digest a sand eel. The female lays a single egg, and the parents take turns bringing fish from the sea to feed the chick several times a day.

By Ray's time, naturalists could afford to be deeply sceptical of hearsay. Ray contradicted the popular notion that swallows hibernate in ponds; he suggested that they migrate to warm regions, like Egypt or Ethiopia. Other naturalists of Ray's generation, such as Charles Morton (1627–1698), suggested that birds migrate to the moon, and this is so fantastic that I present it in some detail. Morton came from Cornwall, attended Oxford for 3 years and then led an uneventful career. Aged 60, he left England for the Massachusetts Bay Colony in Boston, then New England, became vice-president of Harvard and wrote a textbook on astronomy and physics (*Compendium Physicae* 1687). In another treatise he presented his hypothesis that birds migrate to the moon (*An Enquiry into the Physical and literal Sense of the Scripture (Jer. 8.7 [...]. The Stork in Heaven knoweth her appointed times; and the turtle, and the crane, and the swallow observe the time of their coming [...] Written by an eminent Professor for the use of scholars, and now published at the earnest demand of some of them;* before 1686?). No one, he noted, knows the whereabouts of migrant birds during the winter months. The Scripture reads "Even the stork in the sky knows her appointed seasons, and the dove, the swift and the thrush observe the time of their migration" (Jeremiah 8:7). Morton argued that the text reads the stork *in* the sky, not *of* the sky, and hence the skies are their proper place of dwelling. To be more specific, they fly to dwell on the moon. Upon departing, "their cheerfulness seems to intimate, that they have some noble design in hand [...] namely, to get above the atmosphere, and fly away to the other world". Upon returning, they perform most of the journey without any gravitation. Morton suggested that the round trip to the moon takes 2 months each way, and this leaves 4 months on Earth and four on the moon. The distance from the Earth to the moon is 179,712 miles (52 semi-diameters of the Earth), and the birds cover this distance at a speed of 125 miles per hour. As the moon requires 1 month for a revolution around the Earth, birds heading straight for the moon will find it in the same line of direction where it was when they began their journey. During flight, birds encounter no air resistance and are unaffected by gravity, so they fly effortlessly, barely flapping their wings, sustained by excess fat and sleeping for most of the journey.

Morton's strange theory seems to have stemmed from his association with John Wilkins, leader of a scientific group at Oxford and one of the founders of the Royal Society, who had sought to establish the idea of the plurality of worlds. (*The Discovery of a New World in the Moon*, 1638). He had suggested that the moon resembles our Earth in that it has seas, streams and mountains, and he further suggested that there may be inhabitants in this other world, though he was not certain of what kind they are (p.187). Wilkins was also the one who established that the distance between the moon and Earth is 179,712 miles and that on a journey to the moon our bodies will be devoid of gravity, so we would not have to expend muscular energy on the way and consequently would not need to rely on accumulated fat, as do hibernating reptiles and bears.

Another probable source of Morton's theory may have been Francis Godwin (1562–1633), an English historian and Bishop who spun the yarn of an adventurous Spaniard, Domingo Gonsales, who trained 25 swans to draw his chariots up into the skies (*The Man in the Moone: or, a discourse of a voyage thither by D. Gonsales*, posthumously 1638). What seems to have caught Morton's eye in this text was Godwin's mention of birds on the moon. On his flight to the moon, Gonsales saw cuckoos and swallows all flying upwards, and the higher they flew, the less they (and his swans) had to invest in flight until, eventually, they flew swiftly yet without moving their wings, not falling to Earth in spite of their weight, not eating on the way nor sleeping. On the moon, Gonsales found huge herbs, birds and beasts, "all of which, as I now perceived, spend their time, in their absence from us, in that world; neither do they differ in anything from ours, but are the very same kind". After 6 months on the moon, Gonsales prepared to depart. "By the often baying of my birds, he perceived a great longing in them to be gone". Once he let the reins loose, the swans quickly carried him back (Harrison 1954). This story of birds in the moon gives some idea about the theological atmosphere in which Ray was active.

Beyond his *Discovery of a New World in the Moon,* John Wilkins also planned to create a universal language, applicable to all phenomena and to all known ideas and that would be transparently appropriate to the things and notions described. The stimulus for this came from the Royal Society, of which he was a founding member. In his *Essay towards a Real Character and a Philosophical Language* (1668), words were to be created in a top-downwards classification. Concepts were divided into 40 main *Genera*, each of which gives the first two-letter syllables of the word; each *Genus* was divided into *Differences*, each of which added another letter; and *Differences* were divided into *Species*, which add a fourth letter. For instance, Z indicates an animal, Zi identifies the *Genus* of beasts (mammals); Zit gives the

Difference of "rapacious beasts of the dog kind", Zitα gives the *Species* of dogs, say the collie. Ray was then seeking how to classify the living world, to establish method and order in natural history and in Wilkins' *Essay* he contributed the botanical section. Wilkins seems to have expected Ray to fit his classification into the classification system he had just developed, but Ray would not accept it and he distanced himself from Wilkins' scheme. He complained (in a letter to Lister, 1669) that Wilkins had demanded he divide herbs into three squadrons as nearly equal as possible and then split up each squadron into nine "differences", as he called them, that is subordinate kinds, in such a way that the plants ordered under each "difference" should not exceed a fixed number. Ray considered this concept ridiculous and refused to apply it. Thus, during the seventeenth century, while natural history was coping with myths about birds in the moon and with monsters, students classifying the diversity of life were struggling to develop principles that would be true to nature.

We recall the mechanistic views of Descartes, who believed that God can behave only according to mechanical principles, and the mystic views of Kircher, who believed that life came from a universal seed consisting of a seminal vapour made up of salt, sulphur and mercury and possessing a plastic power and a magnetic power. Ray, coming one generation after Descartes and Kircher, was deeply devout. His *The Wisdom of God Manifested in the Works of the Creation* (1691) presented our world and its plant and animal life as evidence for the existence of God (Mayr 1982). To find the deeds of God in nature, he quoted the verse from Psalms 104: "How manifold are Thy works, O Lord! In wisdom hast Thou made them all". Noting the word "manifold" at the beginning of the verse and taking the many organisms in nature, he asked, just how many animals and plants are there? He answered by giving the number of species, as it was then known:

Beasts (including snakes)	150
Birds	500
Fish	500
Fish, including those with shells (i.e. water-dwellers of all kinds)	3000
Insects	20,000
Plants	18,000
	41,650

Forty-one thousand six hundred and fifty species—what strength and wisdom God needed to create them all! Now compare this to Man, who, with all his intense labour, has been able to create only a handful of machines, clocks, pumps and flour mills—and that is it! By creating such a huge number and a variety of species, God reveals the infinity of his Creation. He also reveals his infinite solutions to the same problem. Thus, feathers are essential for bird flight, but God, in his infinite wisdom, has enabled some creatures to fly without them—gliding fishes, gliding lizards, bats and butterflies.

Much of Ray's book focused on the continuation of the verse from Psalms, "in Wisdom hast thou made them all". His opening topic was Creation. Ray was enraged at Epicurus and his theory of falling, swerving and colliding atoms: if the world had been created by random collisions between falling atoms that swerve and collide randomly with other falling atoms—why is there no explanation as to why they fall? Why do they swerve? Which atoms fall and which swerve? How is it possible that atoms colliding randomly can compose something as organized and beautiful as nature, with its animals that are more exact and perfect than any human-created machine? If a random collision of atoms can create a universe, then why can the same fall of atoms not create a temple? A house? A city? Let us look at this question from another perspective. Assume that a person finds a magnificent palace on some remote island, built according to all architectural rules. Would it make sense to claim that an earthquake, or some random shuffling of the materials composing it, had created this palace? If by chance, one were to find an ancient scroll on which wise and significant sayings were written, would it make sense to think that random inadvertent scribblings of a pen had written the text? Likewise, imagine a giant box containing an infinite series of the letters of the alphabet, and suppose that the letters were mixed and then cast; upon falling onto the ground, would they combine to create a coherent text? (This argument, of letters flung randomly and not forming a sensible text, goes back to mediaeval times. The earliest case I know of is that of Bahya ben Joseph ibn Paquda (1050–1120) of Saragossa, Spain, of today. He claimed that there are people who say that chance had created the world without the intervention of the Creator. However, he said, notice that if we accidentally spill ink on a page, no clear writing emerges. We conclude that, if someone claims that a distinct written text was formed by ink that happened to be spilt, we should not believe him. Surely, it would be the deed of a scholarly writer). Beyond Epicurus, Ray also referred to Aristotle, of whom he was remarkably tolerant: though Aristotle had claimed that our world had existed since eternity, he only denied that the world had been created, but he did not deny the existence of God.

Ray's book had another aspect as well. In the Bible, in ancient times and in the Mediaeval, the anthropocentric point of view dominated—Man was the centre of the entire universe, which God had created and now exists solely for him. Maimonides, in his earlier years, held that everything exists only for Man. In later years, he changed his mind: God had created all creatures for themselves, not necessarily for any other purpose. Thomas Aquinas, however, claimed that God supports plants and animals not for their own purpose but for Man.

From the Renaissance onwards, this anthropocentric approach began to falter. Copernicus' calculations dealt one hard blow to the concept that man was at the centre of Creation and Galileo's discovery, with the telescope, of an abundance of invisible stars dealt another blow. Another blow came from Bruno's cosmological vision of an endless universe with an infinite number of worlds populated by other creatures. The possibility of invisible suns shining on worlds and creatures that Man could not see further diminished this approach of Man as the centre of a universe that God had created only for him. It might perhaps seem that an entire generation was distancing itself from anthropocentricity, but the situation was actually more complex because some scientists had difficulties in coping with it. The astronomer Johannes Kepler (1579–1630) accepted the fact that the Earth was not the centre of the universe, but he believed that humans were on the most suitable planet for us, the most noble and logical creature in the universe. We, inhabitants of the Earth, are superior to the inhabitants of Jupiter's moons. Francis Bacon claimed that nature had been created for Man: animals were intended for man, whether for his daily use or for study that would allow their future use. Robert Boyle, one of those who realized Bacon's vision, said that God's plan was that man should rule the land and He gave him all animals. The benefit of a certain creature to man is not immediately recognizable; thus, the silkworm, the sugarcane and cochineal (the insect from which carmine was produced) were once ignored and only eventually were they found to have a use. Henry More (1614–1687) declared that the silkworm came into the world for the sole purpose of providing man with costly clothing. The seventeenth century thus witnessed conflicting views as to how man should see himself relative to the universe.

Within this ideological confusion, Ray attempted to find his way. He believed it ridiculous and illogical to think that God had created enormous bodies, such as stars, only so that they would wink at us; indeed, many of them do not even do so, as they are so distant that we cannot even see them with the naked eye, only by telescope. We may conclude that God had not created distant heavenly bodies just for us. God had, however, created for us the planet Earth. He created it as an infinitely wise designer; indeed, Earth bears the evidence for God's wisdom in designing a planet just for man. Thus, for example, God designed Earth so that in its orbit it tilts towards the Sun at an angle; thanks to its tilt, we humans can enjoy the changing seasons. The mountains are in the centre, and water streams down from them for the benefit of the people on the plain. He created a green, fragrant, flowering fresh carpet with beautiful trees to cover the world and provide us with fruit, medicines and building material. God designed and created all this to provide not only for us Europeans but also for other races of man; thus, for the natives of America, he created the cocoa plant. It, alongside other plants, benefits them as a source of bread, water, wine, vinegar, oil, honey, sugar, needles, thread, cloth, cups, spoons, baskets, paper, masts for ships, sails and house beams. All this requires a wise designer. The number of insect species exceeds 20,000. True, today man has no use for many of these species, but eventually he will use them, whether for exercising his mind or for purposes not yet discovered. Why did God create so many insects, most of which seem to us to be of no value and some even harmful? God created many of them to show and express his great power, lest man underestimate him. The immense number of individuals in each species preserves the species, from one generation to the next and it benefits man indirectly. Birds are useful to us in their tasty flesh that we eat, in their down, which we use to fill our pillows and in their quills, which we use for writing and for dusting our furniture; birds delight us with their songs and bright colours. Birds feed on insects, and this is one reason for insect existence. Another is that many of them (scorpions, spiders) are poisonous and therefore of use to Man in concocting medicines. A third purpose of venomous insects is that God uses them to punish wicked people and wicked nations.

We can conclude God's wisdom in design also from a study of Man's body. Ever so wisely, he placed the senses at the top of our body, like guards in a watchtower, to perceive objects from afar. He placed our sensitive brain in a thick, strong skull and he created our erect walk, which both allows us to see afar and releases our hands to use tools. He also designed and created the functional structure of the eye to perfectly fulfil its specific function. As an organ used by man and other animals so often that they cannot exist without it, it is composed of numerous secondary units whose combined action so well serves its purpose. This perfection in design is also true for all other organs in our body. Nothing in our body is redundant, not even the nipples in males (Ray found documented cases in which men, in the absence of a woman, could produce milk). Also, the great diversity of human faces, a variety so great that there are no two identical faces in the world, is designed with great wisdom, for it enables us to identify forgers, cheats, thieves and robbers. Further, the pleasure derived from those actions intended to ensure the existence of the species (sex) reflects the wisdom of God, and even pain is of value because it warns of illnesses. These and other adaptive traits in our body would not exist if random processes had created it, if chance had located the same organs in the same places, generation after generation.

Beyond Epicurus and his proposal that chance is a major factor in the creation and maintenance of the world, Ray strongly opposed Descartes, his mechanistic approach to nature and his claim that mechanical principles and matter in motion can explain everything. Descartes' views threatened Christianity because in a world governed by mechanical laws, there is no place for miracles, no place for faith, no blessing in prayer and no place for divine inspiration. Ray responded that if Descartes' argument is true, that mechanical principles and matter in motion explain everything, then why do some plants stand tall, whereas others sprawl? Why do all individuals in one species have the same flowers, fruit and seeds? Why do trees not grow up to the clouds? Why do their branches not grow horizontally to infinity? Why are some plant species long-lived and others annual? What laws of motion in matter can determine the number and location of leaves on a stem? The manner of flowering, whether individually or in groups? The ability of a seed to develop a small plant, like an embryo developing in the womb of an animal? Can mechanical forces explain the differences between animal species or why one species cannot create an individual from the seed of another species?

For all this to happen, there must, Ray claimed, be an intelligent force; there must be God's wisdom. God had created each part of a plant to sustain it: roots to stabilize it and to absorb nutrients from the soil, tubes in the stem to transport liquids and air and outer layers to protect the stem from the elements. As for leaves (in those days photosynthesis had not yet been discovered), they protect the flower and fruit from the summer sun, reduce evaporation from the roots and produce the sap to nourish the fruit. Leaves additionally provide cooling shade to Man, in the hot summer ("they shall sit, every man, under his vine and under his fig tree", *Kings* I 5:5, and Abraham, who sat at the entrance to his tent in the heat of the day, and when the angels appeared, he served them "under the tree", *Genesis* 18:8). Moreover, God had created a broad spectrum of plant species for the direct, sole benefit of man: wheat, from which we produce bread, medicinal plants and those designed to serve as hedges.

Ray also rejected Descartes' proposition that animals are mere matter without a soul. We can explain many observations of animals only by assuming that they too have some extent of rationality. Dogs run before their masters and then stop at a road fork, waiting to see in which direction their masters will proceed; having captured prey and fearing their master will take it from them, dogs will run off, hide the prey and then return to their master. A dog wanting to jump on a high table will first climb on a nearby chair and then from there leap to the table. Further, if dogs and other animals were machines or automata, they would not have feelings of pleasure or pain and it would not be possible to be cruel to them; but animals broadcast signals to us when we torment them and people have compassion for tormented animals because their expressions of pain are similar to ours. Finally, the tormenting of animals is against the Scriptures, which say that a good man is merciful to his beast, while the wicked are cruel to it (*Proverbs* 12:10). The Bible would not have mentioned this if animals were machines, as it is impossible to be cruel to a machine. As animals have the same organs of sense as we do, they feel as we do.

It is not mechanical laws that rule the animal world, Ray claimed, but God's will to sustain each species. God, through divine providence, has therefore created male and female animals, and He preserves the correct numerical proportion between the sexes. In viviparous animals during gestation, He has designed, for this purpose, that nourishment should flow to the foetus in the womb, and after birth, the food should flow to the nipple. Birds are all oviparous, and this is another sign of divine providence, for if they were viviparous, the burden of the full womb would make flight difficult, and they would become easy prey. Divine providence for the preservation of the species is also the reason why animals will endanger their lives to protect their young.

The argument of intended design for a specific purpose is sometimes termed the teleological argument (from the Greek *telos* for purpose or reason). This argument stems from the claim that all phenomena fulfil specific purposes. An all-knowing entity designs each phenomenon, and we name this entity—God. Intended design in nature is one of the earliest arguments used to prove the existence of God. During Roman times Rabbi Akiva (first-second century CE) said that "Just as the house attests to the builder and the garment to the weaver and the door to the carpenter—so the world attests to the Lord, it's Creator". Also, later in Roman times, Lactantius claimed that all creatures were created by God, who had carefully planned and provided each animal with the manner of life suitable for it. During the Mediaeval period, Yehudah Halevi claimed that we see distinct purpose in the Creation. God created water so that plants and animals could drink, then covered the Earth with grass so that animals could feed, created high trees so that they would serve as nesting sites for birds, mountains as shelters for some mammals and rock crevices as shelters for others. He created the lion with its claws and the gazelle with its light feet. Also, during the Mediaeval period, Maimonides asked, how are we to know if a land has a king. His answer was that if an elderly moneychanger can sit with a pile of coins in front of him and a young man will not strike and rob him—this is evidence that a king rules the land. Likewise, global laws of order are proof of God's existence. This was Ray's approach as well.

References

Bacon, F. 1620. In *The New Organon*, ed. L. Jardine and M. Silverthornes . Cambridge: Cambridge University Press.2000

Bacon, F. 1626. *The New Atlantis*. Watchmaker Publishing.

Bacon, F. 1629. *Sylva sylvarum.*

Ball, C. S. 1966. The *Apiarium*: An early example of microscopic study. Proceedings of the Oklahoma Academy of Sciences, Section D, Social Sciences.

Breidbach, O., and M. T. Ghiselin. 2006. Athanasius Kircher (1602-1680) in Noah's ark: Baroque "intelligent design" theory. *Proceedings of the California Academy of Sciences, Fourth Series* 57:991–1002.

Cobb, M. 2000. Reading and writing *The Book of Nature*: Jan Swammerdam (1637-1680). *Endeavour* 24:122–128.

Cobb, M. 2002. Malpighi, Swammerdam and the colourful silkworm: Replication and visual representation in early modern science. *Annals of Science* 59:111–147.

Cutler, A. 2003. *The seashell on the mountaintop. How Nicolaus Steno solved an ancient mystery and created a science of the earth*. New York: Penguin.

Des Chene, D. 2004. *Spirits and clocks, machines and organism in Descartes*. Ithaca: Cornell University Press.

Dobbel, C. 1932. *Antony Leeuwenhoek and his 'Little Animals'*. New York: Harcourt and Brace.

Egerton, F. N. 1967. Leeuwenhoek as a founder of animal demography. *Journal of the History of Biology* 1:1–22.

Ford, B. J. 1982. The Rotifera of Antony van Leeuwenhoek. *Microscopy* 34:362–373.

Franklin, K. J. 1961. *William Harvey, Englishman, 1578–1657*. London: MacGibbon & Kee.

Freedberg, D. 2002. *The eye of the Lynx: Galileo, his friends, and the beginnings of modern natural history*. Chicago: Chicago University Press.

Gaukroger, S. 2001. *Francis Bacon and the transformation of early-modern philosophy*. Cambridge: Cambridge University Press.

Hall, T. S. 1951. *A source book in animal biology*. New York: McCraw-Hill.

Harrison, T. P. 1954. Birds in the moon. *Isis* 45:323–330.

Harvey, W. 1651. *Disputations touching the generation of animals*. Translated with Introduction and Notes by G. Whitteridge, 1981. Oxford: Blackwell.

Hiarai, H. 2007. Athanasius Kircher's chymical interpretation of the creation and spontaneous generation. In *Chymists and Chymistry: Studies in the history of Alchemy and early modern chemistry*, ed. L. M. Principe, 77–87. New York: Science History Publications.

Hooke, R. 1665. Micrographia. Royal Society of London.

Karamanou, M., E. Poulakou-Rebelakou, M. Tzetis, and G. Androutsos. 2010. Anton van Leeuwenhoek (1632-1723): Father of micromorphology and discoverer of spermatozoa. *Revista Argentina de Microbiologia* 42:311–314.

Lister, M. 1685–1692. *Historia Conchylorium.*

Lokhorst, G.-J. 2018. Descartes and the Pineal Gland. In *The Stanford Encyclopedia of Philosophy*, ed. Edward N. Zalta, Winter 2018 ed. https://plato.stanford.edu/archives/win2018/entries/pineal-gland/.

Mayr, E. 1982. *The growth of biological thought: Diversity, evolution, and inheritance*. Harvard University Press

Muffet, T. 1658. *The theater of insects, or, lesser living creatures* (as *Insectorum sive Minimorum Animalium Theatrum*). In *The history of four-footed beasts and serpents*, ed. E. Topsell, vol. 3. London: Cotes.

Park, K., and L. J. Daston. 1981. Unnatural conceptions: The study of monsters in the sixteenth- and seventeenth-century France and England. *Past and Present* 92:20–54.

Parke, E. C. 2014. Flies from meat and wasps from trees: Reevaluating Francesco Redi's spontaneous generation experiments. *Studies in History and Philosophy of Biological and Biomedical Sciences* 45:34–42.

Prete, F. R. 1991. Can females rule the hive? The controversy over bee gender roles in British beekeeping texts of the sixteenth-eighteenth centuries. *Journal of the History of Biology* 24:113–144.

Pyle, A. 2006. Malebrabche on animal generation. Preexistence and the microscope. In *The problem of animal generation in early modern philosophy*, ed. J. E. H. Smith, 194–214. Cambridge University Press.

Ray, J. 1691. *The wisdom of god manifested in the works of the creation*. Cambridge: Trinity College.

Rieppel, O. 1985. Ontogeny and the hierarchy of types. *Cladistics* 1:234–246.

Rondelet, G. 1554. Libri de Piscibus Marinis, in quibus verae Piscium effigies expressae sunt. Lugduni: Apud Matthiam Bonhomme.

Schneider, J. R. 2018. The first mite: Insect genealogy in Hooke's *Micrographia*. *Annals of Science* 75:165–200.

Sleigh, C. 2012. Swammerdam's frogs. *Notes and Records of the Royal Society* 66:373–392.

Steno. 1667. *Specimen of Elements of Myology*. Florence.

Swammerdam, J. 1737 (posthumous). *Bybel de Natuure*, Amsteldammer of *Historie der Insecten*. Leiden: Severinus, Vander & Vander.

Thanos, C. A. 2005. Theophrastus on oaks. *Botanika Chronika* 18:29–36.

Topsell, E. 1658. *The history of four-footed beasts and serpents*. London: Cotes.

Tunnacliffe, A., and J. Lapinski. 2003. Resurrecting Van Leeuwenhoek's rotifers: A reappraisal of the role of disaccharides in anhydrobiosis. *Philosophical Transactions of the Royal Society of London B* 358:1755–1771.

Valentin, G. 1837. Über die Spermatozoen der Bären. *Nova Acta Physico-Medica Academiae Caesareae Leopoldino-Carolinae Naturae Curiosorum* 21:239–244. Google Scholar.

West, J. B. 2013. Marcello Malpighi and the discovery of the pulmonary capillaries and alveoli. *American Journal of Physiology - Lung Cellular and Molecular Physiology* 304:L383. https://doi.org/10.1152/ajplung.00016.2013.

Willughby, F., and J. Ray. 1678. *The ornithology of Francis Willughby of Middleton in the County of Warwick*. London: John Martyn.

Wilson, C. 1995. *The invisible world. Early modern philosophy and the invention of the microscope*. Princeton: Princeton University Press.

Chapter 5
Enlightenment

Abstract During the Age of Enlightenment, speculations arose as to the origin, structure and function of "spermatic animalcules": whether they really exist, whether the tail has any function, and whether they are molecules representing a transition from non-living molecules to life, tiny animals with complex organs or perhaps tiny people. Eventually, scientists found that they have no internal organs, cannot multiply and that they are normal constituents of the male organization.

Anatomists could not explain the formation of complex embryos from featureless eggs and semen. How does Matter gain Form? The Epigenesis theory argued that development progresses gradually, from unorganized egg materials to organized structures, by successive building of organs. The power of embryogenesis is in Matter itself, which gains Form. The Preformation theory believed that Form comes before Matter. This theory required no intra-matter formative powers. God had formed all future generations of all living creatures during Creation as miniature versions of themselves. Embryonic development occurs through the expansion and growth of these preformed miniatures, and future potential creatures await the time to develop into actualized creatures when God so wills. Parthenogenetic aphids seemed to prove that the preformed embryo was in the female gamete; other evidence suggested it was in the semen.

Enlightenment naturalists thought that organisms develop from preformed germs consisting of inert, folded nets that expand when activated by fertilization. God can change species' germs so that they would reach higher perfection levels, just like embryos that develop into higher stages. Death is a metamorphosis to a future existence (resurrection), with indestructible soul germs surviving death, rising from the dead and dwelling in bodies of higher animals—repeated rebirth with perfection, up the ladder of nature. Nature's ladder places simple organisms at the bottom and complex ones at the top in a hierarchy of increasing body complexity and perception powers. Individuals reach perfection through ontogeny, as they develop to higher states of complexity and perfection, mirroring this ladder. *Palingenesis* follows global catastrophes that destroy all body germs: the soul germs survive, enabling animals to reappear when Earth is again habitable. Each germ then develops into a new, more perfected organism, and all groups change simultaneously. Every soul has reincarnated several times, in different bodies and epochs. Some

J. Heller, *Nature's Biodiversity*, https://doi.org/10.1007/978-3-031-89790-0_5

naturalists rejected preformation, suggesting that the need to satisfy lust, hunger and security determines diversity. All living creatures have descended from a single ancestor, with diversity rising from different responses to different stimuli.

In classifying diversity, traditional theory maintained that each kind of plant and animal must have a defining essence. New theories claimed that information on essences is not reliable, and such classifications actually rest on indirect knowledge of secondary qualities. As essences are unobservable, we must make do with derived, observable features. As no observable feature is more indicative than another one, our classifications are arbitrary, only possibly conforming to real essences. Even at the real essence level, boundaries between species may be continuous rather than sharp, making it impossible to find boundaries of species on the observable level. Lacking knowledge of the essence, we should replace the traditional principle of using a single feature all the way down by using any of several observable features. These two rival theories resulted in very different, rival classifications.

Linnaeus believed that information gained from the senses is valid and reliable for classifying nature and that essential features, alongside other features, are available for classification. Naturalists must present clear and distinct verbal descriptions—the clearer and more distinct our ability to distinguish and understand, the greater the degree of certainty. Nature forms a natural ladder, rising from minerals through plants to animals, among whom man ranks the highest—God has progressed gradually, from simple to complex. Linnaeus based his classifications on very few external features (even just one). Within a kingdom, he adopted a concise, uniform hierarchy with only four ranks for all minerals, plants and animals: class, order, genus, and species, thereby creating a four-tiered ranking system, as in the church, army and civil service. On his classification scale, he fixed the terms genus and species as absolute ranks, yet he emphasized that ranks were artificial devices to facilitate identification. He imposed on classification the obligatory use of a single pan-European language, Latin, and applied binomials to all three kingdoms, thereby giving nomenclature its simplicity and brevity and expressing the need to distinguish both the general and the particular. Linnaeus erred in assigning fossils to the mineral kingdom.

Linnaeus believed that the mode of generation is a plant's essence. He classified plant diversity by whether sexual organs are exposed or covered, whether stamens and style are on the same flower or not, whether separate from the style or fused and whether all stamens are equal or not. He next created classes by the number, shape and location of the male organs and orders by the female organ. Then he classified genera and species by other traits such as the root, stem and leaves; beyond this, he did not bounce between characters. This was, he believed, a natural classification, in contrast to not-natural, medical classifications. He classified animal diversity into six classes. Many of his orders were unnatural due to convergence. He listed only a few genera, perhaps believing that biblical Adam could memorize only a few names. Species are essential, unchanging entities created during Genesis; they cannot develop from other species and do not form spontaneously. In molluscs, he gave each genus one name characterizing its shell and a different name characterizing the animal inside the shell. Man is highly variable. Beyond mute, hairy savages, there are four races, one for each continent. They differ in colour, temperament and hair.

Linnaeus, finding fossilized marine shells at mountaintops and thinking he had noticed a continuous retreat of the sea, considered this as evidence that land is continuously emerging. This suggested to him that in some past, the land was a small paradisiacal island with all today-existing habitats, and here God had created all creatures. As the island gradually emerged and grew, its creatures began dispersing into the new lands. The fossil beds on mountaintops were, he believed, remains of ancient shorelines of this island. Nature is harmonic and balanced, with every kind having its place in the web of life. God's means of maintaining a balance is that predators daily destroy animals, which would otherwise multiply so rapidly as to deplete their own food sources. Furthermore, if all species inhabited the same place, some would dominate and eliminate the others. To prevent one species from displacing another, God had placed different species in different sites throughout the world. Furthermore, if one species over-multiplies and robs another species of its food, thereby introducing confusion into the harmonic balance of the world, God punishes it with sickness and death. Linnaeus was aware of the significance of biogeography, population dynamics and competition.

Buffon followed the concept of relative time and space in organizing all quadruped diversity relative to a fixed, arbitrary reference point: man. We should study animals by the order in which they appear before us, beginning with the close ones. Accordingly, Buffon grouped all European domestic animals first, then European wild animals; he clustered Asian domestic animals separately. Naming all animals by their French common name, he followed the domestic animal series with a game animal series. His third series consisted of small carnivore pests. Hibernating quadrupeds were in a separate series. Then came species from other continents, broadly from the Americas, then Asia, then Africa (but with exceptions), and finally monkeys. He clustered seals, walruses and manatees by habitat. There were also attempts to erect natural clusters. Natural history became an additional element in Buffon's species descriptions. It relied on documented observations, and these were often irregular, erratic, capricious and difficult to inspect, especially for animals in distant lands. Buffon erred in clustering all elephants into a single species, which he further erred by clustering also the fossil mammoths. These errors had consequences on his views on the origin of life.

Hydra is a transitional creature, between plants and animals. It is green, lacks most animal organs and reproduces by budding; yet it moves, responds to touch and seizes and digests prey, thus suggesting a continuous plant–animal scale, not two distinct realms. When sliced up, hydra pieces regenerate complete individuals. Regeneration ability contradicts the Preformation theory: any claim that every tissue fragment contains a preformed embryo of the next generation is absurd; furthermore, animal regeneration is a new creation, occurring millennia after the biblical Creation. Regeneration also contradicts the belief in individuals possessing a material animal and an immaterial soul. In which of the new parts is the animal's soul? Can a soul divide into many parts? If so, then is the soul merely a function of Matter?

Enlightenment naturalists thought they had found that a "life force" exists, capable of spontaneous generation: new microscopic organisms develop in disorganized material and generate other microscopic creatures. Certain "primitive particles" are present in animals. When animals die, they separate but may rejoin and produce a new living entity. Later naturalists found flaws in these experiments and concluded that life is not only a feature of Matter.

Linnaeus' opponent was Buffon, who offered complete descriptions of animals, not telegraphic descriptions of external morphology. To determine the certainty of knowledge, he distinguished abstract from real knowledge. Abstract, mathematical truths are provable, but as they contain only what we put in them, they are useless in natural history. Physical truths rest only on facts of the real world. They consist of a series of similar facts, such as frequent and uninterrupted repetitions and orderly successions of the same events. Natural events that recur repeatedly are laws of nature. Buffon searched for basic laws in nature that would explain the beginning of the world and how life is generated. To explain the emergence of order out of chaos, he developed a secular history: a comet, hitting the sun, broke off debris, which eventually became the planets. Earth was at first a sphere of hot, melted matter; gradually it cooled and turned into dry land. Clouds then rained to form oceans, within which life spontaneously generated. Earth was then warmer than today, and the oceans once covered mountaintops; eventually, they receded, leaving marine fossils.

Buffon suggested that indestructible "organic molecules", through different combinations, create various species. An "internal mould" positions the molecules throughout the body, creating specific creatures and thereby giving empirical meaning to the species concept. Experiments (erroneously) suggested to him that microorganisms develop spontaneously from organic molecules. Microorganisms, intermediary between living and non-living, resemble natural machines in that they do not feed and do not generate ordinarily. In lower animals, upon death, the organic molecules set free, join other molecules and spontaneously reproduce.

In higher nonsexual animals (plant lice), every organ sends molecules to a place where they reassemble, thus reproducing the parental internal mould. In highly organized creatures, sexual reproduction occurs. The molecules coalesce into a parent-resembling body under a Newtonian field of force, thus passing on the internal mould. The internal mould makes sense of species as genealogical entities, ensuring them as unchanging dynasties. Species fixity is a basic, universal law. If God were to destroy all living creatures, organic matter would form new internal moulds, resembling the old ones; also, on other planets, given proper conditions, the same species would appear. Buffon's species were thus universal types of organic organization.

Buffon rejected the Linnaean classification. Certitude about nature should rely upon real entities, not on abstract essences. In classifying, we need full descriptions, including internal morphology and full natural histories, not just external morphologies representing essences. Genus and order are meaningless ranks, and in practice, naturalists cannot agree on genus-level classifications. Basing life on reproduction, Buffon claimed that in nature, only one category exists, the species: a chain of similar, interbreeding individuals. A species is a system of concrete relationships between real entities. We recall a controversy on the nature of time and space in which Newton had claimed that these are absolute, abstract entities, whereas Leibniz claimed that without concrete things, there is no time or space. Buffon applied this time–space issue to biological classification: species exist only in the time dimension, in the destruction and renewal of creatures, and they are the only living entities in nature. Hybridization can determine whether individuals are of the same species. Beneath species, Buffon recognized varieties: lineages produced in response to changing environments. A permanent variety becomes a "race". Buffon belittled racial differences in man, who forms a single, highly diverse species.

Buffon eventually, encountering fertile hybrids and biogeographic variation, began doubting both fertility barrier absoluteness and species' fixity. Animals change over time, and only the better-fitted survive due to natural selection. Consequently, animals cluster into genera and families, each with a single ancestor that has its internal mould. Ancestors may split into varieties, which sometimes hybridize. His earlier described species are mere races. Families are "expanded species" representing fixed moulds, whose products the environment may modify. Consequently, hybrids no longer ruin the reproductive criterion, domestic animals link to wild animals and fossils become ancient branches of a common root. The family should be the unit of classification, not the species.

Buffon now established a new classification of quadruped diversity: non-splitting families versus splitting ones. A zoogeographic division, in both categories, included families of the Old World, both Worlds and the New World only. The American fauna suggests a continuum that once existed between the Old and New Worlds. Man is superior to all animals, and beyond him come grand, majestic animals. Birds also cluster into families, which he described without describing species. Buffon reached a far more detailed classification than did Linnaeus.

Buffon described the history of nature in light of "historical knowledge", differing from "physical knowledge" in that one reconstructs it. Throughout Earth's slow cooling, oceans appeared and vanished, continents formed and sank and living beings came and went. Animals are perfectly adapted to the climate in which they were born, so as Earth slowly cooled, many animals could not adapt and went extinct. Extinctions release large quantities of molecules, and these, by spontaneous generation, build a new fauna, adapted to now lower temperature. Because of the limited amount of organic molecules, one animal wave must go extinct to enable the next wave to form. Temperature directs the organic molecule combinations, producing some animals rather than others. Animals may migrate, following a cooling climate, toward the equator; here, when the heat is no longer strong enough, they go extinct. In all places where the temperature is the same, one finds the same (or similar) plants and animals. Buffon's history of Earth is, largely, a secular *Genesis*. Man is today transforming nature through land cultivation, domestication, swamp draining, forest clearing and other means. With our help, nature has reached today's magnificent perfection.

5.1 Generation

5.1.1 Sperm, Fertilization and Embryo Formation: Big Controversies

Leeuwenhoek had discovered spermatic animalcules in 1677. Inevitably, many speculations arose as to the origin, structure and biological function of these animalcules over the following decades and for about 120 years; Cole (1930) has admirably summarized the literature, and I have taken much of this chapter from him. Looking back at these sperm speculations, it is impressive that many of those who speculated used reason and thought rationally, bearing in mind the knowledge and concepts then available. Some zoologists, ever so sceptical, denied the very existence of animalcules: when looking through the microscope, what might seem to us as animalcules they considered as merely active portions of the viscid semen. Other sceptics considered them as inert corpuscles with no intrinsic powers of movement, set in motion by the heat of the semen. Several argued that animalcules do occur, but beyond living in semen, they are also in other parts of the body, such as skin, blood, mouth and faeces, where they can hardly be supposed to have any function in a generation. Some suggested that the animalcule's tail impedes locomotion rather than quickening it, in that it produces unstable oscillatory movements. Then, there were mechanistic speculations: the spermatic animalcules prevent the semen from clotting; or, by their movements, they mix male and female seminal liquids or, by their movements, they sweep the spermatic liquid toward the egg. Some speculated that animalcules incite the male to perform the sexual act or that they are responsible for male sexual orgasm. Others thought that the animalcule form fits in with the type of adult into which it is to develop. The sperm of a slug has a long, slug-like structure, and it moves accordingly; the tortoise has a fixed backbone and a mobile neck, and hence, they claimed, its spermatic animalcule has the tail in front and not behind and pulls forwards by hooking movements of the tail.

Some suggested that animalcules are aggregates of organic molecules and represent a continuous transition from non-living molecules to life. They are formed in the semen but also occur in females, and they increase in number, size and activity after having left the generative organs. Discharged semen contains filaments, and from these, the animalcules derive, at first possessing tails broken off from the filaments. Eventually, they detach and move about slowly, encumbered by their tails, which are accidental appendages, not swimming organs; at a later stage, the tails are lost, and motion then becomes quicker. The animalcules now change shape and size and divide by fission. I shall discuss this theory (of Buffon) in the following chapter, in more detail.

Some argued that animalcules are tiny animals dwelling in the semen (hence the term "spermatozoa"). Since the days of ancient Greece, naturalists have defined animals as living entities capable of moving about by their own power. As animalcules answer this definition, they must be minute animals, like so many other minute animals that Leeuwenhoek had discovered. Some classified the animalcules/

spermatozoa together with blood corpuscles as independent organisms that constitute the lowermost group in the animal kingdom. Others considered them minute parasites. By definition, all animals grow and generate, so spermatozoa must also do so. The energy for growth and reproduction must come from food, so the spermatozoa must feed, and for this, they must have an alimentary system. As they occur exclusively inside living bodies, inside the male semen, and the only food inside a body is that body itself, they must be internal parasites. This belief in the animal nature of the seminal animalcules was so deeply rooted that many eminent zoologists believed they could see, under the microscope, animals with complex organizations. In spermatozoa taken from various animals (boar, guinea pig, sheep), they thought they could distinguish a mouth, an anus, a coiled gut, sexual organs and even animalcule embryos (Figs. 5.1, 5.2 and 5.3).

Naturalists considered sperm as parasites also in humans. Felix Pouchet (1800–1872), director of the Rouen Museum of Natural History and the Rouen Botanical Garden and professor at the Rouen School of Medicine, thought he had observed that human sperm possess viscera and also, perhaps, a sucker, as in some biting insects. He further claimed that the spermatozoa, which he considered as

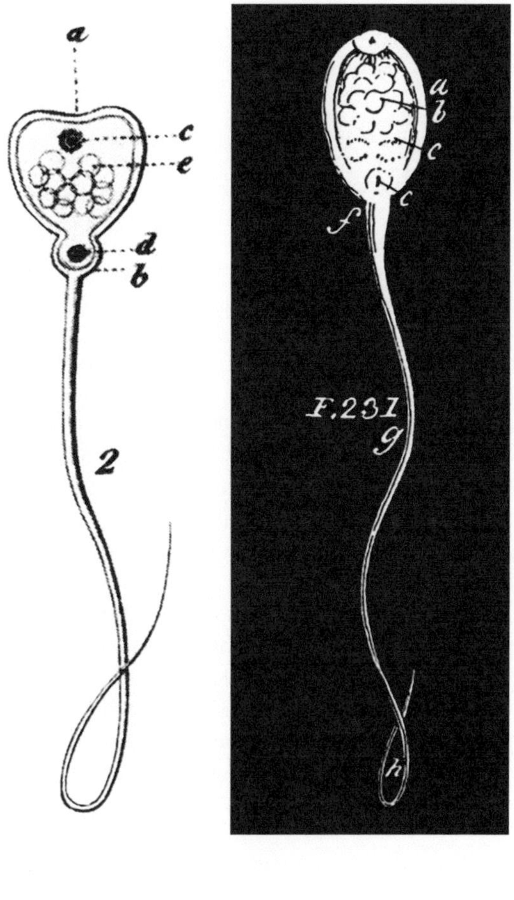

Fig. 5.1 Imaginary organs inside spermatozoa. Left: sperm of a boar, as presented by Valentin. Right: sperm of a guinea pig, as presented by Gerber. (Adapted from Cole)

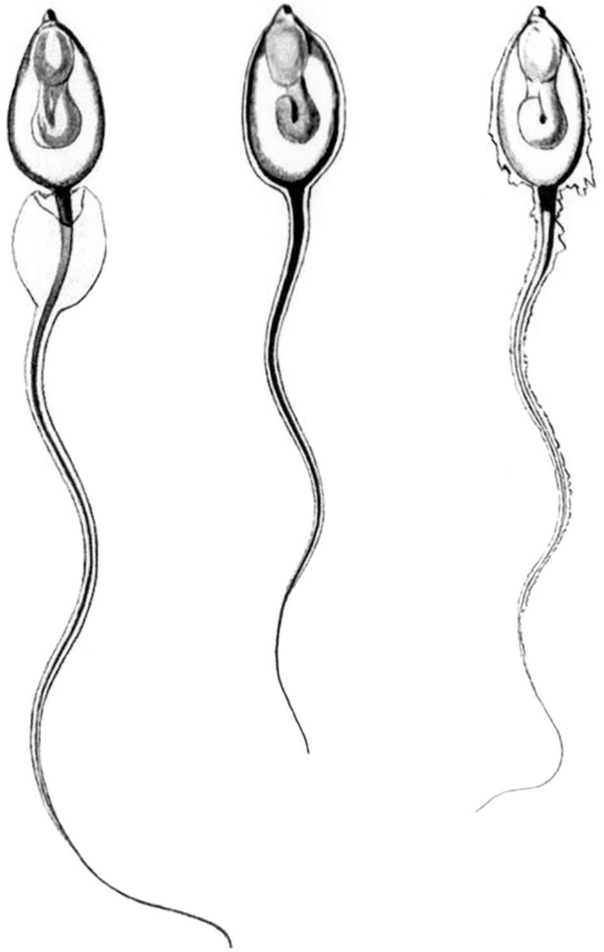

Fig. 5.2 Imaginary organs inside spermatozoa: human sperm as represented by Pouchet. Note the alimentary system in all figures and the moulting epithelial fin in the mid and right figures. (Adapted from Cole)

parasites of the human semen, possess a movable fin, which the creature can control by will. Each sperm is surrounded, body and tail, by an epithelial envelope. One can observe this envelope in moulting individuals, since sperm moult in a process comparable to that of caterpillars.

Then, there were those who thought that sperm contain tiny people. The secretary of the Academy of Montpellier (under the pseudonym Dalenpatius) even claimed to have observed under the microscope the moult of a spermatozoon. He depicted a moulting spermatic animal in which the head, breast, arms and legs still had an outer skin (1679). His figures of spermatic humans were so convincing that, four decades later (1721), a treatise of known spermatozoa reproduced them. Belief in the animal nature of the spermatozoa was so deeply rooted that many eminent experts in microscopy attempted to demonstrate, in a homogenous substance, the presence of a more or less complex organization.

Fig. 5.3 Dalenpatius' figure of the tiny man in the sperm. (Adapted from Cole)

Spallanzani also interpreted spermatozoa as parasites. He termed them 'little spermatic worms' and was sure they played no part in fertilization; he attributed this function to the seminal fluid. Lazzaro Spallanzani (1729–1799, Fig. 5.4) came from northern Italy and was ordained as a priest. He then began studying law at Bologna but switched to science under the influence of a professor of physics at the university. The Reggio College appointed him as a professor of logic, metaphysics and Greek; he then became a professor of physics at the University of Modena, and finally, he accepted Maria Theresa's invitation to become professor of natural history and director of the museum at the University of Pavia, then under Austrian rule. In his studies of sperm function, he enclosed the posterior of a copulating male frog in a bag (made of pig bladder) and collected the ejaculated sperm to verify whether this bag could prevent egg fertilization, which it did (Capanna 1999). Next, taking sperm from the bag, he wetted virgin eggs with it and always obtained complete embryonic development of the egg. An accepted doctrine at that time was that the power of fertilization was in a vapour emanating from the sperm, an *aura seminalis*.

Fig. 5.4 Spallanzani

To study if this is indeed so, Spallanzani placed toad sperm on a watch glass. In the bottom of another watch glass, he stuck a dozen toad eggs, and he then turned it upside down and covered the first. The eggs thus remained a few millimetres above the sperm. The eggs did not develop, and he concluded that spermatic auras do not exist (1776); fertilization takes place only by direct physical contact between egg and sperm. Then, to find which fraction of the sperm is the active part in fertilization, he experimented with sperm diluted with water, which he passed through a series of filters. He found that the fertilizing power of this water diminished progressively in proportion to the thickness and number of filters used, finally disappearing altogether. At this point in his study, had he examined this water under the microscope, he would have seen in it the "little spermatic worms" and concluded that they are the fertilizing element. Unfortunately, he did not do so—a greatly missed occasion in science (Capanna 1999).

Only from the mid-18-20s onwards did scientists first claim that the testis secretes animalcules, and as such, they cannot be parasites. Soon after, scientists denied that any internal organs exist in them or that they can multiply and, thereby, propagate their own "species". Spermatozoan movements were no proof that they are independent creatures since this motion occurs also in other animal and plant structures, the normal status of which is beyond doubt (and the ancient definition of animals as living entities capable of moving about by their own power must be abandoned as an all-encompassing phenomenon). One hundred and seventy-two years after Leeuwenhoek's first description, in 1849, Wagner and Leuckart finally declared that the spermatic animalcules are normal constituents of the male organization, like ova of the female, and that they are as necessary to the semen as blood corpuscles are to blood (of vertebrates). The liquid of the semen holds the spermatozoa in suspension. Of spermatozoa function, science knew nothing certain beyond that, in some way, upon contact, they fecundate the ovum.

We now turn to the embryo, the formation of which has always been a fundamental aspect of life. Anatomists of the seventeenth–eighteenth centuries were incapable, even with the microscopic techniques then available, of discovering a structure in either the egg or seminal fluid, which had any relationship to the creature that eventually emerged. How can one explain the appearance of a highly complex order out of a featureless mass? How does Matter gain Form? Descartes claimed that animals are machines and that we can refer all physiology to mechanical problems, but his mechanical philosophy could not explain generation: "You say that beasts are machines, just like watches? Yet if you put a male dog machine next to a female dog machine, a third little machine could result from them; two watches, however, could rest next to each other for their entire existence without ever making a third watch. … All things that are two but have the power to make three are of a quality well above that of a machine" (Fontanelle 1685). In concept, the generation of life out of a formless fertilized egg and the origin of embryonic order were comparable to the biblical formation of the organized world out of a formless, dark void: God's creative act had created it. However, at the cosmic level, there is only one solar system, created only once in some very distant time, whereas plants and animals are born every day, in front of our eyes. To say that God creates them one by one and at all times does not fit in with the claim that, once God had created the world, He had let the laws of motion govern it. How, then, are human, animal and plant individuals formed?

Two theories sought to explain these basic aspects of reproduction. The classical Epigenesis theory, developed by Aristotle in ancient Greece and held by Harvey in early modern England, argued that embryo development occurs step by step in an orderly manner, gradually progressing from an unorganized, non-functioning body, consisting of undifferentiated materials of the egg, to an organized, functioning one. Since the embryo forms by the successive building up of its parts, the guiding principle of embryogenesis must be inherent in matter itself. Some power inside living matter knows how to organize parental fluids into the physical structure of every organ of every single offspring. In this manner, Matter gradually gains Form.

The rival, the Preformation theory, required no intra-matter formative powers for each generation. It held that at the time of Creation, God had formed together all Matter and Form that will ever be. Accordingly, he created all creatures as miniature versions of themselves—those of the past, today and future generations. These "preformed" miniatures exist in the genitalia of every creature. The development of each offspring occurs in due time as a result of an unfolding, expansion and growth of these preformed miniatures, in accordance with mechanical laws. (We may call to mind the unfolding and expansion of wings in a butterfly emerging from its pupa.) The complete Forms of living things exist, in real terms, prior to their development, and the gradual assembly from parts, as suggested by Epigenesis, is wrong. Every plant was already present, "preformed" in the seed, in the form of a "germ":

extremely small but completely and perfectly organized, needing only to grow. God had created the germ itself at the beginning of time, and it was now waiting for a certain moment to add matter and grow. Form thus comes before Matter. With the origin of all living creatures thus placed at the beginning, time and circumstance have only accidental bearings on them. They account for occasional abnormal productions but leave untouched deep essentialism at the basis of each species.

This theory had philosophical roots in Descartes' theory that God, once He had created the world with its mechanical principles of matter and motion, had become restricted by these laws, and since then, He cannot break them. Now, as mechanical principles cannot explain the (then) mysterious processes of embryonic development, how could an emerging form acquire the capacity to function without some strictly mechanistic factor? To explain this irrelevance of mechanical principles, the theory argued that all generations of all creatures had formed, by God's divine powers, in an era before matter and motion had become the one-and-only governing laws. These miniature organisms, preformed before the rule of mechanical laws, shall expand in due course in accordance with now-ruling mechanical laws.

The Preformation theory presupposed the possibility of infinite smallness in organic beings and rested on an analogy with insect metamorphosis, as represented by Swammerdam's study of late-stage larvae of silkworms. He had demonstrated that wings, legs and other parts of the adult butterfly were clearly visible, as though contained within the larval form, and had concluded that insect metamorphosis consists of the development and growth of pre-existent structures. In insect metamorphosis no parts form anew, as appears at first sight—these parts of the adult have been there from the first, but they then lack visibility. Various stages of an insect can thus coexist simultaneously within an organism. Swammerdam claimed that the egg, larva, pupa and adult are enclosed in that order, as in a series of boxes: the egg enclosing the larva, the larva enclosing the pupa and the pupa enclosing the adult. He did not claim that they all actually exist simultaneously or that the adult as such is present in the egg. He indicated, however, that some mystic entity produces adult form by growth, expansion and accretion, by the gradual accumulation of additional matter. He added that also in the opinion of "a learned friend", all humankind was represented in the loins of the first parents, and thus we can biologically explain the burden of the original sin. Swammerdam also thought he had observed that the black portion of the frog's egg is a little frog, complete in all its organs, and that the rest of the egg is only food for this embryo. This little frog "appears already in the ovary and has the beginning of its growth but is so small as to evade sight. Yet the animal itself is hidden under this minute fundament" (1672, in Cole 1930).

Three years later, Swammerdam disclosed that the "learned friend" was Malebranche (we shall meet him presently). He affirmed that God had comprehended the entire human species in the loins of Adam and Eve, and consequently, humanity will face extinction when the original supply of germs is exhausted. Thus, death, the penalty of the Original Sin, is implicit in humanity.

Swammerdam concluded that generation represents not the creation of a new creature but simply the development of pre-existing structures. Conversely, death does not imply the termination of existence; it merely represents metamorphosis to a future existence. There is no real death or generation in nature. For Swammerdam, metamorphosis became a natural metaphor for the dogma of resurrection. Swammerdam's observations convinced many scientists that Form does not organize itself gradually, as it would under the Epigenesis theory—Form is there from the very beginning. Preformation is thus a theory beyond questions of generation; it is a theory of existence (Wilson 1992). A century ahead, these ideas of Swammerdam would deeply influence another naturalist, Charles Bonnet, as we shall see further on.

A decade after Swammerdam's treatise, Marcello Malpighi was studying the embryonic development of the chick. He (erroneously) claimed that he saw certain structures of the early embryo, such as the head and spine, which pre-exist in a fertilized yet not-incubated egg. Indeed, these structures are there even before the egg is laid. Other structures, such as blood, develop only during incubation. Malpighi described embryo growth as happening gradually, but he believed that all embryonic organs form at once, by a kind of precipitation following fertilization. Upon formation, these organs are small and transparent, and they only become visible as they successively unfold and grow. We can observe the heart beating only 40 h after the onset of incubation; it existed before that but was then invisible because it lacked the red blood content.

The concepts of pre-existence and preformation were compared with insect metamorphosis, the other realm of encased development: the transition from one life stage to another in the lower groups of animals (Wilson 1992). In insects and vertebrates alike, the wormlike appearance of the early foetus, alongside its protective covering, called for a comparison between the silkworm in its cocoon and the foetus in its womb. It was natural to suppose that the emergence of visible form from subvisible form was by the same process occurring inside one as in the other. In the pupa of the caterpillar, Malpighi found the outlines of wings. Legs and antennae of the adult butterfly, just as with the egg. It, therefore, seemed to him that the pupa merely covers the already-formed moth so that it may grow up like a foetus in utero (Wilson 1992).

Building on these studies by Swammerdam on insects and frogs and by Malpighi on chicks, Nicolas Malebranche (1638–1715, Fig. 5.5), a French Catholic priest, combined the mathematical idea of infinite divisibility with the fact that microscopes enable observations on very tiny creatures. These tiny creatures of the microscope suggested to him that there might be creatures that are even tinier than those we currently observe; indeed, the size of some living creatures may be infinitely small (1674). He too advanced the theory that all generations were preformed, each embryo containing an even smaller embryo, like a box within a box, and that this might go on infinitely, like a Russian nesting doll with an infinite number of smaller dolls inside. Each offspring is contained as a preformed entity within the gonads of its parents. As matter is indefinitely divisible, there is no limit to the lower size of an entity. Embryonic development would merely be the unfolding of this preformed being. We recall Descartes's belief that the universe had begun as a chaotic soup of matter in motion and that God had created the world along eternal laws of mathematics and mechanics that now restrict God himself—He can, today, behave only according to mechanical principles of matter and motion. Descartes also suggested

Fig. 5.5 Malenbranche

that an animal's organs develop one after another: first the heart, then arteries and veins, then muscles and so on—the theory of epigenesis. Malebranche, however, thought that whether or not laws of matter and motion explain the creation of the heavens, they—and the theory of epigenesis—cannot explain the creation of plants and animals (Pyle 2003). In these, the functional interdependence of many organs is so complex that no organ can exist and function without a full set of other organs. Consequently, an organism cannot gradually come into existence, organ after organ, as required by Descartes' theory of epigenesis (Pyle 2003). A complex, multi-organ creature can come into being, all at once, only by supernatural creation, by God himself, who had created all creatures during that one great miracle period of Creation ("the body of every man and beast born till the end of time was perhaps produced at the creation of the world"). Malebranche accordingly proposed that all bodies of all humans and animals, those already born and those yet to be born, had formed during the Creation. All future members of each species were present in the ovary of the first female of that species. They were nested inside one another, generation inside generation, from the first moment of creation. God, today, is activating only those laws of nature that account for growth and development—but not those for generations.

In support of his theory of pre-existence, Malebranche noted his observation that in tulips, the bulb encases leaves that are to become green, alongside other structures that are to become flowers, and especially alongside "that tiny triangular part which contains the seed and the six little columns that surround it at the base of the flower". The bulb thus contains an entire tulip that we can see with the naked eye. What is visibly true of the tulip applies also to all other plants and animals. The theory of pre-existence does not exclude a role for maternal imagination in embryonic development, a force in which Malebranche believed and that may explain the specificity of animal generation (Pyle 2003).

It was Malebranche who was Swammerdam's "learned friend", and it could well be that they built up this doctrine together, with Malebranche perhaps crystallizing ideas first developed by Swammerdam. Today, to some of us, it might seem impossible that one organism could hold many millions of tiny miniatures within itself. However, this problem did not worry seventeenth- to eighteenth-century believers in the Preformation theory. The world was thought to be only a few thousand years old, and the Day of Judgement was not so far off, so there did not have to be an endlessly large number of generations.

Malebranche's doctrine was close to that of Augustine, an early Roman Christian theologian and philosopher (354–430) who argued that God had created the complete world by a single act. At the very moment of Creation, the body of every future creature, whether plant or animal, took on full shape. It took the form of a primordial seed (*primordia semina*), waiting for the proper time to unfold and to develop into a fully actualized creature. This primordial seed is not visible to us because of its tiny dimensions. Containing the principles of its subsequent development, it waits for the right time to come into existence and grow when so willed by God. Augustine believed that only God has the power to create life, and nature has only the power to unfold it.

One early criticism of Malebranche's preformation theory was that if embryos are preformed in the egg, then it is impossible to explain the generation of monsters that appear from time to time, say, people with six fingers in each hand rather than the normal five. If, as was then generally believed, monsters are the direct result of the mental state of the pregnant mother, how can such an influence affect a foetus whose form has already been determined? Another criticism is aimed at the theory's suggestion of infinite smallness in living beings. If the first male or female of a species had enclosed all future members of the species, as suggested by Malebranche, then those now inhabiting Earth must have been infinitely small at the time of Creation. If all the rabbits that have been reproduced, from the beginning, had been enclosed in the first male rabbit, then a rabbit existing today would have been, at the time of Creation, about 10^{-100} (not merely10^{-60}, which is about a grain of sand as compared to the whole Earth). To reach this result, naturalists assumed that a seminal animalcule of the rabbit is at least 10000 million times smaller than the rabbit itself; that the creation of the rabbit took place about six thousand years ago and that rabbits begin to reproduce at the age of 6 months (Hartsoeker 1722, 1730). Furthermore, how could this theory be reconciled with the discovery of eggs and spermatozoa in viviparous animals? If the germ was a complete organism, it could not be in the egg and spermatozoon at the same time; also, it could not be half in one and half in the other. One had to choose between two alternatives.

Accordingly, during the seventeenth–eighteenth centuries, the Preformation theory split into two competing models, one that located the preformed miniature in the paternal sperm and the other that located it in the maternal egg ("spermist" versus "ovist"). The spermist model claimed that sperm were actually tiny people that, when placed inside women, grew into children. Offspring developed from a tiny, fully formed embryo contained within the head of a sperm cell. The first to direct attention to the importance of the spermatozoa was Leeuwenhoek, who had discovered them (and had named them animalcules). He believed that in the uterus, the animalcule casts off its tail, and the head then becomes the foetus. He did not accept (1699) Swammerdam's claim that the frog's egg contains a young frog; rather, he stated that parts of the embryo appear gradually in the fertilized egg. Leeuwenhoek was thus a spermist, but not a preformationist. Other spermists thought they had seen miniature organisms inside the animalcules. The Dutch physician Nicolaas Hartsoeker (1656–1725) argued that each animalcule inserts itself into the egg for nutrition and growth. Its tail, he stated, attaches to the uterus and forms the umbilical cord. Hartsoeker even produced an image of a tiny human curled up inside the head of a spermatozoon, which he referred to as *petit l'infant* and *le petit animal* (*Essai de Dioptrique* 1694, a semi-speculative work describing the sorts of new scientific observations that could be done using magnifying lenses; Fig. 5.6). At first, he was very careful regarding the real existence of these preformed embryological beings. He did not claim to have seen them under the microscope; he only suggested that this is what they may look like under a strong lens and careful observation. However, the image began to get its own life, and it soon became the icon of all those believing in preformation. Later, when he copied this picture in other works, the original explanatory function of the image blended with empirical observations and became "real". Today, his figure is one of the most illustrative examples of the preformation theory.

Hartsoeker further discovered that in birds the semen is full of an infinite number of creatures resembling worms that, he argued, the male casts in the female's egg-containing ovary. Here, each worm tries to penetrate one of the eggs to receive nourishment and to grow. Each egg has but one opening, to allow the penetration of just one single worm. As soon as that sole "worm" penetrates that sole opening, the opening closes, refusing passage to other worms. This view of Hartsoeker is close to our current view of the process of fertilization.

One main problem with the spermist model, as applied to humans, was the implied wastefulness of lives or souls. If each human sperm cell contained a potential person, then millions of people would die with each successful conception. In response to this criticism, Hartsoeker formulated a new idea: animalcules not finding an ovum in a female's womb evaporate into the air and become minute particles, which float about until they can be recycled into active sperm. This idea addressed the problem of sperm wastefulness. Hartsoeker abandoned the preformation theory in later years, as we shall soon see. The German philosopher Wilhelm Gottfried Leibniz (1646–1716) speculated that wasted sperm might actually be scattered by the wind and generate life wherever it found a suitable host. Broadly, in line with other believers in preformation, he suggested that at conception, a microscopic,

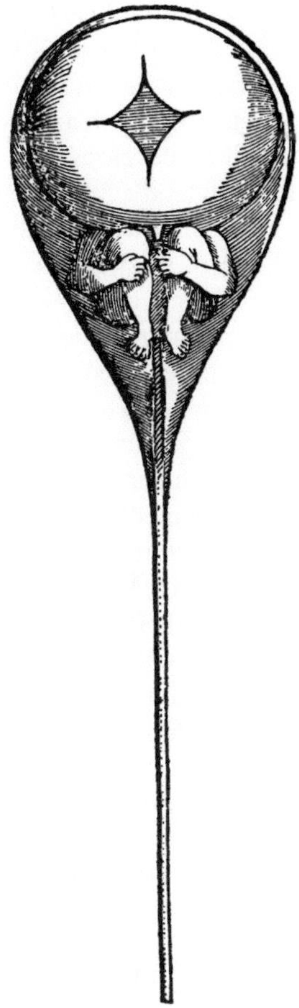

Fig. 5.6 Hartsoeker's *petit l'infant*

preformed entity undergoes a process of growth and unfolding, from which the animal emerges. Leibniz also claimed that what we call "death" is nothing but a process of contraction and infolding of the animal, which nevertheless remains living and retains its own identity, even if reduced to an invisible size. Nature conserves not only the soul but also the animal itself, as its organic machine reduces itself to a tiny, invisible size, as before its birth, minute beyond our senses. Accordingly, death is to be conceived not as a termination of life but rather as the beginning of a new life, as a true resurrection. Thus, not only have organisms always existed in their living form, but they will always exist, body united to soul, even past apparent death. A slow process of growth and unfolding of a tiny microscopic creature would amount to the "birth" of a larger animal, and a similar (but faster) process of contraction and folding of the larger animal into a tiny microscopic creature would amount to its "death". Thus, not only have organisms always existed in their living form, but they will always exist, body united to soul, even past apparent death. Leibniz suggested that of the three souls described by Aristotle (vegetative, animal and rational), the rational soul is a divine endowment restricted to man; its actualization requires God's interference, a process he called trans-creation. He considered the spermatic animalcules, even those of human males, to be animals not yet endowed with a rational soul. Only if a female human conceives a spermatozoon and thus develops into a human being does a trans-creation take place, in that through God's interference, the soul is endowed with rationality.

Back to the spermist model, the trans-creation theory neatly solved the problem of sperm wastefulness. Furthermore, since non-fertilizing sperm had no rational soul, there was no need to assume that God had violated the harmony of nature by creating an inexplicable superabundance of human souls. In 1750, the British Royal Society received a letter titled *Lucina Sine Concubitu* [Pregnancy without Intercourse]. The writer claimed that women could become pregnant without having engaged in sexual activity, due to the presence of microscopic floating sperm in the air. To capture and isolate some of these floating sperm, the author claimed he had invented "a wonderful, cylindrical, catoptrical, rotundo-concavo-convex machine" through which he had observed animalcules floating in the air. As observed under a microscope, these floating sperm were shaped like miniature men and women. This remarkable discovery, he suggested, would enable women to become pregnant without the assistance of men and would help restore the honour of women who could not otherwise explain their pregnancies: it must have been those naughty animalcules floating about. To prove all this, the author proposed a royal edict that would ban copulation for 1 year. If the machine was introduced into use and there were still pregnancies and births, then they must be due to the air-floating animalcules. An engraving accompanying the letter showed a floating sperm approaching a sleeping woman (Fig. 5.7). The letter was signed by "Abraham Johnson, physician and man-midwife", a pseudonym of the botanist Sir John Hill, to satirize the spermist theory.

Fig. 5.7 Pregnancy without intercourse (background faded)

Many naturalists preferred the "ovist" doctrine, which placed the germ in the egg, enveloped in its strongly porous tunics. At fertilization, the most subtle parts of the male semen rush into the egg on all sides through these pores. They then proceed to occupy their appointed stations in an orderly manner, and in this way, they produce the first rudiments and organs of the foetus. The foetus thus forms suddenly at fertilization by several sperm, but it does not pre-exist in the sperm (Drelincourt 1685)—a spermist theory but not a preformation one.

Spallanzani, whom we have mentioned above, observed that under the microscope, amphibian spawns are identical before and after fertilization and that it was never possible to see "the sperm worm", *vermicello spermatico*, inside the egg after fertilization. He concluded that the spermatozoon did not contribute to embryo development (scientists observed sperm penetration of the egg only a century later, in sea urchins). He claimed to have seen preformed tadpoles in unfertilized frog eggs, a clear indication of his commitment to ovist preformation. He prided himself (1767) "in having found that the tadpole, which becomes a frog, pre-exists fertilization". He then concluded that "the unfertilized egg is the actual tadpole, and it, therefore, follows that the tadpole pre-exists fertilization". He was convinced that the egg contained a fully formed tadpole, in potential and not yet alive, which needed only to be fertilized with seminal fluid so as to begin the process of unfolding (1785). The Swiss anatomist Baron Albrecht von Haller (1708–1777) was another advocate of the ovist preformation theory. Observing chick embryos under the microscope, he found that the yolk attaches to the chick's intestine. From this, he gathered that the embryo was created at the same time as the yolk. Now unfertilized eggs also have yolks, and therefore, he concluded, the embryo must exist there prior to fertilization. These preformed embryos, he argued, were simply too small to be seen. Haller believed that God alone possessed creative powers of giving life and formative abilities.

Naturalists of the eighteenth century thought that parthenogenesis studies supported the ovist model, with its belief that the female egg houses the preformed embryo and that the function of the male is only to stir to life the vital processes of the female. Charles Bonnet (1720–1793) of Geneva came from a French Huguenot family driven out of France by religious persecution in the sixteenth century. He was a lawyer by profession, but his favourite pursuit was the study of natural science. To determine whether the generation of offspring without copulation was possible, he isolated a single spindle tree aphid, *Aphis fabae,* kept it from mating with any other aphid from the moment of its birth and found that it gave birth to 95 live young aphids eleven days later (1740; Fig. 5.8). He thus proved that parthenogenesis, namely generation by a female without fertilization by a male, occurs in nature (Egerton 1968). Bonnet's discovery of parthenogenesis in aphids seemed to him to prove that the preformed embryo was contained in the female gamete. In sexually reproducing animals, the offspring must necessarily exist totally within the egg of the mother, and the task of the seminal fluid is merely to open the egg and spur it to start growing. In parthenogenetic animals the mother, he believed, provides this fluid to herself (*Traite d'insectologie*, 1745). We shall come back to Bonnet further on.

Fig. 5.8 The tree aphid *Aphis fabae*. (Illustration by Tuvia Kurz)

5.1.2 Regeneration and Spontaneous Generation

Abraham Trembley (1710–1784), of a prominent family in Geneva, studied at the Calvin Institute (now the University of Geneva). He then became a tutor for the children of a prominent Dutch politician at the family's summer residence near The Hague. It was while teaching the children natural history, during lessons in which they went fishing in ponds outside the house, that he came across a small creature composed of a uniformly gelatinous substance; it had a mouth-like opening at one end of its hollow body, and it reproduced by budding off a series of shoots. This creature, the hydra (*Hydra viridissima* in today's classification; its characteristic green colour comes from symbiotic cells of unicellular alga within its tissues), was first discovered and described by Leeuwenhoek (1703). He classified it as a plant because of its green colour, because it seemed to him that it lacked sense organs, because it lacked most other organs proper to animals (heart, liver, anus) and also because it reproduced by budding (Figs. 5.9, 5.10 and 5.11).

Many had seen the hydra since, without perceiving anything out of the ordinary. Trembley, observing closely, noticed that it had the habits of an animal. It was capable of locomotion like animals (moving much like inchworms); it responded to touch by contraction, like animals, and its eight or ten arm-like projections around the mouth could seize whatever prey came their way, which was then conveyed to the stomach and digested. It seemed, therefore, to be a sort of transition creature between plants and animals. Now, since animals were (in those days) not known to regenerate complete individuals, as do plants, Trembley sought to discover whether these organisms were capable of regeneration. He was surprised to see that no matter how he cut the hydra—lengthwise, crosswise, big pieces or small pieces—each piece always gave rise to an entirely new hydra (Sirks and Zirkle 1964). New individuals formed even when he cut through developing head buds (and this led him to create a seven-headed creature, which he named *hydra*, after the Greek mythological monster; later, Linnaeus would adopt this term as a name for these creatures).

Fig. 5.9 Trembley with his pupils, collecting and observing hydra

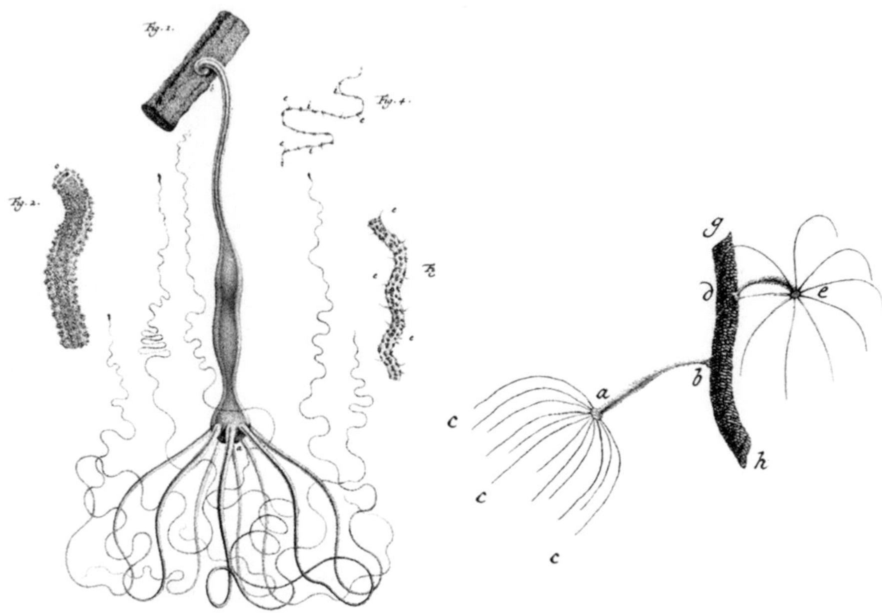

Fig. 5.10 Trembley's figure of a hydra

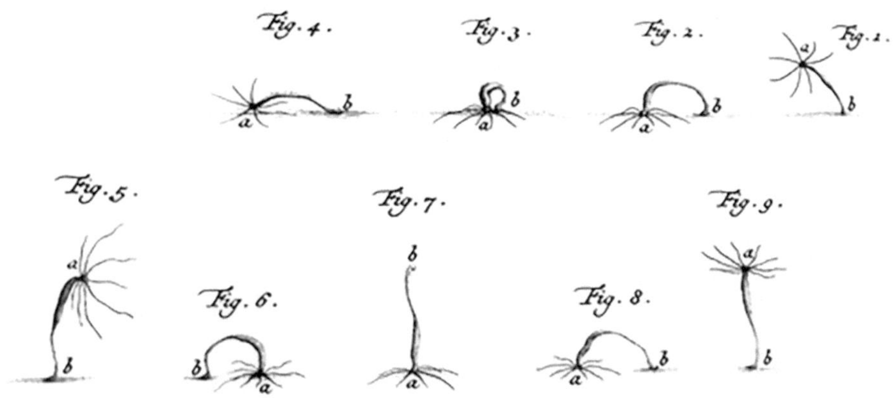

Fig. 5.11 Trembley illustrates hydra movement

Trembley wrote (1741), describing his research, to René Antoine Ferchault de Réaumur (1683–1757), the distinguished French entomologist and authority of the age in such matters. Réaumur had, some decades earlier (1712), described the regeneration of claws, nippers and antennae in crustaceans. Excited by Trembley's findings, he immediately announced them to the Paris Academy of Sciences. In the *Memoires de l'Academie des Sciences* (1741), he wrote that "The story of the phoenix born from its ashes, as fabulous as it might be, offers nothing more marvellous.

[...] From each portion of an animal chopped up into two, three, four, ten, twenty, thirty, forty parts, just as many complete animals are reborn, similar to the first. Each of these is ready to undergo the same division [...] without it being yet known at what point this astonishing multiplication will cease". Trembley's findings took 3 years to be published (*Memoirs for History of a Genus of Freshwater, Horn-shaped Polyps*, 1744, *Mémoires pour servir à l'histoire d'un genre de polypes d'eau douce,*) but long before its publication, talk about Trembley's new discovery had stimulated thought on a variety of subjects (Sirks and Zirkle 1964). Certain animals could propagate from cuttings!

Trembley's study unleashed a succession of theological problems—the hydra's regeneration ability from a fragment of tissue did not agree with the preformation theory, which maintained that the embryo develops from embryos pre-existing since the Creation. To say that every small piece of tissue contains a preformed embryo of the next generation was implausible. Animal regeneration is essentially a new creation, and this is inconsistent with the preformation theory: creation may take place many millennia after the biblical Creation. The hydra regeneration experiments thus threw the preformation doctrine into confusion.

Another theological problem: the hydra's manner of reproduction, by division into many parts without loss of its vital principle, was a severe blow also to those believing in an individual possessing two separate entities, a material animal and an immaterial soul. As every portion of an animal can reconstitute a new animal, in which part is the animal's soul? If it remains in one part to animate the process of regeneration, how does the other fragment, in which it must be lacking, manage to regenerate? Can the soul divide into two to make the two fragments regenerate? Can the soul be divided into very many parts? Should one conclude that every piece of tissue, no matter how small, contains a soul? To say that every small piece of tissue contains a soul was to stretch credulity. If, in certain animals, the soul is divisible indefinitely along with the body, then it is merely a function of Matter.

Further, the idea that plants and animals share the same essential nature, with the hydra representing a connecting link between the forms of vegetative and animal life, would suggest a continuous scale of nature, rather than three distinct realms of nature. Indeed, Bonnet devised a scale of nature. Charles Bonnet was Trembley's nephew, and he too began investigating regeneration, discovering that earthworms, which undoubtedly are animals, can regenerate. He argued that regenerating limbs that replace amputated ones have pre-existed from the beginning and only proceed to unfold themselves. He also claimed that in embryonic development, nature never proceeds by jumps. Every embryonic stage is contained in the preceding stage, and embryogenesis proceeds as a continuous, unbroken chain of events as the various parts are unfolded (Sirks and Zirkle 1964).

Another person to begin studying regeneration was Spallanzani, and he discovered that salamander tails, limbs and jaws could regenerate (Capanna 1999). Questions on regeneration are often linked with those of spontaneous generation, which Spallanzani also studied (1768). Needham (whom we shall meet further on) had claimed that a life-generating force exists in nature, inherent to certain kinds of inorganic matter; this force causes living microbes to generate themselves spontaneously. To prove that life is an inherent feature of matter, he briefly boiled a mutton gravy broth, which was then cooled down to room temperature in an open container and poured into flasks that were then sealed. Needham found, a few days later, that microbes were growing in these flasks. From this, he concluded (1748) that a "life force" exists, capable of spontaneous generation. He thought that he had actually seen new, living creatures forming out of life-lacking matter; that microscopic creatures in the cooled broth were giving birth to further microscopic creatures; and he suggested that a similar process occurs during the embryonic development of higher creatures (1750). Needham gained the support of Buffon, an eminent French naturalist (we shall discuss him further on), who suggested that certain "primitive particles" are present in animals and that when an animal dies, these particles separate but remain capable of rejoining and thus producing a new living entity. Buffon invited Needham to examine fluids extracted from the genitalia of animals, and indeed, under their microscopes, the two naturalists did actually observe globules. Buffon interpreted these globules as "genetic factors" (1749), and he suggested that life is an inherent feature of matter. Spallanzani repeated Needham's experiments with a slight change in protocol: he boiled the mutton gravy for a much longer time, and then he sealed the flasks very carefully. Upon opening the flasks, he found no microscopic creatures whatsoever, and this led him to reject Buffon's theory of life being an inherent feature of matter. As the microscopic creatures did not reappear as long as the flasks were tightly sealed, he suggested both that they fly through the air as tiny eggs and that boiling kills them. There is no "universal seed" floating about in the air that generates living creatures. Only "germs" exist, which some previous generations of creatures had deposited (1765).

In studying the regeneration abilities of the earthworm, Spallanzani sectioned its body at different distances from the head. He found that both the part with the head and the caudal part, and intermediate fragments between the two, regenerated. He even described the regeneration of regenerated parts: Heads could be grown from the neck. Also, some snails can regenerate their heads (*Cepaea nemoralis*), but others (*Helicella ericetorum*) cannot (Capanna 1999).

5.1.3 Organized Life: Birth, Rebirth and Climbing the Ladder of Life

We have briefly mentioned Charles Bonnet, who had discovered parthenogenesis, a discovery that seemed to him to prove that preformed embryos are contained in the female gamete (Fig. 5.12). He was deeply religious, and the issue of generation in its broad sense was, for him, far more than a scientific debate. His main concern was to work out a parallel between scientific ideas on the one hand and beliefs about divine providence with the scriptural account of the resurrection of man on the other. Bonnet believed that God had designed all embryos, their development actualizing His ideas of harmony. All parts of an organism are closely linked like the

Fig. 5.12 Bonnet

parts of a machine, and for this machine to function, they must all be present from the very beginning of embryonic development. Therefore, an organism's parts cannot come into existence separately. All parts are necessary because they act together for a common goal through mutual relations, which they maintain between themselves and with the totality. This is true of a small machine and of a microcosm, just as it is true of the big machine of the universe.

Bonnet's uncle (Abraham Trembley, whom we have already met) had discovered that detached parts of *hydra* undergo regeneration. Now, if detached parts of an animal can grow again into a complete, organized organism with stem and arms, then, surely, there must be an underlying units, much smaller than that organism. Bonnet, in his theory of embryonic development, believed that these basic organizational units consist of fibres [in a way, a forerunner of the cell theory]. Between these germ fibres and the complete organism, there was, he suggested, a broad scale of different organizational levels: from fibre to mesh, from mesh to fibre bundles, then to tissues, organs, systems of organs and finally the complete animal. Each higher organizational unit consists of lower-level organizational units, and thus, the tissue consists of fibre bundles, the system of organs consists of separate organs (say, the digestive system, with its stomach, intestine and liver) and so on. The smallest organized unit that feeds, grows and reproduces is the fibre, which he also named a little machine *machinule*. True, the fibre consists of many sub-units (he named them fibrils *fibrillei*), which in turn consist of many molecules, but it is the fibre that is the basic organizational unit. This is so because, since a sort of envelope encloses it, it is the transition point between external particles floating about without order on the one hand and the organized body in which an internal order dominates on the other hand. The fibre determines the arrangement of nourishment particles entering it. It also determines the patterns of higher organizational levels—meshes, fibre bundles and so on. In this manner, a "network" is formed with germ fibres as its basic unit and arranged in a large number of secondary units (Lawrence 2009). I believe that the structure of striated muscles inspired Bonnet's general hierarchy of units. Each individual muscle cell is sheathed in a thin layer of connective tissue, and neighbouring muscle cells with their connective tissues are wrapped up by a thicker layer of connective tissue (perimysium), thereby forming the fascicle. These are surrounded by an external layer of connective tissue (epimysium), thereby forming the anatomical muscle (say, the biceps), which joins other anatomical muscles (say, the triceps) to form part of the muscles of the arm.

All organisms of all past and future generations develop from pre-existing germs, originally created by God (Sirks and Zirkle 1964; Anderson 1976). Bonnet supposed that successive generations are stored up, encapsulated one within another, so that the first created female of a species contains the whole of the rest of her kind within her ovaries. The germs of all future apple trees were contained within the first apple tree, created on the third day of Creation. All that was necessary for germ growth was the provision of suitable nutrition, which was supplied, first by the male semen and then by the mother. Bonnet envisaged the preformed germ to consist of an inert folded net of elementary fibres, which determine the organism's essential form. Fertilization activates this inert germ, and the net unfolds and starts growing. Embryo growth is thus merely an expansion of elementary fibres already present in the germ, a purely mechanical process resulting from the assimilation of nutritive particles into this net of elementary fibres. The form of a preformed germ does not relate to the form of an embryo. The preformed germ of a chick differs so considerably from the embryonic chick that eventually develops from it that if we were to see this tiny germ greatly enlarged, we would never identify it as a chick. The germ consists of tiny dots, which will later develop into lines, and these will eventually stretch and multiply themselves to form flat surfaces. As the embryo develops, the lower levels of organization (germ fibre, mesh, fibre bundles and tissues) design the species to which it belongs. The three higher levels (organs, sets of organs, and full body) shape the individual (Cheung 2006; Lawrence 2009).

Bonnet's theory of germ pre-existence supports the concept of species stability: All germs of a species are identical, and this stabilizes each species through time. External factors, such as the mother's body and the nourishment she supplies, produce individual variations within each species. Beyond this variation, Bonnet thus had a static, essentialist view of nature: animal forms are unchangeable entities, created and preformed from the beginning (Cheung 2006). Generation, producing endless copies of a given species, permits neither change nor variation. The apple tree growing on the last day of the world and the apple tree that stood in the Garden of Eden will be identical because both are generated from the germ created at the beginning of time. If God chose, however, to create a change in the set of germs and to make them differ, then a species would change within just one generation. It will do so by reaching a higher level of perfection, up the great chain of being, like an embryo changing into a higher stage during its continuous development.

In the Preformation theory from its beginning, the natures of death and of generation were of cardinal importance. In the seventeenth century, Swammerdam believed that there was no real death or generation in nature. Metamorphosis consists of the mere development and growth of existing structures, and likewise, generation represents the development of existing structures, not the creation of something new. Death, on the other hand, implies merely a metamorphosis to some future existence; it does not imply the termination of an existence. Accordingly, metamorphosis became, for Swammerdam, a metaphor in nature for the dogma of resurrection. Bonnet, coming a century after Swammerdam, suggested that the germ from which the material body of each individual develops does not survive death. It returns to become its original condition of a folded net. However, beyond the germ responsible for the body, there is, so Bonnet believed, a tiny preformed germ that is the seat of the soul. He named it the "germ of restitution" and claimed that it is indestructible (Sirks and Zirkle 1964).

Body and soul combine, Bonnet claimed, so that each organism consists of a mixture of a material body and an immaterial soul. These two components act on each other. The outer world stimulates the sense organs, and these stimulations reach the brain. The soul, within the brain, thus traces external impressions. Furthermore, the soul, by its intentions (what we would call will), can exert a "moving force" on the body, causing it to move. Body and soul are constantly changing, mutually modifying and improving each other towards perfection. The eternal germ of the soul, present in every living organism, survives the death of the bodily germ and gains a higher level. In some future, it shall rise from the dead and shall dwell in a future body. Females of a species contain within their ovaries two types of germs. The destiny of one type, the normal one, is to reproduce the species. The destiny of the other type is to determine the future sequence of bodies that her soul would inhabit (Lawrence 2009). Bonnet, in his search for a scientific basis for the dogma of resurrection, thus invented and identified the germ of restitution as the seed that links the bodily and spiritual aspects of humanity between this life and future lives. In this, he was close to St. Paul who had used the metaphor of the seed in his discussion of the resurrection of the dead (1 Corinthians 15):

> What you sow does not come to life unless it dies. When you sow, you do not plant the body that will be, but just a seed, perhaps of wheat or of something else. God, however, gives it a body as he has determined, and to each kind of seed, he gives its own body. Not all flesh is the same: People have one kind of flesh whereas animals have another, birds another and fish another …
>
> This will be so, also with the resurrection of the dead. The body that is sown is perishable, it is raised imperishable; it is sown in dishonour, it is raised in glory; it is sown in weakness, it is raised in power; it is sown a natural body, it is raised a spiritual body.
>
> If there is a natural body, there is also a spiritual body … The spiritual did not come first, but the natural, and after that the spiritual … We will all be changed—in a flash, in the twinkling of an eye, at the last trumpet. The trumpet will sound, the dead will rise imperishable, and we will be changed.

At this point, Bonnet based his theory on the classic theory of the great chain of being, which places all organisms in a hierarchy, with the simplest organisms at the bottom and the most complex and noble ones at the top (man, of course). In ancient Greece, an underlying assumption in Aristotle's zoological works was the concept of *scala naturae*, a ladder of nature often also termed a chain of being, a hierarchical order in which each group of creatures has its relative rank. Above inanimate matter come the plants, above which Aristotle ranked the animals, with the invertebrates lowest, followed by fishes, birds, quadrupeds and finally humans. Intellectuals accepted Aristotle's ladder of life also during the mediaeval and later times. Bonnet, coming twenty centuries after Aristotle, was a staunch supporter of the chain of being (Bowler 1973; Anderson 1976). He differed from Aristotle in that he thought of the *scala naturae* as a hierarchy of successive levels of perfection in terms of increasing powers of perception. The qualities of perception of the soul depend on the perfection of the body, through which the soul obtains sensations from the outer world and through which it acts. The individual state of perfection of an organism potentially existed since the time of Creation, when God had preformed it. Yet the individual reaches its state of perfection only through its ontogenetic development, during which it passes through a series of metamorphoses, which lead from one state of complexity and perfection to the next higher one. What is true of the development of the embryonic body of the individual holds also for animal groups.

The consequence of this view was Bonnet's *Palingenesis*: nature "evolves" up a ladder of successive levels to ever-higher levels of perfection in terms of increasing complexity of the body and correlated with increasing powers of perception, a ladder paralleled by and mirrored in the ontogenetic development of individual organisms. The lowermost steps of the ladder consist of the four classic elements: fire, air, water and earth; further up come metals, crystals and stones, then such minerals as talc and gypsum. The uppermost mineral is asbestos, which consists of flexible fibres. Fibres, we recall, were Bonnet's basic element of life, so he placed asbestos as the highest of minerals and just beneath the plants, which also consist of fibres. Above asbestos come the plants that grow out of rocks (lithophytes); above them come the corals, which seemed to him to grow out of stone tubes (they were then considered to be plants); then mushrooms, lichen and many groups of plants, with trees on the highest step. Above all plants came the various groups of animals. In these, Bonnet determined the precise step by the number of senses and the extent to which they are sensitive. In further detail and taking just one section of the ladder, from bottom to top, animal rungs in the ladder would be (Fig. 5.13): shell-bearing snails, which live in calcareous shells and move about. They grade continuously with slugs, slightly above them in that they are shell-less, elongate, slimy snails lacking legs and crawling about on their stomachs. Slugs grade continuously with aquatic snakes, slightly above them: like them, they are elongate and lacking legs, but they have a backbone. Aquatic snakes are continuous with eels that, slightly above them, are snake-like fish. Eels are continuous with typical fish, swimming about in the sea. Typical fish grade continuously with flying fish, slightly above typical fish in that they are amphibious, living in both water and air. Flying fish are continuous with amphibious and aquatic birds, which are slightly above them in that they can fly much better.

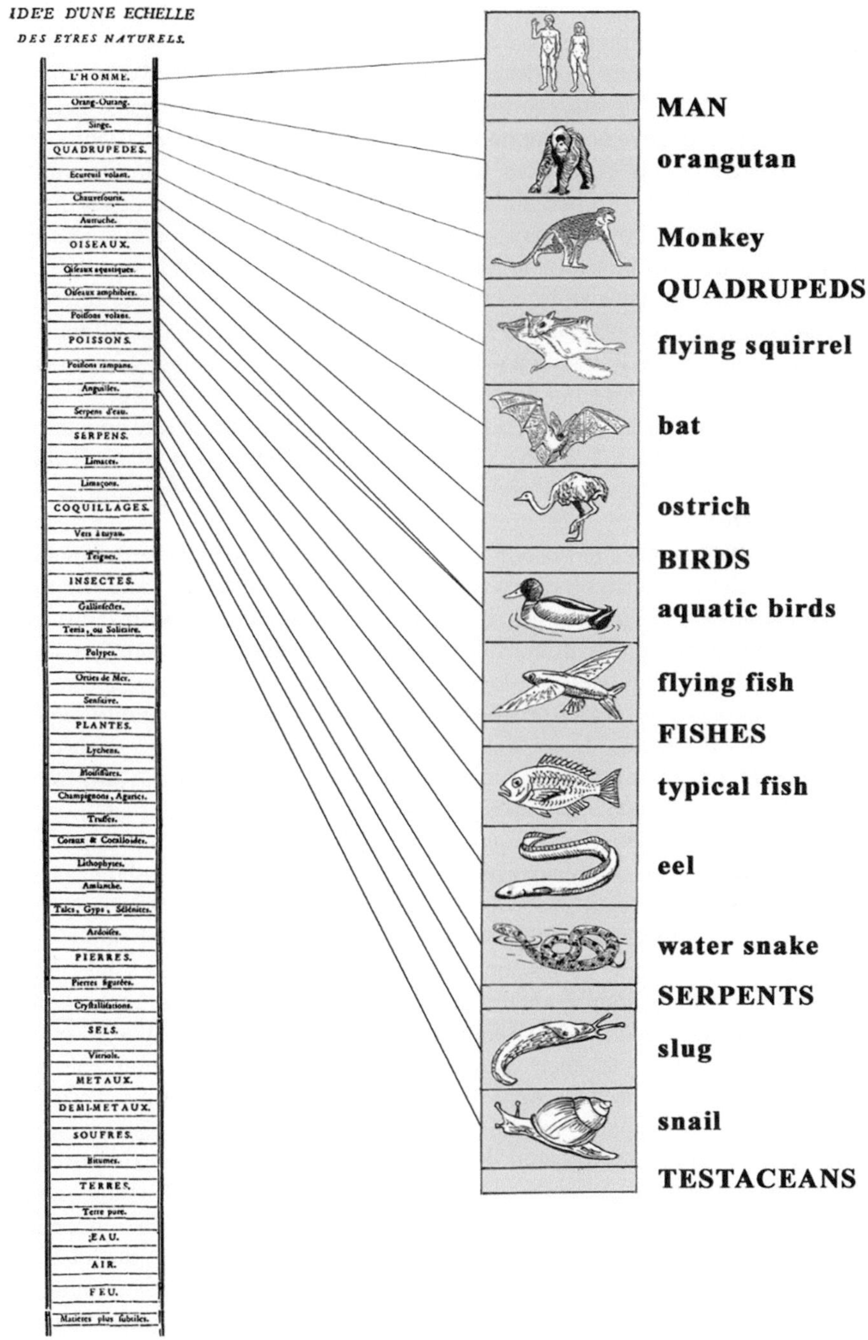

Fig. 5.13 Bonnet's chain of being (1745), as presented by him (left) and an illustrated section (right; see text)

Aquatic birds are continuous with the ostrich, which likewise is a bird but is terrestrial. Ostriches are continuous (via flying birds) with bats, which are slightly above them in that they are furry quadrupeds. Bats are continuous with flying squirrels, which dwell in trees. Flying squirrels are continuous with monkeys, which jump about in trees. Monkeys are continuous with orangutans, which are slightly above them. The orangutan is continuous with man, who is above them in his ability to think rationally. Bonnet in his ladder wisely placed some fishes beneath the term "Fishes"' and others above it; some birds beneath the term "Birds" and others above it and some mammals beneath the term "Quadrupeds" and others above it, thereby further emphasizing that nature is a scale of very gradual, continuous differences from one group to the next. Nature's diversity does not contain gaps. All the manifestations of nature are linked in a single ladder (or chain) of being, with no organism outside this all-embracing continuum. In reading through Bonnet's ladder of life, we recall that Francis Bacon had called for the study of those rare creatures that share two worlds in that they have the morphology of one group but the habits of another: flying fish, bats, apes and moss (Sirks and Zirkle 1964; Bowler 1973; Anderson 1976).

Bonnet's *Palingenesie Philosophique* suggested an innumerable series of earthly lives, in which there is no other reward than the progress effected from life to life. This ran against the Catholic belief that life is a temporary stay in the material world, followed by eternity in the spiritual realm. The book was, therefore, briefly barred from France. Palingenesis does not entail the evolution of new species from antecedent ones. Each species proceeds up the *scala naturae* as an entity separated from other such entities. In all links of nature's chain, it involves a successive improvement not only of the material body but also of the soul. Taking this from the other end, the soul that now animates a high-level creature was once that of the lowest link in the chain of existence. Bonnet's *Palingenesis* broadly calls to mind Plato's theory, by which men who had lived righteously would return to their star after death. Those who had lived cowardly would reincarnate as women, men who were frivolous would reincarnate as birds, men who stopped using their minds would reincarnate as four-legged beasts and men who were the most stupid would reincarnate as mud-breathing aquatic creatures, such as fishes and bivalves. In this context, Bonnet was careful to insist that animals, including humans, change in outward appearance only; their essence remains because essences are immutable. Each entity remains what it is. If it were to change in some essential way, it would no longer be itself; it would be another essentially different entity. The essence, the soul of all created species, remains the same throughout. Organisms simply change their outward appearance, just as an insect undergoes a series of metamorphoses and yet remains the same individual. In embryonic development, God imposes form on matter. In the chain of being, the links are imperceptible. Between the lowest and the highest degree of spiritual or bodily perfection, there is an almost infinite number of intermediate degrees. The succession of these degrees constitutes the universal chain. It unites all beings, links all worlds and embraces all spheres.

While change in essence is impossible, the appearances of species undergo modifications in response to revolutions and environmental changes of the globe—organism and environment change, according to a plan of pre-established harmony. Bonnet's ideas about the transformation of species did not mean a process of Darwinian adaptation. In his theory, both the revolutions of the globe and the changes in the germs of the various species were two sets of preformations, which the Creator coordinates, pre-programming the unfolding of the various animal groups to coincide with the unfolding of the successive catastrophes as parallel but independent timetables; catastrophes do not *cause* changes in species.

To describe the sequence of palingenesis, by which the chain rises toward biological complexity and spiritual perfection, Bonnet used the term *Evolution*. The Latin root *evolvere* means to unroll or unfold; its noun *evolutio* refers to the unrolling of a scroll. Bonnet, retaining the sense of unrolling, applied *evolutio* to the successive unfolding of preformed structures encased in germs: evolution was a synonym for the unfolding of the pre-programmed course of Creation. It was for him as much a theological term as a scientific one (Rieppel 1988).

Various aspects of Bonnet's theories heavily influenced two leading biologists of the next generation, Cuvier and Lamarck, whom we shall meet further on in this

Fig. 5.14 Erasmus Darwin

book, but not everyone accepted them. Erasmus Darwin (1731–1802, Fig. 5.14), the grandfather of Charles Darwin, is one of the most interesting figures in history. The youngest of seven children of a physician, he studied medicine at the University of Edinburgh, and eventually, he settled in the Midlands (Lichfield) and became a successful physician, so highly esteemed that King George III invited him to become his physician; he turned down the invitation. Erasmus had 14 children, including three illegitimate ones, from two different women. In Lichfield he formed a local botanical society (of only three people) to translate the works of Linnaeus from Latin into English, resulting in *A System of Vegetables* (1783) and *The Families of Plants* (1787). He conducted research into cloud formation, experimented with the use of gases to alleviate infections, invented a horizontal windmill, a carriage that would not tip over, a steering mechanism for his carriage, a speaking machine, a canal lift for barges, a minute artificial bird, a copying machine, weather monitoring machines and a rocket engine. He supported the American and French revolutions and established a lifelong friendship with Benjamin Franklin. His most important scientific work was *Zoonomia* (1794–1796). In his *The Temple of Nature* (post mortem, 1803), he expressed in verse many of his biological ideas.

Erasmus rejected the preformation theory. According to his theory, the female produces nutritive particles before impregnation. The male produces a living filament that, when inserted amidst these nutritive particles, is stimulated by them into action. It embraces nutritive particles, adds them to itself and thereby it grows. In doing so, it gains further sensations, to which it responds by embracing further particles. It possesses a selective power of repulsing particles applied to certain parts of it and of embracing others. The embryo thus gradually forms, and the desire to satisfy desires shapes its body. Lust, hunger and security are three major desires that dominate its shape. As all animals generate by means of a living filament, they may all have a similar origin, from one ancestor. Differences in form and quality arise from different responses to different stimuli to which this filament is exposed, and this is how animal diversity has come about. The Earth is millions of years old, and all living creatures, having descended from a single ancestral living filament, are constantly improving their body perfection.

We today might consider parthenogenetic animals as consisting of all-female lines, but this was not how Erasmus saw things. One may perhaps claim that they all reproduce asexually, without a father, but this, Erasmus claimed, is not so. Parthenogenesis is a strictly paternal power of reproduction without copulation. All parthenogenetic individuals are males that produce all-male embryos, which require no mother to supply them with food and oxygen. The Bible allegorizes the generation of animals from a single paternal living filament in the formation of Eve from a rib of Adam. From all this, Erasmus concluded that the embryo is produced solely by the male, the female only supplying food and oxygen.

The developing body gradually grows by the acquisition and addition of new parts. In a newborn baby, all organs of the body seek to add parts to themselves and to grow unlimitedly throughout life, but tissues enveloping the organs prevent them from doing so. If we remove the enveloping skin, the fleshy parts beneath soon shoot out new flesh; if we remove the tissue surrounding bone (the periosteum), underlying bone soon commences growth. In a young embryo, dense confining tissues have not yet formed, and nothing restrains growth. Erasmus argued that this theory of embryo development by new additions explains why some foetuses are deficient at the extremities, lacking a finger or a toe: a deficiency in the number of nutritive particles in the ovum, for the reception of the living filament, results in the extreme parts not forming. This theory also explains why monsters are born with duplicate limbs (four-legged chicks): An excess of nutritive particles in the ovum results in a duplication of some parts, and double yolks in some eggs are evidence that such excess nourishment does exist, sometimes. This theory also explains changes and varieties. Consider the great changes produced in animals and man during their lives: change to butterfly from a crawling caterpillar, to a respiring frog from a tadpole, to a bearded man from a feminine boy and to a lactating woman from an infant girl. Consider, also, great changes introduced into various animals by artificial cultivation. Horses are cultivated for strength in carrying burdens or for swiftness in running races. Dogs are cultivated for strength and courage (bulldog), for acuteness of smell (spaniel), for swiftness (greyhound), for drawing snow sledges (husky) or for children's play (lapdog). Cattle, camels and sheep have undergone, over history, such a total change that we no longer know their origins among the wild animals. Consider the variation resulting from differences in climate and season: sheep of warm climates have hair instead of wool, and hares and partridges of high latitudes become white during winter. Consider also the varieties of man due to physical effort or to diseases, both of which may become hereditary through many generations. ("Those who labour at the anvil, the oar, or the loom, as well as those who carry sedan-chairs, or who have been educated to dance upon the rope, are distinguishable by the shape of their limbs".) Erasmus continues: As labour strengthens the muscles employed and increases their bulk, and as early physical activity becomes hereditary, a few generations of physical activity or laziness may change the form of the body and produce new varieties. Diversity thus forms through the hereditary passing down of acquired characters. Variation may also result from hybridization between species or from differing nourishment to the foetus. Further, many new changes of shape continue as a variety, at least, if not as a new animal species. Peculiar breeds of pigeons had first hatched as monsters, and breeders later cultivated them. He also mentioned a breed of cats with an additional claw on every foot, a breed of poultry with an additional claw, another with wings on its feet and a breed of dogs without tails, perhaps, he suggested, from the long custom of people cutting their tails close off. All these changes and varieties of animal form, and innumerable others, convince us that the embryo is formed by the apposition of new parts, not by the distention of a primordial set of germs.

To these causes of change in the offspring, Erasmus added changes produced by the imagination of the male parent, the father. Pleasurable or painful imaginations affect the secreted fluids of many glands. The imagination of meat, for example, excited in the minds of hungry dogs by their sense of vision or smell, increases their discharge of saliva, as we see in their saliva hanging down in threads from their mouths as they stand around a dinner table (Erasmus wrote this a century before Ivan Pavlov would publish his experiments!). In a similar manner, thoughts of pleasure or pain of peculiar kinds excite a great discharge of tears. Paleness from fear and blushes of shame and joy display the effects of painful or pleasurable sensations on the extremities of the arterial system. Pleasurable or painful thoughts affect all glands and their secretions, in quantity and quality.

Yet no secretion of fluid offers so much pleasurable sensation as that of the semen. The pleasurable sensation is necessary for copulation, and pain, fear or bashfulness easily interrupts it. Even hermaphrodites (snails, worms) that, he thought, can self-fertilize and thereby save the construction of a complex male genital system, prefer to reciprocally receive and give impregnation—surely because of that pleasurable excitation. As evidence for the effect of imagination on the embryo, Erasmus told of a certain gentleman who has a child with dark hair and eyes, though he and his lady have light hair and eyes, and their other four children are like their parents. This gentleman told Erasmus that when his lady lay in labour of her third child, he became attached to a daughter of one of his inferior tenants and offered her a bribe for her favours in vain, and afterwards a greater bribe, but was equally unsuccessful. The image of this girl dwelt much in his mind for some weeks, and his next child, the dark-eyed one, was in both features and colour exceedingly like the young woman who had refused his addresses. As further evidence, Erasmus told of two families, in each of which the father eagerly desired male heirs because of an estate in expectation. In one family seven girls were born and nine in the other family, and only then did each of them have a son. Erasmus concluded that the great desire for a male heir by the fathers produced disagreeable sensations in which, during copulation, they dwelt more on the fear of generating a female than on pleasurable sensations or ideas of their own male organs. Hence, a female character was present in their mind rather than that of a male. Eventually, in despair of generating a male, these female imaginations ceased, and those of the male presided at the genial hour. Erasmus finally noted that the wonderful effects of imagination in parents were there already in biblical times. Jacob had not only placed rods of trees, in part stripped of their bark so as to appear spotted but had also placed spotted lambs before the flocks at the time of their copulation (*Genesis* 30).

The power of imagination, in the act of generation, belongs to the male. The mother's imagination can alter the foetus only through the nutritive particles she imbeds in the ovum before the arrival of the living filament. These particles may vary in size, solidity, colour and form, and this may bear on the living filament as it develops. No such power affects the embryo after it has obtained a placenta and has developed organs that can select and change the food, which it uptakes in the amniotic fluid or in the milk. The female's imagination may affect only minor features, such as, in birds, the colours of the spots on eggs or changes in the colour of feathers in domestic animals.

Erasmus next explained how, in the early embryo, the desire to satisfy pleasures moulds the body. Lust, hunger and security are three great objects of desire of many animals. Starting with lust, there is, throughout nature, a great desire among males for the exclusive possession of females. They combat for a harem of females that, to use Erasmus' words, "like the ladies in the times of chivalry, attend the car of the victor". To combat this exclusive possession, males have acquired weapons, such as the tusks of the boar, which strike obliquely upwards; and the thick, shield-like, horny skin on their shoulder, as a defence against the strikes of other boars. Another example would be the sharp horns of the stag, to stab other stags, and the branching of these horns, to deflect or block incoming attacks by the opponent. Among birds, males have spurs for fighting over females in those species in which they do not care for their young (cocks, quails). In all such cases, these weapons are not for defence against predators, as we can learn from the fact that the females of these species are without them. One result of this lust for females is diversity in structures like tusks, horns and spurs. Another result is that, because of these male contests, the strongest and most active male propagates the species, thereby improving it.

Another great passion is for food, and this too has moulded the bodies of animal species, resulting in a variety of food-obtaining structures. The snout of the swine has become hard for turning up the soil in search of insects and roots; the trunk of the elephant is an elongation of the nose for pulling down branches for food and for taking up water without bending the knees. Beasts of prey have acquired strong jaws or talons, and cattle have acquired a rough tongue and a rough palate to pull off blades of grass. Some birds have acquired hard beaks to crack nuts (parrots); others have long beaks to penetrate moist soil in search of insects (woodcocks) or broad beaks to filter water (ducks). These food-obtaining structures, moulded by the perpetual endeavour to supply the want of food, have gradually evolved over many generations and have been passed on to the offspring, constantly improving the feeding and expanding diversity.

The third great desire among animals is security, which has diversified body forms and colours to escape powerful predators. Hares have acquired great swiftness of foot; flying fish and bats have acquired great lengths of fin or of membrane and tortoises and sea urchins have hard shells. Other animals have acquired cryptic structures and colours to conceal themselves.

Much of the high diversity of form and function in the animal world stems from the need to satisfy those three desires of lust, hunger and security. At a different level, it stems from a living fibre that, during embryonic development, comes under the influence of various forces. All animals generate by means of a living filament, and this suggested to Erasmus that they all have a similar origin, from a living filament. Differences in form and quality arise from different responses to different stimuli to which this filament is exposed and, to some degree, from different particles in the fluid, which first stimulate it into activity; this is how the great diversity of animals has come about. Beneath the great diversity of warm-blooded animals, there is a great similarity in structure. This is so whether in man, this filament develops hands and fingers with a fine sense of touch; in tigers, it develops claws; in seals and geese, toes and an intervening web; in cows and swine, cloven hooves; in horses, whole hooves and in birds, wings. Upon considering the great similarity of the structure of warm-blooded animals, the great changes they undergo both before and after birth, and in how short a time many of the changes have occurred, Erasmus presented his idea:

> Would it be too bold to imagine that, in vast periods of time since the earth began to exist, perhaps millions of ages before the history of man; would it be too bold to imagine, that *all warm-blooded animals have arisen from one living filament*, which God had endued with the power of acquiring new parts, as directed by irritations and sensations; and thus, possessing the ability of continuing to improve by its own inherent activity, and of delivering down those improvements to its offspring, a world without end?

Erasmus thus considered Earth to be millions of years old and all living creatures to have descended from a single ancestor, from one single living filament (we may translate into a single cell). One living filament is the source of all life. From the beginning of the existence of Earth, the animals that inhabit it have constantly improved and are still in a state of progressive improvement. As an example, Erasmus claimed that man may have arisen from a single group of monkeys in which, accidentally, the opposing muscles brought the thumb against the tips of the opposing fingers, and this muscle gradually increased in size by use in successive generations.

In his *Temple of Nature* (1803), Erasmus described the struggle for existence in very clear terms: the ichneumon wasp parasitizing the caterpillar, bee swarms slaughtering one another by the thousands and plants competing for light and, on the way, killing one another.

> The winged Ichneumon for her embryos young
> Gores with sharp horn the caterpillar throng.
> The cruel larva mines its silky course,
> And tears the vitals of its fostering nurse.
> *
> Contending bee-swarms rise on rustling wings
> And slay their thousands with envenom'd stings
> *
> Herb, shrub and tree, with strong emotions rise,
> For light and air, and battle in the skies
> While roots diverging with opposite toil
> Contend below for moisture and for soil
> Round the tall Elm, the flattering Ivies bend,
> And strangle, as they clasp their struggling freind

In later decades Erasmus' grandson, Charles Darwin, would write: "It is curious how largely my grandfather, Dr. Erasmus Darwin, anticipated the views and erroneous grounds of opinion of Lamarck in his *Zoonomia*". The parallelism between Erasmus and Lamarck is striking. Erasmus' *Zoonomia* appeared in 1794. It was not until 1801, 7 years later, that Lamarck published his theory of the mutability of species. Lamarck nowhere refers to *Zoonomia*. He may have arrived independently at the same conclusions, or he may have borrowed freely from Erasmus without giving him credit.

It was in Erasmus Darwin's days that James Hutton (1726–1797), a Scottish naturalist, laid down significant contributions to our understanding, both of Earth-forming processes and of the immensity of deep time. He was born in Edinburgh, where he studied humanities and medicine; in Paris, he studied chemistry and anatomy and finally, he obtained his degree in medicine from Leyden, in the Netherlands. Aged 29, he inherited two farms in Scotland, travelled to Flanders to learn new farming methods and employed them on his own lands. Later he moved to Edinburgh, where, as one of the most influential participants in the Scottish Enlightenment, he co-founded the Royal Society of Edinburgh (1783). Observing processes of erosion and sediment deposition on his farms, he became interested in geology.

Hutton claimed, broadly in line with Buffon (we shall meet him further on), that geological features undergo perpetual transformation over very, very long times. Noting that many solid parts of what at present is land had formed in the sea, he concluded there was once a different world of sea and land, one that maintained different plants and animals. In the seas, at first, loose materials consolidated; and then these consolidated masses were elevated from the bottom of the sea to those positions above the level of the ocean in which we now find them. Furthermore, there must have been innumerable cycles, each involving deposition on the seabed, uplift with tilting and erosion and then undersea again for the deposition of further layers. He believed that this was due to the same geological forces operating in the past as the very slow, gradual forces operating today, and consequently, the great thicknesses of rock layers implied to him enormous stretches of time. He proposed that the interior of the Earth was hot and that this heat was the engine that drove the creation of new rock: air and water eroded land and deposited it as layers in the sea; heat then consolidated the sediment into stone and uplifted it into new lands. He found evidence for his theory in the Scottish Highlands and in nearby islands (Wightman 1953). Hutton, in his book *Theory of Earth; or an Investigation of the Laws observable in the Composition, Dissolution, and Restoration of Land upon the Globe* (1788), concluded that in the history of Earth, "we find no vestige of a beginning, no prospect of an end". He further stated that "from what has actually been, we have data for concluding with regard to that which is to happen thereafter". This restated the Scottish Enlightenment concept of David Hume, that the future will resemble the past (1777). Four decades later, Charles Lyell would rephrase this, as "the present is the key to the past". Hutton claimed that the tremendous displacements and changes that we see in stratification happened very gradually, by processes still happening on Earth today. As these processes were very gradual, the Earth needs to be ancient to allow time for the changes. This opinion of Hutton opposed the then-popular theory of the German geologist Abraham Gottlob Werner, who had claimed that all rocks had precipitated out of a single enormous flood (1774). Hutton repeated his arguments in a later publication, *An Investigation of the Principles of Knowledge and of the Progress of Reason, from Sense to Science and Philosophy* (1794).

Hutton, interestingly, advocated gradual change for living creatures also, and he suggested natural selection as a possible mechanism affecting them. If a creature is not best adapted to its environment, then, bearing in mind an indefinite variety among the individuals of that species, those least adapted will be the most liable to perish. On the other hand, those best adapted will survive and multiply (*Investigation of the Principles of Knowledge*, volume 2). Hutton gave, as an example, dogs that survive through their swiftness and acute sight. Those individuals defective in these features would perish, while those who employed them in greatest perfection would remain to preserve themselves and to continue the race. Similarly, if an acute sense of smell became more necessary to the survival of the dog, the same principle would change the qualities of the animal and produce a race of well-scented hounds instead of those who catch their prey by swiftness. This “principle of variation” would influence every species of plant also, whether in a forest or meadow. Hutton came to these ideas through his experiments in plant and animal breeding, some of which he outlined in an unpublished manuscript. He distinguished between heritable variation as the result of breeding and non-heritable variations caused by environmental differences, such as soil and climate. Charles Darwin arrived separately at the idea of natural selection in his 1859 *Origin*. He may have had some half-forgotten memory from his time as a student in Edinburgh of ideas of selection in nature as set out by Hutton.

5.1.4 Fake Fossils

Dr. Johann Bartholomew Adam Beringer (1667–1738) was appointed professor at the University of Würzburg when 27 years old, just 1 year after achieving his doctorate in medicine there. Aged 28, he was also appointed as keeper of the botanical gardens of the university. Beringer was a passionate collector of fossils, which abounded in the hills about Würzburg, and three lads whom he hired would scan the nearby hills and mountains and collect fossils for him (Jahn and Woolf 1963).

Two colleagues at his university, a professor and a librarian, resented his arrogance and decided to deflate him with a hoax: they hand-carved fake fossils of various living and also of imaginary things and, enlisting the assistance of one of the lads, in 1725 they had these fake fossils planted on Mount Eibelstadt, where Beringer was wont to search for fossils (Jahn and Woolf 1963; Alvis 2019). When the lads came down from the mountain with the first (fake) fossils, Beringer was delighted and remarked that the hillside, which he had “frequented and examined in the past but had never scrutinized very closely, revealed a treasure”. Following the first fake specimens, more and more (fake) fossils turned up. As many of the fossils on the mountain were genuine, this contributed to his belief that these new specimens were also genuine. He published these fake fossils in a book, *Lithographia Wirceburgensis*, that contained over 200 engraved pictures in 22 plates (Figs. 5.15, 5.16 and 5.17). He noted that the animal and plant “fossils” bear the exact consistency, material and colour of the stones in which they lie, with no natural colour, no natural matter and no armour. Furthermore, they very exactly fit the dimensions of their stones, without any further matter. All animals are “in their proper posture which Mistress Nature desires for them, only rarely on their back or side”, and Beringer grouped them as consisting of sand, or of clay, of mud, or of lime. Feeling that Heavens had placed a stone treasury in his hands, he dreamed of a museum that would exhibit his treasury to visitors for them to see what he was showing in pictures (Jahn and Woolf 1963; Alvis 2019). Among the (fake) fossils, I recognize flies (some in copula), snails, slugs, ammonites, scorpions, earthworms, scallops, worms, lobsters, crabs, birds, various plants, spiders in cobwebs, horned beetles, frogs and

Fig. 5.15 Fake fossils, from Beringer's *Lithographia Wirceburgensis*

fish. There were also falling stars and Hebrew words, some with the name of God in Hebrew.

Shortly after its publication, Beringer discovered that he had been the victim of a hoax. Disgraced, he immediately recalled and bought back all printed sheets that

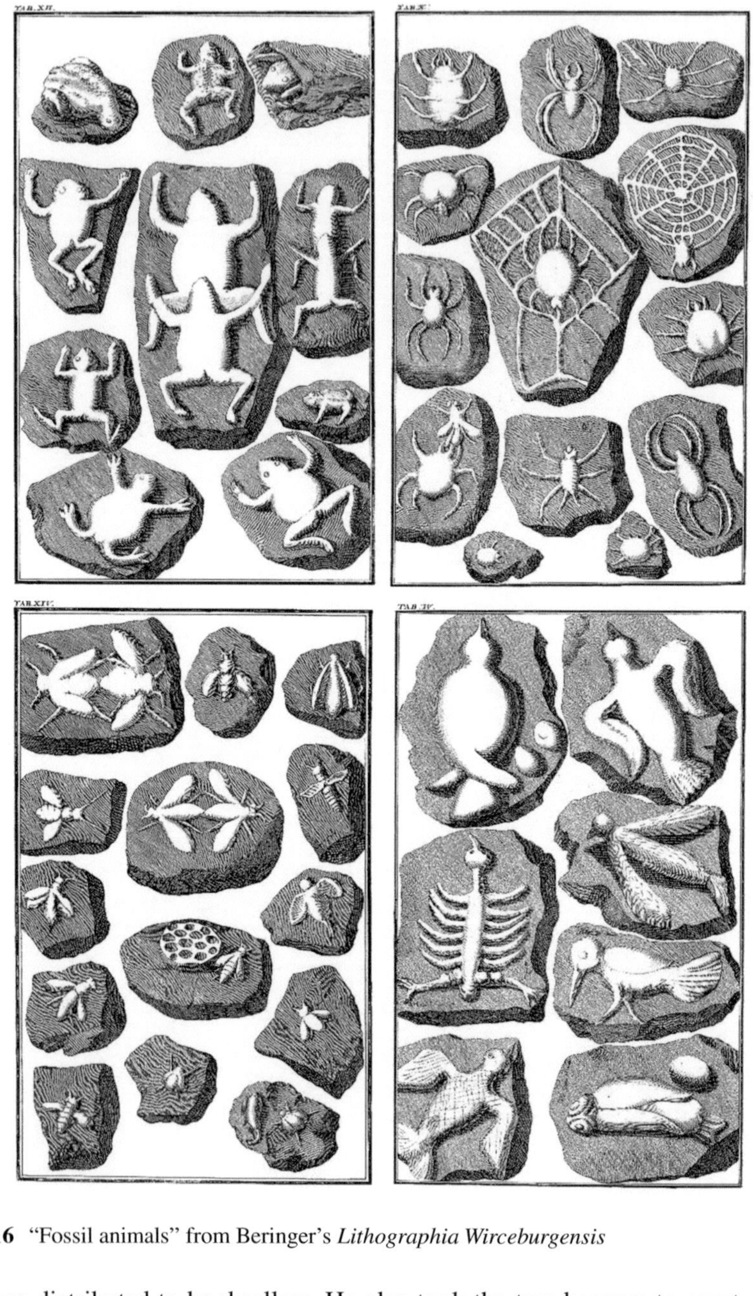

Fig. 5.16 "Fossil animals" from Beringer's *Lithographia Wirceburgensis*

had been distributed to booksellers. He also took the two hoaxers to court and his two university colleagues were found guilty, banished from Würzburg and denied further access to archives of the Duchy (1726; the three lads were found not guilty, Jahn and Woolf 1963). To describe the fraudulent fossils, the term "lying stones" (Lügensteine in German) was used in the court.

Fig. 5.17 Fossil Hebrew scripts from Lithographia Wirceburgensis; the mid "fossil" in the uppermost row and the fossil on right in the second row bear the name of God in Hebrew

We are fortunate in that Beringer, in his book, beyond illustrating the fake fossils, also surveyed the many different opinions as to how fossils come to be. This enables us, coming three centuries after him, to have a panoramic view of how people of the early eighteenth century viewed fossil formation. On this subject, there were at least eight different theories:

- Some considered their formation as yet another of those mysteries of nature, such as magnetism, and considered the question no further.
- Others turned to the influence of the stars or to some intercourse between celestial and terrestrial beings, between constellations and earthly nature. Since the influence of the stars is universal and, for the most part, has a simultaneous effect on all sublunary bodies, it follows that all bodies originating under the dominance of a particular celestial body should be with the same figures.

- Some assigned the formation of fossils to a plastic power [*vis plastica*] of the Earth's salts.
- Some took recourse to vapours rising from the sea and forming rain or mist. These vapours are frequently impregnated with the seeds of various marine animals, whether snails, bivalves or fishes. Depending upon the perfection of the seed and the propensities of the matrix, some of them develop as whole fish, others only into fish teeth, jaws or vertebrae (likewise, among crustaceans, they may develop into whole animals or into their parts, and among the molluscs, into univalves or into bivalves). Rain and mist then transport these seeds to the land.
- Others proposed that marine fossils are generated by seeds dispersed over terrestrial environments. Since the seed of marine animals is tiny and light, it is carried through air and water and eventually is decanted into pores of the earth. Once in subterranean layers, the underground heat, steams and vapours distribute the seed to rock fissures, which are open passages of the underworld. It often reaches the surface of the earth, where it is deposited in a fitting matrix. It clings firmly to this matrix by its innate magnetic powers until, with the coming of the subterranean heat, a latent, plastic force inside it is stirred into action; whereupon, through the medium of a seminal breeze *aura seminalis*, it is shaped and moulded into the same kind of body as that which it produces from the shelly matter in the sea.
- Some suggest that fossils are relics of the Deluge. During the flood, an enormous mixing-up and mingling of things occurred, with vast quantities of originally terrestrial things being hurled into the sea, and things originally in the sea were scattered over the land. After the waters receded, the marine animals, whether molluscs, arthropods or plants, became buried in the soil of mountains and were, over the centuries, permeated by certain salts and by moisture and thereby altered into stony substances. (In 1726, a Swiss natural scientist, Johann Jacob Scheuchzer, claimed to have found a human drowned in the flood, *Homo diluvi testis*; it actually was a tertiary salamander.)
- Some ascribed fossils to errors of nature or to a certain playfulness of nature, *lusus naturae,* resulting in forms that imperfectly imitate the forms of other bodies.
- Fossils are not created by nature. They are the products of an impious art that was customary among the Germans in their pagan state—crude images of plants, trees and animals that were, perhaps, sacred idols.

We now come to Beringer's highly original theory: fossils come to be by light. Light is a flow of minute, fiery solar bodies that permeate, heat and illuminate air. It is a corporeal substance, possibly not distinct from that primordial light that God had created before He had created the sun and the stars. This light takes on the forms and figures of all the things on which it shines, and this is the source of those visible things that we see in still water, in mirrors and in the pupil of the eye. Now, light possesses the faculty of painting and forming such corporeal images as it acquires because it is endowed with an active plastic power of impressing these images on properly disposed matter. The atmosphere is full of worthless leftovers of countless elemental atoms. These leftovers, when hit by the forces of light, would have impressed figures of insects, plants and even stars into soft mud, clay or sandy stones. In this context, we call to mind that visible images of an object suddenly falling on the eyes of a pregnant woman can instantly impress upon the foetus the forms of animal organs. Can we deny the analogy between the womb of a mother and the womb of Earth, the common parent of all? Could not the substance of airy light, thus modified, combined with the activity of the fertile Earth and supplemented by forces of salts and seeds, even of *Panspermia*, be a determining agent in the formation of plant and insect figures in stone or in malleable clay that later hardens into stone? All this would be caused by the refraction, distortion and confusion of the image-producing light, as we see happens to reflections falling upon polygonal, cylindrical, convex or concave mirrors. As to the stones bearing Hebrew characters, an ancient Jewish cemetery is close to our prodigious mountain; it was uprooted during the expulsion of the Jews, not many years ago. The brooding light could absorb letters and even entire words and then transfer them elsewhere (Jahn and Woolf 1963).

5.2 Coping with Diversity

5.2.1 Classification Controversies

The last decade of the seventeenth century and early decades of the eighteenth century witnessed a strong conflict between two philosophies regarding the classification of nature's diversity (Sloan 1972). Should we consider only certain characters and structures as important in constructing a natural classification? Should a single character (or functional system) receive more weight than other characters? Are there any means for deciding which character or structure to weigh more heavily than other ones? Can we ever reach a "natural" system, namely, a system based upon the true "essences" of nature? The alternative would be an "artificial" classification, for example, presenting the various entities in alphabetical order, by habitat or in the form of a catalogue for medicinal or agricultural purposes. Should a natural classification rest upon the resemblance of a few characters or on overall similarity?

The classic tradition, going back to Aristotle (and as we have presented in the first part of this book), had argued that each kind in nature, whether animal, mineral or plant, has an unchanging essence, an inherent, innate nature. By definition, an essential feature of a kind is a feature that a given kind *must* have. This essence makes an entity what it fundamentally is, and it loses its identity without it. The alternative to an essential feature would be an accidental feature, one that an individual of a certain kind happens to have but which it could lack and yet retain its identity (say, blue eyes in humans). To define a kind, one should use only essential characters. The concept of an "essence" originated in ancient Greece. As a brief reminder, Aristotle claimed that each entity in nature has an unchanging essence, which makes it what it fundamentally is. We should use these essences to define a kind by syllogisms. At first, by using the essential feature, one classifies an individual (or kind) into one of several larger, well-known groups (birds, fishes, insects), intuitively known by instinctive knowledge; this larger group would be the genus. Next, by using another essential feature, one distinguishes the individual (or kind) from all other members of that particular genus; this distinguished group will be the species. By first classifying and then distinguishing, we achieve a correct definition. A classification of animals must start with large groups and proceed through subgroups to the species; it should not proceed from the species level upwards. With these theoretical foundations, Aristotle then developed a rather complete, full-blown classification of the animal world. During the Renaissance, Cesalpino, as we have seen, had sought to form a natural classification of plants by following Aristotle's theoretical foundations by reference to Aristotle's two defining, essential functions of the vegetative soul: nutrition and reproduction. His primary division of the plants, based upon root and sprout, was into two major groups: trees with hard medullary substance and herbs with soft matter. Within each of these major groups, he had erected subordinate subgroups by flower and fruit. He divided the trees into those having only a single seed in the fruit versus those having several seeds in a single container and the herbs by the number of seeds produced by the flower or contained in a receptacle of some kind. His classification was "essential" in the Aristotelian meaning of the term.

A rival approach gradually developed in the seventeenth century, in philosophy rather than in biology. An early step in this direction was by Descartes, who deliberated in a completely different direction. In referring to the accepted dogma of his days that our knowledge about the world around us comes through our senses, he maintained that information achieved through our senses might be false, and therefore it cannot be a reliable basis for knowledge. Large distant bodies seem small to us through our senses, and only when we apply our minds and judgement do we conclude that they are large. Since he rejected any knowledge for which there was the slightest doubt, he rejected sense-based knowledge. Descartes' theory undermined that crucial link between what the mind knows on the one hand and what exists in nature on the other hand, the very link upon which Aristotle's claim (and Cesalpino's) that natural essences are knowable rested.

In addition to these theories of Descartes, the mid-seventeenth century saw the revival of atomism. As a brief reminder, in ancient Greece, Epicurus believed that the basic constituents of the world are minute, indivisible atoms and void. All macroscopic things are conglomerations of atoms, and all events are the results of collisions, rebounding and entanglements of atoms. Occasionally and randomly, atoms collide, thereby forming macroscopic bodies that may cluster into organs. If an organ by chance carries out a certain function, then that function sets in. Form thus precedes function. Lucretius, in ancient Rome, echoed Epicurus' philosophy. In the seventeenth century, atomism was largely unacceptable because it excluded God from existence in either the cosmos or the lives of people, because it did not accept a divine purposeful Creation and because it argued in favour of several forever-changing worlds.

The French philosopher Pierre Gassendi (1592–1655) revived atomism as a philosophy of nature. He diverged from the Epicurus–Lucretius atomism by proposing that God had created the atoms and had endowed them with motion. Further, Gassendi considered that *semina*, seeds, are intermediaries between inert atoms and living creatures. God had endowed the atoms with a specific structure and motion so they would combine in a specific order to produce seeds of particular organisms. He had created the seeds of the different plants and animals at the beginning of the world and had cast them over land and sea so that they brought forth all living things, as described in *Genesis*. Gassendi imagined the seeds as containing the rudiments of the organisms they were destined to become. God acts on all parts of the seed at the same time, guiding the development of their rudimentary parts into a complete organism.

A further step was taken by Robert Boyle (1627–1691), the English chemist who discovered that a vacuum exists, that voice does not pass through a vacuum, that fire requires air for burning, that living organisms need air, and that if the volume of a gas decreases, its pressure increases. He explained his discoveries by applying Gassendi's ideas and developing a general theory of chemistry, by which all elements are composed of minute particles of different shapes and sizes, so minute they are beyond our senses (Sloan 1972). Different elements differ by the spatial arrangement of their particles. These tiny particles, the primary qualities of an element, cause secondary qualities that are observable to our senses. Secondary qualities of gold are its glossy yellow colour, specific malleability, specific weight and so on. He rejected the theory that all gases are composed of the four elements of classical Greece (fire, air, water and earth) and any immaterial component (a "soul' or a "spirit") as existing beyond the material component. Boyle took Descartes' concept that our senses can fool us and applied it to chemistry: perceiving the environment is a matter of information penetrating the brain through our senses, where a common sense combines the information from all different senses and processes it to create an image in the brain. The only true, objective existence of a material is by the arrangement of its particles. However, what we humans know about this existence—the colours, tastes, textures and shapes of entities as they are transmitted to us via our senses—is a subjective experience, an imagination connected to concrete reality only via the action of that entity on our senses. This subjective experience does not necessarily represent the actual qualities of the entity. Therefore, all definitions once thought to be so relevant to classification, whether of essential versus accidental characters or of character weighting, are irrelevant. Man's classifications of natural objects into discrete kinds are based on indirect knowledge, as supplied by secondary qualities, all standing on the same footing, and none of which is more essential than another. It is not on direct knowledge of any essences, as Aristotle had claimed (Sloan 1972). This assemblage of secondary qualities, distorted in our minds by subjective processes, is the foundation of all our classifications (*The Sceptical Chymist* 1661, *Origin of Forms and Qualities according to the Corpuscular Philosophy* 1666). In modern terms of today, we would say that oxygen has a unique set of atoms differentiating it from hydrogen or nitrogen. This set of atoms is minute beyond the perception of our senses, which offer only a vague, subjective idea as to what oxygen is as compared to other gases, say that it enables burning.

All this is very remote from plant and animal taxonomy, and Boyle never suggested applying his ideas in chemistry to natural history. It was John Locke (1632–1704, Fig. 5.18) who applied Boyle's theories to biological classification. The son of an English country lawyer, aged 20, he entered Oxford to study medicine, and here he worked with Boyle and with Hooke. He became the personal physician of the Earl of Shaftesbury, and when Shaftesbury became Lord Chancellor (1672), Locke became involved in politics. Twelve years later, he fled to the Netherlands under suspicion of involvement in the Rye House Plot, and there he met Spinoza. He returned home in 1688, after the Glorious Revolution, and there he corresponded with Isaac Newton. Locke was thus in personal touch with quite a few leading scientists, philosophers and politicians of the day. He was also involved in current botanical science, keeping close and ongoing relations with keepers of the Oxford Botanic Garden and distributing seeds and plants to other botanists (Anstey and Harris 2006).

Locke expanded Boyle's ideas in a more biological direction (*An Essay Concerning Human Understanding*, 1690). He accepted Boyle's theory that observable entities are composed of minute, unobservable corpuscles. Along lines broadly similar to those of Boyle, he argued that our knowledge of things surrounding us,

Fig. 5.18 Locke

such as individual plants or animals, are merely mental ideas gained from our senses and elaborated in our minds (Sloan 1972). Our knowledge of entities is comparable to that of a man gazing from afar at a distant clock: he would know only the appearances, as observed by his senses, not the internal mechanical realities that produce its functions. Along further lines broadly similar to those of Boyle, he drew a distinction between real essences on the one hand and derived, observable ("nominal") essences on the other hand. The real essence of an entity is that essential chain of minute, sub-microscopic corpuscles that makes a creature of a given entity what it is. It occurs in all individuals of that entity and only in them. [As a modern example from chemistry, in Mendeleev's periodic table, all and only those members of a particular element share a unique and common atomic structure, and this is their real essence.] Once we know an entity's real essence, we can know its derived, observable qualities without making any observations—the real essence of an entity explains why it has its specific observable features. If we could know gold's real essence, we would know its derived, observable qualities, even if not a single sample of gold existed. We would even be able to deduce an entity's powers to produce effects in other entities. Say, if we knew the real essences of opium and hemlock, then we could deduce that opium produces sleep and that hemlock causes death—we would understand why each entity produces its effects. True, we shall have full knowledge of an entity only when we know its real essence, but this is unobservable, whether due to the minuteness of their particles or due to their remoteness in faraway celestial bodies (Kochiras 2017).

As the real essences are beyond our reach, no reliable, certain knowledge of their entities is possible. In exploring the question of how we acquire reliable knowledge, Locke distinguished among three different kinds of knowledge: intuitive, demonstrative and sensitive. Intuitive knowledge refers to knowledge with high levels of certainty, and it is the most certain of human knowledge because we grasp the truth immediately. There are no intermediate steps, and doubt is impossible because the mind can no more avoid recognizing the truth than the eye can avoid seeing light when turned toward the sun. The mind intuitively perceives that white is not black, that a circle is not a triangle, and that three is more than two. This kind of knowledge is the clearest and most certain. Demonstrative knowledge, though also certain, is less so because it involves intermediate steps. We cannot grasp immediately that the three angles of any triangle are equal to two right triangles and must instead construct the steps of a proof. Upon doing so, and upon grasping connections between the steps of the proof, we reach demonstrative knowledge. Sense-based knowledge lacks full certainty beyond doubt. It is less than knowledge based on intuition or demonstration, but it is far more than ungrounded opinion. Instead of knowing real essences, we can make do with derived, observable ones. All members of an entity share, in addition to their unobservable, real essence, also a set of observable features derived from that real, unobservable essence. By using our senses, we can recognize and classify them. In a modern illustration of this point, the real essence of gold would be its unique atomic structure. Derived, observable features would include its yellow colour, glossy shine, malleability, electricity conductance, melting temperature and so on. These derived, observable essences serve us in providing provisional classifications. We use a set of observable features to define and thereby classify an entity. The naturalist's task, so Locke believed, is to determine observable essences with more accuracy than ordinary people do. Another task is to separate the derived, observable features from accidentally observed ones, which do not derive from the real essence. Entities also have accidental features, not derived from the real essence. Gold on the finger of a princess would be an accidental feature of a particular piece of gold. If located on the wrist of a prince, it would still have gold's atomic structure.

These theories of Locke bear upon questions of classification. Our groupings rest, not on the knowledge of some real essence, as Aristotle would have argued. They rest solely on observed essences. Our classifications only possibly conform to the real essence since the nominal essence only probably indicates it. Therefore, our classifications are largely arbitrary. As far as living entities are concerned, all we can do is put names to what we see as groups of similar individuals (*Essay* III, VI).

This leads to a second problem. It is by observable essences that we classify into large, well-defined groups with distinct boundaries, such as quadrupeds, birds and fishes. Nature, however, does not necessarily divide things into these neat groups with distinct boundaries. Locke believed that the bat, seal and flying fish are transitions between one group of animals and another and argued that in nature there well may be no clear-cut distinctions between major animal groups (Sloan 1972). This is all the more so when we seek clear, well-defined groups further down the taxonomic scale, at the species level. In species defined by observable essences, are the boundaries sharply defined and distinct? Locke tackled this question by taking as a case man the rational and language-possessing animal. He noted that there are creatures in the world that have shapes like ours but are hairy and lack language and reason. Then, some people lack both reason and language. Further, there are "men without beards, women with beards and creatures in other ways like us but with hairy tails". Given this broad range of individual variation, each differing only slightly from one another and sometimes shading into what we would consider another species, nominal essences provide somewhat arbitrary decisions as to boundaries between species.

Another question was whether differing individuals belong to different species also by virtue of differences in their real essences, at a level beyond our reach. Changelings (idiots or simpletons) and mandrills belong to different species, whereas changelings and reasonable men belong to the same species. However, Locke argued that if resemblances and differences between phenomenal features are to serve as signs marking the boundaries of species on the real essence level, then mandrills and changelings, both lacking reason and language, should be considered as not of the same species as man. Now we, of course, hold that reasonable men and changelings belong to the same species, and mandrills do not. It follows that resemblances and differences between phenomenal features do not provide reliable signs of species boundaries at the real essence level. Locke rejected, accordingly, the idea of sharp boundaries between species, even at the real essence level. Continuity of the distribution of features on the real essence level makes it impossible to find non-arbitrary boundaries of species on the real essence as on the observable level. He did not deny the existence of species, but he argued that men determine their boundaries. "I would fain know why a shock [poodle] and a hound are not as distinct species as a spaniel and an elephant" (*Essay* III Ch. VI). The very category of a species, say a man or a horse, seemed to Locke to be an abstraction created by the human mind.

In all animal species, the frequent production of monsters and other strange issues of human birth posed the problem of the limits of a species. The lack of any sharply defined criterion to inform us when real and observable essences are in conformity confines us, Locke argued, to a permanent doubt in our classifications. Given that a gap might exist between the image of a plant created by our senses on the one hand and the real essence on the other hand, the possibility of identity between image and essence is only one of many possibilities. Given this, the only means of classifying lies in an empirical approach: we must simply collect numerous observations, and from the gradual accumulation of information in this way, our knowledge of nature will gradually improve. All observable characters have the same weight in classification, and there is no possibility of weighting one character as being more indicative of the real essence than another one. As all we can know about any entity is through its influence on our senses, the feelings we have regarding the texture, size, shape, taste and odour of a plant each have the same weight in forming the image in our minds as, say, its number of leaves. Lacking, as we do, any knowledge of the real essence, we may replace the traditional principle of classifying plants by one single feature all the way down the classification hierarchy, from kingdom to species, with an approach that permits changing the observable feature at each step as we descend the steps of the classification hierarchy. For example [and using current taxonomic ranks], feather colour to distinguish among classes, diet to distinguish among families, number of eggs to distinguish among genera and feather colour again to distinguish among species. Classification is a matter of the empirical domain, and the only possibility of determining the real essence and constructing a natural classification necessarily rests on a consideration of the total complex of observable features under consideration (Kochiras 2017).

Locke's theory undermined the Aristotelian foundations of taxonomy, in which the key assumption was that knowledge of the real essence was available. There was a serious philosophical disagreement as to whether observable features represent hidden real essences of natural groups and whether a single character implies it. This created a stormy intellectual controversy in the late seventeenth century regarding the classification of plants and how and which key characters to use (Sloan 1972). We shall now descend from lofty philosophies to down-to-earth biological classifications.

We recall the English naturalist John Ray and his *Ornithology* (1676), in which he described and illustrated each bird species. In his *Ornithology,* he presented a key for the reader to find his way easily among the various bird groups. Descending from the highest taxonomic level to the lowest ones, his key "bounced" from one character to another, including every possible feature: habitat (land or water), bill shape (curved or straight), type of food (meat or seeds), time of activity (day or night), character (aggressive or fearful), wing shape (short or long), flesh colour (pale or dark) and clutch size. The mere fact that Ray did not adhere to a single character throughout this key was a statement that natural classifications need not fit rigid norms of essence when descending from high to lower ranks of classification.

A decade later, Ray turned to the classifying of plants (*New Method of Plants* 1682, *Historia Plantarum* 1686). He extensively used dichotomous keys, and, as in his classification of birds, he did not stick to one character throughout (Sloan 1972). First (Fig. 5.19), he divided all herbaceous plants (namely, not including those that are trees) according to one character, the presence of flowers and seeds: the *Imperfectae* (algae, fungi and lichens) lack flowers and seeds versus the *Perfectae* that possess them. He then divided the *Perfectae* according to another character, the seed, into very small or large. He next divided those with large seeds according to a third character, the number of seed leaves in the embryonic plant, into bi-foliate or single-leaved. The single-leaved (our monocotyledons) he then divided according to a fourth character, flower perfection, into not-perfect or perfect (defined by the presence or absence of conspicuous petals on the flower). He then divided those with perfect flowers by a fourth character, the root, into bulbous or tuberous. The rest of the classification focused on the plant's odour, details of the root structure and the arrangement of leaves on the stem. Thus, as in his classification of birds, Ray ignored the approach of a single essential character as being at the heart of the plant classification. Furthermore, he did not attempt to use in his classification only the "number, location and shape" of the parts of reproduction, as Cesalpino had advised. By this, Ray's new system of botanical classification differed significantly from the traditional classifications since the Renaissance days of Cesalpino. Ray's plant classification began, what we may term, a classification war.

It started in 1690 when Augustus Rivinus (1652–1723, Fig. 5.20), professor of physiology and botany at the University of Leipzig and curator of its medicinal garden, published his own version of a classification for the plant world, *Ordo Plantarum* (1690). From top to bottom, it made use of only a single essential character—the flower, with its petals and sepals. By use of dichotomous keys, he divided plants into 18 higher groups, which he called higher genera (*genus summum*), based on the regularity and number of petals and on the fruit (whether bare or surrounded by a dry or fleshy pericarp). He divided all plants (herbs and trees) into those with imperfect flowers versus those with perfect ones. Then, he divided those with perfect flowers into those with compound flowers versus those with single ones. Next, he divided those with single flowers into those with a regular corolla versus those with an irregular one. Finally, he divided each of these into whether they have one, two, three, four, five, six or many petals. This strict use of the structure of the flower only resulted in a classification very different from the classification that Ray had presented (Fig. 5.21).

Criticized this classification of Rivinus by claiming that we should construct natural classifications in which, on the one hand, close genera should not be split, and on the other hand, distant genera should not be mixed. Daffodil and tulip form a natural group of bulb flowers according to root characters; yet Rivinus had split them into two groups: daffodil in his Monopetalia and tulip in his Hexapetalia. Likewise, he blamed Rivinus for creating a mixed artificial group by placing the daffodil together with the bindweed. Further, Rivinus had ignored the traditional

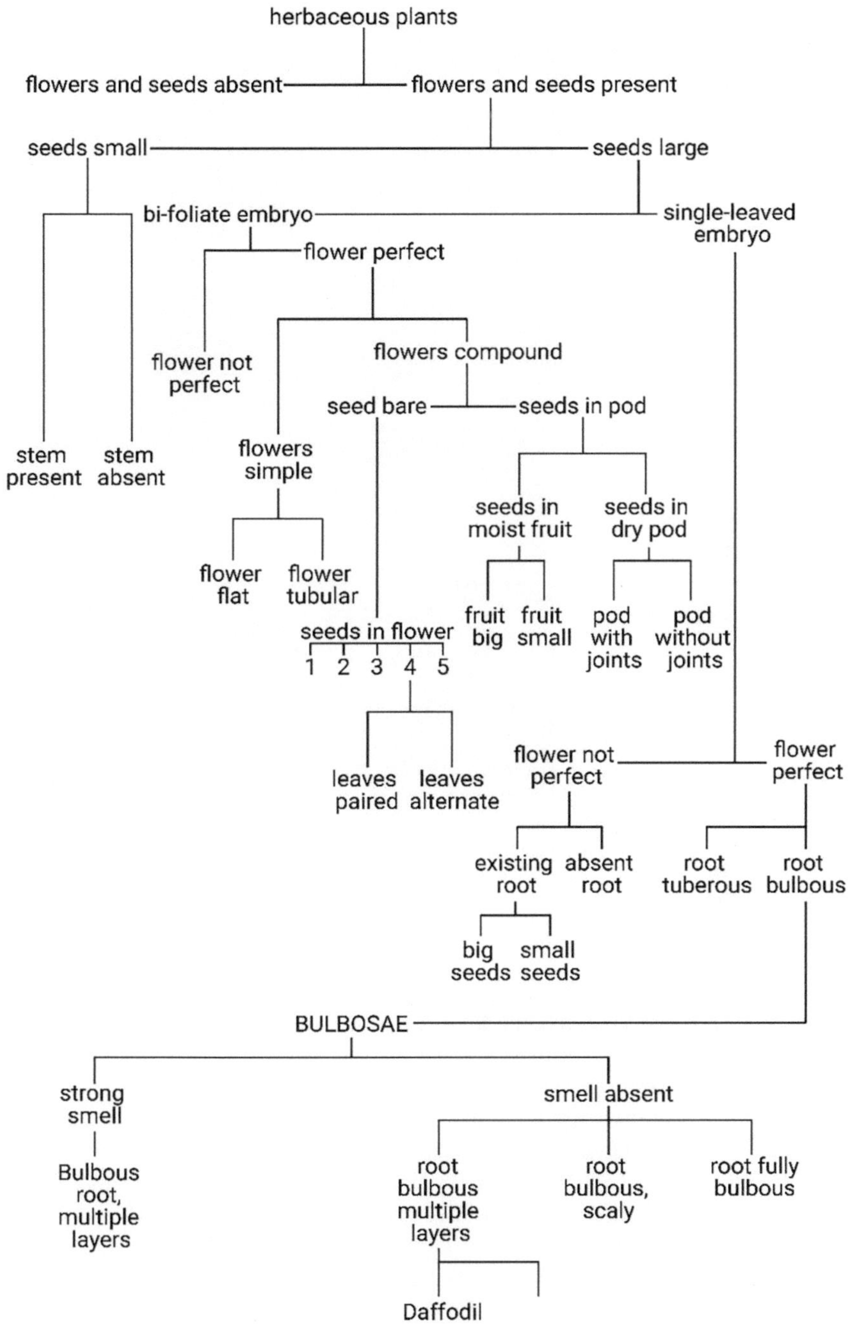

Fig. 5.19 Ray's classification of herbaceous plants, here down to the daffodil. Notice the many different features in use (presence of flower, seed size, number of embryonic leaves, flower perfection, root structure, smell). Elsewhere in this classification, other features are also in use: flowers (compound or simple, tubular or flat), seeds (in moist fruit or dry pod), fruit size and pod (with or without joints)

Fig. 5.20 Augustus Rivinus, dean of the Faculty of Medicine and member of the Royal Society

division into monocotyledons and dicotyledons. Rivinus counter-criticized (1694): it was Ray himself who was mistaken, in that he had grouped daffodils and tulips together—he was forming a similarity based upon the structure of the roots, leaves and cotyledons, thereby following an outdated tradition. He, Rivinus, was the one seeking progress in botany, and he was constant in his use of the essential characters of the flower and the fruit—unlike Ray, who had abandoned essential characters whenever it was convenient for him. Ray responded (*Synopsis Methodica Stirpium Britannicarum*, 1695) that we do not know which of the characters are essential. Therefore, instead of using only essential characters, we must use other characteristic features found in all specimens of a given group. Among animals, for example, essential features in the quadrupeds could be fur, four feet, and terrestrial habits versus fishes with no fur, no feet and aquatic habits. Now the whales, Ray noted, though resembling fishes in shape, in lacking of feet and in being aquatic, we should

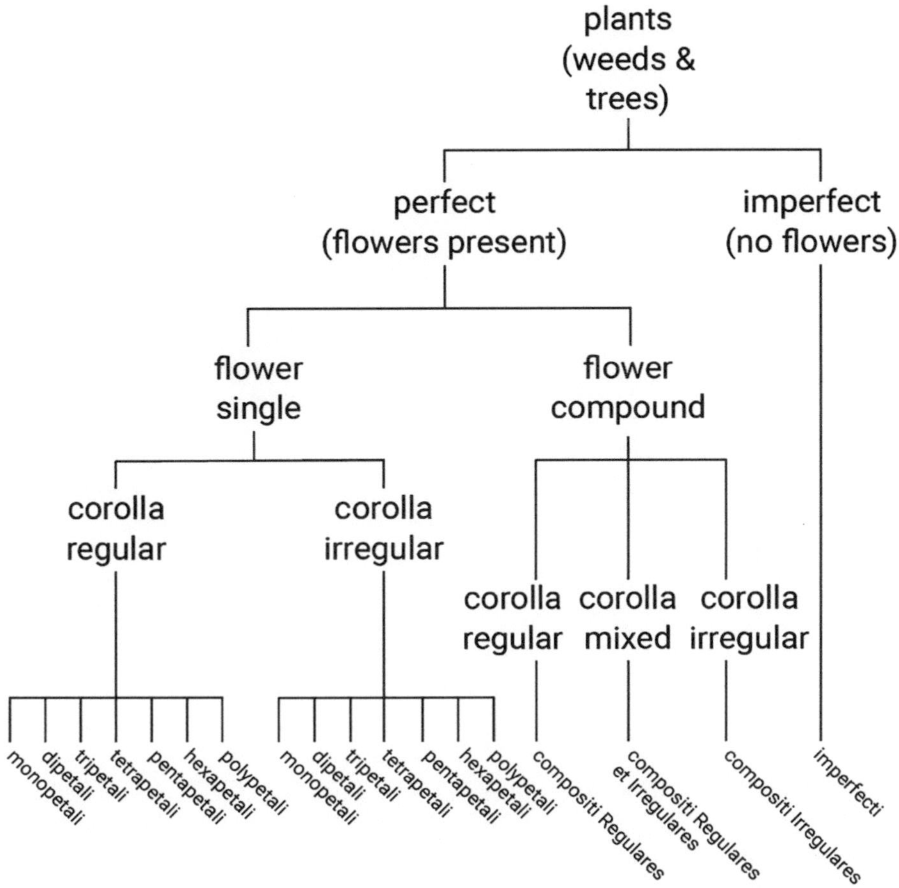

Fig. 5.21 Rivinus' classification of plants. Notice his reliance on flower features only

best place them with the quadrupeds, which they resemble in numerous other features, such as brain structure, respiration organs and mode of generation. By placing the whales near the quadrupeds, the resulting system would have fewer anomalies and exceptions. Back to plants, he went on to say that when he had created the 'tuberous herb' group, he was considering not only the type of root but also the process of germination, the texture and form of the leaves, the shape of the leaf borders and the details of the seed capsule. All these features point to a relation between tulip and daffodil (Sloan 1972).

The conflict escalated when Joseph Pitton de Tournefort (1656–1708, Figs. 5.22 and 5.23), Demonstrator of Plants at the Royal Botanical Garden in Paris, entered the arena. He claimed that we should group similar plants together and separate those not similar (*Eléments de botanique, ou Méthode pour reconnaître les Plantes* 1694). We should deduce plant resemblance from the fact that resemblances in structure between the flowers and fruits of different plants are more perfect and

Fig. 5.22 Tournefort

more striking than those between other parts of plants (Sloan 1972). He accordingly based his system of classification, from top down to the level of the single genus, on **only one** of the plant parts, on the flower: petals and sepals. He disregarded other characters. He described species, though, by characters of their leaves. He argued that the features of flowers, seeds and fruits have received the endorsement and common consent of the botanical tradition for establishing plant groups. As to the possibility that no essential character exists, he relied on a divine guarantee: God has impressed a common character on each of the species, which must serve us as a guide for ranging these in their natural places. Tournefort's approach was thus similar to that of Rivinus, and he regarded Ray's classification as "based on convenience alone and not on penetrating botanical vision, which requires far more precision". What a modest, noble, generous and non-selfish comment.

Ray's response (1696) leaves no room for doubt that he was now recruiting Locke's *Essay* as his philosophical basis. He argued that the essence of things is unknown to us, and if the essence is something not concrete, our senses will not receive it. Consequently, it is not possible to argue that flowers and fruit are the essential organs in a plant. As we do not know the essence, it is necessary to take *several* characteristic features as representative of the essential feature. A collection of as many characteristics as possible, which are common to all the species in the genus, is what should be the criterion for the genus. This may include odours,

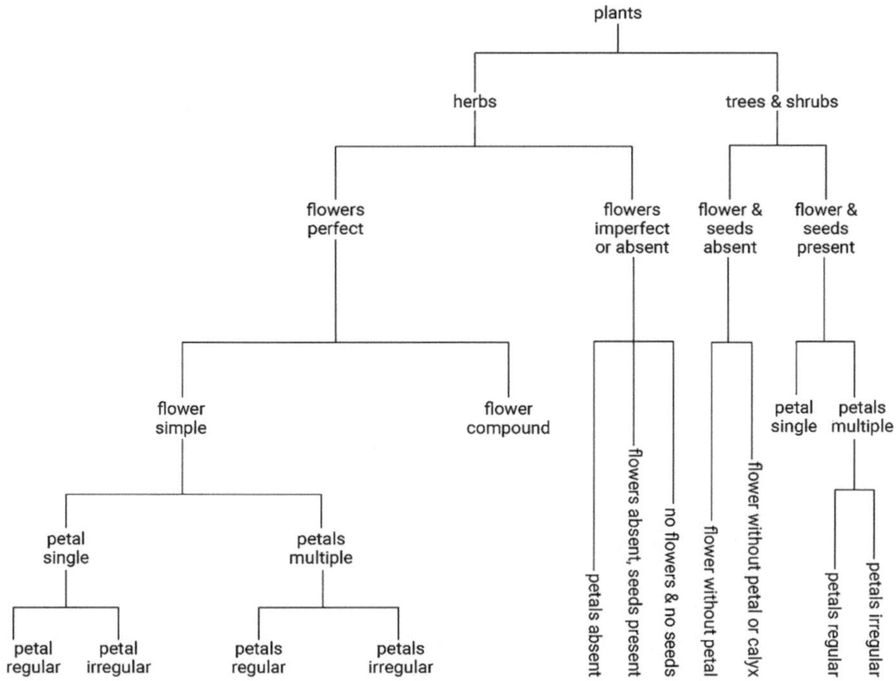

Fig. 5.23 Tournefort's classification of plants

colours, tastes, textures, medical properties and even ecological habitats for defining plant groups—in direct contradiction to Cesalpino's classification.

Tournefort answered Ray broadly by repeating his opinions; within a short time, Ray died, and Tournefort died 3 years later. The study of diversity in the eighteenth century thus began in intellectual confusion, with botanists disagreeing about the basic philosophy of classification and consequently creating very different classifications (Sloan 1972). Whereas Ray presented 38 main groups of plants (in a revised text of 1703), Tournefort presented 22.

In addition to this controversy concerning the essence and key characters in the classification of animals and plants, the early eighteenth century was characterized by considerable chaos also in the number and names of the higher classification ranks, those above the level of (what we today term) genus and species. Some naturalists used numerous high ranks, while others used few, and the terminology for these higher groups was not uniform. In some cases, the terms "order" or "class" were used; in other cases, various other terms were used, such as "kind" (the "duck kind" and the "raven kind" by Ray). In still other cases, a number of genera clustered into a larger group, which was also termed a "genus". The highest groups, such as mammals, birds and fishes, were sometimes termed "gender". Each gender is split into many species, with the term species representing both a single species and a large group of species, such as the "ruminant species" in the mammal gender

(Boyle). Occasionally, relying exclusively on the term "genus" to the point of confusion, groups above the rank of the genus were termed "super genera" *genus sum*, "intermediate genera" *genus intermedium*, and "a genus of close species" *genus proximum*. There were also cases in which classification rested on internal characteristics, which is indeed very reasonable—but this classification was limited to a handful of researchers who could obtain access to those internal features. Classifications by, say, blood colour, the number of heart chambers, digestive organs or reproductive characters—disregarded the fact that for most scholars, foreign animals were available only as skeletons, or in the best case, as a skeleton and a fur (or a plumage or a shell). It was this reality of confusing classifications, taxonomic hierarchies and nomenclatures into which Linnaeus, the founder of modern classification, was born.

5.2.2 Linnaeus

5.2.2.1 Linnaeus' System of Classification

Carl Linnaeus (1707–1778, Fig. 5.24), the son of a plant-loving priest in Sweden, was sent at the age of seven to a school for aspiring priests and government clerks. The curriculum included seventeen weekly hours of Latin, 14 h of theology and 9 h of science (mathematics, physics and logic). Linnaeus spent 13 years at the school, but due to his poor achievements, his teachers prevented his progress to priesthood or government. Fortunately, a local physician believed in the youth and his ability and taught him botany and the principles of herbal medicine. He also acquainted him with the studies of one of the assistants in the King's Garden in Paris, a certain Sébastien Vaillant, who had claimed (1717) that a "vital soul" originating in the pollen grains penetrates the style and reaches the ovules in the plant's ovary. This had a tremendous impression on the developing imagination of the young Linnaeus: plants have a sexuality, comparable to that of humans! In years to come, Linnaeus would elaborate on this comparison extensively, and details of the stamens and the style would be the basis for his sexual classification of plants into orders and genera.

Aged 20, Linnaeus began medical studies at the University of Lund. Here, one of his professors offered the indigent Linnaeus lodging, food and use of his library and his herbarium, the likes of which Linnaeus had never seen before. He soon began a herbarium of his own and attempted to organize and identify plants according to the literature available at the time (mainly that of Tournefort). After 10 months at Lund, Linnaeus moved on to study at the University of Uppsala, where, while he himself was still a student, the professors discovered his botanical tendencies and employed him to demonstrate and lecture on botany to medicine students. Physicians at the time were required to concoct medicines from plants, so botany was a part of the medical curriculum. In addition to lecturing on plants, Linnaeus restored and expanded the neglected botanical garden and attempted to put it into some order. Also at Uppsala, he met Peter Artedi, a medicine student who, like him, was

Fig. 5.24 Linnaeus

interested in nature. The two students became close friends and planned to divide the entire classification of nature between them: Artedi was to classify fishes, amphibians and reptiles, and Linnaeus the birds, insects and plants. Upon completing his medicine studies, Linnaeus rode north to study nature in Lapland. For 5 months he rode from one place to another and studied and collected plants, animals and minerals, thus becoming acquainted with a new world unknown to him. His trip to Lapland was another significant event in his life because he now had to classify and organize the items he had collected. How was he to go about that?

Between one thing and another, Linnaeus fell in love. However, the father of his bride agreed to the match only on the condition that Linnaeus would prove his ability to support a family by completing his medical degree abroad. Therefore, 1 day Linnaeus, together with Artedi, set off from Sweden to the University of Harderwijk (in Holland today), where students could obtain a doctorate in just 1 week and for a very low fee. One needed only to present an original medical thesis and then pass an oral examination. Linnaeus had brought with him from Sweden a thesis he had prepared, "A new hypothesis on the cause of intermittent fevers", in which he hypothesized that some cases of high fever result from the fact that the patients live on clay soil. A mere 6 days after arriving at Harderwijk, he became a Doctor of Medicine.

Linnaeus remained in the Netherlands for 3 years, during which he was responsible for the local botanical garden, the greenhouse and the zoological garden. One night, Artedi accidentally drowned in one of the canals of Amsterdam. He had left a detailed monograph on fishes, which Linnaeus promptly edited and then published (1738). Eventually, Linnaeus returned to Stockholm, where he practiced medicine so successfully that his fiancée's father permitted the wedding. Aged 32, he became president of the recently founded Swedish Royal Academy of Sciences and physician to the admiralty. Aged 36, he received an appointment as Professor of Medicine and Botany at the University of Uppsala. His lifelong record as a university teacher must surely be unique: 23 of his students became professors in branches of natural history or medicine and thereby popularized his methods and nomenclature. Those he inspired to travel far afield reached the Americas, tropical Africa, the Levant, India, Sri Lanka, Indonesia and Australia. When 51, the king of Sweden knighted him as Knight of the Polar Star (a Swedish order of chivalry), and when 54, the king ennobled him as Carl von Linne. Linnaeus, the father of modern taxonomy, was far from modest. Adam had named many of God's creatures, and Linnaeus saw himself as playing a similar role. In his writings, he referred to himself in the third person. He wrote that God himself had given "him" the opportunity to look into his most secret cupboards and, in his infinite mercy, had allowed "him" to see more of his creatures than had any other before "him". God had bestowed upon "him" more insight into nature than upon any other mortal before "him"; God had made "his" name one of the greatest on earth.

Linnaeus had a tremendous obsession to classify, taking the classification of nature as his personal mission: nature is the plan of Creation, and God himself had guided him, Linnaeus, in his search for a *Systema Naturae*—a comprehensive, inclusive system for classifying the world of minerals, plants and animals in nature. Already when travelling to Holland, he had brought with him a manuscript, *Systema Naturae*. In 1735, it was published as a short book of just 11 folio pages, "in which the three great kingdoms of nature are presented, classified into classes, orders, genera, and species".

Linnaeus' belief in an economic role in natural history might have influenced his research agenda. Modern students of the history of biology have claimed that his great classification project was not a purely intellectual enterprise; rather, he was actually seeking medicinal and economical uses of Swedish plants and animals for the benefit of Sweden. Indeed, he lamented about Sweden importing such luxury goods as coffee, tobacco, tea and silk, believing it should be possible to cultivate the same plants in Sweden or, failing this, to find native substitutes. He speculated that one could perhaps grow silk-moth caterpillars, *Bombyx mori*, which feed on white mulberry, *Morus*, in their native countries, on local nettle, *Urtica*, in Sweden; likewise, he attempted to cultivate pearls (Koerner 1999). However, whether national economics was indeed the major motive behind his grand classification project is another question. Most collecting and classifying people I know of do so as an uncontrollable passion. We collect and we classify; we cannot help it. Linnaeus had a tremendous obsession with classifying anything within his reach. Beyond nature, he even classified types of bread.

In here discussing Linnaeus and his system of classification, we refer both to his first, 1735 edition of *Systema Naturae*, where he first presented his concepts for bringing order into nature, and to later editions. Before presenting the minerals, plants and animals, he offered a general outlook on nature in what he termed "observations on the three kingdoms of nature". In observation (§8) he declared that

> Natural objects belong to the field of the *senses* and they are obvious to our *senses*, wherever we turn. This is why the Creator put man, who is provided with *senses* and intellect, upon the earth, where nothing met his *senses* but natural objects, constructed by means of such an admirable and amazing mechanism [my emphasis].

We can rely on our senses to classify the world of nature. We recall Descartes and Boyle and their claim that information retrieved by the senses is not valid and Locke, who had believed in the limited value of knowledge based upon the senses. Linnaeus is here stating that the Creator has given us senses so we can admire his creations and that this is true knowledge. Linnaeus also suggested that if only it were possible to identify *essential* characters, one would need only those essential characters. However, as no appropriate method exists to identify them, one must also use other characters, which "distinguish a genus from all other ones in an artificial order". He believed that a genus' natural definition gathers every possible generic feature, and it, therefore, includes essential alongside other features (*Philosophia botanica* 1751: §186–190). Linnaeus was thus a naturalist of available features that he considered essential (Cain 1958, 1981; Winsor (2006a, b) suggests that Linnaeus, in using the term "essential", meant merely "taxonomically useful").

A point of cardinal importance in Linnaeus' system was his insistence on a verbal description for each group. Some of the previous literature focused mainly on drawings of the genera and species and, lacking the necessary descriptions, could not establish the solid foundation required for grouping genera into higher groups (say, Lister and his book on molluscs). We recall that Linnaeus had studied logic. He might well have derived this basic principle of a clear and distinct description from Descartes, whether directly or indirectly. Descartes had sought to base all human knowledge on unshakeable foundations and argued that for us to know something, this something had to be clear and defined. A clear idea was one on which no doubt could be cast; a distinct idea was one that was separate from other ideas. Anything that is clear and defined is necessarily true, potentially or actually. The clearer and more distinct our ability to distinguish and understand, the greater the level of certainty. The knowledge used to judge something has to be both clear and distinct (*Principles of Philosophy* XLV: 237; clear is the opposite of unclear, and distinct is the opposite of confused). Linnaeus applied this philosophical concept of Descartes to his classification of nature: The description of the different groups must be clear and distinct. He declared that "Distinct ideas, terms, rules—flow from a defining mind. Everything emanating from a confused mind is confusion." Further, he wrote that "I present my request to you: that you should have distinct types, distinct ideas, and distinct names. If you introduce confusion into the names the ideas will be confused, and this is true for the genera as well, and if the genera will be confused, everything will be confused." (*Critica Botanica* §225). He also wrote that "To acquire knowledge regarding the wonders of the world it is necessary to join together a distinct idea and a distinct name" (*Species Plantarum* 1753).

Linnaeus' classification was a descending one: he began at the very top, with the entire universe, which he termed the Natural Empire (*Imperium Naturae* § 5, 6). The universe includes everything existing, everything we know from our senses. The stars are distant, glowing bodies in constant motion. They illuminate, either by their own power (the Sun and stars) or from the light of others (the planets Saturn, Jupiter, Mars, Earth, Venus and Mercury). The four elements (fire, air, water and earth) comprise the atmosphere and fill the spaces between the stars. From the heavens, he descended down to Earth, which had an atmosphere of elements and a crust of natural bodies enveloping it. These natural bodies are classified into three large kingdoms, the boundaries of which meet at the corals (*zoophyta*):

- Minerals: the body is not organized; they have no life and no senses.
- Plants: the body is organized; they have life, but no senses and no ability to move.
- Animals: the body is organized; they have life, senses and mobility.

Having thus declared that the Earth consists of three kingdoms, he next divided each kingdom into classes. Continuing his "'top-downwards" philosophy of systematics, he divided each class into orders, divided the orders into genera and divided generainto species, which were composed of the most homogeneous groups that one could observe.

Linnaeus, like the Greek philosophers, saw the three kingdoms as a natural ladder. It rises from the various mineral groups through the various plant groups to the various groups of animals, with man ranking the highest. The Creator had progressed ever so gradually from the simple to the more complex; nature does not progress in leaps and bounds. To us in the twenty-first century, the transition from mineral to vegetable may seem a giant step, but Linnaeus, in eighteenth-century Sweden, saw things differently. He claimed that minerals are the product of a "marriage" between "paternal" salts and "maternal" sands in a constant state of mating, resembling hybridization between groups of plants and animals. Further, he believed that some minerals (quartz) are parasites since they form and dwell within the spaces of other rocks. He thought that plants and animals come from minerals, with water added. He assigned fossils to the kingdom of minerals, class Lapideum and order *Petrificata*. The *Lithophyta* (corals) belong to the plant kingdom, not that of minerals or animals, because of their shape and structure. Decades after Linnaeus, Bonnet graphically presented the animal kingdom in this natural ladder.

The Linnaean classification was based exclusively on external features, complemented, where relevant, also by teeth, as these were hard structures and therefore durable in collections. He did not use internal features, which can be reached only by dissection. Wherever possible, he based his classifications on variations in a single or a few characters. In plants, this would be some part of the reproductive apparatus; in quadruped mammals, the teeth; in birds, the beak and in insects, the wings. For all three kingdoms of nature, he adopted a uniform hierarchy of only four ranks: class, order, genus, and species. Each of his groups received a name and a definition, thereby distinguishing it from other groups. Linnaeus thus created a fully ordered four-tiered ranking system, resembling the church, army and civil service. The (often not intuitive) order was the link connecting the few intuitive classes to the large number of intuitive genera and species. For example, the intuitive class quadrupeds (our mammals) are linked to the intuitive genera porcupine, squirrel and mouse via the order Rodents (his *Glires*). In this taxonomic hierarchy, every species can be located within a higher-level taxon all the way from genus to kingdom (Ogilvie 2006).

Linnaeus emphasized that his classes and orders were artificial devices, invented to ease identification. This uniform method of four hierarchical ranks, artificial and unoriginal as it may be, when applied to the entire natural world proved very useful for identifying, finding and compiling information. Using this system, anyone seeking to identify any group in the world would easily achieve the same results as Linnaeus did, once he had mastered just a small number of external organs. The terms genus and species, originally having relative meanings and with a shifting reference at all taxonomic levels, were thus narrowed down by Linnaeus and fixed to express only one absolute rank each on his classification scale.

Further, Linnaeus also imposed a uniform language on the classification: Latin, the pan-European ancient language familiar to virtually all intellectuals of his day, no matter what their mother tongue (we recall his intensive study of Latin while at school. I wonder whether he had perhaps considered Latin as the essence of the various modern European languages.) As to the formalistic linguistic aspects of nomenclature, in some animal groups (birds, fishes), centuries before Linnaeus, the name of a species was composed of just two words as it is today, one descriptive term for the genus and one for the species. The use of such two-word binomials is a very old and almost universal manner of naming, rooted in the need to distinguish the general and the particular. Thus, in the native language of Hawaii, the cowry was named *Leho*, a term that included nine subgroups (species): *L. kupa, L. lei, L. paa'a, L. panko, L. opela, L. kolea, L. nuho, L. pisteko* and *L. kiko*—almost all of which correspond to scientific classification. Similarly, the term for a limpet in native Hawaiian was *Opihi*, and four species were recognized: *Oliarus koele*, *O. alanalina*, *O. makai* and *O. awa*. The natives of Hawaii thus used one name for the general, the genus, and one for the particular, the species. In contemporary European classification, this was often the case, but not always. There was one descriptive term for the genus and then, after it, often came a lengthy sentence describing the species. A snail species of England was named *Cochlea citrina aut leucophaea non raro unicolor, interdum tamen unica, interdum etiam duabus, aut tribus, aut quatuor, plerumq; vero quinis Fasciis pullis distincta*—23 words (Lister 1678; probably *Cepaea nemoralis* of today). Likewise, Lister's name for an English spider species was 19 words long: *Araneus cinereus, alvo circiter sensis fasciis transversis, in angulos acutos in medio erectis, argenteis & nigris alternatim dispositis insignita*. In botany, the situation was not much better, and further, some of these sentence-long names were not adequate. Bauhin (1671), in his *Pinax theatri botanici*, named a species as "the other one", thus *Frumentum Indicum Mays dictum alterum*, "the other Indian Grain called Mays". He also named one plant species as true and another one as false, thus *Acorus verus* versus *Acorus adulterus*, leaving the reader to decide how to tell the truth or falsity of a plant by looking at it. Then, there was *Gramini affinis planta ad gutturis inflammationes utilis*, "a plant, related to grass, useful for inflammations of the throat". With such cumbersome and often inadequate species names and definitions, it became difficult to discuss, correspond and publish aspects of nature that concerned the species level (Cain 1994).

To cope with these problems, Linnaeus placed a sharp distinction between a creature's name and its description. Linnaeus was the first to apply binomials to *all* plants and animals uniformly. He (1753) imposed the binomial system on all three kingdoms: we should write the name of each genus with a capital letter, for example, *Hystrix* for the porcupine and *Sciurus* for the squirrel. The name of each species within a genus receives only a single descriptive epithet that follows that of the genus but with an ordinary letter (not a capital one), and one may shorthand the name of the genus to its capital letter. Thus, in the porcupine *Hystrix*, the five species are named *H. cristata, H. prehensis, H. dorsata, H. macroura* and *H. brachyura*; and in the squirrel *Sciurus*, the seven species are named *S. vulgaris, S. niger, S. cinereus, S. flavus, S. getulus, S. striatus* and *S. volans*. As each genus name is unique throughout the animal kingdom, and each species has its unique name within its genus, each species has its unique name in the entire animal classification system, owing to only two words. Linnaeus thereby brought the scientific nomenclature of plants and animals back to the simplicity and brevity of the vernacular nomenclature, out of which it had grown, back to the kind of names used from times immemorial by peasants, woodmen, hunters and herb-gatherers in many lands—folk names. Indeed, thanks to this rule of Linnaeus, which clearly separates the naming and descriptive functions, biologists of today can deal much more easily with biodiversity.

Within three decades of the publication of *Species Plantarum* and *Systema Naturae,* the Linnaean binomials, consisting of a one-word generic name associated with a one-word specific epithet, completely superseded the one- to many-word names. The effect of the Linnaean binomial method has been far-reaching and permanent. Linnaeus also determined further rules for giving scientific names: the genus name must derive from a Greek or Latin root, must not be composed of two words and must not memorialize saints or anyone who is not relevant to science. The separation between name and description, the use of clear, distinct and well-defined descriptions of external characters for each group in nature, the obligatory use of a single language and the use of a concise, uniform hierarchy with only four ranks, each with a unique name in a standard language, were Linnaeus' great contributions to biology. No wonder his method of classifying biodiversity soon became popular.

Linnaeus first published his 11 folio pages of the *Systema Naturae* in 1735. From then onwards he devoted his life to improving it, edition after edition, update after update. Fifty years later, the 13th edition (published posthumously 1788–1793) constituted a ten-volume work of 6,300 pages, a monument to the information upsurge precipitated by the taxonomic methods he had developed. We, today, consider his tenth edition of *Systema Naturae* (1758) as the benchmark of zoological nomenclature: names used in this edition are valid, and any earlier name, if it does not appear in the 1758 edition, is not valid. Likewise, botanists, by international agreement, accept his *Species Plantarum* (1753) together with his *genera Plantarum* (fifth edition, 1753) as the starting point for botanical nomenclature in general. Botanical names published before 1753 have no standing in modern nomenclature unless Linnaeus had adopted them in 1753. He did not aim his Latin, unillustrated writings for a popular audience, but within a few decades, people translated his books into many handbooks, which anyone could use.

Linnaeus' approach to the classification of plants was different from that of animals (we skip his classification of minerals).

5.2.2.2 The Plant and Animal Kingdoms

Summary Linnaeus believed that the mode of generation is a plant's essence. He classified plant diversity by whether sexual organs are exposed or covered, whether stamens and style are on the same flower or not, whether separate from the style or fused and whether all stamens are equal or not. He next created classes by the number, shape, and location of the male organs and orders by the female organ. Then he classified genera and species by other traits such as the root, stem and leaves; beyond this, he did not bounce between characters. This was, he believed, a natural classification, in contrast to not-natural, medical classifications. He classified animal diversity into six classes. Many of his orders were unnatural due to convergence. He listed only a few genera, perhaps believing that biblical Adam could memorize only a few names. Species are essential, unchanging entities created during Genesis; they cannot develop from other species and do not form spontaneously. In molluscs, he gave each genus one name characterizing its shell and a different name characterizing the animal inside the shell. Man is highly variable. Beyond mute, hairy savages, there are four races, one for each continent. They differ in colour, temperament and hair.

The most original classification of Linnaeus, and surely the most important in his eyes, was his so-called sexual method of classifying plant diversity. Indeed, his first edition of *Systema Naturae* (1735) may well have been intended mainly to promote his sexual method of classifying plants and thereby to propose a framework and example of how to classify the entire natural world. Following Aristotle, Cesalpino, Rivinus and Tournefort before him, he too believed that plants have two basic functions: feeding, which preserves the individual, and generation, which preserves the systematic unit, the species. (Naturalists at that time did not know of photosynthesis and of the function of foliage.) He too believed that plant classification should rest upon the reproductive parts, as had been accepted by those earlier botanists. Adherence to this principle was the very mark of those who sought to separate botany from the classifications of medical pharmacists and horticulturists. Two decades later, he would write that orthodox naturalists take their method of classification "from the true foundation in the fructifying parts", as these constitute the very essence of the plant, and all the signifying marks of this essence are to be taken from the "number, figure, proportion and location" of these reproductive parts (*Philosophia botanica* 1751: § 26, 167). However, he had an innovation: In 1717, Sébastien Vaillant, an assistant at the Royal Garden in Paris, had argued that in plants a vital soul from the pollen grains penetrates the style and reaches the ovules, located in the plant's ovary. Linnaeus, two decades later, took Vaillant's observation as the foundation of his plant classification. In order to know to which systematic group a plant belongs, it is first necessary to study its specific generative organs—the stamens and style. These are those "smallest elements in the flower, which previous botanists had never dared observe; and the data were taken from my own observations, having been convinced that these are the most *essential* organs" (my emphasis in bold throughout this section).

To me, it seems that Linnaeus was here following Aristotle's definition of essence as the property (or set of properties) making an entity into what it fundamentally is and without which it loses its identity. He was contrasting essence, the defining character, with accident, a property of the entity without which the substance still retains its identity. Though using the term "essence" as a useful taxonomic character, for Linnaeus this term also had a deeper meaning. Essence was a whole philosophy, from Aristotle, via Cesalpino to Tournefort, and from Boyle, via Locke to Ray. Thus (1735):

> § 6. The *essence of plants* consists in the fructification; of the fructification in the flower and fruit; of the fruit in the seed; of the flower in the stamen and pistil; of the stamen in the anthers; of the pistil in the stigma.
>
> § 7. Each fruit is proceeded by a flower; the *essence of the flower* consists in anthers and stigma, from which I took my method…

Linnaeus was fully aware that a plant is more than just its essence. Beyond the essential character, botanists should pursue the *characteres naturales* (1751: 143):

> Anyone who thinks that he understands botany from the *essential* character and disregards the natural one is therefore deceiving and deceived; for the *essential* character cannot fail to be deceptive in quite a number of cases. The natural character is at the base of plant genera, and no one has ever made a proper judgement about a genus without its help; and so, it is and always will be the absolute foundation of the understanding of plants.

In his classification of plant diversity (Fig. 5.25), Linnaeus largely expanded his own four-rank classification by creating four groups above the class. The "marriage organs" could be either naked, exposed and visible to all, or hidden in a membrane. Regarding first the naked plants, they may be either hermaphrodites with stamens and style in the same flower or flowers in which stamens and style are separate. The stamens can be separated from the style or fused to it. When separate, the stamens may be either equal or different.

Now, at this point, Linnaeus subdivided the plants with equal stamens into *classes* according to a single trait, the number of stamens: *Monandria*, with one stamen (*Andria* is Greek for husband); *Diandria*, with two stamens; *Dodecandria*, with 12 stamens; *Icosandria*, with 20 stamens; and *Polyandria*, with many stamens

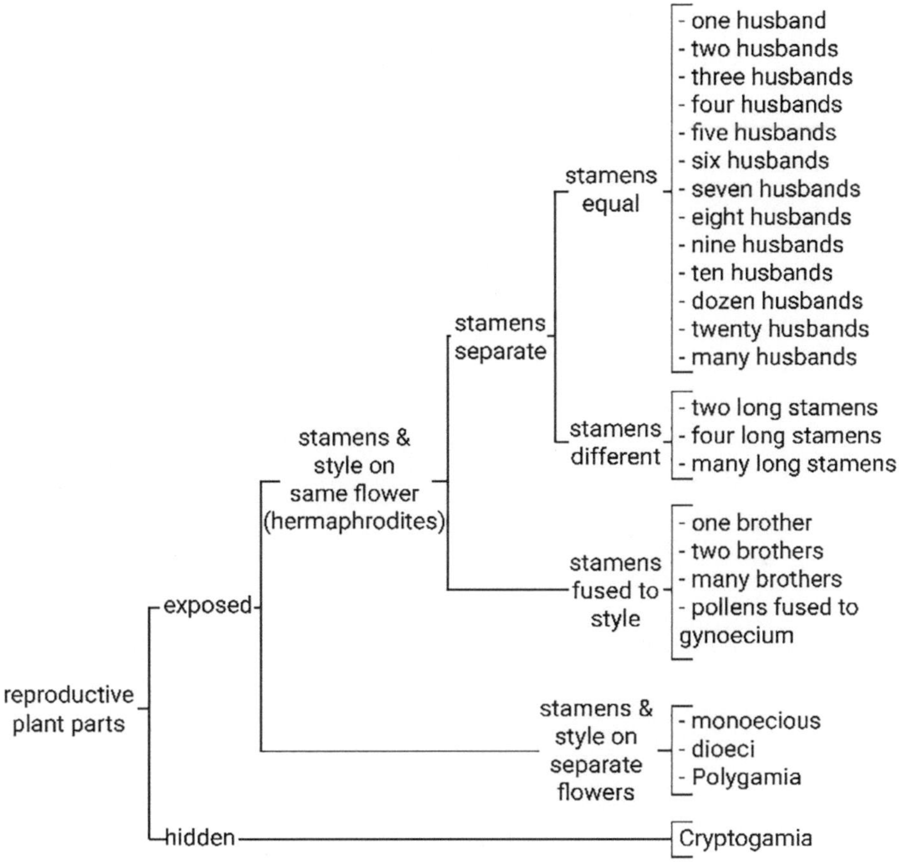

Fig. 5.25 Linnaeus' classification of plants

(=husbands). In those flowers in which the separate stamens are unequal in size, Linnaeus distinguished between *Didynamia*, with two elongated stamens; *Tetradynamia*, with four elongated stamens and *Polydynamia*, with multiple elongated stamens. When the stamens are fused, they form either one body (*class Monodelphia* – one brother), two bodies (*class Didelphia* – two brothers), or many bodies (*class Polyadelphia* – many brothers). If the stamens are fused, they grow together (*class Syngenesia*); if the stamens and styles are mixed, male and female (*Gynandria*). Where there is one flower for the stamens and another for the style (*Diclinia*), Linnaeus distinguished between separate flowers for the stamens and the style on the same plant, *class Monoecia*, or on two separate plants, *class Dioecia*. Cases where some flowers have stamens, others have styles and even others have both in the same flowers were termed *class Polygamia* by Linnaeus.

Flowers with cryptic reproductive organs, *Cryptogamia*, include ferns, mosses, algae, fungi, sponges and corals. In these groups, the stamens and style are irrelevant, and therefore their distinction from the various plant groups with stamens is basic and elementary.

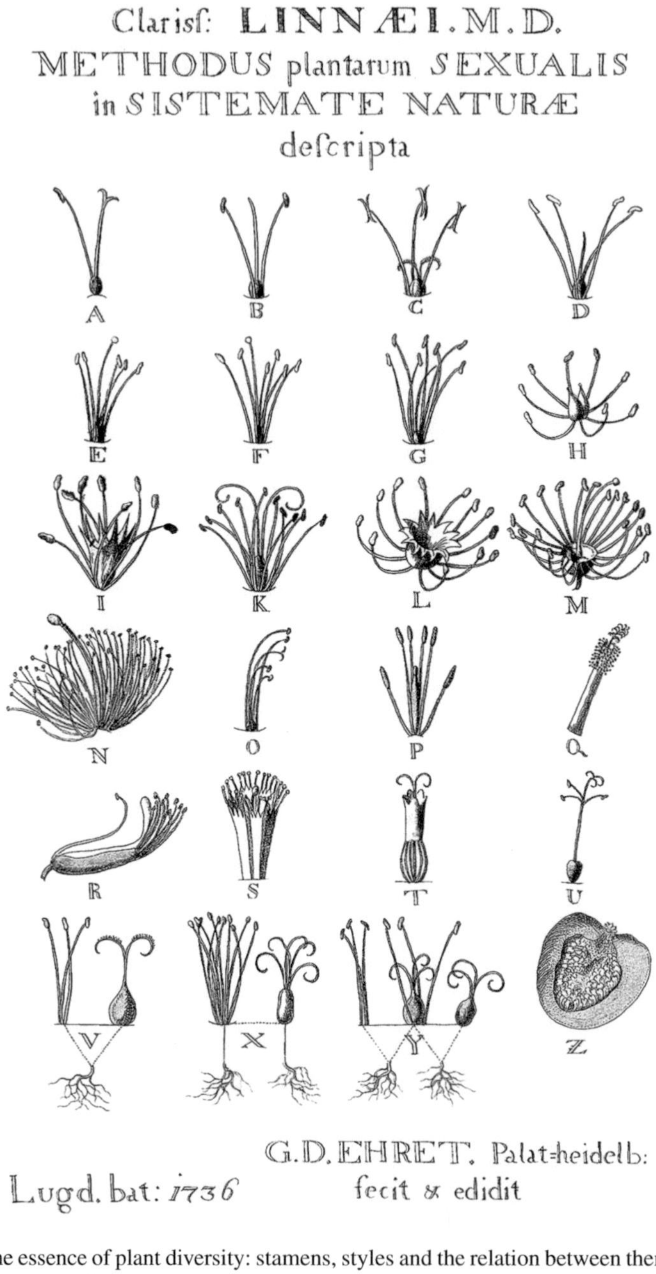

Fig. 5.26 The essence of plant diversity: stamens, styles and the relation between them. Notice the alphabetic arrangement

In this manner, Linnaeus classified all plants into 22 classes according to the placement and number of the male sexual organs (stamens) in the flower. Why 22 and not, say, 27? As his table suggests (Fig. 5.26), a letter marked each *class*, and the number of letters from A to Z was 22 (Fig. 5.26; in those days, J and W were not official parts of the alphabet).

Once Linnaeus had completed this classification of classes according to the features of the stamens (male organs), he subdivided each class into *orders* according to a single trait: the number of styles and the female character. Consequently, in the table, for each *order*, we find a subgroup termed *Monogynia*, which has one wife (style); *Digynia*, which has two wives; *Trigynia*, which has three wives and so forth. In this manner, Linnaeus subdivided the 22 classes into 65 orders based on the number of female sexual organs (styles) in the flower.

Having completed the classification of the orders, Linnaeus went on to classify the *genera*. Here he made use of the other traits of the flower and fruit: flower petals, leaf cup, outer fruit layer, seed, fruit base and so on. Again, he stressed the importance of essential characteristics that give a genus its nature and distinguish it from all other genera in the same natural order. For Linnaeus, God had also determined the number of plant *genera* in nature at the Creation. God had created each genus during the 6 days of Genesis; therefore, one must not split or unite genera, whether arbitrarily or by following one theory or another.

If all characters of the flower and fruit serve to distinguish between classes, orders and genera, what characters shall we use to distinguish between *species*? Here Linnaeus drew from the other vital function of the plant, that of feeding. He used traits of the root, stem and leaves. "The concept of a Species consists of an *essential* feature, by which alone it is distinguished from all others in the same genus [….] A specific definition contains features in which the species differs from those in the same genus. But the specific name contains the *essential* features of the definition [….] Therefore, the specific name is the *essential* definition" (Linnaeus 1751: 219–220).

Linnaeus' sexual classification system of the plant kingdom was very attractive. To reach the right class and order, one needed only to count the flower's stamens and pistils; here, to place it in its right genus, one needed only to examine a few additional characters of the flower and fruit. Linnaeus had thus created a simple and very applicable system for the scientist, the amateur, the gardener and the tourist. Further, rather than use morphological terms such as stamen and pistil, he chose Greek terms for husband (*andria*) and wife (*gynia*) as reflecting the various different types of "marriage" in plants—a large number of husbands, of wives and so on. He described the *Monandria* as like "One husband in a marriage", the *Diandria* as "Two husbands in the same marriage", and so on to the *Polyandria* with "20 males or more in the same bed with the female" (poppy, linden). He paced the marigold *Calendula*, in which the flower head has sterile disk florets and fertile ray florets, in the *Syngenesis Polygamia Necessaria* (Confederate males with Necessary Polygamy), where "the beds of the married occupy the disk and those of the females the circumference; the married females are barren and the concubines fertile".

The classification system of Linnaeus, from the level of the kingdom down to the level of genus, was thus an expression of the marriage relationships in plants—the reproductive system held, for Linnaeus, the essence of the secrets of the blueprint of the Creator. He further noted that the flower petals are not part of the reproductive process. They serve, as he wrote in charming poetic style, only as a "bridal bed", which the Creator had arranged beautifully, decorating it with noble curtains and providing it with pleasant perfumes so the groom can celebrate his nuptials accordingly. Once the bed is set, the groom will embrace his bride and offer his gifts: the anthers open and disperse the reproductive pollen, which falls deep into the tube and fertilizes the ovary (*Syst. Nat.* X ed. Vol. 2: 837).

As a physician, Linnaeus also presented the medicinal virtues of his groups: different plants cause tissue contractions, while others relieve scurvy, increase urine production, promote mucus secretion, soothe the skin, function as aphrodisiacs or are poisonous. I am not competent to go in further depth into Linnaeus' plant classification.

Linnaeus' classification of animals differed from that of plants in that it was mostly not original. In birds, he relied heavily on Ray; in fishes, on Artedi and in molluscs, on Lister. He was not aware of convergence, and he was, of course, not aware of evolution. I am not aware that he used the term "essence" in his classification of animals. Linnaeus designated six classes at the top of the animal kingdom, which he had found in the existing monographs in the literature of the time: quadrupeds, birds, amphibians–reptiles, fishes, insects and worms. Five of these six classes were intuitive, large, natural groups, known to humans for ages. The innovation Linnaeus introduced was that he gave them all a uniform rank—the class—each of which had typical external characters:

Quadrupeds	hairy, four legs, the female bears live young and suckle them.
Birds	feathered, two wings, two legs, bony bill, the female lays eggs.
Amphibia	body naked or scaly, no molars, no external ears [including reptiles]
Fishes	body lacking feet, naked or with scales
Insects	body crustaceous, jointed antennae [actually all arthropods]
Worms	body muscular, antennae not jointed [including molluscs].

In Linnaeus' very short hierarchy of only class, order, genus and species, it was difficult to cram the complexity of animal diversity, which, in the real world, has many more ranks. To solve this problem, sometimes Linnaeus simply omitted certain natural groups. He enumerated the classes of quadrupeds, birds, amphibians–reptiles and fishes but left out their natural grouping into the super-class Vertebrates, known since ancient times as the red-blooded animals.

A distinct and clear description for each group—which features should he use to describe each order? Each order is distinguished according to characters—ideally one, but if not possible, then two or even three. The insects, for which we have classifications preceding Linnaeus, illustrate his use of single characters for defining orders. We recall that during the late Renaissance, Aldrovandi had placed all terrestrial invertebrates into six groups, which he had classified by their legs, wings and readiness to take to the air: bees; dragonflies, cicadas and butterflies; flies and

Fig. 5.27 Frisch in his study. (Adapted from Winsor 1976)

mosquitoes; grasshoppers and beetles; ants, bugs, earwigs, scorpions, spiders, millipedes and centipedes and worms and slugs.

A century and more later, Aldrovandi's classification of invertebrates seems not to have prevailed. Fig. 5.27 presents a woodcarving from 1722, a decade before Linnaeus's first edition: a scientist examining a beetle under a microscope (Winsor 1976). On his table are additional lenses, paper and an inkwell. The carving (6x10 cm) is from a book on the insects of Germany by Johann Leonhard Frisch. It shows the insect classification of the time: the collection specimens were stored in labelled drawers, as insect collectors do to this day. The specimen drawers on the low bench were labelled "large beetles" *Scarabs majores*, "large butterflies" *Papiliones major* and "small butterflies" *Papiliones min[or]*. On the chair were drawers labelled "large dragonflies" *Libella major*, "large wasps" *Vespae major*, "small hornets" *Crabrones*, "large mosquitoes" *Tipulae maj[or]* and "small flies" *Muscae min[or]*. The drawers in the closet were labelled "aquatic insects" *Insecta aquatica*. This classification lacks systematic thinking and has no comprehensive, inclusive outlook.

Thirteen years after Frisch, Linnaeus entered the arena of insect classification. He divided all insects into four orders using a single key character: wings. In his first, 1735 edition of *Systema Naturae*, Linnaeus began his classification of the insects with the largest insect group, the beetles. Those insects with wings enclosed in a hardened sheath he named Coleoptera, "covered wings", a name already given by Aristotle two thousand years earlier. His next step was to group together all insects with well-developed wings that were not beetles enclosed in a hard sheath—butterflies, moths, flies, bees, wasps, mosquitoes and other groups. He characterized them as having fully exposed and well-developed wings lacking a hard sheath, and he named this new order the Angioptera, those with "uncovered wings". All those lacking wings he placed in a separate order, the Aptera, the "wingless" (our acarids, spiders, crabs, scolopenders and scorpions).

Having grouped and named the Coleoptera, Angioptera and Aptera, there remained a jumble of groups of insects that had wings, but they were small and unimpressive or not present in all individuals. He clustered all these "somewhat" winged insects, remaining after he had classified the large groups of winged insects, into a new group that he named the Hemiptera—the "partially winged" or "slightly winged". His Hemiptera included grasshoppers, whose flight wings were concealed under leathery cover wings, and fleas, whose wings were crossed so that they were not covered by cover wings and were not freely visible. It also included ants, whose wings are well developed and not concealed when they exist, but in which many individuals are wingless, and fireflies, in which the males have wings but the females do not.

This was Linnaeus' first step in classifying all insects, and he did so by use of one single character, the wing. Over the following two decades, he compiled a growing list of insects, some of which he and his students had collected, some from the literature and some from museums all over Europe. By 1758, he had already named 1900 species (today the list exceeds one million). Now, in his 1758 edition, he split his original insect order Angioptera and replaced it with four new orders, whose classification he based on the same key character, the wing. However, he now presented a finer resolution:

Lepidoptera ("delicate wings"): wings scale-covered – butterflies
Hymenoptera ("membranous wings"): wings membranous—bees and waspsDiptera ("paired wings"): only two wings—flies Neuroptera ("scaffolding wings"): wings membranous and prominently veined—dragonflies

He added these four orders to the previous Coleoptera (hard wings), Aptera (no wings) and Hemiptera (somewhat winged). Thus, again, Linnaeus based the seven orders and their names on a single key character: wings. Over the centuries since Linnaeus, zoologists have broadly followed in his steps by emphasizing the importance of the wing in insect classification. Storer (1943: 487), in addition to dividing the class Insecta into two large subclasses by presence versus absence of wings (Pterygota for all Holometabola and Hemimetabola, Apterygota for all Ametabola), listed 25 insect orders, of which 17 incorporate the term *ptera* (wing) in their name:

Ortho*ptera*	roaches, grasshoppers
Derma*ptera*	earwigs
Iso*ptera*	termites
Embio*ptera*	embiids
Pleco*ptera*	stoneflies
Hemi*ptera*	true bugs

Homo*ptera*	aphids, scale insects
Thysano*ptera*	thrips
Meco*ptera*	scorpionflies
Neuro*ptera*	antlions, dobsonflies
Tricho*ptera*	caddisflies
Lepido*ptera*	moths, butterflies
Di*ptera*	true flies
Siphona*ptera*	fleas
Coleo*ptera*	beetles, weevils
Strepsi*ptera*	stylops
Hymeno*ptera*	ants, wasps and bees

So far as to insects. In the class Aves, the key character for the orders was the shape of the bill; in the quadrupeds, it was the dental formula.

Because every group has an essence and a key character, many of Linnaeus' orders were exceedingly unnatural due to convergence, of which he was not aware. In his first, 1735 edition, within his class Quadrupedia, in which he based classification into orders on the teeth formula, his order Anthropomorpha included sloths together with monkeys and humans. In his order Ferae, he placed bats, hedgehogs and moles together with bears, lions and seals. In his order Glires, he placed shrews together with mice, and his order Jumenta included horses, hippos, elephants and pigs. His only today-accepted natural quadruped order was the Pecora (camel, deer, goat, sheep and cattle). Likewise, his class Fishes (legless, the body naked or scaly) included the whales. We recall Ray, who, several decades earlier, had used internal features and correctly ascribed the whales to the mammals.

Twenty years later, in his 1758 edition, Linnaeus was already acquainted with many additional animals; he had changed the names of many groups, but he was still not aware of convergence. By now, he already recognized the correct classification of whales, and accordingly, he erected the class Mammalia to replace the Quadrupedia of 1735. He now distinguished between eight mammalian orders, again, by one single trait, the teeth (sometimes also other accessory traits were given). Again, this led him to a ridiculous classification (Fig. 5.28). He placed bats together with monkeys and lemurs in the order Primates (four incisors, one canine; two mammary glands on the breast). He placed elephants and manatees together with sloths, anteaters and pangolins in the order Bruta (no incisors). He placed swine, armadillos, hedgehogs, moles, shrews and opossums in the order Bestiae (an undefined number of incisors; elongated nose). He placed porcupines and mice together with rhinoceroses in the order Glires (two incisors, no canines) and horses and hippopotamuses together in the order Bellua (upper and lower incisors; mammary glands on the rear). Today, we consider only three of Linnaeus' eight mammal orders (in his 1758 edition) as natural groups: carnivores, which he termed Ferae; ungulates, which he termed Pecora and whales, the Cetae. Likewise, among birds, he grouped together raptors and parrots (a curved beak, Fig. 5.29).

Upon descending from the order level and coming now to the genus, it would have been proper for Linnaeus to assign the genera to orders based only on single

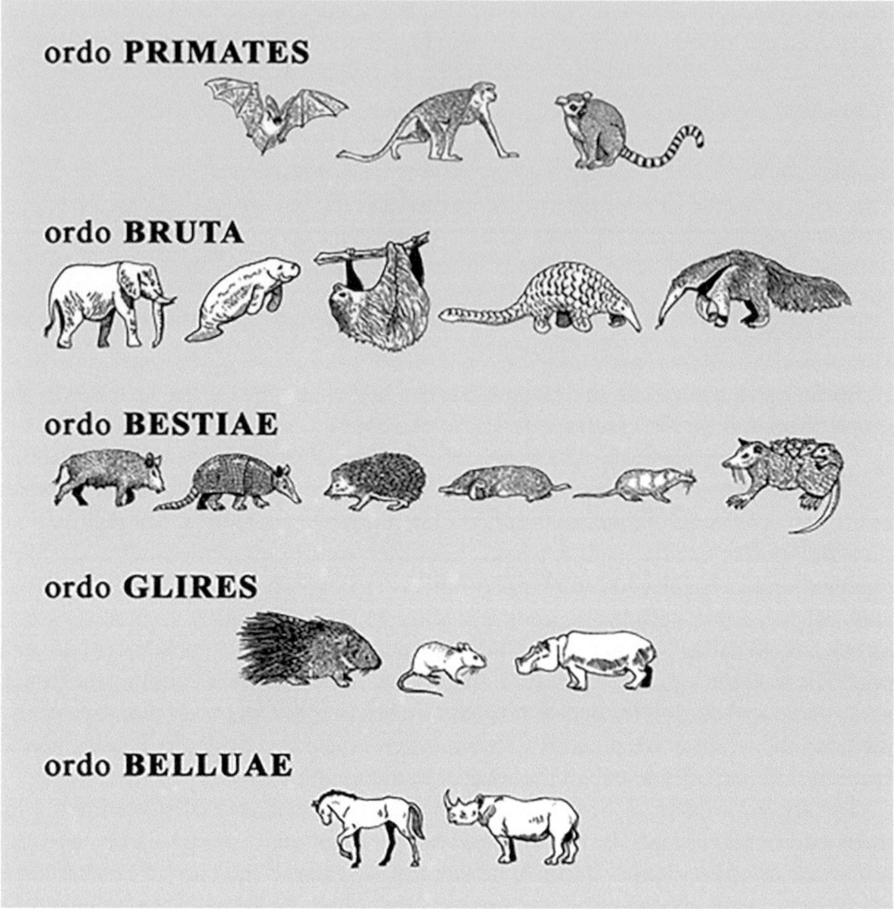

Fig. 5.28 Linnaeus' classification: mammals

key characters, which must always be present, but in fact, he often identified animals also by intuitive impression, and only later would he develop the suitable attribution. When the designation according to characters conflicted with his intuition, Linnaeus preferred his intuition, as in the case of a genus of wingless flies, which he assigned correctly to the Diptera, an order characterized by two wings. Likewise, in birds, in the order Galliformes, the key character was a conical curved beak, and Linnaeus included in it such ground-dwelling birds as pheasants and fowls—and also the ground-dwelling ostrich and cassowary, although their beaks are not curved (Fig. 5.30). In animals as in plants, "It is first necessary to secretly observe the object, as if from under a table, to avoid creating wrong classifications" (*Philosophia Botanica* §168).

Within the animal kingdom in all its classes and orders, Linnaeus (1735) listed a mere 181 genera, giving the impression that only 181 animal genera were known at

Fig. 5.29 Linnaeus' classification: raptors and parrots in the same group because of convergent hooked bills

Fig. 5.30 Linnaeus Galliformes: those with curved beak

the time, which is, of course, not true. In the birds alone, Ray (1676) had recognized and described over 110 genera, yet Linnaeus listed only 47 genera.

We should perhaps understand the paucity of genera noted by Linnaeus in light of his absolute belief in the biblical Genesis and his attempt to link the study of nature with the worship of God. In his eyes, God had commanded Adam to name the animals, and Linnaeus himself, the descendant and personification of Adam, was obligated to name and order all creatures on earth. It is perhaps these religious beliefs of Linnaeus that dictated the approximate number of genera in nature, some 180 animal genera and some 300 plant genera. This would be the number of genera that a single person, Adam, could have coped with and approximately the number of genera that we, Adam's descendants, can memorize.

What if there are many more genera? To solve this problem, Linnaeus (1735) assigned the overflow of genera from the genus column to the last column, in which we would naturally expect him to enlist the species. Here, he distinguished these surplus genera from real species by beginning their names with a capital letter, unlike the "real" species whose names began with small letters. Thus, for the genus

Falco (falcon), which he placed in the genus column, he placed in the species column *Aquilla, Buteo, Lanius, Nissus, Vultur, Falco* and *Milvus* (eagle, buzzard, shrike, hawk, vulture, falcon, kite). In this same column of bird species, he also placed ordinary species, three of magpie (*Pica niger, P. viridis, P. varius*) and two of tit *Parus* (*P. caudatus, P. cristatus*). Likewise, for the genus *Equus* (horse), which he placed in what should be the genus column, he placed in what should be the species column the genera *Equus, Asinus, Onager* and *Zebra* and in this same column of quadruped species, he also placed three ordinary species of mouse (*Mus domesticus, M. brachiurus, M. microtus*).

In the order Testacea, he crammed all bivalves into one genus *Concha*, for which he placed, in the species column, the well-known genera *Mytilus, Pholus, Perna, Chama, Solen, Tellina, Pinna, Ostrea* and *Pecten* (common mussel, paddock, brown mussel, clam, razor shell, pen shell, oyster and scallop)—all with capital letters at the beginning of their names. Here there were no bivalve species. Thereby the significance of the genus rank in his hierarchy of only four ranks lost much of its significance.

By 1758, the number of genera fully recognized by Linnaeus had increased from 181 to 312, each receiving an ordinal number. As in his first edition, also in this edition, many surplus genera appear, but they do so without an ordinal number. For example, in his 1735 edition, he placed *Sepia* (squid) as a genus; in 1758, he gave *Sepia* the genus number 262, followed by the genera *Octopodia*, *Loligo* and *Sepiola*, with no numbers. In this manner, Linnaeus preserved the low number of genera seen by Adam, also in his later editions.

Descending now from genus to species rank, the use of the term "species" was very widespread in Linnaeus' days. It served far beyond the plant and animal worlds of his time, with rock species, metal species, heaven-dwelling angel species and demon species. Also, in the century before Linnaeus, such eminent scientists as Bacon, Boyle and Newton believed that a given species could transform into another species, saying that lead could transform into gold. For Linnaeus, the genus was the effective practical unit in taxonomy, the species often being difficult to determine. If the genus was right, then the species was not far out and could finally be determined later. He adopted Ray's theory of the species as an essential, fixed, unchanging entity, created during the 6 days of Creation. There are no new species (1735, §4), and "Every living being is created from an egg, and every egg creates an offspring very similar to its parents. This leads to the conclusion that new species are not created in our time" (1758). In this, Linnaeus was also following in the steps of Martin Luther (1483–1546, Germany), who had declared, two centuries earlier, that all types of animals were created at one time by God's word. He was also following in the steps of John Calvin (1509–1564, Geneva), who had stated that all animals were created by God in 6 days, and no new species had been created since. All those who think otherwise offend God and should die. Linnaeus also rejected the possibility that from the seed of one species, another species could develop, and he did not believe in the spontaneous formation of species.

In all, Linnaeus named and described 4400 species of animals. Aware of the need to emphasize the usefulness of studying even small animals, he wrote that cockroaches consume bread and clothes, gadflies kill reindeer, shipworms damage ships, fleas trouble women, lice trouble sailors and soldiers, locusts devastate crops, caterpillars eat away tree leaves, spiders and scorpions kill people and worms are parasites. He did not forget to mention that, on the other hand, some tiny animals are useful—in the dyeing industry, in surgery, medicine, weaving or as food.

A most fascinating example of this alternation between observation and intuition is in his classification of molluscs. In his 1758 edition, he divided all invertebrates into two classes: insects, characterized by the presence of jointed antennae on the front of their head, and worms, characterized by the presence of (non-jointed) tentacles. Next, he further divided the worms into several orders. Ever faithful to his method of using external characters alone, Linnaeus placed all shell-bearing molluscs in one order, the Testacea (in which he erred in including also goose barnacles and calcareous tubeworms). He assigned most shell-lacking molluscs, such as terrestrial slugs, sea slugs, cuttlefish and octopuses, to another order, which he named Mollusca (from the Latin *mollis*, meaning soft). This order also included various worms, copepods, sea cucumbers, medusas, sea stars and sea urchins. One shell-lacking genus, the shipworm *Teredo* (a naked wood-boring bivalve) he put in a third order, the Intestina, together with earthworms, nematodes, liver flukes, leeches and hagfish.

Linnaeus seems to have felt, however, that this classification into three orders was not very natural, as it was based solely on the presence or absence of a shell, and that shelled and shell-less snails, bivalves and cephalopods do somehow belong together. Therefore, he fused the two orders in a very remarkable manner: within the order Testacea (the shelled animals), he gave each genus a scientific name characterizing its shell, but he also gave a different scientific name to the animal inside the shell, and this second name was taken from the order Mollusca (Fig. 5.31). Thus, he wrote that the "shell-genus" *Helix,* of the order Testacea, is inhabited by the "animal genus" *Limax* of the order Mollusca, and *Limax* also inhabits the marine testaceous snail genera *Conus, Cypraea, Bulla, Voluta, Buccinum, Strombus, Murex, Trochus, Turbo, Nerita* and *Patella*, and the bivalve *Pinna*. Other testaceous bivalve shell-genera (*Tellina, Cardium, Donax, Venus, Spondylus, Chama, Arca, Ostrea, Mytilus*) were inhabited, according to Linnaeus, by the mollusc animal genus *Tethys*; and similarly, inside the testaceous *Chiton*, he placed the mollusc *Doris*; and inside *Argonauta* and *Nautilus,* he placed *Sepia*. It would seem that Linnaeus was perhaps trying to create a simultaneous classification that was both artificial and natural. We should use the shell only for identification, while the animal inside it denotes the natural systematic position.

Coming now to the diversity of man, Linnaeus placed man in his very first animal order Primates, which he defined as those with four upper incisors and with two mammary glands on the chest. Within the primates, he presented the genus *Homo* and described it only by the phrase *nosce te ipsum* (know thyself), the text that in ancient Greece was inscribed in the forecourt of the Temple of Apollo at Delphi (this is its Latin version). Plato had frequently used this phrase, and during the Baroque, so had Hobbes, in his Forward of *Leviathan* (1651). From the genus *Homo,* Linnaeus went on to the species, which he named *sapiens* (wise, thinking,

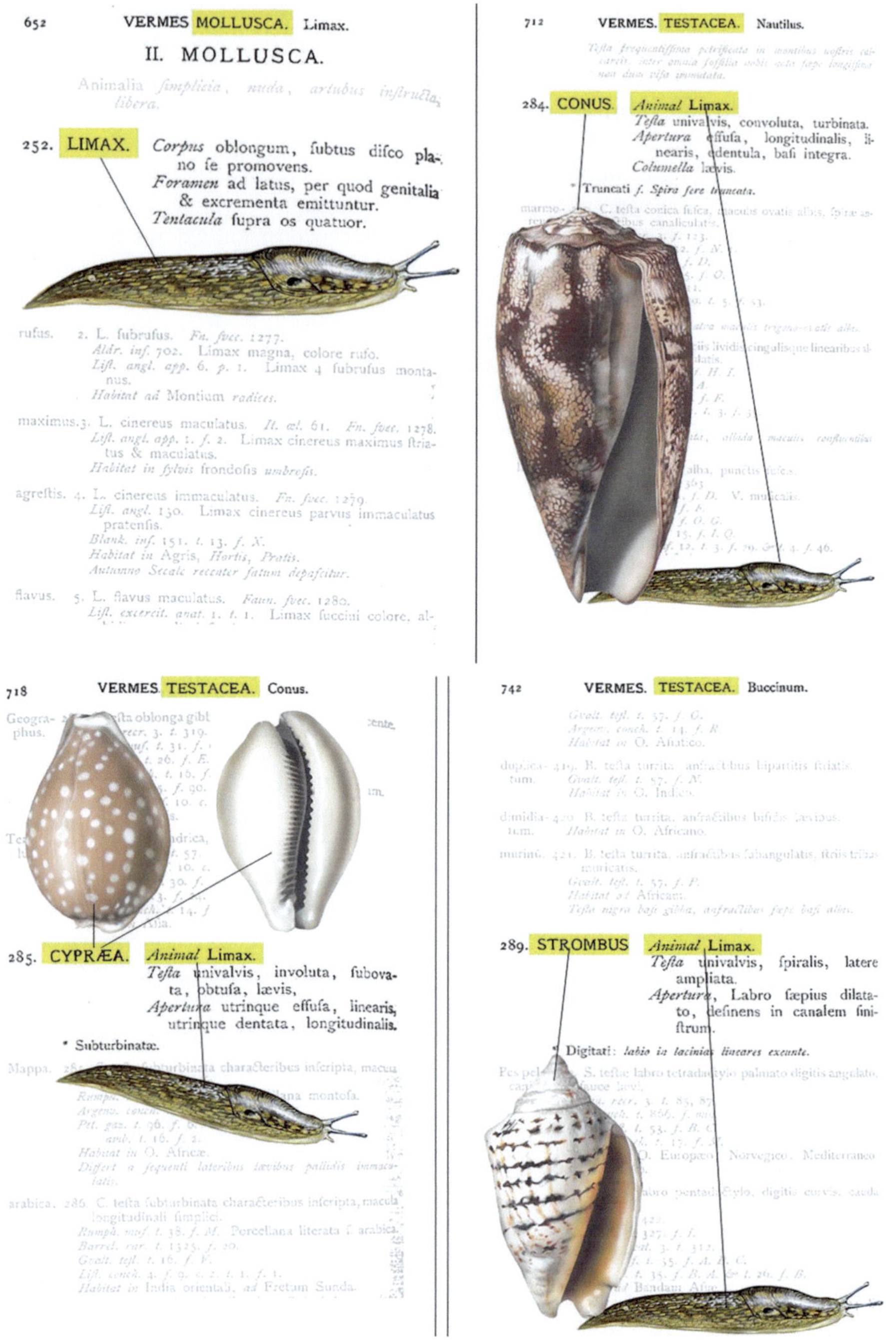

Fig. 5.31 Linnaeus placed shell-bearing molluscs in the order Testacea and shell-lacking molluscs in the order Mollusca. Within the (shelled) order Testacea, he gave each genus a name characterizing its shell but a different name to the animal inside the shell, taken from the (shell-less) order Mollusca. Fourteen examples

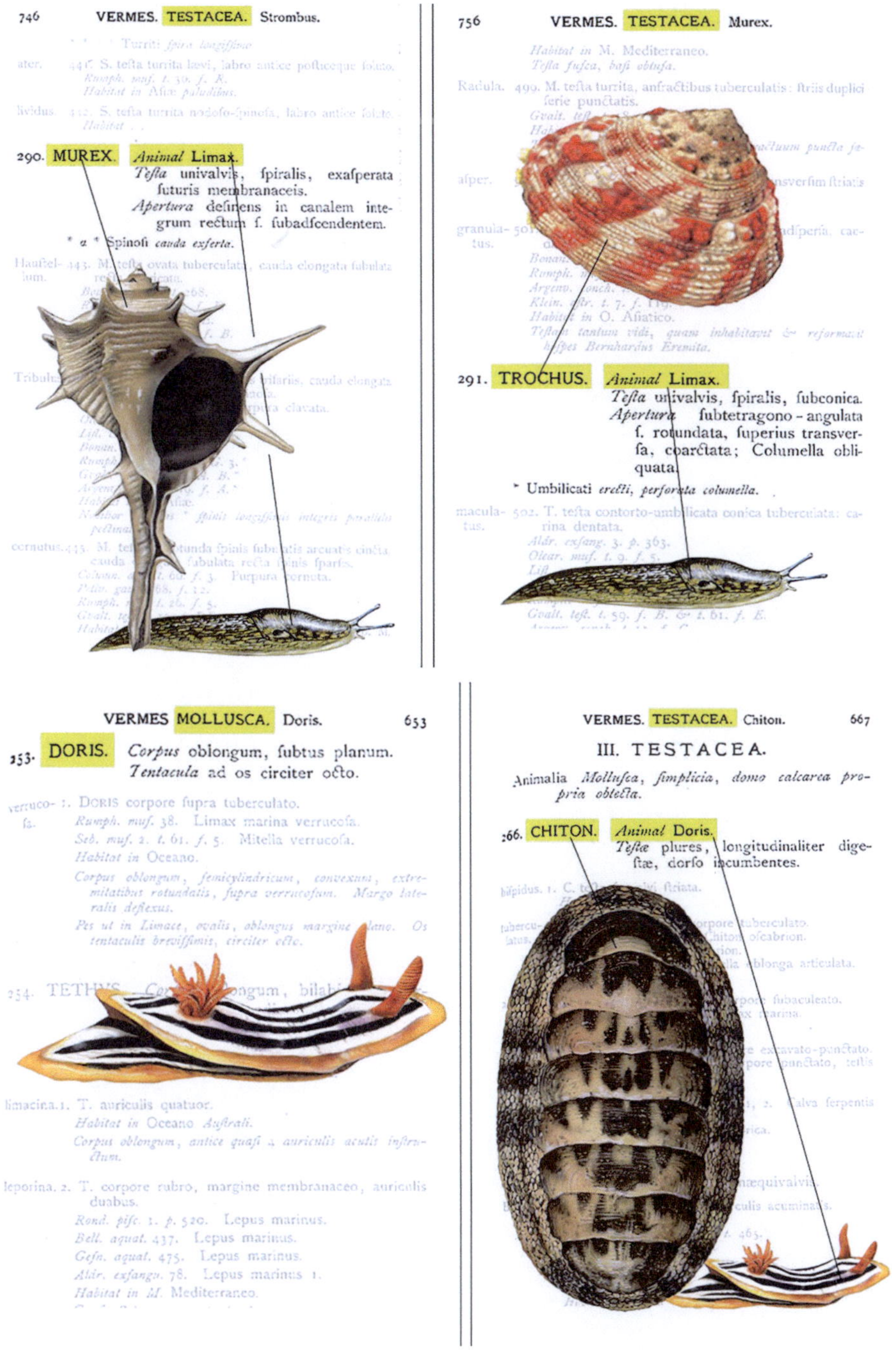

Fig. 5.31 (continued)

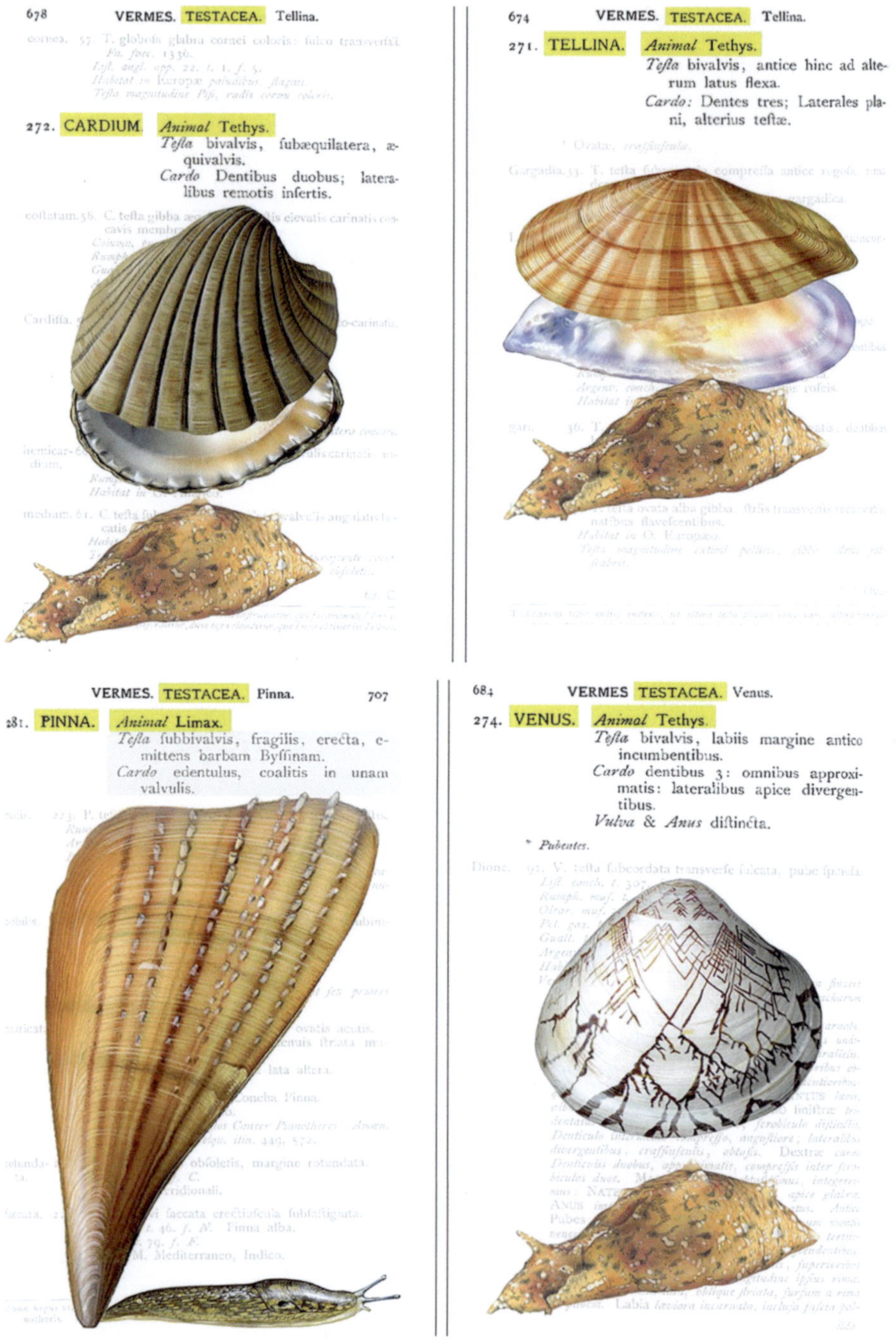

Fig. 5.31 (continued)

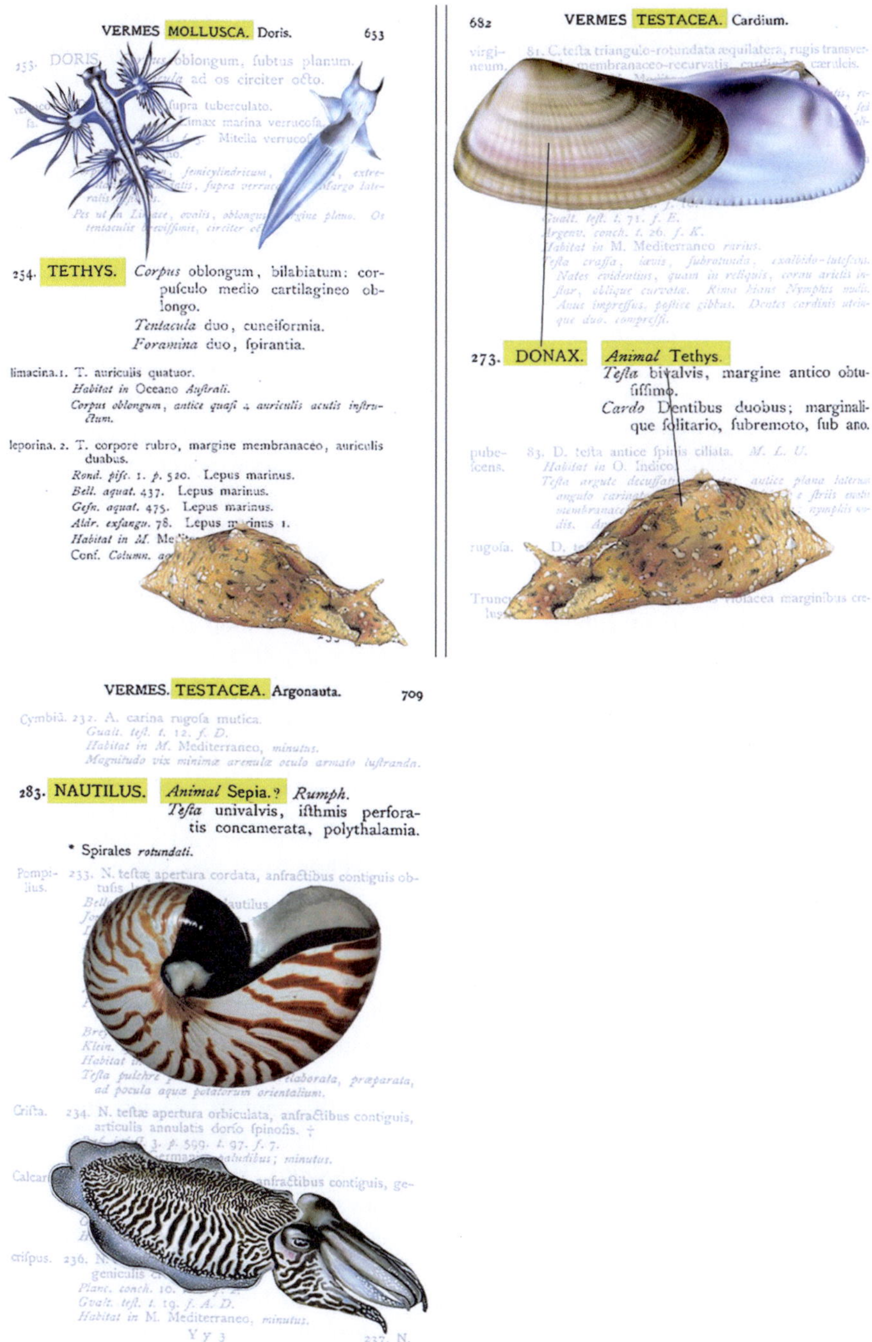

Fig. 5.31 (continued)

rational). He described *Homo sapiens* as diurnal, varying from place to place by his culture. Finally, he described variation within *H. sapiens*, in what we would call subspecies or races. First came *Ferus* (savage man): Walking on all fours, mute and hairy, Linnaeus presented a list of references of observations on men resembling bear, wolf, sheep or bull throughout Europe, whether in Lithuania, Hessen, Hannover, the Pyrenees or elsewhere. Next, he presented the four races with which we are familiar: *Americanus, Europaeus, Asiaticus* and *Afer*, thus one race for each of the then-known continents: America, Europe, Asia and Africa. In the descriptions of each of the races, the first word refers to skin colour: red, white, yellow or black. The second word described the temperament dominant in each race, whether *cholericus, sanguensis, melancholicus* or *phlegmaticus*. These four terms originate in Ancient Greece, where the ancient Greeks were fond of the number four. There were four elements: earth, air, water and fire; there were four basic traits: hot, cold, damp and dry and likewise, the human body contained four basic liquids: blood, phlegm, yellow bile and black bile. These four body liquids were, in the opinion of the ancient Greeks, responsible for a man's temperament. One in whom blood is dominant is cheerful, hopeful and self-confident. One in whom phlegm is dominant is slow, cold and apathetic. One in whom the yellow bile is dominant easily tends to be angry, and one in whom the black bile is dominant tends to be depressed. After colour and temperament, Linnaeus classified the races by their hair, which in the American is black, straight and thick; in the White, it is yellow and flows down in a graceful, orderly manner; in the Asian, it is black and straight and in the Negro, it is curly. We can now present Linnaeus' full classification into four races:

> American *Americanus*: red, easily angered, hair straight and thick, wide nostrils, elliptic face, sparse beard, cheerful, free, paints himself in red stripes, follows his customs.
>
> European *Europaeus*: white, cheerful, muscular, yellow hair falling gracefully and orderly, blue eyes, clever, creative, wears warm clothes and follows his laws and ceremonies.
>
> Asian, *Asiaticus*: yellow, depressed, stiff, black, straight hair, brown eyes, severe, greedy, wears loose clothes, follows his opinions.
>
> Negro *Afer*: black, phlegmatic, lazy, curly hair, silky skin, flat nose, thick lips, female shy, her breasts flowing with milk, dexterous, neglectful, claw-like fingernails, lives by his moods.

Linnaeus also described human monsters, *monstrosus*, which he defined as varying by climate. They included the active, agile and shy *alpini*; the big, lazy *patagonici*; the not-so-fertile *monorchides* (*Hotentots*), who have only one testicle and those with unusual heads, the *macrocephali*. Finally, another species, the cave-dwelling man *troglodytes*, dwells in caves in Africa and the Far East.

A dissertation, submitted by one of Linnaeus' students (Hoppius, 1760), exemplifies the poor state of human classification. The student investigated the borderline, in primates, between the human *Homo* and monkey *Simarium* genera. He defined the human as one who understands, who knows to repair things and to add, subtract, multiply and divide and who can observe and bewilder the wonders of nature. The monkey resembles the man in his morphology and also in these abilities, but to a far lesser extent. The savage *Homo erectus ferus*, say the Hottentot, is a man. He is savage only because he did not receive our education; his customs are different—but he is human, even though, monkey-like, he uses his hands and legs to climb trees in which he dwells. The nocturnal caveman, *H. troglodytes*, with golden iris and pupil, is also human. Probably so is the tail-bearing creature *Lucifer* (=*Homo caudatus*, Figs. 5.32 and 5.33), who cooperates with his fellows in hunting parrots, which they eat raw, and even hunting people of other tribes, whom they eat.

Among the monkeys, he presented what he called *Simia pygmaeus*, of which, he wrote, two preserved specimens were available, both in England. It has a hairy tail and neck, a flat nose as in the Hottentots, elongated fingers and its legs are similar to our hands. It controls hand and foot movements in the manner of monkeys, is born naked, is brown and dwells in the forests of Africa, with customs of its own. Another monkey was the hairy *Satyrus*.

Some features that Linnaeus ascribed to certain human races may seem as racism to our modern eyes, but we should view them in the context of his times. Coming one generation after Linnaeus, Christoph Meiners (1747–1810) of the University of Göttingen believed that each human race had a separate origin. He split humanity

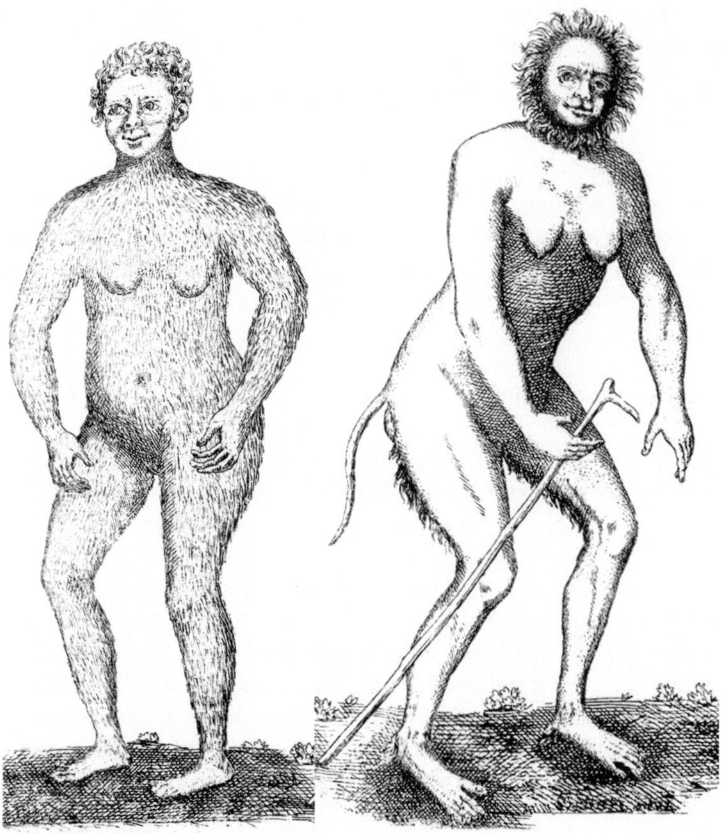

Fig. 5.32 Hoppius: *Homo troglodytes* (left) and *Lucifer* (right)

Fig. 5.33 Hoppius: *Satyrus* (left) *and Pygmaeus* (right)

into the "beautiful White race" and the "ugly Black race". The black ugly Negro has "no human, barely any animal feeling"; has big teeth and jaws because he is a carnivore; his brain smaller than in other races; and he has perverted sex drives (1785, 1815). In comparison to Meiners, I do not suspect Linnaeus of any racism.

5.2.2.3 Nature: Its Beginnings, Its Balance

During his journey in Lapland, Linnaeus found beds of fossilized marine shells at some mountaintops. Also, during that journey, he believed he had noticed an ever so slight but continuous trend for the retreat of the sea amid its cycles of high and low tides. This led him to the conclusion that the continent is continuously growing, and if so, then this is evidence that, at some past time, it was far smaller than it is now; it was merely a paradisiacal island protruding from the sea with a tall mountain at its centre, creating diverse climates over a limited area. This paradisiacal island contained all today-existing habitats: alpine meadow, desert, deciduous forest, marsh, conifer forest and so on. At the time of Creation, God created the inhabitants of this island—a pair of every sexual creature and a single specimen of each hermaphrodite. Later the sea began retreating, and the continent gradually grew, and the creatures of the island began dispersing over the world as the island gradually grew and more land emerged from the sea. The fossil beds he had observed on

Lapland's mountaintops were, he believed, the remains of ancient shorelines, and they are evidence that the environment is constantly changing. Linnaeus needed this paradisiacal island because of the biblical verse "And the man gave names to all cattle, and to the fowl of the air and to every beast of the field" (*Genesis* 2). Adam could have done this only if all creatures occupied a very limited area, which he also inhabited. Linnaeus believed that his classification reflects the concept of Creation as a harmonic, balanced system. Every plant, and every animal, has its special place in the web of life and contributes to maintaining this web. Predators, for example, daily destroy animals that would otherwise reproduce so rapidly that they would deplete their food sources. These complex relations are proof of divine balance in which, he believed, the relation between predators and prey creates harmony.

Linnaeus acquired these ideas, regarding harmony in nature, from William Derham (1657–1735, Fig. 5.34), an English clergyman, natural scientist and Fellow of the Royal Society. Derham's *Physico-Theology* (1713), translated twice into Swedish during the life of Linnaeus, was a doctrine of balance and compensation in nature. It is assumed that there is an appropriate quantity of everything on earth, a quantity sufficient to maintain every species, yet without filling up the world with one species or another. Providence has created various animals to delight in various foods since if all animals were to delight in only one sort of food, there would not

Fig. 5.34 Derham

be sufficient for all. By every kind using different food, there is enough food for all. By these means, also, all provisions of the globe are disposed of and do not rot, thereby causing disease. Furthermore, some animals feed on carrion and rotting matter, and "thus is the world in some measure kept sweet and clean, and at the same time diverse species of animals supplied with convenient food". In this manner, balance is preserved in the world between an animal's ability to reproduce and its lifespan, and thanks to this, the world is not overpopulated. Indeed, Derham was the first to use the term "balance" in referring to nature (1716:171): "Thus the balance of Animal World, is, throughout all Ages, kept even; and by a curious Harmony and Just Proportion between the Increase of all Animals, and the Length of their lives, the World is through all Ages well, but not overstored." Derham thought this to be so regarding human demographics: in some places, a few more people are born than die, and this balances other places, where death outruns life due to the ravages of plagues and war. Consequently, extraordinary losses of lives may be a wise means to keep mankind's balance even. Population surplus also encourages people to seek other, yet uninhabited parts of the earth (Book IV, ch. X). Derham's *Artificial Clockmaker* (1696) was used by William Paley nearly a century later, as we shall see further on.

Linnaeus adopted Derham's approach. He wrote that we might disagree with the idea that God had created some creatures only to slaughter other creatures, but we forget that this is God's means of maintaining a balance between species because nature can maintain only a limited number of individuals. According to Linnaeus, "if thousands of plant species would grow together in the same place, some would dominate and eliminate the others. However, this is where we see God's wisdom. Plant species occur in many different sites in the world: some are in India, others in temperate regions, and some in the poles. Throughout Sweden, there are three hundred plants species, but each species is limited to a different area in the country, and usually there are no more than fifty to one hundred species in a given place. Thus, one species cannot displace the other."

Linnaeus further believed that the balance characteristic of the world, which preserves a steady number of inhabitants, is constantly threatened. There is always the possibility that one species will begin "pestering another species and robbing its food, thus introducing confusion into the order". Therefore, Linnaeus wrote of the terrible punishments that keep every species within its boundaries – "animals that break this rule are punished by sickness and death and in cases where an animal strays from its boundaries it suffers – but this is all for the best, performed by the Creator". Linnaeus was thus aware of the significance, in nature, of population dynamics and of competition.

Linnaeus, with his extensive scholarship, clear, concise expression and methodical organization, brought order into the mass of knowledge of plant and animal diversity. Naturalists held his classifications of the world of nature in very high esteem, and he was widely renowned throughout Europe as one of the most acclaimed naturalists of his time. However, not everybody welcomed every one of the Linnaean names. Johann Siegesbeck (1686–1755), the Prussian-born director of the Medical Gardens at St. Petersburg and Professor of Botany at the Russian Academy of Science, called the Linnaean system "loathsome harlotry". He declared that God would never, in the plant kingdom, have allowed such hateful vice as that several males should possess one wife in common or that a true husband should, in certain composite flowers, have, besides its legitimate flower, a nearby illegitimate mistress. In retaliation, Linnaeus named a small, ugly weed *Siesbeckia* (1737; St. Paul's wort; in traditional medicine, it served to treat syphilis, leprosy and other skin-related diseases). He also relabelled a packet of *Siesbeckia* seeds as *Cuculus ingratus* (ungrateful cuckoo) and had it sent to Siesbeck, who grew the seeds only to find their identity.

Furthermore, Linnaeus' choice of names for shells caused distress among conchologists. At the time Linnaeus was studying molluscs in Sweden, Emanuel Mendes Da Costa (1717–1791) studied them in England. In his *Elements of Conchology* (1776), Da Costa included, aside from facts concerning shells, a heavy personal attack on Linnaeus: "One subject however I shall insist upon; that is, to explode the Linnaean obscenity in his character of bivalves; not only for their licentiousness, but also in that they are in no ways the parts expressed. Science should by chaste and delicate. Ribaldry at times may be passed as wit, but Linnaeus passes it in terms of science. [...] I can compare his terms only to *Spintriae* in a valuable collection of Roman coins" [*Spintriae* - very obscene coins struck by the Emperor Tiberius]. What was it that so infuriated Da Costa? Linnaeus had a lively, flowing imagination in some of the names he gave. He was prone to see, in certain shells, a striking similarity to parts of the human body that, in those days, people did not refer to in public. Linnaeus introduced the names of these body parts into the binomial scientific names of genera and species he described (*Nerita mamilla, Murex anus, Buccinum testiculus, Buccinum glans, Solen vagina, Helix mamillaris, Patella mamillaris, Venus virginica* versus *Venus deflorata* versus *Venus meretrix*). Shell collectors found that, to their embarrassment, they would have to use these terms when labelling their shells and, much worse, when corresponding or discussing them in public. These and other similar names motivated Da Costa to write that "I therefore with due submission recommend to that otherwise great naturalist, to change them, and expunge this reproachable obscenity from his works" (1777, IV).

Not everybody was so severe in his criticism. Ladies in particular found that, thanks to Linnaeus, they were now able to take up the study of botany. A certain Miss Brightwell of Norwich wrote that the easy, pleasant method introduced by Linnaeus so greatly promoted the study of botany "that no wonder the ladies acknowledged with gratitude their obligation to the naturalist, who first originated a method by which this delightful study could be brought within the attainment of those who loved it" (1858).

As with the Linnaean names, not everybody welcomed the Linnaean four-rank system of classification. One of Linnaeus' major opponents was Buffon, the hero of our following pages.

5.2.3 *Against Linnaeus: Buffon*

5.2.3.1 *Histoire Naturelle*: The Grand Picture

Linnaeus' great opponent was Georges-Louis Leclerc de Buffon (1707–1788), a scientist at the French Royal Court and protégé of Madam de Pompadour, the mistress of Louis XV. Georges-Louis Leclerc was the son of an official in charge of the salt tax at Montbard, a small town in Burgundy. He was seven when his uncle died, leaving him a handsome fortune, with which his father purchased an estate containing the nearby village of Buffon. The family then moved to Dijon, where Georges attended a local Jesuit college in which the scientific syllabus included mathematics and physics; he then moved on to study law at a local law school. Aged 22, he left for the University of Angers to study mathematics and medicine, and it was here that he read Newton, who deeply impressed him, as we shall see further on. Here also, he met the young English Duke of Kingston and travelled with him for over a year through France and Italy; during these travels, he added "de Buffon" to his name. Aged 25, he made Paris his winter residence and became a member of the Academy of Sciences, to which he presented a mathematical study introducing differential and integral calculus into probability theory. His permanent residence was, however, in Montbard, with a beautiful new house he erected and with nearby forests that he owned. In his working room, he hung just one single picture, that of Newton. At the request of the French Academy and for the French navy, he began a study of the tensile strength and longevity of wood serving for the building of ships. So as to know which were the best soils for different tree species, he planted seeds in different soils; to find what would increase wood solidity, he stripped bark from standing trees before cutting them; he experimented on the resistance of wood to breaking and he constructed a tree nursery. He found that wood texture was extremely variable and that mathematical extrapolations from results obtained from small logs do not apply to serious predictions on beams of great length—in general, reality prevailed over mathematical generalizations. In his reports to the French Academy of Sciences, he told of his successes but also of his failures, and this gave him credibility for honesty.

As luck would have it, there was an opening for a scientist to catalogue the strange artefacts that had accumulated in the king's cabinet, who would describe the foreign animals wandering around the Versailles Zoo, and who would organize the neglected royal botanical garden along the left bank of the Seine. A century previously, Louis XIII had founded a Royal Garden of Medicinal Herbs in which there were three "demonstrators" to teach students and to study the composition of medicines by herbal and chemical means; drugs and herbs were kept in a special cabinet. Over the decades, this cabinet of drugs became the cabinet of natural history (Fig. 5.35), and decades later, in 1793, it would become the National Museum of Natural History.

Aged 32, Buffon (Fig. 5.36) became Keeper of the King's Gardens in Paris. Over the following decades, he transformed the *Jardin du Roi* into a major international

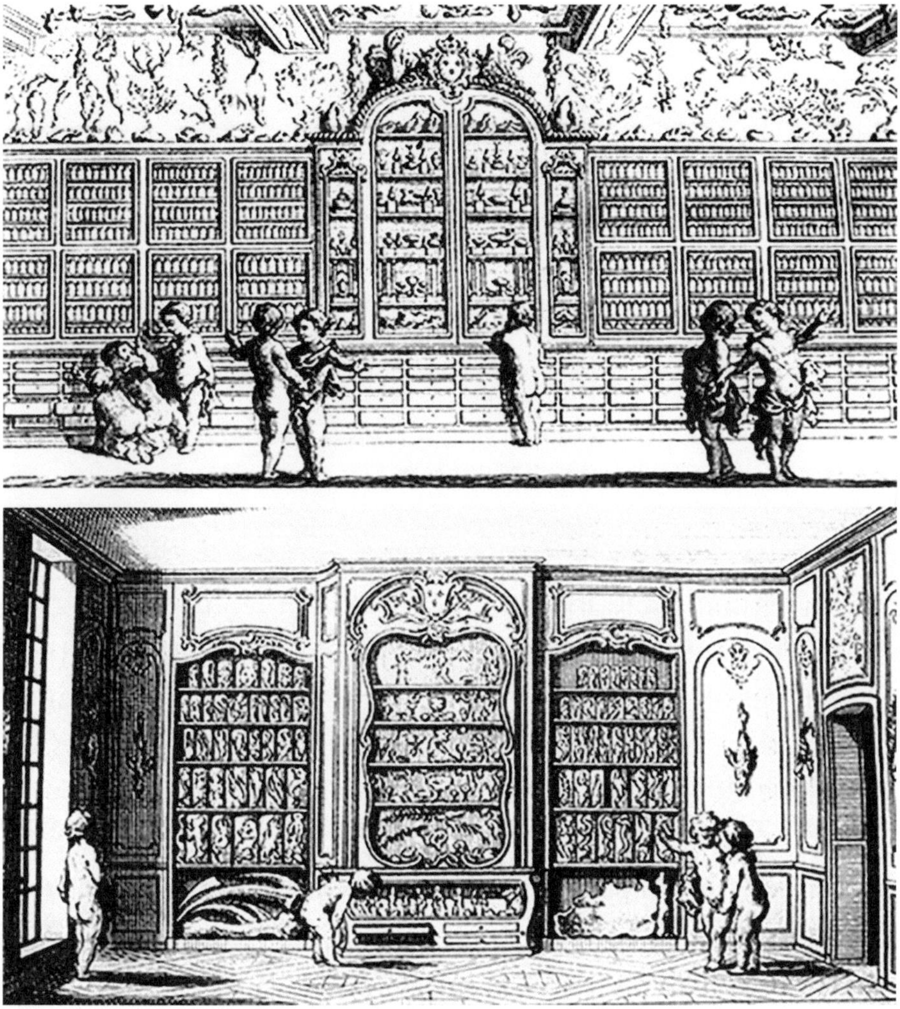

Fig. 5.35 Two views of the King's Cabinet in the mid-eighteenth century. (*HN* 1749, vol. 2)

research centre and museum by enlarging it, arranging the purchase of adjoining plots of land, and acquiring new botanical and zoological specimens from all over the world. With direct access to the king, his chief mistress Madame de Pompadour, and many of the ministers, Buffon was granted generous funds to improve the garden and support the research it conducted. He even convinced the secretary of state to create a special certificate entitled "Correspondent of the Royal Botanical Garden", a purely honorary title that he distributed sparingly. Eventually, the king granted him a noble title, Comte de Buffon. Philosophers of his and later generations described Buffon as "one of those giants, who with bold steps cross the front of their time, as a conqueror carries his sword" (St. Hilaire) and as "the greatest man

Fig. 5.36 Buffon

of his century" (Mirabeau). Buffon himself excelled in flowery speech and fine words. He would use maxims such as "the function of philosophy is not to ask why but how", "we should begin by observing a lot and then return for a second observation", and "genius is nothing but a great capacity for patience". His description of the peacock as having "the plumage of an angel, the voice of a demon, and the stomach of a thief" is adorable.

Buffon believed in a complete description of each animal, as opposed to the Linnaean telegraphic descriptions of only external morphology (and teeth). A description would not be complete without a careful morphological study both external and internal, and lacking the skill and training in anatomy, he sought the aid of someone who could supply him with the anatomical data he required. Shrewd and careful in his choice of new staff, he placed Louis-Jean-Marie Daubenton (1716–1800; Fig. 5.37) of his hometown, Montbard, as Guardian and Demonstrator of the Cabinet of Natural History. Daubenton, then aged 22 and a physician with a strong morphological orientation, joined Buffon and soon became an expert in anatomical dissecting of quadrupeds, in charge of anatomical descriptions, and Buffon's co-author in many publications. His painstakingly exact descriptions supplemented Buffon's philosophical writings; indeed, his supplements were often twice as long as Buffon's article. In describing the anatomical details of 182 quadruped species,

Fig. 5.37 Daubenton

he gained a high reputation as a comparative anatomist. In addition to his anatomical work, Daubenton also wrote brief descriptions of the holdings of the *Cabinet du Roi*, which he inserted after each of his articles. In many respects, Daubenton and Buffon were complete opposites, and they worked well in partnership.

During his long reign as head of the *Jardin du Roi*, Buffon led a very hard-working daily schedule. At his country estate and during 8 months of the year, a servant was engaged to rout him from bed every morning before six, by force if necessary, and he would then work until nine, time for a light breakfast of two glasses of wine and a roll. He would then continue work until two in the afternoon, time for a leisurely meal with his family and with guests. Then came a short walk alone in his gardens and further work until seven, time for 2 h of relaxed talk in his salon. At nine, he would go to bed without supper. From Montbard he would shower his Parisian staff with instructions, yet at the *Jardin du Roi,* he was present in person only during the four winter months.

In Buffon's days, there were no "basic laws" to explain the very first beginnings of life and none to explain the continuous generation of animals and plants. It was a common belief, but only belief that God had created all creatures during the 6 days of Creation. This lack of basic laws in the life sciences was in contrast to the exact sciences, in which the public accepted Newton's theorems as basic laws. Also, in the seventeenth–eighteenth centuries, there was a common attitude for the study of nature to move away from any approach that was overly philosophical and involved too many lofty theories. Many thought that the study of nature should carefully advance the certainty of facts by evidence derived from observation or experiment, not by excessive speculations. People praised giants of the seventeenth century such as Hooke, Leeuwenhoek, Ray and Tournefort for gathering observations and avoiding speculations:

> None of them has thought of anything, but of making us certain of some useful matter of fact, without attempting to explain to us what a globule of air, what the texture of a fibre, the wing of a scarabaeus, or the balsam of an ananas, were. Our most excellent men, in the most celebrated academies, are all of them tired of hunting after evidence which flies from them, and are very well satisfied with experience, which seldom fails to crown their labour with success (Pluche 1740).

Buffon stood up against these approaches and attitudes. He searched for basic laws in the natural world comparable to Newton's basic laws in the exact sciences, basic laws that would explain the great secrets, both of the beginning of life and of how life renews itself, from one generation to the next. In 1749, the first three volumes of his encyclopaedic treatise *Histoire Naturelle, générale et particulière, avec la description du Cabinet du Roi* (*Natural History, General and Particular, with a Description of the King's Cabinet*; an English translation appeared in 1791, and a treatise on birds in 1770) were published; additional volumes were published during the following three decades. The title page presented the royal crown, a globe with three fleurs-de-lis and two trumpeting angels (one male and one female) announcing the forthcoming treatise and noted that the encyclopaedia was printed at the Royal Press (Fig. 5.38).

Surprisingly, on this title page, there was no author's name. Following the title page was a page with the dedication "To the King"; also, it was without the author's name. Only on the next page was the reader introduced to the authors, "Buffon, Keeper of your Botanical Garden" and "Daubenton, Guardian and Demonstrator of your Cabinet of Natural History". Buffon was thus aiming at the very highest persona of the French regime. In this carefully worded presentation, Buffon might perhaps have hinted at a secular biblical story of Adam, keeper of God's Garden of Eden. Throughout the encyclopaedia, Buffon was in charge of the "General" and Daubenton of the "Particular".

Histoire Naturelle, with its descriptions of hundreds of quadrupeds over thousands of pages (and later on of birds and minerals), became a great success. In France, it swiftly became the most widely read work on natural history and was the third most popular piece of literature there in the latter part of the eighteenth century (Farber 1975). Beyond France, many educated men throughout Europe read translated versions, and Buffon came to be considered one of the great scientists of his time, the Age of Enlightenment. His encyclopaedia was to contain everything known in his day about the natural world, but beyond that, Buffon also tried to explain the facts he had amassed. Far from explaining by merely quoting the opinions of others, he tried to explain them in his own way, with overarching theories about the Earth and its inhabitants. He was not afraid to take on big ideas, and his scientific explanations were rich in speculations (Fellows and Milliken 1972).

Fig. 5.38 *Histoire Naturelle*, from the title page

The chronology of Buffon's texts is important since his ideas evolved considerably over the years as he wrote his great work. We may consider the first three opening volumes as a unit since they appeared in the same year, after a decade of research. They dealt with major questions in biology, not with detailed animal descriptions. They turned attention to questions of methodology, the general nature of both man and animals, the species concept, generation, sense perception, anthropology, cosmology and geology; the remaining 12 volumes described the mammalian quadrupeds.

We recall that Locke, half a century earlier and in an attempt to build up a distinction between knowledge and mere opinion (or belief), had distinguished three kinds of knowledge about the external world. He considered intuitive knowledge, saying that a triangle is not a circle, as the most certain of human knowledge because we can grasp this truth immediately. Demonstrative (mathematical) knowledge, say that the three angles of a triangle equal two right triangles, is less certain because it cannot be grasped immediately and involves the construction of intermediate steps of what we, today, would term a deductive proof. Sensitive knowledge, with the least certainty, refers to the entrance of an idea into our mind through our senses; by using our senses, we recognize observable features and thereby classify entities since all members of an entity share a set of observable features. Locke commented that we know the external world only as it appears to us through our senses. We do not know its underlying nature, its essence.

Buffon, surely inspired by Locke, also sought to determine the certainty of human knowledge. In contrast to Locke's intuitive, demonstrative and sensitive knowledge, he emphasized the distinction between abstract and real knowledge (he used the term "truth" rather than knowledge). Science, he wrote, is about finding order in nature, and the disciplinary tool for this must be truth (= knowledge). There are, he argued, two kinds of truth and certitude. Mathematical truths are truths of definition and concern the analytic and deductive relations of ideas. They have the advantage of being always exact and provable, yet they are abstract products of our own intelligence, in that we make suppositions and then combine them in many ways. They contain nothing more than what we have put in them. For all their logical rigour, they are incapable of representing the reality of things; they will never tell us anything about the world we live in, and so in the context of natural history, they are useless. Physical truths differ from mathematical ones in that they rest only on facts. "In mathematics one makes suppositions, in physics one proves. In the former, we find definitions, in the latter facts. One proceeds from definition in abstract sciences, but from observation to observation in the sciences of the real. In the former one arrives at evidence, in the latter at certitude". Natural history must rest on physical truths, on an understanding of organisms in terms of their real relations in the real world. The truth of natural science consists in the direct apprehension of objects in the world (Sloan 1976). Contemplating the mathematical versus physical truths of Buffon, we recall ancient Greece, in which Plato believed in a spiritual world separate from the real world and containing abstract ideas of things that exist in our real world, ideas that we can reach through reason. Aristotle, on the other hand, saw reality as physical objects of the physical cosmos, objects that we can grasp through the senses and through experience. Buffon attached himself to reality, to what could be touched and felt. Beyond touching reality, he also tried to understand it.

How should we consolidate real facts into general laws of nature? Buffon claimed that a series of similar facts, a frequent repetition and an uninterrupted succession of the same events make up the essence of a physical truth: Physical truth includes the orderly succession of empirical events in a series; it is an objective order standing apart from any arbitrary and abstract order as created by the mind in the mathematical truth domain. It is a series of similar facts or a frequent repetition and uninterrupted succession of the same event. If we observe an event in nature, and it recurs 13 or 14 times in the same way, we can be almost completely certain that it will happen that way the 15th time, and we can go as far as to say that this is a law of nature.

Having thus settled the broad philosophical question, namely that the real world was in the domain of physical rather than mathematical knowledge, his next question was: The real world – how was it created? Buffon viewed the study of the Earth as a prerequisite to zoology and he suggested a history of the solar system (1749, vol. 1). The popular theory as to the history of the Earth was the biblical one as presented by Noël-Antoine Pluche (1688–1761), a French priest and professor of humanities and rhetoric at Rheims who argued that modern science (of his days) confirms the biblical narrative. Taking the Deluge as an example, he argued that in today's landscapes, evidence of the flood can be seen in certain folds and fractures of strata; in disorder of the mountains that before had been ordered harmoniously; and, also, in the disappearance of the eternal springtime, which had once dominated the climate. To cause these changes, God had only to incline the axis of the earthly poles, and all the rest then followed. This divine displacement of the axis shook the air and forced the surface of the earth inward, resulting in a changed landscape whereby, instead of delightful valleys and hills carpeted with the greenery that adorned the original earth, the landscape now consisted of rocks tumultuously dispersed, and snow or thick fog appeared on mountain peaks (*History of the Heavens,* 1740).

There were, however, also other opinions. Benoît de Maillet (1656–1738), a nobleman of Lorraine born into a distinguished Catholic family, was the French general consul in Cairo (1692–1708). He was interested in geology and natural history and, after retiring, wrote his main work, *Telliamed* (his name in reverse). He had it published after his death (1748), as a safety measure to protect him, at a time when books contradicting the literal word of the Bible still carried a considerable risk. De Maillet observed fossil shells embedded in sedimentary rocks on mountains high above sea level, and not aware that the land might rise, he concluded that water had originally covered the entire earth; it had since been steadily lost into the depths. He calculated a rate of sea level decrease of about 7 cm per century from sites where former ports were now above sea level. Working back to the highest mountains, he reached a figure of 2.4 billion years for the time that had passed since water had covered the earth. He was thus aware that long periods had formed and shaped the landscape.

De Maillet suggested that life had begun in the shallow sea. Lower forms of life appeared first, and later higher and higher forms arose. When land emerged, marine organisms invaded it, with all terrestrial animals having their origin in marine forms by direct descent. Seaweed developed into trees and shrubs, flying fish into birds, sea lions into lions and man derived from *l'homme marin*, the husband of the mermaid (Osborn 1905; Wilson 1992). Aquatic organisms thus found their way to land, where their new surroundings of air and herbage and their efforts to accommodate themselves resulted in a series of modifications. These modifications took place within the short period of a single life and were transmitted to descendants, who did not revert to aquatic forms. In his account of birds, he described flying fishes as driven out of the water by the ardour of the chase or by pursuit or carried by the wind. They might have fallen some distance from the shore among plants, which, while supplying them with food, prevented them from returning to the water. Here, under the influence of the air, their raised membranes transformed into wings, barbules and feathers; the skin became covered with down; the ventral fins became limbs; the body was remodelled; the neck and the beak became elongated; and the fish discovered itself as a bird. Thus, all organisms formed in two series: first, the aquatic and marine, springing directly from the germs; and second, the terrestrial and aerial, by transformation from the marine.

Yet another opinion was that of Denis Diderot (1713–1784), French philosopher, founder, chief editor and main contributor to the *Encyclopedie*. The secular tone of this encyclopaedia, which included articles sceptical about biblical miracles, angered both the Church and the French government, who had it banned (1758, 1759). Diderot claimed that all organic particles have a rudimentary sensibility that impels them to change their position constantly in search of the most favourable positions (1754). We recall, in ancient Greece, Empodocles' attraction and repulsion theory, applied to organic particles. Diderot's particles made ceaseless blind trials, impelled by their rudimentary senses. All faulty combinations of matter disappeared eventually, and only those favourable individuals who had no major misadaptations and who had the power of supporting and perpetuating themselves survived. In this manner, animals formed. These animals had, at first, only dim sensations, but eventually, the senses refined. Embryos formed in this manner went through a process of organization and development during which they acquired, in succession, movement, sensation, ideas, thought, reflection, conscience, emotions, signs, gestures, articulation, language, laws and finally the sciences and arts. Millions of years elapsed during each of these phases of development.

Buffon, in reconstructing the world's history, also sought to explain the appearance of order out of nothing. Like Diderot, he sought to develop a secular history of the world, a history that would establish science as independent from theology. The biblical Flood was, in his opinion, a moral proverb but not a real event, and we should exclude it from the geological history of the Earth. He designed a cosmogony that would replace the intervention of God by means of a natural phenomenon. Using Newton's physics of matter in motion and restricting himself to the solar system only, he proposed that a comet, hitting the sun tangentially, had broken off debris, which eventually became the planets. These all revolve in the same direction and on almost the same plane, turning on their own axis by virtue of the obliquity of the impact of the comet on the sun. In his secular theory, Buffon was abandoning not only Pluche but also Newton, who had suggested that the only possible explanation of the fact that the planets all revolve in the same direction, in concentric orbits and almost on the same plane, was that God had directly created the planets and their movement. Buffon went on to suggest that Earth, after having been formed in this manner (and not by the Word of God), was a sphere of hot, melted matter; gradually it cooled down until the molten rock turned to dry land. Clouds then rained down to form oceans, and it was here, within these oceans, that life was spontaneously generated, at an age in which Earth was still warmer than today. The oceans once covered mountaintops, but eventually, they receded, leaving marine fossils. Noting the similarity of internal structure in man, quadrupeds, whales, birds, reptiles and fish, he suggested that God, in creating animals, had employed only one idea, which He had diversified in every possible manner—perhaps indeed a common ancestor for all living beings (Wilson 1992).

These secular theories on the evolution of the universe came under attack by Voltaire (1694–1778), the ever-so-witty French philosopher who asked, "Why do we need to submerge the Alps under the sea, just to put a few shells on their peaks?" Pilgrims on their way to Rome had carried snails with them and, after having eaten their fill, discarded the empty shells. Buffon responded by cynically remarking: "Why did he not add that it was monkeys who transported the shells to the top of the high mountains? [...] How can it be that enlightened men who think to know a lot about philosophy, still have such false ideas on the subject?". Réaumur (1683–1757), a leading naturalist in France and one of the most influential personae in the French Academy of Sciences, orchestrated other criticisms. Claiming that Buffon contradicts *Genesis* in everything and that he drives God out of natural history, he recruited many of his disciples against Buffon, among them Charles Bonnet, whom we have already discussed.

Another attack, far more severe, came from the professors of the Faculty of Theology at the Sorbonne, who on religious grounds objected to Buffon's secular history of the world. Buffon was writing when religious orthodoxy held considerable sway, and it was dangerous to disseminate ideas that ran counter to prevailing biblical interpretations. The professors chose Buffon's *Histoire Naturelle* "to be examined and censured because it contained principles and maxims that are not in accordance with those of religion". They brought against Buffon a 14-point indictment because statements taken from his book were contrary to Church beliefs. Buffon had written about mathematical versus physical truths, whereas the only truth is that of the Scriptures. He had written that the Earth and other planets had formed by the collision of a comet into the sun, that the sun will, 1 day, go cold from fuel exhaustion, and that the Earth was once a fiery ball, all of which contradict the Genesis version of Creation. He also wrote that body and soul are but one, whereas the soul is an entity distinct from the body, as known by every devout Christian. The professors demanded Buffon to publish clarifications on all these points.

Buffon, concerned that the Church would blacklist his *Histoire Naturelle*, published the Sorbonne letter, and alongside it, he wrote that:

> I have no intentions to contradict the Holy Scriptures. I believe in everything that has been said as to the Creation and the timetable of its events. I take back all I have said in my previous volumes and books, as to the creation of the globe and Moses' holy book and theory. The idea of the formation of the planets I presented only as a philosophical idea. I retreat from all I have said, in my previous volumes, as to the word truth. I retreat from all what I said as though body and soul are but one. Man consists of body and soul, which are two different entities.

His letter of repentance reminds us that Buffon lived in an age very different from ours in terms of the freedom of scientific ideas.

5.2.3.2 Generation: Now and at the Beginning

Animal generation posed to Buffon a central problem broadly comparable to that of the creation of the world: the appearance of order out of nothing, of form out of matter. Anatomists of the time were incapable of discovering a structure in either the egg or seminal fluid that had any relationship to the extremely complex, organized embryo that eventually emerged. They also could not explain why each species has its specific, broadly fixed form—why offspring resemble parents. How, then, does an embryo develop? Buffon rejected the preformation theory, by which the first germ encapsulated all germs of all future generations of a species, formed by God's will. Theories involving God bring one, he claimed, not only to admit that one does not know how reproduction happens but also to lack the will to think about it. To imagine germs infinitely boxed one inside the other would be, Buffon claimed, like putting the object outside the field of one's vision and then saying that it is not possible to see it.

In seeking to develop a secular theory of generation, Buffon suggested that one of the elements in the world's order is a fixed quantity of what he named indestructible "organic molecules". Beneath all the variety among living creatures, there is a vast stock of different sorts of organic molecules, capable of creating various forms by entering into various combinations among themselves and thus forming diversity. All the apparently unique individuals in nature are nothing but different combinations of the same organic matter. Buffon did not define what he meant by the adjective "organic"; this term clearly did not have the chemical sense of today, which came about only in the nineteenth century. Whatever he may have meant by the term "organic", he believed that these molecules are the primary matter of life and that they are common to all animals and plants. They are absorbed directly from the food, and they enable body growth during an organism's early life.

Having claimed that underneath the great diversity of nature, there is a great reservoir, a constant number of organic molecules that determines form by combining with other organic molecules, Buffon next had to deal with the fact that in every species the combinations are both unique and constant. What determines the creation of a specific, permanent, highly complex living creature from a mass of randomly distributed molecules? Buffon, to fulfil this need for a fixed form in nature, developed the notion of an additional factor: the "internal mould" *moule intérieur*: a guide, unique to every species, which positions the organic molecules. The many sorts of organic molecules differ greatly from each other, and each internal mould admits only those organic molecules that are proper to it. It is the force exerted by the mould that ensures that the right molecules reach the right organs. A power similar to Newtonian attraction attracts molecules to enter the internal mould—"There exists in Nature a power penetrating to the internal parts of bodies that we call gravity, I apprehend the idea of an internal mould as analogous to this power". Buffon did not claim to know what the internal mould was, but he argued that we must recognize it if we wish to understand embryological development into animal species differing from one another. It would be ridiculous, he claimed, to ignore this hidden force just because we cannot see it—we cannot also see the force of gravity. A consequence of the actions of this interior mould is that species are permanent entities.

To us today, Buffon's comparison of his internal mould to Newton's gravity force is problematic: whereas Newton had written of gravity as a universal force common to all matter, Buffon described the internal mould as a collection of individual forces, each influencing some organic molecules and not others. The mode by which the two forces organized matter also differed: gravity was a universal property of all matter that acted uniformly, in contrast to the internal mould that arranged matter according to an intrinsic plan. Finally, Newton's gravity was a constant law of nature, whereas Buffon's internal mould was not; it came into being by reproduction and eventually died out upon the organism's death.

Buffon claimed that an animal takes in organic and inorganic matter indiscriminately. By means of the interior mould, it separates the organic from the inorganic, expelling the latter and utilizing the former to complete its growth. The organic molecules integrate themselves into the internal mould by penetrating it and then assimilating—they become similar in shape and identical to the matter of the mould. The internal mould positions the admitted molecules so that each part of the embryonic body receives only those organic molecules that it can use for its growth. Having thus shaped the form of the embryo, later on, and during growth, the internal mould preserves the body's correct proportions, and at maturity, it supervises the reproductive process of the species. Thus, for Buffon, the internal mould was the mechanism that determined the form of the organism and gave an empirical meaning to the concept of a species. Taking this basic system of attracted molecules back to the early history of the formation of life, Buffon suggested that at first the structures of new forms were the result of chance and that spontaneous combinations then produced an endless number of creatures of any type of structure whatsoever: imperfect, defective creatures were formed alongside perfect ones. Eventually, the imperfect, defective creatures went extinct, while others, who had better internal moulds, lived on and procreated. Survival was thus not random—the organic particles at first produced all possible types of structure, but only a limited number of these remained intact, thereby guaranteeing the absolute fixity of the species at all times. The only change in a species that Buffon would admit, at this early stage of his career, was a diminishing of stature in following generations, resulting from the fact that they would be formed at a lower temperature than the originals, owing to the gradual cooling of the Earth (as we shall see further on). Reading Buffon's theory of the early formation of life, with new forms emerging by chance and with an initially endless number of bizarre creatures, we recall ancient Greece and Empedocles, who had suggested that at first, nature produced only individual limbs and organs. Then, under the attracting power of love, limbs came together and joined in chance combinations. This resulted in monstrous, unviable creatures alongside normal ones. Eventually, those unfit for life died out, and only creatures assembled from homogeneous limbs survived, as the animals of today. We also recall the many reports of severe birth defects in animal and human offspring, from at least the Renaissance and onwards, of "monsters" that never survived.

In line with this idea that in distant times some unfit animals formed, Buffon also suggested that to this very day, some animals are unfit in that their character combinations are non-harmonious. The pig "obviously has useless parts, or rather parts of which it cannot make use, digits in which all bones are perfectly formed and which nevertheless do not serve any purpose" (1755; he was referring to those toes of the pig that do not usually touch the ground; Fig. 5.39).

Buffon also argued that the toucan's enormous, fragile beak (Fig. 5.40), far from serving any functional purpose, only harms the bird whose flight it unbalances, without being useful for crushing food. This beak was, he thought, one of nature's mistakes. He went on to define a mistake of nature. It is a "disproportion joined to

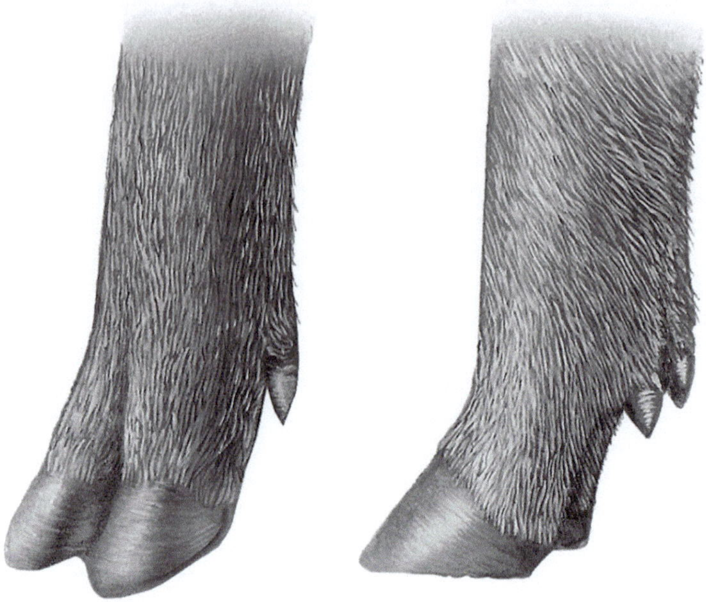

Fig. 5.39 The foot of a pig. The upper toes do not tread the ground, so Buffon claimed they were useless

Fig. 5.40 Toucan. Buffon regarded the enormous bill as a mistake of nature

non-utility: all parts in animals that are excessive, overabundant, placed the wrong way and that are at the same time more harmful than useful, must not be put in the great plan of Nature's direct intent, but in the little list of her whims or, if one prefers, the mistakes". Let us therefore not seek to make everything reasonable; there is no complete perfection of design in nature. (Today we know that the toucan's bill acts like a thermal window or thermal radiator that allows the bird to conserve or release heat, like an elephant's ear.)

Turning to generation, Buffon pointed out two possible methods: spontaneous and sexual. The works of Trembley on hydras seemed to him to have verified spontaneous generation, as also did the studies of John Turberville Needham (1713–1781), an English Catholic priest living mainly on the continent and a promising young expert in microscopy. Needham had published (1745) microscope observations on squid, on "eels" (what we today call nematodes), on blighted wheat and on pollen. In squid semen, he believed he had discovered strange little mobile bodies that act like little machines ("milt vessels") as carriers of the seminal fluid, and he suggested that similar machines occur also in pollen (Fig. 5.41). These observations led him to wonder whether in other animals and plants, spermatic animalcules found in seminal fluids were also small machines, rather than live creatures. He came to Paris in 1746, equipped with recommendations to Buffon by the president of the Royal Society of London, and Buffon recruited him because he sought to submit his reproduction theory to the test of microscope observations. The two set out to test whether all microorganisms (not only spermatic animalcules) are indeed tiny machines formed by the combination of non-living organic molecules. If indeed there were organic molecules in living beings and if these molecules did indeed assemble for reproduction, then they might be observable as distinct units.

In 1748, Needham and Buffon placed an infusion of almond seeds and water in vials, which they stopped up to prevent any contact with the exterior environment.

Fig. 1.

Fig. 5.41 From Needham, *Observations upon the Generation, Composition and Decomposition of Animal and Vegetable Substance* (London 1749): "represents the origin of the spermatic animals"

Fifteen days later, they found in this infusion clouds of small microorganisms moving about, and through the microscope, they thought they could distinguish "organized bodies" formed in the almond infusion. What they described as a coalition of active principles. A distinct particle would detach itself from a cluster of motionless particles and then advance to a distance of about ten times its diameter, or it would move about in circles, fall between other particles and then detach and move on. They believed that something inside that particle caused this movement (Needham 1749; Buffon 1749 vol. 2). Next, they sought to explore whether the particles thus formed were animals. They took hot boiled mutton gravy from the fire and shut it up in a flask, which they tightly closed with a cork. They left this sealed gravy to cool for several days, and when opening the flask, they found that it swarmed with live, microscopic animals of various sizes. They thought they saw certain kinds of microscopic creatures giving birth to other kinds of creatures, and they suggested that in the embryonic development of higher organisms, a similar phenomenon must occur. They argued that these microscopic animals could not have resulted from eggs or organisms previously inhabiting the gravy because the boiling heat would have killed them; they also could not have come from eggs deposited from the air because the flask was sealed. They concluded that the microorganisms must have formed from indestructible organic molecules (observed as globules) that had stood the test of fire and were now combining to form microorganisms. Having thought they had proved that microscopic organisms could develop spontaneously in boiled meat broth, they claimed that there is a "vegetative force" in organic material that can create living organisms; in other words, that spontaneous generation exists.

Some months after having carried out these observations, Buffon left Paris for his home in Montbard, and Needham continued experimenting and observing on his own. He repeated the hot broth experiment but with urine and with blood rather than gravy, obtaining the same results, "and in these I believe no one will suppose that Germs, Eggs or Spawn, are pre-contained, if care is taken to close the phials immediately". He found that even a butterfly's wing, if taken from the embryonic pupa state, would almost immediately vegetate into a microscopic creature. In an infusion of pounded wheat grains, he observed filaments that swell and shed small self-moving globules "due to some inner force"; they came in various forms (spherical, oval, oblong, cylindrical) and died within a day or two, leaving inactive, dead, irregularly formed particles at the bottom of the vial. Then, fourteen-fifteen days later, he thought he had observed that these dead particles united into a big mass from which filaments arose, and from a force lodged within each fibre, various zoophytes then sprung. Some of these zoophytes resembled pearl necklaces, others were uniform throughout their length, and some others had a swollen head resembling the head of a bone at its joint. Zoophytes consisting of four to five globules would advance a little way in concert, by what seemed to him to be a wormlike motion, and then would fall aside motionless, as though exhausted for a while (Roe 1983).

Needham suggested the existence of a vegetative force in every microscopic particle of matter and in every filament of every animal or vegetable. This force, he claimed, enables the filaments to vegetate into numberless zoophytes, which yield many types of microscopic animals. These all eventually become motionless, drop to the bottom and resolve into a gelatinous, filamentous substance. After some time, they give rise to zoophytes of a lesser species. This cycle he believed to be true also of the spermatic animalcules that, after losing their motion, sink to the bottom and are resolved into filaments. We should, therefore, classify spermatic animalcules within the common microscopic animals. He further believed that every animal or vegetable substance seeks to return to one common principle that is the source of all life, "a kind of universal *Semen*, whence its Atoms may return again, and ascend to a new Life". This common element, uniform in its origin, branches out, however, into innumerable species differing in so many ways that the semen specific to one species cannot change into that of another. The once-uniform sperm of all animals has evolved and become species-specific. Needham went on to suggest that microscopic animals constitute a distinct group of creatures, which do not subsist by nutriment as other plants or animals do and do not generate in the ordinary way (Roe 1983).

Buffon, having observed spermatozoa, believed that he and Needham were seeing "globules" progressively detaching themselves from "filaments" to which they stayed attached by a thread for a certain time (Figs. 5.42 and 5.43). As these threads hinder movement, the globules try to rid themselves of them. Eventually, the thread snaps and the globules then free themselves, dragging the thread remnant behind them for some while until, finally, the thread disappears. Thus liberated, a globule now moves faster and more freely in every direction. These globules, which crossed the microscope's field in less than 4 s, line up against each other; they move in a line of seven or eight abreast and follow one another like marching troops. Clearly, Buffon and Needham were lumping together sperm, protozoans, bacteria and cell fragments animated by Brownian movement. What was the nature of these globules?

Fig. 5.42 Microscopic research (*HN* 1749, vol. 2)

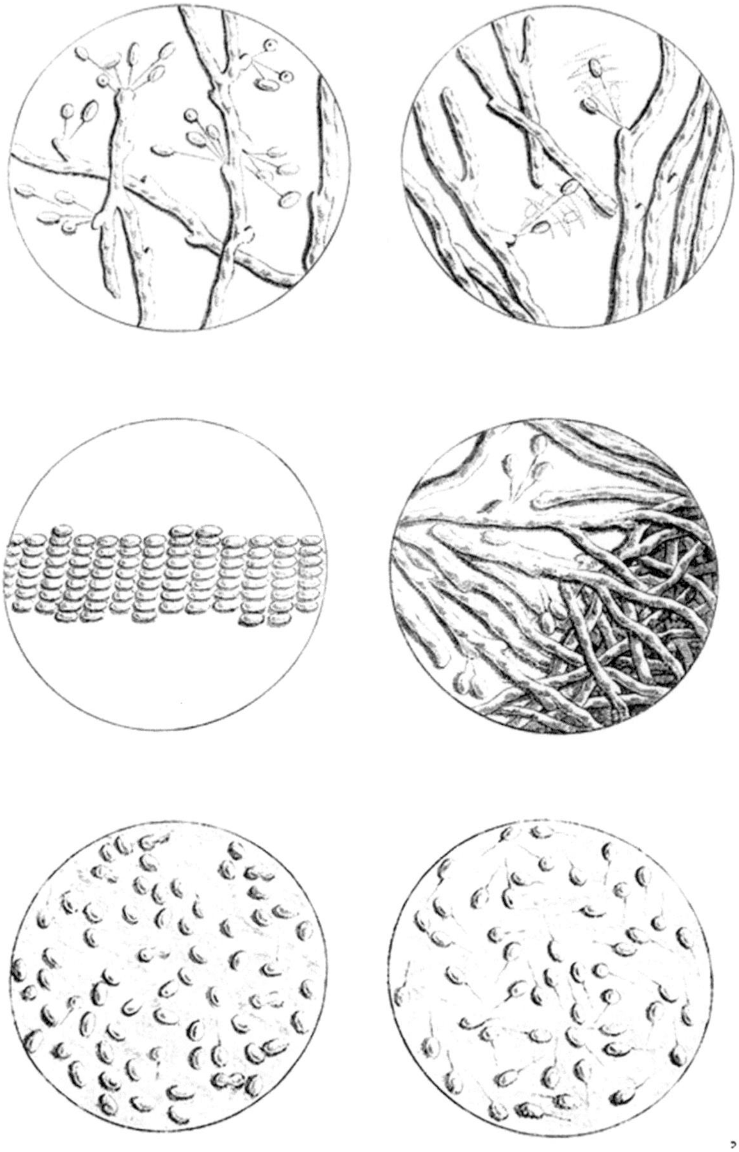

Fig. 5.43 Microscopic observations by Buffon (1749, vol. 2)

Having observed all these many different globules, Buffon began considering the existence of intermediary beings between the living and the non-living beings "that would be no more than the primary assemblage, of organic molecules". These beings might be "organized bodies that are more like natural machines than animals", beings that move about continuously and that constantly change shape. These organic parts in motion, which one can observe in rainwater and in plant infusions, were, in his opinion, the "primary assemblage" of organic molecules. Needham's gravy experiments proved that these moving organic parts were not true animals, as true animals would not have survived the test of fire. They therefore consisted of indestructible organic molecules. This meant the existence, for Buffon, of spontaneous generation, namely the birth of something living from something non-living. In lower animals, such as worms and hydras, Buffon thought that spontaneous generation occurs when a group of organic molecules succeeds, as in death and disintegration, to free themselves from the mother mould. The freed organic molecules then join other molecules and produce primitive organisms. Buffon further suggested that in the case of parasitic worms infesting large organisms, the body of the host could emit organic molecules that then reorganize themselves in the body of another species (*Histoire Naturelle* vol. 7).

Buffon further claimed that in those simple, nonsexual animals composed of uniformly structured miniatures (hydra), each of which can sustain itself as a separate internal mould, reproduction takes place when a part splits off from the whole. In other nonsexual animals (plant lice), after reaching maturity, every part of the organism sends its excess molecules to an appropriate place for reproduction. There these molecules come together and reassemble, by attractive forces, in such a way that a replica, and thus a similar internal mould of its parents, is reproduced.

In more highly organized creatures, such as quadrupeds and men, reproduction is sexual. Buffon argued that the material used for sexual reproduction comes from the surplus of organic molecules not needed for nutrition and development, as in lower animals. He was convinced that he had seen the same basic formations of organic molecules in the female as in the male, and this proved to him that both male and female contribute elements in reproduction. As in so many other cases, Buffon based his explanation of sexual reproduction on the Newtonian analogy of attraction. When two bodies of organic material meet, he argued, a field of force establishes itself that results in equilibrium, and the organic molecules coalesce into a small, organized body resembling its parents. In this way, the form, or internal mould, of the parent passes on to its offspring and on through all the many generations that constitute a particular species. Therefore, it is the internal mould, as the moulding power of the body, that assures the distinctions between the various species. It is what makes sense of a species as a genealogical entity. Accordingly,

> Species are the only beings in Nature; they are perpetual beings, as old, as permanent as Nature herself. In order to judge better of ourselves, we will no longer consider a species as a collection or a continuation of similar individuals, but as one living whole, which is always the same; a whole, which has been reckoned one in the works of creation, and consequently constitutes only one unit in Nature (*Histoire Naturelle* vol. 13).

Coming now to another fundamental question in biology, Buffon asked what determines sex. He answered that both male and female contribute to determining the sex of the offspring. At this point, he (yet again) used the analogy of Newtonian attraction: when two bodies of organic matter meet, a force field forms, and one of the sexes will eventually be dominant. The forces of the sexual molecules in each of the donors determine which sex will dominate—the parent contributing the greater number, whether male or female, will determine the sex of the offspring. If there are more male sexual molecules, then the child would be a boy; if not, a girl. Surplus molecules are not used to determine the sex of the embryo, female or male, depending on whether the child is to be a boy or a girl, from the quadruped placenta. From the moment its sex is determined, the offspring begins developing, and its actual formation takes place. The sexual molecules are the nucleus around which other molecules gather and organize. The force field and principle of attraction ensure that the proper form will result and that all head molecules from the male and female are eventually delivered to the same location where the head is to develop. At the same time, the internal mould is established, and it accounts for the fact that an offspring is an almost exact copy of its parents.

The internal mould ensures the distinction between species and the similarity between members of the same species. It controls the organic molecules and gives significance to the species as a dynastic entity. Thanks to the internal mould, the basic features of each species remain unchanged, no matter how far back we look in time, no matter how much we speculate into the future. All individuals of a species permanently bear the same form as did early generations of that species, back to its first creation. True, they may be subject to variations due to food or climate, and the descendants would inherit these variations, but only to a slight extent that does not threaten to break down the species definition. Buffon considered species fixity a basic law on Earth and throughout the universe. It limits the possible combinations of organic molecules and ensures that all organisms form in the image of a limited number of specific types. The immutable species, an empirical reality, was a basic element in Buffon's order of the world.

Further, his concept of the interior mould was of some form of prototype, which served as a model for the structure of the species. The first horse has been the internal mould and model for all horses since and those that will ever exist. This prototype model, of which today we know only its copies, has, in multiplying its form, the power of slightly varying. Each individual preserves the original impression, but among millions of individuals, not one exactly resembles another nor, of course, the model from which they sprang. "There is a strange variety in the appearance of individuals and at the same time a constant resemblance in the whole species. The first animal, the first horse, for example, has been the exterior and interior model, from which all horses that have existed, or shall exist, have been formed" (*Histoire Naturelle* 1755, vol. 5). Reading Buffon's theory of an interior mould as a prototype upon which each real individual models itself, we recall Plato, who believed that beyond all real creatures in our world, there is an eternal, perfect, abstract model of each group.

To explain the formation of prototypes, Buffon claimed that under proper physical and chemical conditions, an attractive force (yet another variant of Newton's gravitational force) causes the organic molecules to penetrate the malleable matter, and then heat (associated with the expansion of matter) distributes these molecules. Nutrition and growth follow, and the result is a species. Even if God destroyed all living creatures on earth, organic matter, since it has a propensity for organized life, would form new internal moulds that, except for being smaller (heat diminishes with time), would resemble the old ones because the same general plan of organization would exist. Believing that life is an inevitable outcome of environmental conditions on Earth, he further believed that it would appear on other planets too. Given proper physical–chemical conditions, the same species would appear at any time, in any place in the universe. Two identical species, two temporal lineages with the same form, which reproduce in separate times and places, can exist as separate entities without being descended from a common ancestor: they have the internal mould. Buffon's species are not, then, only concrete historical lineages; they are also universal types of organic organization. The laws of nature are universal, and what is true for Earth is true for other planets and their moons: as soon as the temperature was or would be suitable, life appeared or would appear. Buffon argued that life had disappeared from the moon two thousand years ago and that it also had disappeared from Mars, but it was currently in full existence on Mercury, Venus, Saturn and on a few moons that consequently "could be peopled like the terrestrial globe". Jupiter was, however, still too hot and would have to wait yet another 41,000 years for life to appear on its surface. Since the cooling of planets was a continuous process, Buffon also estimated for each planet the time that life still had before it would chill to extinction.

His theory of generation was, for Buffon, a vital part of his natural philosophy. It offered a vital key to the study of nature, which he had to work out in detail before passing on to a study of the species making up the animal kingdom.

5.2.3.3 What Is a Genus? A Species?

Buffon believed that science is all about discovering order in nature's diversity, and yet he firmly believed that this order was not the Linnaean classification. He ridiculed Linnaeus for putting together into the same groups such different plants as blackberry bushes and nettles, tulips and barberry, elms and carrots and roses and strawberries; and for lumping into the same animal groups man and sloth, hedgehog and bat with bear and hippopotamus with horse and elephant. Linnaeus had erred by arbitrarily choosing one single essential part in the plant or animal to be the classifying characteristic; yet if you looked at the entire organism, the seemingly clear-cut Linnaean divisions faded away.

Strongly opposed to the essentialism philosophy behind Linnaeus' classification, Buffon claimed that Linnaeus' philosophical error was in assuming that we can apply a hierarchy of abstract group-category concepts (say, the Linnaean rank order) to a world containing only concrete individuals. Furthermore, the grouping of organisms by one or two so-called essential characters, as Linnaeus had done, results in arbitrary groups imposed by the mind, not in real groups of nature. Buffon's attack on Linnaean taxonomy was thus an attack on the philosophical foundations of Linnaean essentialism. Buffon rejected the ranks of genus and order because there were no empirical grounds, other than resemblance, for claiming them to be more than groups that "we have made only to aid our understanding, and which cannot comprehend the real sequence of creatures". Only if we show that such groups hold empirically could we conclude that they are real at the level of physical truth.

In contrast to Linnaeus and his use of only one or two essential external characters, Buffon believed in giving a full description and a full natural history of each species. One should indeed start by describing a species' external morphology but then go on and include internal morphology. The description further must include such aspects of natural history as generation, pregnancy duration, the number of offspring, parental care, instincts, diet, the way of obtaining food and habits. It would further include the services this kind can render us and "all the uses and commodities that we can obtain from them". Behind each name of a species, there is a living animal with a rich natural history, and we must know this natural history and incorporate it into our definition of that animal. Buffon was thus very far from Linnaeus and from his classification based on external morphologies as representing essences.

Buffon was also against what seemed to him to be a Linnaean overemphasis on the genus category, thereby reducing the species category to a mere supplemental adjective, to a minor phrase consisting of a few descriptive words. Linnaeus had strictly followed a descending classification, from the animal kingdom down to class, from class down to order and from order down to genus. In this descending classification, the final division of the genus into species seemed trivial because it rested on morphological detail that, at this very low level of classification, was rather negligible. Linnaeus had described both horse and ass as of the genus *Equus*, the former as a horse with a tail full of hair *Equus cauda undique setose*, and the latter as a horse with a tail full of hair only at the extremity *Equus cauda extreme setose*. The ass in Linnaeus' description was thus merely a kind of horse. Buffon, in contrast, based his classification on reproduction, so he started with the species category. He could find no meaning in climbing upwards, in an ascending classification from species up to genus and order. He believed that an ass is not a horse with a furry tail. Indeed, Buffon argued, according to this Linnaean outlook, the lynx would be only a kind of cat, and the fox and the wolf, a kind of dog. Would it not be simpler, more natural and truer to say that the ass is an ass and the cat a cat than to believe that the ass is a horse and the cat a lynx? There are only 200 quadruped species on earth, and anyone can remember 200 names. Why, then, arrange them into classes, orders and genera? Furthermore, why employ heavy Latin jargon when we can speak plain language by pronouncing simple names?

Buffon further noted that even at the practical level, those following the Linnaean system do not agree among themselves on genus-level classifications. The mongoose was classified by Linnaeus and his followers once as a badger but then as a ferret, a weasel, an otter or a rat—evidence of how little consistency there is in applying the false, arbitrary, vague and equivocal term genus (Fig. 5.44). To this purely academic argument concerning the justification for using the genus rank, Buffon added a tiny drop of national patriotism. Frederik Hasselquist (1722–1752) was a Swedish pupil of Linnaeus who had travelled to the Holy Land and nearby countries (Asia Minor, Egypt, Cyprus); he died near Smyrna on his way home. Five

Fig. 5.44 Egyptian mongoose (*HN* 1765, vol. 13)

years after his death, Linnaeus published his notes and had them translated into French (1762). Buffon, in reading these notes, found that Hasselquist had insulted France by saying, in his description of the mongoose, that Frenchmen who live in Egypt and who impose imaginary names on all things they have no knowledge of have named it Pharaoh's rat. Buffon responded that had Hasselquist really read the authors he attacked, he would have found that the Egyptians themselves had called it by this name, and "he would have abstained from taking an occasion here to speak badly of our nation. However, we must not be surprised to find the accusation of a pedant in the work of a schoolboy" (1765, vol. 13: 154–5; *The Mongoose*). Further, Hasselquist, in his very lengthy description of the giraffe, had omitted mentioning the nature of its horns. Now, "if horns fall every year, it is of the genus of the deer, and if on the contrary its horns are permanent, it is of that of cattle or goats" (1765, vol. 13.). Actually, in giraffes the horns do not fit into the typical deer-cattle division in that they are solid as in deer and permanent as in cattle.

Further, at a cultural level, a classification like that of Linnaeus that emphasizes the genus category rather than that of the species reflects a primitive approach:

> For primitive peoples and all nascent languages there are almost only general names, that is to say, only vague and undefined expressions of things of the same order. Particular names only came after a detailed examination was made of each sort of thing: the number of these names grew, as Nature was further studied and became better known. [...] Ignorance made genera. Science made and will make proper nouns. (1758 vol. VII: *Le Rat*).

Primitive nations are capable of classifying animals only crudely, down to the genus category. Enlightened nations observe detail and are capable of high-resolution classifications, down to the species level. (With a smile, we recall, at this point, those excellent species-level classifications of "primitive" people.) Also elsewhere, Buffon remained a Frenchman. He wrote that the most temperate climate is from the 40th degree to the 50th. In this zone, the most handsome and the best-built men are found; it is in this climate that the idea of man's true colour takes form; it is from here that the model to which all other nuances of beauty must be compared comes. France so happens to lie mainly between the 40th and 50th latitudes.

In the concrete world, in nature, only one natural category exists, that of the species—that succession of successfully reproducing individuals. Buffon declared that species would be the only units he would present, each separately (Farber 1972). He would not divide groups of creatures in any way differently than what they were in reality; he would not use higher classification categories, as nature does not use these categories either. Accordingly, he proceeded to describe the 200-odd species of quadrupeds known in his day, one after the other. This called for a definition of the term species. In presenting his definition, he emphasized that a species is a chain of similar individuals that reproduce and interbreed. "One must consider as the same species, that which by means of copulation perpetuates itself and conserves the similarity of that species; and as different species, those that through the same means can produce nothing together" (1749 vol. 2). A species is a system of concrete relationships between real creatures at the level of physical truth, not an abstract essence. By presenting this definition, Buffon gave a criterion by which one could decide if an individual belonged to a species, without committing himself to any philosophical position (Farber 1972). Years later, he would point out that

> "[...] these intervals between species are the most uniform and the least variable of all, since one can always draw a line of separation between two species, that is, between two successions of individuals that cannot reproduce with each other. This distinction is the strongest that we have in natural history" (1776 Suppl. 3).

In his definition of species as reproductive units, Buffon was closely following John Ray (1704, *The Wisdom of God*), whose reproductive concept of a species was radically different from the essentialist concept in that it was applicable only to organisms. It was a biological concept since only living organisms reproduce, whether or not two individuals of a (sexual) species are the same. Buffon may probably have developed his ideas on what a species is from a philosophical controversy on a completely different issue—the nature of time and space. Isaac Newton (1642–1726, Fig. 5.45) claimed that time and space are absolute, abstract values. Space is separate and independent of the bodies within it, and time, likewise, is something that passes, unrelated to whatever happens in the universe in the meanwhile. We must

Fig. 5.45 Newton

separate these two absolute, abstract entities from the various ways in which we measure them, measurements that produce "relative space" and "relative time". We should thus consider time and space both as absolute, abstract entities and as relative, concrete ones. Each relative dimension requires a fixed, arbitrary reference point. For any fixed reference point in space, there is a close and a distant; for any fixed point in time, there is a before, and there is an after. The German philosopher Leibniz (1646–1716, Fig. 5.46) and his follower Christian Wolff (1679–1754) disagreed with Newton's concept of an absolute temporal–spatial entity. They believed that temporal–spatial reality exists only in relation to concrete, material objects; without them, time and space have no significance. Space is merely an arrangement of objects existing at the same time; no space exists outside actual existence. Likewise, time has significance only in terms of the order and degree of change. Without concrete things, there is no time or space.

Moving from Newton and Leibniz half a century onwards and into biology, Buffon applied this philosophical issue, of the essence of time and space, to biological classification: he based the concept of the species on a concrete continuum of time and space. Both Newton and Leibniz had defined the reality of the world as composed of the totality of all things that exist in a continuous succession, thereby

Fig. 5.46 Leibniz

defining the concept of time. Buffon placed the species as the ultimate example of a dynastic series of this sort, in which the concrete quality exists in the reproductive continuum. A species is a sequence of similar, reproducing individuals, and it is the only entity in nature. "It is this power of producing its resemblance, this chain of successive individuals, which constitutes the real existence of the species" (1749, vol. 2). A species is not a group of similar individuals. It is the continuous succession and renewal of these individuals from one generation to another—this is what constitutes a species. Species is a term that exists only when we observe nature in a dimension of time, in the continuous procedure of the destruction and renewal of creatures. By thus positioning the study of the species as the study of the concrete and real, with the reality of the species being the reality of time, Buffon differed very considerably from Linnaeus, who had characterized species by structural features representing a hidden, abstract essence. The species, defined as a constant succession of similar individuals that can reproduce, is the only living entity in nature.

> Species is an abstract and general term, to which a corresponding object exists only in considering Nature in the succession of Time, and in the constant destruction and renewal of beings. It is in comparing Nature today to that of previous ages, and existent individuals to those deceased, that we can derive a clear idea of what is called a species, and the [...] resemblances of individuals is only a subordinate and often independent consideration (1749 vol. 2; vol. 4).

The reproductive definition of a species deals well also with the claim presented by Locke that boundaries between species are abstractions made by men, not by nature. ("I would fain know why a shock [poodle] and a hound are not as distinct species as a spaniel and an elephant."). Buffon responded: "The ass resembles the horse more than the water spaniel the hound, but nonetheless the water spaniel and the hound are only one species. This is so since they together produce individuals that can themselves produce others, whereas the horse and the ass are certainly from different species, since they together produce only defective and barren individuals" (1753, vol. 4). Species really exist in nature as elements defined by their very natural limits. Inter-fertility is the criterion that allows recognition as to which animals belong to the same species and which ones belong to another. We may always draw a clear line of separation between two successions of individuals that cannot mix when reproducing, and these intervals are the most well-defined lines of separation occurring in nature. All other resemblances and differences in the comparisons of beings are neither so constant nor real nor certain. The species, defined as a constant succession of similar individuals that can reproduce, is the only entity in nature. One consequence of Buffon's use of reproductive rather than morphological criteria was that he was tolerant of morphological variation within a species.

In different parts of Earth, in similar environments, similar animals occur: elephants in Africa and India, vultures in Europe and America [the condors, an error]. This seems to have led Buffon to generalize that in all the places where the temperature is the same, one finds the same species of plants, insects and animals. (1749, vol. 2: 509).

We recall that Buffon, together with Needham, thought they had observed minute animalcules forming spontaneously in rotting organic matter due to a vegetative force in organic matter that can create living organisms by spontaneous generation. Three decades later, Buffon took this growth force a step further (1777, supplement 4): if God were to destroy all living creatures on Earth, the organic molecules would survive and combine to create a variety of new life. At first, the structure of the new forms would be completely random, and the organic molecules would create an infinite number of organized creatures, of which only a few would survive. Eventually, all imperfect organisms will become extinct, and only the strong, perfect ones will survive. These strong ones will resemble the forms existing today, since live matter is always the same and will yield the same general pattern of organization and the same variations (supplement 4: 365). Survival is not random.

Buffon further speculated that this generalization holds true also at the cosmic level: The very same life was bound to have formed on each of the various planets as they cooled down from their originally molten state and reached temperatures (climates) similar to those prevailing on Earth. This speculation, that the same living structures will appear under the same conditions anywhere in the solar system, led him to believe that universal laws govern life formation. If all organisms in the world were to be destroyed, then organic matter, as it has a tendency to organize itself into life, would rapidly create new internal moulds, which would resemble the old ones, because the same general pattern of organization would still exist: elements that possess the universality of natural laws control the creation of life. When proper physical–chemical conditions exist, identical species will appear repeatedly in any place or time in the universe. There may thus be several dynasties with the same form, which had formed in different places and times and can exist as separate entities with no historical connection. They belong to the same species because they have the same internal mould. Species are potential entities that will form every time the requisite conditions exist for the formation of the internal mould. Buffon's species was a global prototype of organized organic matter.

One consequence of Buffon's reproductive definition of the species was that the term became relevant only to the organic realm. It does not apply to minerals, which Linnaeus had presented as consisting of the same hierarchy of groupings as he had given for organisms. Another consequence was that hybridization experiments could determine, for every two animal groups, whether we are dealing with two varieties of the same species or with two species. Buffon himself carried out several such experiments with the peccary and the pig, dog and wolf, dog and fox, hare and rabbit, zebra and ass—all without positive results (*HN* 1763, vol. 10). He also presented the experiments of others, who had found that European and Siamese pigs interbreed, as do also wild and domestic cats and Indian humped cattle with ordinary European cattle. Some of his breeding experiments yielded confusing results; thus, when a billy goat mated with a domestic sheep, the offspring was a lamb covered with hair, but the ram did not breed with the nanny goat. From this, Buffon concluded that the goat was what he termed the dominant species and the sheep the subordinate one.

Hybridization experiments in which at least one of the participants was a farmyard animal could be carried out with relative ease. For wild animals, such experiments were usually not possible, and one had to explore other evidence. If two species dwelt in the same region without interbreeding, then that in itself was evidence for Buffon that they were distinct species (say, deer and fallow deer). Another clue was gestation. Does carry 8 months, while roe deer carry only five and a half months, a difference so considerable as to prove to Buffon that they cannot interbreed and are therefore of different species. On the other hand, "the goat carries about the same time, and the roe deer can [therefore] be considered a wild goat". I wonder whether the similarity of the Latin names, *Capra* and *Capreolus*, was involved in Buffon's decision to unite goat and roe deer.

Beneath the species level, Buffon recognized the variety: a subordinate historical lineage within a concrete species, produced by the degenerative change in response to environmental conditions. In his view, a permanent variety became a "race": The races in each species of animal are only constant varieties that perpetuate themselves by generation. Buffon's variety, or race, thus differed from Linnaeus' accidental, inconstant variant. A further difference between the two scholars concerned man. Buffon, in describing man, used traits similar to those used by Linnaeus: external morphology (colour, height, hair, lips, nose, face and eyes) and behaviour. He differed from Linnaeus in emphasizing the many gradual gradations between white and black in ever so many shades of brown, suggesting that the range from white to black is continuous rather than discrete. Likewise, hair types are not clearly distinct, as one finds, even in France, people who have hair as short and curly as Negroes. Negroes do have flat noses and large lips, but this is so only because, so he thought, their parents methodically crush them. He claimed that climate and food control racial differences; abundant food can make people stronger and more handsome. He thus thought up anything he could to belittle racial differences and to present mankind as one unified group with gradual transitions from one race to another. Man is almost unique among animals in that he forms a single, highly diverse species that carpet-covers the globe. As this species encounters different climates and foods in different regions, different traits become more emphasized and gradual transitions form. Of the Negro, whom Linnaeus had defined as phlegmatic, lazy, neglectful and capricious, Buffon wrote that they "do not fail to have a great deal of feeling; they are gay or melancholy, hard-working or idle, [...] according to the way they are treated; when well-fed and not mistreated, they are happy, joyful, ready to do anything" (1749, vol. 3). Negroes, like whites, are compassionate and tender with their children, their friends and compatriots, with whom they willingly share the little they have. In a heart-touching text, Buffon wrote that he cannot write their history without being moved by their state.

> Are they not unhappy enough at being reduced to servitude, at being obligated always to work without ever being able to acquire anything? Is it also necessary to exhaust them, beat them, and treat them like animals? Humanity revolts against the hateful treatments that a greed for profits has created [...]. We force them to work, we limit their food, even of the most ordinary kind (1749, vol. 3).

In coming to write his encyclopaedia, Buffon declared that species, namely each succession of successfully reproducing individuals, would be the only units he would present, each one separately. He would not cluster groups of creatures in any way differently than what they were in reality; he would not use higher classification categories, as nature does not use these categories either. Accordingly, he proceeded to describe the 200-odd species of quadrupeds known in his day, one species after the other. In what order should he present all these quadruped species?

5.2.3.4 Species Diversity: By What Order?

Buffon, following the Newtonian–Leibniz concept of relative time and relative space, organized all quadruped species relative to a fixed, arbitrary reference point. The reference point he chose was man, both because he is the best-known animal and because man is a knowing subject, one who makes science. The naturalist should study objects according to the order in which they appear before man's eyes, beginning with the important ones close to him. Therefore, Buffon took man as the starting point for all his comparisons. Not just any man, of course. Man—as represented by the French affluent.

The order of animal species should correspond to what such a man discovers around him in ever-expanding circles. Buffon called upon his readers to imagine a man who wakes up having forgotten everything, one innocent of everything around him. Place that "innocent man" in the countryside where the animals, birds, plants and stones present themselves to his eyes. Gradually, he will arrive at nature's three major divisions: animal, vegetable and mineral. As easily as he distinguished these three great divisions of nature, our innocent man would differentiate, among animals, the quadrupeds from the birds, fishes and "insects" (invertebrates). These are the real groups that "we must respect as a division given by nature herself". We recall Locke and his intuitive knowledge as the most certain knowledge because the truth is grasped immediately. Further back we recall Aristotle, who had claimed that each entity in nature has an unchanging essence, which makes it what it fundamentally is; by using the essential feature, one classifies an individual into one of several larger, well-known groups (such as birds, fishes and insects). Back even further, we recall biblical Adam, to whom Linnaeus liked to compare himself. Buffon's innocent man was a secular Adam.

Buffon's innocent man would next classify quadrupeds according to their relationships with him. Farmyard animals, most necessary to him and most useful, would occupy his first, immediate circle. Next would be wild quadrupeds of his country, and finally quadrupeds of foreign countries. In this way, he would arrange them in his mind because it is indeed both the order according to which he acquired them and those important for him to keep. Of course, that order leads to putting very different animals together. Is it not better to have the horse, which is single-hoofed, followed by the dog, which has many toes, rather than by a zebra, which is rarely known to us and perhaps has no relation to the horse other than being single-hoofed? It would be better to classify by following the order of the real relationship that these quadrupeds have with us. Accordingly, and with the French rural affluent as his reference point, Buffon grouped all domestic animals of Europe in the first volumes—the ass, the cow and the cat—before those wild animals close to them, such as the zebra, buffalo and lion, which he presented in much later volumes. Domestic animals of Asia—the elephant, camel and dromedary, buffalo and Indian humped cattle—he clustered separately and published a decade later.

Then there were matters of nomenclature, an important topic in any encyclopaedia of different animals. Buffon considered name-giving as only a convention, an arbitrary language to be used just as any set of agreed-upon signs are used, so as to improve common understanding. There is no need for a name to consist of a description in Latin, as given by Linnaeus and by others (we are here speaking of the years before 1753, when Linnaeus imposed binomial nomenclature on all three kingdoms of nature). Buffon sought to have, as names, one-word-long terms in a living language, not long Latin descriptions as used by Linnaeus (in his first editions). Why, he asked, should zoologists make up a heavy scientific jargon consisting of long phrases in a dead language when one can speak clearly by pronouncing one simple word, say, by the French common name? Buffon named all animals by their French common name. As to exotic animals not found in France, he gave names used in the country where they live, thus "jaguar", "tapir" and "capybara" as named by the Indians of South America; "desman" for the muskrat of Muscovy as named in Sweden and "caracal" from its name in Turkish. For new species discovered in France, Buffon would coin new names. When Daubenton, on close inspection, had discovered that in France there were seven bat species rather than the two species commonly believed to be, Buffon coined new French names, sometimes taken from the Italian: "oreillar", "noctule", "serotine", "pippistrelle" and "barbastrelle".

Coming now to Buffon's descriptions themselves and starting with domestic animals, the first animal presented by Buffon was the one most important to the French affluent—the horse. His introduction to the chapter on the horse suggests that what mattered to Buffon at least as much as the horse (with, in its illustration, a large rural mansion in the background, Fig. 5.47) was the horse in reference to that fixed reference point he chose, the French affluent landed gentry.

> The noblest conquest ever made by man over the brute creation is the reduction of this spirited and haughty animal, which shares with him the fatigues of war and the glory of victory. Equally intrepid as his master, the horse sees the danger and encounters death with bravery; inspired by the clash of arms, he loves it and pursues the enemy with the same ardour and resolution. He feels pleasure also in the chase and in tournaments; in the course, he is all fire but equally tractable as courageous. He does not give way to his impetuosity and knows how to check his natural and fiery temper. He not only submits to the arm that guides him but also seems to consult the desires of his rider; and always obedient to the impression he receives, he presses on or stops at his rider's pleasure. The horse is a creature that renounces his very being for the service of man, whose will he even knows how to anticipate and execute by the promptitude of his movements: he gives himself up without reserve, refuses nothing, exerts himself beyond his strength, and often dies sooner than disobey.

Was Buffon here describing the horse—or the French affluent horse rider?

Three decades later, he would describe the horse in further anthropomorphic terms: "In each of the horse herds roaming over the plains of Ukraine there is a lead horse, always alert and carefully watching out for possible danger or difficulties. It commands the herd and leads it, whether to walk, gallop or stop. It holds this position for several years, until it weakens. Another strong horse in the herd, one ambitious to command, then attacks the old leader, who keeps its command if it is not vanquished, but retreats in shame into the body of the herd if it is beaten; and the victorious horse places itself at the head of all the others and makes itself obeyed" (1776, supplement III). In those days of the Ancient Regime, affluent men were not

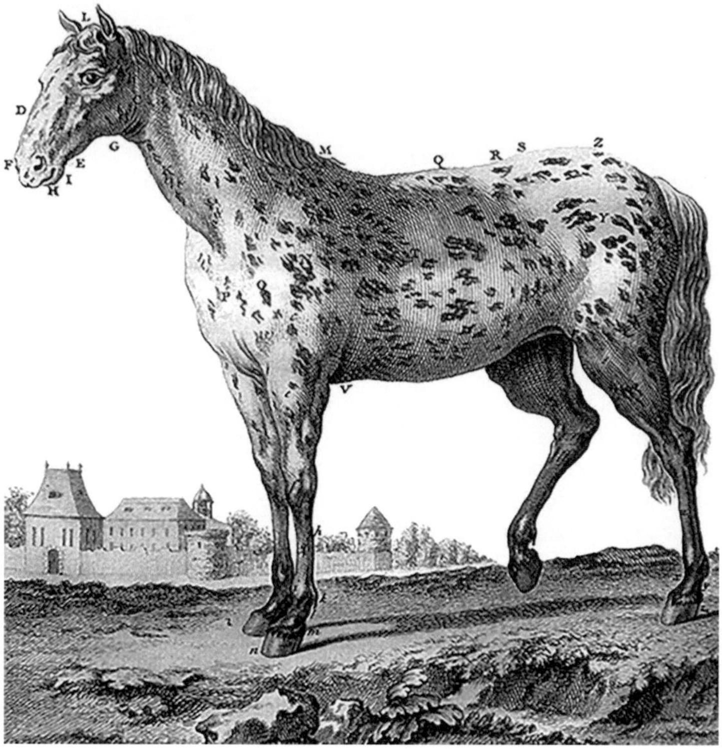

Fig. 5.47 The horse (*HN* 1753, vol. 5)

the only social creatures in which many obeyed the commands of a powerful single. Those who were not affluent would ride a less noble animal, the ass (Fig. 5.48).

The first series in *Histoire Naturelle*, that of domestic animals (1753, 1755; vols. 4–5: horse, ass, cattle, sheep, goat, pig, dog with 27 breeds, cat), was followed by a second series of animals secondary in importance to the French affluent: red deer, fallow deer, roe deer—and then the hare. These were the animals with which the French aristocrats entertained themselves in hunt (1756, vol. 6; Fig. 5.49).

The third series of animals consisted of small carnivores, with an emphasis on some of them being pests (Fig. 5.50). We have here the wolf, which preys on livestock (with a sheep's head and skeletal parts in the foreground of its illustration); the fox (and in the background of its illustration, the poultry it came to snatch); the otter that robs us of fish (dead fish on the riverbank); and the polecat (with a dead chicken; 1758, vol. 7). Then, after hedgehogs, shrews, moles and bats, a series of quadrupeds that hibernate: dormice, marmots, brown bears and polar bears (Fig. 5.51). Then species from other continents, broadly from the New World, then Asia, then Africa, but with many exceptions; and finally, two volumes were dedicated to monkeys—as far as possible from Buffon's fixed point of reference, the French affluent. In this manner, Buffon scattered over five series the carnivores, which we today group into

Fig. 5.48 The ass, side-saddle riding by a peasant woman. Detail from Buffon's figure of the ass skeleton (*HN* 1753, vol. 5)

one order. He placed dogs and cats in the series of domestic animals, together with goats, sheep and pigs. He placed foxes, wolves and otters in the animal pest series, together with rats and mice; bears in the series of hibernating animals, together with dormice, marmots and beavers. He presented lions, tigers, leopards and jaguars in the series of animals from foreign continents and servals and ocelots in yet another series. He grouped together seals, walruses and manatees by their habitat.

There were also other, more subtle layers in Buffon's order of presentation. After the horse, he placed the ass, not the dog, no matter what he had said previously. After the hare, he placed the rabbit, another natural classification. He clustered together small predators and likewise the rat, mouse and water rat. He placed one after the other hedgehogs, shrews and moles, our insectivores and presented in succession lion, tiger, panther, leopard, jaguar, cougar and caracal. He clustered

Fig. 5.49 Animals of game: deer, fallow deer, roe deer and hare. A hunter is hiding in the bushes behind the stone base upon which the skeleton of the hare is erected

Fig. 5.49 (continued)

squirrels, flying squirrels and palm squirrels and presented together all monkeys of the Old World; he clustered monkeys of the New World, though elsewhere. This gives the impression that throughout his fifteen-year-long study of quadrupeds, Buffon was also attempting to erect natural groupings.

Buffon claimed that species consist of living, reproducing animals, each with its specific morphology. He further claimed that species descriptions should include, beyond morphology, also various aspects of natural history. These would include the period of heat, gestation duration, number of offspring per litter, growth duration, parental care, instincts, diet, hunting habits and so on. Natural history thus became an additional element in his descriptions. Now in species descriptions in general, morphology is advantageous in that it consists of everlasting hard evidence.

Fig. 5.50 Pest carnivores: wolf kills sheep, fox chicken, otter fish and polecat large poultry. (*HN* 1758, vol. 7)

Fig. 5.51 Hibernating quadrupeds: dormice, marmot, brown bear, polar bear. (HN 1760, vol. 8)

Skeletons, fur, feathers or preserved entrails remain for many years, so conclusions based upon them can be inspected and criticized by anyone, even centuries ahead, even concerning animals from faraway lands. Natural history, especially its behavioural aspects, is problematic in that it relies only on documented observations concerning animal actions, and these are often irregular, erratic and capricious, and therefore difficult to inspect. Further, observed behaviours might be typical to individuals rather than to an entire species.

Buffon could rely on his own natural history observations. However, how should he cope with descriptions reported in the literature? What if there were several conflicting reported observations? In cases concerning the fauna of France, to reach a decision Buffon could initiate his own observations and experiments. Aristotle (and Linnaeus) had claimed that swallows, when they suddenly disappear from Europe in autumn, swoop down into swamps and stay torpid there until the return of spring, whereas other naturalists claimed that they migrate to Africa. To decide on this question Buffon assumed that if swallows do indeed spend their winter in torpor in swamps, then this was because cold weather produces in them a numbing effect. He experimented on whether swallows are subject to numbing by having a few of them closed up in iceboxes for a certain time. He found that they did not become torpid; indeed, most of them died there, and not one started to move again in the sun's rays. These experiments, when added to observations sent to him about swallows in Senegal during winter, settled the question: swallows in winter migrate from France to Africa, and this aspect of natural history could thus be included in their description.

However, how to cope with animals in distant lands, for which the natural history was known only through stories of travellers? What if these travellers mixed direct observations with fanciful accounts and their reports were exaggerated or even false? In such cases, Buffon could only apply his personal judgement. Jean-François Regnard was a Paris-born playwright, famous for a travel diary he kept of a voyage to northern Europe in 1681 (and published posthumously in 1731). He wrote that in Lapland a certain squirrel crosses lakes by boarding a piece of pine or birch and then using its tail as a sail, and he further noted that "This detail could pass for a fable, if I had not seen it myself". What was Buffon to do with this information? In this specific case, he chose to cite Regnard but without further comment. In another specific case of personal judgement, Buffon wrote that "As for [a certain person named] Kolbe, we count his testimony for nothing, since a man who at the Cape of Good Hope has seen elk and lynx very similar to those in Prussia, could just as well have seen Brazilian anteaters there". In species descriptions, personal judgement concerning natural history is not as strong and hard as morphology.

The elephant (Fig. 5.52) exemplifies the problem of coping with animals in distant lands and travellers' stories about their natural history. Buffon noted that beyond humans, the elephant is the most respectable animal in the world. It is highly sociable and can make itself understood by other elephants, with which it often assembles, disperses and acts in concert. It has great physical strength and deadly weapons (the tusks), which enable it to overcome even lions but further it has courage, prudence and calm, remaining moderate even in its most violent passion. It attacks only those who offend and injure it, never friends. Beloved and respected by all, it serves

Fig. 5.52 Animal groups which had not degenerated: elephant. (*HN* 1764 vol. 11)

man with zeal, fidelity and knowledge. Also, in the wild, elephants are gentle animals. Being social, they march in troops, the oldest keeping the foremost, the next in age bringing up the rear and the young and feeble in the middle; mothers carry their young firmly embraced on their trunks. When the females come in season, the troop separates into pairs of lovers, who sneak off into the bush, elephants being animals of sexual modesty. One never sees them mate in public, as they resent the stares of their fellow creatures and know, perhaps better than we do, the sensual delight of silent enjoyment and of concern only for the loved one (Buffon is here broadly following Pliny). The greatest of animals, how similar to man.

Elephants, Buffon went on, are easily tamed, but when domesticated, they refuse to mate. Being so modest that they can mate only when without witnesses, in their indignation, the male elephants become so violent that their owners must chain them. In refusing to mate when domesticated, the elephants differ from all other domestic animals, which man easily manages because they have no will of their own. Thus, elephants are not a group of animals born into slavery, which man can mutilate or multiply for his purposes. Here, only the individual alone is a slave, but the species as a whole remains independent, uniformly refusing to augment the stock of their human tyrants. Consequently, the elephant has sentiments superior to all other domesticated animals. It is indignant in that it feels the most ardent passion of love, and yet, because it cannot mate in private, it denies its gratification; it experiences all the fury of love and yet does not transgress the laws of modesty.

Beyond these bits of information gathered from travellers and the little known in those days of the natural history of exotic animals, Buffon sought to reconstruct the elephant's natural history also from morphology. From the structure of the mouth and trunk, he speculated that the baby elephant applies his trunk to the teats, sucks the milk and carries it to his mouth. Observing that the female vulva is located away from the anus towards the middle of the belly (rather than adjacent to the anus as in other quadrupeds), he thought he had observed that the male penis is small in comparison to its huge body ("by no means bigger than that of a horse"). He speculated that when copulating, it would be impossible for the male to cover the female in the ordinary quadruped position. Therefore, Buffon speculated that to copulate, the female must lie on her back—she cannot mate in any other manner. To obtain and maintain this position the elephants require more time for copulation than other quadrupeds and more privacy. "The female must not only consent, but solicit the male by a position she never assumes, unless when she thinks herself in perfect retirement. May we not, therefore, conclude that modesty is a physical virtue?" Buffon went on to claim that in nature, modesty is a general and beautiful attribute of the female sex. Some years later, Buffon gathered further information from a naturalist in Sri Lanka who claimed he had actually observed elephant copulation, and Buffon accordingly corrected his ideas as to elephant copulation. "When inclined to copulate, the female bowed down her head and leaned her two forelegs which were also bended, on the root of a tree. The two hind legs remained erect, which gave the male an opportunity of embracing her as other quadrupeds do" (1764 vol. 11). Today we know that the female does differ from most other mammals in that her vulva is somewhat forward to the anus. The male penis, when erect, may reach up to 2 m long, and he covers the female in the normal way of other terrestrial mammals.) As elephants never mate in domestication, all domestic elephants have formerly been wild. To capture a wild elephant, hunters surround a chosen spot in the midst of the forest with strong stakes but for a large gate. To draw him into this enclosure, the hunters bring into the forest a tamed female in season, and her governor makes her utter the cry of love. The wild male instantly replies and hastens to join her. She is then marched into the enclosure, repeating the cry from time to time. The male, following her voice and scent, enters the enclosure, and the hunters then close the gate behind him. His ardour vanishes in the enclosure, never to come back. We recall the way people once used to trap unicorns.

Coming now to matters of elephant taxonomy, to us, today, the differences between African and Asian elephants are obvious. African elephants, *Loxodonta africana,* are large and have relatively big ears; Asian elephants, *Elephas maximus,* are smaller and have smaller ears. Buffon, however, thought that all elephants consist of a single species. For one thing, he observed the very wide distribution of elephants throughout southern parts of the Old World, in all of Africa and in much of southeastern Asia. In Africa, the elephant varied very considerably and yet continuously in size. Buffon, claiming that climate, food and environment influence elephant size, noted that elephants in eastern Africa are the largest, reaching over 4.5 m high, whereas the smallest, in western Africa, reach only about 3.5 m (today the much smaller, forest-dwelling elephants of the Congo basin in western Africa are considered a distinct species, *Loxodonta cyclotis*). In Asia, the elephant also varies considerably and yet continuously in size and shape (today the pygmy elephants of Borneo are considered a separate subspecies, *E. m. borneensis*, with baby faces, oversized ears, plump bellies and tails so long they sometimes drag on the ground as they walk). Further, in Asia the elephant varies also in colour: most individuals are ash-grey to blackish, but others are red, and then there were those rare white ones (we today know that the red colour is due to the red dust with which they cover themselves).

This wide range, combined with local variations in size and colour, reminded Buffon of man, who likewise is widely distributed and variable in colour, with black, red and white peoples, all of the same species. Another point: the mammoths of northeastern Siberia were much larger than elephants and further differed in many morphological aspects. However, Daubenton, the leading anatomist worldwide by then, assured Buffon that mammoths and elephants were of the same species, and Sir Hans Sloane, the highly esteemed British naturalist and collector, shared this opinion. Now if specialists considered mammoth and elephant, with all their many differences, to be of the same species, then differences in ear size between African and Asian elephants become negligible. For these reasons, Buffon considered all elephants and mammoths as one species. This erroneous clustering of all elephants and mammoths in a single species had far-reaching consequences on his views on the origin of life, as we shall see further on.

5.2.3.5 A Change of Concepts, a New Classification

Buffon, in his early volumes of the *Histoire Naturelle*, considered the species as the only living unit in nature. He defined it by a reproductive standard, as a succession of similar, interbreeding individuals, separated from other such successions by fertility barriers. Thanks to the internal mould, the individuals of each species are closely similar to one another and to the earliest, original generation. The only change Buffon would admit was "degeneration"—a diminishing of the stature of a species or its physical or mental characteristics, which he supposed would result from the fact that individuals would form at a lower temperature than the originals, owing to the cooling of the Earth. Low temperatures reduce, he believed, the ability of molecules to pass on the organic identity of the species by means of the internal mould. This results in harmful changes such as making creatures weak and lifeless and, in extreme cases, difficulties in reproducing. Consequently, creatures in a climate different from their natural ones become smaller and less vital, and eventually, they die out: extinction awaits those animals that have degenerated to the point that they cannot reproduce. Indeed, in the end, all species are doomed to disappear due to the long-term cooling of the Earth. On a shorter timescale, one can reverse the perversions of degeneration. Negroes have a perverted skin colour. If removed from that too-hot environment that had altered them and brought them to the environment of origin of the human prototype—Europe—they would slowly revert to proper human skin colours. Alterations are thus either negative, leading to the degeneration and debasement of the species, or they are superficial and do not affect stature and reproductive functions. In neither case can alterations lead beyond the organic form set by the prototype. Above the species level and beneath that of the intuitive classes (quadrupeds, birds and fishes), Buffon rejected the existence of intermediate classification categories of genera and orders. One of his reasons was that, if we admit the existence of genera and orders and that monkey and man are of the same family and have a common origin, we would next accept that every group of plants and animals has only one source of origin. "From here we shall even conclude that all animals are of a single origin; and that its offspring, seeking to gain perfection or degeneration, created all animal groups". So close to the true answer, Buffon rejected this possibility of a common origin to all animal groups because:

- During the entire human history, not a single case of the creation of a new species was recorded.
- Hybrid infertility is an unsurpassable barrier between species.
- If one species had developed from another (such as an ass from a horse), we would have found intermediate stages between them—and this is not the case.

However, as early as 1753, Buffon and Daubenton proposed that several animal species share the same body plan. The idea was, of course, not new; already in the Renaissance, the French naturalist Pierre Belon had presented a human skeleton and a bird skeleton side by side, placed in the same position to show their resemblance. Daubenton, in his description of the horse, meticulously showed the exact resemblance in structure between the skeleton of the horse and that of man, thus establishing the list of what would, a century later, be called "homologies" between these two skeletons. Soon later Buffon took up this theme and gave it a wider generality: The entire skeleton of man can transform into a horse skeleton by modifying the dimensions of the bones and the angles of their articulation, without changing anything in bone order: the same bones occur in the same place. Further, ribs occur in all quadrupeds, in whales, birds and fishes, and even in turtles, where they form ridges under the shell. From this one can conclude that a primitive, general design exists that can be traced in a very wide range of groups.

Another point was that over the years, Buffon gradually began to doubt the absoluteness of fertility barriers and hence the absoluteness of the species concept. He encountered ever-growing reports of the existence of fertile hybrids between wolf and dog (which he erroneously considered as different species) and between goat and sheep. Now if individuals of different species can interbreed and produce fertile hybrids, then this means a collapse of his definition of a species. Moreover, up to this point he had dealt only with quadrupeds. The next group he planned to study was birds, and in his days, naturalists commonly believed that there were more hybrids between bird species than between quadruped ones. Canaries of the Canary Islands mix with siskins, goldfinches, linnets, buntings, chaffinches and perhaps even sparrows. The limits between species, far from being sharp, began to seem somewhat hazy. Furthermore, if birds of so many species can interbreed, then maybe there are dynastic relations between these birds. Could all those birds that we consider as separate species, actually be only variants? Back to quadrupeds, if a dog reproduces with a wolf, then it may well reproduce also with a fox, and that would mean that these three would belong to the same species. Why not erect a mega-species, which would include, beyond the dog, the wolf and the fox, also the jackal and the arctic fox? Yet for such an expanded species, including such a large and varied number of forms, the term Species seems rather inappropriate. Should one replace it with, say, the term genus, or perhaps Family? The idea of a Family could be convenient also when discussing the numerous small quadruped species, all closely resembling one another. "However, we must not forget that we are those who create families, and we do so merely to ease our minds in studying nature's diversity. Nature does not recognise families, only individuals" (*HN* vol. 4).

> It is generation, that older process existing from time immemorial, which seems to have occurred in each *family*, or if one prefers in each of the *genera* within which one can include similar species that differ little from one another. We have, in all terrestrial animals, only a few isolated species that, like man, make up the species and the *genus* at the same time. These include the elephant, the rhinoceros, the hippopotamus and the giraffe, which form *genera* or simple species, which propagate only themselves in a direct line, and have no parallel branches. All the others seem to make up *families*, in which one notices a principal and common ancestor, from which different stems seem to have come and which are much more numerous when the individuals in each species are small and more fertile (1766 vol. 14, my emphasis).

Hybrids were one issue that Buffon had to face. Another issue was the flow of information into Europe coming from Africa, Asia and the New World. Buffon began reflecting on the relations between animals of the Old and New Worlds. He noticed that several quadrupeds of the cold, northern climates of Europe and Asia (bear, martin, squirrel, hedgehog, seal and others) occur also in North America. He further noticed that no quadrupeds of the Old World warm climates (elephant, giraffe) are common to both Worlds. This suggested to him a land bridge that once stretched between the northern parts of the Old and New Worlds and which eventually sank into the ocean. Furthermore, Buffon thought he had found that of those animals common to both Worlds, those of the New World were frequently smaller (bear, deer, roebuck, wolf, fox and others). As another point, when we consider the largest quadrupeds of the Old World in comparison to the largest of the New World, then the elephant, rhinoceros, hippopotamus, giraffe, camel, lion, tiger and so on, are much, much larger than the largest quadrupeds of the New World (anteater, llama, jaguar and others). Another observation: All domestic animals that people have transported from Europe to America (horse, ass, ox, sheep, goat, hog and dog) have, Buffon thought, become smaller over the generations. Not only animals. Buffon also thought that the human native of the New World was a small weakling. He is shorter than the European is, his organs of generation are smaller and his beard less impressive, his senses are not as acute, he is timid and cowardly, his passion for women is feeble and so too are his family bonds (1758 vol. 7). Accordingly, instead of densely populating the land, he has populated the land only thinly, acquired no property, built no empire, domesticated no animal, conquered no sea, redirected no river and cultivated no earth. Nature in the New World has refused him the desire of multiplying. All these observations suggested to Buffon that in the New World, environmental factors prevent a full, flourishing life. This is so because, Buffon argued, the climate of the New World is colder and more humid than that of the Old

World. In Quebec City, at a latitude similar to that of Paris, rivers freeze every year several feet thick, snow covers the land for several months, and the air is so cold that the birds fly off during winter. Likewise, in Peru, at a latitude similar to Senegal, temperatures are far more temperate and cool. Accordingly, the local savages are only tawny brown, whereas the Negroes of hot Senegal are black. For a similar reason within the New World, the natives of Peru are less brown, red or tawny than those of Brazil. Turning from temperature to humidity, Buffon claimed that in the New World, there are more waters than in the Old and that this water is short of drains or outlets. As the natives have neither stopped the torrents nor directed the rivers nor drained the marshes, stagnating waters cover immense tracts of land, increasing air moisture and diminishing its heat. As the earth never dries, trees, shrubs and grass herbage densely cover it. The transpiration of this vegetation produces large quantities of moist, noxious exhalations. In these melancholy regions, Nature, uncultivated by man, never opens her beneficent womb. The earth here has never seen its surface adorned with those rich crops that demonstrate her fecundity and constitute the great wealth of civilized nations. The humid air and tree-covered frigid earth are unable to profit from the warmth of the sun and can afford nourishment only to cold, feeble animals and men, whereas large quadrupeds require the warmth of the sun for their activity. Consequently, the sun's powers are limited to the production of only moist plants, frogs, insects, and other animals that wallow in the mire and whose blood is watery. These are large and numerous in the low, moist, marshy lands of the New World. All these climatic differences contribute to the diversity between the faunas of the Old and New Worlds.

Buffon, on reflecting on all these differences, further believed that the New World is more recent than the rest of the globe; it had continued longer under the waters of the ocean. For if this is not so, if the New World was as ancient as the Old World, then why is it so thinly peopled? Why are its inhabitants wandering savages? Why do they lack writing? Why did they not domesticate the llama and other animals? Why are their crafts so primitive? Why is their language so barbarous? On this last point, Buffon presented a list of animals, whose names are so difficult to pronounce that, he commented, it is surprising how Europeans could submit to the trouble of writing them:

Native American name	**European name**
Pelon ichiatl oquitli.	Llama.
Tapiierete, or maypoury.	Tapir.
Tamandua-guacu, or ouariri.	Great anteater.
Ouatiriouaou.	Little anteater.
Ouaikaré, or ai or hai.	Sloth.
Aiotochtli, tatu or tatupeba.	Armadillo.
Tatu-ete, or tatou-kabassou	8-banded armadillo.
Macatlchichiltic or temamacama.	Antelope.
Jiya or carigueibeju.	Brazilian otter.
Quauhtla coymatl, or casigoara.	Mexican hog.
Tlacoozclotl, or tlalocelotl.	Mountain cat.

Cabianara, or capybara.	Cabiai, thick-nosed tapir.
Tlatlaubqui occlotl, or jaguara.	Jaguar, Brazilian cat.
Cuguacu arana, or cuguacu ara.	Cuguar or brown cat.
Tlaquatzin, or aouaré.	Opossum.
Hoitzlaquatzin.	Porcupine.
Cuandu, or gouandou,	Brazilian porcupine.
Tepe-maxtlaton, or maraguao.	Cayenne cat.
Quauhtechallotl thliltic, or tlilocotequillin.	Black squirrel.
Quinichpatlan, or assapanick.	Flying squirrel.
Tzquieptal.	Moussette, stiffling weasel.
Xoloitzcuintli, or cuetlachtli.	Mexican wolf.

All evidence indicates that America is a new land in which the temperatures are low, the humidity is high, its animals are small, and its native men are weak. The future, however, is bright: some centuries ahead, when the lands shall be cultivated, the forest cut down, the river courses properly directed, and the marshes drained, Buffon envisioned that this New World will become the most fertile and rich land in the whole world. Indeed, it is already so in those of its parts that have experienced the industry and skill of the Europeans. In some not-too-distant future, the native fauna will not become larger, and the tapir and capybara will never acquire the size of the elephant or hippopotamus; but domestic animals brought over will no longer become smaller, as they have formerly done.

Now, what if the literature suggests that once, there were many natives in the New World? Well, then these data are wrong. The first historians of the Spanish conquest, to augment their glory, had exaggerated the number of enemies they had to encounter. Travellers assured Buffon that they would travel 400 leagues without seeing a single family of savages, evidence that the New World is underpopulated. What if the literature suggests an American fauna that is different from that which Buffon had described? Well then again, these data are wrong. Among Old World animals which previous naturalists had erroneously ascribed to the New World, Buffon enlisted the pangolin of Java (West Indies), claimed to be an American animal of the East Indies; and the civet of southern Asia, which some naturalists had wrongly claimed to be in Central America. Likewise, previous naturalists had denied America some of her animals. They placed the opossum in southeastern Asia and the sloth in Ceylon, though both belong solely to the New World. Buffon added to this list the many blunders in the taxonomy of previous authors, such as that the hyena and glutton are the same; that the arctic fox occurs in the Cape of Good Hope and that the aardvark of Africa is the anteater of the New World. Some naturalists had indeed claimed that the reindeer is an antelope, that the genet is a raccoon of the New World, and that the skunk is a badger. Buffon, in his notorious self-confidence, was here thus putting theory before facts and claiming that facts contradicting theory are erroneous.

Man, Buffon emphasized, is the only living being on whom Nature has bestowed sufficient strength, genius and ability to enable him to subsist and multiply in every climate of the earth. (This is a secular version of the biblical narrative in which God told Adam and Eve to "Be fruitful and increase in number; fill the earth and subdue it"; *Genesis* 1). No other animal has obtained this great privilege of multiplying everywhere; animals are limited to certain climates, even to particular countries. Those of one continent almost never occur in another. Noteworthy, in those few exceptions that are widespread, the animals often vary so much that we can barely recognize them as the same species. Furthermore, Buffon stated that species change, whether by degeneration or improvement, in response to the influence of favourable or hostile climates. Past changes might have been to such an extent that the descendants are now no longer recognizable as the parent creatures. Hence, it is possible that many animals of the New World were originally the same as their ancestors in the Old. Afterwards, immense seas or impassable lands separated them from the Old World. Subject to the effects of a new, different climate, they degenerated to such an extent that by today they are different species.

In addition to the flow of data from all continents and the issue of hybrids, a third issue facing Buffon was the practical problem of organizing extensive amounts of data. In his time, there were only about 200 known species of quadrupeds, but there were about 2000 known species of birds. It had taken him 20 years to describe quadrupeds—it would take him a much longer time to deal with the description of birds, a group ten times larger.

These three factors (additional information on hybridization, reflections regarding the faunal relations between the Old and New Worlds and the upcoming work on birds) began to modify Buffon's opinions on the nature of the species. An early indication of this change in concepts appears in his chapter on the sheep. He had treated domestic sheep already in 1755. Nine years later, he had the opportunity to study also wild sheep (which he named *Le Mouflon*). Buffon presented all varieties of domestic sheep as degenerations from the first wild type (Fig. 5.53). In doing so, he explained that although there are no genera in nature, man could create genera by domestication.

> We must use *genera* in a natural history of domestic animals. In nature, there are only individuals and series of individuals, i.e., species. [...] We have altered, modified, and changed domestic animals. We have created, then, real physical *genera*, quite different, to be sure, from the arbitrary and metaphysical *genera* that exist only in the mind [a sarcastic reference to the genera constructed by Linnaeus]. These physical *genera* are composed of species that we have worked on, modified, and changed. All the species, although differently altered by the hand of man, have a common and unique origin in nature. The entire *genus* must have been formed from one species. (1764, vol. 11: 369; my emphasis).

Here, for the first time, Buffon proposes the use (albeit limited) of a unit beyond the level of the species. We should group domestic animals into genera, each with a single point of origin.

The turning point was in 1766. Having almost completed his study on quadrupeds, Buffon wrote a chapter on degeneration, *Degeneration des animaux*. This chapter presented a declarative change in conception, in that he introduced the term *family*: the animal world may well consist of families, formed over time by successive variation or by degeneration, of the original species. Each family has only a

Fig. 5.53 Domestic sheep (above and middle) and wild sheep (*Mouflon*, below; (*HN* 1755 vol. 5, 1764 vol. 11)

single ancestor. The ass is of the same family as the horse, differing from it only in that it had degenerated over time; it is thus a degenerate horse.

The French botanist Pierre Magnol (1638–1715), director of the Royal Botanic Garden of Montpellier, had coined the term family half a century previously. He had sought to establish a natural classification of groups of plants that have features in common, and he clustered all plants into 76 groups, which he termed families. Buffon, coming half a century after Magnol, used Magnol's term family and the classic term genus interchangeably. He went on to develop an idea in which he emphasized a broader concept of the species. The term *genus*, which in 1749 he had totally rejected and in 1764, he still attributed only to domestic animals, he by now considered as a natural unit of nature. One could thus interchangeably use the terms genus and family for both wild and domestic animals. Using the strictly reproductive species concept, Buffon began referring to his earlier defined species as races, a *race* being a "constant variety" within a species. If animals "came from the same ancestor, if they were indeed from the same family, one could bring them closely, blend them once again, and undo with time what time had done" (vol. 4).

Buffon further argued that each genus (= family) had a common root *premier souche*, which may degenerate over time to several varieties (we would consider them as species), all offspring of a common ancestor. These varieties all resemble one another morphologically, and they are able, sometimes, to hybridize. The common root of each family may still exist, and we might be able to identify it; thus, the horse is the original root of the horse family, with the ass and zebra being units that had degenerated over time. Each common root has an internal mould of its own, and the different variations (what we call different species) are different expressions of the same internal mould. Buffon's natural family was thus an expanded species, whose members are morphologically similar and who may hybridize. He no longer regarded the species as distinct since he was convinced that the species contained within a single family could interbreed, but the families themselves represented distinct, fixed moulds whose products the environment could modify to a certain extent.

The concept of natural families allowed Buffon to explain away many of the issues he had faced before when he had still held his old species concept. Fertile hybrids no longer crushed his reproductive criterion; on the contrary, they supported his claim of a dynastic relation between members of the same family. Likewise, we can easily link domestic animals to wild animals. Finally, fossils resembling modern animals are no longer a riddle: they are remains of some ancient branch of some common root of today's species that lived in another time and another environment. So, it is the family that should serve as the main unit of classification, not (what we today would call) the species. We should arrange species in genera and families as a function of their relative similarities and differences. How are we to arrange all these families?

Buffon outlined a new classification of quadrupeds, in which he reduced to 28 families all 200 quadruped species he had described over the past 15 years. His basic division was between families that had never split into different species versus those that had done so, in that they had degenerated. A further division in both categories was zoogeographic: Families were in the Old World, in both the Old and the New World or only in the New World.

Quadruped Classification

(From Roger 1997: 332, slightly modified).

Isolated Species Which Make Up a Genus by Themselves

SPECIFIC TO THE OLD WORLD: elephant, rhinoceros, hippopotamus, giraffe, camel, lion, tiger.
COMMON TO BOTH CONTINENTS: bear, mole.
SPECIFIC TO THE NEW WORLD: tapir, water cavy, llama, peccary.

Genera, Each with Several Species ("Degenerated")
Old, Old and New World

WHOLE-HOOFED: horse, zebra, ass, mule. *Old World*
CLOVEN-HOOFED, LARGE, HOLLOW HORNS: cattle, buffalo, bison. *Old and New World*
CLOVEN-HOOFED, SMALL, HOLLOW HORNS: sheep, goat, gazelle, musk deer. *Old World*
CLOVEN-HOOFED, SOLID ANTLERS: elk, moose, reindeer, caribou, deer, fallow deer, axis, roe deer. *Old and New World*
CLOVEN-HOOFED, EQUIVOCAL: boar, pig and its varieties. *Old World*
FISSIPED [divided paws] CARNIVORES, HOOKED, RETRACTABLE CLAWS: Panther, leopard, cheetah, snow leopard, serval, cat and its varieties. *Old World but connected to New World*
FISSIPED CARNIVORES, *NON*-RETRACTABLE CLAWS: wolf, fox, jackal, arctic fox, dog and its varieties. *Old World*
FISSIPED CARNIVORES, *NON*-RETRACTABLE CLAWS, POCKET BENEATH TAIL: hyena, civet, zibeth (Indian civet), genette, badger. *Old World*
FISSIPED CARNIVORES, ELONGATE BODY, FIVE DIGITS ON EACH PAW: stone marten, marten, polecat, ferret, mongoose, weasel. *Old and New World*
FISSIPEDS, TWO LARGE INCISORS ON EACH JAW, NO QUILLS: hare, rabbit, squirrel, dormouse, marmot, rat. *Old and New World*
FISSIPEDS WITH QUILLS: porcupine, hedgehog. *Old World*
FISSIPEDS WITH SCALES: pangolin, phatagin. *Old World*
AMPHIBIOUS FISSIPEDS: otter, beaver, desman, walrus, seal. *Old and New World*
QUADRUMANES: monkey, baboon, guenon, lemur, loris. *Old World*; American Quadrumanes differ
WINGED FISSIPEDS: flying fox, bat. *Old and New World*

New World

SAPAJOUS: eight species.

SQUIRREL MONKEYS: six species.
OPOSSUMS: marmoset, *kaipollon* (a small possum), phalanger, tarsier.
JAGUARS: cougar, ocelot, margay.
COATIS: three-four species.
SKUNKS: four-five species.
AGOUTI: acouchi, paca, aperea, tapeti.
ARMADILLOS: seven-eight species.
ANTEATERS: two-three species.
SLOTHS: unau, ai.

Buffon's group of families, which had never split (degenerated) into different species, included Man, who is absolutely superior to all animals in that only he has domesticated other animals, only he speaks and only he thinks rationally; mankind consists of one single highly diversified species, found in all climates with only minor, superficial alterations. Beyond man came animal groups that, like man, had not degenerated—the grand, majestic animals of the Old World: elephant, rhinoceros (Fig. 5.54), hippopotamus, lion, tiger and camel. These, he claimed, had always lived in the same climate ("Elephants are faithful to their country and never leave it") and are often large. In this group of families, Buffon also included the tapir, water cavy, llama and peccary of the New World and the bear and mole in both Old and New Worlds.

Buffon's next group consisted of families that had degenerated, each with several genera and often several species, whether only of the Old World or of both the Old and the New World. Many of these are natural families to modern eyes. Among the hoof-possessing quadrupeds, he correctly placed together all the single-hoofed (horse, zebra, ass, mule—Equidae of today), all cloven-hoofed with solid antlers (elk, moose, reindeer, caribou, deer, fallow deer, roe deer—Cervidae of today) and the large cloven-hoofed with hollow horns (cattle, buffalo, bison—Bovidae of

Fig. 5.54 Animal groups which had not degenerated: rhinoceros. (*HN* 1764 vol. 11)

today). He correctly placed some of the cloven-hoofed quadrupeds (boar, pig) in a separate group of uncertain nature. Among the carnivores, Buffon correctly distinguished cats with their retractable claws (panther, leopard, cheetah, snow leopard, serval, cat) from dogs without retractable claws (wolf, fox, jackal, arctic fox, dog) and from carnivores with elongate bodies (marten, polecat, ferret, mongoose, weasel). He correctly placed together all monkeys of the Old World (baboon, guenon, lemur and others) and then all bats. Almost all these are natural groups by modern classification. Not aware of convergence, Buffon erred in creating a family of quadrupeds with quills (porcupine, hedgehog). True to his view that we should group animals also by habitat, Buffon erred in forming a family of amphibious quadrupeds (otter, beaver, desman, walrus and seal). I do not understand his very strange group of those quadrupeds with "a pocket under the tail" (hyena, civet, genet, badger).

Buffon's third group consisted of families confined to the New World and that had "degenerated", namely, they had split into sub-taxa. Of these, he noted that many do not relate to any group of the Old World (opossums, coatis, armadillos, sloths, anteaters, tapirs and water cavies). However, several other quadrupeds seemed to him to have distant relationships with those of the Old World, perhaps indicating some common source of origin (llama, vicuna, peccary, jaguar, cougar and ocelot). Then, there were the New World monkeys, differing from the monkeys of the Old World in their flat noses and long prehensile tails. Buffon suggested that the most reasonable way to explain both the differences and resemblances was by claiming that in early times the Old and New Worlds were continuous, and that quadrupeds confined to the New World were separated by the flooding of waters that eventually divided Africa from America (1766). We shall return to this point in the following chapter in further detail.

We can compare Buffon's classification with that of Linnaeus, who, by the time of his last, 1788 edition of the *Systema Naturae*, had placed whales together with quadrupeds into one new, big group, which he named Mammalia. In the mammals, Linnaeus erected six orders, by their teeth formula:

PRIMATES: man, monkey, lemur, bat
BRUTA: rhinoceros, elephant, sloth, manatee, anteater, pangolin, armadillo
FERAE: seal, dog, cat, genet, weasel, bear, opossum, mole, shrew, hedgehog
GLIRES: porcupine, guinea pig, beaver, mouse, squirrel, dormouse, jerboa, hare, coney
PECORA: camel, musk deer, giraffe, deer, antelope, goat, sheep, cow
BELLUAE: horse, hippopotamus, tapir, pig
CETE: narwhal, baleen whale, toothed whale, dolphin.

Linnaeus thus, even in this very late edition, combined man with bat, elephant with armadillo, dog with opossum and hedgehog and horse with hippo; only his Pecora and Cetae would today be considered natural; Buffon, in his ruminants Pecora and carnivores Ferae, reached a far more detailed classification than Linnaeus did.

Having completed the quadrupeds, Buffon began applying his idea of the species as a complex of similar species to a new group of animals, the birds. In *Histoire*

Naturelle des Oiseaux, which he began publishing in 1770, he applied the concept of higher groupings. Why continue studying the comparative anatomy of each variation in each family? It would be sufficient to provide the features of the family and to omit morphological descriptions of the individual species. His ambition was "to compose a genus that is not strict, rigorous, and in consequence imaginary, but a real and true genus", one which brings together beings of the same origin. Consequently, he grouped similar birds into genera and families (again, he did not differentiate between these terms and used them interchangeably) and presented individual species only by illustration. He selected, tentatively, an original root and dealt with close species as variations from the standard; thus, the original root for the birds of prey was the great eagle (Fig. 5.55). In this manner, Buffon organized the study of bird diversity by families without describing each distinct species separately and without describing their anatomy, thereby shortening the text to only nine volumes. Distinct species closely resembling each other were considered as branches of a common stem with a common root (a common ancestor) and therefore retained many common characteristics. Climate, nutrition and time separated these related species from one another, as we shall see in the following chapter.

Fig. 5.55 Eagle, the common root of raptors

5.2.3.6 Seven Epochs

Late in life, Buffon wrote *Epochs of Nature* (*Époques de la nature,* 1778), an epic treatise that elaborated and reorganized his earlier sketch of the creation of the solar system and of life. In 1749, in his introduction to natural history, Buffon differed between mathematical (abstract) and physical (observed) types of knowledge and claimed that only physical knowledge is relevant to the study of nature. Now, in describing the history of nature, he presented a new type of knowledge, a historical one, which differs from physical knowledge in that it is reconstructed, not observed. He claimed that Earth had a beginning and that it then went through a succession of changes, with animal species coming and going. In this, he differed from Linnaeus, who had argued in favour of a divine, unchangeable, steady state of the living world, with a fixed number of unchangeable species. To reach Earth's history, Buffon used evidence referring to its shape, raised at the equator and flattened at the poles; its interior heat, which, he thought, was the main source of warmth for living nature, rather than the sun; its ancient rocks, which, he thought, were either "glassy" or "chalky" and fossils. *Époques* main biological aspects concern animal and plant fossils and recent zoogeography as explained by geology.

To reconcile his secular history of Earth with the religious history of the world, Buffon divided the history of Earth into seven epochs, which he openly compared to the Days of Creation. Also now, as three decades earlier, he would not have God intervene either in the formation of the earth and planets or in the appearance of life. Therefore, he coped with the secular-religious conflict by offering alternative explanations to the *Genesis* text. The Bible, he claimed, spoke in popular language to people of three thousand years ago, people who were ignorant and incapable of understanding scientific truths. The biblical "days" of Creation were meant as epochs of many years, not as 24-h spans of time as we understand them today and which correspond to solar days. We can know this from the fact that three of these "days" had passed before the sun had been placed in the sky. Therefore, we must consider these seven "days" as seven spans of time, not as solar days. In addition, he claimed the first biblical sentence, "In the beginning God created Heaven and Earth" does not mean to say that at the beginning God created Heavens and Earth as they are now. It was only at a later time that he made them as they now are, by giving form to Earth and by placing the sun, moon and stars in the heavens. To understand that first sentence, it is necessary to add three words that reconcile the whole. We should read this sentence thus: "In the beginning God created *the matter of* the Heaven and Earth"; in later days, God would shape them into today's Form.

The theological ground thus laid, Buffon started his secular history by stating that natural history encompasses all of space and all of time and has no limits other than those of the universe, not merely that of Earth. He started his history at the time when Earth chipped off from the sun. Reminding ourselves that in those days, the world was considered to consist of four elements (fire, air, water and earth), we shall now follow Earth's history.

The Birth of Earth (first epoch). A heavy comet hitting the sun obliquely tore off some of its hot, molten matter, which divided into rapidly rotating spheres. In this manner, Matter achieved Form at the cosmic level. Of these spheres, one was Earth. Buffon calculated the time needed for the cooling of a hot, blazing sphere as big as Earth. He experimented on the cooling rates of hot molten iron balls of various diameters and, later on, of glass, sandstone, calcareous rock and marble. From these data he concluded that to reach its current temperature, Earth needed 74,832 years; this was its age since it had been chipped off the sun, over ten times more than the six thousand years as calculated from the Bible and later books of history.

Cooling down (second epoch). Left to itself, Earth was a globe of molten matter spinning rapidly on its axis. This rotational motion, with its centrifugal force, shaped it as a sphere flattened at the poles. Gradually it began cooling and solidifying. Cooling produced effects similar to those that Buffon had observed when he had experimentally left a hot molten iron ball to cool: holes, waves and wrinkles formed on its surface, and below the surface, hollows, cavities and blisters appeared. Likewise, but on a much larger scale, that of Earth, Buffon claimed that the cooling molten matter produced huge "wrinkles"—ancient, primitive mountains all over the globe (Alps, Andes, Caucasus, Himalaya). Buffon believed that Earth's matter consists of materials that are either glassy or chalky, depending on whether, when subjected to great heat, they are reduced to glass or lime (glass minerals: quartz, crystalline rock, sand, granite, shale, clay, metals; lime minerals: limestone, marble, alabaster, calcite spar). As the glassy mountains formed when Earth was still scorching hot, no life could then inhabit them, and today their peaks are void of fossils. Also, like those molten iron balls he had left to cool, the hot molten Earth, when cooling, produced huge "cavities" and "blisters"—gigantic subterranean caverns, to which we shall refer presently. Solid but still hot, Earth's primitive landscape consisted of glassy rocks, enveloped by steaming hot water vapours. In this manner, Matter on Earth achieved Form.

Water world: Life begins (third epoch). An atmosphere rich in water vapours, volatilized by heat, surrounded Earth, which was then a globe of hot matter. Gradually it cooled enough for water vapours to condense and for rainfall to gather on its surface without being instantly vaporized. The element of water was thus separated from that of air. In this manner, a vast ancient, primitive ocean came into existence, its level reaching four thousand metres higher than that of today. Earth was then a water world, with a few very high mountains towering above it. Next, the water element began attacking the earth element, breaking down the glassy rocks of the ancient mountains to produce salt, which dissolved into the ancient sea; and produced sand, mud, silt, shale and slate, which accumulated on the sea floor.

It was in this third epoch, broadly together with the formation of sedimentary strata, that life spontaneously appeared in Earth's warm waters. In the Philippines, certain plants and fish live in fresh waters so hot that one could barely plunge one's hand in them, and this suggested to Buffon that in the primitive ocean, life started soon after the earth had cooled to such an extent that it could be touched without burning oneself. As soon as thermal conditions became tolerable for life, in the rather hot waters of the primitive ocean, the organic molecules organized themselves into complex living organisms. The first living organisms, thus spontaneously formed, were large and had complex forms. These included giant ammonites up to several feet across, belemnites, stony corals, starfish and fishes. Buffon believed that these creatures were large because nature was then in its prime, and it worked with a more active power and in a warmer temperature upon the organic molecules, which then united and combined into great masses. This resulted in the gigantic animals of these first living animals of the world. Noting that they today do not occur in nearby lands or adjoining seas, he suggested that they had since died out.

Buffon next addressed the question as to where, on Earth, life began. He thought that as the polar zones were the first to cool, it was there that rain first fell and where the ancient ocean first gathered, and so the (then moderately cool) polar regions were the first in which the organic molecules first combined spontaneously to give life; equatorial regions were then still scorching hot and inhospitable. As the cooling of the earth continued and gradually proceeded from the northern lands toward the equator, the watery, habitable zone gradually followed. Meanwhile, plants formed on the emerging land. These were tropical plants, which rains uprooted and swept down from the mountains onto muddy, boggy ground, there to die and decay; successive layers of plant debris and mineral sediment accumulated to form coal.

Water world sinks (fourth epoch). We recall that when cooling down, Earth's surface formed enormous subterranean caves and caverns, as bubbles beneath the cooling crust. The element of water now attacked these caverns, which the element of fire had produced. During the fourth epoch, the roofs of these huge caves collapsed, and waters of the ancient ocean rushed underground into the eroded caverns. Thereby, the level of the ancient, primitive ocean sank, and vast new lands emerged, thus exposing the "secondary mountains" with their fossil-bearing strata. The days of the vast water world were over. Now the water element began attacking the earth element, carving deep valleys into the secondary mountains and toppling those peaks that were less solid (Kircher had offered very similar ideas over a century earlier).

Volcanism, the element of fire, began when the sea level dropped. Noting that today several active volcanoes are near the sea (Stromboli, Etna, Teide, Santorini and Vesuvius, which during his lifetime had erupted four times), Buffon reasoned that volcanic eruptions had something to do with the proximity of rock, air and water. Volcanoes, he suggested, had then erupted because something spontaneously ignited some combustible matter, upon contacting water. Knowing of the extinct volcanoes in France, which were inland and on high ground in the Auvergne, he suggested that further retreat of the primitive ocean extinguished many volcanoes. During all this epoch, only fish and shelled animals in the sea and vegetation in the mountains inhabited Earth; this was a fauna and flora adapted to warm environments. Eventually, Earth cooled further, and this ancient fauna and flora died out. It was only after this epoch had entirely run its course that terrestrial animals appeared.

Giant Elephants and Man (fifth epoch). After the extinction of that ancient, third-epoch fauna, in Buffon's fifth epoch (corresponding to the fifth day of *Genesis*, on which animals were created), with the emergence of the continents from beneath the falling levels of the oceans, there appeared a new fauna suited to cooler climates. Giant elephants appeared and roamed the northern lands. Buffon supposed that the first fauna went extinct before the second could form since he thought that the spontaneous generation of such large new forms required large quantities of matter. It was by assuming that species are limited to a particular temperature range that he was able to envisage a major extinction, allowing large quantities of free organic particles to build up, ready for the generation of a new fauna. One entire set of species had to go extinct, by global cooling, to enable the next set to form. Accordingly, Buffon did not imagine a continuous replacement of species. The organic molecules of the first wave of animals, those of the third epoch, were left without function, so they recombined, in the north, to spontaneously generate a new wave of animals, adapted to the now lower temperature.

Buffon wrote of enormous fossil skeletons found in North America, in swamps adjoining the Ohio River. The bones and ivory tusks resembled those of an elephant, but the teeth were quite different, terminating in five or six blunt points and without complex grinding surfaces. Because an animal of elephant dimensions cannot hide elsewhere on Earth to the point of remaining unknown, he concluded that it had not survived to the present. Buffon's "elephant" was the mammoth, which had cusps on its molars, thereby differing from elephants, which have ridged molars. Although he did not separate the mammoth as a species different from the elephant, Buffon was aware of the differences between the two and noted them.

The many bones resembling elephant, rhinoceros, hippopotamus and oxen found in North America, in the frozen lands of Siberia and elsewhere in Eurasia were, for Buffon, further evidence of a once warmer Earth—still unbearably hot in the tropics, but with tropical-style animals inhabiting those northern lands that are now frozen wastes. Those northern lands might then have been even hotter than the tropics of today because some of the Siberian bones were much larger than the bones of modern elephants and hippopotami, and the environment (temperature), he believed, controls size. With further cooling of the northern lands in Eurasia, many of these archaic animals migrated to warmer regions, and their remains occur in temperate lands. Eventually also here they were doomed to extinction by the increasing cold.

In the New World this migration southwards by the first and largest terrestrial animals encountered a major obstacle: high mountains near the Panama Isthmus blocked their way to South America. Elephants and other quadrupeds could not break through this barrier because the cold at these mountain heights was by now too unbearable for them. Thus, isolated from the rest of terrestrial Earth, South America was beyond the reach of the northern animals—the elephant, the rhinoceros, the hippopotamus, the giraffe and others. South America produced animals on its own, which are feebler and much smaller than those animals in the north, inferior to them in grandeur and in strength. This fact, that none of the terrestrial animals of South America resemble those of the southern Old World, is however only part of the story. In the seas too, Buffon noted, the whales and narwhals of the north are larger than the manatees of the south. Accordingly, he suggested that in general, the quantity of organic molecules in the South was less than in the North. Also, all the "aqueous, oily, and malleable parts" that enter animal composition were in much smaller quantity in the south than in the north, and it is in these aqueous and malleable parts that the organic molecules begin to exercise their power to shape and to develop organized bodies. In South America, with its small quantity of malleable materials, only small, feeble and generally inferior species formed.

The reindeer represents a third generation of animals slowly shifting southwards. It today occurs in temperate regions, and with the passage of time, as the earth gradually cools even further than today, there may well appear new species in the temperate region. To conclude this point, Buffon suggested that the northern regions, whether marine or terrestrial, were the first to become fertile, and here, life reached its greatest dimensions.

Man, the great and last product of Creation, also formed during the fifth epoch. He differs from animals by the divine ray that God had bestowed upon him so that his soul and spirit lead his matter. By finding means to make and control fire and to provide clothes, he moderated the ravages of nature, and climate effects became all the same for him, so he traversed the globe and settled everywhere, as the master of Earth. In each region, climate and nourishment have caused slight variations, but all populations remain of the same species because they can reproduce. Buffon placed all men into one single, highly diversified species.

Continents separate (sixth epoch). Since elephant remains were scattered across Europe, Asia and North America, Buffon deduced that there must have been free movement between these continents. He proposed various connections between North America, Greenland, Scotland, Scandinavia and Spain, and that the Mediterranean once did not connect to the Atlantic and not to the Black Sea.

Human actions assist nature (seventh epoch). Man is superior to animals in that he has domesticated animals, whereas no animal has ever domesticated another; in that he speaks, whereas animals do not and in that he thinks, whereas animals merely follow their instincts. Man has emerged as a species apart from all other animals, in that he has begun to transform the Earth. This was not always so. Humans started as a bunch of savages who were terrorized and fearful of volcanoes, earthquakes and ferocious animals. Gradually they learnt to use tools, to make fire and weapons and to fell trees; clans were formed and then small nations, in which individual interests became part of national interests, law and order followed and society assumed power. Crafts, such as land cultivation, weapon making, clothes weaving and so on, spread all over the world and were further perfected. Through man's crafts, the Earth increased its fertility, and metals were discovered; by his intelligence, animals were tamed; by his works, swamps were drained, river cataracts smoothed and forests cleared; by his reflection, space was measured and time was counted, celestial movements calculated, heaven and earth compared, knowledge of the universe enlarged and the Creator respectfully adored. Through crafts derived from science, people crossed seas and mountains, discovered a new world and extended their domain over all isolated lands. The power of man, though subordinate to that of nature, has often created more than she did, and it is with the help of our hands that nature has reached the point of perfection and magnificence that we see today.

Buffon saw human influence on the Earth as a good thing also because it could warm the Earth and stave off the final, terminal freeze. It may seem impossible to oppose the progressive cooling of the Earth and to raise its climate, but man has done just this by felling forests (he believed that trees attract cold) and by draining swamps—measures that warm the climate. This good human influence on the Earth does not refer to every man. As compared with those great civilizations of Europe, the savage nations of America and the half-civilized peoples of Africa are either stupid or lazy; they starve the Earth without nourishing it, destroy without building and use everything yet renew nothing. The neglected state of their lands reflects their small worth. Buffon's emphasis on the disasters of a cool climate might perhaps reflect the period of severe climatic cooling in Europe during the Little Ice Age, which reached its minimum at around 1650. Its encroaching glaciers destroyed farms and villages in the Alps; it froze canals, rivers and even the sea to such an extent that a Swedish army marched across the frozen sea between Sweden and Denmark to attack Copenhagen. Throughout Europe, famine claimed the lives of many. In the decades after the climatic minimum and during much of Buffon's time, temperatures increased, and farming was more successful.

As Buffon divided the history of Earth into seven epochs that he openly compared to the Days of Creation, we can reflect on his comparisons. Buffon's water world calls to mind *Genesis*, in which God on the second day created Earth's waters by separating them from the waters in the sky, and on the third day, He let Earth's waters gather to one place, and dry ground appeared in other places. God then created vegetation, both the grain-bearing plants and various fruit-bearing trees. The late appearance of mammals, as compared to aquatic animals and plants, as presented by Buffon, calls to mind also *Genesis*, in which God only on the fifth day created mammals, having created plants and aquatic life on the third day. Buffon wrote that man was the last living being to have appeared, and that he appeared in the same epoch as the mammals; in the Bible also, he was the last living being to appear, and he appeared on the same day. Buffon's emphasis on clothes as a means to moderate the effects of the harsh climate in various parts of the world calls to mind *Genesis* 3, in which God made garments of skin for Adam and Eve and clothed them. Buffon emphasized that man has crossed mountains and seas and inhabited every corner of the globe, and this calls to mind God's command to Adam and Eve to fill the earth and subdue it and to rule over every living creature. Buffon's history of Earth is, largely, a secular *Genesis*.

References

Alvis, A. K. 2019. A heavy hoax: The "lying stones" of Johann Beringer. In *Unbound Smithsonian Libraries and archives, July 25, 2019*.

Anderson, L. 1976. Charles bonnet's taxonomy and chain of being. *Journal of the History of Ideas* 37:45–58.

Anstey, P. R., and S. A. Harris. 2006. Locke and botany. *Studies in History and Philosophy of Science Part C: Studies in History and Philosophy of Biological and Biomedical Sciences.* 37:151–171.

Bowler, P. J. 1973. Bonnet and Buffon: Theories of generation and the problem of the species. *Journal of the History of Biology* 6:259–281.

Buffon, G. 1749–1767. *Histoire naturelle, générale et particulière, avec la description du Cabinet du Roy*. Paris.

Buffon, G. 1770. *Histoire Naturelle des Oiseaux*. Paris.

Buffon, G. 1791. *Natural history, general and particular, by the Couint de Buffon*. Translated from French into English by W. Smellie. Third edition. London.

Cain, A. J. 1958. Logic and memory in Linnaeus's system of taxonomy. *Proceedings of the Linnaean Society of London* 169:144–163.

Cain, A. J. 1981. The development of systematic ideas of variation illustrated by malacology. In *History in the Service of Systematics*. London: Society for the Bibliography of Natural History.

Cain, A. J. 1994. Rank and sequence in Caspar Bauhin's Pinax. *Botanical Journal of the Linnean Society* 114:311–356.

Capanna, E. 1999. Lazzaro Spallanzani: At the roots of modern biology. *Journal of Experimental Zoology* 285:178–196.

Cheung, T. 2006. The hidden order of preformation: Plans, functions and hierarchies in the organic systems of Louis Bourguet, Charles Bonnet and Gorges Cuvier. *Early Science and Medicine* 11:11–49.

Cole, F.J. Frs. 1930. *Early theories of sexual generation*. Oxford: Clarendon. (*Romantic Review* 28: 369–378).

Darwin, E. 1794. *Zoonomia*. Full text via Project Gutenberg.

Darwin, E. 1803. *The Temple of nature*. Full text via Project Gutenberg.

Egerton, F. N. 1968. Leeuwenhoek as a founder of animal demography. *Journal of the History of Biology* 1:1–22.

Farber, P. L. 1972. Buffon and the concept of species. *Journal of the History of Biology* 5:259–284.

Farber, P. L. 1975. Buffon and Daubenton: Divergent traditions within the *Histoire naturelle*. *Isis* 66:63–74.

Fellows, O. E., and S. F. Milliken. 1972. *Buffon*. New York: Twayne.

Jahn, M. E., and D. J. Woolf. 1963. *The Lying Stones of Dr. Johann Bartholomew Adam Beringer, being his Lithographiae Wirceburgensis*. University of California Press.

Kochiras, H. 2017. Locke's philosophy of science. In *The Stanford encyclopedia of philosophy*, ed. Edward N. Zalta, Winter 2017 ed.. https://plato.stanford.edu/archives/win2017/entries/locke-philosophy-science.

Koerner, L. 1999. *Linnaeus: Nature and nation*. Cambridge, MA: Harvard University Press.

Lawrence, C.R. 2009. *Charles Bonnet (1720–1793)*. Embryo Project Encyclopedia (2009-06-10). ISSN: 1940-5030. http://embryo.asu.edu/handle/10776/1745.

Linnaeus, C. 1758. *Systema Naturae Regnum Animale*. 10th ed. Lipsiae: Guilemi Engelmann. (Facsimile).

Linnaeus, C. 1735. *Systema Naturae*, S.J. Engel-Ledeboer and H. Engel (Translators). Hes abd De Graaf, Amsterdam.

Linnaeus, C., 1751 (2003). *Philosophica Botanica*, S. Freer (Translator). Oxford: Oxford University Press.

Lister, M. 1678. *Historiae animalium Angliae tres tractus. Unus des Araneis. alter de Cochleis tum terrestribus tum fluviatilibus. Tertius de Cochleis Marinis*. London: The Royal Society.

Needham, T. 1749. *Observations upon the generation, composition and decomposition of animal and vegetable substances*. London.

Ogilvie, B. W. 2006. *The science of describing: Natural history in renaissance Europe*. Renaissance Quarterly. Chicago: The University of Chicago Press.

Osborn, H. F. 1905. *Tyrannosaurus and other Cretaceous carnivorous dinosaurs*. Proceedings of the Academy of Natural Sciences of Philadelphia.

Pluche, N. A. 1740. *The history of the heavens, considered according to the notions of the poets and philosophers, compared with the doctrines of Moses*. London: J. Osborn, etc.

Pyle, A. 2003. Malebranche. London: Routledge.

Rieppel, O. 1988. The reception of Leibniz's philosophy in the writings of Charles Bonnet (1720–1793). *Journal of the History of Biology* 21:119–145.

Roe, S. A. 1983. John Turberville Needham and the generation of living organisms. *Isis* 74:159–184.

Roger, J. 1997. *Buffon, a life in natural history* (Translated from French by S.L. Bonnefoi). Ithaca: Cornell.

Sloan, P. R. 1972. John Locke, John Ray, and the problem of the natural system. *Journal of the History of Biology* 5:1–53.

Sloan, P. R. 1976. *The Buffon-Linnaeus Controversy*. Isis 67.

Sirks, M. J., and Zirkle, C. 1964. The evolution of biology. https://www.cabidigitallibrary.org/.

Wightman, W. P. D. 1953. *The growth of scientific ideas*. Yale University Press.

Storer, T. I. 1943. *General zoology*. New York: McGraw-Hill.

Wilson, E. O. 1992. *The diversity of Life*. Penguin Group.

Winsor, P. M. 1976. The development of Linnaean insect classification. *Taxon* 25:17–67.

Winsor, M. P. 2006a. The creation of the essentialism story: An exercise in Metahistory. *History of Philosophy of the Life Sciences* 28:149–174.

Winsor, M. P. 2006b. Linnaeus' biology was not essentialist. *Annals of the Missouri Botanical Gardens* 93:2–7.

Chapter 6
Paving the Road to Evolution

Abstract The decades before Darwin's *Origin* were rich in hypotheses and speculations regarding diversity and classification and rich in new zoological material from expeditions. Zoos, collections and popular books increased public awareness of nature's diversity. Cuvier followed Buffon in claiming that to classify animal diversity, we must know all features, not just external ones. To classify animals that fit one group by one trait and another group by another, he outlined his *Principle of Character Subordination*: the more significant a functional organ is, the more it should weigh in classification. Accordingly, he distinguished animals from plants by the presence of a nervous system, and within animals, four major branches by nervous system complexity. He expanded the Linnaean taxonomic hierarchy by establishing additional optional ranks (Branch, Section, Tribe and Subgenus). He viewed organisms as integrated wholes, whose organs function in harmony; consequently, no organ can change without impairing this functional integration. Cuvier used this *Correlation of Parts* to predict the structure of complete animals from a few remaining fossils. He rejected theories of animals forming a single Great Ladder of Life, with a rising order of complexity.

Cuvier created and established modern Palaeontology by presenting extinction as a natural, global process. By proving that mammoths differed from elephants, he rejected Buffon's theory of the Earth gradually cooling down and of animals migrating from slowly cooling poles to the warm tropics. He identified an extinct giant sloth, an extinct giant deer and an extinct tapir, evidence of once giant creatures that no longer exist. Reptiles came into existence before mammals and were once the dominant fauna and there were flying fossil and swimming ones. From proving that a fossil with a skull and teeth of a marsupial must also have other marsupial bones, Cuvier formulated the *Principle of Correlation between Organs*. As all organs of every animal function together harmoniously and are interdependent, and the existence of a species depends on the manner in which these organs interact, no organ can change without the other organs also changing. Hence, species cannot change; as fixed entities from their moment of creation, they are so well adapted that only catastrophic events could exterminate them. Cuvier suggested a series of sudden disastrous catastrophes that hit Earth, destroying many animals. New fossil forms

J. Heller, *Nature's Biodiversity*, https://doi.org/10.1007/978-3-031-89790-0_6

appear abruptly after each catastrophe and persist unchanged until their sudden extinction by the next catastrophe.

Lamarck believed in a Principle of Life that drives life towards ever-greater perfection. Animals and plants form two series of increasing perfection. Plants synthesize simple compounds that animals assimilate to create complex compounds. The Principle of Life expresses itself as the movement of fluids within the plant, as determined by the environment. In animals, the movement of fluids is by internal forces. Lamarck rejected the claim of extinction as a natural process, with previous animals going extinct by massive catastrophes, as Cuvier had suggested. He claimed that nature changes gradually, constantly balancing itself, so no species is ever lost. Earth's climate has changed very slowly, and so too have plants and animals. Many seemingly fossil forms actually live today, slightly changed but still closely resembling the fossil. Ammonites and belemnites are still alive in unexplored deep seas. Marine fossils occur on mountaintops because of a slow, continuous westward movement of the Atlantic Ocean basin over Earth's surface, the sea ceaselessly eroding eastern coasts and abandoning western ones. This displacement has left fossil marine shells far from the sea.

Lamarck at first believed that animals formed a single Ladder of Life, of increasing complexity. Classes, Orders and Genera interlink via interim animals, as Lamarck's (remarkably poor) mollusc classification demonstrated. Nature began (and begins) by spontaneously creating simple animals from molecules that spontaneously assemble to form infusoria, which proceed to form some major groups. Intestinal worms also appear spontaneously and proceed to form other ones. The Principle of Life builds nervous system complexity, with animals lacking nerves being insensitive; animals with simple nervous systems following instincts; and intelligent animals following ideas. Adaptive change also determines structure. Amphibious worms adaptively diversify into air-breathing amphibious insects, which change to terrestrial insects that become arachnids, which change into crustaceans. Other worms become annelids, barnacles and molluscs. In originally amphibious mammals, plant browsers became terrestrial herbivores, flesh-eaters became terrestrial carnivores, and aquatic dwellers became whales. Species-level diversity, in response to environmental diversity, involves only external features. Species changing habits change shape and convert into new species as lateral ramifications from the linear ladder. Use inherently strengthens and enlarges organs; disuse weakens them to disappearance, so inheritance of acquired characters creates animal diversity, in contrast to theories of unchanging species that perish during catastrophes.

Geoffroy suggested that all animals of a group share the same basic set of organs, which never change their relative locations. Animals, during embryonic life, go through phases recalling their ancestors, and this, with vestigial organs, indicates the derivation of animals from a common archetype; hence, species transform over time. He erroneously extended the vertebrate archetype to include insects as vertebrates that live inside their vertebrae and walk on their ribs with their bellies facing upwards and later claimed that molluscs also share this archetype. He thus unified all bilateral animal groups under one basic plan, thereby suggesting to reject Cuvier's theory of four animal branches.

Goethe suggested that vertebrates have basic units that repeat themselves throughout the body. This calls for an ideal vertebrate, from which diversity develops by modifications of such units.

Others believed that each such segment is a *zoonite*, a miniature animal. Zoonites are identical throughout the animal kingdom, the animals they form differing in number and grouping ways. Animals thus consist of multiple organisms that were once capable of living independently; gradually, they coalesced and modified to conform to the functions of the body, thus resembling a civilized human society evolving from a savage one.

Blainville classified biodiversity by external, sensing features that enable movement. The more an animal feels and moves, the higher its place on the Ladder of Life. Forming a strange classification that linked vertebrates to arthropods and molluscs to arthropods, he rejected Cuvier's four branches: animals form a single Ladder.

Oken suggested five main steps of animal progress: invertebrates, fish, reptiles, birds and mammals, corresponding to the five senses of touch, taste, smell, hearing and sight. Animals reach higher steps on the Ladder by adding senses; as no senses are ever lost, organs in primitive groups may become different organs in advanced ones: bivalve valves become bird wings. The skull consists of modified vertebrae; jaws correspond to arms and feet and teeth to claws or nails. Plants, animals and man first formed from a primitive slime by unicellular organisms agglomerating into multicellular organisms. Semen consists of single-celled creatures and generation repeats that primitive, slime-formed creation. In developing from a fertilized egg to an adult, animals pass through stages representing the groups below them. The joint manifestation of nutrition, digestion and respiration is self-movement.

Some authors strictly divided each vertebrate class into five series, each with precisely three orders. This, too, resulted in erroneous classifications.

Humboldt travelled extensively in Spain's realms in the Americas. He viewed nature holistically, and the unity of nature meant for him the interrelation of all physical sciences. Life is a web of interactions, and the conjoining between biology, meteorology and geology determines where specific plants shall grow. Altitude determines where plants on mountains live and also determines temperature and humidity. Furthermore, vertical plant stratifications reflect global vegetation patterns. Elevation gradients compress life zones that, near sea level, are thousands of kilometres apart. Mountains, by lumping contrasting climates onto a single slope, host more diverse life than other terrestrial systems. By interconnecting climate, geography and nature, Humboldt laid the foundation of plant geography. He also introduced the isotherm, a line connecting points on a map that receive equal quantities of heat. The geographical distribution of temperatures on Earth correlates with the geographical distribution of life on Earth, so he drew curves of equal mean temperatures over the globe, thereby devising means of comparing climatic conditions of various countries. Humboldt collaborated to produce the first comprehensive physical world atlas. It included maps of plant distributions in both latitudinal and altitudinal perspectives.

In the early 1800s, the leading embryological concept was that ontogeny recapitulates phylogeny, with stages of embryogenesis mirroring steps of the Ladder of Life, a single Ladder and a single unified body type. Baer challenged this concept. He distinguished four types of embryonic development, corresponding to Cuvier's four Branches. Embryogenesis advances from the centre to the periphery (Radiata); creates identical forms along a longitudinal axis (Articulata); spirals around a conic space (Mollusca); or forms identical organs along the sides of an axis (Vertebrata). These distinct embryonic types reject the existence of a single Ladder. Early embryos of all animals are similar, but changes accumulate as development proceeds and diverge towards four specific adult forms. General features of a large animal group (a Class) appear earlier in the embryo than specific characteristics (a Genus). Accordingly, Baer classified the vertebrates (with emphasis on mammals) by extensive use of embryonic structures: allantois, umbilical cord and yolk sac. Other zoologists followed Baer and proposed a classification based upon the vitelline sources and from which sides they entered the embryo. By now, scientists had begun abandoning the preformation doctrine; this enabled them to accept animal change through time, as did Baer. New studies of small, low organisms demonstrated their highly complex organ systems and their diversity and that infusorians do not appear spontaneously. This, consequently, rejected theories of a Ladder of Life, of spontaneous creation and of low organisms transforming into highly organized creatures. A new classification based on the nervous system: vertebrates with a spinal cord versus invertebrates with ganglionate nerves. Vertebrates subdivide into those that care for their young (mammals, birds) versus those that do not. Invertebrates split into those with a heart versus those without. The former split into arthropods, with similar ganglia throughout their bodies, versus molluscs with different types. Within Cuvier's Radiata, the anatomy of starfish and sea urchins differs entirely from that of medusas, corals and sea anemones. The newly established cell theory, that the cell is the basic unit of the living body, inspired studies of the single-celled creatures. Scientists now considered these creatures as consisting an entirely separate Kingdom, the Protozoa, with no digestive, vascular or secretion systems, no muscle or nervous tissue and no egg laying. It consists of five Classes. Gradually, the classification of animal diversity became close to that of today, as regards the major groups.

Some naturalists arranged biodiversity in circles, not ladders. Essential to this method were homologies leading from one form to the next, connections linking each member of one series to the corresponding member of another and the homologue series being circular. Consequently, each series must have the same number of members, which happens to be five. Each member in a major circle consists of five minor sub-circles. Biodiversity thus falls into circles within and next to circles, each with five members. Each member of a five has typical features. The first, *typical* member has the most complete, general features. The second, *sub-typical* member consists of powerfully armed, destructive animals that subsist on prey. The third is aquatic, bulky and partakes in destructive deeds of the *sub-typical*. The fourth feeds by suction, has a small head and a weak body, lacks offensive protection and (often) is a parasite. The fifth comprises domestic animals; gentle and familiar with man, they have strong limbs and many head ornaments. Many naturalists experimented with this "Quinarian" classification. Its strictly regular structure was, however, irreconcilable with the idea of change in form over time, following environmental changes. Change would break down this neat five-in-a-circle design into irregular patterns.

Environmental pressures may modify pre-existing structures but cannot give rise to new ones; species do not change into other species, and all major groups have always been present. Closely similar environmental conditions scattered over the globe may harbour different species, and this is so because of separate creation events: God had created each species separately in a given area. Slight but continuously changing environments might cause considerable changes in animal distributions, with species going locally extinct and others invading and replacing the annihilated groups. Competition, migration, food shortage, climatic changes or high population densities may also change distribution, and even minor-scale invasions might influence an entire local fauna. Man also changes the environment by transforming natural habitats into agricultural regions.

The world's birds and mammals fall into six major regions that are separate centres of creation: Australian, Neotropical, Ethiopian, Oriental, Nearctic and Palaearctic. Submerged land bridges and continents may account for populations of terrestrial species now separated by barriers of water.

Paley suggested that animal adaptations bear evidence of God's design. There is no order without someone making order, no means towards a purpose without someone thinking of the purpose. Only a wise, all-potent entity can cause this miracle of animal structure corresponding to function. Animal design calls for a designer; the designer is God, and animal adaptations are evidence of His skilful design. Adaptations occur in all creatures and all organs, with some being prepared for future function (milk teeth). Thanks to God's goodness in creating adaptations, nature is full of happiness and harmony. Evidence rejects claims of organs changing by over- or underuse, of environments determining shape or of animals having achieved their shape by gradual evolution. Evidence also rejects claims that creatures achieve adaptations by constraints during embryogenesis, the shaping of one organ linking to that of another. Over a broad taxonomic front, homologous organs occur—God shows His greatness in that He purposefully constrains His power to a small number of well-defined types. All the endless adaptive aspects of nature point to the existence of God. This argument assumes an analogy between nature and human contrivance: since the latter involves intelligence and purpose, so must the former.

Robert Chambers, in his *Vestiges,* sketched an evolutionary history of life. Fossils progress over time, from simple to complex animals, mostly within existing classes but some as inter-class creatures. Environment and life advance side by side, with life going extinct when environments change. The unity beneath diversity suggests one creation event of potentiality, but potential animals did not all appear at one time. During embryonic development, chemical-electric operations produce cells that may remain as animalcules or start foetal development. Monstrosities might result in transitions from one class to another. There is thus a threefold parallelism:

the order of appearance of animals in the fossil record mirrors the order of appearance of their morphological novelties and the order of embryonic development. Harmonious adaptations prove that design presides in all forms of animal life. Similar forms of life occur on stars because matter everywhere is the same; those elements of matter that build animals occur throughout the universe, as do electricity, gravitation and light, all involved in life's basic procedures and enabling Life to rise spontaneously. Where there is light, there will be eyes, and these, in the stars, will be the same as on Earth. Star inhabitants, therefore, resemble us.

6.1 The First Modern Zoologist

6.1.1 The Layout of the Animal Kingdom

The seven decades before Darwin's *Origin* were very rich and adventurous in hypotheses and speculations regarding animal diversity, with naturalists searching for a plan and consistency that would explain the rich diversity of animal form and for a way to integrate this plan into their classifications. During the last decade of the eighteenth century and the first two or three decades of the nineteenth, England experienced a deep enthusiasm for natural history. Many middle-class homes had fern cases, butterfly cabinets, seaweed albums or shell collections; many newspapers ran natural history sections; many popular books appeared; and when the London Zoo opened (1828), people flocked to see its wonders. In 1853, a "Marine Vivarium" opened as well, with crabs, molluscs and fishes.

It was Paris, however, that became the European centre for the scientific study of animal diversity. One reason for this was the great success of Buffon's *Histoire Naturelle,* which sold out within 6 weeks after its first three volumes had appeared (1749). An immense popular enthusiasm for natural history arose, in France as in England, among naturalists, apothecaries, businessmen, military officers, lawyers, ministers of state and members of the royalty who, following the Royal Natural History Cabinet, sought to possess natural history cabinets of their own. A host of merchants emerged to supply these many collectors with specimens for their cabinets, so much so that at the end of the century, some 600 merchants were making their living in Paris alone by selling shells to the clientele. The public attended courses in natural history offered at the *Jardin du Roy*, and there were brief, popular excursions to the countryside to study plants in their natural habitats.

A second reason was that when the French Revolution began, the professors of the Royal Garden and Cabinet wisely portrayed themselves as willing and useful citizens of the Republic. They initiated a large-scale reorganization of the Royal Garden and Cabinet of Natural History, previously a royal institute (directed by Buffon), and reformed it as a republican institute, now to be named the Museum of Natural History (*Muséum National d'Histoire Naturelle*). When the newly named Museum opened its doors (1794), its faculty consisted of 12 professors, whose job was to oversee the organization and development of the collections under their charge and to give public lectures on natural history; they were also collectively responsible for the museum's administration. The spirit of the French Revolution also seeped into the Museum and it was Daubenton who, in line with the revolutionary sentiments of his students, lectured (1795) that the word "kingdom" was altogether out of place in natural history, it being improper and unintelligible. He had already drawn an enthusiastic response from his listeners by observing, in an earlier lecture, that "The lion is not the king of animals. There is no king in nature". By changing its name and spirit, the institute survived the Revolution's Reign of Terror.

The Museum expanded its collections by claiming objects from the cabinets of the aristocracy, from other institutes such as the Royal Academy of Sciences and from spoils of war. Professors of the Museum, sent by the National Convention, accompanied French armies that passed beyond their country's borders into conquered areas (1794) to seek natural history objects and send them back to France. Prominent among these was the superb natural history cabinet of William V, the ruler (*Stadthauser*) of Holland. Three ships loaded to the brim with boxes of specimens left Hague on their way to France, and when the first 150 cases of the cabinet arrived in Paris, the French authorities displayed the zoological specimens in the museum amphitheatre: skeletons and skins of elephants, rhinoceros, orangutans, birds, insects, shells and many other treasures.

Beyond these spoils of war, the Museum also benefitted considerably from French exploratory expeditions. During the latter half of the eighteenth century, maritime expeditions were a means of expanding colonial empires, establishing new trade routes and extending diplomatic and trade relations to new territories. With the Enlightenment, scientific curiosity became a new motive for exploration, to add to the commercial and political ambitions of the past. Hans Sloane (1650–1753), a member of the British Royal Society, travelled to Jamaica as a physician to the Governor of Jamaica. He collected numerous specimens, which he bequeathed, with his vast library of over 50,000 books, to the nation, prompting the establishment in 1753 of the British Museum. His travels also made him a wealthy man as he patented a recipe that combined milk with the fruit of cocoa, which he saw growing in Jamaica, to produce milk chocolate. By the eighteenth century, maritime exploration had become safer and more efficient, with technical innovations that vastly improved navigation and cartography (precision clocks, the compass, the telescope and general shipbuilding techniques). Scientific expeditions mapped newly discovered regions, brought back to Europe newly discovered faunas and floras, made hydrological, astronomical and meteorological observations and improved the methods of navigation. This stimulated great advances in natural history, contributing to the sense of improvement and progress that characterized the Enlightenment. One English naturalist-traveller-collector was Joseph Banks (1743–1820), the son of a wealthy member of the House of Commons who had a strong passion for plants. As a student at Oxford, he was so determined to receive botanical instruction that he paid a Cambridge botanist to come over to Oxford and deliver a series of lectures. When 21, he inherited a large estate in Lincolnshire, became the local squire and magistrate and, dividing his time between Lincolnshire and London, he became a member of the Royal Society. Aged 23, he went aboard a light warship to study the flora of Newfoundland and Labrador and aged 25, he went on another expedition, this time to the South Pacific and on board HMS *Endeavour*, with James Cook as captain, on a joint project of the Royal Society and the Royal Navy. Its official mission was to reach Tahiti and from there to observe a rare event, the movement of the planet Venus across the face of the Sun; this would enable them to determine the distance to the Sun and thereby aid in British naval navigation. Another secret mission was to continue south and west to find that rumoured but yet undiscovered Great South Land, *Terra Australis Incognita*, and claim it for England. Cook was to take with him a team of naturalists that would study and collect natural materials, beasts, fishes and minerals found in these exotic lands. Banks generously brought to the project £10,000, somewhat more than the £4000 donated by the King. He also brought with him two other naturalists, two artists and four servants. One of these naturalists was the Swedish botanist Daniel Solander, a pupil of Carl Linnaeus who had first-hand knowledge of the new Linnaean method of plant classification. Solander had travelled to England 8 years earlier, had begun cataloguing the natural history collections of the British Museum and had become a Fellow of the Royal Society.

The three-year voyage (1768–1771) went from South Africa to South America, then to Tahiti, then to New Zealand, where Banks and Solander discovered many new species of tree ferns and noted the potential economic significance of the New Zealand flax *Phormium tenax*, which the Maoris were using for clothing. The expedition discovered Australia in 1770, and the Endeavour first anchored offshore at a site that Cook later named *Botanist Bay,* in recognition of "the great quantity of plants Mr Banks and Dr Solander found in this place"; the name eventually changed to Botany Bay, the core area around which the future town of Sidney would develop. Banks and Solander here discovered and collected many new plant species. Later, proceeding northwards along the east coast of Australia, the *Endeavour* ran aground amidst the corals of the Great Barrier Reef. It took the crew 7 weeks to repair the ship, during which Banks and Solander collected further specimens. Upon returning to England, Banks intended to join Cook on his second voyage, but the Admiralty regarded his scientific requirements on board Cook's new ship as unacceptable and withdrew his permission to sail. Banks immediately arranged an alternative expedition: on board another ship and taking Solander with him again, he visited the Isle of Wight, the Hebrides, Iceland and the Orkney Islands, returning to London with many plant specimens. Solander, upon returning, resumed his duties at the British Museum and became Keeper of the Natural History Department.

Banks and Solander, on their three-year Endeavour voyage, had collected over 3000 plant specimens. They planned a momentous series of 14 volumes under the title *Florilegium,* which would describe and illustrate all the Australian plants collected during the expedition. They intended some 900 illustrations, and as the artist on board the ship had died of fever and dysentery on the return voyage, Banks employed several other artists to complete the task of drawing. Unfortunately, Solander himself, who had supervised the whole project, including the finishing of the drawings, died suddenly (1782). Banks still hoped to publish his *Florilegium,* but then, when aged 35, the Royal Society elected him as President, and he could no longer spare the time. Furthermore, during much of his time as adviser to King George III on the Royal Botanic Gardens, Kew, he dispatched explorers and botanists to many parts of the world. Through these efforts, Kew Gardens became the pre-eminent botanical gardens in the world. Carolus Linnaeus the Younger, the son of Linnaeus, named one of Bank's Australian plants *Banksia* (1782), placing the genus in Class Tetrandra, Order Monogynia of his father's classification. Fortunately for us, the *Endeavour* drawings and copper plates still survive today in the Natural History Museum and Bank's *Florilegium* was eventually published, two centuries after the expedition (between 1980 and 1990).

Books of distinguished figures like Buffon and scientist-travellers like Joseph Banks increased the desire of European governments and of the general public for accurate information about the newly discovered distant lands. In 1792, the French government sent a scientific expedition to the Middle East to collect animals and plants, which doubled as a diplomatic mission. The French Republic, then under political attack from the rest of Europe, decided to re-establish relations with Persia, which was at war with Russia at the time. The two scientists chosen for this dual mission were Antoine-Guillaume Olivier (1756–1814), a physician whose main interests were in botany and entomology, and Jean Guillaume Bruguière (1749–1798), a physician interested mainly in molluscs. Olivier's abilities as a doctor greatly facilitated the progress of the journey. In Baghdad, Suleiman Pasha was bedridden with illness, and his physicians believed him to be on the point of death. The vice-Pasha seized the opportunity and assumed his master's title, intending to overtake the reign. When Olivier arrived at the Palace, he thwarted the vice-Pasha's plan by curing Suleiman of his illness within just 3 days. When the Pasha arose from his sickbed, he executed the vice-Pasha. The two Frenchmen eventually reached Persia, and the Persian Prime Minister granted them an interview. The return journey took some time because Olivier was wounded in Kurdistan by bandits and later, when once again in Baghdad, the chief janissary kept him there to cure his venereal diseases. Bruguière died on the journey home, and Olivier returned to France after 6 years of adventure with rich collections.

Other expeditions, initiated by the Museum and approved by the French government, were to the West Indies (1796–1798) and to the South Seas: Peru, Chile, the Pacific Ocean, Tahiti and Australia (1800–1804). These expeditions added over 18,000 new specimens to the Museum's zoological collections, representing over 3800 species, 2500 of which were previously unknown. In those distant days of wooden sailing ships, a corvette was a small, fast naval ship with only about 20 guns on the top deck, which served mainly to convey messages between the battleships of a fleet. The *Coquille* was a French Navy corvette that sailed to the South Seas (1822–1825) with a zoologist and two illustrators on board that brought animals and plants from the coasts of South America, the archipelagos of the Pacific and New Zealand, New Guinea and Australia. A year after returning, the *Coquille,* now renamed *Astrolabe* (Fig. 6.1), sailed to the Pacific Ocean again (1826–1829). These two expeditions brought back illustrations of live animals, published as "atlases", which consisted of illustrations and names only, without further text (D'Urville

Fig. 6.1 The *Astrolabe*

1833). These on-board illustrations advanced knowledge of the natural history of marine animals of distant lands, far beyond the knowledge gained from museum specimens, as their illustrations portrayed the live animal, a point so significant that I briefly present some examples of gastropods, the group I am acquainted with (Fig. 6.2). Careful recording by highly accurate and detailed images meant that even if specimens were not available to study at a distance, because the images travelled, the zoologists and botanists did not have to.

The cowry shell is unusually smooth, symmetrical and domed, with a glistening glaze (Cypraeoidea), and in museum collections, this is all that we see. The live cowry, however, has a magnificent mantle with two huge flaps that spread over the shell and cover it, thereby preventing shell-boring organisms from settling on it. These flaps, which may vary among species, have different colours than the shell beneath, and they may be smooth or tuberculate or have small finger-like papillae that branch or fuse at their bases to a series of tufts; some are glandular and secrete an acidic liquid when the snail is disturbed. The *Astrolabe* illustrations, true to nature, present living cowries with their mantle flaps partly extended—very different from the empty shells of museum collections. The violet shell (*Janthina*, Epitonioidea), as can be seen in museum collections, has a purple-blue shell. The live snail dwells near the surface of the sea, where it builds a raft of bubbles from which it hangs upside down and floats. The female places her eggs within a special capsule that she attaches to the lower part of the raft. Again, true to life, the *Astrolabe* illustration presents a violet shell hanging upside down from a floating raft, with egg capsules on its lower surface. Moon shells (Naticoidea), as seen in museums, have a thick, almost spherical shell. In the live snail, as the *Astrolabe* illustration shows, there is a huge foot; a fold of skin rises from the front of the foot and covers the entrance to the mantle cavity, preventing silt and sand from entering. Moon shells glide over muddy-sandy sea floors with a foot distended by seawater, capturing buried bivalves by digging and covering them with mucus and then boring through their valves. The surf whelk (*Bullia*, Buccinoidea) has an ordinary, spiral shell and that is all we see in museum collections. In the live snail, as the *Astrolabe* illustration shows, it has a huge, thin, flattened, supple foot. Today, we know that this foot allows it to surf up and down the shore. It burrows into the sand at low tide, and as the tide comes in, it emerges from the sand, turns on its back and spreads its foot, which now functions as an underwater sail that "catches the wave". Foot above and shell below, it now drifts up and later down the sandy shore to reach the stranded animals on which it feeds. Furthermore, the *Astrolabe* illustration shows that the surf whelk's shell is partly green. Today, we know that in addition to carrion, the surf whelk also grazes on plant material. Green algae grow on those parts of its shell that receive abundant light, and the surf whelk regularly crops these algae by extending its long mobile trunk over its head and onto the algal garden, moving its trunk over it and plucking the algae. The *Astrolabe* illustration shows a cone (Conoidea) with an extended breathing tube, with a black ring on a white background. Today, we know that cones kill benthic creatures with a venom-filled tooth that triggers continuous muscle spasms or paralysis (or both); their sting may be fatal to humans;

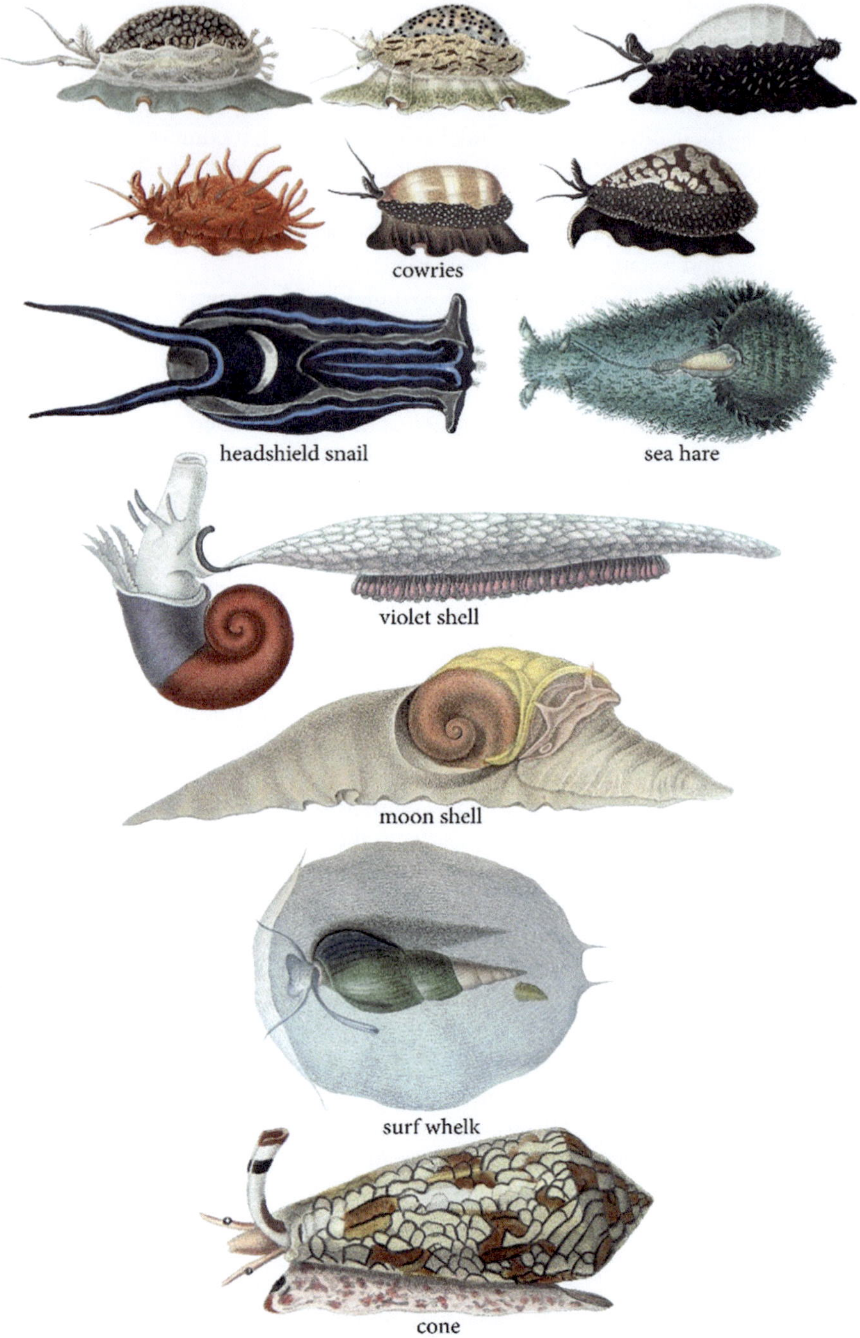

Fig. 6.2 Figures from the *Astrolabe* expedition. Cowries, their mantle partly or completely covering the shell. Headshield snails and sea hares are without external shells and colourless in preserved specimens, but the illustrations depict their rich colours. Violet shell, live snail with bubble raft and egg capsules beneath; moon shell (Naticidae) with foot extended. Surf whelk *Bullia*, live snail with foot extended. Cone, with a white breathing tube and a black stripe as warning colouration

the conspicuous, strongly contrasting colours of the breathing tube may be warning colouration. The *Astrolabe* on-board illustrations indicate all this natural history.

Furthermore, in 1819, the Museum established a programme to train young travelling naturalists in collecting, preserving, labelling and classifying specimens. The programme proved successful, and Museum professors sent trained collectors to areas poorly represented in their collections. Only later did other countries follow suit, and representatives from Berlin, Vienna and Leiden began scouring the remote corners of the colonial world. Many of these collectors died of disease, accidents or at the hands of locals, but the lure of uncovering more of the globe's riches did not diminish.

We have mentioned two reasons why the museum in Paris became the centre of European research in zoology. A third reason was Cuvier. Georges Jean-Léopold Nicolas-Frédéric Cuvier (1769–1832, Fig. 6.3) was born into a French-speaking, Protestant family, the son of a lieutenant in the Swiss Guards in Montbéliard, then in the Duchy of Württemberg (and annexed to France in 1793). By the age of 12, he had read Gessner's *Historiae Animalium* and Buffon's *Histoire Naturelle*, which kindled his interest in zoology. The Karlsschule Academy in Stuttgart, where he later studied, aimed to educate future civil servants. Cuvier took administration, law, chemistry, mineralogy, botany and zoology; it was here that he also learnt to

Fig. 6.3 Cuvier

dissect animals and to document his dissections by drawing them. Here too, a professor presented him with Linnaeus' *Stystema naturae*, which for the following 10 years would comprise his entire natural history library. The education he gained at Stuttgart would serve him well later and throughout his career. At 19, he took an interim job as a private tutor to the son of a noble Protestant family in Normandy, and it was here, along the beaches of Normandy, that he discovered the rich world of marine animals, many of which he carefully dissected, thereby finding an abundance of new features and types of animal organization. His stay in Normandy was fortunate in that it kept him away from the violent Reign of Terror that then raged in the streets of Paris (1792–1794). Eventually, he was invited to the Museum as (assistant and then) professor of comparative anatomy. Coming from the snails and fishes of Normandy, animals were the focus of his research, and he had never dissected a human body. He was, however, in close touch with men of medicine, and physiological-functional aspects of the human body were to shape some of his theories in zoology.

Soon after reaching Paris, Cuvier became a member of the Academy of Sciences and, some years later, its permanent secretary. Beyond his academic career, he was appointed, time and again, to public positions on matters dealing with higher education and interior administration. He presided over commissions charged with examining the state of the higher educational establishments in the districts beyond the Alps and the Rhine that France had annexed. Napoleon appointed him to the council of the Imperial University, admitted him to the Council of State, made him administrator of the non-Catholic religions in France and appointed him as president of the Committee of the Interior. For this, Napoleon appointed him Grand Officer of the Legion of Honour, knighted him as a Baron and honoured him as a Peer of France. Cuvier was very fond of all the honorary titles showered upon him and would often present them in the title pages of his zoological publications, together with the scientific societies and academies that had elected him as a member. In 1817, Cuvier published *The Animal Kingdom* (*Le Règne Animal*), ten large, richly illustrated volumes that present a broad—and yet deep—view of the entire animal world. With a panoramic view of the entire animal world and only the animal world, *Le Règne Animal* marks the birth of modern zoology.

A devout Protestant, he regularly attended church services and was a founder of the Parisian Biblical Society. Alongside his rich public career, he carried out intensive research, with the ability (somewhat like Napoleon) to see the broad, panoramic picture of zoology, gaining control over several fields of study within zoology simultaneously and not missing details, all at the same time. His studio at the museum, in which he placed a plaster bust of Linnaeus, whom he admired (Eigen 1997), and which nobody was permitted to enter, was furnished with 11 desks, each with pens, ink and a bell to call servants. He would move between the desks, from one study to another, and research assistants would bring him any scientific publication he needed, whether from his library of 8252 volumes or from his collections of 5000 species and 15,000 individual specimens (Eigen 1997); he did not permit his assistants to enter the room, nor to speak. Benches in the room enabled him to rest when tired. A towering figure, Cuvier died at the age of 62, during an outbreak of cholera.

Cuvier, like Buffon before him, sought to promote comparative anatomy as a discipline as rigorous as the physical sciences by establishing basic principles in zoology that would be comparable to Newton's basic principles in physics. He was heavily influenced by the German philosopher Immanuel Kant, who had suggested that living beings are powered from inside, through an intimate interlocking of internal functions (Outram 1986; Wilson 2006). Kant, following the scheme of Epicurus, had believed that the universe is constantly unconstant, forever shifting between chaos and order by two opposing universal forces, attraction and repulsion (Wilson 2006; we recall Empedocles of ancient Greece, who had believed that two opposing forces, Love and Strife, govern the universe.) From these forces, the emergence of order in terms of simple, general laws expresses the wisdom and power of the Creator. There must be a God, Kant says, because nature proceeds even in chaos, in a regular and orderly fashion (Wilson 2006). Kant had further suggested (1790), echoing Maillet, that many animals share a certain scheme in their bone structure and in the arrangement of their other parts, a basic outline that is simple yet able to produce the great diversity of species by shortening some parts and lengthening others, by the involution of some and the evolution of others (we recall Aristotle's words of the more or the less). Despite all the variety among these forms, they have been produced according to a common archetype, and this reinforced Kant's suspicion that they are actually akin and have been produced by an ancient, common mother (Wilson 2006). Kant further believed in the existence of a vital, immaterial principle, distinct from the power of generation, an organizing force outside space but active in matter. This principle, which is beyond the limits of our insight, extends through all parts of the world. It transforms bodies and replaces dead ones with new formations; it is the *anima mundi,* the world spirit that builds individual bodies and also builds the world as an integrated system (Wilson 2006).

Cuvier was also acquainted with the strange theories of the French physician Marie Francois Xavier Bichat (1771–1802). The son of a doctor in eastern France, Bichat studied surgery at Lyon (Nordenskiold 1935). During the Revolution, he moved to Paris, where he lectured on surgery, was appointed to a professorial chair and took part in the life of the medical faculty. Bichat suggested that animals are assemblages of 21 different kinds of tissue, each executing a different function and thereby contributing to the preservation of the individual, somewhat like several separate machines all coordinated in one general machine. Each organ of the living animal is thus composed of several elements (we would call them tissues). Chemistry has its elements (such as hydrogen or oxygen), of which compound molecules are formed; and in the same way, anatomy has its simple textures, which, by their combinations with other textures, make the organ. Bichat was thus searching for the elements of living animals and found them, as he thought, in the tissues, whose various combinations formed the familiar organs of the body (Wightman 1953). He further distinguished between "organic" organs and functions of the body responsible for the vegetative, unconscious life of an animal and "animal" organs and functions, controlled by the will of the individual. He claimed that animal organs are always symmetrical and organic ones are unsymmetrical and went to great efforts to prove that lungs and kidneys are actually unpaired. Furthermore, he claimed that animal functions are always "harmonious", while the organic ones are "discordant", meaning that it makes no difference if one lung functions more or less than the other, whereas dissimilarities of the visual or auditory organs cause serious disturbances. Thus, the lack of a gift of music is due to the fact that the ears possess different powers of hearing (Nordenskiold 1935).

Cuvier, seeking basic principles of natural history, followed Bichat in broadly defining the concept of Life as the ensemble of functions resisting death. Living beings exist in a continual defiance of those laws of chemistry and physics that reduce the living body to its constituent elements. These forces are, like machines, mainly internal to an organism's body; they are not external features as used by Linnaeus in his classifications. Cuvier thus considered life as a rebellion against physical and chemical rules by creating a harmony of functions. Organisms are integrated wholes whose organs function together in harmony to maintain them, and this harmony dictates that none of the organs can alter without impairing this total functional integration (Outram 1986).

One of the central questions in the classification of plants and animals was whether to use a single external key character that reflects the inner essence of a group, as practised by Linnaeus, or all its characters, as claimed by Buffon (Outram 1986). On this question, and in line with his view that an organized being is an ensemble of organs acting on each other reciprocally, Cuvier's opinion was clear. Every group has many characters that identify and distinguish it from other organisms "that, to distinguish one organism from all the rest, we need its full description".

True to this claim, Cuvier set out to dissect animals to incorporate their internal anatomy into their classification (Outram 1986). Whereas Buffon and Daubenton had dissected mainly large mammals, Cuvier dissected a wide range of vertebrates, including the small ones and invertebrates (Figs. 6.4, 6.5, 6.6, and 6.7). His dissection (and sketching) abilities were admirable, and the range of organs and animals he dissected was truly amazing, from the inner ear of marine sunfish to that of freshwater carps, from the digestive system of porcupines to that of dolphins, from the genitalia of kangaroos to those of spiny anteaters and from crocodile hearts to hedgehog muscles. Mammals were of a size convenient for dissection, but the molluscs he dissected included small organisms (dorids, sea butterflies, *Lymnaea*). Modern comparative anatomy was born with Cuvier.

Cuvier's concept of an animal was of an ensemble of organs acting on each other reciprocally and therefore of distinguishing among groups by a full description of all its organs. This concept is easily applicable when all distinguishing characters point in the same direction. However, what should we do when an animal fits into one group according to one trait and into another group according to another trait? How are we to decide? To solve such contradicting data, Cuvier outlined his *Principle of Character Subordination*: the more significance a functional organ has, the more it should be the leading character for classification (Mayr 1982; Eigen 1997). We must weigh some characteristics of an organism more heavily than other ones. Therefore, we should rank characters by their order of importance, and within

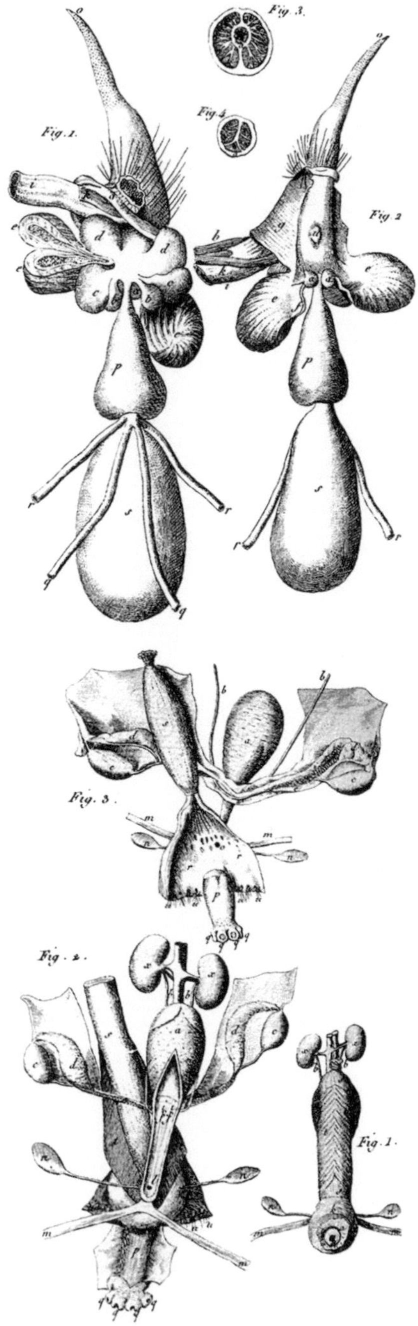

Fig. 6.4 The genitalia of a kangaroo and a spiny anteater (Cuvier 1805)

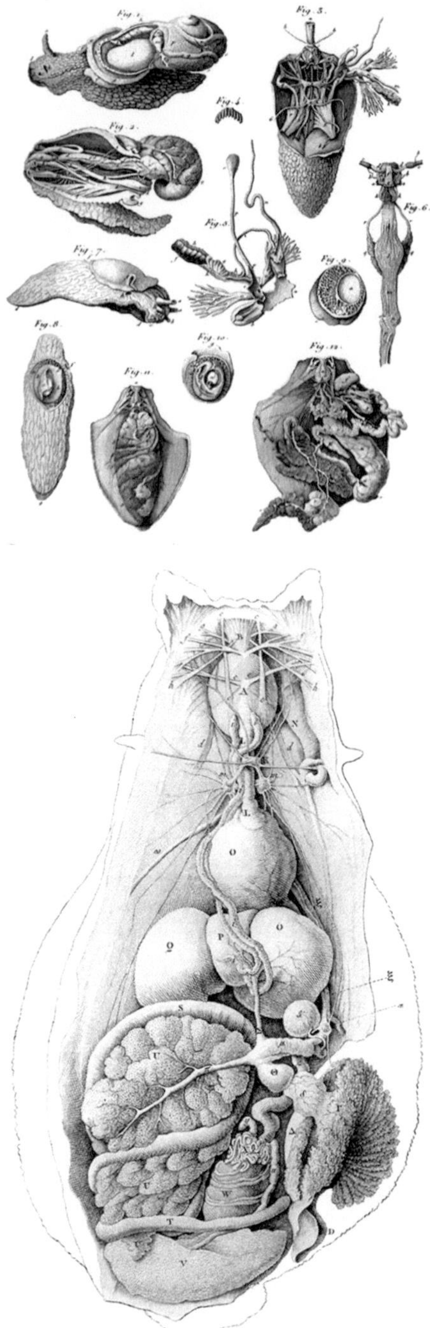

Fig. 6.5 General anatomy of a land snail and a sea hare (Cuvier 1817)

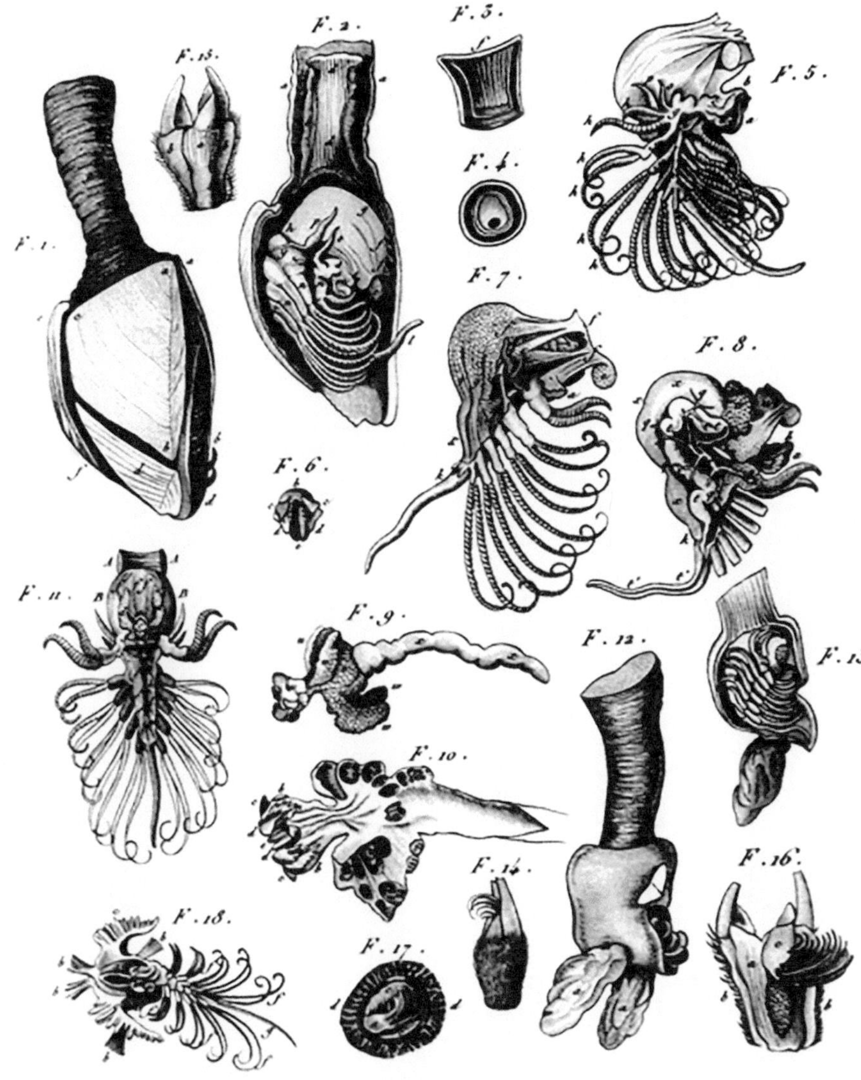

Fig. 6.6 Anatomy of a goose barnacle (Cuvier 1817)

the animal kingdom, classification into groups must follow this leading character (Eigen 1997). (We recall that Aristotle had believed that for any given natural group of living creatures, some features are essential: they appear in all its members and in no other group and they define the group. We further recall that Cesalpino had sought to establish a classification of plants based on essential features of the plant, rather than on extraneous characters.) Following his Principle of Character Subordination, Cuvier distinguished Animal life from Plant life by the presence of

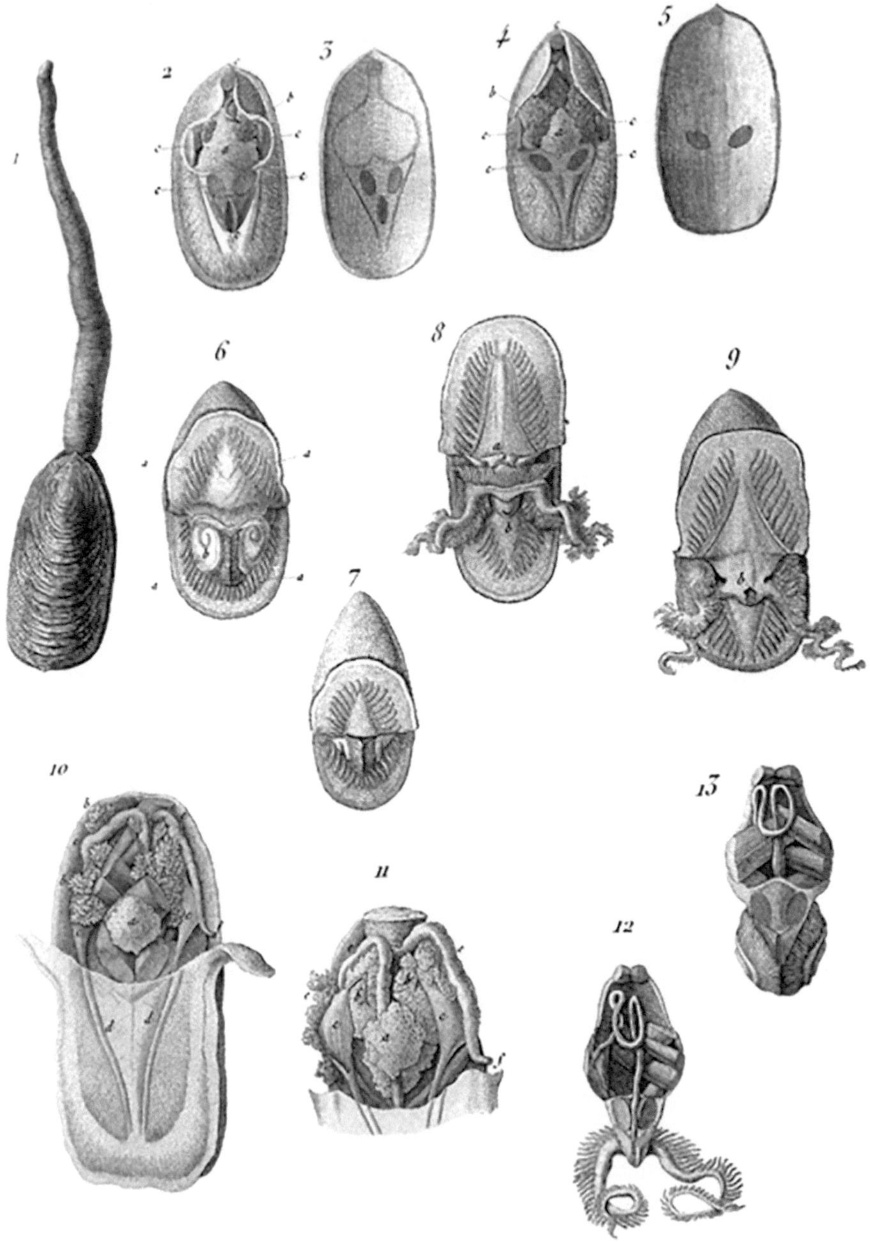

Fig. 6.7 Anatomy of brachiopods (Cuvier 1817)

a nervous system, which functions both to channel sensations and to enable voluntary movement. Next, he used the nervous system as the dominant characteristic of distinction between major animal groups (Eigen 1997; Outram 1986). According to the complexity of the nervous system, he divided the entire Animal Kingdom into four large major groups of equal rank, which he named *embranchements* (Branches):

Vertebrata	a developed brain, a dorsal nerve cord (and a segmented body, bilateral symmetry, axial skeleton)
Articulata	a small brain (and paired ventral nerve cords, a segmented body, bilateral symmetry, external skeleton; largely arthropods and annelids)
Mollusca	a primitive brain (and a nervous system consisting of separate ganglia, muscles attached to the skin and shell, bilateral symmetry)
Radiata	no distinct nervous system (sensitive to touch; radial symmetry; echinoderms, cnidarians, worms, sponges, corals, rotifers and unicellular creatures—"waste-basket" groups)

Cuvier insisted that these four major Branches are so fundamentally different from each other that they represent large natural groups in the animal kingdom, above the Linnaean Classes; and that individuals of one Branch can never transform into another (Mayr 1982). Indeed, he argued, even within each Branch, between the different Classes, there are no intermediate creatures. Consequently, he rejected Bonnet's theory that the animal kingdom falls into a single Great Ladder of Life in which all beings fit in a rising order of complexity of organization.

Cuvier also formulated another principle of classification: the same character may vary in taxonomic importance within different animal groups. The teeth formula determines an Order in mammals, only a Genus in reptiles and only a Species in fishes; the form of the hind foot has Order-level importance in mammals, in which it is the major organ of locomotion, but only Family-level significance in birds, in which the wing is the main organ of movement. By following these principles to some extent or another, Cuvier classified the entire Animal Kingdom.

Three of Cuvier's four major Branches—vertebrates, articulates (our arthropods) and molluscs—are still accepted, two centuries later. The fourth, Radiata, was weak because it lumped together echinoderms, worms, cnidarians, rotifers and protozoans, among which there is no connection. Cuvier placed 14 classes in his three correct branches, 11 of which we accept as natural groups today as well. His classification differed from today's phyletic classification, mainly in that he assigned goose barnacles, brachiopods and tunicates to the molluscs, close to the bivalves. These three groups resemble bivalves in that they filter-feed, are sessile, have a degenerate sensory system and blood system and have a calcareous armour or tough tunic, which encases their body. In a classification lacking the dimension of evolution over time, Cuvier was thus classifying them by their common function, filter feeding (Fig. 6.6).

In the mammals, Cuvier put primary emphasis on the function of movement as expressed by the feet, and accordingly, he divided all mammals by their feet into those with claws (or nails), with hooves or with fins (whales, seals; his Unguiculata Ungulata and Mutilata). This resulted in his placing the rabbit, bear and ape closer to the kangaroo than to the hooved elephant, horse and sheep. This error of Cuvier emphasizes the importance of choosing the most significant function when grouping by a single diagnostic function, because had he chosen reproduction as the prime function in mammals rather than movement, he would have gained today's division into placentals, marsupials and monotremes. The principle of the Subordination of Characters is thus weak in that it opens classification to subjective decisions (Outram 1986). Furthermore, Cuvier placed secondary emphasis on the function of obtaining food as expressed by the teeth and, unaware of convergence, he placed the spiny anteater, a monotreme, together with the giant anteater and the aardvark, all ant feeders (Fig. 6.8). He further erred in lumping together all those with a thick hide into a single group, the Pachydermata (elephants, conies, hippopotami, pigs, rhinoceros and tapirs); today we place these into three distant groups, with elephants closer to conies, hippopotami to pigs and rhinoceros to tapirs.

Cuvier, in his classifications, considered the adult organism as the main focus of natural history. In his search for a pattern in the animal kingdom, he did not integrate aspects of embryology (Outram 1986).

One of Cuvier's important contributions to classification concerns the expansion of the taxonomic hierarchy. Linnaeus had allotted a fixed number of only four (to five) ranks in all three kingdoms of nature, whether animals, plants or minerals: Kingdom, Class, Order, Genus, Species and, sometimes, Variety. With the sharp increase in the number of animals discovered, it was inevitable that additional ranks would come into use and that most of Linnaeus' ranks would undergo secondary

Fig. 6.8 Cuvier placed together the spiny anteater together with the giant anteater and the aardvark, all ant feeders. (Illustration by Tuvia Kurz)

splitting. Cuvier's splitting of the Animal Kingdom into four major groups required a new rank between the Kingdom and the Class, which Cuvier named "Branch" (today a Phylum), and the many new genera demanded a new rank between the Genus and Order, the Family (Buffon's term). Cuvier also created ranks above and beneath the Genus: Section, Tribe and Subgenus. His scale was thus:

Kingdom

Phylum

Class

Order

Family

Section

Tribe

Genus

Subgenus

Species

(Variety)

Not every group required all those new ranks, in his opinion. In some groups (mainly insects), he made use of the full, detailed hierarchy; in some groups, he would use only some of the new ranks (mainly vertebrates). Thanks to this flexibility in expanding the range of classification ranks by use of compulsory versus non-compulsory ranks, Cuvier well represented the natural classification of the animal diversity known in his time. His new additional ranks were also important because they stabilized nomenclature, as naturalists had already begun using the terms inconsistently and for different ranks (in a study of the insects of Paris, the term Family was used throughout as a rank beneath the genus, not above it; Fourcroy (1785), see Fig. 6.9). Cuvier considered most of these ranks as subdivisions created merely for convenience in classification and did not ascribe them any further importance (Eigen 1997). His one exception was the Species, which he defined as "individuals of the same origin or common parents, and they resemble parents just as they resemble each other". To allocate species to genera and genera to families, Cuvier developed the concept of a type species as a sign for a Genus and a type genus as a sign for a Family; and he derived the family name from the best-known genus within it. A species was to be assigned to a given genus by whether or not it resembled the type species of that genus more than that of another type species, and along similar lines, a genus was to be assigned to a given family by whether or not it resembled the type genus of that family more than that of another type genus of another family (Eigen 1997). In his opinion, classification could be arrived at either

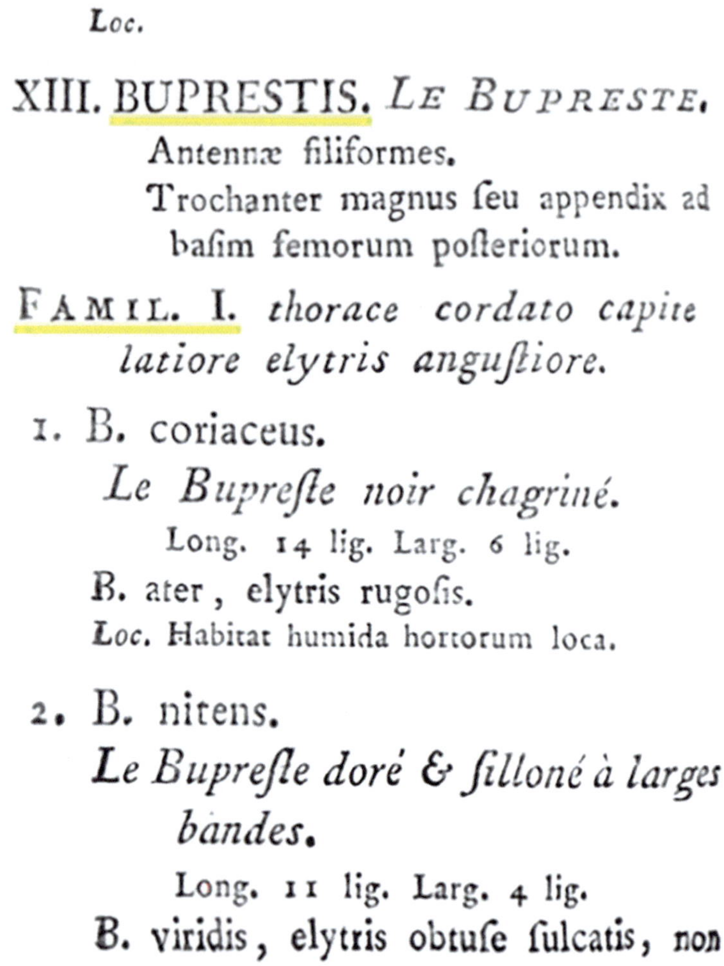

Loc.

XIII. BUPRESTIS. *Le Bupreste.*

Antennæ filiformes.

Trochanter magnus ſeu appendix ad baſim femorum poſteriorum.

FAMIL. I. *thorace cordato capite latiore elytris anguſtiore.*

1. B. coriaceus.

Le Bupreſte noir chagriné.

Long. 14 lig. Larg. 6 lig.

B. ater, elytris rugoſis.

Loc. Habitat humida hortorum loca.

2. B. nitens.

Le Bupreſte doré & ſilloné à larges bandes.

Long. 11 lig. Larg. 4 lig.

B. viridis, elytris obtuſe ſulcatis, non

Fig. 6.9 Fourcroy 1785, divides a Genus (*Buprestis*) into Families; *Entomologia Parisiensis*

by going upwards, from species level to the genus, family, order and class, or downwards, from the higher levels to that of the species (Outram 1986). He defined a species in terms of a reproductive and a morphological unit: "a collection of all organised beings born from each other or from common parents, and those who resemble them as much as they resemble each other" (Outram 1986).

In his *The Animal Kingdom* (*Le Règne animal distribué d'après son organisation, pour servir de base à l'histoire naturelle des animaux et d'introduction à l'anatomie compare*; 1817, 1829–1830), Cuvier was, we could say, the first modern zoologist. He created a broad, general and yet deep view of the entire animal kingdom and only of the animal kingdom. He based his treatise, of course, not only on his own direct studies. The 50 years following the tenth edition of Linnaeus' *Systema*

Naturae (1758) were years of considerable expansion in animal classification. With the discovery of more and more animals, more and more people began studying the specimens piling up in museums and in private collections. Some of these people defined these specimens so sloppily that Cuvier would not rely on them. Johann Friedrich Gmelin (1748–1804), a professor of medicine, chemistry and botany in Göttingen, had published the 13th edition of Linnaeus' *Systema Naturae* (1788–1789) with descriptions and names of many new species. Cuvier complained (Preface, pp. 2–3): "what can one say of Gmelin's *Trichechus manatus* [a manatee] who, under one species, had placed three species and two genera, which differ in almost everything? What name shall we use when we come to discuss *Velella* [a free-floating cnidarian], placed by him twice among the medusas, and once among the sea cucumbers? How are we to reconstruct *Biphorae*, of which some are named *Dagysa* [a tunicate] and most *Salpa* [a tunicate], which he places among the sea cucumbers?" Cuvier thus carefully inspected the studies of other naturalists to determine which were reliable.

In classifying fishes, for example, he commented that he had examined all the material of Bloch and that he supported his identifications (Eigen 1997). Marcus Eliezer Bloch (1725–1799), the son of a Biblical scribe in southern Germany, knew only Hebrew and Yiddish until he was 19. Eventually, he studied German and Latin, qualified as a physician and practised in Berlin. A devoted amateur fish lover, he amassed a large private collection and an extensive library by buying from other collectors and from scientific expeditions. His comprehensive, 12-volume, beautifully illustrated treatise on fishes, *Allgemeine Naturgeschichte der Fische* (1782–1795), contained, in addition to well-known groups, the description of 19 new genera and 176 new species and it ran rapidly through five editions, both in German and in French (Fig. 6.10).

Cuvier also relied, and very heavily at that, on the entomologist Pierre André Latreille (1762–1833). Born in south-central France as the illegitimate child of a baron who never recognized him and an unknown mother who abandoned him at birth, at the age of 16, Latreille came to Paris to study for the priesthood, but throughout his studies, he was also interested in natural history. With the fall of the *Ancien Régime*, the Civil Constitution of the Clergy required priests to swear an oath of allegiance to the state (1790). Latreille failed to do so and was imprisoned (1793) under threat of execution. The prison's doctor, upon inspecting the prisoners, was surprised to find him scrutinizing a beetle on the dungeon floor and explaining that it was a rare insect, the red-necked ham beetle *Necrobia ruficollis* (Fig. 6.11). The impressed physician managed to obtain his release, whereas all the other inmates were executed within the month; it is nice to find that one beetle had once saved a man's life.

Latreille was then employed at the Natural History Museum, where he curated the arthropod collections, eventually became a member of the French Academy of Science, was appointed professor of entomology (then including all terrestrial arthropods, thus also spiders, isopods and others) and was instrumental in founding the Entomological Society of France. Besides describing countless genera (*Genera Crustaceum et Insectorum*, 1806), he also named many higher taxa, with which

Fig. 6.10 Fish species described and illustrated by Bloch

many modern zoologists are familiar (Thysanura, Siphonaptera, Pycnogonida, Ostracoda, Stomatopoda, Decapoda, Amphipoda, Isopoda, Xiphosura, Melipona, Myriapoda). Furthermore, Latreille introduced the concept of a "type species", a species to which the name of a genus is attached and the method of naming a family after one of its genera, thereby designating a type genus. Latreille also contributed the section on entomology to Cuvier's *Regne animal*. In his own textbook of entomology, dedicated to Cuvier, he disagreed with Cuvier's emphasis on internal organs, commenting that it was impossible to group insects into families without recourse to external characteristics, such as the form of the antennae, wings or legs (Outram 1986; Fig. 6.12).

Cuvier emphasized the function of animal structure. An organism, he claimed, was a functional unit whose structure was determined by its relationship to its environment. According to Cuvier, a functional unit signified that all organs relate

Fig. 6.11 *Necrobia ruficollis*

functionally to one another and operate together in harmony—they cannot vary independently of the entire system because variation in an organ would render it unable to function. Cuvier called this broad view of functional interdependence *the Correlation of Parts* (Eigen 1997; Mayr 1982; Outram 1986). As an (erroneous) example, the size of a liver seemed to Cuvier to be inversely related to the size of the respiratory system. Completely absent in insects, whose entire outer surface was given to the action of respiration, it was far larger in mammals, where the business of respiration was the function of a single, concentrated system of the body. From this, Cuvier concluded that in mammals, the liver aids the respiratory action of the lung (Outram 1986). The animal body, according to Cuvier, displayed intricate integration and precise coordination with its surroundings (we shall soon return to this correlation).

6.1.2 The Time Dimension: Extinction

One of Cuvier's major contributions to biology was the establishment of extinction as a solid fact. In those times, many people could not believe that God, having created all things and having pronounced them good (*Genesis* 1), would allow any of his creatures to be wiped out. They interpreted fossils as remains of living species, whether already known to us or yet waiting undiscovered in some remote places on Earth. Furthermore, scientists of that time did not know enough about the skeletal anatomy of fossil versus modern groups to answer the question as to whether fossil

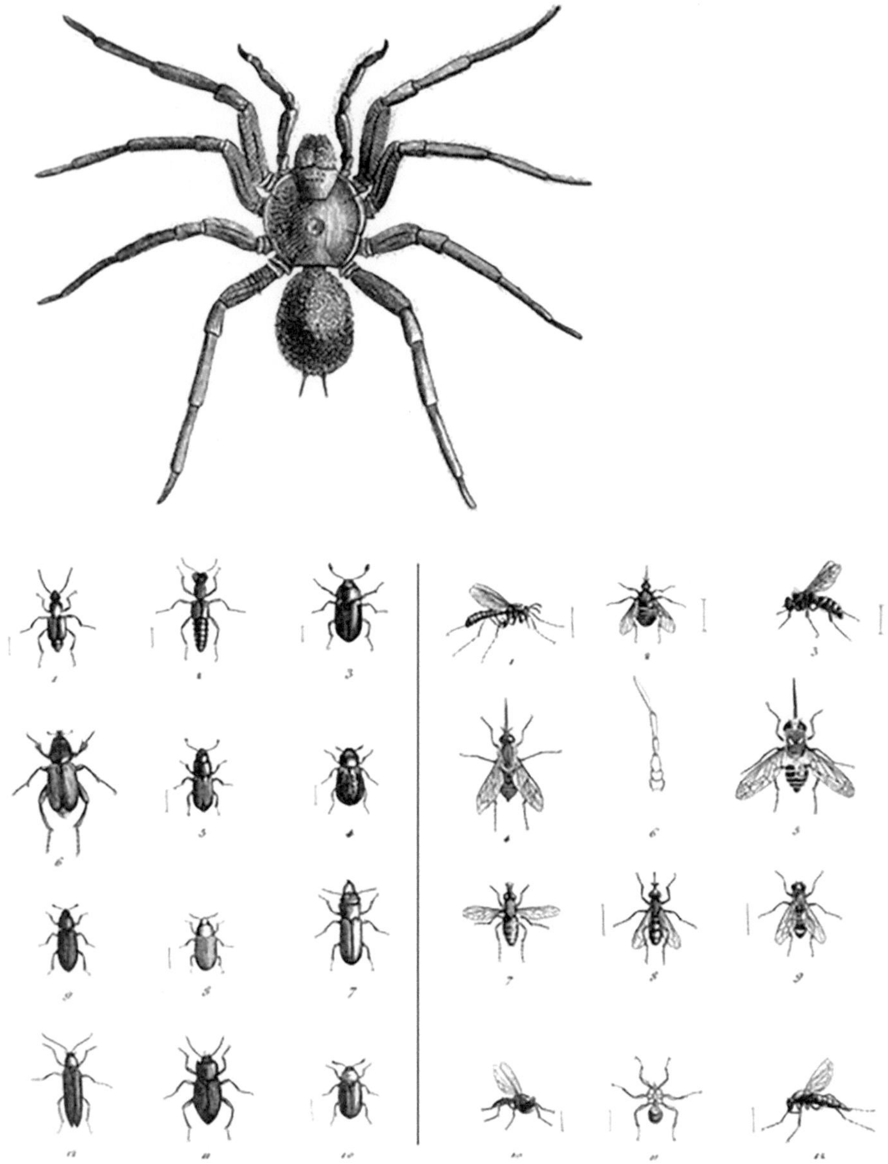

Fig. 6.12 Insect classification based on external features (antennae, wings, legs) and including spiders (Latreille 1806)

remains were of extinct or modern creatures. In cold regions of northern Europe, layers close to the surface contained fossil bones and teeth of large mammals, and Buffon had identified these fossils as the remains of elephants and rhinoceros. From their very northern distribution, Buffon had concluded that in the past, tropical warm climates prevailed in the far north; with the gradual cooling of the globe, these giants migrated southward to the warm tropics where they live today. Cuvier, coming some decades after Buffon, was fortunate in that, by the time he began studying these fossil bones, the Revolution's spoils of war contained rich collections of the previous ruler of Holland (William V), conquered by France in 1795. An extensive collection of treasures of natural sciences thus reached Paris. It included specimens that enabled Cuvier to prove, by observing skeletal differences, that the African and Indian elephants were different species. Furthermore, the fossil bones of the cold northern regions differed so considerably from both species that he included them in a separate, extinct species adapted to cold climates. ("The inhabitants of Siberia believe that these bones come from a subterranean animal like our moles, which never lets itself be taken alive; they name it 'mammoth', and mammoth tusks, which are similar to ivory, are for them quite an important item of commerce", 1796, in Rudwick 1997). The mammoth differs from the elephant at least as much as a dog differs from a jackal and a hyena. The dog tolerates the cold of the north, whereas the other two live only in the warm south, and the same applies to mammoths versus Indian and African elephants. As the mammoth was not the same species as either of the living elephant species, it could well have been adapted to a quite different environment, namely to the cold climates in which its bones were now found. Cuvier, therefore, rejected the theory of the gradual cooling down of Earth and of the migration of animals from slowly cooling poles to warm tropics. He further found that another fossil elephant, Buffon's "Ohio elephant", represented a distinct, extinct species that differed even more from living elephants than mammoths, in that it had "enormous teeth armed with points" (in 1806, he would name it *Mastodon;* Fig. 6.13). All elephants are very large and therefore easily seen, and yet no human had ever seen these fossil elephants—evidence, he claimed, that extinction was a fact. Extinction, he claimed, is a natural process in nature, and the mammoth and *Mastodon* bear witness to an animal world preceding the present world, which went extinct because of some sudden catastrophe. As evidence for this abruptness, he noted that many remains of mammoths had their skin, hair and flesh so well preserved that when discovered, nearby dogs ate the flesh. Their high state of preservation pointed, in his opinion, to a sudden extinction: had they died off gradually, they would have decomposed rapidly by putrefaction, and their bones would have been scattered.

Cuvier studied other massive fossils as well. Near Buenos Aires, then a Spanish colony, Spanish colonialists had discovered an almost complete skeleton of a single individual of some huge, elephant-sized mammal (1789). They shipped it to the royal museum in Madrid, where curators assembled it, and illustrations were drawn and later engraved in preparation for future publication. Seven years later, a French official in Madrid obtained these still unpublished engraved illustrations and sent them to Paris, where Cuvier reported on them to the Academy. From a study of the illustrations alone and without ever seeing the fossil itself, he announced that the skeleton was of an extinct animal belonging to the Edentata (today the Xenarthra), a largely South American order characterized by its reduced teeth, which includes sloths, anteaters and pangolins. Cuvier named the fossil *Megatherium*, the "huge beast", and concluded, by comparing its skull with living species of tree-dwelling

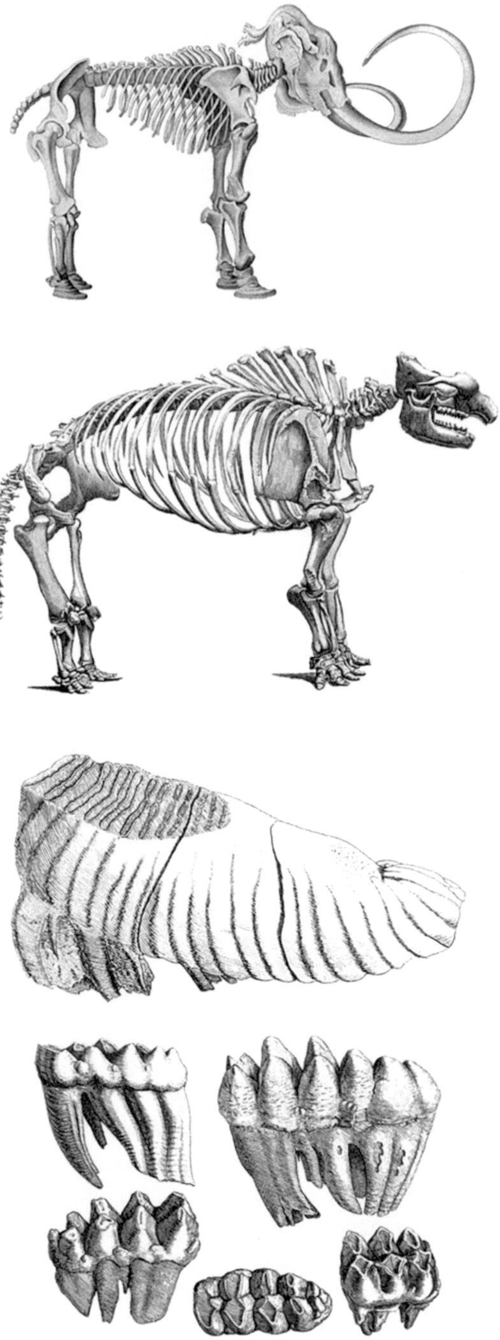

Fig. 6.13 Cuvier's (1825) illustration of the skeleton of an elephant (above) and of *Mastodon* (centre), and of the tooth of an elephant versus teeth of *Mastodon* (below)

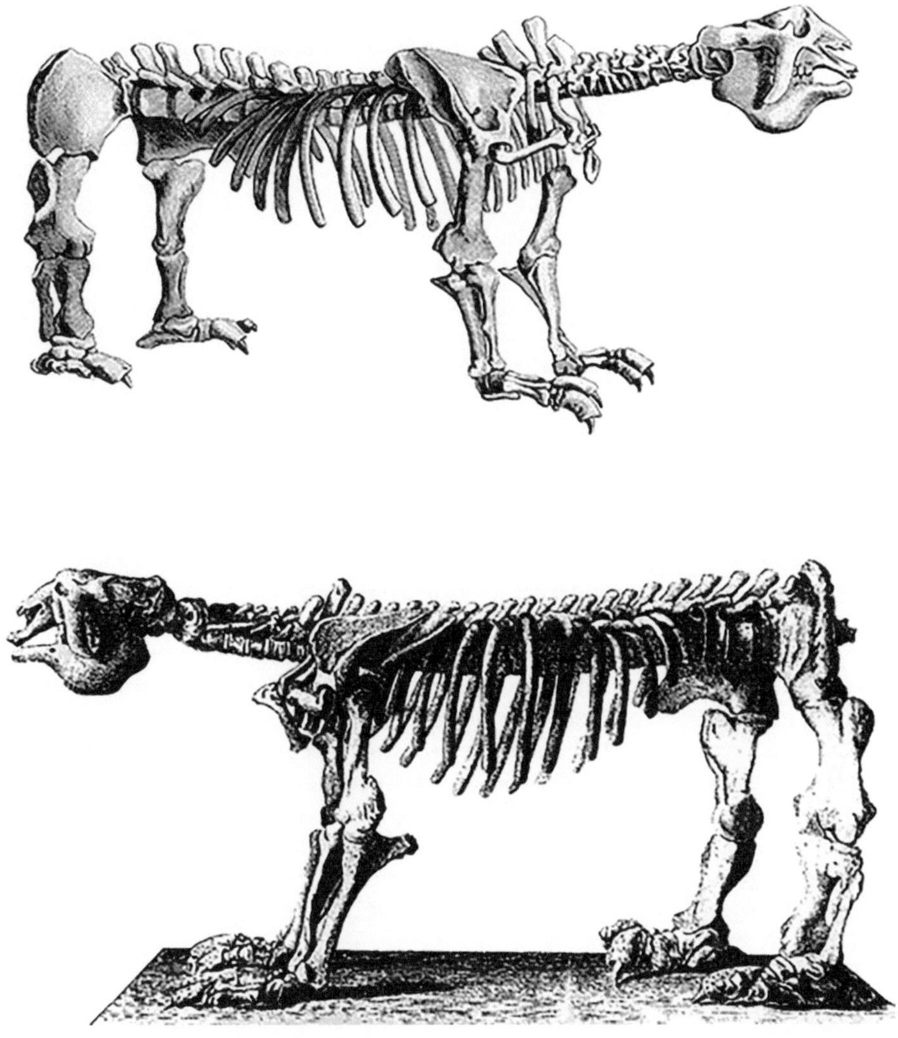

Fig. 6.14 Bru's reconstruction of the skeleton from South America, redrawn to illustrate Cuvier's treatise in which he named *Megatherium*

sloths, that it was a ground-dwelling giant sloth (Fig. 6.14). This was a great achievement for Cuvier because only a handful of scientists in the world at that time would have had the ability and experience to understand the skeletal anatomy of the little-known sloths, and only the museum in Paris could offer the broad range of comparative material necessary for this conclusion. The huge beast added further evidence to the concept that there once were all kinds of giant creatures, which no longer exist today. Cuvier claimed that *Megatherium* too went extinct because of a sudden global catastrophe.

(Today we know that *Megatherium* was an elephant-sized ground sloth of South America that lived from the Pliocene through the Pleistocene in lightly wooded areas of South America and fed on leaves and grasses. Rising on its powerful hind legs and using its tail to form a tripod, it could pull down branches with leaves. It walked on the sides of its feet because its sharp claws prevented it from placing them flat on the ground.)

In Ireland, antlers of a giant deer were discovered as early as 1588 and described as those of the American moose (Molyneux 1697). In 1812, two centuries after its description, Cuvier compared skeletons of various animals and showed that these huge antlers and bones belonged not to any living deer but to an extinct giant deer, which he named *Megaloceros giganteus*. It had antlers spanning 4 m, was 2.5 m tall at the shoulder, weighed about 1 tonne and its antlers weighed about 40 kg. (*Megaloceros*, an open-country deer feeding on grasses and shrubs in open birch and juniper shrubland, lived from 400,000 to about 10,000 years ago, across northern Europe and Siberia.)

In 1798, Cuvier began a study of fossils exposed in gypsum quarries (of Eocene age) on the outskirts of Paris. He found another extinct animal that, he argued, had perished due to a sudden catastrophe. He described it as *Palæotherium,* and because its retracted nasal bones suggested a small tapir-like trunk, he suggested it was a tapir (Fig. 6.15). With short, stocky, three-toed limbs, it lived in tropical forests of Europe some 45,000,000 years ago and some of its species were almost as large as a horse. (Recent studies suggest that *Palaeotherium* did not possess a trunk and that it related more closely to horses than to tapirs.)

One specimen among the quarry fossils was exceptionally important from a methodological aspect. From details of its skull and teeth, Cuvier suspected that it was a marsupial. If so, then the pelvis of this specimen must have marsupial (epipubic) bones to support the marsupium. Chiselling into the gypsum in the presence of witnesses, Cuvier did indeed find the marsupial bones. He also found that this fossil is an extinct species (*Didelphis gypsorum*, Fig. 6.16), close to the American opossum of today but distinct from it.

Then, there were the fossil reptiles. A very peculiar fossil had been described and illustrated (1784) from the Jurassic of Bavaria. Some had proposed that this creature

Fig. 6.15 Cuvier's reconstruction of the tapir *Palaeotherium*

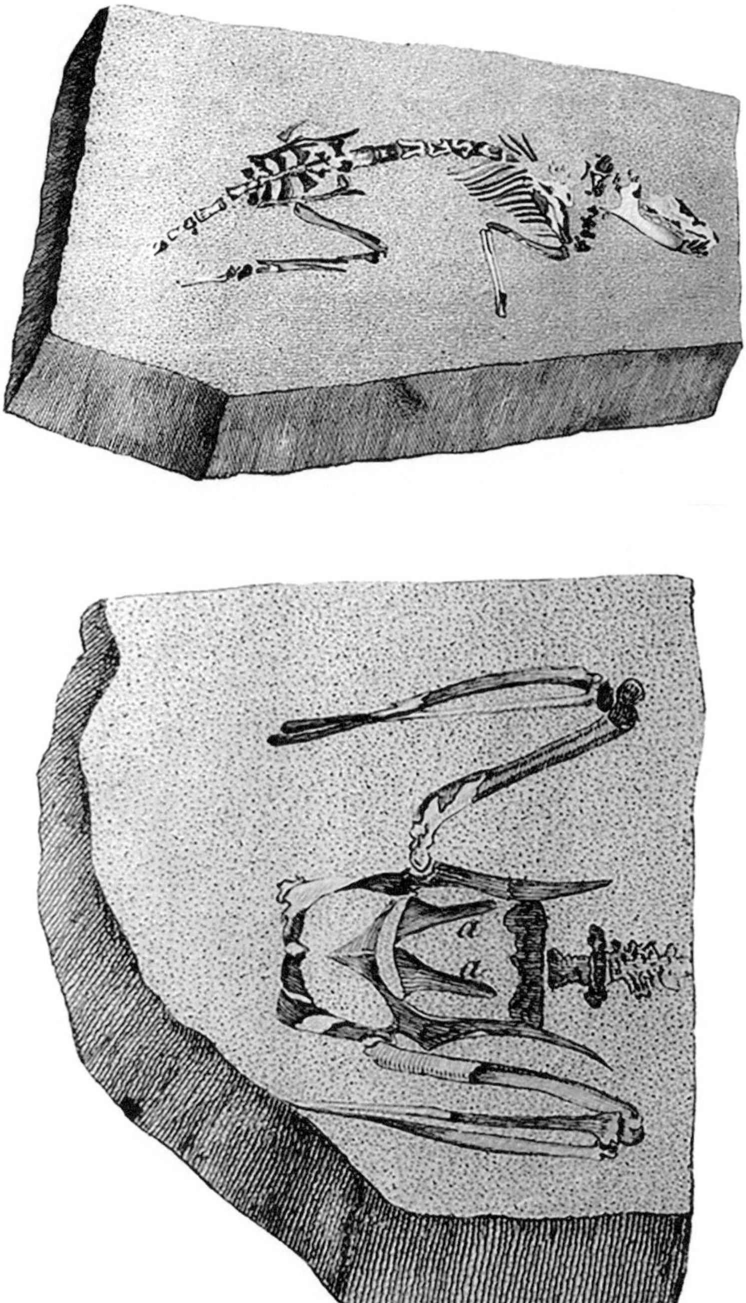

Fig. 6.16 Cuvier's *Didelphis gypsorum,* Cuvier 1825. Above: the complete skeleton as found; below: the two epi-pubic bones as Cuvier excavated them (marked by him as *a*)

lived by swimming in the sea, while others suggested that it was intermediate between birds and mammals, somewhat like a bat. Cuvier, again working only from the published illustration and without ever seeing the original specimens, correctly identified the mysterious animal as a reptile that could fly (1800). He named it *Ptero-Dactyle* (later Latinized as *Pterodactylus*); it was the first known pterosaur (Fig. 6.17). Another strange fossil skull was discovered in a quarry in the Netherlands and brought to Paris and Cuvier confirmed that it too was a reptile, but one that swam in the sea; he named it a Mosasaur (1808). He also reconstructed an ichthyosaur and a plesiosaur (Fig. 6.18, described by others in 1821), and he speculated that reptiles came into existence before mammals and were once the dominant fauna. He also described fossil species of hippopotamus, cave hyena, rhinoceros, the cave bear, manatee, seals, crocodilians, chelonians, fish, birds and molluscs. Cuvier, indeed, created and established modern Palaeontology.

The opossum discovery led Cuvier to formulate another basic principle in anatomy, the Principle of Correlation between Organs, which we have already briefly mentioned. As all features and organs of every animal function together harmoniously and are interdependent, and the existence of a species depends on the manner in which these organs interact, no organ can change without the other organs also changing—they correlate. Indeed, these correlations between functions (and, consequently, between organs) were, he believed, the basis for the order of living entities. If an animal is a predator, then its gut must be specifically adapted to digest meat: it

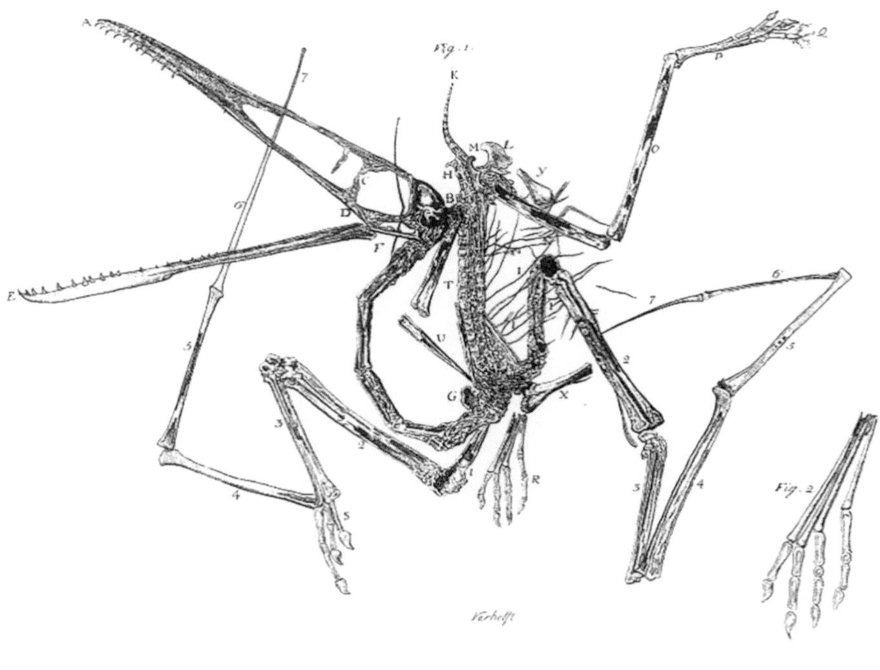

Fig. 6.17 *Pterodactylus antiquus*

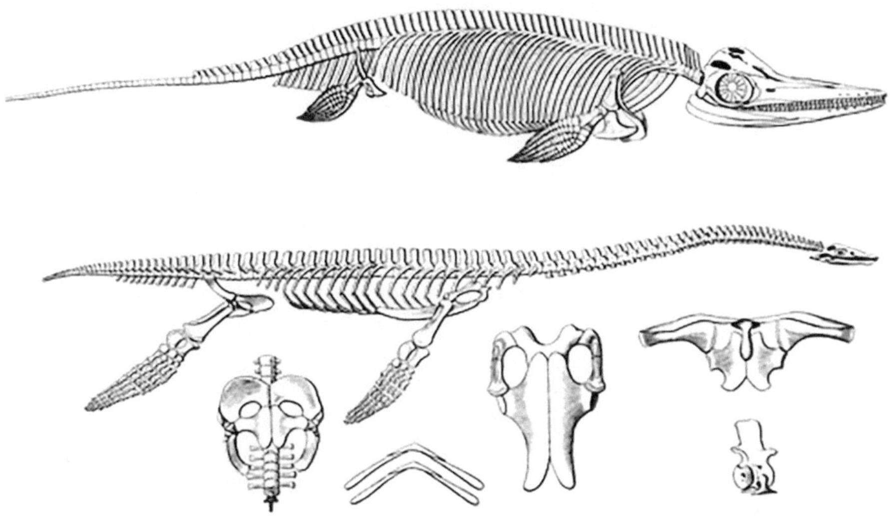

Fig. 6.18 Ichthyosaurus and Mosasaurus

must have sharp carnassials to slice flesh, jaws adapted to bear these teeth and claws to grasp the prey and to tear apart its flesh. It would also need long legs for speedy hunting and keen eyesight to see the prey from a distance. A predator will never have cloves or flat molars, as in herbivores. These constant correlations between organs enable us to identify an animal by observing just one of its parts in a fossil assemblage, such as a tooth or a claw. Today, the Principle of Correlation is central to comparative anatomy and palaeontology as a rational way to make predictions and reconstruct fossils. In emphasizing correlations between organs, we may recall Aristotle, who had maintained that the parts of an animal must be coordinated in such a manner that it is in harmony with the environment. He also noted that viviparous quadrupeds always have fur, whereas egg-laying ones have scales; that animals with large canines do not have horns; that birds with talons never possess spurs; and so on.

Cuvier's studies demonstrated conclusively that extinction is a natural, global process, to the extent that most fossils are the remains of extinct species. He suggested a series of sudden disastrous catastrophes, of "terrible events" that hit the Earth, exterminating so many animals that most species did not survive, from one period to another (*Recherches sur les ossemens fossiles de quadrupeds*; *Researches on quadruped fossil bones,* 1812). These catastrophes were sudden and violent; they hit every part of the globe and long, tranquil periods separated them. They may have occurred because of some sudden chemical change in the seawater, which animals could not tolerate (we might think of hypoxia). Alternatively, they might perhaps have occurred because of sudden massive waves violently hitting the globe and killing many species; or because the continents were flooded, drowning many terrestrial animals; or perhaps because the sea suddenly receded, destroying many marine creatures. Catastrophic events could explain the completely different faunas found in fossil records. New fossil forms appear abruptly in the geological record and then persist unchanged in overlying strata until the time of their sudden extinction. (Most English translations use the term "catastrophes", but Cuvier himself, in most cases, used the term "revolutions"). Cuvier demonstrated that the fossils occurring in the highest strata belonged to species or genera that still have living representatives, but that the fossils become increasingly dissimilar to modern forms the deeper we penetrate into the geological sequence. He did not deal with the issue of where the new fauna came from.

We recall Buffon's claim that extinctions release large quantities of molecules, and these, by spontaneous generation, build a new fauna and that because of the limited amount of organic molecules, one animal wave must go extinct to enable the next wave to form. Cuvier repeatedly emphasized that one fossil form does not gradually change into a succeeding one (Outram 1986). His commitment to the idea that, in each species, the body is an interacting system whose organs function together in harmony to maintain that body caused him to reject the idea that any organ could ever gradually change independently of all the other parts. For him, a species was such a well-adapted animal machine that only a catastrophic event could cause it to go extinct: under the impact of more gradual changes in the environment, it would simply migrate and survive. Hence, he insisted, a species is a fixed entity from its moment of creation. Species only become extinct; they do not change over time. Fossil mammoths differ from modern elephants, and likewise fossil remains of lions, bears and other animals belong to species different from those of today. Intraspecific variation is too limited to allow the transformation of one species into another. Viewed from the opposite side, Man belongs to the modern age, and therefore it is impossible for his remains to appear in fossil assemblages. As evidence that each epoch has its own fossils, Cuvier noted that in our modern period, mummified cats and ibises found in the pyramids during Napoleon's invasion of Egypt and brought to the Museum did not differ from their living counterparts (Fig. 6.19). They all belonged to modern species, and this, he claimed, was evidence that within an epoch, species do not evolve over time. To the claim that evolution happens much too slowly to be observed over just a few thousand years (as advocated by Lamarck, whom we shall soon meet), he argued that one may judge what a long time would produce only by multiplying what a shorter time produces. Since a shorter time produced no changes, neither, he argued, would a much longer time. Cuvier, to conclude, emphasized sudden changes in the animal kingdom, revolutions caused by violent catastrophes that befell the Earth time and again.

Cuvier's zoology had an enormous impact on the study of animal diversity because it combined the inclusiveness of Linnaeus with Buffon's quest for underlying laws of order in diversity. Two naturalists at the museum in Paris did not accept Cuvier's classification, according to which there are four animal branches, and did not accept his theory of catastrophes. One of these was Lamarck. He found the bivalve *Trigonia*, an ancient saltwater clam, which Cuvier claimed became extinct by a sudden catastrophe, very much alive in Australia. He wrote that "catastrophes

Fig. 6.19 Cuvier: Ibis, a modern skeleton and a mummified skull

which do not regulate anything, which mix and spread everything, is a convenient way for those naturalists who seek to explain everything without bothering to observe and study natural history". Lamarck was 15 years older than Cuvier was, and he had been in the Royal Gardens years before him.

6.2 Ever-Changing Animals

6.2.1 Animals Never Go Extinct

Jean Baptiste Pierre Antoine de Monet, Chevalier de Lamarck (1744–1829, Fig. 6.20), of a poor noble family in northern France, was the youngest of 11 brothers. At the age of 17, he left home to join the French army, at the time engaged in Germany in the Seven Years War (France against Prussia). He placed himself in a company of grenadiers that, during the battle of the following day, suffered heavy fire from the Prussian artillery. In the confusion of the retreat, the high commanders had forgotten the company and, with all its officers killed, only 14 men survived. The oldest grenadier, seeing that there were no longer any French troops in sight, suggested to the young volunteer, who so speedily became a commander, to withdraw the little troop. To this, Lamarck answered: "This post has been assigned to us. We must not quit it unless we are relieved". The unit remained on its post until, eventually, it received the order of retreat. Soon after, the army nominated Lamarck as lieutenant. Later, the army garrisoned his company on the Mediterranean seashore, where he became interested both in the surrounding vegetation and in seashells. Later on, his company served in the Fortress of Mont Dauphin, where he had the opportunity to compare the Mediterranean flora of Toulon with the rocky alpine flora of his new surroundings. At the age of 22, he resigned from the army, lived in Paris off occasional jobs, began studying medicine and, in the mid-1770s, joined the circle of botanists and students at the King's Garden and Cabinet, then under the direction of Buffon. It was here that he wrote a three-volume dichotomous Guide on the flora of France (1779), in which, as author, he presented his title of nobility, "Par M. Chevalier de Lamarck" (Stafleu 1971). Buffon, impressed by non-Linnaean aspects of Lamarck's book, arranged to have it published at government expense. His book appeared at a time when botany had become a popular science, and the success was rapid. He also obtained a commission from the king for Lamarck, as a correspondent on behalf of the Royal Garden, in charge of visiting foreign botanical gardens and museums, to establish communications with the French institute. Lamarck travelled to Holland, Germany and Hungary, where he visited museums, botanical gardens and mines, together with Buffon's son. Shortly later, the Academy of Sciences chose him to fill a vacancy in its botanical section (with Buffon's considerable support: the votes were in favour of another botanist, but Buffon recruited the King himself to overrule in favour of Lamarck). An active botanist, he described new plant genera, discussed plant classes, wrote about arboreal flowers and

Fig. 6.20 Lamarck

contributed botanical items to the *Encyclopédie méthodique* (published between 1782 and 1832 in over 200 volumes). Eventually, he received a post as Botanist of the King and Keeper of the Herbarium of the King's Garden (with Buffon's help, once again).

Following the French Revolution and in its true spirit of freedom, equality and fraternity, Lamarck volunteered to give free public courses in botany and conchology and undertook several tasks for the Committee of Public Instruction (established in 1791 to reorder the education system). Further, he dedicated one of his scientific papers "to the French people", asserting that he had resisted all those who had urged him to dedicate his book on the flora of France to the King or to anyone of the nobility and that he had never bowed down to anyone in the old regime. All he now wanted was to be useful to "my kind, my brothers, to my equals". As Lamarck himself was of the nobility, it took a special dispensation from the Committee of Public Instruction to allow him to stay in Paris in 1794, when a law was passed forbidding all ex-nobles to reside in the city. The former chevalier (knight) now became Citoyen Lamarck, the Citizen Lamarck. By turning his back on the old regime that had so favoured him and by so enthusiastically contributing to the new one, Lamarck survived the Revolution's Reign of Terror. From 1795, he was a professor at the Natural History Museum and a member of the *Institut de France*, the reconstituted academy. The Museum's faculty was to include two chairs of zoology, one for mammals, birds, reptiles and fishes, and the other for "Inferior Animals". (This brings to mind Democritus, who, in fifth-century BCE Greece, divided all living creatures into animals with red blood, the vertebrates and, those without red blood, the invertebrates.) Geoffroy Saint-Hilaire (whom we shall meet further on) received the first chair. For the second chair, consisting of those parts of the Animal Kingdom only vaguely known, there were no suitable candidates. Lamarck was by then a botanist of 25 years standing, but the chair of botany had passed to someone else, and there now seemed nothing suitable remaining for him except this chair of zoological remnants, to which he was accordingly appointed as professor of "insects, worms, and microscopic Animals". Lamarck and Cuvier thus became zoologists at the museum at about the same time, but whereas Cuvier was a promising young man of 25, Lamarck was a deposed botanist in his forties; the relations between the two were bound to be complex.

The Invertebrates collection at the Museum was immense. Within a few years, Lamarck established several new orders and families of marine invertebrates, separated the Insects from other groups of terrestrial arthropods and became an authority on molluscs. His major work on the classification of invertebrates (*Histoire naturelle des animaux sans vertèbres*, seven volumes, 1815–22; it was Lamarck who coined the term Invertebrates) considerably expanded the Linnaean classification of invertebrates (the words "*histoire naturelle*", in his title, comes closest to our "systematics"). Lamarck also devoted his first 4 years as curator to physics, chemistry, geology and meteorology, to questions far beyond invertebrate classification. He tried to predict the weather by publishing long-range forecasts at a time when meteorological prediction was still impossible because of the absence of instantaneous communication over large areas, and his predictions discredited him in the eyes of his contemporaries, who considered him a ridiculous speculator (Stafleu 1971). Lamarck's last years were sad, as he lost his eyesight and was struggling with severe poverty. He was buried in a pauper's grave when he died and, after 5 years, his remains were dumped into some unknown communal grave. Cuvier wrote, for the Academy of Sciences, a eulogy that was a bit nasty. It opened by declaring that, among the men devoted to the noble employment of enlightening their fellows, a small number have both a lofty imagination and a sound judgement and can embrace the entire field of sciences, laying before the world nothing but the truth, men of genius whose immortal writings will shed everlasting light on the paths of science. Cuvier was here referring to himself. The eulogy went on by declaring that others have mingled many fanciful conceptions and have laboriously constructed vast edifices on imaginary foundations, resembling the palaces of old romances, which soon vanished. Cuvier was here referring to Lamarck.

In many ways, Lamarck was a follower of Buffon. Like him, he introduced the time element in biology and, like him, he drew attention to biological change, suggesting the possibility of an open-ended concept of life. Like Buffon, Lamarck's approach was "cosmic" in that, alongside biology, he gave much attention to physics, chemistry and the Earth sciences (Stafleu 1971). He planned a grandiose enterprise that would provide theoretical foundations of three related sciences: meteorology (the study of the Earth's atmosphere), hydrogeology (the study of the Earth's surface and the effects of the movement of water on it) and biology (the study of the origin and development of living things; he was the first, in 1802, to coin and define the term Biology; Stafleu 1971). Beyond his classifications of plants and animals, Lamarck's thoughts centred on broad processes of nature and on general theories. (*Hydrogéologie* 1802, *Philosophie Zoologique* 1809, see Elliot 1964). He criticized Linnaeus for forming three kingdoms, as he was thereby placing minerals as a kingdom at the same level as those of plants and of animals. Rather, Lamarck claimed, the fundamental difference is between living and not-living beings: not-living entities by adding external matter at their periphery (we may think of crystals), while living beings do so by an internal "Principle of Life", which enables the assimilation of matter. It is only once this basic living versus non-living division has been made that we can move ahead and distinguish between animals and plants. This criticism was of prime importance because all through the ages, including Linnaeus and Bonnet, people had believed in a single ladder (chain) of nature, from mineral via vegetable to animal and then to human (sometimes even further upwards, to angels of diverse levels). Lamarck, in claiming that inorganic and living beings grow in completely different manners, was claiming that they should not be placed on one continuous ladder. He further broke the old chain of being by claiming that plants and animals form two different, parallel series. He at first admitted only a single unbranched series of plant taxa, parallel to one of the animals (Fig. 6.21). In later years, he admitted the possibility of further branching and of the occurrence of dead-end series in the course of evolution. His concept of two series of animals and plants was at first only a descriptive picture of the order of complexity of organization, which lacked the dimension of time. Lamarck's awareness of a difference in principle between the inorganic world and that of living beings signalled the birth of biology. His table of "organic beings subject to death, and which have the faculty to reproduce themselves" (Fig. 6.21) shows an intentional parallel between the animal and plant series, what we may call a double ladder ("a perfect counterpart with the six classes of the animal kingdom"; *Botanique* 1786, 2). Both have six classes, arranged in order of decreasing complexity of organization—a forced plant-animal symmetry. Each class is divided into two–four sections, which, for the plants, have much in common with the groups distinguished by Tournefort. His within-class order partly resembled Bonnet's Ladder of Life, in that terrestrial birds lead down to wading birds and these down to swimming ones; four-legged amphibians (then a group that also included reptiles) lead to those with no legs; and insects with four wings lead down to those with two, which lead to those with none (Fig. 6.21). In later years, when studying invertebrates more closely, he detached himself from this close comparison between animals and plants and

* *Etres organiques vivans, aſſujettis à la mort, & qui ont la faculté de ſe reproduire eux-mêmes.*

ANIMAUX.	VÉGÉTAUX.
1. *LES QUADRUPEDES.*	*LES POLYPÉTALÉES.* 1.
1. Terreſtres onguiculés.	Thalamiflores. 1.
2. Terreſtres ongulés.	Caliciflores. 2.
3. Marins.	Fructiflores. 3.
2. . . . . *LES OISEAUX.*	*LES MONOPÉTALÉES,* . . . . 2.
1. Terreſtres.	Fructiflores. 1.
2. Aquatiques à cuiſſes nues.	Caliciflores. 2.
3. Aquatiques nageants.	Thalamiflcres. 3.
3. . . . . . *LES AMPHIBIES.*	*LES COMPOSÉES.* 3
1. Tétrapodes.	Diſtinctes. 1
	Tubuleuſes. 2.
2. Apodes.	Ligulaires. 3.
4. . . . . *LES POISSONS.*	*LES INCOMPLETTES.* . . . . 4.
1. Cartilagineux.	Thalamiflores. 1.
	Caliciflores. 2.
2. Epineux.	Diclynes. 3.
	Gynandres. 4.
5 . . . . *LES INSECTES.*	*LES UNILOBÉES.* . . . . 5.
1. Tétraptères.	Fructiflores. 1.
2. Diptères.	Thalamiflores. 2.
3. Aptères.	
6. . . . . *LES VERS.*	*LES CRYPTOGAMES.* . . . . 6.
1. Nuds.	Epiphylloſpermes. 1.
2. Teſtacés.	Urnigères. 2.
3. Lithophytes.	Membraneuſes. 3.
4. Zoophytes.	Fongueuſes. 4.

Botanique. Tome II.

Fig. 6.21 "Organic living beings subject to death and which have the faculty to reproduce themselves". Lamarck's table of the order of gradation in the Plant and Animal kingdoms. (From Burkhardt 1995)

increased the number of animal classes. However, he did not completely give up the idea of a general arrangement of the animal kingdom in a series sequence of increasing (or decreasing) complexity.

According to Lamarck, two powerful opposing forces are in perpetual combat in living beings, and each force constantly destroys what the other produces (*Hydrogéologie* 1802). One force is the natural tendency of all compounds to decompose. Holding to the traditional view of fire, air, water and earth as the four basic elements, he assumed that each element tends to disengage itself from combinations with other elements to decompose from chemically complex compounds. Plant remains and also those of animals, being chemically complex, decompose and gradually turn, first into minerals and then into the basic elements. These ideas on the origin of minerals, which are broadly similar to those of Cuvier, seem to have stemmed from his observations in mines of Saxony and Hungary, where newly formed soil on the surface was complex and soft. Deeper in the mines, there were more anciently formed soils that had changed over time, and by now they were harder, denser, less complex and with a significant component of glassy quartz. This suggested to him that dead organic matter passes through successive mineral stages, as the different elements in organic substances slowly disintegrate. As the elements disintegrate, the Earth element of the organic substance, masked initially by other elements present in the soil compound, reveals itself more and more until what was once an organic substance becomes quartz and then finally pure rock crystal. Plant remains give rise to one series of minerals; animal remains give rise to another. In both cases, the final product is the same: the Earth element. "One sees from this table the infinite distance between living beings and inorganic substances, and the need not to present beings of such different character in a single continuous chain". The mineral chain moves away from life, the only force that can synthesize complex compounds; inorganic nature destroys complexity (Fig. 6.22; *Hydrogéologie* 1802). This theory, of the organic origin of all materials, is obviously untenable: to state that metallic ore, granite and basalt derive from organic substances is simply too far-fetched, and the sweeping generality of this thesis detaches it from reality. There was, in Lamarck's days, ample evidence of the direct volcanic origin of many minerals and rocks.

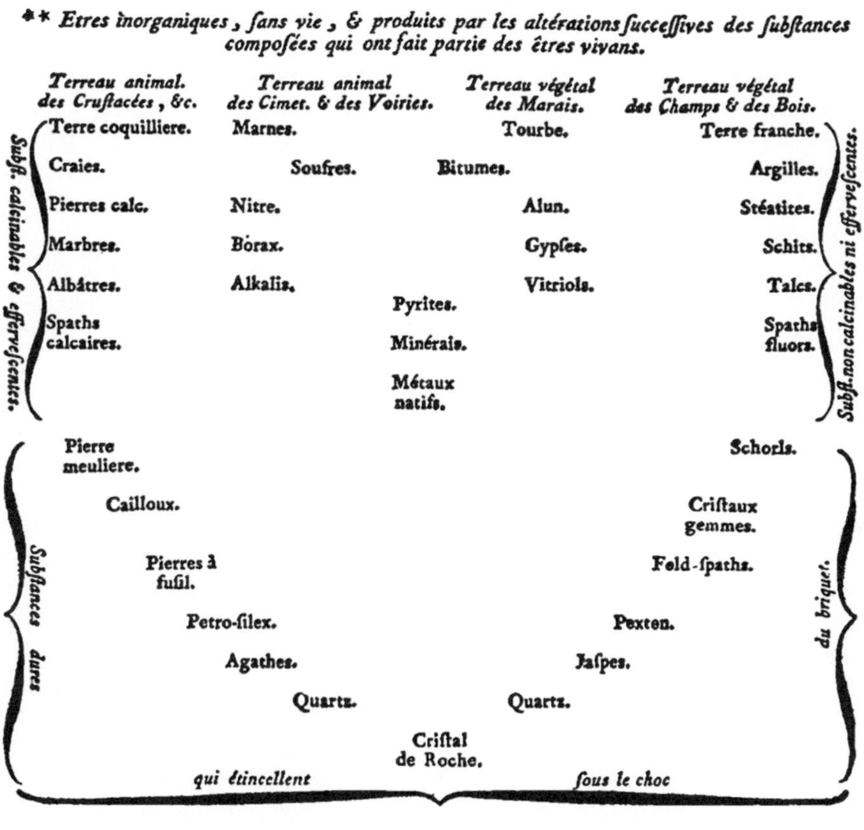

Fig. 6.22 "Inorganic beings without life, and produced by the successive alterations of the composite substances which have been part of living beings". Lamarck's illustration of his idea that the different minerals are formed through the successive breakdown of organic matter. (From Burkhardt 1995:70)

The counter-force to decomposition was what Lamarck called the Principle of Life, a non-physical force that drives the living world upwards towards ever-greater perfection. By this Principle, matter assimilates in living beings. This, and only this principle, can form compounds. Were it not for life, there would be no minerals and also no other compounds. Living organisms ceaselessly create complex compounds that would never have existed otherwise and which the destructive power of Nature continuously tries to annihilate. Plants, feeding (as he believed) only on water but benefiting from air and light, develop and increase in size. In doing so, they combine free elements and produce such compounds as gum, resin, sugar, oil, starch, gluten and lignin. These compounds do not exist in the soil, evidence that plants do not extract them by means of their roots: they synthesize them. Plant growth produces carbon, potash, clays and iron. Consequently, wherever plants occur, these elements exist. Animals, unlike plants, do not form basic compounds. They use the basic compounds produced by plants as food and modify them into complex compounds by use of their digestive organs. They therefore depend, directly or indirectly, on plants for their food. Horses convert grass into flesh, blood, vessels, glands, sinews, cartilages, bones, hair, bristles and hooves. When the power of life no longer maintains these complex compounds, and the plant or animal dies, these compounds decompose, undergo many successive changes and finally become minerals. Bitumen, coals, clays and some salts are minerals of plant origin; chalk, sulphur and marine salt come from animal remains. Whether of plant or animal origin, all minerals in Earth's crust, so Lamarck proposed, result from the remains of living organisms.

Lamarck further believed that the Principle of Life expresses itself in physical terms, as the movement of fluids within more solid tissues of the body, in response to a stimulus. In plants, an external stimulus causes this internal movement. Sap rises in the plant during the day, as the heat dilates the plant's tubes and pores in the leaves serve as excretory ducts, excreting the more fluid nutritive juices of the sap. At night, when temperatures drop, the plant deposits unexcreted nutritive juices, which assimilate in the interstices of the plant fibres and the pores of the leaf receive the "atmospheric juices", which descend all the way down to the roots. Annual plants die, he believed, when the first winter frosts cause their tubes to shrink, thereby preventing sap movement. Animals differ from plants in that the stimulus causing the movement of solids and fluids is internal, due to the sensitivity of the animal fibres. At this point, Lamarck noted an exception to the generalization that plants differ from animals in their lack of sensitivity. The touch-me-not *Mimosa* responds to touch by folding inward, drooping and reopening a few minutes later. Accordingly, Lamarck (Fig. 6.21) placed the touch-me-not in the plant class of greatest complexity (his *Polypetala*): "the principle of life renders itself more manifest in these plants, and brings them near, in a way, to other organic beings in which the irritability is joined to a more prefect quality called sensibility" (*Encyclopédie méthodique* 2:30. We recall Erasmus Darwin and his theory of the first living filament being irritable).

Slow, gradual change was a basic, general principle in Lamarck's thought. All entities in nature undergo change, whether by decomposition or by the Principle of

Life. In minerals, nature's power to decompose organic compounds is the main force; they evolve from the complex to the simple. In living entities (plants and animals), the Principle of Life is dominant and they evolve from simple forms of life to complex ones. The production of form, in each case, is gradual (*Hydrogéologie* Chap. 4). A destroying power governs the world of minerals; a building power governs the world of life.

Extinction became an acute issue in the academic circles of Paris following Cuvier's claim that extinction is a natural process and that there once were previous animal worlds, which went extinct by some sudden, massive catastrophe. This question, whether or not sudden mass extinctions had occurred, leads to a more basic question: does the fact that many fossils have no known living counterparts suggest species extinctions? Lamarck became involved in this issue through his studies of fossil and living shells. Do fossil shells also suggest large-scale extinctions during Earth's past? He believed, as Linnaeus had suggested, in a balanced world in which no species prevails to the point of destroying another. Because of this concept of nature constantly balancing itself so that no species is lost, he was reluctant to acknowledge extinction. True, a few large quadrupeds might have become extinct, but this was because of man, rather than an unbalanced nature. For many fossil forms with no living counterparts, he could point to living species of the same genus that closely resembled the fossil species. Furthermore, he could not think of any natural means by which organisms could ever become extinct. What could bring about the destruction of any of those insects that multiply so prolifically? What could cause the demise of those many marine invertebrates, which the ocean bottom shelters from unfavourable climatic conditions and from hunting by man? Lamarck concluded that no grand-scale extinctions had ever occurred.

To explain the widespread occurrence of marine fossils on mountaintops, Lamarck developed his own theory on the history of Earth (*Hydrogéologie* 1802). Looming in the background of this theory was his belief in Earth's vast age. How much time did Lamarck have in mind? As he saw it, time was essentially unlimited. He never offered a precise estimate of how old he believed Earth might be, but he was thinking in terms of thousands or even millions of centuries (We recall Erasmus Darwin and his belief in millions of years of evolution). Lamarck suggested that the ocean basin, throughout its vast existence, constantly shifts over the globe to such an extent that it has already covered all places on Earth several times. Indeed, he believed that even in his days, the basin of the Atlantic Ocean was undergoing a westward movement at an estimated speed of three leagues per 24 hours (about 12 km). During this displacement of its basin, the sea ceaselessly erodes the eastern coast of continents while abandoning the western ones. This continuous displacement of the ocean basin over the entire surface of the Earth has left the remains of marine shells far from the sea, as fossils. Some of these fossil shells are of littoral molluscs; others are of deep-sea bottom-dwelling ones.

Living deep-sea shells would include, besides auger shells, also ammonites, belemnites and orthoceratids. True, Lamarck noted, no one has ever found living representatives of these three groups, and therefore some naturalists consider them extinct. However, he claimed, we do not know enough about life in the depths.

Ammonites are still alive in yet unexplored parts of the deep sea. Indeed, the deep sea in those days yielded living forms of life then known only as fossils. The prime example was the marine bivalve order Trigoniida, characterized by the complex dentition of the hinge articulating its two nacreous valves. It originated in the Devonian, underwent adaptive radiation in the Jurassic, peaked in the Cretaceous and since then, geologists of the day considered it extinct. Lamarck himself had described several such fossils (Fig. 6.23). In 1802, however, a French explorer discovered a living trigoniid off the coast of Tasmania, which Lamarck described and named (*Trigonia margaritacea* 1804, today *Neotrigonia*).

Shoddy scientists, Lamarck commented ("Persons who are careless in their search for exact knowledge"), have claimed that marine fossil shells, strewn over the continents, bear evidence of a sudden universal catastrophe. If so, he asked, how could a catastrophe have left perfectly horizontal marine layers on land? How can the hypothesis of a universal catastrophe explain the preservation of so many delicate shells, which the least movement would break, but which, nevertheless, appear in perfect condition? How could bivalves have retained the two valves of one individual, one close to another in the fossil bed, had not their corresponding animals lived in those places? There is, Lamarck claimed, no need for a great catastrophe hypothesis. Severe, violent geological events did admittedly happen in the past, but these were all restricted events: local volcanic eruptions, earthquakes, floods, strong storms and so on. The effects of these local catastrophes were as limited as their locations. The past processes that shaped the Earth's face were gradual, not sudden, large-scale catastrophes. As the Earth's climate was subject to a continuous, though very slow, environmental change, plants and animals correspondingly changed, but they did so slowly and continuously ("nature has done everything little by little").

6.2.2 A Theory and Overview of Evolution

Early in his career, Lamarck had closely followed Bonnet in arguing that animals assemble on a single natural Ladder of Life, on which internal organization gradually increases in complexity of the different organs. There may seem to be missing rungs along the ladder, but these are due to groups yet undiscovered. Between birds and mammals, the gap once seemed so wide as to be impassable, but the platypus (described in 1800) and the spiny anteater (in 1811) possess characteristics of both mammals and birds. They seemed to Lamarck to confirm his claim that gaps in the linear series of animals merely represent the existence of hitherto undiscovered living animals in unexplored lands or deep seas.

In Lamarck's opinion, the three largest units of Linnaean classification, the mineral, vegetable and animal kingdoms, should be replaced by a scheme of only two "branches": lifeless entities (minerals), whose motion is passive and depends on transmission, versus organized, living entities, whose motion depends on stimulus (plants and animals). They absorb food, develop, reproduce and die. Bonnet had proposed a single ladder from mineral via vegetable to animal and human. Lamarck

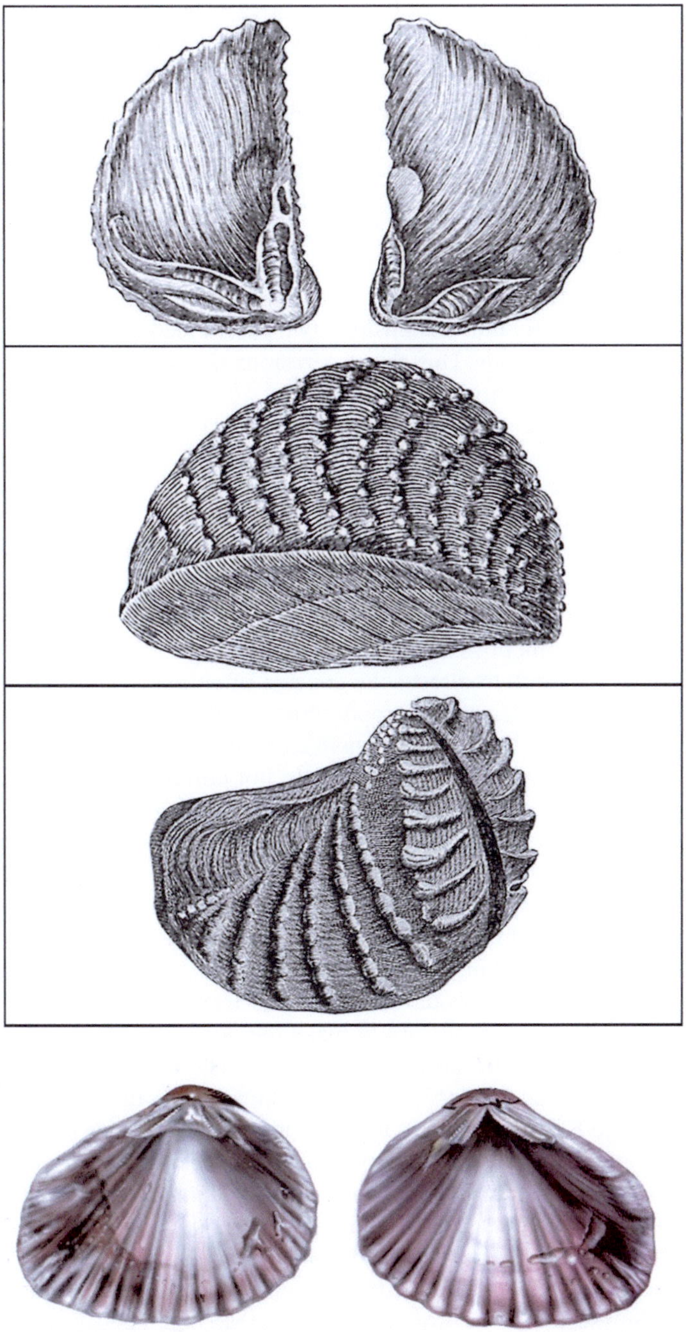

Fig. 6.23 Fossil *Trigonia* (Lamarck 1797) and a modern *Trigonia* from southern Australia

proposed to divide it: one ladder for living entities and another for minerals. For the science dealing with both animals and plants, he proposed (already in 1802) the term "Biology": the science of living organisms. It was Lamarck, the botanist turned zoologist, who observed that plants and animals, so different as to comprise different lineages, also have so much in common as to be treated as one. This was, perhaps, the birth of Biology as a scientific discipline. Lamarck followed Linnaeus in dividing the living entities into two clear-cut kingdoms. Animals possess irritability (a contraction that takes place instantly throughout an irritable fibre on contact with a foreign body); the ability to digest food, on which they live and the ability to move. Plants lack irritability, do not digest food and do not move (*Philosophie Zoologique* Chap. IV). By thus establishing two parallel ladders of living beings, he broke Bonnet's single ladder. This recalls Erasmus Darwin, who also defined animal life in terms of irritability.

During 1800–1803, Lamarck abandoned his belief in the fixity of species and developed his main thesis, by which organisms change with time (Stafleu 1971). Within the animal kingdom, at the level of large, high-ranking groups such as Classes, Lamarck believed that the Principle of Life develops body features that are internal, as these are the features most essential to survival and the most distant from direct environmental influence. This power constantly seeks to complicate the body's internal organization by gradually adding more organ systems to carry out new or more specific functions.

This view led Lamarck to discuss the concept of animal groups. He maintained that, as nature makes no leaps, it is impossible to find absolute lines of separation between groups. Classes, Orders and Genera are artificial aids that man has invented merely to assist in grouping animals. They do not correspond to anything real in nature, "Let me repeat, that nothing of the kind is to be found in nature". Groups that might seem to us as distinct actually link with one another via ever so many interim animals, and any proposed borderline between them is bound to be artificial. Our classification into these groups merely serves "to furnish points of rest for our imagination", by means of lines of demarcation that enable easy identifications. The primary reason for establishing genera is to reduce the number of principal names to remember, a number that would be enormous if every species had a different single name. It is therefore necessary to make the genera neither too large nor too small. Further, as the classification ranks are artificial entities that we accept only as a necessity, we must keep them to a minimum. Lamarck was therefore against dividing Kingdoms into Branches, Orders into Suborders and Families into Subfamilies, Tribes and Subgenera and so on, as Cuvier had done. These ever so many subdivisions were, he thought, an overuse of non-natural rankings (Stafleu 1971).

What made Lamarck change his mind? We, at this point, turn to another French naturalist of the Museum, Lacépède (Bernard-Germain Étienne de la Ville-sur-Illon, comte de Lacépède, 1756–1825). A protégé of Buffon, who appointed him sub-demonstrator in the *Jardin du Roi* (1785), Lacépède continued Buffon's *Histoire naturelle* under other titles such as *Histoire naturelle des quadrupèdes ovipares et des serpents*, *Tome premier* (1788) and *Histoire naturelle des serpents, Tome second* (1789). As an active revolutionist of the French Revolution, he later became a member of the Legislative Assembly. During the Reign of Terror (1793), as one who disapproved of the massacres, he fled Paris. Later that year, he returned, however, and, in the newly organized Museum, he was appointed to the chair of Reptiles and Fishes (1793). Six years later, he became a senator, then president of the senate, perpetual secretary of the French Academy of Sciences, grand chancellor of the Legion of Honour, minister of state and a Peer of France. We return to his years as a naturalist because in 1800, in the Introduction to the second volume of his *Histoire naturelle des poissons,* he concluded that a species may undergo so many modifications that, ultimately, it may be more distant from its original state than from another species; a species may thus change ("undergo metamorphosis") to another species (Stafleu 1971). Lacépède did not further elaborate on these ideas, essentially those of Buffon, but they may well have influenced Lamarck.

We can demonstrate Lamarck's concept of groups linking with one another by his classification of the molluscs (1795 and later). On his Ladder of Life, on the level just beneath the class Mollusca and leading to it, he placed a shelled group with no head (barnacles); and in the molluscs, his first order was likewise that of shelled animals with no head ("Acephalic molluscs"). Here, he placed first the brachiopods, linked to barnacles in that they too are shelled and headless; each has two valves that cover the dorsal and ventral surface of the animal. Second came the bivalves, linking with the brachiopods in that they too have no head, differing in that they have two shells that cover the lateral surfaces. Finally came the ascidians, linked with the bivalves in that they too have no head, differing in that they have no shell, only a leathery jacket; one of them floats (*Salpa*). Lamarck next led this last headless, shell-less "mollusc" to the order "cephalic molluscs" (gastropods). Here he included first those that link with the ascidians in that they too have no shell and can float, differing in that they have a head and fins. Second came those linked to the previous group in that they have a head and no shell, differing in that they have no fins; some float, others crawl on the sea bed. The next three groups link to the previous group in that they crawl and all have a head; they differ in that they have a shell. In some, the shell is small and internal; in others, it is small but external; others have a hard shell; and in one, it is fragile, thin and slightly spiral. The next group resembles the previous one in that its members crawl and have a head and a shell; they differ in that the shell is always spiral and so on, from those with a spiral shell and a syphon canal. Lamarck's third molluscan order was the Cephalopods. Here, first came floating animals with spiral, multi-chambered shells. They link to floating animals with straight, multi-chambered shells. From here, he went on to present a group of those with a shell but no chambers, which float; that link to floating animals with an internal shell (squid); that link with burrowing-floating animals with an internal shell and no chambers (cuttlefish); and these finally link with crawling animals with no shell (octopus).

This classification of molluscs by Lamarck is simply terrible. He put a sea slug, a limpet and a chiton into one family, foraminifers with cephalopods (different phyla! Fig. 6.24) in another and in a third case, he lumped a floating octopus with a floating snail into one family (*Argonauta* with *Carinaria*). He seems to have first developed a concept of linking and then constructed a classification that would

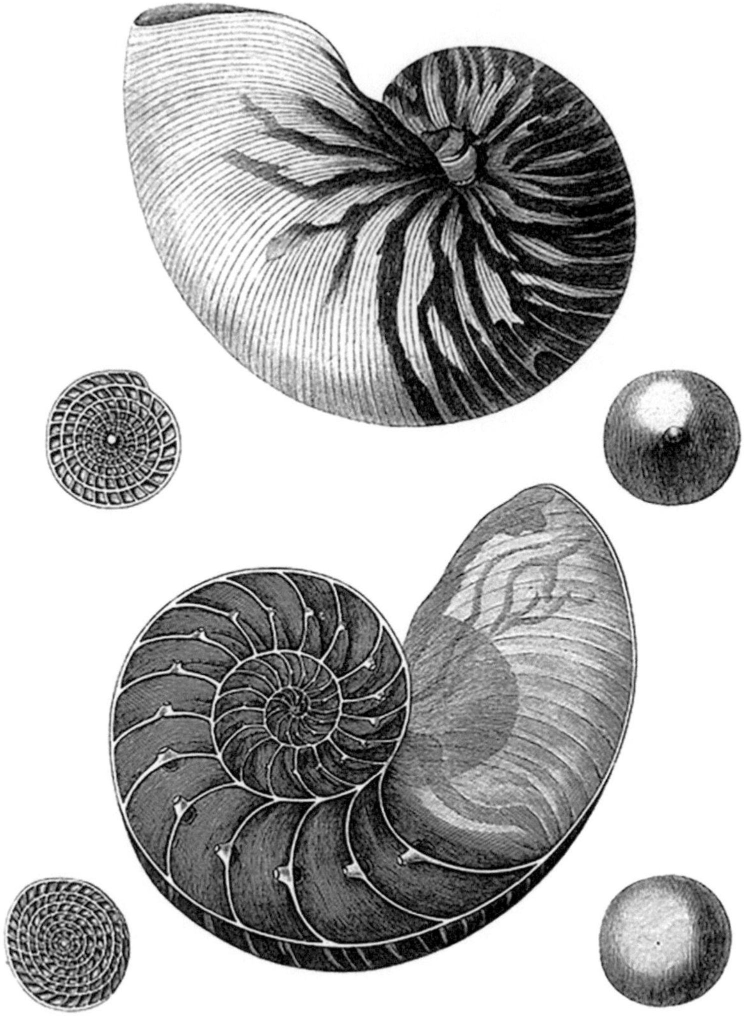

Fig. 6.24 Lamarck's group of floating cephalopods, with spiral, multi-chambered shells: nautilus and foraminifera

support it. His belief that ammonites and belemnites still live today in some deep sea may have stemmed from this very faulty classification, which lumped these fossils together with living *Nautilus, Spirula* and Foraminifera. As fossil "cephalopods" broadly resembled living representatives of this group, there was no reason why they too should not be still alive today.

Lamarck saw the origin of life and the process of increasing complexity as closely connected with the diversification of organs. Animal classes, Lamarck claimed, form a ladder of increasing organizational complexity. In classes of simple organisms, there are fewer or no specialized organs, and additional complexity means further morphological specialization and the division of functions. The ten invertebrate classes, together with the four vertebrate ones, form a ladder of perfection in which there are six stages of complexity of the nervous system (Table 6.1).

The Principle of Life is responsible for gradually transforming simple organisms into more complex ones over time. Hence, the origin of complex organisms is in simple organisms, and these form by spontaneous generation. Yet if a spontaneously generated creature transforms into a more complex one, and that in turn develops according to the driving progressive force Lamarck described, then the final product would be just one species, the most advanced, man. To explain the existing plurality

Table 6.1 "An arrangement and classification of animals, by the order most conforming that of nature"

Invertebrates
1. No nerves, vessels or specialized organs except for digestion.
I. Infusorians. Amorphous, reproduce by fission or budding, gelatinous, transparent, minute, no tentacles, no internal organs.
II. Polyps [rotifers, hydrozoans, anthozoans, bryozoans, sponges]. Reproduce by budding, body regenerates, only one internal organ, the alimentary canal, with a single aperture.
2. Nerves present, no ganglionic longitudinal cord, no vessels.
III. Radiarians. [Coelenterata exclusive of anthozoans, echinoderms]. Regenerating body, no head, eyes or jointed legs, alimentary canal arranged radially, mouth on underside.
IV. Worms. Regenerating body, no metamorphosis, no eyes, no legs, internal organs not radially arranged.
3. Nerves terminate in longitudinal cord, tracheal respiration; circulation absent or imperfect.
V. Insects. Metamorphosis, eyes present, six jointed legs, many tracheae, single fertilization during life.
VI. Arachnids. No metamorphosis, jointed legs, few tracheae, primitive circulation, several fertilizations during life.
4. Nerves terminate in the brain or longitudinal cord; respiration by gills; arteries and veins for circulation.
VII. Crustaceans. Jointed body and limbs, crustaceous skin, eyes, four antennae (usually), respiration by gills, ganglionic longitudinal cord.
VIII. Annelids. Elongated, ringed body, no jointed legs, rarely any eyes, respiration by gills, a ganglionic longitudinal cord.
IX. Barnacles. A mantle, jointed arms, horny skin, no eyes, respiration by gills, ganglionic longitudinal cord.
X. Molluscs. Body soft and moist, unjointed, mantle variable, respiration by gills, no spinal cord, no ganglionic longitudinal cord, nerves terminate in a brain.
Vertebrates
5. Nerves terminate in a brain that does not fill the skull; heart with single ventricle, cold blood.
XI. Fishes. Respiration by gills, two or four primitive limbs, locomotion by fins.
XII. Reptiles. Respiration usually by lungs, which exist either throughout life or in the latter part of it; four, two or no limbs.
6. Nerves terminate in brain that fills the skull; heart with two ventricles; warm blood.
XIII. Birds. Four jointed limbs, two of which are wings, respiration by lungs, feathers.
XIV. Mammals. Viviparous, with mammae, four jointed limbs or only two, hair.

From pp. 131–133, slightly abridged

of living species, Lamarck added another hypothesis: animals are *constantly* coming into existence and travelling up the course of evolution. Lamarck argued that nature began, *and still begins*, by creating the simplest of organized bodies, those that have only rudiments of organization, which spontaneous generation directly creates. Continuous spontaneous generation was, as we have just mentioned, an essential feature of Lamarck's theory of life (Stafleu 1971). The Earth, especially near water surfaces and even in the atmosphere, "is populated in some way by animated molecules, the organization of which, however simple, is sufficient for their existence". These animalcules then reproduce. At the bases of both the plant and the animal scales, tiny, transparent, unorganized masses of matter are organized and brought to life through the action of subtle fluids. Plants and animals form from different materials: "If the tiny mass in question is gelatinous, animal life will be established there, but if it is simply mucilaginous, then only plant life will be established there". Whereas among animals only infusoria appear by spontaneous generation, among plants, Lamarck believed, moulds, mushrooms and lichens form in this manner. (Infusoria was a collective term for minute aquatic creatures such as ciliates, euglenoids, protozoa, unicellular algae and small invertebrates living in ponds. In modern classification, the term is obsolete.) The origin of the infusoria, he explained, was too recent for them to have greatly diversified: these creatures had not been subject to change for long enough since the appearance of the spontaneously generated forms from which they originally derived. Erasmus Darwin held similar views of life, beginning with the spontaneous generation of simple organisms (in modern eyes, one drastic implication of this hypothesis would be that no living species is actually a blood relation of any other; their relationship lies merely in similarities resulting from their parallel histories).

In later years, the simplicity of some intestinal worms led Lamarck to suspect that they, too, like simple infusoria, appeared by spontaneous generation. Consequently, he retracted his previous statement, as though all animals of all groups fall on one single linear ladder, as suggested by Bonnet. The animal scale began, he believed, with two animal branches (*Philosophie zoologique* 2). Nature formed infusoria in some instances and intestinal worms in others, in the same way it forms plants in some instances and animals in others: it gradually produces different kinds of life when working with different materials. In his two major branches of animal descent, Lamarck proposed that one, starting with the infusoria, proceeded via cnidarians to echinoderms and via ascidians to molluscs (bivalves and snails). The other major branch, starting from worms, proceeded to annelids and, via insects, to arachnids or to crustaceans and barnacles (Fig. 6.25). Lamarck's overall classification was not, in the end, that very different from that of Cuvier, who classified the animal kingdom into "radiates", arthropods, molluscs and vertebrates.

Lamarck further believed that throughout the animal kingdom, the Principle of Life gradually built up the complexity of the nervous system. He thought that the simplest animals lack a nervous system, and he considered them insensitive, passive machines whose drivers of activity are fluids in the environment, which invade the body from the outside, thereby causing mechanical movement. These simplest,

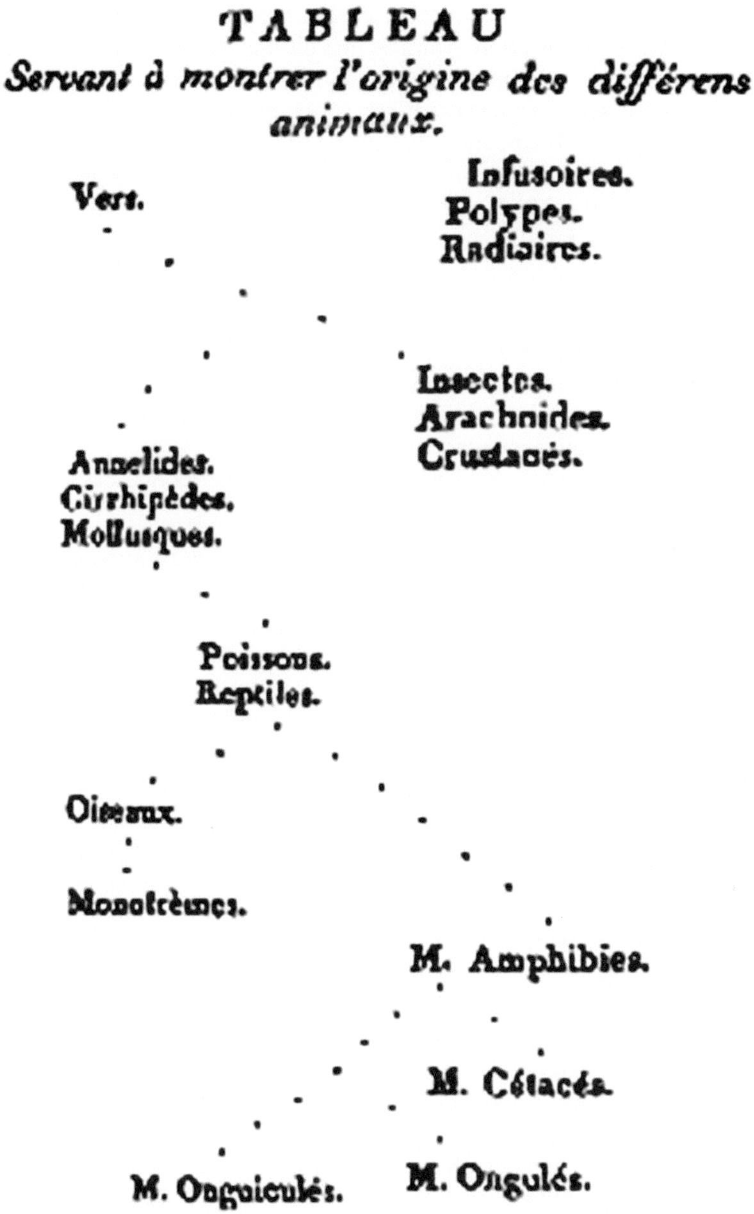

Fig. 6.25 The order of production of animals, as presented by Lamarck in his *Histoire naturelle des animaux sans vertèbres*

insensitive animals do not develop their habits in response to perceived needs, for they lack the ability of sensation. Animals that have a nervous system are sensitive, and their primary cause of behaviour is internal instinct. By this term, Lamarck meant an automatic set of responses set into action by an inner sensation in response to environmental stimuli, an action that does not require any will on the animal's part. The environment no longer influences them directly (*Philosophie zoologique* 2: 279; it calls to mind a reflex, as that of an automatic knee-jerk). Ideas, in addition to instinct, motivated intelligent animals with complex nervous systems (vertebrates). Accordingly, Lamarck classified the animal kingdom also into three primary levels (Fig. 6.26):

Insensitive animals, *les animaux apatheques*, the simplest of invertebrates, lacking any nervous system (infusoria, polyps, ascidians, worms).
Sensitive animals, *les animaux sensibles*, the higher invertebrates (barnacles, molluscs, insects, arachnids, crustaceans, annelids).
Intelligent animals, *les animaux intelligents*, the four vertebrate classes (fish, reptiles, birds, mammals).

In addition to the Principle of Life that gradually built up the complexity of internal functions of the body, another force was dominant in determining the structure of the body: adaptive change. An environmental change leads to changes of habit, and consequently, the more frequently used organs become stronger and bigger, whereas those less frequently used degenerate. The result is variations on a common theme (Erasmus Darwin once again). Though referring mainly to variation within Classes, this force applies to some extent also to differences among Classes. Lamarck supposed that aquatic worms had diversified over time. Those in the habit of exposing themselves to air gave rise to "amphibious insects" (midges, mayflies), and these in turn gave rise to many terrestrial insect groups. Some of these insects, forced by circumstances to change their habits, gave rise to the arachnids, some of which developed the habit of frequenting water, and these gradually changed into the crustaceans. Those aquatic worms that never exposed themselves to the air, on the other hand, led to the formation of the annelids, barnacles and molluscs.

Beyond adaptive diversification between classes, Lamarck also described adaptive group origins within classes. He suggested that the major mammalian groups had originated from "amphibian mammals" that lived along the shore. Some of these, the plant browsers (walruses, manatees), gradually led to the formation of such terrestrial herbivore groups as the ruminants and the "pachyderms" (thick-skinned mammals, including elephant, rhinoceros, hippopotamus and others; today the group is obsolete). Others, with the habit of feeding on flesh (seals), gradually led to the formation of terrestrial carnivores. Others still developed the habit of never leaving the water and only coming to breathe at the surface, and these gave rise to whales. The adaptive derivation of these three groups from the "amphibian mammals" was portrayed in Fig. 6.26, which Lamarck presented, to show "the origin of the different animals".

Within-class morphological change due to changes in habit was, in Lamarck's opinion, noticeable in some birds that, attracted to water to find food, developed the

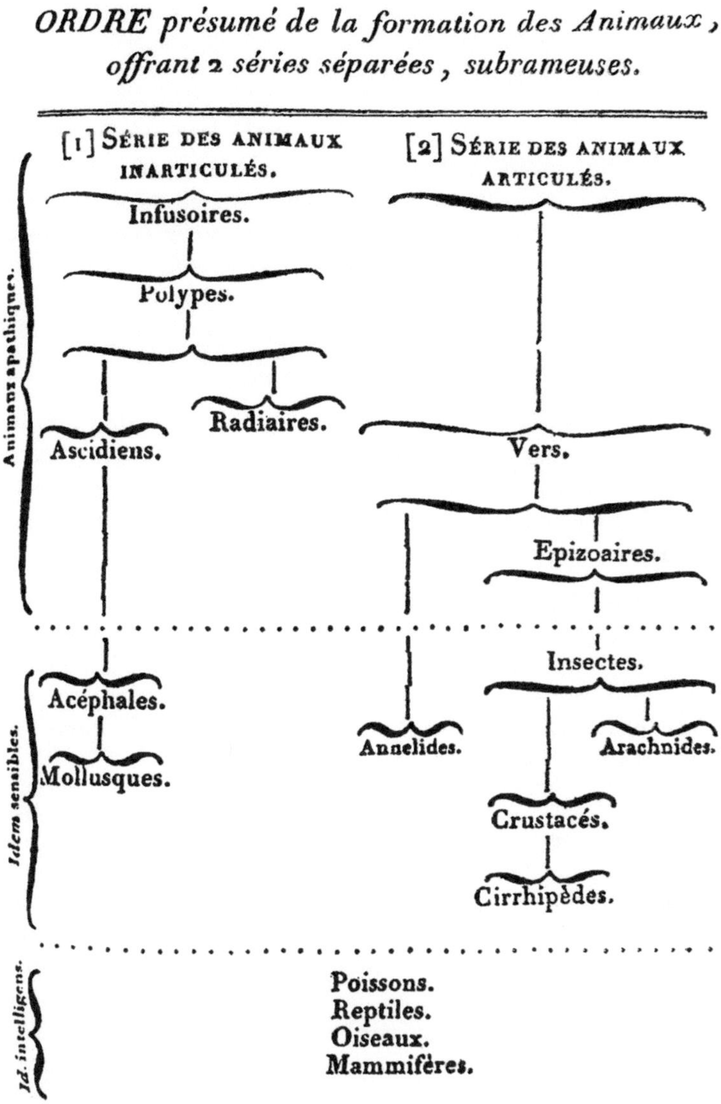

Fig. 6.26 The order of production of living things, as presented by Lamarck. (From Burkhardt 1995)

habit of spreading the digits of their feet when striking the surface, thereby stretching the interdigital skin that, over time, became large swimming membranes (ducks). As further evidence that the frequent use of an organ strengthens it, Lamarck cited the by-now-famous neck and forelegs of the giraffe:

> This animal, the largest of the mammals, is known to live in the interior of Africa in places where the earth is nearly always arid and without pasturage, obliging it to browse on the

> leaves of trees and continually to strive to reach up to them. This habit, maintained for a long time by all members of the race, has resulted in the forelegs becoming longer than the hind legs and the neck being so lengthened that the giraffe, without standing on its hind legs, with its head raised reaches a height of six meters. (*Philosophie zoologique* 1: 256–257; Elliot 1964).

As evidence of the opposite trend, that failure to use an organ causes this organ to become weaker and eventually disappear, Lamarck cited the absence of teeth in whales and anteaters, the rudimentary eyes of moles, leg absence in snakes and the lack of wings in many insects. Habits might also result in the displacement of organs, as the placement of both eyes on the upper side of the flatfishes. Furthermore, an animal's behaviour could bring into existence wholly new structures, such as the horns and antlers of ruminants that serve for inflicting head blows in combat. Thus, behaviour too determines form.

In this manner, external circumstances responsible for departures from the regular gradation in the animal scale account for the fact that many groups do not fall on a linear line. They form "lateral ramifications" around the major classes. Stressing the importance of the use and disuse of organs and the idea of the inheritance of acquired characters, Lamarck formulated two rules (Wightman 1953):

1. In every animal, frequent use of an organ will strengthen this organ little by little and develop and enlarge it. Disuse of an organ will weaken it, lead to its deterioration, progressively diminish its faculties, and finally cause it to disappear.
2. Everything that nature has caused individuals to gain or lose, over a long time, by constant use or disuse of an organ, is conserved, over generations, in individuals descending from them.

Animal groups gradually change due to the continued use of an organ, which strengthens and enlarges it, or by permanent disuse of an organ, which causes its deterioration and eventual disappearance. Therefore, when environmental changes lead to changes of habit, the more frequently used organs become stronger and bigger and the less frequently used slowly disappear. Lamarck's second rule implied the inheritance of acquired characters. To put this idea into the words of Cuvier, in his eulogy of Lamarck, wants and desires, produced by the circumstances, will eventually produce organs. Organs do not determine the habits of an animal. It is the desire and the attempt to swim that produces membranes in the feet of aquatic birds; it is the desire to avoid wet that, in wading birds, has lengthened the legs; and it is the desire of flying that has converted the arms of all birds into wings and their scales into feathers (Burkhardt 2013).

The heredity element in this mechanism is clearly untenable, but to regard it as the main feature of his thought would be an injustice to his much more important general theories, such as the long span of time available to the animal kingdom and the existence of gradual adaptive change. Lamarck never displayed much interest in heredity as such. He presented the idea of the inheritance of acquired characters as self-evident, never claiming it as his own. ("The law of nature, by which new individuals receive all that has been acquired in organisation during the lifetime of their parents, is so true, so striking, so much attested by the facts, that there is no observer who has been unable to convince himself of its reality"). His innovation was not in the formulation of the idea as such but in his claim that the inheritance of acquired characters was an agent of unlimited change—it led to the diversity of animals.

At the level of species, Lamarck noted that changes occur only in external, less essential features of organisms, those subject to direct environmental influences. Challenging the assumption that species are constant entities that do not vary much, he claimed that individuals of a species reproduce without variation only as long as the conditions of their existence do not vary. Species undergo changes, arising from the diversity of conditions in which they live and not because of any increasing complexity of organization. As individuals of a species change their habits, they little by little change their overall shape and will convert, after many generations, into a new species distinct from the original. For example, in plants, if the seed of a meadow grass germinates on a dry, arid, stony, windy mountain, it will be small and thin, clearly not resembling its parents. Likewise, in domestic animals, if we imprison in stables a wild horse running swiftly over plains, after some generations it will lose its swiftness, its body will thicken and the strength of its limbs will diminish. Furthermore, some fertile species may come from different parent species as fertile hybrids. For these reasons, genera and species form lateral ramifications of the series of large groups. These views were a further breakdown of Bonnet's Ladder of Life. Lamarck's theory also contrasted with Cuvier's theory of species as unchanging units, some of which perish during global catastrophes. (Erasmus Darwin, as we have seen, had claimed that man might change due to physical effort or to diseases, both of which may be hereditary; a few generations of physical activity or of laziness may change the form of the body and produce new varieties.)

Lamarck noted that "some naturalists" claim that among individuals of a group, the proportions between organs are fixed, so that, by examining detached parts of an individual, we can decide to what group it belongs (Cuvier's rule of The Correlation of Parts). This generalization, Lamarck declared, applies only to modern animals and to recent fossils; it is invalid for older fossils since animals may slowly change the relation between their parts in response to slowly changing environments. These slow changes are imperceptible over brief periods, such as the time since the Egyptians mummified ibises and placed them in pyramids. Cuvier had considered the mummified ibis as evidence that species do not change. To this, Lamarck replied that the mere few thousand years since the days of the pyramids are not nearly long enough to observe change. Furthermore, no environmental change had occurred in Egypt since the days of the pyramids, so there was no need for a corresponding change. Indeed, among fossils in general, those closely resembling species living today are the least ancient. This is so because the species to which they belong did not have enough time to undergo considerable changes. In line with these views, Lamarck defined a Species as a group of individuals perpetuated by reproduction without change, so long as their environment does not alter enough to cause variations in their habits, character and shape (*Philosophie zoologique*). Thus, in his later years, Lamarck viewed the living world not as a ladder but as a continuous central stream with side streams.

In December 1809, the members of the Institut de France attended the emperor in the palace to present Napoleon with their new publications. Lamarck, now 65, presented him a copy of his *Philosophie zoologique*. Napoleon, who in the Lamarck-Cuvier enmity sided with Cuvier because of his administrative ability, merely glanced at it and growled: "What is that? Is it your absurd meteorology with which you are disgracing your old age? Write on natural history and I will receive your works with pleasure; this volume I only accept out of consideration for your grey hair. Give it here!" Napoleon handed the book to an aide-de-camp, while Lamarck protested in vain that it was a work of natural history (Gershonowitz 1980).

6.2.3 All Animals Are the Same

Étienne Geoffroy (1772–1844, Fig. 6.27) was the youngest of 14 children of a procurator in a small town near Paris. While still a child, his parents added Saint-Hilaire, and he later added this name to his family name. At first, he studied law, and later he turned to medicine, but included botany, zoology, mineralogy and crystallography in his studies as well. During the Reign of Terror, the revolutionary authorities imprisoned his teacher of crystallography, a priest who refused to take an oath accepting the Civil Constitution of the Clergy. Geoffroy risked his life to release

Fig. 6.27 Geoffroy Saint-Hilaire

him from the guillotine just days before the large-scale clergy massacres of September 1792. In gratitude, when the Museum of Natural History opened its gates, its lecturers saw to Geoffroy's appointment, a 21-year-old crystallographer with no experience in zoology, to the chair of vertebrate taxonomy. They likewise appointed Lamarck, his elder by almost 30 years and with only slight experience in zoology, to the chair of invertebrates. Geoffroy joined the team of scientists accompanying Napoleon on his campaign in Egypt and collected both the Egyptian fauna (including the primitive, lung-breathing fish *Polypterus*) and many animal mummies from within the pyramids (1798–1801). A short time after returning to France, he entered the Academy of Sciences and received the chair of zoology at the Faculty of Sciences. His major treatise was *Anatomical Philosophy* (1818, 1822). Starting around 1834, his writings became increasingly vague, and thereafter, the Academy published only the titles of his communications. He went blind in 1840, resigned his chair the following year and died 2 years later (Le Guyader 2004).

Geoffroy, like Buffon and like Cuvier, took Newton and his universal laws of celestial movements as his inspirational model in that he sought to discover a grand basic law that would explain almost everything in zoology. Observations of the wings of the ostrich during the expedition to Egypt made Geoffroy aware of the potential importance of rudimentary organs. He noted that this bird, though flightless, has a flat breastbone and this caused him to suggest that even when superfluous, vestiges of an organ will remain if that organ is of importance in other species of the same family. Like the ostrich, the cassowary also has small wings that cannot lift it into the air and yet it has a breastbone; and as a more distant case, humans have a bloated flesh at the inner corner of the eye, a rudiment of the nictitating membrane of many lizards and birds. These observations led him to suggest that it is rare for organs to disappear completely. They exist, though often as rudiments, in the same parts of the body as the more fully developed organs to which they correspond. This fact demonstrates a fundamental principle in zoology, the Unity of Plan, that he later termed the *Unity of Composition*: all animals of a group share the same basic organs in their bodies. His criterion for deciding when structures in two different organisms were variants of the same type (homology, in modern terminology; he used the term *analogie*) was by their position relative to other organs. An organ may change in shape or in size, but it will never move to another location relative to its neighbours (his *Principle of Connections*). Taking his example, the anterior part of vertebrates consists of four sections: the shoulder, the arm, the forearm and a terminal section, forming the hand of a man, the claw of a cat and the wing of a bat. This fixed order was, he thought, a basic law of Nature. Some proportions of some components may change to fit new functions, but their order is fixed; nature does not make nor arrange new organs. This was an entirely new approach to organs because, until then, naturalists had thought of an organ as designed to perform a particular function. Geoffroy was now claiming that an organ simply exists, independently of any specific function that it may serve. His Principle of Compensation dictated that, if one of the organs is excessively developed, the others would be correspondingly reduced—there was a limited amount of material from which an individual animal could develop. He was not searching for differences between animals but for resemblances that they might share.

Geoffroy first proposed his unity of plan in 1796 for mammals. A decade later, he extended this theory to all vertebrates by comparing the bony heads of mammals, birds, reptiles and fishes and establishing that they all presented the same plan (1807). As he claimed that every structure in one animal must have its homologue in another, he attempted (1818) to find, in the skull of a mammal, the homologues of the apparent extra bones in the skull of a bony fish. Now, the head of an adult, a bony fish, clearly has many bones with no corresponding bones in the head of an adult mammal, and this seemed an obstacle to his principle of a Unity of Composition. At this point, Geoffroy compared the skulls of fish with those of mammals in different embryonic stages, not only with those of adults. From this study, he declared that he had found the same number of centres of bone formation in fish and in mammals and noted that at the end of embryonic development, in the adult, secondary fusions conceal the original number of bones and obscure the basic plan of the skull. In this way, he thought that he could identify not only the bones in these animals but also the centres of ossification and how they relate to one another. This allowed him to show that, despite apparent differences, there are unquestionable resemblances between the bones of the heads of fish, reptiles, birds and those of mammals. Having established, so he thought, that the tympanic cavity of mammals was the homologue of the gill chamber of fishes, he erroneously equated the middle ear bones of mammals with the operculum bones of fishes, thus:

Mammals	Fishes
Stapes	Operculum
Malleus	Inter-operculum
Lenticular and Incus	Sub-operculum
Tympanic	Pre-operculum

Pursuing this approach, he also wrote that he had discovered rudimentary teeth in the embryos of birds and in the jaws of very young whales, which lack teeth as adults. These embryological observations led Geoffroy to an additional aspect: vertebrates, in the course of their embryonic life, go through phases recalling their ancestors. The human embryo, for example, goes through preliminary forms typical of fish. In those days of the early nineteenth century, the term Evolution, which then meant the gradual development of something from a simple to a more complex form, was for Geoffroy the sum of transformations undergone by an embryo. Embryonic recapitulation was the idea that embryos of animals high on the Ladder of Life developed successively through the forms of animals lower on the ladder. Embryological evolution of an individual thus follows the path of the evolution of the animal group itself (Panchen 2001). Geoffroy (1831) adopted this meaning of evolution, which implied a transformation of species in the course of time. An animal of high rank passes, in the course of its development, through stages that correspond to adult stages of inferior animal groups. By establishing a connection between transitory embryonic stages of certain animals and the adult stages of others, Geoffroy reinforced his idea of the unity of vertebrate structure: lower animals were, so to speak, permanent embryos of higher animals. Furthermore, abnormal embryonic development ("monsters") suggested to him that morphological change occurred in bursts, not as the slow and gradual process as Lamarck had proposed. Vestigial organs and embryonic transformations indicate the common derivation of an animal from its archetype, whether or not they serve any functional purpose. Geoffroy's theory of homology among vertebrates was not entirely new. Aristotle had sketched it in ancient days; Belon had found homologies between birds and a man in the Renaissance, and his (then) friend Cuvier had considered vertebrates as one of his four major branches, though claiming that fish fins bear no relation to mammal paws.

After yet another decade, Geoffroy proposed to extend the vertebrate Unity of Composition to include insects as well (1820). The studies of French entomologists had shown that the seemingly highly varied mouths of the beetles, fleas, bees, flies and butterflies always consist of the same parts, which are in similar places in the most diverse species. He also found that all arthropod appendages, including the wings of insects, are modified legs, thus relating them to the respiratory appendages of crustaceans and aquatic arthropods. The moment had come to try to show that these two groups distinguished by Cuvier, the articulated animals, that is to say, the arthropods and the vertebrates, were really only one. He noted that in both groups, the body consists of a single, segmented tube that contains the intestines and spinal cord. Carried away by his imagination, Geoffroy reached the strange conclusion that the vertebral column, an internal skeleton, is homologous to the insect's cuticle, an external skeleton. He maintained that every animal lives either within or outside its vertebral column and that "insects live within their vertebral column, like molluscs inside their shell" (1822). Further, he homologized the vertebrate ribs with the legs of insects, claiming that insects were vertebrates that walked on their ribs. The hind wings of insects were, in his opinion, homologous to the swim bladder of fishes and air sacs of birds; during insect metamorphosis from pupa into adult, they serve as a bladder. Finally, he homologized the series of openings of the respiratory system on the abdomen (spiracles) in insects to the lateral line openings in fishes.

In this grand insects-are-vertebrates festival, a major problem facing Geoffroy was the nervous system, which is dorsal in vertebrates yet ventral in arthropods. He explained this difference by claiming that insects are upside-down vertebrates who walk about with their backs below and abdomens above (1822). As evidence, he noted that backswimmers (Notonectidae) swim "wrong side up", following the vertebrate position. Accordingly, insects are among the vertebrates (*Troisième Mémoire* 1820). To summarize Geoffroy's theory, insects are vertebrates that live inside their vertebrae and walk on their ribs with their bellies facing upwards (Fig. 6.28). In claiming that insects have the same body plan as vertebrates, Geoffroy united two of the four great animal Branches that Cuvier had claimed to be completely distinct.

Geoffroy's final stage in promoting the concept of the unity of all bilateral animal groups under one basic plan came when two obscure young naturalists wrote to the Academy of Sciences of Paris (1829). They proposed a transition from cephalopods (squid, cuttlefish, octopus and others) to fish, claiming that the organs of a cuttlefish

Fig. 6.28 Illustrating Geoffroy's theory: vertebrates and insects share the same body plan

Fig. 6.29 Illustrating Geoffroy's theory: vertebrates (above) and molluscs (below) share the same body plan

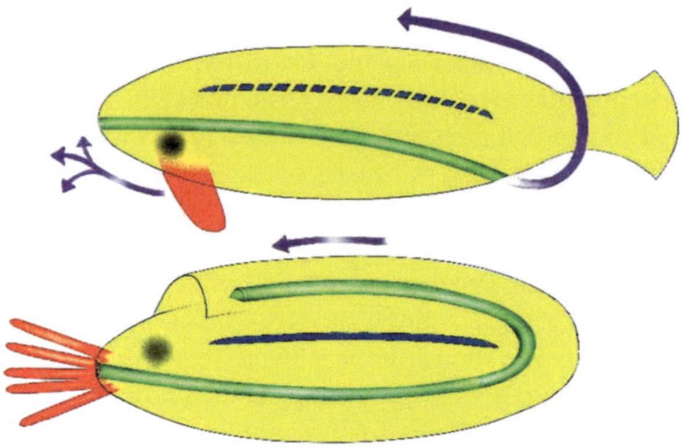

are arranged like those of a vertebrate. The Academy appointed two senior naturalists to referee this paper, Geoffroy and Latreille. Geoffroy was enthusiastic, as the paper put forward the possibility of a unified plan for vertebrates and molluscs, a plan that would now unify three of Cuvier's animal Branches. His review to the Academy appealed to "the universal law of nature, the unity of organic composition" and suggested connections between various organs of a cephalopod and those of a vertebrate (1830; Latreille removed himself at once). Geoffroy assumed, in the name of the two young naturalists, a vertebrate doubled over at the level of its navel so that the face of the ventral side remained outside and the two halves of the back were welded together (Fig. 6.29). The pelvis was close to the neck, and the limbs were on the extremity of the body, so that when the animal was walking on these limbs, it took "exactly the same position as an acrobat who turns his head back over his shoulders and hands". Eight or ten limbs around the head now served as arms and legs, the cephalopod intestine doubled over and pieces of cartilage in the rear of the neck became connected to the funnel. The parrot-like beak of the cuttlefish and its large, complex eyes made these comparisons even more convincing. Geoffroy enthusiastically adopted this claim as proof of his theory of the Unity of Plan shared by all bilateral animals. Homologous organs exhibit similar positions in different animals, regardless of their shapes or functions and they are evidence of an ideal pattern present throughout the animal kingdom. "Nature has formed all beings according to a unique plan. Although the details of their relationships may vary in a thousand different ways, they all rest on the same fundamental principles". The molluscs could now combine, with the vertebrates and arthropods, into one unified plan of composition, and Cuvier's system of four branches of the animal kingdom could be rejected.

This theory provoked a fierce reaction from Cuvier, who could not reconcile Geoffroy's speculations with the results of his own careful anatomical research. The controversy resulted in eight public debates between Cuvier and Geoffroy (February–April 1830) at the Collège de France. It was a debate about whether or not the structural plans of all animals can fall back onto a single archetype. In this debate, Cuvier convincingly showed that many of Geoffroy's supposed examples of Unity of Composition were not accurate and that similarities between vertebrates and cephalopods were superficial. Geoffroy had stretched the Unity of Type much farther than was warranted, according to the available evidence.

While Geoffroy was playing with the idea that all animals have the same basic units, others were playing with the idea that within every (plant and) animal there is a basic unit that repeats itself throughout the body. Johann Wolfgang Goethe (1749–1832, Fig. 6.30) was a German scientist, writer and statesman, close in age to Lamarck. The son of a wealthy family in Frankfurt, he had private tutors to teach

Fig. 6.30 Goethe

him languages, dancing, riding and fencing. He became interested in literature and developed a lively devotion to theatre. At the age of 25, he published his first novel, *The Sorrows of Young Werther,* which brought him such fame that the Duke of Saxe-Weimar invited him to his duchy. Goethe went to live in Weimar, where he remained for the rest of his life, becoming the friend and chief adviser of the duke. He became a member of the Duke's Privy Council, sat on the war and highway commissions, oversaw the opening of silver mines, implemented a series of administrative reforms at the University of Jena, managed the Weimar theatre, and published the first part of his most celebrated drama, *Faust* (1808). As a politician, he sympathized with the American Revolution, but at the time of the French Revolution, he remained sceptical of the ability of the masses to govern. During his meeting with Napoleon (1808), the two discussed politics and literature, and Goethe was deeply impressed with Napoleon's enlightened intellect and his efforts to build an alternative to the corrupt old regime. He always spoke of Napoleon with the greatest respect, confessing that "nothing higher and more pleasing could have happened to me in all my life" than to have met Napoleon in person.

Goethe was keenly involved in studies of natural science. In 1790, while walking through the Jewish cemetery in Venice, he accidentally kicked a sheep's skull, breaking it into pieces. These few pieces gave him the idea that the skull, rather than being a skeletal part completely different from the vertebral column, actually consists of a number of modified vertebrae: the same organ, in repeating and modifying itself, could form parts of an organism with a very different appearance. The skull is a modified part of the spinal column, in which the vertebrae have been enlarged, transformed and partly welded to make a protective case for the brain. Goethe sought to construct an ideal vertebrate, from which the entire vertebrate diversity develops by simple modifications of certain parts. For this, it was necessary to observe which parts are common to all vertebrates and which differ, and "The appropriate assemblage is then used to deduce a hypothetical form that could serve as a model for various types of natural creatures". Stemming from this abstract model, the features of living creatures may change and modify over the course of time. All forms are products of a single plan. He established (1795) the concept of "types" of a plan, to which all kinds of animals could be fitted and classified, thus providing a scheme in which every animal could be included and described.

Horace Benedict Alfred Moquin-Tandon (1804–1863) was Professor of Botany at the University of Marseilles (and later, director of the *Jardin des Plantes* in Paris). He found that all segments of the body of a leech are identical, and he thought that each contained all it needed to live independently, as a distinct organism. He considered each such segment to be a miniature animal, which he termed a *zoonite* and he believed that all of Cuvier's articulated animals could be broken down in a similar way; they are no more than a linear series of simpler animals, of zoonites (*Monographie des hirudinees* 1827). Antoine Louis Dugès (1797–1838, Fig. 6.31), professor of pathology and operative medicine in Montpellier, expanded on this idea and claimed that it applied to all animals. Individual zoonites always have the same basic structure and composition; the animals that they form differ only in their

Fig. 6.31 Dugès

number and in that they group in different ways. Zoonites can group in a linear series, or as a double chain, or radially around a central point, or they can branch out. The zoonites, or the organs that they make up, can fuse together, partially or totally, to the extent that it becomes impossible to discriminate one from another. Transitional forms of different structural arrangements might link animals that, at first sight, seem completely unrelated. These abilities allow for an infinite number of intermediate forms and prevent sharp demarcations between members of the animal kingdom.

Zoonites are the segments that build the animal body. Sensibility and mobility cause them to arrange themselves and distribute their functions in a manner that achieves their common purpose. An organ may form in some segments and not in others, according to where its function in the total organism is most effective. Such organs may form by a partial coalescence that draws all elements of a similar nature towards a common centre. This reorganization and reciprocal interaction of the individual components aim at perfecting the general vitality of the organism, and the association of the organs resembles a human society. Civilization brings together individuals who, through their different contributions, augment the resources and benefits shared by the community as a whole, unlike a primitive tribe of savages, who live a crude and simple way of life. In this, the savage is comparable to a tapeworm, with a limited number of functions, "just as the animal itself is organized in segments that differ little from one to another".

Unlike Geoffroy, Dugès no longer tried to explain insects by reference to vertebrates, nor did he try to find the equivalent of the mammal vertebrae in the segments of the bodies of arthropods. In Dugès's view, all zoonites are identical throughout the animal kingdom, which accounts for the uniformity we find among the animals. "Although there is no uniform plan on the scale of animals, there is conformity in the way the component elements are always the same, and their disposition does not vary enough to make a clear distinction between the animals they constitute". He searched for intermediate types to establish a transition from worms to vertebrates, and he was convinced that he saw affinities (what we now term homologies) between leeches and lampreys. He believed that each species has its own particular form and style. At times, he sought to show that very different animals always have the same parts in identical numbers, without offering any reason why the preferred number should be constant. For example, he attempted to show that three vertebrae form the neck of vertebrates, just as the thorax of insects has three parts. He also thought that there was a connection between the five pairs of feet of the crustacean decapods and the five primitive appendices, which, when welded together, constitute the limbs of the higher vertebrates. In sum, Dugès proposed (1832) that all animals are composed of multiple organisms that, in the early stages of development, were capable of living independently; gradually, they coalesced and became modified to conform to the functions of the body.

Henri Milne-Edwards (1800–1885; Milne, which he added to his name, was the name of his godmother) was born in Bruges. When Belgium became independent, he chose French citizenship, studied in France and eventually joined the staff of the Museum of Natural History. He studied invertebrates of coastal regions, organizing excursions along the shores of the English Channel where he collected, classified and observed animal behaviour. In his comprehensive three-volume *Histoire naturelle des crustacés* (1834–1840), he suggested that the crustaceans consist of 21 homologous metameric zoonites (segments) that are variously fashioned according to the functions they fulfil and the mode of life of a species, whether free, fixed or parasitic. Later, (*Introduction a la zoologie générale, ou considérations sur les tendances de la nature dans la constitution du règne animal* 1858), he further developed this idea into a great fundamental principle, the law of the division of labour within organisms. In lower animals, the same tissue can adapt to different functions; thus, in coelenterates, a single fragment can regenerate the entire animal. In higher animal groups, however, this ability tends to disappear, and tissue specialization progressively replaces it, with body systems becoming individualized in order to carry out precise and exclusive functions. We find a reproductive system, a respiratory system, a digestive system and so on. Within each such system, each organ has a well-defined role. The digestive system divides into a digestive tube and the attached glands, and the digestive tube itself consists of a first region into which food is introduced, a second in which the nutrients undergo the action of the digestive juices and a third where substances useful to the organism are absorbed and where waste products are eliminated. Such specializations, which become more and more precise, determine the rank of an organism in the animal series.

6.2.4 Senses Dictate Biodiversity

Henri Marie Ducrotay De Blainville (1777–1850) was born in the French Flanders, a member of the lesser Norman nobility. He went to Paris to study music, art and literature; moved on to study medicine; and eventually attracted the attention of Cuvier, who offered to make him a collaborator on a work of comparative anatomy for an annual wage. Cuvier also authorized Blainville to teach his courses when he was away on administrative assignments for the Napoleonic regime. Most importantly, he aided him in acquiring the position of assistant professor of anatomy and zoology in the Faculty of Sciences (1812). Blainville soon became an outstanding comparative anatomist, lecturing and publishing on invertebrate zoology, comparative osteology, the history of science and the principles of natural history. Eventually, relations between the two men soured and ended in open enmity.

Blainville was quick to take offence, difficult to get along with and his extreme pride led to bitter relations with contemporaries. When Lamarck died, he received the chair of molluscs and worms only through the interference of the Minister of Public Instruction. Two years later, on the death of Cuvier, he obtained the chair of comparative anatomy. During the 1830s, he was the leading zoologist in France. One of Blainville's important contributions was the separation of amphibians (frogs and salamanders) from the reptiles (lizards, crocodiles and tortoises) into a separate class, between reptiles and fish (1816). Frogs and salamanders are more primitive animals than lizards, crocodiles and tortoises because they pass through a larval stage in which they breathe by gills, like fish.

Blainville's general classification of the animal kingdom appeared in 1816, the year of his break with Cuvier; it later appeared in 1822 and in 1839. He defined life as the ability to sense and respond to the environment. Organs of sense and of movement, therefore, assumed primacy and allowed Blainville to base his classifications upon those external features that sensed the environment and enabled movement. In place of Cuvier's classification by internal characters, taken from the nervous, circulatory and respiratory systems, he introduced classification by external characters—external covering, number of limbs, digits and so on. In place of Cuvier's four major branches, he divided the animal kingdom into three major groups, which he termed "subkingdoms": animals with an irregular form (corals, infusoria, sponges), those with radiate form (coelenterates, bryozoans, echinoderms) and those with bilateral form (all the rest). The more an animal felt and the more it moved, the more it expressed itself as an animal and the higher its place on the scale. Whereas Cuvier had placed the molluscs above the arthropods because of their complex circulatory system and sense organs (mainly in cephalopods), Blainville placed the arthropods higher on the scale because their senses and locomotor organs are more perfectly developed, and their mode of reproduction is more often via individuals of separate sexes. Whereas Cuvier had insisted on the absolute discontinuity of his four major Branches, Blainville claimed to have detected continuities between them. He linked Cuvier's vertebrates to the arthropods (then named Articulata) by assuming that both were composed of a series of homologous segments. He linked the molluscs to the arthropods by establishing a strange new intermediary group that he pointedly called Malentozoa ("articulo-molluscs"). It included the barnacles, encased in a calcareous shell, like molluscs and the chitons (coat-of-mail shells), whose calcareous shell is articulated (Fig. 6.32). Blainville thus deliberately attempted to overthrow Cuvier's principle, that there were no homologies across his four branches, and tried to show that the animal kingdom forms one single, gradated series. He completely removed Man, whom he classified as a separate order of mammals in 1822, from the animal kingdom in his 1839 publication.

Blainville did not accept Cuvier's doctrine of successive catastrophes followed by new creations. He also did not accept Lamarck's ideas of evolution, with species changing over time. He believed that there had been but one Creation and that all animals fall into one single linear sequence, the Ladder of Life, leading up to man. In the beginning, God had created a perfect series of animals, from sponges to the highest primates. In the course of time, this animal series gradually became imperfect through extinctions due to natural causes and, especially, to man. The dodo was proof that an animal might become extinct in recent times, though not through a catastrophe but through the intrusion of humans into its limited range (Appel 1980). Blainville argued that fossils tend to fill gaps in the series of living forms. He placed the pterodactyls (Mesozoic flying reptiles) in a separate class between birds and reptiles and the ichthyosaurs (Mesozoic swimming reptiles) in a class between reptiles and amphibians. The theory of the Ladder of Life was thus still very much alive in the mid-nineteenth century. Whereas Cuvier had claimed that there were no fossil primates, by the 1840s, naturalists had found primates as far back as the Eocene. Blainville concluded, from this, that extinct forms occur in the same geological layers as forms still existing on Earth. Together, extinct and extant organisms testify to the original fullness, and hence rightness, of God's creation (Appel 1980). He based his claim on his own, sometimes controversial, identifications of fossil and living

Fig. 6.32 Blainville's Malentozoa, a group in which he grouped chitons with barnacles. (Illustration by Tuvia Kurz)

species. For example, he believed that the mammoth of Siberia was of the same species as the Asian elephant.

To forward his theory of biodiversity classified by external features, Blainville also insisted on replacing familiar names with a new terminology, based exclusively upon external characters. The mammals were to be named *Pilifera* ("fur bearing"), the birds *Penifera* (feather bearing), the reptiles *Squamifera* (scale bearing), the amphibians *Nudipellifera* (nude skinned) and the fishes *Pinifera* (fin bearing). Beyond vertebrates, he named the spiders *Octopoda* (eight legs), the insects *Hexapoda* (six legs), the corals *Dendrolitharia* (stones with roots) and so on. We see this progressive order of naming by external features particularly well in his classification of arthropods:

Hexapoda	Six legs	Insects
Octopoda	Eight legs	Spiders
Decapoda	Ten legs	Crabs
Heteropoda	Different legs	Shrimps
Tetradecapoda	Fourteen legs	Woodlice
Myriapoda	A thousand legs	Milllipedes
Chaetopoda	Arrow legs	Ringed worms
Apoda	No legs	Leeches, parasitic worms

In Germany, during the first half of the nineteenth century, a somewhat different school emerged in the study of diversity. It is believed in a world in which everything can also be anything else, with electrical, chemical or magnetic properties of Earth recurring in the sensory equipment of man, who, consequently, is an abbreviated Earth and the Earth is but a drawn-out human being. One representative of this school was Lorenz Oken (1779–1851, Fig. 6.33), who was about the same age as Geoffroy. Coming from a peasant family in Baden and orphaned at an early age, he studied medicine and natural history at the universities of Freiburg, Wurzburg and Gottingen. Eventually, he joined the faculty of the University of Jena, where he founded the German association of scientific and medical researchers and promoted the idea of annual general meetings of German naturalists; his idea soon took form and indeed, he held the first meeting in 1822. Oken also founded a journal (*Isis*) in which he declared that he would publish everything submitted. This was a bit too liberal for its time; the university dismissed him, and he spent the following 8 years as a private researcher. Eventually, he joined the faculty of the University of Munich and, after 5 years, moved to the recently established University of Zurich, where he became the rector. My direct knowledge of Oken's theories is from his *Naturphilosophie* (translated by Stallo as *General Principles* of *the Elements of Physiophilosophy*, 1848).

Oken disliked Cuvier's classification, which he described as resembling a miserably crippled, crooked, hunchbacked child and criticized its "horrible lack of symmetry" where one category could have only two members and another equal category could have hundreds. He believed that man was a microcosm of the entire world, and each different kind of mineral, plant or animal represented, in some

Fig. 6.33 Oken

manner, a specific part of man. Thus, there are five main kinds of animals: invertebrates, fish, reptiles, birds and mammals, corresponding to the five senses found in humans: touch, taste, smell, hearing and sight. In his textbook (1812), Oken proposed the following classification:

"Half animals"

1. Element animals
2. Mineral animals
3. Plant animals

"Whole animals"

A. *Feeling animals*

I. Sexual animals

Sperm-animals: infusoria
Egg-animals: corals
Germ-animals: zoophytes

II. Torso animals

Artery-animals: jellyfish

Gut-animals:	molluscs
Skin-animals:	insects

B. *Intellect-animals*

III. Head-animals

Tongue-animals:	fish
Nose-animals:	reptiles
Ear-animals:	birds

IV. Eye-animals mammals

Oken, considering himself a philosopher of zoology rather than a zoologist, believed that the entire animal kingdom is a huge organism, *Weltorganismus*, held together by the interplay between similarity and difference, in which each organ is in itself a complete entity. The various taxonomic groups are analogous to organs of this huge organism. Man is the most supreme creature in the animal kingdom, the highest, most complete and perfect realization of God. As a microcosm, he bears within him all that was before him, just as Oken wrote, the fruit contains all the early stages of the plant. Hence, the entire animal kingdom merely represents the activities of man's organs. Zoology is the science of turning Man into the entire animal kingdom, the groups of which we should classify as an analysis of the human microcosm. Consequently, Oken described classification as presenting a kind of anatomy of nature. We recall Plato and his claim that the universe (macrocosm) is a huge, living, gigantic animal, shaped as a completely smooth sphere, with a body, a soul and reason; and that whatever man (the microcosm) has, the cosmos must also have.

Oken argued that on our planet, when only two elements combine, they are still part of the planet. When, however, three Greek elements combine—life forms of Earth, water and air—the combination is no longer a complementary part of the planet but a miniature planet complete in itself; it is an organism. Every such organism is alive with three processes, corresponding to those of the planet: a formative process (earth), a chemical process (water) and an electrical, oxidizing process (air). When carbonic acid mixes with water and air, it forms a primitive slime, *Ur-Schleim*, which creates a soft mat in the places where earth, water and air make contact. Hence, primitive slime forms in shallow waters near the shore, and it is here that plants, animals and even man were created (Wightam 1953; Breidbach and Ghiselin 2002; this, as noted by Kocandrle and Kleisner 2013, brings to mind Anaximander, in ancient Greece, who had suggested that animal life originated from a primordial moisture that had covered the Earth in some distant age).

Some of these unicellular organisms agglomerated in some distant past, forming the multicellular organisms of today. Slime thus creates unicellular organisms, and larger creatures develop from these existing elements. Existing advanced organisms do not transform into other advanced groups ("A finished or perfect organism cannot gradually transform itself into another"). Nutrition, digestion and respiration correspond to the terrestrial activities of magnetism, chemistry and electricity. Their joint manifestation is self-movement, which discriminates between them and the inorganic substratum. When it dies, every plant and every animal decomposes and then, it reverts to a mass of unicellular organisms. Generation, on the other hand, is the agglomeration of infusoria to higher forms of life. Animal semen consists, accordingly, of single-celled creatures and generation is a repetition of that primitive creation, first formed in slime. Indeed, sperm is the total organism, reduced to the primordial menstruum (Oken 1848; the term menstruum, which Oken used, was also used in alchemy, as a solvent).

Oken suggested that animal diversity is maintained by the addition of successively higher-ranking organs, culminating in man, who possesses all organs in their highest forms ("The greater the number of cooperative organs in an animal, the more perfect its organisation and the higher its rank on the organic scale"). In ranking the major animal groups, he believed that the sense organs provide the primary criteria: every major taxonomic group of animals corresponds to one of the five senses. By adding one sense to another, animals reach a higher step on the Ladder of Nature and come closer to perfection. Consequently, there are five major animal groups. Oken believed, however, that each animal consists of a double series of organs—the external organs of the senses (including the sexual parts) and the internal anatomical systems. As they correspond to each other, one may classify according to either of these series. His five sense groups were:

1. **Skin-animals**, that he termed *Dermatozoa*, which have only primitive organs of sense located in the skin, and these are mainly the sense of touch (invertebrates; Oken commented that they may also be named intestinal animals, body animals or touch animals). By means of the skin, the animal becomes an individual that is distinct from nature in general; the animal distinguishes itself from everything extraneous by its sense of feeling, of touch. In these animals, we find mainly the digestive, circulatory and breathing systems.
2. **Tongue-animals,** *Glossozoa*, in which a true tongue and a well-developed sense of taste first appear (fishes; they may also be named bony animals, as they have many skeletal elements). There are still no internal nares, the ear is only internal, the eyes lack eyelids and the muscle fibres are still white.
3. **Nose-animals**, *Rhinozoa*, in which the nose opens into the mouth to enable the passage of air for improving the sense of smell (reptiles, which may also be termed muscle animals). The muscles are already red, the ear opens externally, but the eyes lack the ability to move and the eyelids are not fully developed.
4. **Ear-animals**, *Otozoa*, in which the ear possesses a cochlea (birds, which may also be termed nerve animals), in addition to a tongue and a nose with internal nares. The eye still lacks the ability to move, and only the lower eyelid is "perfect".
5. **Eye-animals,** *Ophthalmozoa*, in which all sense organs are fully developed, the eyes are able to move and have lids (mammals, which may also be termed sense animals or fur animals). All body systems and sense organs are fully developed, the tongue and nose are fleshy, the ear has an external component and the eyes are movable with fully developed eyelids.

Oken's classification thus created a single ladder, in which the entire animal kingdom fell into five major groups. Groups that were poor in senses and had only a few features were low on the ladder; those with many senses and many features were higher. Features also accumulated in the subdivisions of each class. In his Skin-Animals, he first placed the "Intestinal Animals" that have no major organs beyond skin and intestines (Cuvier's Radiata). After these, he presented the "Blood Animals" that possess a circulatory system, in addition to skin and intestines (Cuvier's Mollusca). Finally came the "Breathing Animals" that possess a breathing system, in addition to skin, intestines and a circulation system (Cuvier's Articulata). Furthermore, within these Breathing Animals, he placed first the segmented worms that breathe through the skin; then the crustaceans that breathe via external gills arising from the skin; and finally, the insects that breathe via internal tracheae (their wings, he claimed, correspond to the crustacean gills). Oken thus divided the animal kingdom into groups and subgroups that correspond to one another in a manner that places the lowest animals of a higher class below the highest animals of a lower class. Sadly, his knowledge of invertebrate functional morphology was inadequate. For example, he wrote that "most of the annelids, mollusca, snails etc., breathe through the skin", an incorrect observation even for his time.

In Oken's single Great Ladder of Life, organs only accumulate, and none is ever lost. Consequently, every organ in a primitive group may well become a different organ in an advanced group. The bivalves' two valves correspond to the gill-covers of fishes; the structure locking the two valves (umbones) corresponds to the neural arch of vertebrates; the two muscles locking the valves correspond to the shoulder and thigh muscles of mammals; and the tentacles surrounding the mouth correspond to the feet of insects (Fig. 6.34).

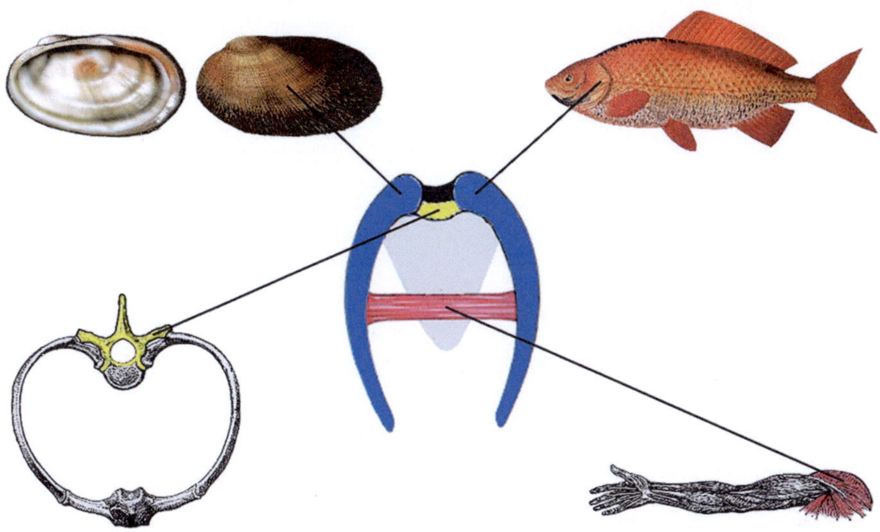

Fig. 6.34 Oken: Organs of a bivalve become different organs in other, more advanced groups

Furthermore, Oken claimed that in the course of its development from fertilized egg to adult, an animal of a given type passes in turn through the stages representing each of the classes that rank below the class to which it belongs. The foetus is a representative of all animal classes in time. At first, it is a single vesicle, stomach or vitellus, as in the Infusoria. Then, the vesicle doubles inside the albumen and shell and develops an intestine, as in the Corals. Next, with a muscular heart, a testicle and a penis, it becomes one of the class of Snails. With the appearance of the bony skeletal system, it passes through the class of Fishes, with the evolution of muscles, through the class of Reptiles and with the advent of respiration through the lungs, into the class of Birds. In this theory of recapitulation, which is actually that of Geoffrey, stages in embryonic development recapitulate the panorama of animal diversity, and "Animals are but foetal forms of man" (Oken 1848: 303–304).

The five senses also determined, for Oken, the number of vertebrae that form the vertebrate skull (Oken 1848). The vertebrae envelop the nerves and protect them, and the number of nerves determines the number of vertebrae. He further believed that the number of nerves depends on that of the senses they serve, and it follows that the number of vertebrae equals that of the senses. There are consequently vertebrae of touch, of taste, of smell, of hearing and of sight. The last four belong to the head and form a system distinct from the vertebrae of the sense of feeling, which is distributed all over the body and served by the nerves of the vertebral column. As the brain is but a voluminously developed spinal cord, the braincase is but a more voluminous spinal column. Accordingly, Oken suggested that the vertebrate skull is the anterior continuation of the vertebral column and that it consists of four vertebrae, corresponding to the four senses in the head. The vertebrate jaws correspond to the arms and feet further behind; the teeth correspond to their claws or nails. Oken elaborated his speculations without any support from observation.

Oken's concepts of classifying animal diversity were followed and further developed by Leopold Fitzinger (1802–1884, Fig. 6.35), of the Vienna *Naturhistorisches Museum* (and, eventually, director of the zoos of Munich and Budapest). He divided each vertebrate class into five series, each with precisely three orders. He subdivided the mammals, for example, (*Systema Reptilium* 1843) into precisely five groups by their senses, each consisting of precisely three orders:

1. Cetacea, emphasis on the sense of touch
   (a) Balanodea (whales)
   (b) Delphinodea (dolphins)
   (c) Sirenia (dugongs)
2. Pachyderma, emphasis on the sense of taste
   (a) Phocina (seals)
   (b) Obesa (?)
   (c) Ruminantia (ruminants)
3. Edentata, emphasis on the sense of smell:
   (a) Monotremata (spiny anteater)

Fig. 6.35 Fitzinger

   (b) Lipodontia (giant anteaters)
   (c) Tardigrada (sloths?)

4. Unguiculata, emphasis on the sense of hearing:
   (a) Glires (rodents)
   (b) Bruta (edentates, elephants and walruses)
   (c) Ferae (carnivores)
5. Primates, emphasis on the sense of sight:
   (a) Chiroptera (bats)
   (b) Hemipitheci (monkeys)
   (c) Anthropomorphi (apes)

The strictly structured classification of Fitzinger thus lumped together seals with ruminants, rodents with elephants and lions and bats with monkeys.

Early Biogeography

We have seen how Buffon, back in 1788, had explained the creation and distribution of animals in terms of Earth's heat. Some three decades later, another naturalist, Alexander von Humboldt (1769–1859), who was a German explorer, geographer and naturalist, likewise placed emphasis on Earth's heat, this time in seeking to explain the distribution of plants. He was born in Berlin into a wealthy family. His father, an officer in the Prussian Army, was rewarded with the post of royal chamberlain for his services in the Seven Years' War (he died when Alexander was nine); his mother, wealthy from a previous marriage, was emotionally withdrawn. When she died, Alexander's inherited money contributed most of his private income and funded most of his travels and publications. Enlightened tutors educated him, one among them to become an important botanist. When 18, he studied finance at the University of Frankfurt, and when 20, he matriculated at the University of Gottingen. He then travelled to England, where he met Sir Joseph Banks, president of the Royal Society. Banks showed Humboldt his huge herbarium, with specimens of the South Sea tropics. Humboldt then studied commerce and foreign languages at Hamburg, geology at the Freiberg School of Mines, anatomy at Jena and astronomy and the use of scientific instruments. Aged 23, he gained a Prussian government position as a mine inspector from the Department of Mines, and he studied and published on the vegetation of mines (1793). Eager to explore the world, at age 26, he toured Switzerland and Italy on a geological and botanical expedition. He then travelled to Marseilles in the hope of joining Napoleon, who was then in Egypt. However, the French refused him travel permission, so he moved on to Madrid.

In Spain, the monarchy had already authorized and funded several botanical expeditions to Peru, Chile (1777–88), New Granada (1783–1816), New Spain (Mexico, 1787–1803) and around South America (1789–94). These lengthy, state-sponsored enterprises gathered information about plants and animals from the Spanish realms, assessed economic possibilities and provided plants and seeds for the Royal Botanical Garden in Madrid. Crown officials had already issued, in 1779, and systematically distributed instructions concerning secure and economic means to transport live plants by land and sea from distant countries, with illustrations including one for the crates to transport seeds and plants. Humboldt, now in Madrid, sought royal authorization to travel to Spain's realms in the Americas, with his own financing. He gained a positive response, and further, he gained access to crown officials and documents on Spain's empire. While in Madrid, he visited the Natural History Museum to examine specimens of previous botanical expeditions.

Having achieved authorization from the King of Spain, Humboldt finally realized his dream of a scientific expedition, and he hastened to sail off (1799) to Latin America. Obsessed with the importance of measuring whatever one could measure, he took with him 43 measuring and observation instruments. There were telescopes and microscopes, barometers, thermometers, a pendulum clock, compasses and even a "cyanometer" to measure the intensity of the blue sky—instruments necessary for all the observations he intended to carry out. He landed in Venezuela some months later.

This was not the ship's original destination; an outbreak of typhoid on board had forced the captain to change course from Havana to northern South America. Humboldt started his South American research with the valley of Aragua, where export crops of sugar, coffee, cacao and cotton were cultivated, and here he predicted harmful human-induced climate change. When studying the rapid fall in the water level of the valley's Lake Valencia, he credited the desiccation to the clearance of tree cover and to the inability of the exposed soils to retain water. With their cutting of trees, the farmers were removing the woodland's moderating influence upon temperature, such as cooling shade, evaporation and radiation. From there, Humboldt crossed the Amazon jungle, sailing along the Orinoco and its tributaries for three months, venturing deep into the rainforest and then crossing the Andes. He discovered dangerous electric eels, whose shock could kill a man. To catch them, locals suggested he drive wild horses into the river, which brought the eels out from the river mud and resulted in a violent confrontation of eels and horses, some of which died. Humboldt captured and dissected some eels, which retained their ability to shock; he received potentially dangerous electric shocks during these investigations. In Cuba, he collected a vast collection of the island's flora and fauna, which he later published (*Essai politique sur l'îsle de Cuba*). In Ecuador, he climbed the Chimborazo, which, until the beginning of the nineteenth century, people thought to be the highest mountain on Earth. Humboldt tried to reach the summit. He collected and documented plants all the way up, from the tropical rainforest at its base to lichen above the treeline. He reached an altitude of 5878 m but had to return, suffering from altitude sickness. This was 300 m short of the summit, which is 6234 m high (we shall return to his climb of the Chimborazo some paragraphs ahead).

After an expedition to the sources of the Amazon, he headed for New Spain (Mexico), where, having been officially welcomed via a letter from the king's representative, he was given a special passport to travel throughout the country, with access to crown records, mines, landed estates, canals and Mexican antiquities. Then, leaving Spanish America, he took a short visit to the United States. Knowing that the US president, Thomas Jefferson, was himself a scientist, Humboldt wrote to him saying that he would be in the United States. Jefferson warmly invited him to visit the White House. In his letter, Humboldt had gained Jefferson's interest by mentioning that he had discovered mammoth teeth near the equator. Jefferson had previously written that he believed mammoths had never lived so far south. Humboldt had also hinted at his knowledge of New Spain. In Washington, Humboldt held intense discussions with Jefferson on both scientific matters and on his stay in New Spain. Jefferson had recently concluded the Louisiana Purchase, which now placed New Spain on the southwest border of the United States. The Spanish minister in Washington had declined to furnish the US government with information about Spanish territories, and access to the territories was strictly controlled. Humboldt was able to supply Jefferson with the latest information on the population, trade, agriculture and military of New Spain, information that would later be the basis for his *Essay on the Political Kingdom of New Spain* (1810). After 6 weeks in the USA, Humboldt set sail back to Europe (1804), laden with 60,000 plant specimens representing 6000 species—2000 of which were previously unknown in Europe. He settled in Paris, where he became the nucleus of a thriving scientific community.

Throughout his expedition to Spanish America, Humboldt kept a detailed diary in innumerable notebooks, with tens of thousands of annotations, observations, ideas and measurements, and from it he drew and published, immediately and continuously, in multiple volumes over 20 years. He used his personal fortune to produce densely illustrated books that gained him an international reputation in scientific circles and, through popular, condensed versions of his work in many languages, among the broad public. He soon became one of the most famous men in Europe, with academies throughout the world eager to elect him to their membership. Furthermore, King Frederick William III of Prussia conferred upon him the honorary post of royal chamberlain, without exacting the duties, with a pension of 2500 thalers. This stipend became his main source of income in later years, when he had exhausted his fortune on the publications of his research. Humboldt spent his later years in Berlin. His only additional expedition was to Russia, at the invitation of the Czar.

Humboldt viewed nature holistically, and the unity of nature meant for him the interrelation of all physical sciences. Life on Earth is a web of interactions; everything is interconnected, and it is the conjoining between biology, meteorology and geology that determines where specific plants grow. Upon returning to Europe, he based his *Essay on the Geography of Plants* (1807) on this idea of studying the distribution of organic life as affected by various physical conditions. He presented a colour pictorial cross section of Chimborazo (*Ein Naturgemälde der Ande,* "a painting of nature", known today as "the Chimborazo Map", Figs. 6.36 and 6.36a). This meticulously rendered diagram shows plants distributed according to their altitudes on mountains (Munzar 1967; Wulf 2015; Korner and Sphen 2019). To the left and right of the mountain, Humboldt placed several columns that provided information from temperature, gravity and humidity to the blueness of the sky, again all related to the height of the mountain. Thus, when choosing a certain height up the mountain, one could draw a line through the illustration to see the temperature,

Fig. 6.36 Humboldt's *Naturgemälde,* the "Chimborazo Map". (a) Humboldt's Chimborazo Map: detail

Fig. 6.36 (continued)

humidity or atmospheric pressure, as well as the plant and animal species observed at that altitude. It thus depicts the relationship between elevation and vegetation zones.

Over the globe, we have rainforests in the tropics, deciduous forests and grasslands in temperate zones, and tundra at high latitudes. Humboldt showed that a vertical stratification reflects these global vegetation patterns. On mountains, temperature falls with altitude, not because less heat is received from the Sun (the Sun's rays burn intensely at high altitudes) but because the air becomes thinner and therefore colder. Steep elevation gradients compress life zones into 4–5 km of elevation, which otherwise, near sea level, would be separated by thousands of kilometres of latitude. Mountains, by lumping contrasting climates onto a single slope, host more diverse life than any other terrestrial system. Not only do mountains harbour a high concentration of endemic species in specific regions, but their ruggedness also creates contrasting exposure to sun and wind. The result is a myriad of different niches for life, which explains why mountain ecosystems are so rich in biodiversity. The Chimborazo Map, by interconnecting climate, geography and nature, laid the foundation of plant geography as a scientific discipline (Nordenskiold 1935; Munzar 1967; Wulf 2015; Klein 2018; Korner and Sphen 2019; Miller 2019). Furthermore, its richness of information, combined with the simplicity of the scientific visualization, was part of Humboldt's general insistence on presenting data in an easily understood way. Much of his success in gaining a general readership was his understanding that charts and graphs, which speak to the senses without tiring the intellect, can bring attention to a large number of important facts.

Humboldt (1807) also introduced the concept of the isotherm, an invisible line connecting points on a map that receive an equal quantity of heat (Fig. 6.37; in Greek, "isotherm" means "of the same heat"). Until then, scientists would present meteorological data in long tables of geographical places and their climatic conditions. These gave precise temperatures but were difficult to compare. Following his research on the distribution of plants, and with a feeling that the geographical distribution of temperatures on the planet closely correlates with the geographical distribution of life on Earth, he had the idea to draw curves of equal mean annual temperature over the globe. By his delineation (in 1807) of isothermal lines, he devised the means of comparing climatic conditions of various countries. At sea level, the isotherm 30 connects Florida, Madeira and Cairo, and also at sea level, the isotherm 45 connects Boston, Paris and China. Furthermore, these sea-level isotherms correlate with mountain heights. The sea-level isotherm 30 corresponds to a mountain height of 1500 metres; sea-level isotherm 15 (Naples and further away from the equator) corresponds to a mountain height of 3000 metres; that of Stockholm (sea-level isotherm 5, even further away from the equator) to a height of

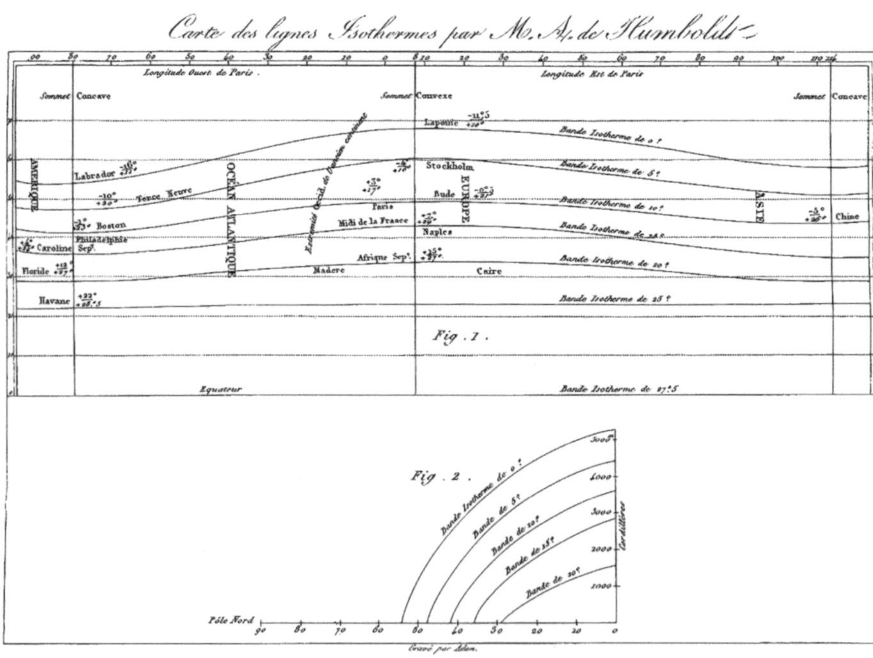

Fig. 6.37 Humboldt's isotherm map

6000 m; of Labrador (isotherm 0) to 4000 m. Climate thus does not depend solely on latitude (Nordenskiold 1935; Munzar 1967; Wulf 2015; Klein 2018; Miller 2019).

Humboldt's isotherm concept (1807) was a sketch containing only a few sites and their data. Five years later, William Channing Woodbridge (1794–1854), a schoolteacher from Connecticut, produced the first world isothermal chart (1823). His work with the deaf had led him to discover maps as an ideal visual vehicle for delivering information. Harbouring an interest in geography and natural history, he was keen to introduce the latest geographical information into his maps, and he travelled to Europe, both to teach and to gather ideas for future textbooks and atlases. When in Paris, he met Humboldt, who introduced him to the concept of isotherms and demonstrated the ability to illustrate them diagrammatically. Upon returning to Connecticut, Woodbridge expanded on Humboldt's findings to publish his own world isothermal chart, in which he depicted the relationship between mean annual temperature and place and illustrated climatic regions in colour (1830). Following the isotherms, he divided the world from the northern pole to the tropics into Frozen, Winter, Cold, Temperate, Hot and Torrid regions, with corresponding regions in the southern part of the globe, towards the southern pole. He entitled this map the *Isothermal Chart, or View of Climates and Production, Drawn from the Accounts of Humboldt & Others* (Fig. 6.38).

Some years later, Humboldt collaborated with the German publisher Heinrich Berghaus to produce the *Physikalischer Atlas*, the first comprehensive physical

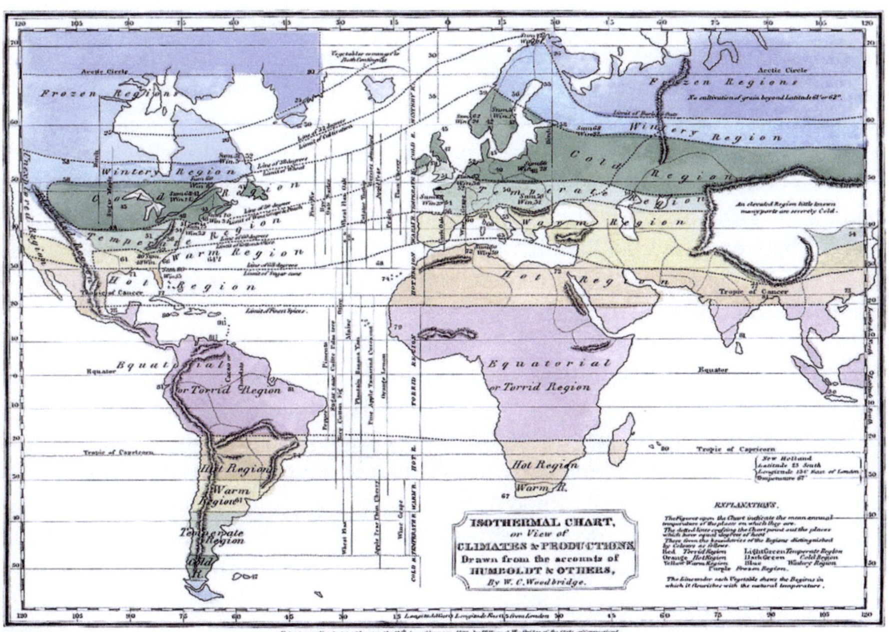

Fig. 6.38 Woodbridge's *Isothermal Chart, or View of Climates and Production, Drawn from the Accounts of Humboldt and Others*

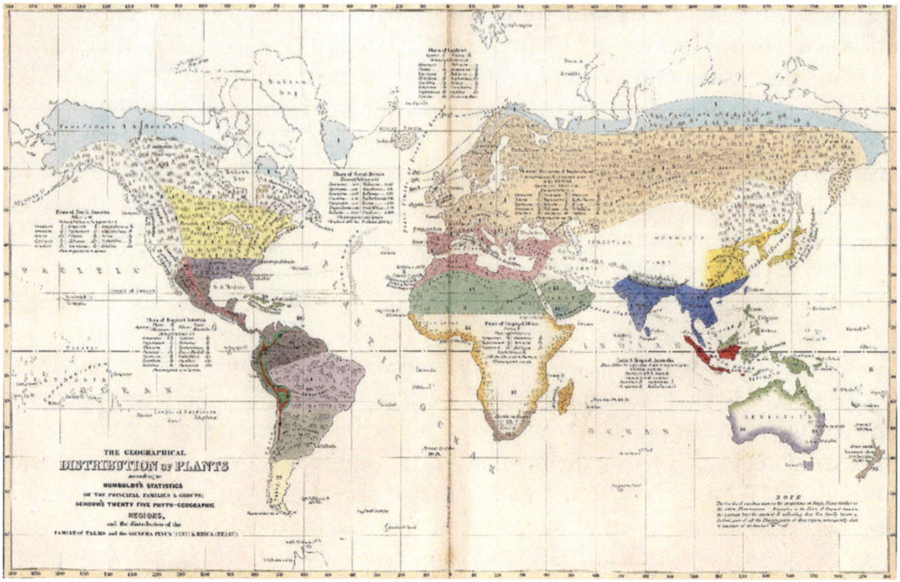

Fig. 6.39 Berghaus's *Outlines of Botanical Geography. The Distribution of Plants in a Perpendicular Direction, in the Torrid, the Temperate and the Frigid Zones; with Indications of the Mean Temperature of the Year and of the Coldest and Warmest Months*

world atlas (1838). The atlas accompanied Humboldt's bestselling book, *Cosmos*, and included maps showing the distribution of plants from both latitudinal and altitudinal perspectives (Fig. 6.39).

Humboldt's research on biodiversity changing with latitude was seminal. He proposed both a specific causal factor, winter temperature, and a mechanism, loss of fluidity. In stating that fluidity is essential to life, he pinpointed the dynamic relationship between water and temperature so crucial to life on Earth: biological processes, such as photosynthesis and metabolism, require water to be in a liquid state, and this is controlled by temperature.

In 1869, on the 100th anniversary of his birth, 25,000 people gathered in New York's Central Park to listen to speeches extolling his accomplishments and witness the unveiling of a large bronze bust of Humboldt, who had died 10 years earlier. Flags and enormous posters showing Humboldt's face lined the streets of Manhattan. Similar celebrations took place around the world—in Berlin, Humboldt's birthplace, 80,000 admirers gathered in chilly rain to listen to eulogies and songs sung in his honour. Today, his name lives on in the Latin names of nearly 300 plants and more than 100 animals and in many geographic features and place names, from a glacier in Greenland to a mountain range in Antarctica. Humboldt was the most famous scientist of his age (Schrodt et al. 2019).

6.2.5 Enter Embryology

Karl Ernst von Baer (1792–1876, Fig. 6.40), of a German noble family, was born in Estonia, then a Baltic province of the Russian Empire. Having studied at the universities of Tartu, Vienna and Wurzburg, at the age of 30, he joined the faculty of the University of Königsberg as professor and director of the anatomy museum and of the botanical gardens. He later moved to Saint Petersburg, as a member of the Imperial Academy of Sciences, roamed the vastness of the Russian Empire, travelling to such diverse regions as Novaya Zemlya in the Arctic Ocean, Lapland and the Caspian Sea. Being interested in physical anthropology, he organized the first international congress of anthropology, with the purpose of standardizing skull measurements (1861). Late in life, he retired to Estonia, the country of his childhood.

Baer discovered the mammalian ovum (1827) as a minute cell inside the follicle that de Graaf, two centuries previously, had erroneously supposed to be the ovum itself. He also introduced the term "spermatozoa" for what had previously been referred to as "animalcules" in the seminal fluid, and which he erroneously believed to be parasites (Brauckmann 2008, 2012). Baer was among those who believed that species change over time. He (1828) placed embryology in a central position in biology.

Fig. 6.40 Von Baer

As a brief reminder, Aristotle had developed the concept of a great Ladder of Nature, with all organisms arranged in a hierarchy based on the complexity of their form, with humans at the top. After Aristotle, various naturalists restructured the Ladder to reflect God's place in creating the hierarchy of nature. In this new framing, humans maintained their position at the top of the earthly creatures because of their likeness to the Deity, just under God and the Angels. During the late eighteenth century and the early 1800s, many accepted the Ladder of Life as the main explanation for the variety of creatures on Earth, and many sought evidence, mainly in the developing embryos of vertebrates, for the idea that embryos pass through the morphological stages of lower animals on the Ladder.

Coming now to the early 1800s, two leading physicians, Meckel and Serres, both suggested that stages of ontogeny parallel the stages of the Ladder of Life. The embryos of higher animals progress through successive stages in which they resemble lower-level forms ("recapitulation").

Johann Friedrich Meckel (1781–1833), of Halle, grew up helping his father prepare specimens for the family's anatomical collection. He graduated from the local university and went to Paris, where, as an assistant of Cuvier, he translated Cuvier's five-volume *Leçons d'anatomie Comparée* into German (1810). In the Cuvier-Lamarck disputes, however, he adopted Lamarck's evolutionary beliefs. At the age of 27, he became a professor at the University of Halle, where he was meticulous in his comparative study of abnormal embryological specimens, which he collected from all over the world; he considered these malformations as the results of early terminations of development. In his *Contributions to Comparative Anatomy* (*Beyträge zur vergleichenden Anatomie*, 1811), he outlined the idea that embryonic stages of higher forms recapitulate the forms of animals lower on the great Ladder of Life. He compared the arrested organs from organisms high on the Ladder to the normally developed organs of organisms on lower rungs and argued that the organs of higher life forms develop through the stages of lower morphologies. Embryonic development thus mirrors the hierarchy of organisms on the Ladder of Life.

Étienne Serres (1786–1868) was a physician at a hospital in Paris; aged 55, he became president of the French Academy of Sciences, and during 1850–1868, he held the chair of comparative anatomy at the Museum of Natural History of Paris. In his comparative study of vertebrate brains, he argued that the developing human brain progresses through the hierarchy of nature (1821). At first, it looks like the brain of a fish, then a reptile, then a bird, then a general mammalian brain and finally it settles into the form of a human brain. This order of brain development appeared, to him, to mirror the Ladder of Life. Serres argued, broadly in the footsteps of Lamarck, that a Formative Force propels the development of species. The individuals of lower groups have too little of the formative force to have their organs develop into complex organs, as in higher groups. Humans are the most complex life form because they have the greatest amount of formative force. On this scale, animals become perfect by the addition of organs, and the embryo successively adds those organs that characterize the animal groups in the ascending scale. The human embryo, when but a single vesicle, is an infusorian; when it has gained a liver, it is a mussel; with the appearance of bones, it enters the class of fishes; and so forth, until it becomes a mammal and then a human being.

Meckel and Serres did not work together, but they had similar approaches. They believed that within the entire animal kingdom, there was a single unified body type and that during embryonic development, the organs of higher animals match the forms of comparable organs in lower animals. Higher animals go through embryological stages analogous to the adult stages of lower animals in the course of their development. This idea, of an animal passing through the stages of each of the groups ranking below it during embryonic development, depends on the assumption of a single linear scale of nature. The leading embryological concept of Baer's time, thus, was that ontogeny recapitulates phylogeny.

Von Baer denied this claim of different grades of embryonic development in the animal kingdom, forming a single series that recapitulates the animal panorama. He distinguished four types of embryonic development, corresponding to Cuvier's four Branches. Embryogenesis either advances from the centre to the periphery (Radiata); creates identical forms along a longitudinal axis (Articulata); forms a spiral shape around a conic space (Mollusca); or forms identical organs along the two sides of an axis formed by the notochord, neural tube and gill slits (Vertebrata). The existence of these distinct types of organization rejects the existence of a linear scale of organic perfection. Baer noted, however, that the infusoria and entozoan (parasitic worms) are not natural groups, because size or habit defines them, not their structure. Entozoans have a simple structure, but this is a consequence of the demands of their parasitic lifestyle.

Baer further proposed that in all animals, throughout the animal kingdom, an identical mechanism regulates only early embryonic development (1828, 1834). The development of all animals starts from the same point, the germ, which is the common primary form for the entire animal kingdom, and proceeds to an initially hollow embryonic sphere (which he himself had discovered and named a blastula). During the next gastrula stage, the embryo separates into primary germ layers (ectoderm, endoderm, usually also mesoderm), and differences appear between internal and external layers. From these early common embryonic forms onwards, the paths of development of the different animals diverge towards specific adult forms. In the vertebrate embryo, for example, an elastic rod, which constitutes the primitive axial skeleton, develops beneath the nerve cord. This rod is the notochord (which Von Baer had discovered), and when it differentiates, the embryo is first recognizable as a member of one of Cuvier's four types: it is a vertebrate (with a notochord). It is never a mollusc or an arthropod, as Oken, Meckel and Serres had claimed. Now, at the stage in which only the notochord differentiates, the embryo is merely a general vertebrate. Subsequently, it becomes a specific type of vertebrate embryo, which may or may not possess an allantois (an embryonic pouch formed as an outgrowth of the embryo's gut). Some vertebrates lack the allantois (fishes); others have it (reptiles, birds, mammals). Thus, at this stage, the embryo can differentiate into either a fish or, alternatively, into a reptile, bird or mammal. At a still later embryonic stage, in birds and in reptiles, the allantois surrounds the embryo; in mammals, it (usually) forms part of the umbilical cord that connects to the placenta. The embryo thereby further differentiates, for example, into a mammal. At an even later embryonic stage, the umbilical cord falls off, whether early, as in marsupials, or much later, all the while nourishing the embryo, as in placental mammals. In this manner, as early embryonic stages are similar throughout the animal kingdom but changes accumulate as development proceeds, general features of a large group of animals (say, a Class) appear earlier in the embryo than more specific characteristics (say, of a Family or a Genus). Consequently, the embryos of higher and lower animal forms may be similar, but this is never true of adults. Table 6.2 presents Baer's classification of vertebrates (with emphasis on mammals). The most noteworthy point in this classification is his use of embryonic structures: the allantois, umbilical cord and yolk sac.

Other zoologists followed Baer in recruiting embryology for the study of animal classification. One was Pierre-Joseph van Beneden (1809–1894), professor of zoology in Louvain, Belgium, the person who had discovered the complex life cycle of parasitic worms. Beneden proposed a classification based upon the vitelline sources of the embryo (1845). In vertebrates, vitelline matter enters the body from the ventral side; in arthropods, from the dorsal side and in molluscs, worms, echinoderms, tunicates, bryozoans, coelenterates and various protozoans, it enters neither from the ventral side nor from the dorsal side.

Beyond such changes in the embryonic development of the individual, Baer argued that species also change over time. This, he claimed, makes preformation no longer plausible. Preformation, which claimed that each individual had been created when the first female of that individual's species had been originally created, denied the possibility of individual variations from generation to generation, and this was the basic condition to account for species change. It was impossible to give up the theory of species fixity as long as the preformation doctrine remained valid. Only when naturalists abandoned preformation could they accept change within a species. Baer was the person who placed embryology in a central position in biology.

6.2.6 Diversity: The Small and the Tiny

During the first half of the nineteenth century, major contributions were made to the lower steps of the Ladder of Life: corals, medusas, sea anemones and microscopic animals. In the infancy of microscopic knowledge, naturalists had lumped together all those creatures that they could discern only under a high magnification and whose internal structure was not clear with the microscopes then in use. These animals were termed the Infusoria, animals with infusions of organic matter. The group was a ragbag assemblage of plants, bacteria, single-celled creatures, cnidarians, rotifers and minute crustaceans (such as water fleas), together with larvae of worms and of molluscs.

Table 6.2 Baer's classification of vertebrates, with emphasis on mammals

The embryo:	
1. Acquires gills	
(a) No development of lungs	
Skeleton not ossified	Cartilaginous fish
Skeleton ossified	Bony fish
(b) Lungs develop	
Gills remain	Sirens
Gills disappear with age	Uredela, Anura
2. Acquires an allantois	
(a) No umbilical cord	
No wings, no air sacs	Reptiles
Wings and air sacs	Birds
(b) Has an umbilical cord	Mammals
Which disappears early	
Unconnected to mother	Monotremes
Brief connection to mother	Marsupials
Which persists longer	Placentals
Yolk sac continues growth for long time	
Allantois grows slightly	Rodentia
Allantois grows moderately	Insectivora
Allantois grows extensively	Carnivora
Yolk sac increases slightly	
Allantois grows slightly, umbilicus very long	Monkeys, man
Allantois grows extensively, placenta simple	Ruminants
Allantois grows extensively, placenta complex	Pachyderms, whales

From Agassiz (1859)

Christian Gottfried Ehrenberg (1795–1876, Fig. 6.41) was the son of a Lutheran minister and municipal magistrate in Delitzsch, a small town then in Saxony. After the Napoleonic wars, the Vienna congress granted Delitzsch to Prussia, and Ehrenberg moved to Berlin, the Prussian capital. Here, at the University of Berlin, while studying medicine, he also undertook research on fungi, demonstrating the constancy of the species, their sexual generation and their origin from seeds, thereby rejecting Lamarck's claim that fungi originate directly from mud by spontaneous generation. From these early years onwards, throughout his career, he placed considerable emphasis on demonstrating that lower organisms resemble higher ones in that they arise only from living ancestors. In 1820, he participated, as a scientific *attaché,* in an archaeological expedition to the Nile, Libya, Sinai, Lebanon and the Red Sea, sending home boxes packed with specimens of animals, plants, seeds, rocks, fossils and mummies (the expedition yielded 4000 animal species). Seven members of the expedition died by the time it had ended. When, 6 years later, he

Fig. 6.41 Ehrenberg

returned home, the Berlin University appointed him a professor, and he applied himself with unceasing zeal and diligence to his rich collections, publishing a series of elaborate scientific works with splendid illustrative plates (1828–1830). In 1829, he undertook another scientific expedition, financed by the Russian Czar Nicolas I, to explore mineral resources in the Ural and Altai regions. The inscription on Ehrenberg's grave reads *Also the small in the world are wondrous and great, and worlds are built from the small* (*Der Welten Kleines auch ist wunderbar und gross, Und aus dem Kleinen bauen sich die Welte*).

Throughout his career, Ehrenberg consistently disputed Lamarck's theories, both of the Ladder of Life and of spontaneous creation, namely, of the lowest organisms emerging spontaneously from inorganic matter and transforming into more highly organized creatures. These lower organisms were all in Cuvier's little-investigated radial animals: echinoderms, intestinal worms, medusas, polyps and Infusoria, and were to be the animal groups that Ehrenberg would study in detail to demonstrate that, unlike the basic argument of the Ladder of Life, they too have complex organ systems. He offered careful, detailed descriptions of the organs of medusas and employed galvanic currents, as stimuli, to ascertain their nervous capacity (Fig. 6.42; Williams and Huxley 1998).

Ehrenberg paid special attention to the corals of the Red Sea, which he could observe alive, unlike his predecessors who worked mainly with the calcareous skeletons. He studied how a coral grows by budding, rather than by egg production, and

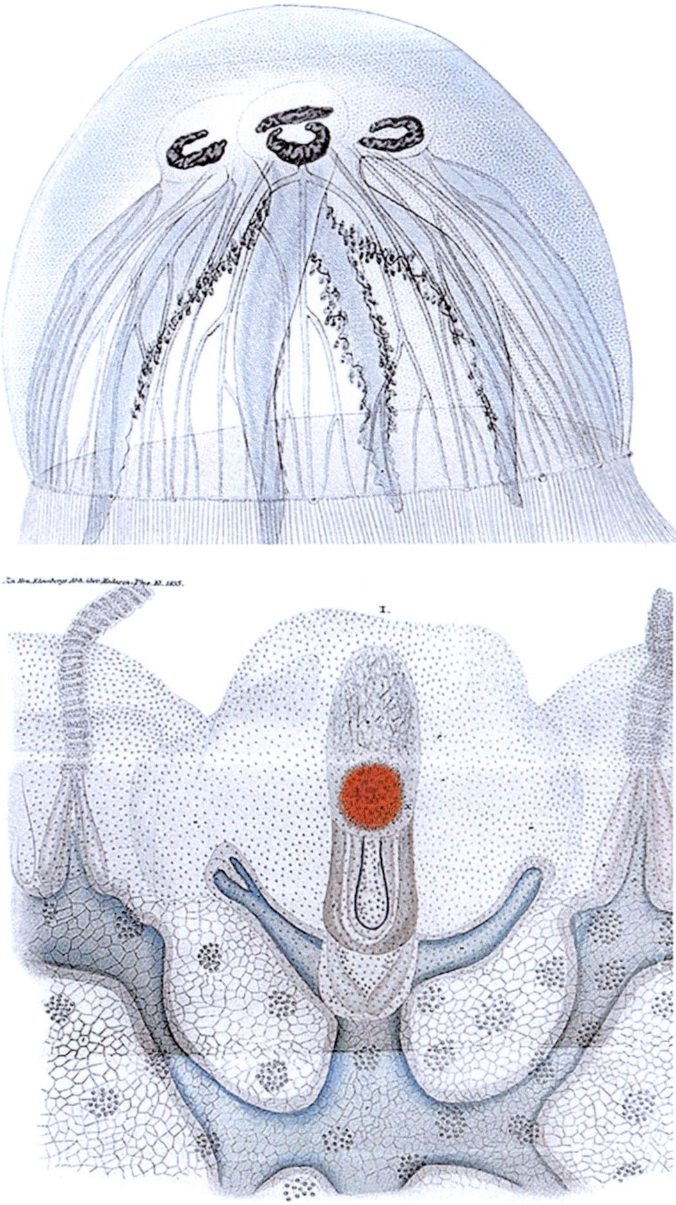

Fig. 6.42 Ehrenberg's illustrations of a medusa. Above, a general view; below, the structure he believed to be an eye. (Courtesy Hartmann Library, Hebrew University)

he noted that the animals clearly have senses, though he could find no traces of a nerve system. Some, which he named Bryozoa, have a separate mouth and anus, while others, the Anthozoa, have just one opening into their gut cavity and are distinctive in having radially arranged chambers in their body cavity. Having demonstrated their structural complexity, he concluded that corals are not simple animals, as suggested by Oken.

In his studies of jellyfish (1835; Fig. 6.42), he combined observations from the Red Sea expedition with studies of *Medusa aurita* (today *Aurelia aurita*) from the Baltic. He found that the jellyfish disc is a highly organized arrangement of membranes and tiny vessels; it is not a formless jelly. The ovaries are separate from the stomach cavity, the digestive system has eight excretory pores, there are muscle fibres, eyes and nerves, and it has globules that move about. By watching the movement of coloured particles through live specimens, he discovered that there were anal openings at the end of these intestines. He thought that eight dark spots at the edge of the disc were eyes after noting a tiny red spot on the top of each, and he noted that in starfish, likewise, there is a tiny red eye on the underside of the tip of each arm.

Over the years, Ehrenberg's research emphasis shifted to the infusorians (Williams and Huxley 1998). To investigate whether these lower animals arise spontaneously out of water, he observed drops of dew and found that they do not contain single-celled (or other) creatures, thereby proving that they did not appear spontaneously (1839). He rejected Oken's theory, popular at the time, that the smallest unit of life was the infusoria, and growth meant the addition of further infusoria. Eager to prove that even small organisms are complete, he endowed them with organs such as those found in complex animals—nervous, digestive and sexual systems (he erroneously considered the chloroplasts as ovaries)—tiny but "complete" creatures, with minute organs comparable with those of more familiar animals. They have a nervous system, have a strong chewing apparatus with teeth in their mouth, have a very complex digestive system and reproduce both by self-dividing and by egg laying. They also possess perfectly distinct intellectual abilities, like other animals. Whether they have a more personal soul, as has been suggested, is open to question.

These observations, he suggested, refute the theories of Oken: if all animals are essentially of the same complexity, then there is no gradation from simple to complex. Furthermore, eager to demonstrate that species do not change over time, that they are constant not only in every area but also across the ages, he studied modern alongside fossil diatoms and foraminifers in soil and rock formations, distinguishing between the shells and skeletons of freshwater and marine animals (1839, 1840; Fig. 6.43).

Ehrenberg also proposed a classification of the entire animal kingdom (1836). Rejecting Cuvier's classification of man among the mammals, he set man as a separate systematic category, unlike the animals, because of his "capacity for mental development". Man is the perfect animal because of the unique harmony of the components of his organization. Ehrenberg, therefore, proposed a primary division of the animal kingdom into the circle of man as opposed to the circle of animals. This aspect of his classification seems influenced by Oken's idea that the animal kingdom represents man, each group of animals embodying, in some sense, an organic system found in man. He retained Lamarck's distinction between vertebrates and invertebrates, but he based his own division on the spinal cord versus

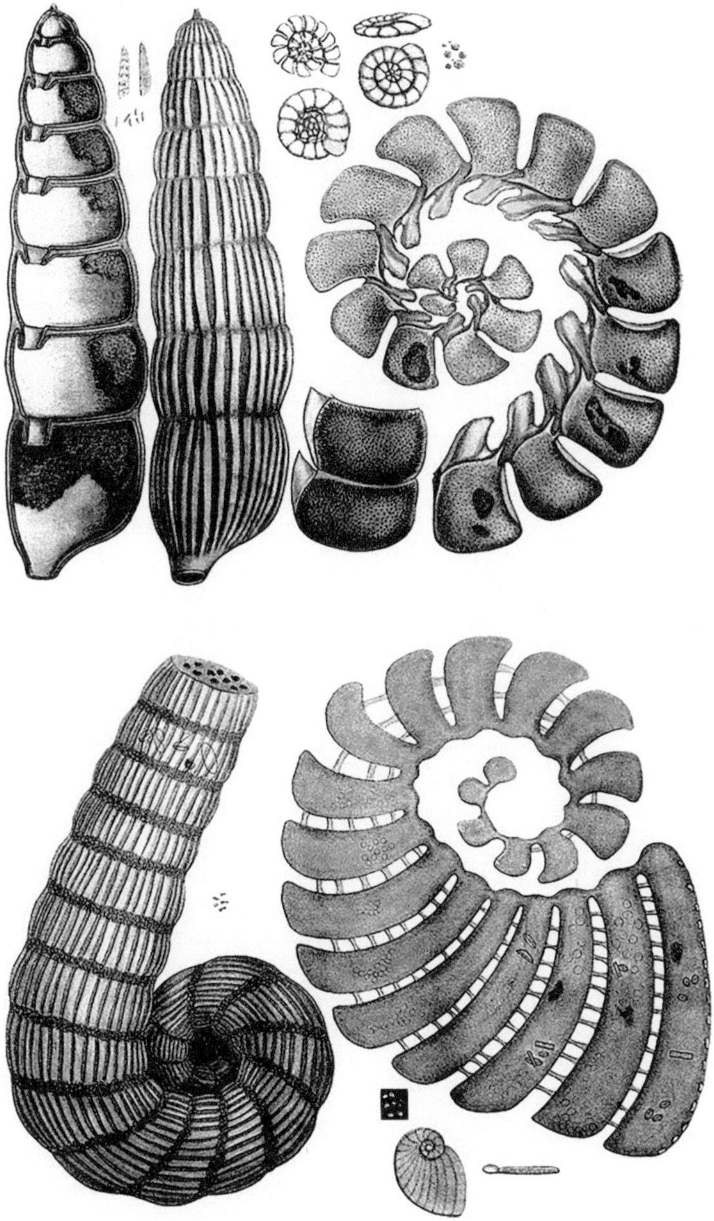

Fig. 6.43 Ehrenberg: Fossil foraminifers

ganglionated nerves, thereby giving the invertebrates a positive rather than a negative definition. Accordingly, he divided all animals into those with nerve marrow (he termed them Myeloneura, = vertebrates) versus those with ganglia ("Gangloioneura", = invertebrates). He further subdivided the vertebrates into those that care for their young, which he named Nutrientia (mammals, birds), versus those that do not, which he named Orphanozoa (amphibians, presumably also many reptiles and fishes). He split the invertebrates into those with a heart or pulsating vessels versus those with non-pulsating vessels. Those with a heart he split into the arthropods, with similar ganglia all along their body ("Homogangliata"), versus the molluscs, with different types ("Heteroganglia"). A major innovation was that he split up the Radiata, or pulseless invertebrates, into several subgroups. In the Infusoria, he split off the rotifers (Williams and Huxley 1998). The following list (taken from Winsor 1976: 42 with modifications) compares Cuvier's Radiata with those of Ehrenberg, two decades later.

Cuvier 1817	Ehrenberg 1835	
Echinodermata	Echinodoidea	Sea urchins
	Asteroidea	Starfish
Entozoa	Turbellaria	Turbellarian platyhelminthes
	Nematoidea	Nematodes
	Trematodea	Trematode platyhelminthes, flukes
	Complanata	?
Acaleph	Acalephae	Jellyfish
Polyps	Anthozoa	Sea anemones, stony and soft corals
	Bryozoa	Moss animals
	Dimorphaea	?
Infusoria	Rotatoria	Rotifers, wheel animals
	Polygastrica	Various protozoans

After Ehrenberg, further studies were into those of Cuvier's radiated animals that are large-sized and multicelled: corals, medusas and sea anemones, all today in the phylum Cnidaria, with four classes. We now know that, broadly speaking, cnidarians carry out a life cycle that alternates between asexual polyps and sexual, free-swimming medusoids, which differ markedly in appearance. The medusoids alone bear sexual organs whose gametes, when released into the water, unite and form an embryo, which develops into a polyp. However, this is an idealized case. The anthozoan cnidarians live only as polyps; the scyphozoan and cubozoan cnidarians live most of their life cycle as medusoids; and the hydrozoan cnidarians live as polyps or as medusoids or as species that alternate between the two.

Cuvier, when he presented (1817) his Polyps and Medusae (under the name Acalepha) as two classes within his Branch Radiata, did not know of these complex life cycles. Only gradually, in the two decades following 1830, did naturalists come to recognize that the asexual (polyps) and sexual (medusoid) individuals are not distinct groups. Rudolph Leuckart (1822–1898) was Professor of Zoology at Giessen and then at Leipzig. He demonstrated that the internal anatomy of medusas, corals and sea anemones is entirely different from that of the starfish, sea urchins and their allies, who have more highly developed internal organs, such as the alimentary canal that exists separately within the body cavity and their peculiar water vascular system. He therefore split Cuvier's Radiata into Coelenterata (today

Cnidaria, broadly speaking) and Echinodermata (1848; the discovery of the complex life cycle of cnidarians is beautifully described by Winsor 1976).

The first half of the nineteenth century teemed with political theory, and it was inevitable that political thought would penetrate biological thought. In 1848, two German philosophers and revolutionary socialists, Karl Marx and Friedrich Engels, defined and popularized the term Communism (*The Communist Manifesto*). We now turn to the Siphonophora, a small order of marine animals within the Cnidaria, of which the Portuguese-man-o-war *Physalia* is well known (it superficially resembles a jellyfish). Most siphonophores are long, thin, transparent floaters. A siphonophore "individual" appears to be a single organism, but in fact, each specimen is a colonial organism composed of small, individual, connected animals ("zooids"), each with its own unique function, thus forming a community. Some individuals are adapted to be floaters, others to be pulsating swimming bells, others specialize in sensing the environment, others as feeders with long tentacles for catching prey, others for reproducing and so on. The integration with other zooids is so strong that the colony attains the function of one large organism, in a manner somewhat analogous to the construction and function of social organisms (Wilson 1992). Leuckart was the first to show (1851) that every zooid and every organ in the siphonophore is a single, modified individual. The seemingly "individual" is actually a community with division of labour, in which each individual specializes and, accordingly, changes its morphology (Wightman 1953). He suggested the term polymorphism for this phenomenon. Having thus described the siphonophores and believing he had to go beyond the scientific world, he added a political view:

> What the individual acquires becomes the property of the community, and serves the welfare of every member in the community. Likewise, damage to one individual causes damage to all in the community. As in a communist state, there are here no poor next to the rich, no hunger beside surfeit, no lazy member next to the industrious. Each individual contributes his part to the existence and welfare of the whole group, each according to its powers (Leucart 1851).

Politics thus came into natural history. Decades later, with Darwin's theory of natural selection, natural history would come into politics.

Theodor Schwann (1810–1882) came from a town near Düsseldorf. He studied at the universities of Bonn, Würzburg and later Berlin, where he worked for 5 years as an assistant in the museum. He was eventually appointed professor of anatomy at the University of Louvain (Belgium), and at the age of 38, he moved to the University of Liège (also in Belgium), but most of his research was conducted during those 5 years in Berlin. There he studied, by microscope, the structure and function of nerves, muscles and blood vessels. His research included digestive physiology (he discovered pepsin), spontaneous generation (he proved the idea was false), fermentation (he discovered that yeast caused fermentation), muscle contraction and the structure of the nervous system (he identified the sheath of cells surrounding peripheral nerve fibres, "the Schwann' sheath"). In the early decades of the nineteenth century, people believed that fibres, or globules, or fibres consisting of globules, were the basic unit of the living body. In 1839, Schwann (in a treatise translated to English as *Microscopical Researches on the Similarity in the Structure and the*

Growth of Animals and Plants, 1847) demonstrated that both plants and animals consist of cells and their secretions and that cells of the body exist independently, subject to the organism's life. With this generalization, that the cell is the basic unit of the living body, the cell theory was established. The theory was central to modern embryology, for it described embryonic development as a succession of cell divisions. Schwann further elaborated that the continued division of the egg gives rise to tissues, which he classified into five groups:

(a) Tissues in which the cells are independent and separate (blood)
(b) Tissues in which the cells are independent but pressed together (skin)
(c) Tissues in which the cells have well-developed walls that have coalesced to a greater or lesser degree (cartilage, teeth, bones)
(d) Tissues in which the cells are elongated (fibres, tendons, ligaments)
(e) Tissues generated by the coalescence of the walls and cavities of cells (muscles, nerves)

Today, this classification is largely outdated. We classify tissues into four basic types:

(a) Epithelium, which lines and covers surfaces
(b) Connective tissue, which protects, supports and binds together
(c) Muscular tissue, which produces movement
(d) Nervous tissue, which receives stimuli and conducts impulses

The establishment of the cell doctrine closely affected the classification of single-celled creatures. Carl Theodor Ernst von Siebold (1808–1894) of Bavaria was a professor of zoology in Erlangen, then in Freiburg, in Breslau and finally in Munich. Building on Schwann's cell theory, he suggested that the so-called Infusoria are actually a Kingdom of single-celled organisms, completely separate from the Animal and Plant Kingdoms. This called for a definition of the existing kingdoms: in what ways do plants and animals differ from minerals? What is a plant, and what is an animal?

Plants and animals differ from minerals in that they consist of cells, in that they have a definite form and in their ability to reproduce. A plant has neither mouth nor stomach, exhales oxygen and is built of tissues consisting of cellulose. An animal differs from plants in that it receives nutritive matter through a mouth, inhales oxygen and exhales carbon dioxide, consists of tissues such as muscles, nerves and has organs such as vascular and secretory systems; it is also capable of voluntary movement from one place to another. Siebold defined his new Kingdom Protozoa with admirable conciseness: "Animals in which the different systems of organs are not sharply distinguished, and whose irregular form and simple organisation may be reduced to one cell". In the Protozoa, there are no digestive, vascular or secretion systems, no muscle or nervous tissues and no egg laying (Siebold 1848 in English translation, 1854).

The kingdom Protozoa consists of two Classes, one in which the individuals have a definite form and move by "vibration" (= cilia or flagella), to be termed Infusoria; the other, with no definite form and moving by retractable, ramifying projections of the body, the Rhizopoda. The class Infusoria (ten families, over 30 genera) splits into two Orders, those with a mouth and those without one (Stomatoda versus Astoma). The class Rhizopoda also splits into two Orders, those which may have a single shell (amoebas, he termed them Monosomatia) versus those which produce a shell that can have multiple chambers (foraminifers, he termed them Polysomatia). This newly created Kingdom did not include rotifers, as these are multicellular organisms. It also did not include single-celled organisms containing chlorophyll. Over the decade following Siebold, the study of microscopic animals made considerable progress, and by 1856, Carpenter recognized several groups of microscopic animals (Fig. 6.44):

Protozoa: isolated cells, or aggregations of cells, with no differentiation of parts. *Amoeba* and *Actinophrys* with contractile tentacle filaments often radiating from the centre (today placed in the class Raphidomonadea).
Rhizopoda: the body extends into root-like processes. *Difflugia, Arcella.*
Infusoria: a ragbag group of one-celled animals that may undergo encysting by secreting a gelatinous case that hardens, as a coffin for the creature living inside (*Desmidia, Diatoms, Paramecium, Volvox, Chilodon, Coleps, Trichoda, Stentor, Palmella*).
Foraminifera, with three groups: those with calcareous shells, those with siliceous shells and sponges. Forams: *Rosalina, Nodosaria, Dentalina, Cristellaria, Faujasina.*
Rotifera: wheel creatures (*Rotifer, Stephanoceros, Floscularia*).

With the downfall of Cuvier's Radiata, on the eve of Darwin's *Origin* (1859), the classification of the major groups of multicellular animals had reached a point at which it broadly resembles what we accept today. Or had it?

6.2.7 Biodiversity Is Circular

The first half of the nineteenth century witnessed an attempt to arrange the major groups of animals and plants in circles instead of in ladders. William Sharp Macleay (1792–1865), the son of the secretary of the Linnaean Society, was a civil servant, amateur entomologist and member of the Linnaean Society. He served as an attaché in the British embassy in France, and it was here that he met Cuvier and Lamarck. Upon returning to England, he published a two-volume text, *Horae Entomologicae* (1819, 1821), about how chains of affinity within the animal kingdom run back to themselves, forming circles. Five years later, he left for Australia, but a discussion group, "the Zoological Club", was formed within the Linnaean Society, and throughout its 6 years of existence (1823–1829), it served as a forum for papers and debates about Macleay's method of classification. Winsor (1976) offers a very good summary of Macleay's system, and I follow her closely.

Three ideas were essential to Macleay's system of classification, in which he emphasized the difference between "affinity" (our term for homology) and analogy. Firstly, within a group, homologies (his "affinities") lead from one form to the next

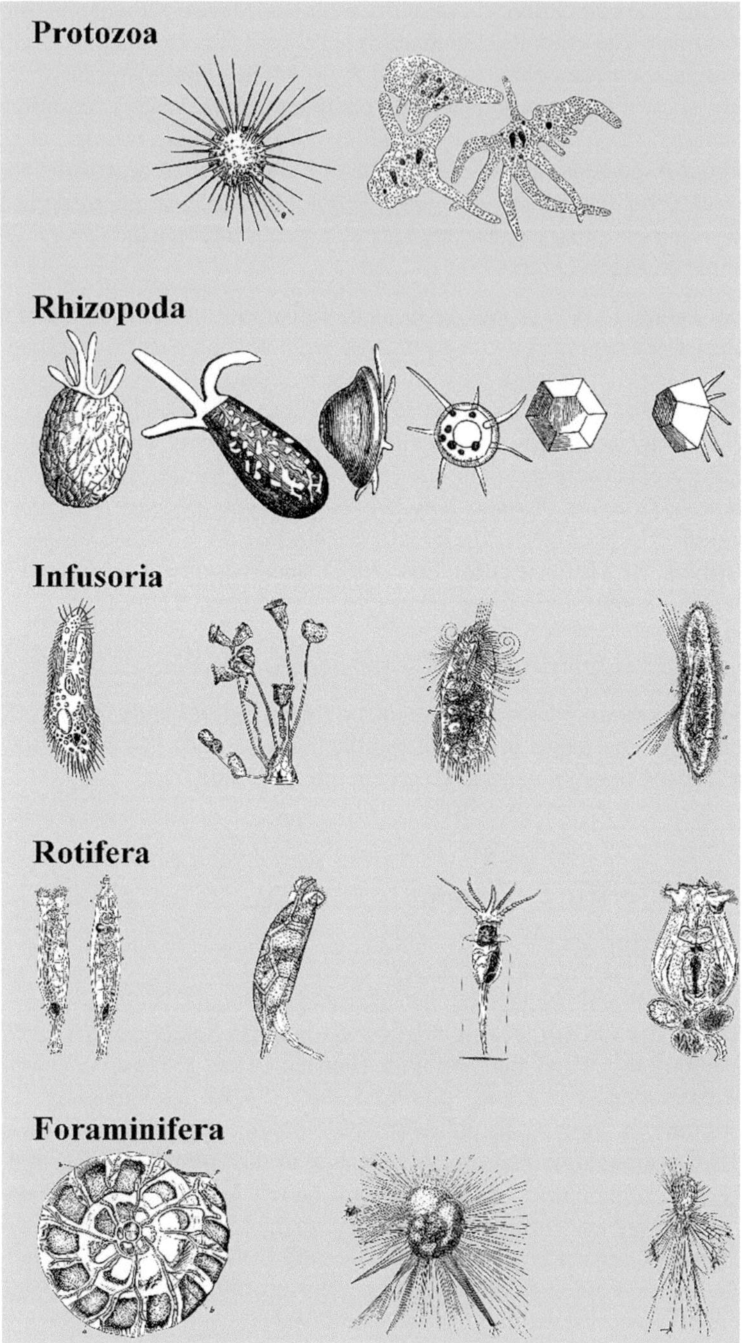

Fig. 6.44 Unicellular diversity. (Assembled from Carpenter 1856)

in a linear fashion. Secondly, different series of homology (different groups) may run parallel to one another. Thirdly, in any natural group, the series of affinities is circular. When following the affinity of A to B, to C, to D, to E, the next link is the affinity of E to A. From the point where you start, if you follow structural modifications and go through all shapes, you return to the starting point. As different series are parallel and correspond member for member, it follows that each series must have the same number of members. This number happens to be five. Viewed from a different point, an affinity is the close tie linking the five members of one circular series; an analogy is the relationship between corresponding members of parallel series. Each of the five-in-a-circle members of any natural group form consists, in turn, of five subgroups that also form circles, which express their relations of affinity and which parallel the five subgroups in other circles. The entire diversity of nature thus falls into circles within circles and circles next to circles, each consisting of five members. This was the Quinarian system (Fig. 6.45).

Macleay's favourite example came from insects (Table 6.3). The major group Mandibulata, with biting mouthparts, consists of five orders (Hymenoptera, Coleoptera, Orthoptera, Neuroptera, Trichoptera) connected, each to the one above and below it, and the last back to the first, by their affinity. Another major group, the Haustellata with sucking mouthparts, consists of five other orders (Diptera, Aptera, Hemiptera, Homoptera, Lepidoptera), also connected, each to the one above and below it, and the last back to the first, by their affinity. In addition, however, the first order among the Mandibulata (Hymenoptera) shares with the first order among the Haustellata (Diptera) the similar features of the larva being without feet and that a pupa encloses the larva during metamorphosis. Likewise, the fifth order of the Mandibulata (Trichoptera) shares with the fifth order among the Haustellata (Lepidoptera) a caterpillar-shaped larva that undergoes complete metamorphosis.

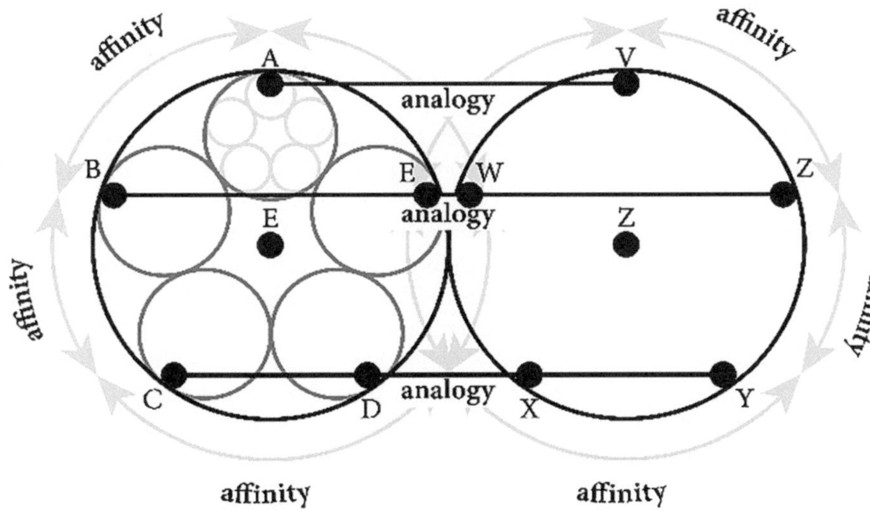

Fig. 6.45 The Quinarian view of order in biodiversity

Table 6.3 Macleay's analogy between two groups of insect orders

Mandibulata		Haustellata
Biting mouthparts		***Sucking mouthparts***
Hymenoptera *Wasps, bees, ants*	Larvae without feet, the body is enclosed in a puparium shell during metamorphosis	Diptera *Flies*
Coleoptera *Beetles*	Larvae varied or unknown, metamorphosis incomplete	Aptera *Fleas, lice*
Orthoptera *Grasshoppers*	Larvae resemble adults, semi-complete metamorphosis	Hemiptera *Bugs*
Neuroptera *Dragonflies*	Larvae with six feet, sub-semi-complete metamorphosis	Homoptera *Cicadas, aphids*
Trichoptera *Caddisflies*	Larvae caterpillar-shaped, complete metamorphosis	Lepidoptera *Butterflies*

Adapted from Winsor (1976: 84)

This similarity of features is an analogy. Macleay argued that such a complex and regular pattern of relations could not have arisen by chance; it exists in nature, as part of the original plan of Creation. Indeed, the fact that these arrangements were orderly and parallel was his reason for believing them to be natural. It was inconceivable that animals would fit a pattern so well if it were not correct, and it was consistent with God's creative intellect (Macleay did not create these orders, nor their grouping into two divisions).

Macleay made no claim to have discovered the proper arrangement for all groups. He did suggest that there were five major Branches in the animal kingdom, not four as Cuvier had claimed. He did so by separating the simplest radiates (polyps and worms, which he named Acrita) as a Branch distinct from higher radiates (sea urchins, starfish). He proposed that these major Branches formed a chain of affinity running from Acrita to molluscs, to vertebrates, to arthropods, to echinoderms, with the last leading back to the first. He likewise proposed five classes for each branch and showed how they, too, might form circles. Thus, he developed a uniform system of classification for the entire plant and animal kingdoms, based upon an idealistic view of the number five. Each of these five major groups consists of five groups that merge into each other. Macleay did not progress beyond the level of a tentative sketch, but others, in his wake, noted that in vertebrates, for example, reptiles merge into birds, which merge into mammals, and these into fishes and these into amphibians, who bring us back to the mammals, to the point where we started. Vertebrates are thus a circular group and therefore a natural one.

One enthusiastic supporter of the Quinarian theory was William Swainson (1789–1855, Fig. 6.46), who soon took the lead in circular diversity. The son of a customs collector at Liverpool, he joined the army but retired at the age of 26, went on an expedition to Brazil, and returned after a year with a rich collection of plants, insects, fishes and bird skins. Swainson elaborated Macleay's general sketch of Quinarian biodiversity by suggesting that in a Class consisting of five Orders, the

Fig. 6.46 Swainson

first Order is the most complete with regard to the general character of the class and therefore its best representative; this is the *typical* Order. The second, *sub-typical* Order has a less complete balance of qualities than the typical one, and it always consists of the most noxious and destructive animals of the circle to which it belongs. This sub-typical Order comprises those that are most powerfully armed, that live by rapine and subsist on the blood of other animals. These are, in Swainson's words, the symbol types of evil. The evil is most noticeable at the level of the Order. In the Class reptiles, these evil sub-types are the serpents, many of which are particularly noxious. In the mammals and birds Classes, we have the carnivores and raptors, the great butchers of terrestrial animals. As we descend the taxonomic hierarchy through tribes, families, genera and species, the extent of evil becomes fainter, but it never vanishes altogether. In the genus *Bos*, the sub-typical group comprises the bison, "wild, revengeful, and shewing an innate detestation of man". In *Equus* it is the zebra, which has the stripes of a tiger and is as remarkable for its wildness as the horse and ass are for their docility and usefulness.

The other three groups he termed *aberrant*, as they depart much more from the typical standard, though the last of the three recovers somewhat and joins the typical group to complete the circle. The first of the aberrant groups is remarkable for dwelling in water and being bulky. Animals in these groups have large heads and weak feet. They partake in the destructive deeds of the adjoining *sub-typical* group, but they prey by mouth alone, not by use of their forelimbs. In the primary division of the animal kingdom, we find this type in the Radiata, all of which are aquatic. In the secondary circle, the vertebrates, it occurs in the fishes; in both of these, feet are totally lacking. In the tertiary circle, these would be the swimmers among birds, *Ichthyosaurus* among reptiles and whales among mammals.

The second aberrant group is that of what he considered to be the meanest and least complete organization, the lowest of all groups. It feeds by suction. Animals in these groups have a small head and mouth, a weak body, lack an active means of offence, lack organs of mastication and (often) lead a parasitic mode of living. In the primary circle of the animals, Swainson included here Macleay's Acrita, or polyps. In the secondary circle, and taking the arthropods, he included the intestinal worms; in the tertiary circle and among the reptiles, he included the tortoises; among mammals, the armadillo, scaly anteater, pig, mouse, jerboa and kangaroo; among birds, the waders; among fishes, the hagfish *Gastrobranchus*; and among insects, the bug, louse and flea.

The third aberrant type comprises the domestic animals. These include the chicken among the birds, the dog among the carnivores, ruminants and others. Leading features of groups belonging to this type are gentleness, familiarity with man and "an approach to human intelligence". They usually have strong limbs and abundant ornaments on the head (tufts, crests, horns or bony excrescences). In the animal kingdom, molluscs belong to this type because of their soft and sluggish character and their general edibility. Among the winged insects, the ants and bees belong here, "the most social, intelligent, and in the latter case, most useful to Man, of all the annulose animals".

Swainson was aware that speculations on representation were still imperfect because of the novelty of the doctrine and of defective knowledge. The doctrine has, however, been so fully proved in birds and traced so clearly in other parts of the animal kingdom that its universal existence is beyond doubt. I offer two examples of Swainson's Quniarian classification, birds and molluscs. In Birds (a class), the major circle consists of five Orders:

Perching birds *Incessores*
Birds of prey *Raptores*
Swimming birds *Natatores*
Waders G*rallatores*
Scrapers *Rasores*

Each group within this major circle of birds consists of its own sub-circle; thus, the perching birds (the typical Order of Birds) consist of five subgroups:

Conirostres (finches, weaverbirds), the most perfect of the perching birds Order, they have a small notch on their bills.

Dentirostres (thrushes, flycatchers, shrikes) have a strong, tooth-like notch, making them similar to the Raptors.

Fissirostres (swifts, swallows, nightjars) have great powers of flight, and their feet are only slightly developed. In this, they are analogous to the swimming birds.
Tenuirostres (hummingbirds, sunbirds) have the small mouths and long, soft bills of the waders.
Scansores (parrots, woodpeckers and toucans) resemble scrapers in their superior intelligence and docility and in having strong limbs.

In a tabular form:

Bird order	*Characters*	Incessores *tribe*
Incessores	Most perfect of their circle; bill notch small	Conirostres
Raptores	Bill notch resembling a tooth	Dentirostres
Natatores	Slightly developed feet; strong flight	Fissirostres
Grallatores	Small mouths; long, soft bills	Tenuirostres
Rasores	Strong feet, short wings; docile and domestic	Scansores

Swainson presented his quinarian arrangement of mollusc diversity in his *Treatise of Malacology* (1840; Fig. 6.47). In molluscs (a phylum), according to existing classifications, the number of major groups reached only four: gastropods, bivalves, nudibranchs (today included in the gastropods) and cephalopods. To achieve five major groups, Swainson brought into the molluscs an outside group, Cuvier's Parenchymata, a group (today obsolete) of worms that included the common liver fluke *Fasciola* (Trematoda) and the flatworm *Planaria* (Platyhelminthes). Swainson believed that parasitic worms led, via Cuvier's (erroneous) *Hectocotylus*, to the floating, largely naked snail *Firola*. Free-living worms crawling on their belly and with a rudimentary head led to nudibranchs, on the other. *Firola* leads to the naked nudibranchs, thus forming a circle.

In arranging the five major groups of molluscs as against those of vertebrates, Swainson (1840) thought he saw a pattern of analogous similarities between the two groups.

Orders of the *Testacea*	Circle of the *Vertebrata*
Gastropods	Quadrupeds
Dithyra [Bivalves]	Birds
Nudibranchs	Reptiles
Parenchymata [planariae, liver flukes]	Amphibians
Cephalopods	Fishes

He noted that "the best modern zoologists", in presenting the molluscs, place the gastropods first and then the bivalves; in presenting the vertebrates, they place the quadrupeds first and then the birds. Now, quadrupeds have teeth-bearing jaws, and they chew their food, whereas birds have toothless bills and swallow their food whole. In a similar manner, gastropods have fleshy lips and corneous jaws and teeth, whereas bivalves have no lips, jaws or teeth, and they too swallow their food whole. The two "most perfect" groups of molluscs thus have the same mode of feeding as

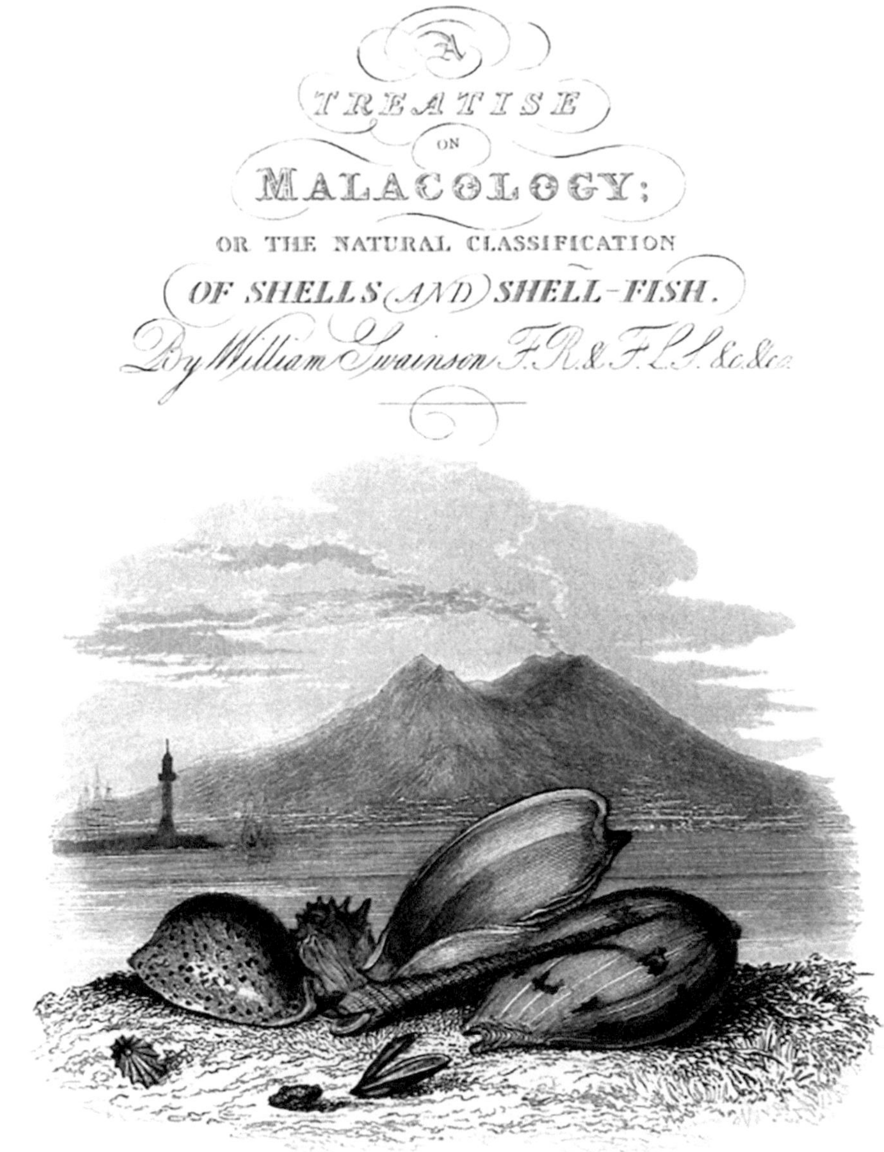

Fig. 6.47 Swainson, 1840. *A Treatise on Malacology.* (Courtesy Molllusc Collection, Hebrew University)

the two "most perfect" groups of vertebrates. Swainson thought he could see further resemblance between birds and bivalves: in birds, two broad and expansive wings envelop the sides of the body, and in bivalves, likewise, ample folds of the mantle envelop the sides of the body (Fig. 6.48).

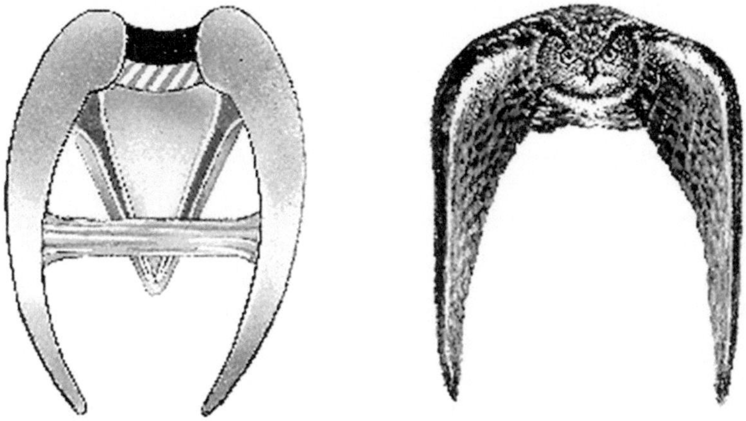

Fig. 6.48 Analogy between bivalve valves and bird wings

Another analogy that he noted was that nudibranchs and reptiles (did he mean snakes?) both crawl on their bellies and lack any limbs. Somewhat swept away at this point, he wrote that "The tritons, in fact, have no real feet". He went on to claim that the amphibians are the most imperfect of vertebrates, and likewise, the Parenchymata are the most simply constructed molluscs. Furthermore, cuttlefish (cephalopods) resemble fish in containing large individuals, in swimming with fins and in having very similar eyes. Beyond resembling this group of vertebrates, cuttlefish also resemble a certain group of arthropods, the barnacles, in that both the mouth is surrounded by long, slender arms to catch prey and both contain animals whose softer parts are protected by shells. Nevertheless, barnacles have nothing to do with molluscs. Swainson presented many such comparisons within the molluscs in search of resemblances.

Moving down to an inner five-circles-in-a-circle within the molluscs, Swainson established a new Order (Phytophaga) of grazing snails in the gastropods, consisting of five families, one terrestrial and the others marine (he erred in including the carnivorous moon-snails Naticidae).

1. Helicidae
2. Turbidae
3. Naticidae
4. Trochidae
5. Haliotidae

The first of these five grazing families (Helicidae) consists of those that breathe with a lung, and Swainson claimed that it consists of five subfamilies (Limacinae, Lucerninae, Helicinae, Achatinae, Limnacinae). One of these subfamilies (Achatinae) consists mainly of snails with spiral shells that, he claimed, consist of five genera, each with five subgenera:

1. Clausilia: *Pupella, Eruca, Balia, Clausilia, Macrodontes*
2. Bulimus: *Auricula, Bulimus, Bulimulus, Leptospira, Goniostoma*

3. Achatina: *Macrospira, Cochlycopa, Achatina, Achatinella*
4. Cyclostoma: *Cyclophora, Cyclostoma, Cyclotus, Megalomastoma*
5. Helicina: *Helicina, Pachytoma, Oligyra, Trochatella*

In the genera *Achatina* and *Helicina*, he explained the lack of a fifth subgenus by gaps in taxonomic knowledge and by not having yet elaborated his theory in full detail.

Among marine snails, Swainson even illustrated affinities (=homologies) versus analogues (Fig. 6.49).

The Quinarian system was a prominent feature of the intellectual landscape from 1820 to 1840 and possibly later as well, and several of the best naturalists used it, more or less enthusiastically (the entomologist Rev. William Kirby, the mammologist G.R. Waterhouse of the British Museum, the marine biologist Edward Forbes). Chambers (1844; we shall meet him further on) even applied it to dog breeds, with the bulldog and mastiff representing the ferocious sub-typical group, the greyhound with its speed and long muzzle the suctorial group and the shepherd dog and spaniel representing gentle and serviceable character. The system had the great advantage of predictability; it could predict what relations of affinity would turn up among material not yet classified by it, and it could determine how many forms remained to be discovered in imperfectly known groups. The regularity in the structure of the Quinarian classification was, however, totally irreconcilable with the idea of change in animal form over time, in response to changes in the environment. Change would break down the neat five-in-a-circle design into irregular, arbitrary patterns.

Swainson failed, of course, in his search for an order in animal diversity that would reflect a strict numerical pattern. Three decades later, however, such an order emerged in another field of science. Dmitri Mendeleev, a Russian chemist, presented a tabular display of the chemical elements then known, arranged by atomic number. He listed the elements in rows and columns by order of atomic weight, starting a new row or column when the characteristics of the elements began to repeat. The table consisted of seven rows, which he called periods. He left gaps in the table when it seemed to him that the corresponding element had not yet been discovered, and he predicted some properties of those missing elements that would eventually fill the gaps within his periodic table of elements. Most of his forecasts

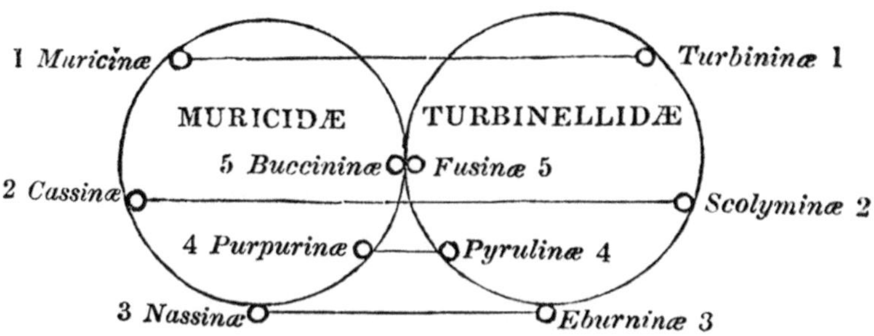

Fig. 6.49 Homology (within circles) versus analogy (connecting lines) in two sea snail families. (From Swainson 1840)

proved to be correct. A numerical order in diversity that fails in one science may thus be very relevant in another.

Macleay's and Swainson's ill-conceived quinarianism represents a sincere endeavour to reconcile the seemingly chaotic diversity of nature with the then prevailing conviction that there has to be some "higher" order of nature. Other naturalists believed that such a high order does exist but that it consists of a Ladder of Life, arranged by modes of reproduction. William Clark, of Bath, divided molluscs into six groups by their generation (1855). Brachiopods and bivalves are on the lowest two steps of the ladder, as lowly hermaphrodites in which each individual fertilizes itself, *Hermaphrodita sine concubita*. Next comes a group of sea-tusks, chitons and limpets, hermaphrodites in which each individual fertilizes another individual, but it does so in discretion, not in groups *Hermaphrodita sinne congressu.* Then come the sea slugs and relatives, together with land snails, in which each individual fertilizes another individual in a social reality (sex orgy), *Hermaphrodita congressu*. The remaining two groups, marine gastropods and cephalopods, are highest on the ladder in that the sexes are separate (bisexual). It is noteworthy that in five of the seven groups in which Clark claimed that hermaphrodite reproduction is dominant, this is simply not true.

6.2.8 The Space Dimension: Biogeography

Charles Lyell (1795–1885, Fig. 6.50) was the son of a prosperous Scottish landowner who was an enthusiastic botanist and familiarized his son with the study of nature; the young lad did indeed develop an avid interest in insects. At Oxford, he attended geological lectures but took up law as a profession after graduation. At the age of 33, he left law and turned to fossil and biogeographic aspects of natural history (then loosely termed geology) as a full-time profession. His respected *Principles of Geology* (1831–1832) went through many editions. He was highly religious, with a strong belief in the special status of human reason. Lyell was knighted at the age of 53, made a baronet (a hereditary honour) at 69 and when he died, he was buried in Westminster Abbey.

Lyell studied shells of the Caenozoic era (Age of Mammals). By the study of Caenozoic shells, he divided the entire Age into four periods based on the proportions of surviving species that tertiary shells exhibit.

Of the Eocene shell species, 3.5% still exist.
Of the Miocene shell species, 18% still exist.
Of the Old Pliocene shell species, 35–50% still exist.
Of the New Pliocene [Pleistocene] shell species, 95% still exist.

Lyell believed, like Hutton before him, that the same natural processes, which are still in operation today, had shaped the surface of Earth at similar intensities. In this, he differed from Maimonides, who, in mediaeval times, had argued that from the present we cannot automatically conclude about the past, when Life was

Fig. 6.50 Lyell

created, and from Cuvier, who believed in global catastrophes. In Lyell's opinion, and in line with that of Lamarck, the steady accumulation of minute changes over enormously long spans of time was the cause of geological change. Regular physical forces, by their very gradual operation over an immensely long time, slowly erode mountains, gently deposit layers of sediment upon the ocean floor and occasionally throw up small volcanoes. Such slight but continuous geological forces could, he believed, eventually cause considerable faunal changes. There is no need to imagine catastrophic events as Cuvier had in seeking to explain changes in the history of life.

We recall that Lamarck had believed that as a species changes its habitat, climate or manner of living, it gradually changes its structure and behaviour, and thereby, through many intermediate forms, it becomes a new species. Lyell asked whether species do indeed change slowly under the pressure of the environment and whether the fossil record presents evidence of such changes in the animal world. If indeed species do change in this manner, then the existence of a Creator would become superfluous. He accepted that environmental pressures might indeed modify the proportions of certain **pre-existing** structures but claimed that such pressures do not give rise to any completely **new** ones. A changing climate may have induced elongation of the giraffe's neck, but it had not formed any giraffe (nor any neck). Furthermore, in many recent cases, changes in the environment do not result in a change in shape. The cat has remained a cat and the ox an ox, despite transportation throughout the globe by man into so many different environments. Lamarck had not discovered any source of novelty, of creation.

Another of Lamarck's theories was the successive development of organic creation, leading to a Ladder of Life. We recall Lamarck's claim that Life had started with simple creatures and that it gradually increased in complexity over time by adding new organ systems, which carry out new functions. This gradual increase has resulted in a ladder of all living entities, with plants leading to primitive animals, these to more advanced ones and so on, with man at the highest rung. As nature makes no jumps, groups that may seem to us as distinct actually merge into one another via ever so many interim animals.

In response to this theory of Lamarck, Lyell claimed that palaeontological evidence does not support any Ladder of Life (*Principles of Geology* 1: 167). There is no succession from plants to animals in the fossil record and no progression from primitive to advanced animals. Representatives of all major groups have, he claimed, been present at all times, somewhere on Earth. In the earliest fossiliferous deposits, fish, reptiles and snails all occur together with plants (reptiles, in Lyell's days, also included amphibians, which were present in the Palaeozoic). Birds and mammals are admittedly absent from early deposits, but this is due, he claimed, to their more delicate bones, causing difficulties in their preservation and fossilization. Change and extinction are indeed omnipresent, but the large animal groups have always been there. The recent appearance of man on Earth might perhaps seem to support the Ladder of Life theory, on which the most supreme creature made its appearance last. In Lyell's opinion, however, man is wholly set apart from the remainder of creation by the possession of reason ("the superiority of man depends not on those faculties and attributes which he shares in common with the inferior animals, by which he is distinguished from them"). By dissociating man from other animals, Lyell rendered the supposed progressive tendency of nature meaningless. One might perhaps argue that hybridization, by producing intermediates, demonstrates the instability of species. To this argument, Lyell responded that fertile hybrids are not the offspring of interspecific crosses; they are merely the offspring of races of the same species. Lyell was thus confident that his analysis had removed all foundation from the arguments that one species may change into another (Coleman 1962). The absence of an evolutionary mechanism, denial of the successive appearance of organisms and the inability of hybrids to generate new organs all supported his view that a species was a discrete and unvarying unit. He concluded that each species "was endowed, at the time of its Creation, with the attributes and organisation by which it is now distinguished".

Lyell then moved on to problems of distribution to clarify to what extent the endless changes of the non-living world influence the animal world. Knowledge of the distribution of animals had increased in his days, and therefore, Linnaeus' idea of one single distribution centre for the entire animal and plant kingdom was by now outdated. Why, then, do species occur only where they are? One might have expected a given plant or animal to occur in all places in which there are environmental conditions suitable for their survival, namely in all places with the same climatic conditions (as expressed by the same latitude). In reality, however, different species may occur in the same climatic conditions (same latitudes), such as in the tropics of the Old versus those of the New Worlds. To this question, of why animals are only where they are, Lyell replied that the present distribution of species on Earth is the result of separate creation events: God had created each species separately within a given area. When creating the first pair of individuals of each species, He had endowed it with all the features that enable its existence in that particular area. Consequently, many species are today limited to regions separated from the rest of the globe by natural barriers. Each species has a sharply circumscribed range that it must occupy. One species within a genus may occur in China, another near the Black Sea, a third along the Mediterranean coasts and yet another on the Siberian uplands, all with no intermediate forms, each having its own special adaptations to its own specific environment. Different but similar species are usually isolated from one another by an intervening sea, mountain chain, broad river, desert or some other impassable boundary. Some birds "are so local, that in the same archipelago, a single island frequently contains a species found in no other spot on the whole earth" (there is no reference here to the Galapagos Islands, visited 2 years later by Darwin). Local distribution is the basic order of nature. The present distribution and number of species on Earth are the result of special creations, each species in a specific time and at a small specific place, where it survives for an appointed period.

In the following decades, Andrew Murray also suggested a multiple origin of mammals (1866). "May they not have sprung from different stocks—the bat from the pterodactyls; the duckbill, or ornithorhynchus, from the birds; the whale from the ichthyosaurus, and the general mass of mammals from terrestrial reptiles?"

Creation is one aspect of the distribution of life on Earth; dispersal is another, much more complicated one. In the beginning, Lyell argued, each species did indeed populate its own, well-defined territory. As environments changed over time, however, so too did animal distributions, with one species displacing another. Often, an entire fauna would displace another one. Such displacements occur now as well. In slowly flowing rivers, suspended matter settles and buries bottom-dwelling animals beneath it but at the same time creates a platform for other shallow-water creatures that require more amounts of light for their survival. In the sea, waves pound the rocks and gradually crush them into stones, which will become the habitat of shore-dwelling animals. As a pond gradually fills with sediment and becomes a swamp, its pond fauna goes extinct and swamp fauna replaces it; in the next shift, from swamp to meadow, the swamp fauna becomes extinct and meadow species replace it. Thus, changing environments might cause changes in animal distributions. Natural environmental forces constantly destroy and demolish, but they also build and create new environments, into which surviving creatures disperse.

Because of the close interaction of species with their environment, a slight change in the environment may provoke a significant change in the range of a species. Imagine, Lyell suggested, that a slight earthquake would lower the flatlands between the Sea of Azov and the Caspian Sea, even just a little. That would be enough for the Caspian Sea to flood the sunken region and cause a "revolution" in the flora and fauna of the area. Similarly, a slight rise of the shelf of the Gibraltar Strait would revolutionize the fauna of the Mediterranean basin. (Lyell deliberately retained the term "revolution" to emphasize that drastic faunal changes may be due to slight physical causes, and there is no need to suggest revolutionary environmental catastrophes.) Following such slight physical changes, some species might go locally extinct while others, from adjoining areas, might invade the disturbed areas and replace the annihilated groups, resulting in new distribution patterns. Lyell thus did not accept Cuvier's opinion that species become extinct through sudden violent catastrophes. Because of the close interactions between a species and its environment, a slight change in the environment may cause a significant change in the distribution of many animals, and the fossil record would express this faunal change as an apparent revolution in the fauna. Slight geological changes may have considerable influence on the redistribution of species.

Beyond physical forces, biological forces may also change distribution patterns. Competition between species for resources may cause one species to displace another, leading to great alterations in animal distribution. Biogeography is, in general, a dynamic process, and changes in distributions result from the motivation to wander and migrate, a motivation stemming from food shortage, climatic changes or high population densities, following changes in the non-biotic environment. (Lyell's concept of competitive struggle was of a process that occurred between species and would lead to local extinctions. It was unlike the within-species competition, which Darwin would later associate with natural selection, as we shall see in a subsequent chapter. Lyell's competition is a mechanism that leads only to extinction or to a reduction in the distribution of a species. It cannot create new species. The creation of a species, according to Lyell, is the privilege of God.)

Lyell was also aware of the fact that minor-scale invasions, for example, of just one new species into a region, might influence an entire local flora or fauna. Polar bears sometimes float on ice, from Greenland to Iceland. If a population of polar bears were to establish itself in Iceland, a catastrophe would occur. The populations of deer, fox, seals and birds that bear prey on would dwindle rapidly. The plants on which the deer fed would flourish and provide more food for insects and snails, which would multiply, and supply food for other insects and birds, the numbers of which would rise. The decrease of seals would result in an increase in the numbers of certain fish. Waterfowl, whose eggs and young foxes had once been devoured, would now increase in numbers, and this would cause a drop in the numbers of those fish on which waterfowl feed. "Thus, the numerical proportions of a great number of inhabitants, both of the land and sea, might be permanently altered by the settling of one new species in the region; and the changes caused indirectly might ramify through all classes of the living creatures, and be almost endless". Minor events can determine nature's course, not Cuvierian catastrophes. Lyell claimed that man also plays an important role in changing the natural environment. Man has exterminated the dodo and has transformed many natural regions into agricultural regions, thereby reducing the number of wild animals. Furthermore, within less than 300 years, the handful of horses and cattle brought to the Pampas of South America grew to over 15 million, and the local fauna diminished accordingly.

The ornithologist Philip Lutley Sclater furthered Lyell's idea of independent centres of creation. He argued that different regions of the flora and fauna of the world have been the result of distinct creations. Therefore, he suggested, naturalists should determine how many centres of creation really exist. He found that the world's birds fall into six major regions ("divisions"): Australian, Neotropical, Ethiopian, Oriental, Nearctic and Palaearctic. Three decades later, he found that terrestrial mammals also fit into these six regions. Each region was well defined, both by the mammals present in it and those absent from it (Table 6.4). It is indeed remarkable that in spite of his erroneous basic theory of separate centres of creation, Sclater reached a pattern of biogeographic regions that is correct to such an extent that we still use it today in modern biogeography. This suggests that the taxonomy of birds and mammals was by then sufficiently sound for the formulation of broad distributional patterns. It was, of course, not as perfect as it is today; thus, Sclater noted the abundance of porcupines in the Neotropical region, whereas we today place the New World porcupines in a separate rodent family. Three decades later, Wallace adopted Sclater's faunal regions, and they are now in general use in our approach to the distributional patterns of the world's fauna. Today, some researchers united Sclater's Nearctic and Palaearctic into one region, the Holarctic.

Some of Sclater's ideas in biogeography are no longer accepted. In studying "lemurs" (which included related primate groups), he was puzzled by the presence of their fossils in both Madagascar and India, but not in Africa. Unaware of the possibility of continental drift, he speculated that Madagascar and India had once been part of a now sunken continent that had once occupied much of the Indian Ocean. This continent was later broken up into islands, of which some merged with Africa, some with Asia and some remained as islands. Madagascar has living relics of this

Table 6.4 The six biogeographic regions of mammals

Region	Characteristics
Australian	Absence of nearly all Eutherian mammals, but for a few rodents and bats. Presence of six families of marsupials, with 100 species; and the only known monotremes.
Neotropical	Absence of insectivores, civets, elephants and frugivorous bats; no ruminants except deer and lamas. Presence of two peculiar families of monkeys, of vampire bats; sloths, anteaters and armadillo; one family of marsupials (opossums), tapirs, abundance of the porcupine family.
Ethiopian	Absence of bears and deer. Presence of chimpanzee and other monkeys, lion, African elephant, hyrax, rhinoceros, hippopotamus, warthog, numerous antelopes, giraffe, pangolin, ant bear, general richness of highly organized ungulates.
Oriental	Presence of orangutan, gibbon, peculiar monkeys, flying lemur, tiger and other cats, Indian elephant, rhinoceros, Malayan tapir, pangolin (*Manis*).
Nearctic	General mammal fauna, very like that of the Palaearctic region (bears, beavers, sheep, deer), but mixed with endemic forms (prongbuck, pouched mice) and intruders from the south (raccoon, opossum).
Palaearctic	Absence of monkeys, lemurs, frugivorous bats, elephant, hyrax. Presence of ounce (snow leopard), lynx, wolf, fox, bear, weasel, marmot, beaver, pika, ungulates sheep, deer, chamois, musk deer.

Adapted from Sclater and Sclater (1899)

great continent, for which he proposed the name Lemuria (some decades before that, Geoffroy, also looking at the relationship between animals in India and Madagascar, had suggested a southern continent). Biologists of the nineteenth century frequently postulated the existence of submerged land bridges and continents to account for populations of terrestrial species now separated by barriers of water. Today, we know that other sunken continents, such as Zealandia in the Pacific, Mauritia and the Kerguelen Plateau in the Indian Ocean, do exist.

Carl Bergmann (1814–1865) was a German professor of anatomy and physiology at the University of Rostock who sought to explain how heat production and heat dissipation in animals vary with size. He also sought to reveal whether there is a minimal and/or a maximal possible body size for warm-blooded animals. All animals produce heat, yet the difference between warm- and cold-blooded animals is not, he claimed (1847), in the total amount of heat produced. It is in that warm-blooded animals maintain a constant internal temperature, unlike cold-blooded animals. This difference in warm- and cold-blooded animals' ability to maintain a constant temperature inspired Bergmann to propose changing the terms cold- and warm-blooded to homoeotherm for an animal that keeps its body at an almost constant temperature and poikilotherm for one that does not.

Noting that in general, the more the production of heat is increased and the dissipation of heat is limited, the smaller the animal, Bergmann wondered whether there are body size limits because of the extremes in the relationship between heat production and heat dissipation. Small animals need less muscular activity for their movements, and they would therefore find it difficult to increase their heat production. With a reduction of body size, the volume that produces heat decreases faster than the surface area that dissipates heat, and this leads to a lower size limit for homoeothermic (warm-blooded) animals. Hummingbirds are close to this limit, and therefore, for the animals smaller than hummingbirds, homoeothermy is not possible. He further noted that these smallest of homoeotherms live in hot tropical America (most species) and that, in addition, they are extremely active, leading to an extraordinary amount of muscular activity. Many small birds have extra ornamental feathers that increase load and drag, and in turn, a proportionally larger muscle mass is needed.

Terrestrial homeotherms do not reach the largest potential body size with respect to heat dissipation. Since an increase in muscular activity leads to increased heat production, conditions should be prevailing in large animals that reduce muscular activity. Bergmann noted that large mammals such as elephant, rhinoceros and hippopotamus live in a hot climate. This, he thought, indicates that the largest possible sizes are not realized in the cold climate of the far north, where some small animals live. The large mammals of the tropics spend ample time in the water to control their temperature, have hardly any fur, and tend to be active at night. In the tropics, animals thus reach or even exceed their "thermic" size limits. We must also consider body shape. The round form of large mammals results in a small surface area in relation to volume, meaning that they do not have ideal shapes for heat dissipation because of their mechanical constraints. In contrast, bats have a large surface area/volume ratio because of their extended wing surface. This is possible because they hibernate, and their wings have low blood flow and therefore probably a low temperature.

Bergmann was aware that though the surface area/volume ratio is important for maintaining an internal temperature, other factors related to heat production (say, the quality and quantity of food) and dissipation of heat are also important (say, skin evaporation, fur structure, mode of life, the environment). Homoeotherms of different sizes live in the same climate, and this is so because of differences in the many factors influencing the production and dissipation of heat. Animals differ in morphology and behaviour, and this will consequently mask the predicted "thermal" pattern. These differences become smaller the more similar different animals are. If two species differ only in size (say, two in the same genus), the smaller would need a warmer and the larger a colder climate, proportional to their size difference. Birds in Iceland inspired Bergmann as an expression of the rule: in raptors and in ravens, the smallest species leave Iceland in winter, whereas the largest stay. Among birds, indeed, as there are more similar species within a genus than in mammals, Bergmann examined the geographical distribution of body sizes in relation to climate. He found data in the literature for over 300 species of 86 genera of birds. For each, he described whether the species distribution in relation to its size is as expected by his "thermal rule", and, in cases of deviations from the expected pattern (in that smaller species in a genus live farther north than larger species), he explained why this may be the case. Species listed as belonging to the same genus may not be as closely related to each other as previously thought; or there may be some influence of altitude; or the proportionally longer wings of some species may mask real size; or we must consider the quantity of food consumed or the quality of plumage. The only species for which Bergmann mentioned intraspecific size variation was the white-tailed sea eagle, *Haliaeetus albicilla*, in which the larger individuals live mainly in the north and the smaller ones in the south.

Bergmann also studied intraspecific comparisons in races of domesticated animals. Focusing on the development of feathers and fur, which influence energy budgets, he thought that young chicks lose their down feathers quickly in tropical America and that sheep in this region lose their wool and develop goat-like hair. In dogs, which often live with humans, the differences in size are especially pronounced. However, horses in Iceland and the Shetlands, though having well-developed fur that could aid in heat retention, have smaller races in the north. As a reason for this size reduction in the northern horses, he mentioned the "miserable life" and lack of care when breeding in Iceland and that horses there must look after themselves during the winter. He then explained how young animals live in the same climate as larger ones: young animals are often not yet homoeothermic; they acquire heat from their parents in warm nests and dens. They also grow up during warmer times of the year and are voracious, which aids in heat production.

Since large animals can maintain a constant body temperature with less effort, it follows that poikilotherms (cold-blooded animals) are often much smaller than homoeotherms (especially when animals that live in the same climate and medium are compared). Also, small poikilothermic animals cannot be homoeothermic because of their body size. Bergmann thus sought to establish a biophysical rule that animals, to raise their body temperature above that of the environment, produce less heat the larger they are.

Bergmann did not study latitudinal size clines. Indeed, he never formulated an explicit rule. He assumed that the surface area of an animal is a measure of heat dissipation and that an animal's volume is a measure of its heat production. As body size increases, an animal's surface area increases less than its volume. When everything but size is equal, the smaller animals should live in warmer areas. From his text and from a distance of nearly two centuries, Salewski and Watt (2017) formulated Bergmann's rule as he would have seen it: within species and among closely related species of homoeotherms, animals often achieve a larger size in colder climates than in warmer ones, and this is linked to the temperature budget of these animals.

6.2.9 Evolution, Yes Indeed!

Owen ardently believed in evolution since, in the fossil record, primitive groups appear before advanced ones; since some organs have degenerated in some animals to rudiments; and since we can follow evolutionary change from fossil to modern groups. Evolutionary change occurs by abrupt modifications, not by accumulating minute changes. Extinction is a natural phenomenon, with much evidence in the fossil record. Animals become extinct because of catastrophic events or competition. Homology among animals calls for an underlying common structural plan, a basic "archetype" possessing the most basic, generalized structural arrangement common to all members of a group. This was the body plan of the primitive ancestor, from which later descendants diversified by modifications and adaptive divergence. Diversity is also built up by the alteration of proportions after birth from infancy to maturity, not only by divergence during late embryonic stages. As animals grow, different structures begin and end their growth at different times, resulting in an adult differing from the juvenile; such are the changes in the brain structure of primates. Accordingly, Owen classified mammals into four subclasses, corresponding to four levels of brain development. By their brain, humans are a distinct subclass, which did not evolve from apes.

Richard Owen (1804–1892, Fig. 6.51), the son of a merchant in Lancaster, was apprenticed to a local surgeon as a youth and from there he went on to study medicine. He became assistant conservator, went to Paris, where he attended the 1831 debates between Cuvier and Geoffroy and, at the age of 32, was appointed professor and conservator of the museum of the Royal College of Surgeons in London. At the age of 54, he became superintendent of the Natural History department of the British Museum and worked tirelessly for his great dream of an independent natural history museum that would be a British counterpart to the *Muséum National d'Histoire Naturelle* of Paris. Eventually, the natural history collections of the British Museum indeed moved to a new building in South Kensington, and the "British Museum (Natural History)" was established. It opened its doors in 1881, and today it is the

Fig. 6.51 Owen

Natural History Museum. It houses some 70 million objects from every realm of life and every corner on Earth.

To appreciate the importance of this step of Owen, we must put it in historic perspective (and I am closely following Barber 1980). An Act of Parliament created the British Museum in 1753, and its Natural History Department consisted of collections bequeathed by Sir Hans Sloane (1660–1753), which were important and extensive enough to ensure the Department's respectability for the remainder of the eighteenth century. By the early decades of the nineteenth century, however, most of Sloane's collections had disappeared. Some of the items had simply crumbled away from a lack of any effort to preserve them, but the museum also sold off many specimens. Furthermore, the museum's Keeper of Zoology in the years 1813–1822 despised the taxidermy of Sloane's age to such an extent that he made periodical bonfires of Sloane's specimens (he called these deeds "cremations") and, by 1823, of the 5500 insects listed in Sloane's catalogue, not a vestige remained. The inability of the Natural History Department even to conserve its specimens, let alone display them, was so notorious that the Treasury frequently refused to entrust it with specimens that had been collected at the government's expense. Purchasing was also impossible since the Natural History Department's acquisitions allowance in the first half of the century was generally under 1000 pounds per year, and this had to cover geology, mineralogy and palaeontology, as well as botany and zoology. Entomology, for instance, had just 10 pounds a year to spend on new acquisitions (about 1000 pounds in today's value). Many important private collections went to other buyers because the British Museum could not afford them. In 1856, the director of the museum declared that he "would at any time, give three mammoths for an Aldus". Another problem was a lack of staff. Until 1856, there were only three Keepers (of Zoology, Botany and Geology-Mineralogy) with a handful of assistants, and appointments went by patronage rather than by merit. Consequently, for the first three decades of the century, the natural history collections were unpreserved, uncatalogued, unlabelled, unarranged and therefore unused. When Owen became Head of the Natural History Department (1856), besides giving it the prestige that it so badly needed, he began a campaign to move the Natural History Department to a separate building. Many scientists opposed the move, among them Darwin and Huxley, but by 1869 Parliament formally accepted Owen's idea of a separation of the natural history collections.

Owen was one of the greatest British anatomists. He described the anatomy of molluscs, sponges, worms, brachiopods, lungfish, the kiwi, dodo, great auk, monotremes, marsupials, orangutans and gorillas, along with many extinct vertebrates. He became known as "the British Cuvier" when, presented with a bone fragment from New Zealand, he noted that it resembled an ostrich bone and stated (1839) that giant flightless birds had once lived in New Zealand. This was the extinct moa *Dinornis*, the tallest bird ever (3.6 m). He also examined reptile-like fossil bones from southern England and concluded that they were not of lizards but of a distinct suborder of Reptiles, for which he coined the name Dinosauria (Padian 1997). He described an unusually long-tailed, fossil, toothed bird from the Jurassic Period in Germany, *Archaeopteryx* (1863; Fig. 6.52), finding missing links in the horse dynasty (1866: 789), thereby proving that animal forms may change over time. During 60-odd years of academic activity, he wrote approximately 800 papers, books and monographs, of which the two major ones were *On the Classification and Geographical Distribution of the Mammalia* (1859) and *Anatomy of Vertebrates* (1866).

Owen also had a gift for reaching the highest levels of society. When Sir Richard Owen, the zoologist, went to stay with Sir Robert Peel, the prime minister, he took his microscope along with him, and all the guests gathered after lunch to examine the remains of the joint they had just eaten. By mid-century, many naturalists were thinking of Owen as the person most likely to carry comparative anatomy to new heights and to synthesize existing knowledge into a unified system. On a personal level, naturalists described Owen as a malicious, dishonest and hateful individual, driven by arrogance and jealousy, with an almost fanatical egoism and a callous delight in savaging his critics and a treacherous liar with a streak of sadism. The only other English villain I can think of who enjoyed so many such compliments was King Richard the Third.

In his scientific thought, Owen was deeply convinced that animals undergo evolutionary change. As evidence for this, he noted that in the fossil record, primitive groups appear before more advanced ones. Cold-blooded vertebrate groups (fishes and reptiles) already appeared in the Palaeozoic, whereas the more advanced, warm-blooded groups (birds and mammals) appeared much later, in the Mesozoic; and the

Fig. 6.52 *Archaeopterix*, as reconstructed by Owen (1866)

most advanced, man, the reasoning, appeared only by the end of the Caenozoic. Owen was convinced of evolutionary change from comparing modern to fossil fauna as well. In modern horses, the digit on each side of the one supporting the hoof is a concealed rudiment. Earlier, in the Miocene horse *Hipparion,* these digits reached their full length and supported small hooves, like the spurious hooves of an ox; and earlier still, in the Eocene horse *Palaeotherium*, these hooves were so developed as to form a functional three-toed foot. Moreover, certain teeth, which are rudimentary in the modern horse, were persistent and functional in *Palaeotherium* (Fig. 6.53). Evidence for evolutionary change also emerges, he argued, from comparative anatomy. Organs that occur throughout most of the animal kingdom have degenerated over time to mere rudiments in certain animals, such as the loss of eyes in inhabitants of dark caves. Owen was thus an ardent believer in evolution. He differed from Darwin, his contemporary (and rival), mainly in that **he did not believe in the evidence that natural selection was the driving force of evolution**. Natural selection was, in his opinion, merely a hypothesis of what might perhaps have happened. It lacked observations of what did indeed happen (1866). Furthermore, Owen believed that evolutionary change occurs by abrupt, noticeable modifications, not by the endless accumulation of minute changes. A few years after Darwin published his *Origin* (1859), Owen published *Anatomy of Vertebrates* (1866–1868), which aimed to comprehend the unity that underlies the diversity of vertebrate structures (Camardi 2001; Amundson 2007).

Owen (1859) was fully aware that extinction was a natural phenomenon, but he asked whether extinction follows a specific pattern. He claimed that large-sized mammals are more prone to extinction than small ones because, if a dry season gradually lengthens, the large mammals will suffer from the drought sooner than the small ones. If a climatic change also affects the quantity of vegetable food, the bulky

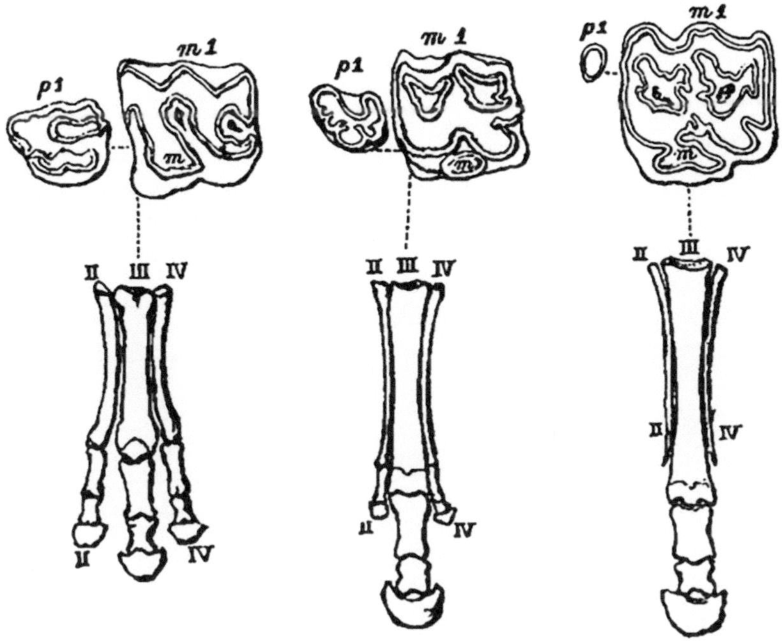

Fig. 6.53 Evolution of fossil horses, as illustrated by Owen

herbivore will be the first to feel the effects of a deficient diet. Finally, if new enemies appear, the large and conspicuous mammals will fall prey while the smaller kinds will conceal themselves and escape. Small mammals are, moreover, more prolific than large ones, which produce only a few offspring each time.

Is there evidence of species extinctions today? There is abundant evidence, but to discuss fossils as well, we need extinctions not caused by man. Owen mentioned the great auk, *Alca impennis*, a flightless bird up to 70 cm long, with small wings well suited to paddling underwater. It inhabited the Atlantic coast of Europe, North America and Iceland and fed on fish. Owen believed that man did not hunt it down selectively, compared to other aquatic birds. In Owen's days, it used to breed on just three rocky islets, of which one sank during a volcanic disturbance in 1830, resulting in its decline towards extinction, evidence that extinctions occur today as well, without human involvement.

Looking from the modern era to the past, Owen noted that in the sea, the successful whales displaced the marine reptiles, and on land, the extinction of Mesozoic mammals was followed by the more numerous, varied and highly organized forms of the class during the Tertiary Period. Animals may thus become extinct because of catastrophic events or because of competition.

In trying to understand the basic structure of the vertebrate body, Owen sought, first, to find a distinct difference between analogy and homology. He defined analogy as "a part or organ in one animal which has the same function as another part or

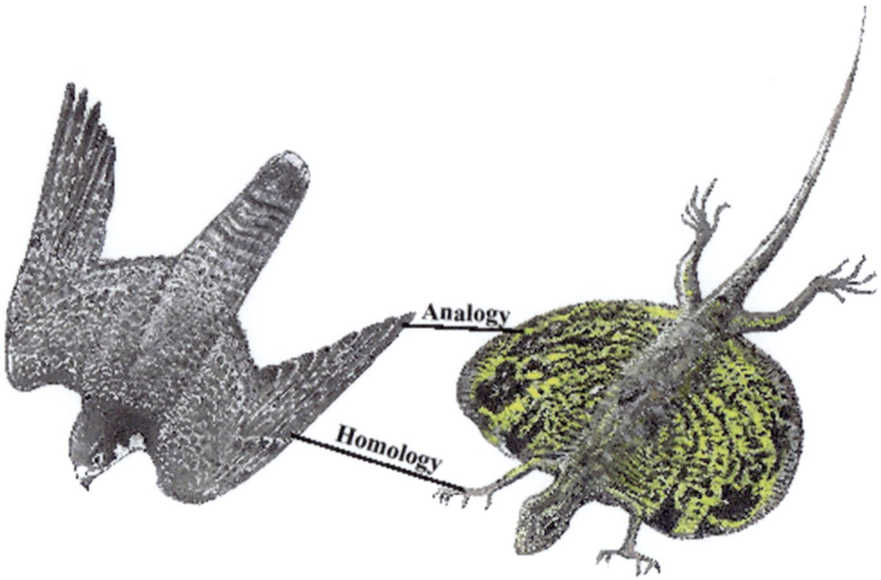

Fig. 6.54 Owen's example of homology versus analogy: bird versus gliding lizard

organ in a different animal", and homology as "the same organ in different animals, under every variety of forms and functions" (Padian 1997). To illustrate these terms, he suggested the gliding lizard *Draco*. Its forelimb, composed of the same basic parts as the wings of birds, is homologous with them; its parachute, composed of different parts yet performing the same function as the wings of a bird, is analogous to them (Fig. 6.54).

Homology calls to mind an underlying pattern. Owen reasoned that there must exist a common structural plan for all vertebrates, a single, basic ground plan that he called the "archetype" and which, he thought, underlay all adaptive modifications (Fig. 6.55; from the Greek, *arkhe*: original and *tupos*: imprinted image). The archetype was an abstract model that summed the most basic, most generalized structural arrangement common to all the members of a given group of organisms; it was a divine plan that existed in the mind of the Creator, and it allowed naturalists to interpret the concept in a variety of ways. Some saw it as an ideal plan, a blueprint in a Divine mind that guided the creation of vertebrates. Others considered it the body plan of a primitive ancestral vertebrate, from which later vertebrates developed. Owen favoured the latter idea, which he believed the fossil record supported.

He may have gotten this concept of an archetype from Pierre Belon, who, during the Renaissance (1555), had presented a human skeleton and a bird skeleton side by side, placed in the same position to show a resemblance. He might also have gotten it from Buffon and Daubenton, who had proposed (1753) that many animals have the same body plan. They had pointed out the close resemblance in structure between the skeleton of the horse and that of man and had remarked that, since the

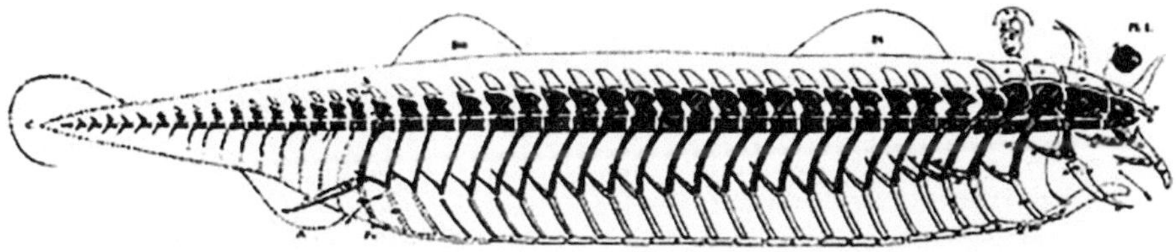

Fig. 6.55 The ideal archetype skeleton of a vertebrate, as suggested by Owen (1866)

same bones are in the same place, by modifying the dimensions of the bones in a human skeleton, one can transform it into that of a horse. This suggested to them that "a primitive and general design exists that can be traced very widely". Similarly, Denis Diderot (1713–1784), a philosopher of the French Enlightenment, had suggested that first, there was a single animal that served as a prototype for all others, and that all nature had done was to lengthen, shorten, alter, multiply or eliminate certain organs. Geoffroy also believed that all animals of a group share the same basic organs and that Dugès had claimed that individual zoonites, of the same basic structure, compose all animals. Owen was thus not the first to address that broad field of animals consisting of a basic structure.

Owen suggested that each group of animals has a unique archetype, a unit of organization that is progressively lost as the individual proceeds towards its specialized form. All vertebrates share some basic structures that originate from a common archetype, from which they gradually diverge. In seeking to add a palaeontological aspect, he suggested that the mammalian archetype ancestor was hornless, with 28 molars, four incisors and two canine teeth in each jaw, and five toes on each foot. Modifications of these characters, such as the development of horns, a reduction in the number of toes or a loss in the number of teeth, represent later adaptive divergence from the ancestral mammalian type (Gould 1977; Camardi 2001; Ghiselin 2005; Amundson 2007). Progressive improvements ensue from within the organism, with the archetype slowly advancing from that first embodiment and gradually diverging until, at its peak, "it became arrayed in the glorious garb of Human from".

Owen closely accepted Baer's opinion that divergence in embryonic development separates different animal groups. Aware of the cardinal importance of individual development in the study of morphology, he took Baer's divergence principle a step further. To understand the diversity of the animal kingdom, we must consider, he claimed, not only divergence during late stages in the development of the embryo but also the alteration of proportions after birth, from infancy to maturity. As animals grow, different structures begin and end their growth at different times, resulting in an adult that differs from the juvenile. (We recall Harvey, who, two centuries previously, had noted that in the rooster, crests, combs, wattles and spurs form only after hatching, resulting in an adult very different from the chick.)

Owen studied this aspect in the structure of the brain (Fig. 6.56). In apes and in monkeys, the brain reaches its full size at about the same time that the first set of teeth completes its growth. At this infancy stage, before having developed many of

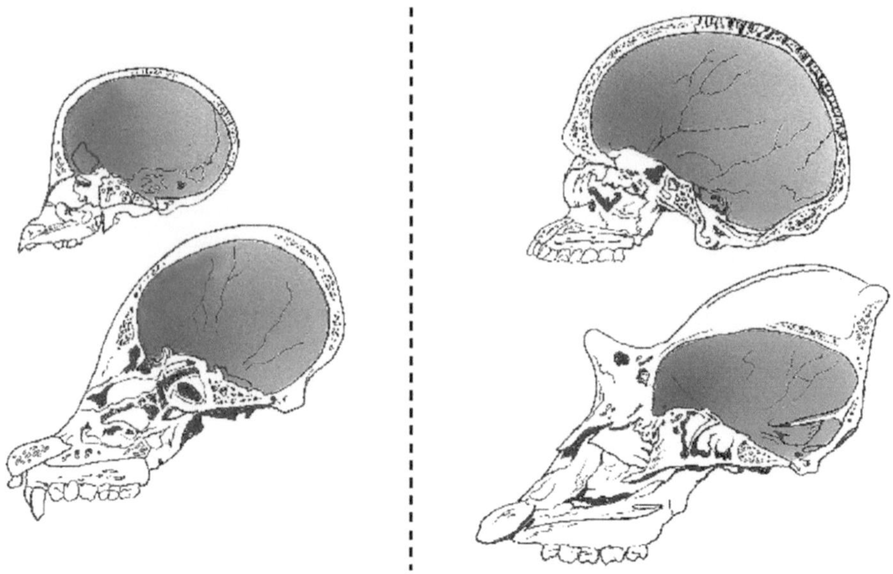

Fig. 6.56 Left: Cross section of an immature orangutan skull (above) contrasts with the skull of an adult. Right: Cross section of a human skull (above) and a gorilla skull (below). (After Cosans 2009)

their species-distinct characters, apes and monkeys appear deceptively similar to humans. Later, after the first set of teeth develops, the brain of maturing apes stops growing, whereas the jaws expand considerably, affording ample space for the large canine teeth, which serve as weapons. Consequently, the relative size of the ape's brain decreases compared to its body and face, and it loses much of its resemblance to humans. In humans, the skull of the adult has the proportions of an infant ape because human brains continue to grow well after the development of the first set of teeth, thus keeping pace with the rest of the body. Meanwhile, the jaws expand only modestly, as they house only more equally developed teeth; large canines are unnecessary, as man can build tools for his defence. Humans thereby achieve both a large stature and a relatively large brain. Divergence between groups thus occurs also after embryonic stages, in the transition from infancy to adulthood, and can explain structural differences between two species (Fig. 6.56).

Owen (1866) classified all vertebrates into two large groups, those with cold blood (fishes and reptiles, which at the time included the amphibians) versus those with warm blood (birds and mammals). In classifying mammals, he emphasized what he believed to be at the core of what it is to be a mammal, the brain. His mammals are classified into one of four subclasses, corresponding to four levels of brain development (Fig. 6.57). Monotremes (platypus, spiny anteater) and marsupials (kangaroo, opossum and others) made up the most primitive subclass. Their cerebral hemispheres are smooth and of such small size that, when viewed from above, they do not cover the olfactory ganglions, cerebellum and most of the optic lobes.

Since the two cerebral hemispheres are not connected, Owen named this subclass Lyencephala (loose-brained). At the second level, in rodents, shrews, bats and sloths, the two cerebral hemispheres are connected, but they have no (or only a few) convolutions, Owen designated the subclass Lissencephala (smooth-brained). The brains of ungulates, cetaceans, carnivores, monkeys and apes are included in the third level of development, in which there is such an increase in the cerebrum's relative size that it extends over much of the cerebellum and the olfactory lobes. Owen named this subclass Gyrencephala because most of its members have numerous gyri or convolutions on their cerebral hemispheres. In humans, (Archencephala, the ruling brain) the cerebral hemispheres are so large that beyond merely covering the olfactory lobes and cerebellum, they "extend in advance of the one, and further back than the other". To illustrate the extent of the human brain's growth, Owen cited the prominence of three structures: the posterior lobes, the posterior horn of the lateral ventricle and its floor the *hippocampus minor* (now termed the *calcar avis*). While all other parts of ape and man are strictly homologous, this unique feature of the human brain, together with its large size, indicated, Owen believed, that humans could not possibly have evolved from apes, and he placed them in a distinct subclass. In later years, he claimed that unique features in humans include the very large size of their brain relative to their body size.

Owen's classification of man in a separate, higher subclass of mammals calls to mind both the separate Creation of man in the Biblical narrative and the claims of Cuvier, Oken, Ehrenberg and Lyell that man is the supreme creature in the animal kingdom and is in a group far above all other animals.

6.3 Theology

6.3.1 God Designed a Happy World

From Owen, we step back several decades in time, to the very beginning of the nineteenth century. William Paley (1743–1805, Fig. 6.58) was an English clergyman. He studied philosophy at Cambridge, became Archdeacon of Carlisle at the age of 39 and, for his services in defence of the faith, was given a stall in St. Paul (London) and made sub-dean of the Lincoln Cathedral. In his widely read *Natural Theology or Evidences of the Existence and Attributes of the Deity* (1802), his key argument was that animal adaptations bear evidence of God's design of the world. There is no order without someone making order, no means towards a purpose without someone who thinks of the purpose. This old argument goes back to John Ray and, many centuries earlier, to mediaeval scholars (Lactantius, Yehuda Halevi). Paley, like these and others over the centuries, reached the conclusion that only a wise, all-potent entity can be responsible for this mechanical miracle of animal structure so closely corresponding to animal function.

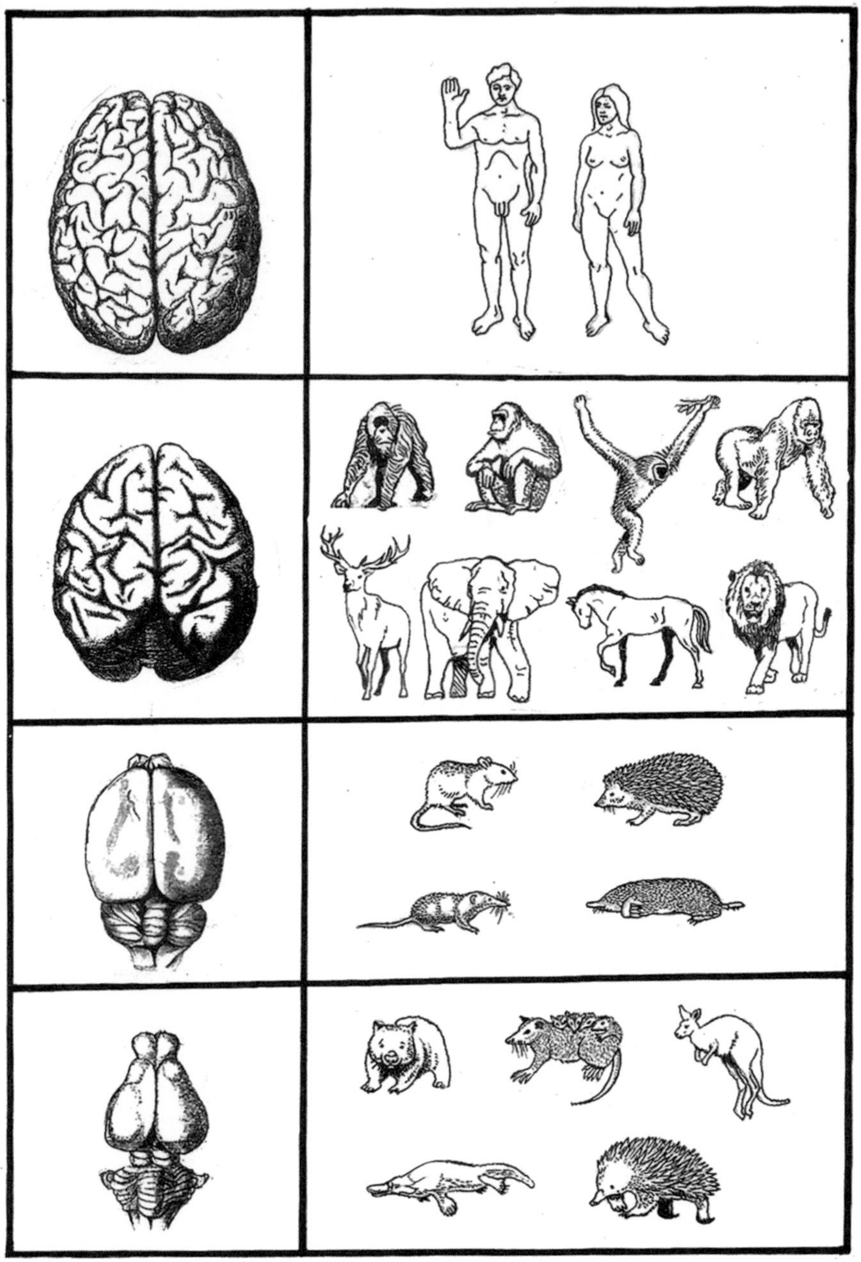

Fig. 6.57 Mammal classification by brain structure, according to Owen

Fig. 6.58 William Paley

Natural Theology opens with Paley's famous watch analogy. If we come upon a watch while strolling across a heath, we perceive that someone had formed its separate parts and had assembled them for a specific purpose: to produce movement that would point to the hour of the day. From the very complexity of its parts and the manner by which they fit one into another, we conclude that some intelligent entity had designed and made the watch for a specific purpose. Two features of the watch bring us to this conclusion. One is its complexity; the hand of chance could not have created something so complex. Chance can form simple things, say a wart or a pimple, but never something as complex and sophisticated as a watch. The second, and more important, feature is that the watch was designed and created for a specific purpose. This requires a designer with foresight.

These conclusions for watches apply, all the more so, to those entities in the heath, which are far more complex, each serving its own destiny, such as to animals. Animal design calls for a designer; the designer is God the Creator, and the adaptations we find in nature are evidence of His skilful design. Every hinge in every wing of an insect is intricately wrought, with no signs of any decrease in devoted care because of the numerous individuals that He must care for or any distraction because of the diversity of animal shapes that He must care for. Paley's model example for perfection in adaptive design was the human eye. He noted that eye and telescope structures follow the same principles, as both comply with the laws that regulate the transmission and refraction of rays of light. The lenses of a telescope resemble parts of the eye in their shape, position and ability to converge light rays at a precise distance from the lens. Rays of light refract by a more convex surface when transmitted through water than when passing out of air into the eye, and accordingly, the eye of a fish is much rounder than the eye of terrestrial animals. What plainer manifestation of design can there be than this difference? What could an instrument maker have done better to show his knowledge of this principle? Paley further pointed out that the lens of the eye is transparent, very different from the opaque materials of which the rest of the body is composed, and the retina is the only membrane in the body that is black. It spreads out behind the lens to receive the image formed by the transmitted light and is at the precise distance at which, and at which alone, a distinct image can form. The eye also possesses a large nerve, communicating between the retina and the brain, without which the sensation of light upon the membrane would be lost. Could this complex, sophisticated eye, with its unique functional anatomy, have come about without design for a preconceived purpose?

Adaptation occurs not only in the eye; it occurs in all creatures and in all their organs, and even if sometimes difficult to identify, careful observations eventually uncover it. The wild hog *Babirusa*, of eastern Asia, has strangely long, bent tusks growing out of its upper jaw (Fig. 6.59). They grow upwards from their bases so that they pierce the skin of the muzzle and curve backward, eventually almost touching the forehead. *Babirusa* is a retiring, nocturnal animal of dense jungles. These tusks do not serve in defence because two other tusks, issuing from its lower jaw, perform that function, as in the common boar. What purpose, then, do these upper tusks serve? Paley answers: this hog sleeps while standing, and, Paley claimed, while doing so, it hooks these upper tusks onto nearby branches to support its heavy head. Upper tusks are therefore adaptations. As a general conclusion, whatever at first sight may seem a purposeless structure shall eventually be found as purposeful. Even pain is an adaptation, in that it signals some distress, to which the individual must pay attention. It further expresses God's glory in that, when it disappears for a while, we become very happy. After a fit of gout, people briefly feel levels of happiness far higher than healthy people do. As further evidence of a divine designer, Paley noted that some structures are prepared for future function, such as the milk teeth of a baby, which are already present inside the gums at birth.

Thanks to God's overall goodness in creating these adaptations, nature is full to the brim with happiness and harmony, and even baby creatures, newly born into nature, already express the pleasure of entering this wonderful world. Swarms of newly born flies fly about, and their sportive movements and continual change of place without purpose testify to the triumphant joy that they feel in their just-discovered abilities. By the seaside and during the ebbing tide, clouds of young shrimps bounded into the air from the shallow margin of the water or from the wet sand. "If any motion of a mute animal could express delight, it was this; if they had meant to make signs of their happiness, they could not have done it more intelligibly". If each individual shrimp in this cloud is in such a high state of enjoyment, we can imagine how much gratification and pleasure the whole cloud of shrimps has here before our eyes!

Some people may perhaps claim that it is not providence but the overuse or underuse of organs that determines the shape, both of specific organs and of entire creatures. Is this indeed so? Men, Paley claimed, have never suckled babies; they have no use for their nipples, and yet, their nipples have not disappeared. Similarly, Jews have been circumcising the foreskin of their newly born boys for centuries, and yet their foreskin has not shortened (Paley's modesty led him to envelop this classical

Fig. 6.59 *Babirusa*: in its environment (above); skull with long, bent tusks (below)

case in untranslated Latin: *nec curtorum, per multa saecula, Judaeorum propagini deest praeputium*). Some may perhaps claim that it is not providence but the environment that designs the shape of a creature. Is this indeed so? The Sun might shine forever upon animals and plants, without the smallest tendency for them to produce an additional sense of sight. Light in itself has never succeeded in creating an eye. Those endless rays of light, falling upon so many animals, do not "breed" eyes.

People might perhaps claim that God had not created all creatures instantly, as they are today. Animals were once different, and they achieved their current shape by very gradual evolution. The elephant once had a very short proboscis, which gradually, over many generations, grew longer to reach the high branches. Is this indeed so? How did elephants survive during the intermediate stages until the elongation of the snout was completed? On what did they feed while perfecting? Some people might perhaps argue that the whole approach, as though God had shaped animal organs for a specific purpose, is not true. Shape forms for reasons not related to any specific aim, and the use of an organ took morphology over only secondarily, as animals discovered the possibilities of various shapes and organs. Is this indeed so? We might have accepted this argument if animals had only simple organs. We cannot believe, Paley claimed, that a structure as complex as an eye was created without any regard to vision and that, one day, the animal itself happened to discover that the eye, though not created for sight, happens to serve for sight.

Some people might perhaps claim that creatures achieved their high levels of adaptations, not by Creation but because of developmental constraints. During embryonic development, the growth and shaping of one organ may be linked to that of another organ. As a result, if one of these linked organs develops, for any purpose whatsoever, we automatically also get the other organ. Is this indeed so? Paley called our attention to the swans that have both a long neck and webbed feet, which might lead us to claim the two features are linked developmentally. We can conclude that this is not so from the many other waterfowl in which long necks and webbed feet do not occur together. How does it happen, therefore, that they meet only when a particular design demands the aid of both? Paley's answer: only God can combine the two. As another case, the mole has both tiny eyes and large, shovelling hands, and we might perhaps think that these two features develop while linked. This is not so. Small eyes often go together with small hands, and large eyes with large ones. How, then, do these two happen to be together only when a special design demands a combination of the two? Just like man, who has both made and combined the axes, springs, gears and weights in the watch, God the Creator purposely designed, made and combined the mole's small eyes together with its shovel-hands.

Some people might perhaps argue against the concept that nature was created by God, that over a very broad taxonomic front, we find the same [= homologous] organs. Within a large, highly diverse animal group, each species uses the same building blocks, merely altering them one way or another, and thereby adapting them for its own needs. If God had indeed created the world, why did He not create each species separately from the beginning, from building blocks that would suit it and only it? Why did God create a common type when each of the species participating in this type functions in a different manner? To this claim, Paley responded: God shows His greatness in that He purposefully constrains His own power to a small number of well-defined types to work His ends within these limits. He prescribes limits to his power to demonstrate his wisdom.

Paley saw the Divine Plan of God everywhere in nature and could not reconcile himself to a theory that did not invoke design. His argument regarding design assumed an analogy between nature and human contrivance: since the latter involved intelligence and purpose, so must the former. It also assumed a vast difference between natural and artificial things, between the skill of nature and the imperfection of art. In this way, it hoped to prove God's omnipotence.

Paley's biological knowledge was extensive and deep. He discussed the fish's air bladder, the viper's fang, the heron's claw, the camel's stomach, the woodpecker's tongue, the elephant's proboscis, the bat's wing hook, the spider's web, insects' compound eyes, the glow-worm, bivalves and on and on, with accuracy and as much detail as was known to the best zoologists of his time. In each case, he concluded that the organized complexity and purposeful function of organisms and their parts reveal an intelligent design and that only the omnipotent Creator could be this Intelligent Designer. All the endless adaptive aspects of nature point to the existence of God the Creator. To know more of Him—of his features, his intentions—we must further investigate the secrets of adaptation.

Paley's argument for creation by an intelligent designer is the most common and widespread philosophical argument to convince people of the existence of an all-powerful God, who had created man and who is interested in the fate of humans, and it comes up time and again in the sermons of different religions. However, from this argument, we cannot conclude that Moses, or Jesus, or Mohammed had any special relationship with Him who had created Man and animals. We also cannot conclude that we are forbidden to eat one food or another or, indeed, any other religious custom—the argument from intelligent design does not offer support for most laws of the historic religions.

A few years after the publication of *Natural Theology*, the eighth Earl of Bridgewater endowed a large sum of money to the President of the Royal Society of London to promote the publication of treatises on "the Power, Wisdom and Goodness of God as manifested in the Creation". Eight treatises appeared during 1833–1840, several of which had considerable influence on the public and on scientists. Charles Darwin, on completing *On the Origin of Species* (1859), told a friend: "I do not think I hardly ever admired a book more than Paley's *Natural Theology*: I could almost formerly have said it by heart". Years later, Darwin would shatter the claim of an intelligent designer by suggesting natural selection as a mechanism to reach adaptive perfection (Ayala 2007).

6.3.2 *Vestiges of Natural History*

Robert Chambers (1802–1871, Fig. 6.60) was a Scottish publisher, geologist, author and journal editor, highly influential in scientific and social circles. The son of a cotton manufacturer, he was born with six fingers on each hand and six toes on each foot, and the operations to correct this abnormality left him partially lame; while other boys roughed it outside, he stayed indoors and studied his books. At the age of 16, he began a business as a bookstall keeper and, in later years, joined his brother, who had a printing press: Robert would write, and his brother would print

Fig. 6.60 Robert Chambers

magazines and books. The two brothers eventually united as partners in a publishing firm. Robert had 14 children. A member of the Royal Society of Edinburgh, he took a keen interest in geology and was a fellow of the Geological Society of London. While recovering from a mental depression, he dictated *Vestiges of the Natural History of Creation* to his wife and published it anonymously (1844; I wonder whether, in using the term *vestiges,* he was inspired by another Scotsman, James Hutton, who had used the term six decades earlier). The book became an influential international bestseller with many editions, provoking widespread discussions on evolution; even Prince Albert read it aloud to Queen Victoria over several days.

Chambers sketched an evolutionary history of life. He was not very original, and we have encountered many of his ideas in previous sections of this book. He found that, in the fossil record, animals (and plants) evolved from the simple to the more complex. Sponges, cnidarians and sea lilies were the earliest on record, together with a few molluscs and arthropods. Only much later did the "most humble" first vertebrates appear, all of which were cartilaginous fishes. Bony fishes, more advanced than cartilaginous ones, soon followed. Some freshwater fishes (e.g. *Megalichthys,* a lobe-finned fish of the Carboniferous) were of a "reptile character" in that they breathed air. Then came the reptiles themselves, a terrestrial class just above fishes, whose imperfect respiratory system suited them for enduring an atmosphere not yet suitable for birds or mammals. Later came the birds and finally the mammals, the highest of all vertebrate classes: first the pouched (marsupials) and later the placental ones, superior to marsupials in their more perfect brain and more advanced mode of gestation. Finally, man arrived on the scene. The fossil record thus suggests progress over time, what we today call evolution. The progression from low to high forms, in any taxon, parallels the order of appearance in the fossil record.

Viewing this fossil record from a broad aspect, Chambers noted that the earliest creatures belonged to Phyla and Classes that exist today as well, but to Species and Genera that no longer exist, sometimes even to Orders that no longer exist. These early creatures soon gave way to new species and genera, which also do not exist today. In this manner, the fossil record consists of many extinct groups, but all within existing branches, belemnites and ammonites within the molluscs, trilobites within the arthropods and so on. Viewing it from yet another aspect, he found that the biotic world diversified similarly to the diversification of the abiotic environment. For example, the abiotic habitat diversified, and land eventually emerged, and the biotic world diversified with terrestrial plants and animals appearing soon later, while corals and crinoids remained in the sea. When vast, low, muddy coasts and broad sea margins formed between sea and land, huge reptiles appeared that were adapted to these habitats. When the Earth was generally a flat, imperfectly variegated state, as in Australia today, marsupials appeared. Physical conditions and animals thus advanced side by side, with life invariably "pressing in" (Chambers' term) immediately when suitable conditions arose. No niche that could support any form of life remained unoccupied for any length of time. Various animal groups went extinct, on the other hand, when conditions for their existence changed. This was the fate of trilobites, of ammonites, of Tertiary mammals and of many other groups. Chambers did not say which changing conditions caused extinction.

In the great Cuvier-Geoffroy dispute, regarding whether there are four distinct branches with no intermediates between them or only one basic plan throughout the entire animal kingdom, Chambers noted that, as an exception to the observation that most fossils belong to existing phyla and classes, some fossils were intermediate between two of Cuvier's major branches. The first fishes differed from modern fishes in that they formed a link between arthropods and true fishes (Fig. 6.61). *Cephalaspis* only slightly advanced beyond the arthropod type, with its longish tail-like body inserted within the cusp of a large crescent-shaped head, strong, enamelled plates of bone covering its body, a large plate shielding the head from above and only small fins, suggesting weak swimming powers. *Coccosteus* marks the next step towards fish. Its thick, rounded body, covered with strong bony plates, ended in a long tail, which was the sole organ of movement. While the tail classifies this creature among the fishes, its mouth opened downwards, like those of the crustaceans, is evidence that *Coccosteus* formed yet another link between these two great branches of the animal kingdom. *Pterichthyodes* had strong bony plates over its body, much like those of a tortoise, and it had a long tail; its most remarkable feature was a pair of long, narrow, wing-like appendages attached to the shoulders, which it, so Chambers believed, raised for its defence when attacked. *Holoptychius* was a fish furnished with fins, whose body ended in a long tail; its body was covered with strong overlapping plates, and its head formed only a slight rounded projection from

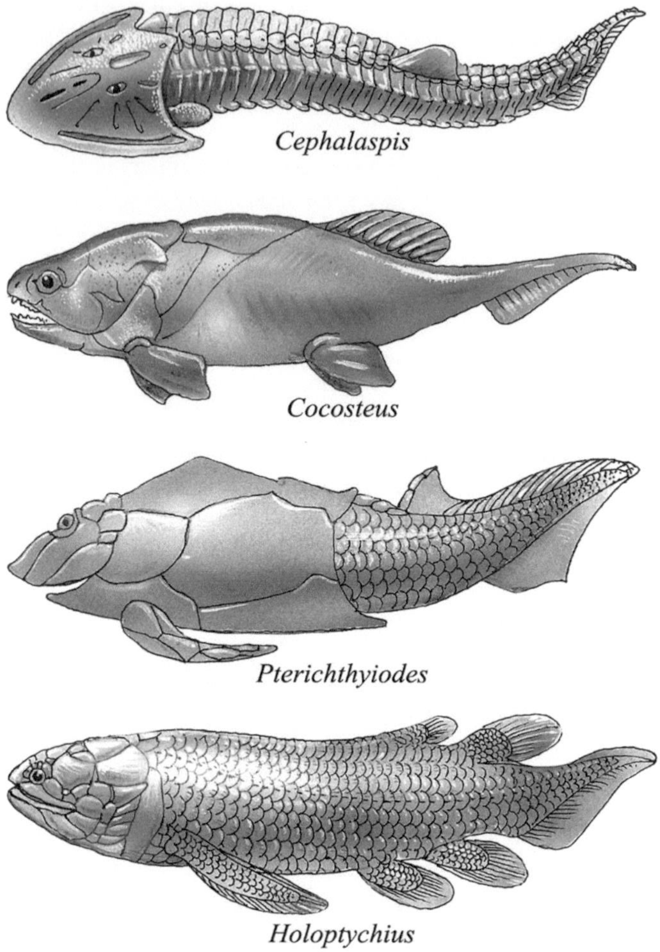

Fig. 6.61 Early fishes linking arthropods to fishes, in Chambers' opinion. (Tuvia Kurz)

the general shape. [Modern Zoology classifies *Cephalaspis* as an agnathan vertebrate of the Devon. *Coccosteus* was a freshwater placoderm with a joint between the armour of the body and skull, its jaws forming a beak. *Pterichthyodes* was a placoderm of the Devon, with heavily armoured head and unarmoured scaly tail, its name referring to their odd wing-like appendages that correspond to the pectoral fins of modern fish. *Holoptychius* was a carnivorous lobe-finned fish of the Devonian and mid-Carboniferous, with fang-like teeth on its palate and smaller teeth on its jaws. Its asymmetrical tail had a caudal fin on its lower end.] Other fishes of the period were more like modern fishes but covered with enamelled bony scales and bony-tipped spines, and their tail had a fin only on the lower side, and the internal skeleton was cartilaginous (*Osteolepis, Glyptolepis, Dipterus*). They thus combined characters of cartilaginous fishes with characters peculiar to themselves, in which

Chambers saw clear vestiges of the pre-existent arthropod form. [Again, modern zoology classifies these three fishes as lobe-finned fish and a lungfish.] We recall that Blainville had also claimed that he detected intermediate groups between arthropods and molluscs in his attempt to overthrow Cuvier's claim of four distinct branches.

Chambers' next step was to show that the formation of life does not require any supernatural, miraculous, divine intervention; it closely follows basic laws of nature that form basic compounds, many of which man can synthesize. Thus, for example, take the overall shape of plants, with their trunks, stems and branches pointing upwards and roots downwards. If we take an amalgam of silver and mercury dissolved in nitric acid and suspend in this solution a small piece of silver, then particles of the silver amalgam will crystallize in a form that precisely resembles a shrub. Likewise, in other solutions subject to electric current, tree ramifications with leaves appear near the positive pole and spreading roots near the negative one. Electric energy is thus involved in determining plant form to such an extent that Chambers noted that germination would not proceed in water charged with negative electricity, whereas water charged positively greatly favours it; and a garden will increase in luxuriance if conducting rods are placed above it. Furthermore, in nature, the atmosphere has a positive charge, the Earth a negative one and a plant thus gains shape as a result of a simple natural electrical current. As another example, this time concerning the chemical composition of animals, Chambers claimed that the living body of animals is composed mainly of albumen, fibrin, urea and allantoin, some of which man can produce synthetically, further evidence that the creation of life does not require any supernatural, divine intervention. Shell production is another natural operation that man can imitate. In bleaching workhouses that rinse cotton cloth free of the lime employed in its purification, encrustations with all characters of a shell gather on the wheel, displaying a highly polished, beautifully iridescent surface and a foliated texture, just like mother-of-pearl (we recall that his father was a cotton manufacturer).

Chambers now came to his main thrust: Life can, did and does arise spontaneously. Place lime on a waste moss ground, and a crop of white clover will sprout; mix cow and horse dung, and mushrooms will grow; take organic matter inside other animals, and internal parasites will be born. Furthermore, certain tapeworms (*Echinococcus*) infest only domestic pig, not wild boar, and pig domestication is a recent event, how can we explain the first progenitor of this parasite other than by spontaneous creation? Larvae of the fly *Oinopota cellaris* live only in wine and beer, beverages manufactured only by man, how can we explain the first progenitor of this fly, other than by spontaneous creation? [*Oinopota* is Greek for a wine drinker; this is surely the fruit fly *Drosophila*.] The fish *Pimelodes cyclopum* (Fig. 6.62) exists only in South American subterranean cavities formed by very recent volcanic activity, how can we explain the first *P. cyclopum*, other than by spontaneous generation? Chambers further alluded to a certain Mr Crosse, who, a few years before, had conducted experiments on crystallization with a powerful battery operating on a saturated solution of potassium silicate. Suddenly and unexpectedly, insects were present above the solution. Crosse subsequently tried a solution

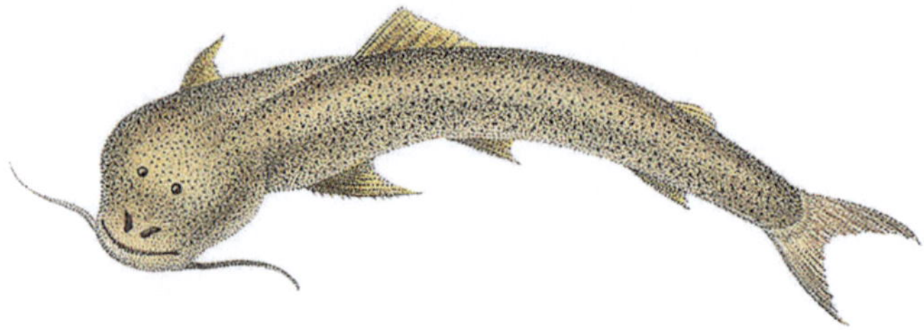

Fig. 6.62 *Pimelodes cyclopum.* (Tuvia Kurz)

of copper nitrate, which is a deadly poison, and above that fluid, too, he found live insects. Some years later, a certain Mr Weekes got similar results by employing potassium ferrocyanide: insects emerged from this solution as well and climbed to the surface of the fluid. Some of them, soon after their existence had commenced, began to reproduce, returning to the fluid to feed and occasionally to devour each other. The insects produced by both men were minute, semi-transparent mites with long bristles of the genus *Acarus* [the flour mite, a pest of stored grain and flour, about half a mm long, with a life cycle of 2 weeks]. All this amounts to evidence supporting spontaneous generation under certain specific conditions. Chambers believed that *Acarus crossii* was a being ordained from the beginning as a potential creation of the Almighty, who destined it to become a reality but only under very specific physical conditions. Humans, occasionally and without being aware of it, happen to create those specific conditions under which the creative laws work. Now, if this is so for mites, fish, flies and tapeworms, then why not accept that also pterodactyls appeared by spontaneous generation when Earth had become suitable for their existence?

Moving on to diversity, Chambers declared that while animals may differ in their external features, they are all variations on a single fundamental plan. (At this point, he was vague: he wrote of all animals, as suggested by Geoffroy, but in his examples, as we shall see, he restricted himself to vertebrates, as suggested by Cuvier.) Mammals breathe by lungs and fishes by gills, two distinct organs, yet gills exist in mammals at an early embryonic stage; later, they disappear, and lungs then develop. Whales have baleen, and terrestrial mammals have teeth, two different organs, yet the whale embryo has rudiments of teeth; these, being unnecessary, do not develop further, and baleen develops instead. Thus, in many instances, a particular structure advances to a certain point in a particular group of animals, although not required there. These, what we term "rudimentary organs", include the useless mammae of males, the unrequired marsupial bone in the male opossum, the imperfectly developed wings of the ostrich, foot rudiments in some serpents, and the undeveloped tailbone of man (*os coccygis*). As another aspect of unity beneath diversity, the giraffe has, in its tall neck, the same number of vertebrae as the pig, which scarcely has a neck at all; and the paddles of the whale have the same bones as in the extremities of the terrestrial mammals. All animals on Earth thus share a fundamental unity that pervades and embraces them all. On a theological note, Chambers declared that the entire animal creation depended upon one single decree of the Almighty, that formed all animals in potentiality; these potential animals do not all come forth at one time. He rejected ideas of a separate creation for each group because if so, then we must regard the rudimentary organs as blemishes or blunders, and this is irreconcilable with the idea of Almighty Perfection. When we accept animal creation as a general law, however, we see nothing in these abortive parts but harmless peculiarities of development.

Embryonic development was a key issue in Chambers' evolutionary theory. He noted that each animal passes through a series of changes in the course of its embryonic development that resemble the forms of the various animal groups below it on the scale of life. The frog, upon hatching, is a fish with external gills suited for an aquatic life, but it changes as it advances to maturity and becomes terrestrial. Man gradually passes through forms generally resembling a fish, a reptile, a bird and a lower mammal before attaining his specific form; at one of the last stages of his foetal career, he has an intermaxillary bone, as in apes; this is later suppressed, and he may then "take leave of the simial type" and become a true human. So also, the heart in mammals consists of four cavities, in reptiles of three and in fishes of two; in articulated animals, it is merely a prolonged tube. In the mammal foetus, at an early stage, the developing heart is a prolonged tube (the insect heart). Subsequently, it shortens and widens and divides into two parts, a ventricle and an auricle (the fish heart). Subdivision of the auricle makes a triple-chambered form (the reptile heart); and lastly, the ventricle also subdivides, and it becomes a full mammal heart. These cases led Chambers to consider the embryonic stage as the main one involved in evolutionary change. The entire series of animals is a succession of advances of embryonic development, from the simplest and most ancient up to the highest and most modern. In the first step in the creation of life, a chemical-electric operation produces a simple, germinal vesicle, the nucleated cell. This basic form of all living organization is the meeting point between the inorganic and the organic; it is the end of the mineral kingdom and the beginning of the vegetable and animal kingdoms. It may be an independent being, as in microscopic animalcules, but in higher creatures, it is the starting point of foetal progress. The next (vertebrate?) step, Chambers suggested, was an advance from the simplest forms of being to the next more complicated through generation. In the reproduction of the higher animals, the new being passes through stages in which it is successively fish-like and reptile-like. The resemblance is not to the adult fish or adult reptile but to the fish and reptile at a certain point in their foetal progress. The gradation from low to high forms parallels the order of appearance in the fossil record and the order of the stages in the organism's embryonic development. There is thus a threefold parallelism, the order of appearance of animals in the fossil record largely mirrors, the order of appearance of their morphological novelties, and the order of embryonic development.

Sex determination is also a matter of embryonic development, at least in some cases. Bees can modify a larva so that it will develop into either a queen or a worker by manipulating the size, temperature and food of the cell. The queen is, to cite Chambers, destined to enjoy love, to burn with jealousy and anger, to turn to vengeance and to pass her time without labour. The worker would be one zealous for the good of the community, a defender of public rights, enjoying an immunity from the stimulus of sexual appetite and the pains of parturition; laborious, industrious, patient, ingenious and skilful; incessantly engaged in the nurture of the young, in collecting honey and pollen, in elaborating wax and in constructing cells and the like. A mere modification of the embryonic process produces all these differences between the two adult types, queen and worker.

Another embryonic aspect of evolutionary change is embryonic monstrosities. We know of cases in which, Chambers claimed, in a human foetus, the heart goes no farther than the three-chambered form, so that it is a reptilian heart; indeed, in some cases it remains two-chambered, a fish heart. We can imagine (although we do not know of) favourable conditions that reverse the phenomenon and make a fish mother develop a reptile heart or a reptile mother develop a mammal one. A goose might perhaps give its progeny the body of a rat and thus produce a platypus; a platypus might perhaps give its progeny the mouth and feet of a rodent and thus complete the passage from birds to mammals. Such monstrous cases would result in the transition of animals of one class into another, higher one. In the history of life, as reflected in the geological record, the simplest and most primitive type gave birth to the type next above it; this again produced the next higher and so on to the very highest. The stages of advance were, in all cases, very small, from one species only to another. Fifteen years later, Darwin would repeat this argument, that changes in the animal world are slow.

Chambers believed that the harmonious adaptations of animals to the physical environment in which they are destined to live clearly prove that design presides in the creation of different forms of life. Design implies a designer, a Creator and Chambers fully accepted Paley's argument on this point. He, too, like Paley, emphasized that joy is the proper attendant of animal existence. We perceive symptoms of joy even in the lowest creatures, whose whole being is a system of needs that, when satisfied, supply gratification and of deeds that, when exercised, grant pleasure. True, there is misery in our world, but it is only an uncommon side effect of something good in the main. Consider war and all the misery, hardship, waste of human life and misspending of human energies that it causes. War emphasizes such positive tendencies of human nature as keen assertion of a right, resentment of injury, acquisitiveness, a desire for admiration, combativeness and love of excitement. These tendencies benefit us daily; man would be a tame, indolent, unserviceable being without them, doomed to starvation. War is but an uncommon side effect of these tendencies, an occasional misdirection of essentially good properties and powers, which God has given us for a benevolent purpose. Consider sexual passion, which likewise leads to the great miseries of overpopulation; again, these are uncommon side effects of something that is, basically, good. Providence has made ample provision for the preservation of all species, and sex aims to diffuse existence as widely as possible, to fill up every vacant space with some sentient being, to be a vehicle of joy. In discussing misery resulting from sexual overactivity, Chambers brings in Malthus: the relation between the number of beings and the means of supporting them must balance. There may be occasional discrepancies between the forces operating on the multiplication of individuals and those operating to supply them with the means of subsistence. Misery will follow in consequence. (Darwin, 15 years later, would also refer to Malthus and would make this the central point of his theory of natural selection.) Yet against all miseries from the deeds of sex, there is the happiness we derive from it. Sex is the basis of all domestic affections, the sweetening principle of life, the prompter of our most generous feelings and of our best moral qualities. In the balance, the misery that it causes is negligible. As yet another case, consider the sufferings of disease. Many diseases are the result of effluvia from decaying organic matter, which is extremely useful when mixed with Earth, in growing vegetation as food; such diseases intend to signal to us to put away decaying organic substances into the Earth, where they are so useful. Other diseases result from mismanagement of our personal economy, such as eating and drinking to excess. Further, miseries may bring blessings. When a child is born lacking arms, or eyesight, or hearing or intellect, there will always be, in the parents, in relatives and in the surrounding public, a sympathy with the sufferer; this would make up for his deficiency and make it lighter. Further, in the defective child itself, a defect in one feature is compensated for by an extra power in another. The blind have a sense of touch more acute than those who see; persons born without hands acquire the power of using their feet. Whatever the miseries in our world, the purpose of life is joy and happiness.

In seeking the origin of Life, Chambers soared far beyond the fossil record of our little planet and gazed at the universe, with its endless number of planets and stars, in infinite space. He suggested that numberless celestial bodies are either already a theatre of life or on the way to becoming so and speculated as to what the (plant and) animals of these remote celestial bodies look like. Starting from the formation of the solar system from a nebular cloud, he concluded that all matter was originally in one mass; hence, matter is everywhere broadly the same. Next, he noted that on Earth, the animal body is composed, basically, of only four elements of inorganic matter, carbon, oxygen, hydrogen and nitrogen. These would also be the elements composing life on distant celestial bodies. These elements compose flesh and blood that are, therefore, bound to be the components of all living beings on all celestial bodies throughout the universe that are the seats of life. All celestial bodies share such universal phenomena as electricity, which is involved in all basic procedures of life; they share the principle of gravitation, and they share light. Light, on Earth, is so important that the structure of one important organ, found throughout the animal kingdom, is in direct relation to it. Where there is light, there will be eyes, and these, in all celestial bodies, will be the same. Thus, as one conspicuous organ of our animal kingdom is universal, parity in all other organs is highly likely. The inhabitants of all celestial bodies must therefore be of one general character and are bound to resemble us, at least in broad terms.

The reaction to Chambers' *Vestiges* was colossal. Adam Sedgwick, Professor of Science at Cambridge and President of the Geological Society, was utterly outraged. "… if our glorious maidens and matrons may not soil their fingers with the dirty knife of the anatomist, neither may they poison the strings of joyous thought and modest feelings by listening to the seductions of this author, who comes before them with … a false philosophy". The critics shredded Chambers to pieces. With such publicity, the *Vestiges* had magnificent sales, with 11 editions between 1844 and 1860 (24,000 copies).

By way of summing up the study of biodiversity in the decades up to the mid-nineteenth century, some scientists of those days viewed classification as circular, whereas others believed in a straight Ladder of Life, with mystic forces pushing life upward. Some thought that all fossils had gone extinct, whereas others believed that fossils never go extinct; they merely change into modern creatures. Some thought that there once was only one centre of creation; others thought that there were several such centres. Naturalists did not know how to cope with convergent versus divergent evolution, and they did not offer a natural mechanism by which the diversity of life changes over time. Into this zoological landscape, Darwin was born and grew up.

References

Agassiz, L. 1859. *An essay on classification*. London: Longman, Brown and others.

Amundson, R. 2007. Richard Owen and Animal form. In *Richard Owen: On the nature of limbs: A discourse*, ed. R. Amundsen. Chicago University Press.

Appel, T. A. 1980. Henri de Blainville and the animal series: A nineteenth-century chain of being. *Journal of the History of Biology* 13:291–319.

Ayala, F. 2007. Darwin's greatest discovery: Design without designer. *PNAS* 104 (supplement 1): 8567–8573.

Baer, K. E. 1828. *Über Entwickelungsgeschichte der Thiere. Beobachtung und reflexion.* [On the Developmental History of the Animals. Observations and Reflections]. Königsberg: http://www.biodiversitylibrary.org/item/28306. Accessed May 3, 2012.

Barber, L. 1980. *The Heyday of natural history. 1820–1870*. New York: Doubleday.

Bergmann, C. 1847. Ueber die Verhältnisse der Wärmeökonomie der Thiere zu ihrer Grösse. *Göttinger Studien* 1:595–708.

Brauckmann, S. 2008. The many spaces of Karl Ernst von Baer. *Biological Theory* 3:85–89.

Brauckmann, S. 2012. Karl Ernst von Baer (1792–1876) and evolution. *International Journal of Developmental Biology* 56:653–660.

Breidbach, O., and M. Ghiselin. 2002. Lorenz Oken and 'Naturphilosophie' in Jena, Paris and London. *History and Philosophy of the Life Sciences* 24:219–247.

Burkhardt, R. W., Jr. 1995. *The spirit of system, Lamarck and evolutionary biology*. Cambridge, MA: Harvard University Press.

Burkhardt, R. W., Jr. 2013. Lamarck, evolution, and the inheritance of acquired characters. *Genetics* 194:793–805.

Camardi, G. 2001. Richard Owen, morphology and evolution. *Journal of the History of Biology* 34:481–515.

Carpenter, W. R. 1856. *Microscope and its revelations*. Philadelphia: Blanchard and Lea.

Chambers, R. 1844. *Vestiges of the natural history of creation*. London: Churchill.

Coleman, W. 1962. Lyell and the 'reality' of species: 1830-1833. *Isis* 53:325–338.

Cosans, C. E. 2009. *Owen's Ape and Darwin's Bulldog*. Bloomington: Indiana University Press.

Cuvier, G. 1800–1805. *Leçons d'anatomie comparée* (5 volumes.) Paris.

Cuvier, G. 1817. *Mémoires pour servir à l'histoire et à l'anatomie des mollusques*. Paris.

Cuvier, G. 1825. *Discours sur les révolutions de la surface du globe et sur les changements qu'elles ont produits dans le règne animal*. Paris.

D'urville, J. D. 1833. *Voyage de la Corvette L'Astrolabe, execute pendent les annees 1826–1827–1828-1829, sous le Commandement de M. Jules Dumont D'Urville, Capitaine de Vaisseau*. Paris: Atlas. Tasti.

Ehrenberg, C. G. 1835. *Die Akalephen des Rothen Meer und den Organismus der Medusen der Ostsee*, 181–260. Berlin: Abhandlungen der Königlichen Akademie der Wissenschaften zu.

Ehrenberg, C. G. 1840. *Uber noch Zahlreich Jetzt Lebende Thierarten de Kreidebildung*. Leipzig: Koniglichen Akademie der Wisenschafte.

Eigen, E. A. 1997. The overcoming first impressions: Georges Cuvier's types. *Journal of the History of Biology* 30:179–209.

Elliot, H. 1964. *Introduction to zoological philosophy by J.B. Lamarck*. New York: Hafner. (Originally published in 1914).

Fourcroy, A. F. 1785. Entomologia Parisiensis: sive, Catalogus insectorum quae in agro parisiensi reperiuntur.

Gershonowitz, H. 1980. Napoleon and Lamarck. *Indian Journal of the History of Science* 15:204–209.

Geoffroy Saint-Hilaire, É. 1831. Letre sur quelques points du mémoire ayant pour titre: de la conformite organique dans l'échelle animales. Gazette médicale de Paris 2: 381–382.

Ghiselin, M. T. 2005. Homology as a relation of correspondence between parts of individuals. *Theory in Bioscience* 124:91–103.

Gould, S. J. 1977. *Ontogeny and phylogeny*. Belknap: Harvard.

Humboldt, A. 1807. Essay on the Geography of Plants.

Klein, M. 2018. *The first isothermic world maps*. Geography and Map Division, Library of Congress.

Kocandrle, R., and K. Kleisner. 2013. Evolution born of moisture: Analogies and parallels between Anaximander's ideas on origin of life and man and later pre-Darwinian and Darwinian evolutionary concepts. *Journal of the History of Biology* 46:103–124.

Korner, C., and E. Spehn. 2019. A Humboldtian view of mountains. *Science* 365 (6458): 1061.

Lamarck, J. B. 1795. *Tableau Encyclopedique et Methodique des Trois Regnes de la Nature. Mollusques Testaces*. Paris: Agasse.

Lamarck, J. B. 1802. *Hydrogéologie*. Translated by Carozzi A.V. 1964, as *Hydrogeology by Lamarck*. Urbana: Illinois University Press.

Lamarck, J. B. 1809. *Philosophie Zoologique*. Translated by H. Elliot 1963, as *Zoological Philosophy*. New York: Hafner. (Originally published in 1914).

Latreille, P. A. 1806. *Genera Crustaceum et Insectorum*. Paris.

Le Guyader, H. 2004. *Geoffroy Saint Hilaire 1772–1844. A visionary naturalist*. Translated from French to English by Grene, M. Chicago: University of Chicago Press.

Leuckart, R. 1851. Ueber den polymorphismus des individuen oder die erscheinungen der arbeitsheilung in der natur. Ein beitrag zur lehre von generationswechsel. *Geissen*.

Mayr, E. 1982. *The development of biological thought. diversity, evolution, and inheritance*. Belknap: Harvard University Press.

Miller, G. 2019. The Pioneering Maps of Alexander von Humboldt. *Smithsonian Magazine*, October 2019.

Molyneux, W. 1697. A discourse concerning the large horns frequently found under ground in Ireland, concluding from them that the great American deer, call'd a moose was formerly common in that island.

Munzar, J. 1967. Alexander von Humboldt and his isotherms. *Weather* 22:360–363.

Nordenskiold, E. 1935. *The history of biology, a survey*. New York: Tudor.

Oken, L. 1848. Isis von Oken.

Outram, D. 1986. Uncertain legislator: Georges Cuvier's laws of nature in their intellectual context. *Journal of the History of Biology* 19:323–368.

Owen, R. 1859. *On the classification and geographical distribution of the Mammalia*. London: Parker.

Owen, R. 1866. *Anatomy of vertebrates*. London: Longman, Green & co.

Padian, K. 1997. The rehabilitation of Richard Owen. *Bioscience* 47:446–453.

Panchen, A. L. 2001. Étienne Geoffroy Saint Hilaire: Father of "evo-devo"? *Evolution and Development* 3:41–46.

Rudwick, M. J. S. 1997. *George Cuvier, fossil bones and geological catastrophes*. Chicago University Press.

Salewski, V., and C. Watt. 2017. Bergmann's rule: A biophysiological rule examined in birds. *Oikos* 126:161–172. https://doi.org/10.1111/oik.03698.

Schrodt, F., M. J. Satos, J. J. Bailey, and R. Field. 2019. Challenges and opportunities for biogeography - what can we still learn from von Humboldt? *Journal of Biogeography* 46. https://doi.org/10.1111/jbi.13616.

Sclater, W. L., and P. L. Sclater. 1899. *The geography of mammals*. London: Kegan, Trench & Trubner.

Siebold, C. Th., and H. Stannius. 1854. Anatomy of the Invertebrata. In *Comparative anatomy*. Translated from the German by W.I. Burnett, with Notes and Additions Recording the Recent Progress of Science. Boston: Gould and Lincoln.

Stallo, J. 1848. *General principles of the philosophy of nature. With an outline of some of its developments among the Germans. Embracing the philosophical Systems of Schelling and Hegel, and Oken's system of nature*. Boston: Cosby and Nichols.

Stafleu, F. A. 1971. Lamarck: the birth of biology. *Taxon* 20:397–442.

Swainson, W. 1840. *A treatise in malacology. On the natural classification of Shells and Shell fish*. London: Longman, Orme, Brown, Green and Longmans.

Wightman, W. P. D. 1953. *The growth of scientific thought*. New Haven: Yale University Press.

Williams, D. M., and R. Huxley. 1998. Christian Gottfried Ehrenberg (1795-1876): The man and his legacy. An introduction. In *Christian Gottfried Ehrenberg (1795–1876): The man and his legacy*, ed. D. M. Williams and R. Huxley, 1–13. London: Linnaean Society of London/ Academic Press.

Wilson, E. O. 1992. The diversity of Life. Penguin Group.

Wilson, C. 2006. Kant and the speculative sciences of origins. In *The problem of animal generation in early modern philosophy*, ed. J. E. H. Smith, 375–401. Cambridge University Press.

Winsor, M. P. 1976. *Starfish, Jellyfish, and the order of life. Issues in nineteenth century science*. New Haven: Yale University Press.

Wulf, A. 2015. *The invention of nature: Alexander von Humboldt's New World*. England: John Murray.

Chapter 7
Darwin

Abstract Darwin found that birds and plants of different Galapagos Islands were distinct species, yet slightly different between islands. Why did God create a separate species on each island? He also found fossil mammals resembling modern forms. Why do animals go extinct? He developed a hypothesis that in nature, hereditary variation occurs among offspring. Populations multiply beyond their resources, and this results in a struggle for existence among descendants, in which the strongest prevail and the weakest fail and die out. Individual variation—ever so slight—determines the winner. This natural selection of the fittest explains the origin of all species. Diversity forms by descent from common ancestry, with modification piling up in adaptive directions by non-random survival of random hereditary varieties. Darwin claimed that basic morphology, comparative embryology, rudimentary organs, fossils, classification, fertile hybrids and human selection of domestic breeds all support his hypothesis. Variation is a basic feature of all species. The individual is important, not the type, because there is no "typical" individual of a species. Organisms form an ever-branching tree, not a Ladder of Life. Evolution through natural selection explains the diversity of nature, and this is the basic law of Life.

Darwin's theory of natural selection as the force behind evolution was not readily accepted. Agassiz denied variation within a species, thereby dismissing Darwin's theory. Individuals differing even slightly are separate species, not variants. Nature is the product of God's Creation and symbolizes divine intelligence, and therefore species are immutable. Animals had been created wherever they are found, land animals on land, sea creatures in marine habitats. The Creator wipes out one fauna and starts anew, with models different from those cast aside. There is no evidence of a relationship between distinct species, and hence, there is no proof for evolution. Some fossils, intermediate in that they possess a combination of characters of different groups, offer evidence that the plan of the creation had been considered long before it was executed. To further disprove Darwinian evolution, Agassiz claimed that a huge ice sheet had once covered South America, destroying all life. The current flora and fauna of South America were created anew after the ice had receded,

J. Heller, *Nature's Biodiversity*, https://doi.org/10.1007/978-3-031-89790-0_7

with no relationship between animals that had lived there prior and after this glacial epoch. This, he claimed, is evidence of the mistaken concepts of Darwinian evolution.

7.1 Darwin, Barnacles, Wallace and the Origin of Species

Charles Darwin (1809–1882; Figs. 7.1, 7.3, and 7.12; his son Francis published his autobiography) came from a family, which included wealthy doctors who also devoted themselves to philosophy on the one side and consisted of wealthy industrialists on the other side (his father had married the daughter of Wedgewood, of the porcelain industries; she died when Charles was eight). As a youth, he loved to collect, to fish, to shoot birds and to read, and he was an orthodox Christian who believed in a created world, inhabited by constant species. At school, his education did not include natural sciences—the subject did not even exist at the time. As he was doing no good at school, his father took him away and sent him, before he was 17, to the Medical School in Edinburgh. Here, finding the lectures in medicine, geology and zoology intolerably dull and, unable to bear the sight of live dissections, he neglected his medical studies, spending his time investigating marine invertebrates and discovering that in oysters, black spores are actually the eggs of a skate leech. He also classified plants and assisted with work on the collections of the University Museum. Most important, it was here that he became aware of his grandfather's *Zoonomia* and of Lamarck's evolutionary ideas. His disappointed father next sent him to Cambridge for a degree that would assist him in becoming a country parson, but the young Darwin preferred riding, shooting and beetle collecting. Indeed, nothing at Cambridge gave him so much pleasure as collecting beetles. One day, upon tearing off some old bark, he saw two rare beetles, and seized one in each hand; then he saw a third and new kind, "which I could not bear to lose, so that I

Fig. 7.1 Darwin, early years

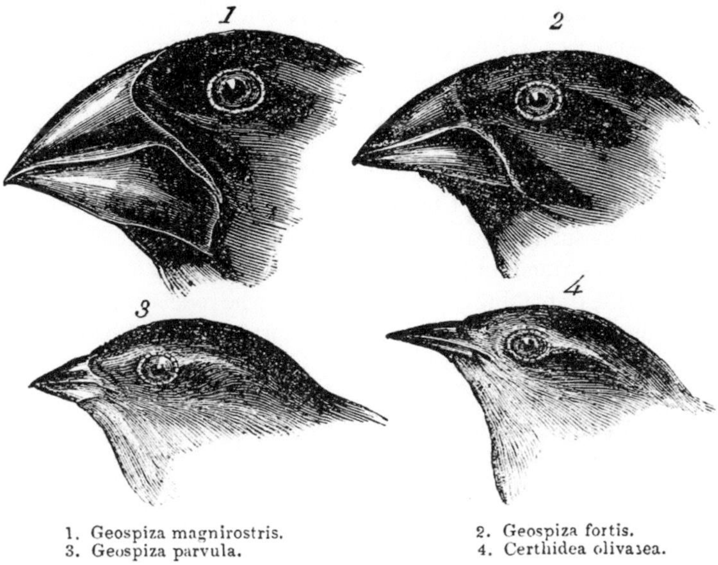

Fig. 7.2 Galapagos finches. (From Darwin (1845))

Fig. 7.3 Darwin in his middle age

popped the one which I held in my right hand into my mouth. Alas! It ejected some intensely acrid fluid, which burnt my tongue so that I was forced to spit the beetle out, which was lost, as was the third one". While at Cambridge, he ardently read both Paley's *Natural Theology* and Humboldt's *Personal Narrative*, which fired in him with a desire for travel, and from which he copied out long passages to read them aloud to his friends.

A life in a rural vicarage seemed to await him but, at the age of 22, he accepted an invitation to serve as a self-funded naturalist on the research vessel *HMS Beagle*, on a 5-year expedition that was to chart the coastline of South America and to make longitude measurements around the world (1831–1836). The captain, Robert FitzRoy, was looking for someone who would make good use of the expedition's opportunities for researching natural history, would share his scientific tastes, and provide friendly companionship. Charles was eager to accept the offer, but his father strongly objected, as such a voyage would interfere with his career as a clergyman. Fortunately, Charles's grandfather on his mother's side (Josiah Wedgwood) believed that this would be the right thing for the young lad, so he rode 30 miles to Shrewsbury and convinced the father. Then, another obstacle turned up: FitzRoy believed that he could judge a man's character by the outline of his features, and he doubted whether anyone with Darwin's nose could possess sufficient energy and determination for the voyage. Eventually, Darwin accepted the position, and years later, he would write that the Beagle voyage, by far the most important event in his life, "depended on so small circumstances as my uncle offering to drive me thirty miles to Shrewsbury, which few uncles would have done, and on such a trifle as the shape of my nose". In general, FitzRoy and Darwin got on well together during the voyage, but there were inevitable moments of strain. The captain had a violent temper (his outbursts had gained him the nickname "Hot Coffee"), and this resulted in quarrels sometimes "bordering on insanity", as Darwin later recalled.

Darwin, upon leaving England, took with him a copy of the first volume of Lyell's *Principles of Geology*. While other shipmates chartered the coasts, he investigated geology and collected animals and plants on land, leaving the ship for extended periods. Reading Lyell's *Principles of Geology* when on board, Darwin began to see geology through Lyell's eyes: expanses of land rising and falling ever so slowly, over immense stretches of time. He saw layers of seashells high up in the mountains, he experienced an earthquake, and he saw local mussel beds stranded above high tide, the mussels all dead, evidence that the land had just risen. Lyell had argued that geological formations were the result of steady cumulative forces, and Darwin had actually seen that the continent was thrusting itself up, a few feet at a time. Finding fossilized trees in sandstone (once seashore mud) at altitudes of above 2000 m, he imagined the aeons it had taken to raise them to such heights and began thinking in terms of deep time.

Turning from geology via palaeontology to zoology, during the *Beagle* voyage, Darwin found fossil bones of huge extinct mammals beside modern seashells, indicating recent extinction with no signs of drastic change in climate or of a violent catastrophe. He found bones of what he assumed were rhinoceroses, mastodons, cow-sized armadillos, anteaters and giant ground sloths (*Megatherium*). These fossils led Darwin to thinking about primeval animal worlds, and about what had caused those giant beasts to die out. Accepting Lyell's view of centres of species creation, on the geologically new Galapagos Islands, he searched for evidence that would attach wildlife to an older centre of creation. He found birds allied to those in Chile, but differing from island to island, with a remarkable diversity in beak form and function, highly adapted to different food sources (Fig. 7.2; in his *Naturalist's*

Voyage (1888), this is the only illustration). "Seeing this gradation and diversity of structures in one small, intimately related group of birds, one might really fancy that, from an original paucity of birds in this archipelago, one species had been taken and modified for different ends" (1845). In Australia, the marsupial rat-kangaroo and the platypus seemed so unusual that Darwin thought it was almost as though two distinct Creators had been at work. While sailing, Darwin's plankton-filled tow-net left him wondering why such beautiful creatures teemed in the ocean's vastness, where no human could appreciate their beauty.

In his ornithological notebooks of 1935 Darwin wrote: "When I recollect the fact, that from the form of the body, shape of scale and general size, the Spaniards can at once pronounce from which [Island] any tortoise may have been brought;—when I see these Islands in sight of each other and possessed of but a scanty stock of animals, tenanted by these birds but slightly differing in structure and filling the same place in Nature, I must suspect they are only varieties. [...] such facts would undermine the stability of species" (Darwin, in De Beer 1958a). His visit to the Galapagos Islands in 1835 was thus a critical period for his theory of evolution. He was fully aware of the implication of the revolution in thought on which he was embarking. He wrote, in his diaries of 1837: "if we choose to let conjecture run wild, then animals, our fellow bretheren in pain, disease, death, suffering and famine—our slaves in the most laborious works, our companions in our amusements—they may partake [of] our origin in one common ancestor—we may be all melted together" (Darwin 1837, in De Beer 1958a).

Aged 27, Darwin returned to England, having been away for 5 years and 2 days, and he then married his cousin (on the Wedgewood side). Her riches added to his, and that allowed him not to work for any research establishment for the rest of his life. He could carry out whatever research he wished, at his own pace. Also, on his return to England, he fell into a strange long-term illness, the bouts of which would leave him chronically listless with severe nausea, palpitations, migraines, exhaustion, shortness of breath and depression. Often, he could work for no more than 20 minutes at a time, sometimes even not that. The cause may have been Chagas's disease, a tropical malady that he might perhaps have acquired from the bite of a bug in South America. Kissing bugs (so-called for feeding around the mouths of people), the Triatominae, feed on vertebrate blood, and transfer this disease. The parasite *Trypanosoma*, which induces anaphylaxis, infests these bugs. Indeed, Darwin in his *Voyage of the Beagle* (1835) wrote that (while in western Argentina of today),

> At night I experienced an attack (for it deserves no less a name) of the *Benchuca* the great black bug of the Pampas. It is most disgusting to feel soft wingless insects, about an inch long, crawling over one's body. Before sucking, they are quite thin, but afterwards they become round and bloated with blood, and in this state are easily crushed. They are also found in the northern parts of Chile and in Peru. One which I caught at Iquiniqe [northern Chile of today], was very empty. When placed on the table, and though surrounded by people, if a finger was presented, the bold insect would immediately draw its sucker, make a charge, and if allowed, draw blood. No pain was caused by the wound. It was curious to watch its body during the act of sucking, as it changed in less than ten minutes, from being as flat as a wafer to a globular form. This one feast, for which the *benchuca* was indebted to

> one of the officers, kept it fat during four whole months; but, after the first fortnight, the insect was quite ready to have another suck.

At first, Darwin did not doubt, either his religious views or the stability of species that were all, so he then believed, created by the Lord. A series of findings that began emerging from the taxonomic studies of his collections on the Galapagos stirred his doubts as to the stability of species. In parcelling out the material he had collected to leading specialists in England, he entrusted his birds to John Gould, the most eminent ornithologist of the Zoological Society of London. Gould informed him that most birds from the different Galapagos Islands were distinct species, slightly different between one island and another, and yet overall similar, not only one to another but also to those of the nearby continent, South America (Wightman 1953).

Darwin entrusted his Galapagos plants to the botanist Joseph Dalton Hooker, a 27-year-old botanist (8 years Darwin's junior), the son of the director of the Royal Botanic Gardens at Kew). Hooker found that in plants as in birds, the flora on each island consisted of slightly different plant species. If God had created each of these species separately, why then, Darwin wondered, did God find it necessary to create a separate species on each separate island, differing only in minute details and with such a limited distribution? Why did God bother to create so many species in the Galapagos when he had created far fewer species in the entire adjacent continent of South America? The islands were all in sight of each other and all had the same geology and climate. Why did different but closely related species occupy them?

Darwin handed the large fossil mammals he had collected in South America over to Owen, who ascertained that they virtually all resemble the smaller, present-day forms that now characterize the continent: llamas, sloths, armadillos and so forth. Why did these animals become extinct if they had the same body plan as mammals living today? Another observation was that similar abiotic conditions do not result in similar fauna. The fauna of Cape Verde resembles that of Africa, whereas the fauna of the Galapagos, with similar abiotic conditions, resembles that of South America. Viscachas (Chinchillidae) are South American rodents that closely resemble rabbits, but are not closely related. Why, Darwin wondered, are these South American, rabbit-like rodents based on the special South American type, and not on the rodent type of North America and of the Old World?

In the years following his return, Darwin began to develop a hypothesis of his own: Species are mere variations that can change throughout their long existence and split into several distinct species. This hypothesis, if indeed correct, would answer all his questions. The birds of the Galapagos resemble each other (and those of mainland South America) because they are all of the common descent, having originated from a common centre—they were not created separately, each where it presently exists. They differ one from another because the flora and fauna of the islands had diverged from the parent types, each local population becoming adapted to life habits unique to its particular island, one to feeding on seeds on the ground and the other to feeding on insects on trees. The volcanic nature and abiotic conditions of Cape Verde resemble those of Galapagos, but the birds of Galapagos differ from those of Cape Verde, evidence that the abiotic conditions of the islands are not the only factors that determine differences in faunas. These differences are there because the birds of Cape Verde share a common ancestor with those of South Africa, and those of Galapagos with South America. The rabbit-like rodents of South America have the structure of the South American rodents because all South American rodents share a common ancestor. The fossil *Glyptodon* resembles an armadillo because it shares a common ancestor, and if modern species resemble fossil ones, then there is no reason to believe that extinct types of animals did not leave modern offspring. It could well be that they have living representatives, and that means that the fossil evidence is available to science, as a tool to study problems of evolution (De Beer 1958b). With this theory in his hands, gradually during the years, Darwin's religion dissipated (Fig. 7.3).

Darwin now had a *hypothesis*, that species undergo modification over time, from a common ancestor species, that they share with other species. He lacked *evidence*. His next step was to plough through all the information on zoology and botany he could lay his hands on, in search of facts that would be relevant to his hypothesis—one of the most thorough, broad attacks ever, on any scientific problem (De Beer 1958b). He noted the extensive variation in domestic plants and animals, and he learned from breeders that in one litter of domestic animals, there could be considerable hereditary variation. The breeder selects the individuals that possess a desired trait, and when these selected individuals breed among themselves, they often give birth to offspring, which may further emphasize the desired feature. In this manner, purposeful accumulative selection creates new races and breeds that differ considerably from the original ancestor race. Darwin reviewed the extent of intra-specific variation in 20 domestic animals, from horses via fowls and goldfish, through bees and silk moths (1868, 1884). He presented data that he collected from the literature, and from private correspondence, as well as the data of his own research on pigeons (Wightman 1953). Breeding fancy pigeons was then a popular pastime in Britain, with enthusiasts spanning the entire social spectrum. Darwin's pigeon research was very thorough in that he raised many breeds, prepared skeletons and received skins from as far away as Persia and India (Figs. 7.4 and 7.5). He also joined two pigeon-breeding clubs, corresponded with amateurs, reviewed the collections in the British Museum and delved into the history of pigeon breeding, back to 1600. Classifying the pigeon breeds into races, sub-races, varieties and sub-varieties, he distinguished between the wild rock pigeon, *Columbia livia*, those semi-domesticated races allowed to mate freely in captivity, and those fully domesticated races, subject to purposeful matching and selection of the offspring. He found variation in very many features, which I list here in full, both to present what Darwin had in mind when he thought of variation and as evidence of his very detailed, thorough research:

Beak and accompanying bones (length, breadth, curvature)
Skull shape (shape)
Lower jaw (curvature)
Mouth (gape)
Tongue (length)

Fig. 7.4 Five of the pigeon breeds in Darwin's study of variation within a species (1884): English carrier, English pouter, English fantail, African owl, short-faced English tumbler

Wattles around nostrils and eyes
Eyelids (length)
Nostrils (length)
Oesophagus and crop (capacity of inflation)

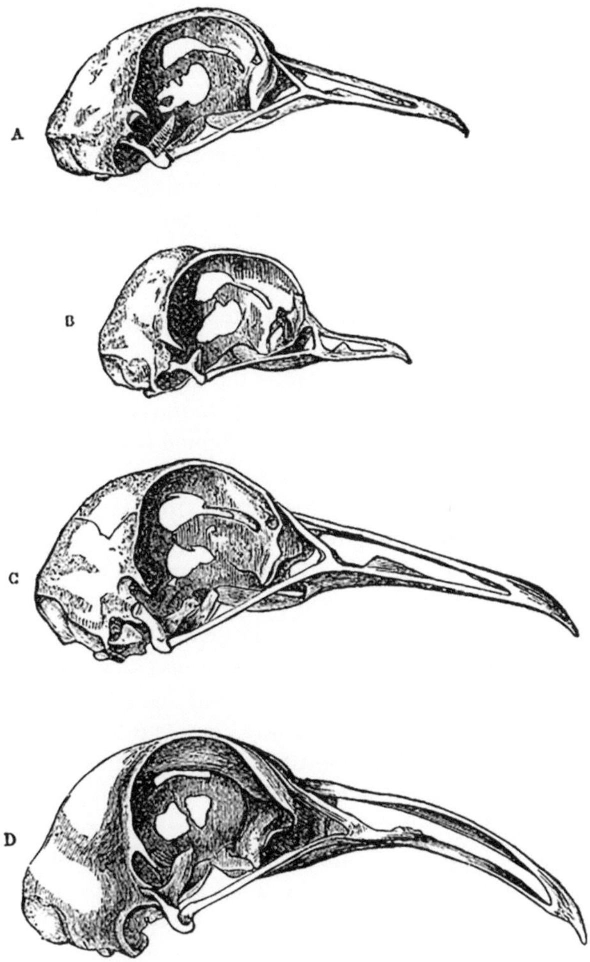

Fig. 7.5 The skulls of four pigeon breeds in Darwin's study of variation within a species (1884): wild rock pigeon *Columbia livia*, short-faced tumbler, English carrier, Bagadotten carrier

Neck (length)
Rib number and breadth, presence of rib processes
Sacral vertebrae (number)
Sternum (length, perforations)
Furculum size and divergence of arms
Oil gland (development, sometimes absent)
Feather (direction, length, colour)
Wing feathers (size and number)
Tail feathers (size and number)
Leg (size)
Foot (size)
Body (size)
Egg (size and shape)

Down cover (extent)
Manner of flight
Song.

These features may vary, in Darwin's opinion. Over the generations, nature creates slight variations and man, by intentional artificial selection, steers them in directions that are useful to him, thus forming new races. Darwin agreed with Leibniz that *Natura non facit saltum* (there are no jumps in nature), concluding that "it would take a great deal more evidence to make me admit that forms have often changed *per saltum*".

If man can, by selection, create so many races of pigeons, dogs and horses, can nature not create species in a similar manner? The difference between a poodle and a bulldog is much greater than the differences between the diverse species on the Galapagos. Could it not be possible that in nature, too, there is a selective force, comparable to a human breeder, which selects new forms of animal races? What could that force be?

Two years after having returned from the Galapagos, on 28 September 1838, Darwin began reading the studies of Malthus on the dynamics of populations, and here he found what he sought. Thomas Robert Malthus (1766–1854) was not a biologist. The son of a rich landowner in England, he had no interest in classifications or in evolution. He was a Professor of Economics at the University of London. The decade after the French Revolution saw such topics as "rights of man" and "natural justice", and people believed that a day was dawning when, amidst universal peace, all men would enjoy complete liberty and complete equality. Malthus pointed out the difficulties of such ideal dreams in states of over-population. During discussions with his father about connections between riches and happiness, he developed a theory that he published as *An Essay on the Principle of Population as it affects the Future Improvement of Society* (1798). Its main point was that every population could multiply itself at a geometric rate, whereas the means of living can multiply at only an arithmetic rate. Among plants and animals, as new individuals are born in numbers far more than the means for their existence, many perish by want of room and nourishment. Man differs from animals in that reason interferes and asks whether one should indeed produce offspring, for whom one cannot provide means of support. Consequently, certain human population constraints will be taken at will (marriage to one wife only, at not too early an age, moral sexual life). If these checks of reason fail, other checks will come into action: starvation, disease, epidemics and wars. Inequality appears in a population growing beyond its ability to survive, with poverty-stricken people whose children lack education, a life of crime and a violent competition for survival. Just as if we eat too much, our health will fail, so also if we reproduce too much, many of our descendants shall die in poverty and disease. The one way to maintain happiness is by living in restraint, and by not establishing a family before having beforehand the economic ability to feed children and educate them. Malthus was aware of variation among domestic stock and of artificial selection as a means to improve races. He emphasized, however, that there were limits to variation (Book III chapter 1): Farmers breed the

Leicestershire sheep to produce offspring with small heads and legs, but sheep will never have the tiny heads and legs of a rat. Claims that progress has no limits are absurd, the error lying not in supposing a small degree of improvement impossible, but in not discriminating between a small improvement, the limit of which is undefined, and an improvement unlimited (Wightman 1953). Malthus' theories of population control largely resemble those of Sir Matthew Hale (1609–1676), an influential English jurist whom Cromwell, during the Civil War, had appointed as a Justice of the Common Pleas, and whom Charles II, during the Restoration, had appointed as Chief Baron of the Exchequer and as Chief Justice of the King's Bench. Hale had recognized the struggle for existence and had listed checks to excessive population increase: plagues, epidemics, famines, wars, floods and disastrous fires. Such checks, happening once in only 5, 10 or 20 years, result in an overall consistency of animal and man population numbers.

Malthus theory of population control gave Darwin the idea: in nature, all organic beings tend to increase at so high a rate that the descendants of just one single pair, if not checked, will cover the entire surface of the earth, or the entire ocean, within a few generations, far beyond what the Earth can support. The inevitable result will be a perpetual struggle for existence among the descendants, a struggle in which the strongest shall prevail, whereas the weakest shall fail, and die out. In this struggle of existence between individuals of a species, variation—even if ever so slight—determines the victor. Slight variations occur among individuals of a species, and these determine which shall be the stronger individual. The gazelle with slightly longer legs will flee the lion faster than the slightly more short-legged gazelle and shall survive to reproduce long-legged offspring. "This preservation, during the battle of life, of varieties which possess any advantage in structure, constitution or instinct, I have called Natural Selection" (1884, Introduction). In the wars of survival under conditions of over-population, those that do not succeed in adapting better than others to existing conditions perish, whereas the strong, best-adapted individuals survive and multiply. That is how the species change, and that is how new species are formed. The struggle for survival is that which creates natural selection, which operates in a manner similar to the selection of races of domestic animals, by man (with the difference that nature has at its disposal a much longer time). "Owing to this struggle, variations, however slight, if they are in any degree profitable to the individual of a species, tend to the preservation of such individuals, and will generally be inherited by the offspring". Elsewhere, he wrote that this is the doctrine of Malthus, applied to the whole Animal and Vegetable kingdoms. Given (virtually) endless variation, given extinction and given adaptation, enormous pressure operates in nature, in favour of the individual that is more adapted, as against the less adapted. From these components, of unrestrained multiplication, of adaptation, of extinction and of ecological niches, Darwin distilled his theory of evolution by natural selection. It was based on seven points:

- All organisms vary; no two are ever alike; hereditary variation is random.
- More offspring are born than can grow to maturity.

- Consequently, there is competition between the individuals of every species (a struggle for existence).
- The weakest, or the least fit among the varying offspring, are eliminated in this competition; the fit or better adapted to the environment will survive. This is natural selection.
- The hereditary features that give the survivors an advantage will carry on to succeeding generations.
- Over time, in ever so slight steps, selected variations thus modify a species into a new one.
- This new species will form a basis for new variations and so on.

Darwin's theory of natural selection accounted for the origin of species neither through special creation nor through the inheritance of acquired characters but through selection by survival of the fittest. For new species to arise, it was only necessary for the old species to vary and for the variations to be heritable. Whereas hereditary variation is random, evolution via natural selection is directional. Populations that are too large and a diversity of offspring followed by a struggle for life result in natural selection of the fittest—these explain not only the origin of all species in nature but also the entire hierarchy of the animal and plant kingdoms. The boundaries between species and varieties are vague; species are varieties that have evolved to a higher rank by natural selection, while varieties that were defeated in the struggle perished. A species is, according to Darwin, a category created merely for convenience, for several individuals that resemble each other and interbreed, and in this, the term "species" does not differ essentially from the term "variety", which serves for less well-defined forms. "There is no clear line between a species and a subspecies […] or again, between a subspecies and well-defined variations; or between secondary variations and individual differences. These differences merge one into another, in a continuous series" (1859). Randomness is an integral part of the evolutionary process in that hereditary varieties appear, whether or not they bring any benefit to their bearers. Natural selection cancels this random process. Random variation on the one hand, and directional selection on the other, together lead to evolution. Species extinction is common in the evolutionary process; indeed, the vast majority of all species that have ever existed on Earth—have gone extinct. Many diverse animals are of common descent, species are not constant, and in some cases, are impossible to define.

Natural selection operates ever so slowly, by tiny variations that accumulate, over an almost endless number of generations, each slightly improving the survival chances of the individual. "It may be said that natural selection is daily and hourly, scrutinizing throughout the world, every variation, even the slightest; rejecting that which is bad, preserving and adding up all that is good; silently and insensibly working, whenever and wherever opportunity offers, at the improvement of each organic being in relation to its organic and inorganic conditions of life. We see nothing of these slow changes in progress, until the hand of time has marked the long lapse of ages, and then so imperfect is our view into long past geological ages, that we only see that the forms of life are now different from what they formerly were"

(1859). Usually, it is the environment that operates natural selection, and hence its great importance. However, not the environment but natural selection is the force that determines evolutionary change. On islands, for example, some groups of insects have a limited ability to fly compared to their relatives on the continents: The well-flying individuals flew into the sea and drowned; the variety of weakly flying ones remained on the island, living on and reproducing, and thereby evolving into a new, weakly flying species.

Darwin's theory of common descent was revolutionary due to the forces he involved in the evolutionary process: The continuous production of an inexhaustible supply of genetic variation, and the differential survival among this oversupply of individuals, through natural selection. In 1838, Darwin had worked out his theory of natural selection, and 4 years later, in 1842, he committed his ideas to paper in the form of a short, rough sketch. In 1844, he wrote down a pencil sketch of 231 folio pages, a complete presentation of the arguments familiar to us in the *Origin*. If, as he feared, he would die before his study would be finished, his wife was to see to the publication of this draft, and his great theory would thus reach the world; for this purpose, he set aside £400 (£42,500 in today's currency; Colp Jr. 1986).

We have, some paragraphs above, mentioned Hooker, the botanist who investigated Darwin's collection of Galapagos plants. Darwin and Hooker exchanged letters about the plants, and Darwin revealed to him that "I am almost convinced (quite contrary to opinion I started with) that species are not (it is like confessing murder) immutable. Heaven [distance] me from Lamarck's nonsense of 'tendency to progression', adaptations from slow willing of animals" &c, but the conclusions "I am led to are not widely different from his—though the means of change are wholly so—I think I have found out (here's presumption!) the simple way by which species become exquisitely adapted to various end" (Darwin, in Colp Jr. 1986). Darwin thus disassociated himself from Lamarck.

Darwin revealed to Hooker his two main revelations, that he had discovered a theory of transmutation and that his theory differed from Lamarck's (Colp Jr. 1986). He did not tell Hooker (or to any other correspondent) that he was writing a 231-page sketch of his transmutation theory, of variation in nature and natural selection, of the gradual production of complex organs and of evidence for the concept of descent from common stocks, from palaeontology, distribution, classification, morphology, embryology and rudimentary organs. This sketch was his first coherent presentation of his theory (Colp Jr. 1986). Throughout 1846 and 1847, Darwin steadily extracted facts and views from Hooker about the classification and distribution of plants—through letters and through meetings, at Down or at the Kew Gardens. Hooker would sometimes stay for several days at Down, and later he recollected those visits. "It was Mr. Darwin's practice to ask me, shortly after breakfast, to retire with him to his study for twenty minutes or so, when he brought out a long list of questions put to me on the botanical subjects then engaging his attention. These questions were sometimes answered offhand, others required consideration, and others research in the Herbarium or in the gardens at Kew. The answers were written on slips of paper, which were deposited in bags or pockets which hung against the wall within reach of his arm, each of them a receptacle devoted to a special object of inquiry. … These morning interviews were followed by his taking a complete rest, for they always exhausted him often producing a buzzing noise in the head, and sometimes what he called 'stars in the eyes', the latter too often the prelude of violent eczema in the head, during which he was hardly recognisable" (Hooker in Colp Jr. 1986). Darwin's attacks of violent eczema in the head which occurred during his morning discussions with Hooker suggested to Colp Jr. (1986: 29) that they might have been psychosomatic symptoms caused by anxiety over his disagreements with Hooker about the ability of species to change, and the past migration of species. Hooker read Darwin's sketch and offered many comments that aided Darwin in improving his text, but nowhere did Hooker indicate that Darwin's theory had won his approval or influenced his thinking. Hooker would later reflect that at the time he failed to grasp the full significance of the theory and did not assimilate it at once and apply it to his own work on the distribution of plants (Colp Jr. 1986).

Darwin revealed his theory also to Jenyns, that species are mutable, and that allied species are co-descendants from common stocks (Colp Jr. 1986). Leonard Jenyns (1800–1893), an English clergyman and naturalist, was a founder member of the Ray Society and of the Zoological Society of London, a member of the Society of Entomologists of London and of the Geological Society of London; in Cambridge, he set up what would become the Cambridge University Museum of Zoology. John Stevens Henslow was his—and Darwin's—mentor. Jenyns was the original choice for a naturalist on the *Beagle* voyage, but he turned down the offer. His diary entry for 1831 records "This year I had the offer of accompanying Capt. Fitzroy, as Naturalist, in the Beagle, on his voyage to survey the coasts of S. America, afterwards going round the globe: [I] declined the appointment, that was afterwards given to Charles Darwin Esq. of Xts' College Cambridge". A decade later, he wrote the chapters on Fish, in Darwin's (1845) *The zoology of the voyage of H.M.S. Beagle, under the command of Captain Fitzroy, R.N., during the years 1832–1836*.

Darwin revealed his theory also to Lyell, whom he told (1938) of the "delightful number of new views, … bearing on the question of species" that he was writing up. Lyell, however, looked at the *Beagle* observations and did not see any changing of one species into another. In the successive editions of his *Principles of Geology*, Lyell claimed that species had been separately created at different times and by unknown means and that the evidence did not support any transmutation theory, especially that of Lamarck (Colp Jr. 1986).

Having completed the sketch, Darwin now put aside all further writing of evolution by natural selection, in favour of an eight-year study of barnacles (Winsor 1969). He might perhaps have put his ideas on evolution aside because in England of the 1830s and 1840s, the concept of one species changing into another had been publicly found to be faulty by William Whewell and Adam Sedgwick, leaders in the scientific establishment, as a fallacy associated with such anathemas as spontaneous generation, lack of design, and the animal origins of man. Transmutation was also associated with the theory of Lamarck, which was held to have deleterious religious, social and moral consequences (Colp Jr. 1986). Whatever his reasons, Darwin now turned to barnacle classification (Newman 1993).

Linnaeus, a century earlier, had grouped barnacles and goose barnacles among the molluscs because of their calcareous shell and lack of segmentation. Lamarck (1809) had placed them in a class to themselves, the Cirripedia (from the Latin *Cirrus*—fringe, and *Pes*—foot), considering them as a transition group between the class of Articulata (annelids and crabs) and the class of Mollusca (snails and bivalves). Some years later, Cuvier (1812) denied, however, the possibility of any intermediate between the Articulata and Mollusca, so he classified the barnacles and goose barnacles as the sixth class of molluscs. A decade later, Blainville (1822) placed them in a separate group with chitons, the "Malentozoa", thus upholding Lamarck's view. These early naturalists believed that barnacles differ profoundly from crustaceans in that they are sessile, not mobile; in that their body does not consist of segments; and in that they are hermaphrodites, not bisexual.

It was John Vaughan Thompson, a British military surgeon by profession and a marine biologist by inclination, who discovered that the larval form of barnacles is the *naupilius*, a form typical only of crustaceans. Thompson had settled in Cork (Ireland) as an inspector of hospitals, and on the ferry crossing the harbour he would cast towing nets overboard, collect plankton and later observe them under magnification. In 1823 he observed, to his amazement, that some plankton with stalked eyes had "thrown off their excuvia and wonderful to say, were firmly adhering to the bottom of the vessel and changed into young Barnacles! ... Another individual was seen in the act of throwing off its shell, and attaching itself as the others, to the bottom of the glass". ("excuvia", in arthropods, are the remains of a moulted exoskeleton). This observed metamorphosis bore evidence that barnacles are crustaceans, together with shrimps, crabs, isopods and amphipods, and Thompson published his results in 1827 and 1835. Most naturalists from now recognized barnacles as a group (Cirripedia) belonging to the Crustacea, and define them as sessile, predominantly suspension feeders, with two active swimming larval stages, the *nauplius* and the *cypris*. The *nauplius* stage is characterized by the use of the appendages of the head (the antennae) for swimming, and by the appearance of a simple, unpaired eye. It develops into a *cypris*, a nonfeeding larval stage that has a bivalved shell of chitin, cement glands on the antennules (first antennae), and a series of swimming legs on the chest. The *cypris* eventually cements itself to a hard substrate and undergoes metamorphosis to the sessile, adult form (Winsor 1969; Newman 1993).

This, broadly, was the state of the art when Darwin began examining several invertebrates that had remained undescribed from his *Beagle* collections, among them some tiny, shell-less barnacles that he had found off the coast of Chile, buried within a hole they had drilled in the shell of a living sea snail. He immediately recognized them as cirripedes, initially named them *Arthrobalanus*, later renamed them *Cryptophialus* (Fig. 7.6), and he originally intended to describe only this single abnormal barnacle.

Comparative aspects led him, however, to examine the anatomy of many other genera, and eventually, the keeper of the zoological collections at the British Museum (John Edward Gray, himself a barnacle expert) suggested that he prepare a monograph of the entire group. This resulted in eight "barnacle years" that culminated in two volumes of living Cirripedia (1851, 1854), and two fossil ones (1851,

Fig. 7.6 *Cryptophialus*. (From Darwin plate XXIII, Fig. 5)

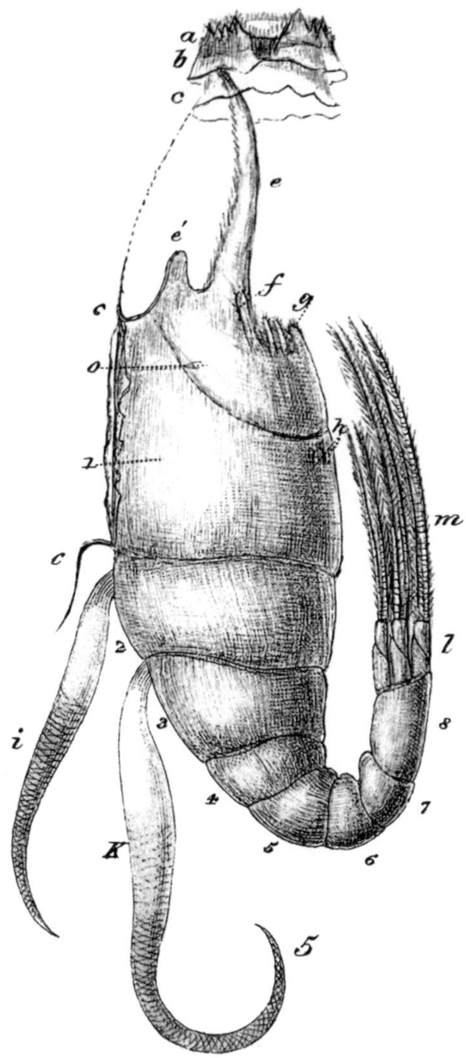

1855). Knowledge of barnacle systematics and morphology was then in an early stage of development, and Darwin based his classification on the number and forms of the plaques composing the shell, and of the muscles attached to them. He coined a specific name for each plaque, and this detailed terminology, still used today, allowed him to compare shell morphology across species. Going about the study of barnacles in a very thorough manner, he provided detailed accounts of larval development and metamorphosis, of the nervous system and sense organs, of feeding and of growth. He called the cypris a *pupa*, likening it to the pupa of insects. The metamorphosis of a *cypris* into a juvenile barnacle is indeed as profound as that of a caterpillar into a butterfly, in that it involves much cell and tissue lysis and, though a case of convergence, this reflects Darwin's panoramic view of the animal kingdom. Darwin correctly homologized the chest limbs of the *cypris* to the cirri of the adult and, countering previous beliefs that barnacles are always hermaphrodites, he described several species in which there were males, living in close relation with females or with hermaphrodite adults. He viewed his discovery of males and complemental males in these two genera as a remarkable illustration of a gradual change in nature, within a few members of a group, from one condition to another, in this case, from hermaphroditism to separate sexes (1854; Winsor 1969; Newman 1993).

A major feature that distinguishes barnacles from other non-parasitic crustaceans is attachment throughout life, and Darwin studied in detail how barnacles accomplish this attachment. Here, in describing how the *cypris* larva settles, he blundered. He thought he saw a connection between the cement glands and the ovaries ("ovarian tubules"), he missed the connection between the oviduct and the genital aperture and oviducal gland ("acoustic organ"), and he postulated a connection between the oviduct and the salivary gland ("ovarian gland"). Thus, to Darwin, the female genitalia had no openings to the exterior. He explained this by postulating that eggs went from inside the body into the mantle cavity during a moult. He further suggested that in the larva, the cement escapes through the prehensile antennulae; and that in adults, the cement gland forms from a part of the ovarian tube. Darwin erred also in interpreting some limbs of the *nauplius* (Fig. 7.7). Cirripede *nauplii* bear two pairs of appendages in the front of their heads. The more anterior "frontal filaments", do not occur in any other crustacean (except in some peculiar recently discovered crustaceans living in marine caves). Darwin erroneously thought that these frontal filaments were articulated (Fig. 7.7b), and their position and supposedly articulated morphology led him to erroneously interpret them as the first pair of antennae. The second pair of appendages in the cirripede nauplius comprises the "horns" (Fig. 7.7c), chemosensory organs specific to cirripedes. Darwin erred in interpreting these hollow tubes as the second pair of antennae. Consequently, he mistakenly took the other three pairs of appendages (Fig. 7.7e–g) to be appendages of the chest. These three pairs of appendages are actually those found in almost all crustacean nauplii: the first and second antennae and the mandibles. Further errors concern the non-segmented posterior end of the arthropod body (the telson), which he misidentified as the abdomen (Winsor 1969; Newman 1993).

Beyond his classification, Darwin also sought to construct a cirripede archetype, a basic body plan, an ideal model that does not represent any real animal, extant or fossil. We recall that Owen had proposed this archetype concept a few years earlier, as an ideal model to which the anatomist can refer in order to draw homologies between species, despite anatomical differences. Owen had erected an archetype of vertebrates, and Darwin now sought to erect an archetype of barnacles, that he would compare to a general archetype of crustaceans, to which the barnacles belong. Accordingly, he presented a figure that "gives at a glance the homologies of the external parts" between a cirriped and a prawn (Fig. 7.8). As a representative of the cirripeds, he chose the goose barnacles (genus *Lepas*), which attach to the substrate by stalks. They cement the antennules of the *cypris* larvae to the substrate and then elongate that region of attachment into a stalk. As a representative of all other

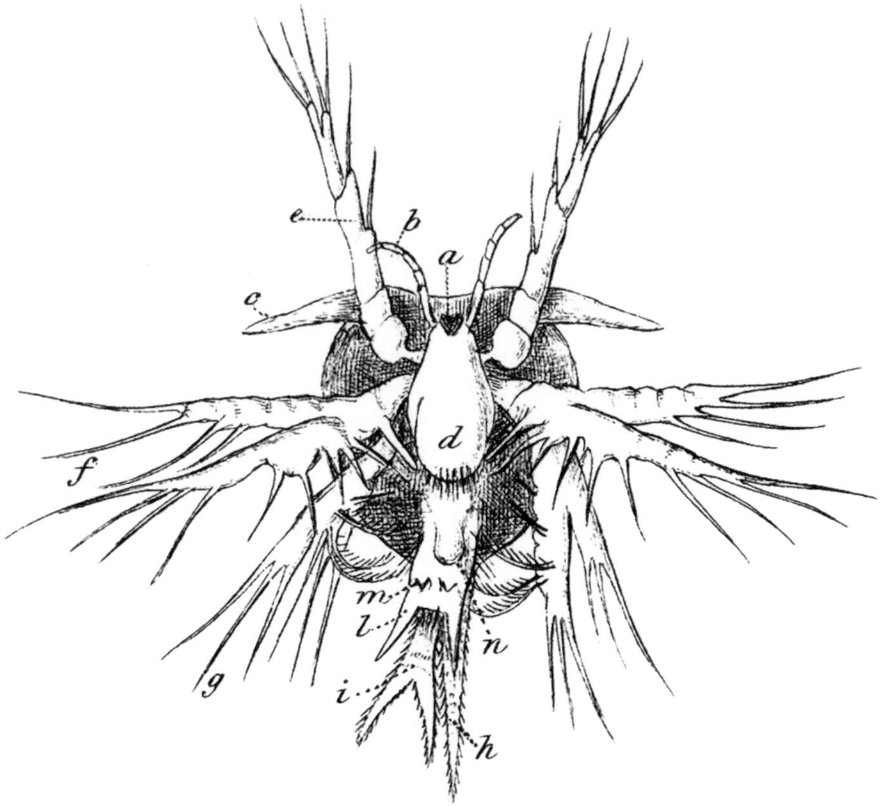

Fig. 7.7 Darwin's interpretation of naupliar appendages. Nauplius larva of *Chtamalus stellatus* (a sessile barnacle). (**a**) Eye; (**b**) first pair of antennae; (**c**) horns, including the second pair of antennae; (**d**) mouth; (**e**) first, uniramous natatory leg; (**f**) second, biramous natatory leg; (**g**) third, biramous natatory leg. (From Darwin)

crustaceans, he chose the prawns (Malacostraca), the largest of crustacean classes. His model prawn was the genus *Leucifer* (now called *Lucifer*), a degenerate prawn with a long body but fewer appendages than other prawns, without claws, without gills and with long eyestalks (p. 28). Darwin, following Milne-Edwards, wrote (p. 562) that the archetype crustacean consists of 21 segments, and this fits well with the body plan of prawns, which is composed of six head, eight chest and six to seven abdominal segments. He illustrated the abdomen of *Lucifer* only in faint lines, because "the abdomen, which we know becomes in cirripedes after the metamorphosis, rudimentary, does not fairly enter into the comparison". We, today, would find it hard to reconstitute the malacostracan body plan from this sketch of *Lucifer*.

Darwin believed in the origin of sessile barnacles from a stalked ancestor. His drawing of the archetype cirripede presents seven pairs of cirri, suggesting seven chest segments. For the anterior end, Darwin wrote that "the lower figure is a mature

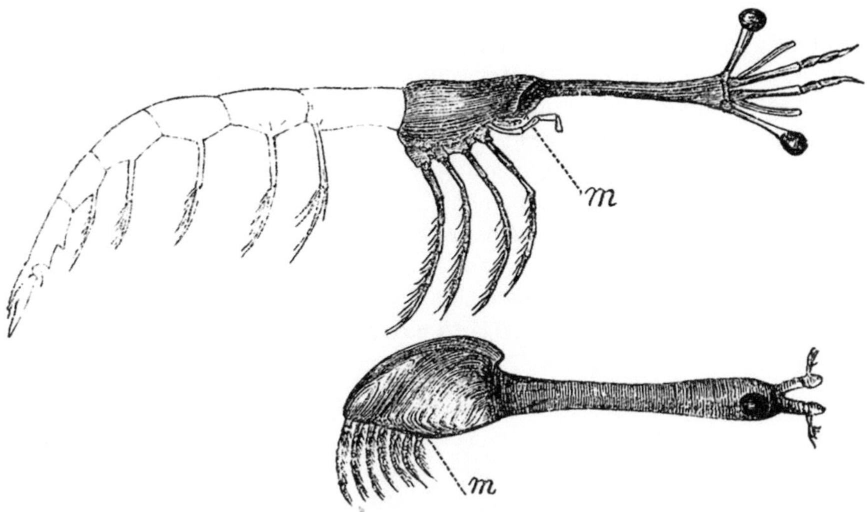

Fig. 7.8 Darwin's comparison between a decapod crustacean and a cirripede. (From Darwin, p. 28)

Lepas, with antennae and eyes, which are actually present in the larva, retained". Thus, the picture shows a chimera with mixed features, some belonging to the adult, some found only in the larva. Darwin did not find antennae at the base of the stalk in the adult. What he actually found is that, in the larva, the cement gland is located within the antennulae. While this is correct, it does not imply that the antennulae are located at the tip of the stalk in the adult. The adult cement glands, which derive from the larval ones, are in fact located at the base of the stalk. Furthermore, the eyes of the *nauplius* degenerate during the metamorphosis of the *cypris* into the juvenile barnacle, yet Darwin put them near the tip of the stalk, because he thought that this represents the anterior of the animal, where the eyes would be, had they remained. Barnacles have six chest segments, while prawns have eight. To homologize the body plans of these two different crustaceans, segment by segment, Darwin suggested that in some cirripedes, "there must be two segments missing between the outer maxillae and the first thoracic pair of legs". He thought he had found these two lacking segments in a curious cirripede that he named *Proteolepas bivincta*, for which he created a special order, the "Apoda", comprising only this species, which Darwin described from a single specimen. Adding its two additional "thoracic" segments and three abdominal ones (see above), Darwin brought the number of segments in the cirripede archetype up to 17, as compared with the 21 of the "archetypal crustacean", "the four missing being abdominal, and, I presume, the four terminal segments". A recent re-examination of Darwin's specimen has revealed that it is not a cirripede, but an isopod that parasitizes cirripedes (Winsor 1969; Newman 1993). With all these drawbacks in Darwin's cirripede archetype, it formed a classification that, beyond being descriptive, was explanatory in an evolutionary perspective, of descent with modification.

Another notable aspect of Darwin's study of barnacles was the large amount of variation he found in a species, causing considerable difficulty in deciding whether such variations indicated different species or merely varieties (*Living Cirripedia* 1854: 155). Darwin had difficulty in deciding where within-species variations stopped and distinct species began, and he became aware at first hand of the variation that occurs within a given species. This reality of nature's diversity was very far from the type concept, so popular among naturalists since the times of Linnaeus, of each species having a type, to which all individuals of a species are closely similar.

Darwin, in his eight barnacle years, placed the classification of barnacles on a firm footing (Fig. 7.9). Towards the end of his study, he became deeply fed up with the group and wrote to a friend (1852): "I hate a barnacle as no man ever did before,

Fig. 7.9 Barnacles. (From Darwin)

not even a sailor in a slow-sailing ship". In later years, most of his interpretations of cirripede anatomy proved wrong, but he gained, from his barnacle years, an awareness of the large amount of variation that might occur in a species.

In 1856, 2 years after having completed his barnacle studies and 12 years after having written his sketch, Darwin began to push forward his ideas about a book on natural selection. Earlier, though, he wrote of his theories to friends, in private letters, and already in 1837, he opened a notebook on the subject—the first point in his slow, 12-year-long journey towards his *Origin of Species*. Darwin noted that in animal groups, comparative anatomy reveals the presence of common archetypes. Vertebrates are a large group in which the different skeletal parts of the front limbs fit bone-by-bone, from the top of the arm to the last digit of the last finger, whether they serve for walking, running, swimming or flying, whether in a frog, lizard, tortoise, hare, seal, bat, bird or human (De Beer 1958a). The basic morphological similarity is, therefore, evidence of hereditary relationship, and the assumption that all vertebrates have descended from a common ancestor could explain this homology. Beyond comparative anatomy, comparative embryology also reveals an amazing similarity in structure between the young embryos of all vertebrates, whether a fish, a lizard, a bird or a human. This similarity goes down to such details as the manner in which the blood vessels turn from the heart towards the aorta. Clearly adaptive in a fish that breathes through gills, it has no significance in a lizard, a bird or a human, in which gill chambers form in the embryo, but rapidly transform into new, different structures, while other structures (lungs) serve for breathing. How can we explain this similarity between the embryos, if not by descent from a common ancestor? Embryology also supplies evidence about rudimentary remains of organs. Once, in some common ancestor, these organs carried out important functions; but today they carry out other functions or none at all. Such rudiment structures would include the limbs of snakes, or the embryonic remnant of an egg tooth, which still appears in certain marsupial mammals—even though 75 million years have passed since their ancestors needed it to break through the eggshell and emerge from it. Again, common ancestry explains all these cases.

In the fossil record, Darwin noticed that the deeper (older) a geological layer, the fewer fossils of recent species it contains. Palaeontology thus shows that new species appeared rather late and that old species gradually died out (and not all at once, as we might expect from Cuvier's catastrophic extinction theory). This can be so only if we accept that new species form by descent and modification from other species.

Classification also fits in. We classify animals by the extent of their similarity, placing them in a few large groups such as molluscs, mammals and so on. We subdivide each of these large groups into smaller, secondary groups—the mammals, for example split into rodents, carnivores, ungulates, primates and so on. We further subdivide these secondary groups, and so forth. Our classification always places species in groups, which are within larger groups. Why do we classify species in this manner? Why do they not all form a single Ladder of Life? The arrangement of groups within groups is a natural classification that reflects the progress of evolution. It is the result of descent from common ancestors, and evidence of close relationships, with the differences between groups resulting from splitting and modifications of descendants. This generalization applies to all animals and also to Man: "… animals, our fellow brethren in pain, disease, death, suffering and famine—our slaves in the most laborious works, our companions in our amusements—they partake [of] our origin in one common ancestor—we may be all melted together" (Darwin 1902: 170). The links that bind species to genera are of the same nature as those that determine the relationship of subspecies and varieties. The relationship between the various genera of a family is between the species of a genus. Descent with modification is the chief law of nature throughout the living world. Upon launching the first volumes of the *Zoology of the Voyage of HMS Beagle*, Darwin was elected Fellow of the Royal Society (1839). We recall the Quinarian, circular system of classification that Macleay had developed, and brought into practice for insects in 1819–1821, and Swainson for molluscs in 1840 (Part 6 in this book). Darwin argued against the circular classification system. In his opinion "**classification consists in grouping beings according to their actual relationship, i.e., their** consanguinity, or **descent from common stock**" (1843; Colp Jr. 1986).

Darwin also studied hybridization. He noticed that some distinct species form fertile hybrids, and consequently, hybrid sterility was in no way absolute. From this aspect, species behave like varieties within a species. Why, then, can species not develop from the beginning as varieties, by descent and modifications from other species? Speaking of intermediate forms, he remarked: "Opponents will say—show them to me. I will answer yes, if you will show me every step between bulldog and greyhound" (Darwin 1902: 170).

Fifteen years after having written the sketch draft of his book (1857), Darwin received a friendly letter and the draft of a paper from an English hunter who was roaming remote wild places in the Far East, in search of animals that he would sell to zoos and museums. Alfred Wallace (1823–1913, Fig. 7.10) came from a modest social rank (his father made a living from teaching a few young scholars), and he had to leave school at 14 for financial reasons. He could not afford university education but as an avid reader, he educated himself in public libraries, focusing on natural history and geology. Later, he decided to be an animal hunter and travelled to South America on a collecting journey; a disastrous fire on board the ship during the return voyage burnt most of his collections. His second expedition was to the Far East, to which he set off in 1854 and returned 8 years later, with a collection of over 125,000 specimens, from vertebrates through molluscs to insects. He had found orangutans and gliding frogs. He had seen also how the natives climb trees into hides, and from there shoot down, male birds of paradise in a lek with bows and arrows (a lek is an aggregation of male animals gathered to engage in competitive displays, to entice visiting females that are surveying prospective partners to mate with). He had seen ever so many, ever so strange insects (Fig. 7.11). Also, during his long period of travelling in the Malay Archipelago, he discovered that there was a host of plant and animal species found only on the western side of a line on a map (today's Wallace line) and another host of species found only on the eastern side of that line. This represents the line joining the points of contact of Asian and Australian species, advancing towards each other.

Fig. 7.10 Alfred Wallace in Malaya

When immobilized by an attack of Malaria, he had time to re-read Thomas Malthus's essay on populations, and this gave him too the idea that natural selection could be the active agent causing evolution. He put his ideas in an essay for publication, *On the Tendency of Varieties to depart indefinitely from the Original Type*, and sent it to Darwin, in England, requesting that Darwin forward the essay to Lyell. (Wallace and Darwin were already corresponding and Wallace, generously, had several times sent Darwin specimens that he thought might be of interest.)

In his essay, Wallace argued that the life of wild animals is a continuous struggle for existence and for providing for offspring. Animal populations are broadly stable in nature, and nearly all the immense numbers that are born perish. As the survival of each individual depends upon itself, those who succumb in the struggle for existence are the weakest—the very young, the aged, the diseased. Those that survive are the most healthy and vigorous—those that are best at obtaining food and avoiding enemies. This applies both to individuals and to species: Those best adapted to obtain food and to defend themselves against enemies become more abundant, while others diminish in numbers and eventually become extinct. The survivors are variants that have an ever-so-slight advantage in habits or abilities. A slight change in colour may conceal one individual more than the others; a slight increase in limb length may enable one individual to run slightly faster. Such a better-adapted variety may slightly increase its chances of survival. Over time, this variant would become more numerous, and eventually, it would become the typical variety of the species. It would replace other varieties that would go extinct. Later on, this new, improved

Fig. 7.11 Some of the strange animals Wallace saw on his trip to Malaya. (From Wallace (1894)). Orangutan, a gliding frog, insects and natives hunting lekking male birds of paradise from behind shelters

Fig. 7.11 (continued)

and populous race might itself eventually give rise to yet newer varieties, which might replace it. The result would be continued divergence from the ancestral form, by minute steps.

Wallace further noted that whereas high systematic groups, such as classes and orders, are usually widely distributed, low systematic groups such as genera and species frequently have very small localized distributions. Also, when a group is confined to one region and is rich in species, usually the most closely allied species are in the same locality or in closely adjoining localities and, therefore, the natural sequence of the species by affinity is also geographical. Also in the fossil record, high groups extend through several geological formations whereas low groups do not (De Beer 1958a). From these observations, Wallace concluded that "every species has come into existence coincident both in space and time with a pre-existing closely allied species".

Thought out about 1835, written in Sarawak in 1855, and published in the same year, Wallace wrote that the natural series of affinities will also present the order in which the several species came into existence, each one having had for its immediate ancestor a closely allied species existing at the time of its origin. It is possible that two or three distinct species may have had a common ancestor, and that each of these may have become the ancestor from which other closely allied species were created (De Beer 1958a; Wallace used the term antitype for the ancestor).

All the ideas that Darwin had developed and elaborated over the years were here in the essay of an unknown animal hunter. In forwarding Wallace's essay to Lyell, Darwin sadly wrote that "If Wallace had my MS. sketch written out in 1842, he could not have made a better short abstract!" Darwin felt that he could no longer honourably publish a paper claiming evolution be natural selection as his own idea, and yet he was reluctant to give up the results of his many years of painstaking study. Will Darwin lose priority? Lyell and Hooker, Darwin's two friends, came to his help, with a unique suggestion. They told him to seek anything that he had ever written on the matter. Eventually, they found evidence that long ago, Darwin had **intended** to write things. They decided on a joint communication at a meeting of the Linnaean Society of London that they would present in such a manner that Wallace could not claim priority of publication (Spearling 2002). Their communication consisted of an introduction, explaining that Darwin and Wallace had, independently and unknown to one another, conceived the same theory to account for the appearance and perpetuation of varieties and species. Neither of them had published his views, though Mr. Darwin had for many years past been repeatedly urged to do so. In order of dates, their communication consisted of an extract from an unpublished text by Darwin (1844); an abstract of a letter from Darwin to a professor in Boston (1857); and, lastly, a recently submitted text by Wallace (Darwin and Wallace, 1858). By thus listing unpublished texts with a submitted one, all in chronological order, Wallace lost priority and Darwin gained it. Some 30 people were present at the Society's meeting, but not the authors, as Darwin was then attending the funeral of his son, who had just died of scarlet fever, and Wallace was still in Borneo. There was no discussion, perhaps because the audience was fatigued. A few magazines reviewed the paper, and Darwin, years later, recalled one review. It was of a professor from Dublin who had responded that "all that was new in them was false, and what was true was old". The president of the Linnaean Society, in his presidential report for 1859, wrote that "The year which has passed has not, indeed, been marked by any of those striking discoveries which at once revolutionize, so to speak, the department of science on which they bear".

Darwin, within a year, completed his book *On the Origin of Species by Means of Natural Selection, or the Preservation of Favoured Races in the Struggle for Life* and sent it to a publisher, John Murray, who also happened to publish the *Quarterly Review*. Murray cautiously trusted the editor of the *Quarterly*, a certain reverend Whitwell Elwin, to review the book. Elwin passed a thumbs-down verdict, claiming that Darwin was putting forth a theory without any evidence. He further complained about the book's style: whereas Darwin's *Journal of Researches* (later known as *Voyage of the Bea*gle) was "one of the most charming books", Darwin had written this new work in a "much harder & drier style". Having consulted Lyell and to be on the positive side, Elwin advised that the book should focus on observations of pigeons, kindly commenting that everyone is interested in pigeons. Every journal in the kingdom would review such a volume and, he suggested, it would soon be on every table. Very fortunately for us, Murray ignored this advice. On 24 November 1859, he published *The Origin*, pricing it at 15 shillings. All 1256 copies sold within a day.

Never did a book have such a huge influence, far beyond the borders of biology. Darwin's idea that species are exposed to change had been suggested by several naturalists before Darwin, as he himself had commented (Buffon, Lamarck, Geoffroy, and others). None of them had suggested, however, promoting natural selection as the central force in evolution, and none had phrased the issue in such a flawless manner; none had brought such a thick body of evidence; none of them had placed the idea as a central theme for the entire animal world. Darwin's claim of evolution by natural selection undermined Paley's claim that nature's complexity is evidence of the existence of God, a purposeful designer, who had created all creatures, equipped with all their adaptations, directly out of nothing. True, natural selection does not "create" the entities on which it operates. It does, however, favour and forward functional features that would not have existed otherwise, by increasing the chance of a bearer of these features to survive. In these terms, it is a creative process. The existence of a process that is creative but not conscious, that forms design without the need for a designer, was Darwin's basic discovery.

His claim of natural selection also undermined Paley's belief (echoed decades later by Chambers) in God who had created a joyful nature, full to the brim with happiness and harmony. Darwin noted that parasitic wasps (Ichneumonidae) inject their eggs into the body of a caterpillar or a spider. The hatching larva then feeds on the host's living flesh, and the host dies only after some weeks; the wasp larva then emerges and pupates. "I do not succeed in convincing myself that an ever-loving, omnipotent god, purposefully created the Ichneumonidae with the clear intention that they would feed inside the living body of caterpillars" (Darwin in a letter to Asa Gray, 1860). Blindness to suffering is a built-in consequence of evolution by natural selection. Nature, as Darwin understood it, is neither merciful nor cruel. It is indifferent. The concept of natural selection explains both the evolution of adaptation and the evolution of new species. Paley, like Ray a century before him, was convinced of purposeful design and concluded that God exists as an omnipotent and purposeful designer. There is, in nature, however, much imperfection and non-function, evidence of faulty design; Man's jaw has too many teeth for its size, to the extent that we must remove wisdom teeth. Would we want to blame God for such an imperfect design? Evolution gives a good answer to this imperfect case: In our ancestors, the brain grew over time, and the re-design of the skull to fit the growing brain involved a reduction of the jaw. In another case, the birth canal of women is too narrow to enable the easy passage of a baby's head, and therefore thousands of babies (and women) die during childbirth. The females of other primates do not experience this difficulty. Would we want to blame a God for such a faulty design, which kills babies? Science explains this difficulty of human birth by the fact that it results from the evolutionary growth of our brain, which, in turn, makes the baby's head so large. Nature consists of much imperfection and non-function that Paley, in his claim for a purposeful designer, failed to explain. Darwin claimed that the accumulation of innumerable slight hereditary variations, each good for the individual, perfected organs (and instincts) and not divine power.

Darwin's theory of evolution by natural selection put an end also to the view, that existed since the days of Plato and was followed closely by Linnaeus, that every species has an essential ideal of itself; the flesh-and-blood individuals of a species vary and differ slightly from one another in that they are imperfect expressions, bad copies of this ideal. Essentialism, with its emphasis on discontinuity, constancy and typical values ("typology") had dominated thought in the Western world for two millennia, but now Darwin showed that intra-specific variation, in one feature or another, is a basic feature of all species and that thanks to it, species undergo evolution, by natural selection. Evolution means gradual change, as explained by Darwin, and this is quite incompatible with essential thinking. Plato's essentialism had dominated Western philosophy for more than 2000 years. It was not until Darwin that a new, different way of thinking about nature began to spread, which stresses the uniqueness of the individual, not the type. Every individual, in a sexually reproducing species, is uniquely different from all others. There is no "typical" individual.

Darwin's theory also contradicted the classic theory of the Great Ladder of Nature that placed all organisms in a hierarchy, with simple organisms at the bottom and complex organisms at the top, as a ladder of successive levels of perfection, in terms of increasing powers of perception. Darwin showed that organisms form an ever-branching tree, not a ladder. Hereditary variation occurs among offspring, and individual variation—ever so slight—determines the winner, among descendants, in the struggle for existence. Diversity forms by descent from common ancestry, with modifications piling up in adaptive directions by non-random survival, of random hereditary varieties. Natural selection, in which the strongest prevail and the weakest fail and die out, explains the origin and diversity of all species. Variation is a basic feature of life; there is no "typical" individual of a species, and one can no longer define a species as having a typical essence. Furthermore, many species are subject to change and to extinction; they are not timeless entities.

In the seventeenth century, Newton laid down the basic laws of Physics, and he convinced his contemporaries that natural laws rule the whole universe. In 1859, Darwin laid down one of the most basic laws of Life: Evolution through natural selection forms nature's diversity. The demonstration that the members of the plant and animal kingdoms are as they are because they have become what they are and that change, not immutability, is the rule of living things, is one of the most important contributions ever made to knowledge, and its effects have been felt in every

Fig. 7.12 Darwin, later years

field of human thought (De Beer 1958b). The *Origin of Species* ended one era in the study of biodiversity and introduced another (Fig. 7.12).

Here we have two major competing philosophies: that of Paley, who explained biodiversity as a purposeful design by God, an omnipotent and purposeful designer (Argument from Design), versus that of Darwin, who explained it by evolution, through natural selection. One basic fault in the Argument from Design is that, in seeking to explain the formation of biodiversity, it suggests an intelligent entity, a Creating God who is as diverse in features as the organisms he has created (at least). Therefore, if biodiversity requires an explanation, so too does God's existence, and for the same reasons. We are left with the same problem with which we started. Having asked "how come that there exist complex, purposeful entities", we have answered "because there is something even more complex and determined than them". Clearly, the problem has not been solved.

One advantage of the theory of evolution by means of natural selection over the Argument from Design is that it explains biodiversity by processes that are within nature, without assuming forces beyond it. Biodiversity has evolved gradually, by the accumulation of random changes, with organisms more fitted to their environment surviving because they produce more offspring; their process of selection does not involve any planned choice. The theory of evolution does not assume the presence of anything in addition to the world of nature. The theory of evolution by means of natural selection is advantageous also in that it can explain the presence of fossils. The Argument from Design by a Creator cannot explain why we find fossils that differ from the creatures living today.

The theory of evolution by means of natural selection explains also the distribution of animals over the globe. Why we do not find elephants in America? Why are there no penguins at the North Pole? Why are there no frogs on oceanic islands? These and other animals could live also in places in which they are not found naturally, as proven by the fact that they have flourished in places to which they have been brought by man. If organisms have been formed by a design creator, it seems that he was whimsical in distribution over the globe. If, however, creatures have evolved in an evolutionary process, then creatures that have evolved in one site, even if they were pre-adapted to another site, should have reached it somehow. Frogs, for example, should have found a way to cross the ocean to reach the oceanic islands. But the big, salty sea they could not cross. The distribution of creatures, today and in the past, supports the theory of evolution, as Darwin had already said.

Another important advantage of the theory of evolution by means of natural selection is that evolution is not all-powerful. Not everything can evolve in an evolutionary process, and not everything that evolves must necessarily be the best we can think of. The theory that creatures were formed by Design does not explain why organisms are far from perfection in various aspects.

Let us discuss various aspects of this lack of perfection. One such aspect is continuity. Evolution is a gradual process. A complex organ is not formed all at once but by the accumulation of simple changes in existing organs. For a complex change to survive, every separate simple change on the way to it must give an advantage to the creature that bears it. If the way to a better change passes through less successful changes in the existing organ, then the successful organ will not evolve.

We may ask, for example, why birds do not have, beyond feet and wings, also a pair of hands? Such a bird, that would be able to pick fruits and peel them, would have been far more successful than a similar bird without hands. But such a bird could evolve only through intermediate birds that would have been less successful than the existing birds: All birds have four limbs. As an additional pair of functional limbs cannot evolve immediately, what should have evolved first, usually would have been limbs that make it difficult for the bird. Therefore, such a bird would not evolve via natural selection. The Argument from Design, on the other hand, would not be able to offer an explanation for this absence. On the contrary: in many myths, we find "birds" with hands: the angels. The Design theory cannot explain why the existing creatures are actually found, and not other successful creatures that we can think up.

Another aspect of non-perfection in evolution is the presence of features that have remained in certain species of animals and that were beneficial to one of the species from which it evolved, though for it itself, it is of no advantage. As a feature that was beneficial to one species might not be so or even harm another species that had evolved from it, it might remain, even if it degenerates to a certain extent. One example of such a feature in humans is the ability that some of us have to move their ears and even the skin covering the upper part of the head, the scalp. As far as we know, there is no benefit for this in humans. How, then, can the theory from Design explain it? Maybe God is not just wise and can do everything, but also with a sense of humour? The explanation of the evolutionary theory, on the other hand, is that such a feature is a remain of the creatures with fur to move their skin so as to get rid of pests—we can see how cats and dogs and horses do it in response to a light touch of their fur. Indeed, this ability has remained in the last remnant of our fur, in man. This feature, by the way, is transferred through heredity, as one might expect from the evolutionary theory. Darwin already, in his *Descent of Man* (1871), discussed this feature as evidence of the origin of man from a more ancient species.

Another aspect of non-perfection is improvisation. In the process of natural selection, a new species is formed as a variety of an older one. In natural selection, we do not rub away everything; we merely add a little, stretch a little, shrink a little and so forth. Therefore, so many of the organs of creatures that serve a specific function were formed by the adaptation of an old organ, or old structure, to a new purpose, and this so as not always in a very successful manner. This fact can explain the non-functional similarity between organisms. All mammals, for example evolved from a common ancestor, by a limited variation of his features. Therefore, we may expect points of similarity in the structure of mammal structures, even if this structural similarity is not functional. One example would be the similarity between the hand of a human, the wing of a bat and the fin of a whale. The three of them have the same basic similarity, and the difference is mainly in the proportions of the different bones. This similar basic plan has no engineering justification, but we can understand it if we see in these three creatures the product of natural selection.

Beautiful examples of non-perfection in evolution in nature that result from variation of an existing structure are to be found in Gould's *The Panda's Thumb*. I give another example: birth among humans, or the reason for "I will make your pains in childbearing very severe; with painful labour you will give birth to children". During the evolution of man from ape, there was an advantage to intelligent individuals. Intelligence depends on the size of the brain, and therefore, there was an advantage to individuals with a large head. But the birth of a baby with a large head is difficult and dangerous, to the mother and to the baby. Therefore, there was an evolutionary advantage to women who gave birth to babies at an earlier stage of development of the embryo, in which the chances of the embryo to survive are bigger, since in this manner the chances of their children to die at birth were lessened. As a result, the human baby is helpless at birth as compared to the babies of other apes: he is born proportionally younger. But one cannot have given birth to a baby at a stage that is early enough of its development so that the passage of his head through the vagina would not be problematic: at this stage, the newborn would surely not survive. Therefore, the passage of the big head through the vagina is traumatic and dangerous to the mother and her baby. The percentage of deaths at birth of the mother and the baby among humans was, up to the twentieth century, among the highest among primates. Only in modern times and in modern society did death at birth stop being a significant danger. Birth in man is actually a very not-successful process, also from the aspect of the helplessness of the baby and from the aspect of danger to the mother; but we can understand it if we see in man a variation of apes, that along history gave birth to new-borns with a small head and had to because of the change in their life conditions, give birth to babies with a large head.

These aspects show how much the theory of evolution is a much more successful answer than the theory of Design, to the question as to how organisms were formed.

7.2 Against Darwin: Species Do Not Vary

Darwin's theory of natural selection as the force behind evolution was not readily accepted. Variations that natural selection favoured could have been divinely guided. Agassiz denied variation within a species, thereby dismissing Darwin's theory. Individuals differing even slightly are separate species, not variants. Nature is the product of God's Creation and symbolizes divine intelligence, and therefore species are immutable. Animals have been created wherever they are found, land animals on land and sea creatures in marine habitats. The Creator wipes out one fauna and starts anew, with models different from those cast aside. There is no evidence of a relationship between distinct (and separately created) species, and hence, there is no proof for evolution. Some fossils, intermediate in that they possess a combination of characters of different groups, offer evidence that the plan of the creation had been considered long before it was executed. To further disprove Darwinian evolution, Agassiz claimed that a huge ice sheet had once covered South America, destroying all life. The current flora and fauna of South America were created anew after the ice had receded, with no relationship between animals that had lived there prior and after this glacial epoch. This, he claimed, is evidence of the mistaken concepts of Darwinian evolution.

When the *Origin of Species* was published, there was no great immediate uproar and no historic clash of science and religion, perhaps because the Victorian public that had first read Darwin's book consisted largely of people who already believed that the Biblical Book of *Genesis* should be read in metaphorical terms (van Wyhe 2018). Darwin's theory inspired a broad scale of reactions, from rejection to support, some feeling (and writing) that his theory was an attack on the role of a Creator in nature, others writing that he had uncovered a new law by which God governs nature, and his theory thus fits into religion. Darwin's broad ideas on the common descent of morphologically similar groups were soon accepted, but his theory of natural selection as the primary force behind evolution was less welcome. Many suggested that the variations that natural selection picked out were themselves divinely guided or caused. The bottom line seemed to be—was there a meaning or intention behind how life changed? (Van Wyhe 2018).

Louis Agassiz (1807–1873), the son of a Protestant pastor in Switzerland, loved natural history as a child (Lurie 1960). In later years, he studied, first at Heidelberg, where Leuckart and Oken were among his teachers, then in Munich. In 1817–1820, a German expedition to Brazil had yielded a large collection of fishes and Agassiz, then 22 years old, was invited to study it. His investigation yielded a publication, *Brazilian Fishes*. He travelled to Paris to meet Cuvier, whom he admired, and following closely in Cuvier's footsteps, he came to believe that species are immutable and do not evolve; that living forms bear no relationship to fossil forms; and that there was a series of past catastrophes and faunal successions on earth, under the will of the Creator. Cuvier entrusted Agassiz with the study of several fossil fish and when he suddenly died (1832) this fact, that the great naturalist had entrusted him with the study of important fossils, led Agassiz to view himself as Cuvier's disciple.

In Paris, Agassiz also met Humboldt, who taught him the social and institutional support needed, for intellectual activity. At 25, Agassiz returned to Switzerland, to the small town of Neuchâtel, where he began intensive research on fossil fish; this culminated in *Recherches sur les poisons fossils* (1843)—a description of over 1700 species, in which he devised a fish classification by features of their scales: Placoid scales resemble teeth in that each scale consists of dentine covered by an enamel-like substance; they occur in sharks and rays. Ganoid scales resemble teeth overcovering bone; they occur in extinct groups and in primitive modern bony fish. Ctenoid and Cycloid scales are thin, transparent and lack hard structures; they occur in most living bony fishes, the ctenoid scales differ from cycloid ones in that their hind end splits into a comb-like fringe.

Agassiz studied also glaciers and moraines in the Alps, and he concluded that in a catastrophic event during the Pleistocene, a gigantic ice sheet had covered the world, from the North Pole to the shores of the Mediterranean and Caspian seas. To investigate the ice sheet's impact, in the following years he travelled through England and Scotland and performed careful measurements (*Études sur les Glaciers* 1840, *Systeme glaciare* 1847). Sadly, the scientific community felt that his Ice Age studies were not entirely original, and Agassiz lost quite a few friends, who had feelings of plagiarism (Lurie 1960).

Nevertheless, many considered Agassiz as a highly capable, universal naturalist. He became involved in urban matters of Neuchâtel, the school in which he taught became an academy with him as rector, he rented a large house and hired a private secretary. To publish his books he set up his own publishing house, and by 1839, he had 12 people working for him, some of whom lived in his house. His ambitious drives resulted in heavy debts that were unbearable for his wife, and she left him, with their three children.

Amidst these domestic tensions and when at his lowest ebb, Humboldt came to the rescue: he informed him that Fredrick William IV of Prussia would grant him 3000 dollars to undertake a 2-year journey to the United States, to study the natural history of the New World. In 1846, aged 39, Agassiz embarked for America, where he was warmly welcomed by both Academic circles and the general public. He went on a (well paid) lecture tour, became a professor of Zoology and Geology at Harvard, founded the Harvard Museum of Comparative Zoology, was editor of the *American Journal of Science* and held an additional professorship in Charleston; he also remarried and brought his children over from Europe.

Agassiz launched a grand research plan to study the entire fish fauna of the United States (then about half its size of today). To this purpose, he recruited a collecting force of amateur naturalists, sportsmen, fishermen, government officials and professional scientists, who sent him specimens from all over the country. In describing the many specimens that he received, he rigidly stuck to the concept that species are rigidly fixed in their morphology, with no variation whatsoever. Consequently, if a single batch of specimens contained individuals differing from one another even just ever so slightly, he would describe them as separate species, not as variants. In another grand project, he intended a series of volumes that would cover the entire American fauna: *Contributions to the Natural History of the United*

States (coming out from 1857 onwards). In his *Contributions* he interpreted nature as the product of God's Creation and as symbolic of divine intelligence, and therefore species (and larger taxonomic units) are immutable. There exist specific zoological provinces in nature, each with its distinct flora and fauna, and it is impossible to believe that all animals were created in a single place, as the Bible suggests. *Genesis*, Agassiz believed, refers only to those animals placed near Adam and Eve by the Creator, in just one of these provinces, yet there had been many separate, successive creations that had occurred in many different parts of the world at many different times, giving rise to many separate and distinct animal forms. Animals and plants had been created wherever they were found, land animals on land and sea creatures in marine habitats. The Creator, for some reason, wipes out one fauna and starts anew, with models different from those cast aside (Lurie 1960).

Palaeontology provided Agassiz with empirical demonstrations of cosmic assumptions in natural history. He identified the existence of what he called "prophetic types" in palaeontology, intermediate forms like sauroid fishes, that possessed a combination of both ichthyic characters and the reptilian characters that came after them, in geological time. These forms were defined as a prophecy, in earlier times, of an order of things not yet possible with the combinations then prevailing in the animal kingdom. They afforded the naturalist with the most unexpected evidence that the plan of the whole creation had been maturely considered long before it was executed (Lurie 1960; Did Agassiz have in mind lizard fish, Synontidae, whose heads superficially somewhat resemble those of lizards?)

In 1859, Darwin's *Origin* came out, with its emphasis on intra-specific variation and on common descent, two points deeply opposite to Agassiz's view of nature. Agassiz was by then a symbol of scientific authority, and Americans expected their grand interpreter of natural history to inform them of the validity of such views and to assess Darwin's work. At a meeting of the American Academy, Agassiz denied evolution: There were no genetic relationships between shells of the Tertiary period and those of modern times; and varieties do not exist, in the animal kingdom (Lurie 1960). There is no such thing as random variation in the wild state of nature and Darwin's emphasis on variation in species is ridiculous. The only variation occurring in nature is that of different stages of individual growth when, during growth to maturity, it occurs in terms of transitionary phases. In dismissing Darwin, Agassiz informed readers of the *American Journal of Science* (1860) that the arguments of the *Origin* had not made the slightest impression on his mind, nor modified in any way his already existing views. With Darwin's intellectual framework in mind, he asked: "If species do not exist at all, as supporters of the transmutation theory maintain, how can they vary? And if individuals alone exist, how can differences which may be observed among them prove the variability of species?" The Darwinian idea might be true only if evidence from all parts of the world would show a genetic relationship between species that Agassiz had, previously, considered distinct and separately created. There is no proof of evolution. Even if the Galapagos species had indeed originated in South America, why did their transmutation require such very long periods, as the evolution advocates claim necessary? Furthermore, among domestic animals, new variations return to their original, wild type, not to new forms. Why does Darwin's evolution concept require such a great amount of time to accomplish changes in species, when change occurs suddenly and very rapidly, in individuals growing from egg to adult? Change, in embryology and in geology, is dynamic and catastrophic (Lurie 1960; Agassiz even disputed the common derivation of language (1861). He was certain that later languages were not derived from earlier ones; languages, like species of plants and races of men, were created separately and bore no relationship to one another.)

To further prove the fallacy of Darwinian evolution, Agassiz claimed that a huge ice sheet had once covered also South America, destroying all life in its path. The current flora and fauna of South America had been created anew after the ice had receded, and there was no hereditary relationship between animals and plants that had lived there prior and after this glacial epoch. This, he thought, was undeniable evidence of the mistaken concepts of evolutionists. Agassiz travelled to Brazil (1865–1866), announcing beforehand that he expected to discover glacial remains. He did not find reliable evidence for glaciation in terms of furrows, scratched rocks and moraines, so he interpreted as glacial drift some loose rocks and pebbles in some unstratified clay deposits, and upon this meagre evidence, he claimed that an Ice Age had occurred in Brazil with huge ice sheets covering vast regions, as in Europe and in the United States. His motive was plain: the more ice he could discover all over the world as marking the earth's history at the close of the Tertiary and beginning of the Quaternary periods, the greater the barrier he could construct between the animals and plants that had immediately preceded the present creation; and the more authoritative would be his affirmation that no hereditary affiliation exists between life in the present and the past. Pleistocene glaciation, then, was the great catastrophe inspired by supernatural power that disproved Darwin's transmutation of species (Lurie 1960).

References

Colp, R., Jr. (1986). "Confessing murder". Darwin's first revelations about transmutation. *Isis, 77*, 9–32.

Darwin, C. R. (1845). *The zoology of the voyage of H.M.S. Beagle, under the command of Captain Fitzroy, R.N., during the years 1832–1836*. John Murray.

Darwin, C. R. (1851). *Living Cirripedia, A monograph on the sub-class Cirripedia, with figures of all the species. The Lepadidæ; or, pedunculated cirripedes* (Vol. 1). The Ray Society.

Darwin, C. R. (1854). *Living Cirripedia, The Balanidæ, (or sessile cirripedes); the Verrucidæ* (Vol. 2). The Ray Society.

Darwin, C. (1859). *On the origin of species*. John Murray.

Darwin, C. (1888). *A naturalist's voyage. Journal of researches into the natural history and geology of the countries visited by the H.M.S. Beagle on its voyage round the world, under the command of captain Fitz. Roy, R.N.* Murray.

Darwin, F. (1902). *Charles Darwin: His life told in an autobiographical chapter, and in a selected series of his published letters*. Murray.

Darwin, C., & Wallace, A. R. (1858). On the tendency of species to form varieties; and on the perpetuation of varieties and species by natural means of selection. *Journal of the Proceedings of the Linnaean Society, 3*, 42–62.

De Beer, G. (1958a). *C. Darwin and A.R. Wallace: Evolution by natural selection*. Cambridge University Press.

De Beer, G. (1958b). The Darwin-Wallace centenary. *Endeavor, 17*, 61–76.

Lurie, E. (1960). *Louis Agassiz: A life in science*. University of Chicago Press.

Newman, W. A. (1993). Darwin and cirripedology. In F. Trueseday (Ed.), *History of carcinology* (pp. 349–434). Balkema.

Spearling, J. (2002). Book review, Alfred Russel Wallace – A life (by P. Raby). *The Linnean Newsletter, 18*, 43–47.

Van Wyhe, J. (2018). *Darwin: The man, his great voyage, and his theory of evolution*. Rosen Publishing Group.

Wallace, A. R. (1894). *Malay Archipelago. The land of the orang-utan and the bird of paradise. A narrative of travel with studies of man and nature*. Macmillan.

Wightman, W. P. D. 1953. *The Growth of Scientific Ideas*. Yale University Press.

Winsor, M. P. (1969). Barnacle larvae in the nineteenth century: A case study in taxonomic theory. *Journal of the History of Medicine and Allied Sciences, 24*, 294–309.

Chapter 8
After Darwin: Classification

Abstract Darwin's emphasis on the species created a rush to search for new species, and to extreme species splitting. Taxonomists, to overcome problems of convergence in external structure, combined studies of external and internal structures, often of the genitalia. Darwin's legacy called for a definition of what a species is. Some suggested that species definitions should involve reproductive criteria: A species is an isolated group of interbreeding populations. This definition is relevant only to local situations and to sexually reproducing groups. Another evolutionary definition: A species is a lineage evolving separately from others. This ignores the causation and maintenance of discontinuities between contemporary species, does not refer to why phyletic lines do not interbreed and raises any race to the rank of species. In practice, the classifying of species seeks discontinuities among groups. On a geographical scale, sharp discontinuities over short distances suggest strong reproductive isolation and indicate different species; gradual differences over large distances suggest one species, with interbreeding populations.

Scientists of the century following Darwin argued that diagnostic definitions of species relate not to adaptive characters but to discontinuities, with no relation to specific advantages, and that they do not necessarily have survival value. Also, alongside the slow accumulation of hereditary advantages, evolution may proceed by mutations producing a completely new body plan in a single step. New species evolve by the drastic re-patterning of the gene sequences along the chromosome.

The discovery of variation in chromosomes took taxonomy beyond the level of external morphology and internal anatomy, to that of chromosome numbers, which may be group-specific. Furthermore, at every taxonomic level, some proteins are highly specific. The proteins of one organism will show a stronger antibody reaction to those of a closely related organism than they will to those of a more distantly related organism. When one introduces foreign cells or particles (antigens) into an organism, highly specific protein molecules (antibodies) are produced in the immune system of that animal. The degree and rate of blood reaction offer an index of relationship; closely related blood react more powerfully and more rapidly than do distantly related blood. There is a general correlation between differences in blood reaction and broad lines of zoological classification. Serology, by detecting

J. Heller, *Nature's Biodiversity*, https://doi.org/10.1007/978-3-031-89790-0_8

homologous proteins in the serum, can therefore be a taxonomic marker that contributes to classification.

When classifying biodiversity by a few features, the resulting classification may not be repeatable with other features. Furthermore, natural groups may not share any single defining character, as members of a group might lose their diagnostic features. We should classify by overall similarity in many features, rather than by over-weighting a few. Phenetic systematics clusters individuals into groups by overall similarities, in many numerically encoded features. Computers determine overall similarities in quantitative, repeatable terms and an objective classification automatically emerges, that is theory-free and does not attempt to reflect evolution.

Phylogenetic systematics (cladistics) differs from phenetics in that it classifies organisms by descent from a common ancestor, not by overall similarity. It reflects the evolutionary history of a group and corresponds to all descendants of an ancestral lineage. Features evolve, from ancestral to derived states. We classify only by shared derived features, not by ancestral ones. We reach the ancestral character state by consulting an additional group that is phylogenetically close, but not within the study group. Features occurring within both the study group and the outgroup are ancestral to the study group; those in the study group that are absent from its outgroup are derived.

8.1 Origin of Species—What Is a Species?

The century and a half since Darwin witnessed a surge in so many aspects of the study of biodiversity that both knowledge and theory have split into many disciplines. These include, beyond the refining of classifying techniques, also the rise of new disciplines such as genetics, evolution, ecology, biogeography and species-specific reproductive behaviours. The historic treatment of these new disciplines is far, far too broad for any one book. In describing the post-Darwin era, I focus, therefore, on just two disciplines, Systematics and Biogeography. Many excellent books cover numerous day-to-day principles and methodologies of these two dynamic scientific fields. My two following chapters emphasize historic perspectives, they are not up-to-date summaries of everything in modern Systematics or modern Biogeography.

Systematics seeks to discover, define, describe and classify biodiversity (Mayr 1963). Until recent decades, the terms Systematics, Classification (in the context of biodiversity) and Taxonomy were broadly synonymous. Today we use "Systematics" for the study of organic diversity in broadly defined contexts, and restrict "Taxonomy" to the theory and practice of classifying, as more narrowly defined. Systematics, from the Latinized Greek *systema* meaning together, is the theory, principles and practice of ordering the diversity of organisms into general patterns. It studies the diversity of organisms, including their past and present relationships. Taxonomy, from the Greek term *taxis*, arrangement, and *nomos*, law, is the theory and practice of classifying organisms. It includes classification and nomenclature, and it leans heavily on systematics for its concepts. Classification, Systematics and Taxonomy are broadly overlapping terms.

In the decades after *The Origin*, literature in botany and zoology bloomed. One reason for this was the innovations in printing, from the mid-nineteenth century onwards. Paper making from lumber by machines, instead of hand-dipping individual pages, printing processes by steam instead of completely manual, and book binding with cloth covers instead of leather—reduced publication costs, and this, in turn, contributed to an increase in printed matter. The invention of lithography also encouraged publication because of its low cost and the great accuracy that it allowed. Furthermore, colour printing became available, enabling the publication of very beautifully illustrated books in popular natural history (Fig. 8.1).

Fig. 8.1 Colour printing: from Haeckel's book on natural history (1904)

Some popular books of those decades tended to illustrate natural history in over-dramatic tones. *Nature's Wonderland, or Short Talks on Natural History* (1890), a three-volume book "profusely illustrated with scientific accuracy", showed a grizzly bear carrying off a deer (Fig. 8.2), in a posture that reminds me of paintings of The Rape of the Sabine Women, rather than of bear and deer in nature. Another figure, of "The Lion in Captivity", shows a lion behind bars, fed by a beautiful woman from the colonies, with naked breasts (Fig. 8.3).

Other zoological literature catered to a narrow audience of scientific specialists. A new scientific genre sprang up, the monograph, which offered extensive studies on a single topic. Owing to its inherently limited appeal, the monograph could become a basic scientific tool only thanks to affordable printing. Low printing costs also resulted in specialized scientific journals, enabling communication that is faster than lengthy books. The first, overall division of natural history was into journals of zoology versus those of botany, with the *Proceedings of the Zoological Society of London* appearing as early as 1830. Eventually, however, as the monographs and articles narrowed down to restricted animal groups, such as a particular family of birds or snails, the broad field of zoology sub-divided into specific branches, giving rise to specific journals. Journals devoted to the study of shells appeared as early as 1850 in France (*Journal de Conchyologie*), and three decades later in England

Fig. 8.2 Grizzly bear (*Ursus horribilis*). (From *Nature's Wonderland* (1890) III: 640, Fig. 517.)

Fig. 8.3 Lion in captivity (*Felis leo*). (From *Nature's Wonderland* (1890) III: 662.)

(1879, *Journal of Conchology*). This specialization in research and publication made sense since expeditions to exotic countries often yielded very many new species, but mostly of a single group. As an outcome of these tendencies, zoology began splitting up.

Darwin's legacy caused a great species rush, a frantic search for more and more new species and, also, to an extreme species splitting. Every slight difference became sufficient to divide a group into multiple groups. One species describer, among the extreme splitters (Jules-René Bourguignat, 1829–1892), created a long trail of 2500 described mollusc species.

Darwin's legacy called also for the need to define what a species is. The rediscovery of Mendel's laws of inheritance encouraged ideas that, in addition to close comparisons of external and internal structures, in defining species, we should also involve a genetic element. It is the concurrence of heredity factors and genotypic resemblance, which determines affinity, not phenotypic resemblance in structures. We can assess genotypic resemblance only by experimental means. Johannes Paulus

Lotsy (1867–1931), director of the herbarium at Johns Hopkins University and eventually of the Herbarium at Leiden, was a very active botanist who carried out expeditions to several remote countries. He claimed (and here I am closely following Nordenskiold 1929) that a species in the Linnaean sense is not really a species, but merely the product of our imagination, and consequently, we should no longer call them species but "linnaeonts". Rather, he wrote, a species is a group of all the homozygous individuals having the same hereditary character. Consequently, not even pure reproductive lines are true species; they are so only if they are at the same time homozygotes. Regarding organisms with asexual reproduction, we can never know whether they are species, for we can answer this question only through the analysis of hybridizing experiments. In these circumstances, it becomes doubtful whether any species at all exists in nature. Lotsy thus suggested that species are strictly identical genetic entities, whether or not species in this sense of the word exist at all in nature.

Darwin, in his *Origin*, had claimed that the only distinction between species on the one hand, and well-defined varieties, on the other hand, is in that the latter are connected by intermediate gradations, whereas species were formerly connected. This opinion, that there is no clear line between species and varieties, probably represents the opinion of most biologists of today (Mayr 1963). Furthermore, from the early twentieth century onwards, biologists felt that the definition of a species should contain elements of diagnosis, of reproduction (in sexual organisms) and of common descent (Poulton 1908). Concerning diagnosis, discontinuity in structure is a basis for differentiating a species, and it has always lingered around the species concept. However, as Robson (1928) had already pointed out, there is no universal criterion, no absolute standardized method by which to distinguish species one from another and to attribute species rank to some assemblages of individuals but not to others. Groups that we recognize as species may differ considerably in the number of their unique features, in their distinctness, and in the very criteria of distinction. We group certain organisms because they resemble each other in certain attributes, but the basis of resemblance may be any of countless features that we cannot define in comprehensive terms. To broaden our diagnostic basis, we seek correlations of discontinuity between unrelated structures, say discontinuities in the structure of genitalia that correlate with discontinuities in external features. The land snail *Cepaea nemoralis* differs from *C. hortensis* both in that it has a brown lip on the aperture of the shell rather than a white one; and in details of its love dart; and this strict correlation between external skeleton and internal genitalia broadens our basis for distinguishing between these two species (Robson 1928).

A further problem in defining species, in addition to variation in the number of differentiated characters, is the great variety in the degree of distinctness of individual characters (Robson 1928). We may separate two species by, say, five characters; but of these five, two may be very distinct in the two species, the remainder less distinct though still discontinuous. Discontinuity may be in one character or in many, and we cannot specify how many characters must be involved. No hard-and-fast line exists between species and variety, and we can formulate no absolute criteria for the two. To sum up, at the structural level, no rigid definition of the species is available; there is every degree of differentiation in the number of characters affected, and we can apply no absolute criterion. We are unable to standardize a particular degree of structural differentiation as "specific", as we have no absolute criterion by which to recognize species as standardized classificatory units. Nevertheless, well-defined groups do occur, of which individual representatives having the same structural constitution occur in places distant from each other and after long intervals of time.

Reproductive isolation is another criterion in the definition of species: We view species as breeding units. De Vries (1903, in Robson 1928) attempted to distinguish between species and varieties by the occurrence of segregation in crosses between the latter and its absence in species-crosses. In the capacity of producing normal and healthy offspring, we thus have what seems at first sight to be a valid test, as to the status of two forms that we bring into a cross: two forms that do not have this degree of affinity are different species. Such a test seems about as rational and natural as one could hope to obtain. However, taxonomists today reject the inability to obtain fertile offspring as a sure index of species level. There are cases in which individuals refuse to pair in the lab. In an early experiment in which scientists caged up a male poplar hawk-moth *Laothoe populi* (Sphingidae) with its own female and that of an eyed hawk-moth *Smerinthus ocellatus*, the male paired with the latter. Then, there are cases in which two species with overlapping distributions seldom interbreed in nature, but readily yield hybrids in captivity. Two ducks, the pintail *Anas acuta* and the mallard *Anas platyrhynchos*, easily form hybrids under artificial conditions but only rarely do so in nature, where their distributions overlap (Sturtevant 1920; Robson 1928; Capy and Gibert 2004). There is, indeed, every degree of gradation in the reproductive capacity of morphologically differentiated species, from absolute fertility to complete sterility. Stages in this gradation series include hostility between male and female, inability to pair, coitus without reproduction, F1 obtained but with a disturbed sex ratio, ill health or other abnormality, F1 normal and healthy but sterile, F2 obtained but weak or unhealthy; and finally, fertile and viable F2. On the whole, then, the breeding test is a consistent index of morphological differentiation only in a very wide sense, and it does not give a sharp criterion. Conditions in captivity may alter the psychological and physiological conditions of animals to such an extent that certain species will not mate with their own species in captivity, animals will pair with another species to which they would show repugnance in nature, and closely allied species may display animosity and unwillingness to pair. The concept of a species as a breeding unit has only partial validity when brought to the test.

Today, we have two major concepts as to what a species is. One modern concept views a species as a reproductive community protected by isolating mechanisms, as a large, inter-communicating, collective gene pool in which individuals recognize each other as potential mates and seek each other for reproduction. Reproductive isolation results in a discontinuity both in a species' genotype and in its phenotype. This concept expresses itself in a clear-cut definition:

A species is a group of interbreeding populations that is reproductively isolated from other such groups.

We term this the "biological species concept" because the definition is biological: The terms "interbreeding" and "populations" are meaningless in the inanimate world, of metals or of books, for example. As taxonomists, we must ask ourselves if differences, in and among the groups that we study, are of such amplitude that we assume they have accumulated because of reproductive isolation.

The biological species concept has several weaknesses, one of which is that it applies only to plants and animals living in the same area, whose distributions overlap. The house mouse in Italy is reproductively isolated from the black rat in Italy, but also from the house mouse in Iceland. We can only speculate whether the morphological differences between house mice of Italy and Iceland suggest reproductive isolation, and yet there is no reason why the sexual choices of mice should follow our speculations. Organisms can develop reproductive isolation without (or with very little) equivalent morphological change, and they can develop strong structural differences without reproductive isolation. Furthermore, isolating mechanisms may break down occasionally, resulting in hybridization. Finally, this concept does not apply to organisms that practice uniparental reproduction, in which, the terms "interbreeding" and "population" are meaningless. Such organisms include self-fertilizing hermaphrodites, parthenogenetic individuals or groups reproducing by vegetative reproduction (budding, as in corals).

From concepts to actual practice, in some animal groups, such as birds, all these weaknesses are of negligible importance. In over 600 North-American birds, the biological species concept failed to clarify only a single case (Mayr and Ashlock 1991). In other animal groups, however, these weaknesses are very meaningful. In snails, among stylommatophorans, self-fertilizing is uncommon, but among basommatophorans it is common. Furthermore, within the same superfamily, some snail groups reproduce sexually whereas others, living in the very same aquatic habitats, reproduce asexually (*Melanopsis* versus *Melanoides*). In *Melanopsis*, some species seem to have acquired reproductive isolation without morphological differentiation, and in *Cerion* of Cuba, one species has acquired strong morphological differences without reproductive isolation.

A different concept of what a species is involves evolutionary criteria: A species is a lineage (an ancestral-descendant sequence of populations) evolving separately from others and with its own unitary evolutionary role and tendencies (Simpson 1961; Mayr 1961). In its modern version and as formulated by Barraclough (2019), a species is an independently evolving group of organisms that is genetically and phenotypically distinct from other such groups. Barraclough defined independent evolution as the limits to which a new mutation can spread. This definition encapsulates both the pattern of species, as genetic clusters of variation separated by gaps, and the causes of that pattern, namely cohesive processes that limit variability within species, while independent evolution leads to divergence between species. One weakness of this definition is that two of its components, "an independently evolving group of organisms" and "genetically and phenotypically distinct from other groups", are slightly circular: We define independently evolving groups by their being genetically and phenotypically distinct.

8.2 Studying Biodiversity: The Classification Tool Kit Expands

In Darwin's days, the toolkit of the taxonomist contained mainly external structures such as feathers, fur characters, scale counts and shell shapes; alongside internal anatomical structures such as the morphology of the genitalia, of prime importance in that they overcome taxonomic problems stemming from convergence in external structures. Close resemblances in external structure can occur in entirely different groups as a result of convergence and, therefore, it is the concurrence of internal structures, not outward resemblance, that proves true affinity. For example, two intertidal limpets, *Patella* and *Siphonaria*, closely resemble one another in their kneecap-shaped shells, to such an extent as to suggest that they might be closely related (Fig. 8.4). Internal structure reveals, however, that in *Patella*, males and females occur in different individuals and there are about a dozen teeth in each row of the radula, some of which contain iron oxide. *Siphonaria*, on the other hand, is a hermaphrodite and has about eighty teeth in each row, which lack iron oxide and are therefore much weaker. The structure of the sperm is often highly specific as well.

In the decades after Darwin, the toolkit of the taxonomist began expanding, as biology progressed beyond internal anatomical structures, into the cell nucleus.

Fig. 8.4 Convergence in external structure: the limpets *Patella* (left) and *Siphonaria* (right). (From Heller.)

Upon finding variations in nucleus contents, students of biodiversity were quick to search whether intracellular features bore any relevance to classification. By the mid-nineteenth century, scientists had already understood that hereditary information was located in the nucleus, but the structural nature of the hereditary material was unknown. Walther Flemming (1843–1905, Fig. 8.5) was a German anatomist who studied medicine, served as a military physician in the Franco-Prussian War (1870–1871), and, at the age of 30, became a professor of anatomy at the University of Kiel. A noble-souled philanthropist, he would weekly feed those who were homeless, yearly donate 20% of his salary to homeless shelters, and, beyond his lectures at the university, teach young children who were too poor to attend school.

Using aniline dyes to enhance the contrast of cell contents, Flemming found a structure that strongly absorbed basophilic elements and which he named chromatin, a "stainable material". He further identified that chromatin was correlated to a network of threadlike, fibrous structures within the cell nucleus—what we would name the chromosomes (meaning "coloured bodies"; these were named as such only some years later). The chromatin forms threadlike bodies during cell division, and he deduced the process of cell division and the distribution of chromosomes to the daughter nuclei, in a process that he called mitosis (Greek for thread). He also

Fig. 8.5 Flemming

observed that chromosomes split along their length during mitosis, and he suggested that, at the end of mitosis, the split chromosome parts into different daughter cells. From these discoveries, he concluded that all cell nuclei came from other, previous nuclei, and he coined the phrase *omnis nucleus e nucleo* (after Virchow's *omnis cellula e cellula*, all cells come from cells, a rejection of the concept of spontaneous generation, which was after Harvey's phrase *ex ovo omnia*, all life originates from the egg). Unaware of Mendel's 1886 work on heredity, Flemming did not connect his observations to genetic inheritance. Nevertheless, by recognizing that chromosomal movement during mitosis offered a mechanism for the distribution of nuclear material during cell division, and by providing a detailed description of the initial mechanisms underlying the process of cell division, he paved the way for the discovery of hereditary mechanisms.

A few years later, Theodor Boveri (1862–1915), another German zoologist and comparative anatomist, took Flemming's findings to the next level by providing evidence that chromosomes of germ cell lineage provide continuity between generations (1888). In sea urchins, for proper embryonic development to take place, it was necessary to have all chromosomes present in the egg cell; he also discovered the importance of the centrosome for the formation of the spindle during mitosis, and that the various chromosomes possess different genetic makeups. He also studied embryos of a nematode (*Parascaris*, a parasite of horse intestines), in which the cells are large and have only two pairs of chromosomes, and the embryos develop distinct somatic cell and germ cell lineages during the first few cleavage divisions. He could thus trace the fate of the chromosomes in individual cell lineages, with considerable precision (Fig. 8.6), and he observed that only the germ cell lineage, the source of future gametes, retained the full complement of chromosomes. By contrast, the chromosomes in somatic cells, the source of all other adult tissues, underwent fragmentation and elimination. Knowing that chromosomes are also present between two nuclear divisions in the cell nucleus, he developed the idea that chromosomes retain their individuality during interphase.

Walter Sutton (1877–1916) was an American biologist whose most significant contribution to present-day biology was his awareness of the fact that the Mendelian laws of inheritance apply to chromosomes. He grew up on a farm and enrolled at the University of Kansas in engineering but following the death of his brother from typhus, he switched to biology with an interest in medicine. For his M.Sc. thesis, he studied spermatogenesis in a grasshopper indigenous to the farmlands of Kansas. He then moved to Columbia University for further study in zoology, and it was here that he wrote his two significant works in genetics, "On the morphology of the chromosome group in *Brachystola magna*" and "The chromosomes in heredity". After obtaining his doctorate (1907), he shifted away from biology into medicinal surgery.

Having discovered that one could distinguish individual chromosomes in cells undergoing meiosis in the testes of a grasshopper (1902), Sutton identified 11 pairs of chromosomes—and an unpaired chromosome, which he correctly presumed to be a sex chromosome. Due to its lack of a partner, this chromosome sorted into only half of the sperm cells, while the remaining half never received a copy. The unpaired chromosome replicated and entered stages of mitosis in the same manner as all

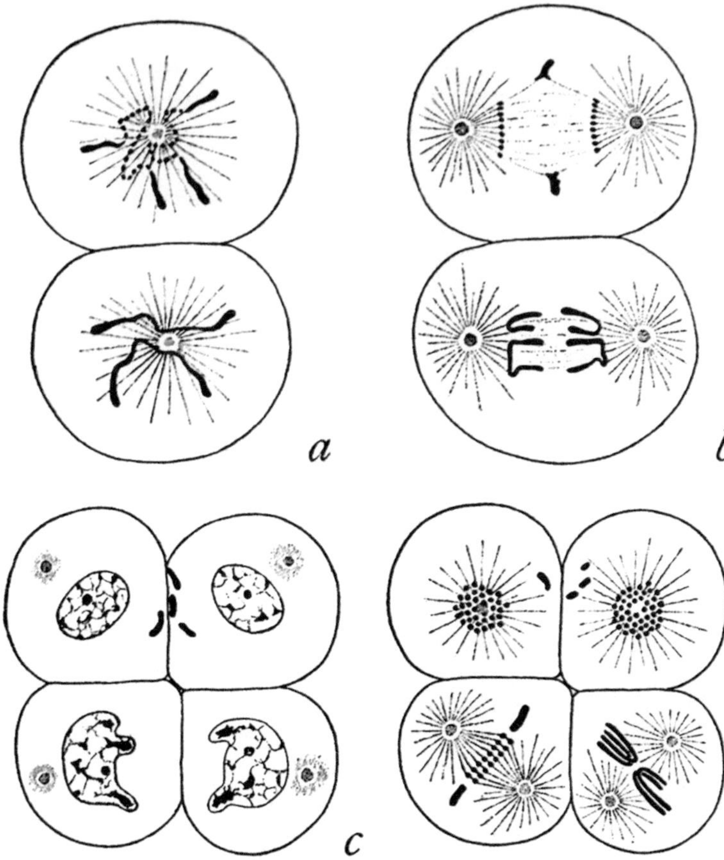

a b c d

Fig. 8.6 First and second cleavages of the egg of *Parascaris*, with two chromosomes. Figures *a* and *b* show the fragmentation of the two chromosomes: *c* and *d* show the germ cells. (After Boveri, in Morgan 1926.)

other chromosomes. Sutton declared that all chromosomes have a stable structure that they maintain between generations, and he used this feature to follow the behaviour of individual chromosomes through the various stages of meiosis. Most notably, he recognized that his observations were consistent with the recently rediscovered work of Mendel. Believing that Mendel's factors are chromosomes, he proposed the chromosomal theory of inheritance. He next suggested the basis for variation among heritable traits. Noting that the position of each chromosome at the midline was random during metaphase, he concluded that each chromosome was independent of the others. This chromosomal independence during meiosis means that one could calculate the number of possible chromosomal combinations for each gamete, by the number of chromosomes in an organism: There are 2^n possible combinations of chromosomes in gametes, with "n" representing the number of chromosomes in the gamete. Furthermore, considering all possible pairings of one gamete with another, the variation in zygotes is $(2^n)^2$, a fairly large number. Sutton also suggested that there was more than one trait present on each chromosome, so the actual total variation was even higher. This observation of potential variation added strength to his assertion that he had discovered the physical basis for Mendel's principle of independent assortment. Genes are sections of chromosomes. During reduction division (meiosis), the chromosome number of a cell halves. The chromatids in each pair of chromosomes separate, just like Mendelian factors, segregate and each gamete receives one chromosome of each pair. Fertilization restores the former number of chromosomes in the zygote.

Still lacking, however, was definite proof linking the inheritance of genetic traits to chromosomes. Such a proof turned up soon thereafter, with a mutation in the fruit fly *Drosophila melanogaster*. Thomas Hunt Morgan (1866–1945) was an American evolutionary biologist, geneticist and embryologist. He grew up in Kentucky and studied zoology at Johns Hopkins University, where he completed a thesis on the embryology of sea spiders. At the age of 34, he became an associate professor at Bryn Mawr College, where he produced descriptive studies of sea acorns, ascidians and frogs. In 1894, he spent a year in the marine laboratory of Naples, where he shifted from descriptive to experimental embryology, seeking physical and chemical explanations for organismal development in ctenophores. He also found that by adding magnesium chloride, he could cause sea urchin eggs to divide without fertilization. Upon returning to Bryn Mawr, he switched to experimental work on regeneration and larval development in tadpoles, fishes and earthworms. In 1904, he moved to Columbia University.

Morgan accepted Darwin's theory of evolution as the common descent of similar species, but he rejected natural selection acting on small, constantly produced variations as its mechanism. Natural selection, he argued, is a theory that Darwin had erected to explain the process by which evolution takes place, to explain the manifold adaptations of an organism to the surrounding world. However, the definitions of a species relate not to adaptive characters but to discontinuities, to differences with no obvious relation to specific advantages. The diagnostic differences that taxonomists use to separate species do not necessarily have any survival value. In this argument, Morgan sought to support the findings of De Vries.

Hugo De Vries (1848–1935), a Dutch botanist, was among those who, in 1900, had rediscovered the work of Gregor Mendel. He doubted whether slight selective advantages, by which species slowly change into a new type, as Darwin had believed in, could explain evolution. Many species in nature remain constant, and no one has ever observed those slight changes whereby one species is supposed to convert ever-so-slowly into another species. He developed an alternative, "mutation theory", in which he suggested that new species and varieties form by sudden, short, large-scale leaps in existing species during which the parent type itself remains unchanged, and may repeatedly give birth to new forms (De Vries 1906). Furthermore, he believed that he had found evidence of such large-scale changes in the evening primrose *Oenothera*, a hermaphrodite plant capable of self-fertilization. Of American origin, this plant had spread, by De Vries' days, over various countries in Europe, grew in masses around Amsterdam, and was highly variable, with both the typical form and a few forms that were very different (Fig. 8.7). De Vries transferred some plants to a garden and allowed them to multiply by self-fertilization. Over just a few years, seeds of the typical species developed forms that were similar to it, as well as some specimens with well-marked features of new species. There was a dwarf form, a leafy form, a giant form and some other forms, which he retained for further cultivation. Having raised 55,000 plants, he counted 8034 mutant plants among this very large number. The mutations he discovered—he introduced the term—consistently recurred, and he classified them into seven species, which he proved to be species by breeding them true. Here, then, we have, De Vries claimed, a species that suddenly "explodes" (his terminology) into several new species, each with definite features.

This case of the evening primrose shows, he claimed, how species arise in general. Species originate through old species suddenly shooting off new forms. Some of these new forms survive unchanged until the next mutation, while others cannot sustain the struggle for existence and succumb to natural selection. Fresh features form by a sudden outburst of dramatic mutations, and between such outbursts, a

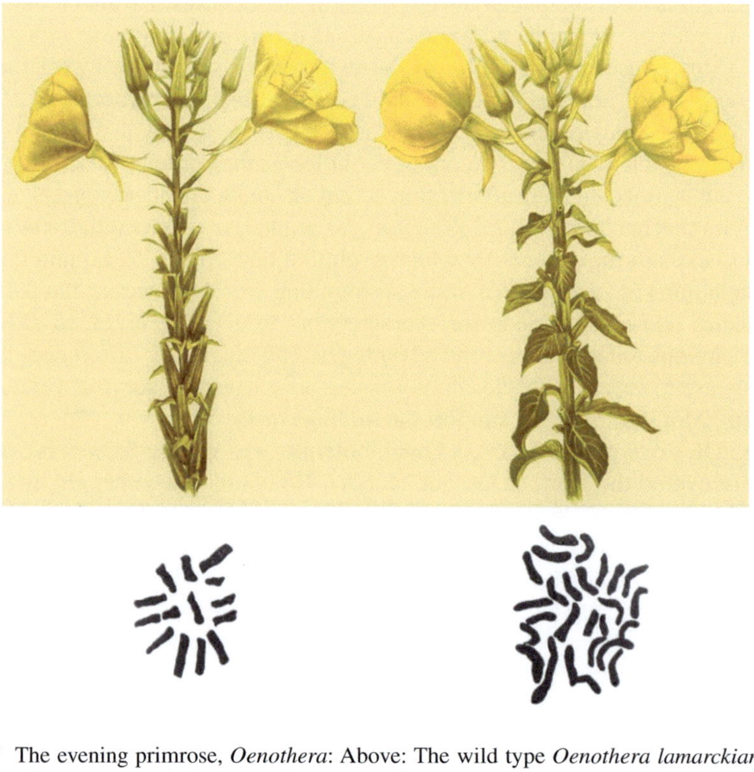

Fig. 8.7 The evening primrose, *Oenothera*: Above: The wild type *Oenothera lamarckiana* (left; *today O. glazioviana*) and *O. gigas* (right), one of the mutant species that De Vries suggested developed from the wild type. (From De Vries 1901). Below: The 14 diploid chromosomes of *O. lamarckiana* (left) and the 28 chromosomes of *O. gigas*. (From Morgan 1926.)

species survives with its characters unchanged. Darwin's slight, minor variations present in the daily life of a species have no significant effect on evolution.

De Vries thus claimed to have seen evolution before his eyes, and it was quite different from Darwinian evolution. His new species came into being by saltation, not by any struggle for existence among individuals of a continuously varying population. Natural selection is therefore not the omnipotent designer of a species; it does not eliminate the unfit from a population of continuous, slightly varying individuals. Forces at work in the genetic material of a species are the ones that play a directing role, and evolution occurs through a process that bypasses the need for Darwin's natural selection ("Natural selection can explain the survival of the fittest, but cannot explain the arrival of the fittest"; De Vries 1906, 1977).

Back to Morgan, he presented evidence that one of De Vries' mutant species, *O. gigas*, differed from the wild *O. lamarckiana* in that it was a polyploid (Fig. 8.7). Seeking to prove De Vries' mutation theory by experimental heredity research, he began (around 1908) investigating the fruit fly *Drosophila melanogaster*, mutating it through physical and chemical means and by radiation, and setting up cross-breeding experiments, to find heritable mutations. He and his colleagues identified hundreds of genes. His cytological investigations showed that *Drosophila* possesses only four pairs of chromosomes, including a pair of sex chromosomes. Female flies normally have two identical X chromosomes, whereas males have a single X chromosome and a smaller, gene-poor Y chromosome. Unlike humans, however, sex in fruit flies is determined by the number of X chromosomes, rather than by the presence of the Y chromosome (Fig. 8.8).

In 1910, a heritable mutant appeared that displayed Mendelian inheritance patterns: Among red-eyed wild types, a white-eyed mutant male suddenly appeared (Fig. 8.9). Later on, breeding experiments established that the white eye colour was caused by a recessive mutation in the gene responsible for normal red eye colour, and furthermore, that the gene was probably located on the X chromosome. When Morgan bred these white-eyed flies with red-eyed females, all the progeny were red-eyed; but the second-generation cross, produced white-eyed males—a sex-linked recessive trait. He concluded (1911) that some traits were sex-linked, and that specific chromosomes probably carried other genes as well. The observation of a miniature-wing mutant, also on the sex chromosome but sometimes sorting independently of the white-eye mutation, led him to the idea of genetic linkage, and to

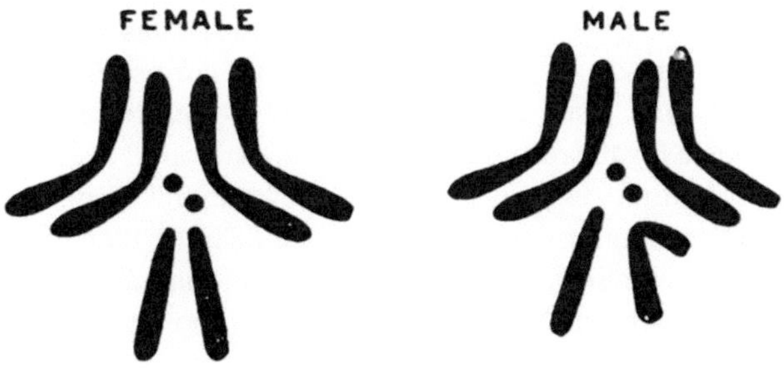

Fig. 8.8 Female and male chromosomes of *Drosophila melanogaster*. (From Morgan 1925.)

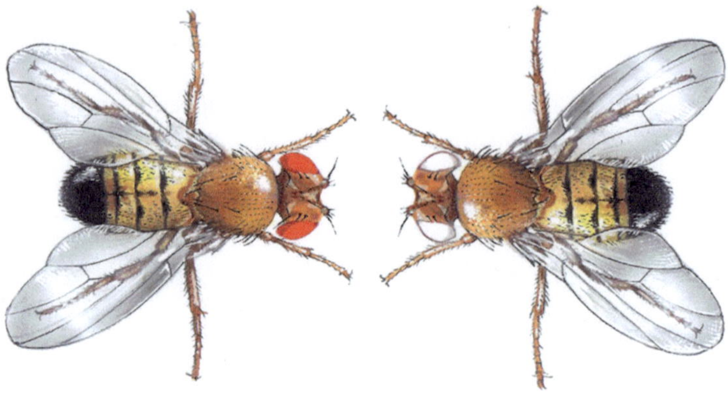

Fig. 8.9 *Drosophila melanogaster*: the wild, red-eyed type and the white-eyed mutant. (Tuvia Kurz.)

the idea of crossing over. Morgan (1916) concluded that the characteristics of wild animals and plants, as well as those of domesticated races, are inherited both in the wild and in domesticated forms according to Mendel's Law. In 1933, Morgan, who by now had joined the faculty of the California Institute of Technology, received the Nobel Prize.

Other biologists viewed chromosomes from a different angle. Richard Goldschmidt (1878–1958) was a German-born Jew. He studied the embryology and chromosome behaviour of nematodes and of other worms, became a professor at the University of Munich, and later became head of the genetics section of the Kaiser-Wilhelm Institute for Biology (1914), where he investigated different races of the gypsy moth *Lymantria dispar*. Two decades later, he fled to the United States (1936) when the Nazi regime began applying its Nuremberg racial laws, and joined the staff of the University of California, Berkeley. The Nazis, during World War II, published a propaganda poster entitled "Jewish World Domination", in which they displayed the Goldschmidt family tree.

Goldschmidt rose against the strict Darwinian gradualism in which most biologists of his day believed. In the 1930s, when chromosomes were among the frontlines of genetics, he suggested that in nature, two modes of evolution were in operation (1940). One is by the gradual accumulation of small "micro-mutations", and this mode results in evolution within a species. It brings about the formation of clines, of geographic races and of subspecies—of intra-specific units whose features are adaptive, in that they follow corresponding variations in environmental conditions. Such races and subspecies are, however, neither incipient species nor models for the origin of species; they are blind evolutionary alleys within a species.

The first step towards "Macro-evolution", namely the step from one species to another, requires an evolutionary mode other than that of the sheer accumulation of micro-mutations. Goldschmidt claimed that, at the point of emergence of separate species, chromosomes are of prime importance in that they undergo large-scale re-patterning. This may be due to a translocation between non-homologous chromosomes (segmental interchange), or to the inversion of part of a chromosome—a segment of a chromosome is reversed end-to-end, changing the serial order of the parts (genes) of the chromosome. Inversions tend to be additive, in that a second inversion may occur that has one break inside the first one and the second one outside, say (1940, p. 189):

Start A B C D E F G H I
1st inversion: A – E D C B – F G H I
2nd inversion: A – E – H G F – B C D I

Within species, the chromosomal pattern may change in a single step, or in a series of brief steps, without any visible effect on the phenotype and without any accumulation of gene mutations. The overall result of such major chromosomal re-patterning would be the existence of different, non-interbreeding species that differ mainly in their chromosome patterns, but not (or very little) in their morphology. Such would be the case of *Drosophila pseudobscura* and *D. miranda* (and here Goldschmidt drew on Dobzhansky's studies), so similar in appearance that taxonomists would hesitate to separate them on the basis of morphological differences alone, yet hybrids between these two species are sterile and the males are abnormal. The two species differ considerably in the site of their homologous genes along the same chromosome due to inversions, or they occur on different chromosomes, the result of translocations. Pattern changes within the serial structure of a chromosome may thus change the genotype, without any change of any individual gene. Inversions, translocations, deficiencies and duplications may produce a change in pattern that would lead to reproductive isolation, with the individual genes having nothing whatsoever to do with the process. The change from species to species is not a change involving more and more atomistic changes, but a complete change of the primary pattern into a new one, which afterward may produce intra-specific variation by micro-mutation.

Goldschmidt called this type of change in the intra-chromosomal pattern a "systemic mutation". Whatever gene mutations may be, they do not enter this picture at all. This new pattern emerges in a series of rapid consecutive steps that at first may be without any visible effect. Separated from its parent species by a sudden, rapid, bridgeless gap and an incompatible intra-chromosomal pattern, a new phenotype might emerge, a new species. Rapid, sudden re-patterning may also enable the expression of genes that in the parent species only rarely reach expression. As an example of this type of quick evolution, Goldschmidt offered the faunas of the geologically recent Pacific islands. The closest relatives of such faunas, living on the nearest continent, have undergone very little diversification, but explosive evolution has taken place upon the islands (Darwin's finches). In summing up. He noted that an intra-chromosomal pattern might exert a considerable phenotypic effect independent of genetic changes. It is possible to produce immense phenotypic changes of a macro-evolutional order by relatively small systemic mutations, not involving the creation of any new genes.

Another mechanism of rapid macro-evolution would be by mutations producing, though very rarely, a completely new body plan, what Goldschmidt termed a "hopeful monster". A genetic monstrosity appearing in a single step might permit the occupation of a new environment. Small (genetic) rate changes in early embryonic processes can produce a large effect on an adult phenotype. A fish undergoing a mutation that makes for the distortion of the skull and the carrying of both eyes to one side of the body is a monster. The same mutant in a very compressed form of fish living near the bottom of the sea is a hopeful monster, as it enables the fish to live on the sandy bottom of the ocean, as exemplified by the flounders (Goldschmidt 1940).

Over 50 years, a series of researchers thus developed the chromosome theory of inheritance. It began with Flemming's discovery of chromosomes and description of their behaviour during mitosis; Boveri and Sutton later strengthened the idea of a connection between chromosomes and heredity, and Morgan offered experimental proof that chromosomes were indeed the physical carriers of hereditary information. Goldschmidt suggested that new species evolve because of drastic re-patterning of the gene sequences along the chromosome, and also by the rare appearance of hopeful monsters. The biological significance of chromosomes, however, emerged with no deep connections whatsoever to the classification of animals.

What brought chromosomes into taxonomy was their diverse variation in number, structure and behaviour, to such an extent that up to the 1950s, the inference of species relationships was often based on the assumption that closely related species share similar characteristics in their chromosomal setup (their "karyotype"; Mayr 1963). Chromosomes consist of DNA; variation in DNA leads to a change in the chromosome, and this may cause variation among individuals, species, genera and higher ranks. By analysing chromosome similarities and differences, one can reconstruct the evolution both of the karyotype and of the animal or plant groups studied. The character most frequently used in chromosome-based taxonomy was chromosome number. In plants, speciation often relates to changes in ploidy, and furthermore, long chromosomes often form by the fusion of smaller chromosomes. As herbaceous plants possess longer chromosomes than woody plants, this suggests their position is more advanced than woody ones. Among animals, the jack jumper ant *Myrmecia pilosula* of Australia has two chromosomes (for females, with males being haploid and thus having one)—the smallest number possible. At the other end, the catfish *Clarias batrachus* of Southeast Asia has 104 chromosomes, the lamprey has 174, and the atlas blue butterfly *Polyommatus atlantica* of North Africa has 448–452 chromosomes. In some cases, the number is species-specific, thus humans have 46 chromosomes, whereas other primates have 48; in other cases, it may be family specific, as in mosquitoes, all of which have six chromosomes. The platypus has 52 chromosomes, of which ten are sex chromosomes: males have XYXYXYXYXY and females have XXXXXXXXXX. Crocodiles (Crocodylidae) have no sex chromosomes and sex, in this group, is determined by ambient temperature (Werner 2016). The mule has 63 chromosomes, an odd number between donkey (62) and horse (64).

Chromosome-based taxonomy (cyto-taxonomy) is today useful as an indicator of genetic relationships, but the fact that badgers, jellyfish and dolphins all have 48 chromosomes is irrelevant, as is the fact that both fox, starfish, earthworm and clawed frogs have 36 chromosomes. At the opposite extreme, some animal species are polymorphic in their chromosome numbers (the gecko *Hemidactylus flaviviridis* of Africa and Asia has either 46 or 40 among interbreeding populations, Werner 2016). Today, we do not commonly use chromosomes in classification.

One important breakthrough after Darwin's *Origin* was the discovery that at every taxonomic level, some proteins are highly specific. The proteins of one organism will show a stronger antibody reaction to the proteins of a closely related organism than they will to those of a more distantly related organism. When one introduces foreign cells or particles (antigens) into an organism, highly specific protein molecules (antibodies) are produced in the immune system of that animal. George Nuttall FRS (1862–1937) was an English bacteriologist who in 1904, by using serological methods that detect chemical (not morphological) differences, showed that Old World primates relate more closely to Man than to New World primates, in exact accordance with Darwin, who reached the same conclusion by relying on morphological similarities. He suggested that the degree and rate of blood reaction offer an index of relationship; closely related blood react more powerfully and more rapidly than do distantly related blood (Nuttall 1904: 409). The data in Nuttall's book show a general correlation between differences in blood reaction and the broad lines of zoological classification. Serology, by detecting homologous proteins in the serum, can therefore be a taxonomic marker that contributes to classification. Serological studies using plant protein extracts have been widely used in elucidating the taxonomy of a wide variety of groups and in estimating phylogenetic relationships, mainly around the mid-twentieth century. In animals, their use has been less common.

With different groups having different molecules, one early technique to separate and identify different molecules in a mixture uses capillary action on filter paper (paper chromatography). In this method, the investigator places a sample (a tissue or body fluid) of the taxonomic group under investigation on a spot on a paper, and then dips it into a solvent. Movement of the solvent carries different molecules at different rates so that different plant and animal groups produce different patterns of spots, representing various components of the sample. A later technique, electrophoresis, uses the different electrical potential of molecules instead of capillary action, and hydrolyzed starch or polyacrylamide instead of filter paper. Most biological molecules carry an electrical charge and, therefore, they will migrate through an electrical field at a rate that is proportional to their charge density. An electrical field in a solvent can, therefore, separate differently charged proteins (in addition, small proteins will migrate more rapidly than large proteins). Electrophoresis can therefore separate different kinds of proteins, some of which are specific to a given taxonomic group. In groups that have "featureless" external morphologies and internal anatomies, such as parasitic nematodes, parasitic insects and bacteria, electrophoresis has been a particularly useful tool for classification.

Today, taxonomic research at the intra-cellular level has shifted from differences in proteins to differences in DNA. Every living organism contains DNA, and its structure is more similar in closely related organisms than in distantly related ones. The DNA nucleotides (adenine, cytosine, guanine and thymine) have been known since 1910. In the 1950s, scientists worked out that adenine always pairs with thymine and cytosine always pairs with guanine; that DNA is the molecule that carries the genetic information; and that its three-dimensional structure consists of a double helix. In the 1960s, scientists found that letters in DNA are read in blocks of three, called "codons". Each codon specifies one amino acid that, during its synthesis, joins a protein.

An early method of obtaining taxonomic information from DNA was by DNA hybridization: Upon heating, the bonds between corresponding DNA nucleotides break down and the double-stranded DNA dissociates. Upon cooling, the disassociated strands recombine and bonds form, but only at the corresponding sites. When putting together single strands of DNA from different taxonomic groups (say, human and frog), they form hybrids that, when heated, separate at a temperature much lower than the one required to separate the original DNA pair because of the formation of much fewer bonds. This temperature difference is a measure of genetic similarity, between the two groups. Sibley and Ahlquist (1990) used DNA hybridization to reconstruct an overall phylogeny and classification of birds. Their study accepted, with minor adjustments, many of the groups recognized by traditional tools. It also yielded a new group that today is widely accepted, that of waterfowl together with landfowl (today termed Galloanserae).

DNA hybridization would perhaps have been more widely accepted, were it not for Sibley's studies in hominoid phylogeny. In these studies, he concluded that humans and chimpanzees are more closely related than chimpanzees and gorillas (Sibley and Ahlquist 1984, 1987; Sibley et al. 1988, 1990). However, Marks and his colleagues pointed out (1988, 2011) that Sibley and Ahlquist's *results* were that the chimp-gorilla DNA hybrids were more thermally stable than chimp-human hybrids. The conclusions they *reported*, however, were that chimp-human was more thermally stable than chimp-gorilla. Marks thus accused Sibley and Ahlquist of falsifying data. Consequently, many felt that DNA hybridization is tricky to do and tricky to interpret, and refrained from using DNA hybridization in classification and phylogenetic studies.

Other molecular techniques in the study of classification make use of small sections of nuclear DNA and of mitochondrial DNA. In the nucleus, in a small subunit of 18S ribosomal RNA, DNA mutations are unlikely to survive natural selection, and consequently, these genes are highly conserved in their structure, having changed slowly during evolution. Ribosomal DNA thereby provides clues about changes that occurred very early in evolution, such as the splitting up of classes within a phylum, or the origin of new phyla.

The mitochondria carry a small amount of DNA originating from the bacteria engulfed by the early ancestors of today's eukaryotic cells. Discovered in the 1960s, this DNA is located on a small chromosome (that is linear in many cnidarians but circular in bilateral animals) and its size varies from 80,00 base pairs in the anemone *Isarachnanthus nocturnus* to less than 6000 in the parasite *Plasmodium falciparum*; human mitochondrial DNA consists of under 17,000 base pairs, and it encodes 13 proteins. In most animals and plants, it is inherited from a single parent, the mother; double inheritance occurs in bivalves. Mitochondrial DNA changes fast in evolution as compared to nuclear DNA and it is therefore useful in determining differences that have accumulated less than 15 million years ago, as among many species. By comparing mitochondrial DNA sequences among different species within a genus, and sometimes between different genera of a family, biologists can build evolutionary trees for the taxa examined. For example, whereas humans and chimpanzees have nearly identical nuclear genes, their mitochondrial genomes are 10% different. Human and gorilla mitochondrial genomes are 12% different, suggesting that we may be more similar to chimps than to gorillas (Xu and Arnason 1996).

A more advanced method of obtaining information on the detailed molecular structure of nucleotides is by sequencing, which allows the identification of each single nucleotide in a whole sequence of a DNA molecule. The number of differences in the base pairs of two nucleic acids broadly reflects the number of mutational steps that separate them and, therefore, by comparing the sequence of base pairs of two homologous DNAs, one can determine the number of mutational differences between them. The measurement of such differences is of prime importance to the (evolution and) classification of plants and animals.

Sequencing produces the exact sequences of nucleotides, or bases, in DNA segments. Over the last 50 years, research has produced many techniques to sequence DNA (and RNA) molecules. Innovations have increased the technological capabilities of sequencing, allowing the reading of hundreds of base pairs in length and producing giant bases of data in one run (Heather and Chain 2016).

As molecular differences between groups of animals or plants can indicate how close they are in evolutionary terms, nucleotides can serve in the study of evolutionary history and hence in classification. Molecular characters are superior to morphological ones in the extraordinary detail they can supply. They might produce unequivocal conclusions in cases in which morphological evidence is ambiguous. In a classification, typical molecular systematic analysis requires the sequencing of around a thousand base pairs. At any location within such a sequence, the bases in a given position may vary between organisms. We term the particular sequence found in a given organism a "haplotype". In theory, as there are four base types, with a thousand base pairs, we could have 4^{1000} distinct haplotypes. In practice, however, for organisms within a particular species or in a group of related species, only a minority of sites show any variation at all, and most of the encountered variations correlate, so the number of distinct haplotypes found is smaller. When comparing the sequences of different haplotypes, one assesses the difference between two haplotypes by counting the number of locations where they have different bases, and express the difference between organisms as a percentage of the number of base pairs analysed. Molecular biology has indeed had a revolutionary impact on classification.

From a historic perspective, the 1990s witnessed the first completion of the genomes of a bacterium, a yeast and a nematode worm; and soon later, the full genome sequence of a fruit fly was completed. UK Prime Minister David Cameron launched the Human Genome Project, an international research effort that aimed to sequence all 33 billion letters of the human genome within 15 years. By 2003, the Human Genome Project was completed (at a cost of over $300 million), confirming that humans have 20,000 to 25,000 genes. In the 2000s, scientists completed the full genome sequence of the mouse; they found it smaller than the human genome by 14%, and that is similar to ours by 95%. The rat genome is smaller than the human genome but larger than that of the mouse. Today, one can sequence a complete human genome in one day, for under $1000.

Molecule sequences bear another, most important aspect. Mutations in a particular phyletic lineage occur at a broadly constant rate. Consequently, they provide a "molecular clock" for dating divergence. The clock is calibrated by reference to the fossil record or to biogeographic events whose age we know, allowing us to calculate the number of amino acid replacements per 10 million years. The resulting timeframe permits the reconstruction of the evolution of closely related groups, in the dimension of time: The probable time of the splitting of phyletic lines and sometimes even the geography of the splitting event. The assumption of a completely constant molecular clock is, of course, unrealistic. The same molecule may evolve at different rates in different groups. The less affected the selection of a certain molecule is, the more likely it is that it will change at a constant rate. The molecular clock is thus only a first approximation, and its results must be fine-tuned by using additional methods.

Molecular taxonomy measures differences in molecules that contain genes alongside junk material. It is excellent in assigning an individual to one of two known species with overlapping distributions; say, to return to our previous example, to the house mouse of Italy or the black rat of Italy. It does not answer questions concerning reproductive isolation—whether or not the house mouse in Italy is reproductively isolated from the house mouse in Iceland. We can only speculate whether the molecular differences between house mice of Italy and Iceland suggest reproductive isolation, just as in morphological data. Molecular taxonomy does not differentiate between variety- and species-level ranks.

Another weakness of molecular systematics is its assumption that classification must correspond to phylogenetic descent and that all valid taxa must be monophyletic. The discovery of extensive horizontal gene transfer among lower organisms complicates molecular systematics, as different genes within the same organism can have different phylogenies. Horizontal gene transfer is the movement of genetic material between organisms, rather than by the "vertical" transmission of DNA from parent to offspring (reproduction). Frederick Griffith (1877–1941) was the first to suggest, in the 1920s, that bacteria are capable of transferring genetic information. Born in Merseyside and having attended Liverpool University, at the age of 33, he began working in the lab of the Royal Commission on Tuberculosis. During World War I (1914–1918), the laboratory became the Ministry of Health's Pathological Laboratory, and he became a medical officer who examined pneumococci samples taken from patients throughout the country and classified each, in search of patterns of pneumonia epidemiology.

Pneumococci have two general forms: smooth and rough ones. The smooth form bears a capsule outside the cell wall that prevents phagocytosis by the host's immune system and is highly virulent. Mice injected with the smooth form succumb to pneumonia and death within days. The rough form lacks a capsule, its outer surface is the cell wall, it is not as virulent, and does not cause pneumonia as often. Griffith injected heat-killed Smooth bacteria into mice, and as expected, no disease ensued. However, when he injected mice with a mixture of heat-killed smooth and live rough forms, pneumonia and death ensued: the live rough had transformed into the smooth form, and replicated as such. More accurately, rough tended to transform into smooth if a large amount of live rough alone was injected, and adding masses of heat-killed smooth made the transformation much more reliable. Griffith's report that smooth had been produced from rough and that rough had been transformed into smooth (1928), was the first demonstration of bacterial transformation in which a bacterium distinctly changed its form and function: *Streptococcus pneumoniae* transformed from one strain into a different one. Griffith attributed his observation to an unidentified Transforming Principle (later identified as DNA). Subsequent research, in the 1930s and 1940s, revealed that isolated DNA was the material that communicated this genetic information, and in the 1950s, scientists demonstrated the transfer of antibiotic resistance between different species of bacteria. Horizontal gene transfer, so common among bacteria, also occurs from bacteria to fungi, to plants and to insects. In some rotifers, 8% of the genes are of bacterial origin. Additionally, horizontal gene transfer commonly occurs from plant to plant, from plant to fungus, from fungi to insects and from one fish to another.

Horizontal gene transfer bears on the classification of lower organisms and on those phylogenetic trees based on the sequence of a single gene. When using single genes as phylogenetic markers, in the presence of horizontal gene transfer, it is difficult to trace organismal phylogeny. Given two distantly related bacteria that have exchanged a gene, a phylogenetic tree including those species will show them to be closely related, because that gene is identical, even though most of the other genes are dissimilar. For this reason, phylogenetic analysis should rely on as wide a range of genes as possible. A gene frequently used in constructing phylogenetic relationships of prokaryotes is the 16S ribosomal RNA gene, since its sequences are sufficiently variable among members with close phylogenetic distances to allow us to measure differences. However, because of the horizontal transfer of this gene, its validity in constructing phylogenetic trees must be re-evaluated. Horizontally transferred genes are typically concentrated in only ~1% of the chromosome, in regions called hotspots, which diversify by rapid gene turnover.

8.3 Handling Many Features: Phenetics and Cladistics

In the 1960s, with the many breakthroughs in genetics, taxonomists felt that the classification of biodiversity must recognize that we are not classifying genetic entities ("genotypes", the set of genes an organism carries)—as we do not know what the genotype is. We are classifying by observable physical features of organisms, namely by their "phenotypes" (from the Greek *phainein*, to appear). These are determined by the genotype, as well as by environmental influences on the genes. Furthermore, in constructing systems of high-level groups (say, of orders or classes rather than species) we do so by subjectively choosing just a few features. These may not be repeatable in that, say, bird classifications by feathers might differ from bird classifications by bill, which might differ from classifications by their feet. A more objective way of classifying would be by taking into account both feather bills and feet and considering overall similarity (Mayr and Ashlock 1991).

Coming now to definitions, a character is any feature that the taxonomist uses to study variation within and among species—whether at the morphological, chromosomal, or molecular level. A character state is a particular form manifested by a variable character in a specific individual or group of organisms. A character may be flower colouration and its character states may be red or blue (Schuh 2000).

Numerical phenetics, as the name implies, consists of applying various mathematical procedures to numerically encoded character state data for the organisms under study (Sokal and Sneath 1973). Computers, by analysing similarities in a large number of features (preferably well over 60) shall objectively determine overall similarities in quantitative terms, and these, as they rely on many features, would be repeatable. By ordering many raw data by their similarity, an objective classification will automatically emerge. By relying on far more features than in traditional methods, numerical phenetics can achieve classifications that are more natural than those relying on a few features. Because computers produce repeatable results, any beginner can make classifications as good as those that a specialist can. Phenetics, in its classifications, does not attempt to reflect evolution. We expect the descendants of a common ancestor to be similar and therefore, bringing the most similar species and higher groups together may well produce a phylogenetic classification, as a side effect. We should completely exclude evolutionary considerations from taxonomy, however, because in the vast majority of cases, phylogenies are unknown. Furthermore, classifications based on phylogeny are by their very nature designed for a purpose. Phenetics suggests incorporating the maximum number of observations into our classifications, rather than being disposed toward any particular scientific theory, such as evolution. The more characters included, the more trustworthy the result must be. Numerical phenetics is the methodology of assembling individuals into groups by estimating overall similarity, regardless of phylogeny or evolutionary relation. This strictly theory-free approach is advantageous in that it requires no previous knowledge of the group under study. It demands only the ability to observe and quantify. To free their taxonomy of any theoretical implications and any a priori assumptions, pheneticists rejected any reference to the formal categories of species, genera or families. They replaced them with the concept of the operational taxonomic unit (OTU) as representing individuals, populations, species, genera and so on. The product of the clustering process is a "phenogram", which various procedures convert into a classification. In practice, character states are each assigned a numerical value, and the resulting sets of values (data matrices) obtained for several characters of several groups are then processed to obtain unbiased taxonomic inferences. Natural groups may not share any one single defining character, as a member of the group might lose the diagnostic feature (Sokal and Sneath 1973). For example, flies Diptera are characterized by two wings, but some flies are wingless. We should classify groups as such if they possess at least a minimum number of characters, but not all.

The measuring of very many variables and then presenting them as two-dimensional graphs requires clustering, the task of grouping a set of objects in such a way that objects in the same group are more similar to each other than to those in other groups. We can achieve clustering by various algorithms that differ significantly in their perception of how to cluster, and the appropriate clustering algorithm and parameter settings depend on the individual dataset. To illustrate numerical phenetics, an example in Fig. 8.10 presents three groups: A, B and C; and the states of five binary characters, coded as 0s or 1s, to present ancestral and derived states, respectively. A and B are more similar to each other than either of them is to C in that they share three out of five character states (0, in characters 3,4 and 5).

Numerical phenetics avoids errors that result from over-emphasis on, or over-weighting of, one feature or another. One of its weaknesses is that it lacks any theoretical justification for choosing among different clustering methods, which may yield different results for the same set of data. Scientists chose clustering methods as a matter of personal preference, and this imparts an inability to meet the claims of objectivity and repeatability. Another weakness of the method is that by converting the totality of the similarity values of all individual characters into a single overall similarity value, the clustering method results in an enormous loss of information, mainly regarding the character direction and origin of differences. Finally, phenetics does not distinguish between convergent evolution and adaptive radiation can mislead it. Today, Cladistics has largely superseded numerical phenetics, in classifications.

characters	taxa A	B	C
1	0	1	0
2	0	1	1
3	0	0	1
4	0	0	1
5	0	0	1

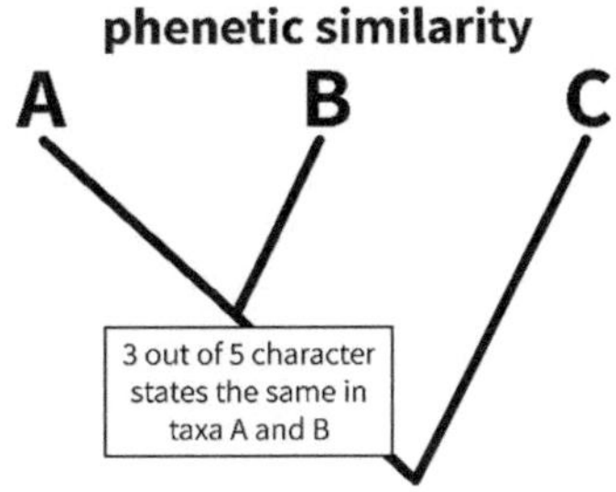

Fig. 8.10 A simple data matrix with three taxa and five characters, to demonstrate classification by overall phenetic similarity. (Modified after Quicke 1993, p. 6, Fig. 1.1.)

Willi Hennig (1913–1976), the founder of phylogenetic systematics (cladistics), was a German biologist. The son of a rail worker and a factory worker, while at school he volunteered at the Dresden State Museum of Zoology, went on to the Leipzig University and from there to the German Entomological Institute of the Kaiser-Wilhelm Gesellschaft, where he began research on the larvae of flies and mosquitoes. During World War II, he served in the German infantry in Poland, France, Denmark and Russia, was injured by a grenade shrapnel (1942), and then became an entomologist at the Institute for Tropical Medicine and Hygiene in Berlin. On a mission to Italy to fight malaria, the British captured him (1945), and during his brief time as a prisoner of war, he drafted his important contributions to systematics (which he published in 1950). After the war, upon returning to the German Entomological Institute in Berlin, he qualified in zoology and became Head of the German Entomological Institute he was in the Soviet Sector of East Berlin while living in the American sector. While on a trip to France, he heard of the building of the Berlin Wall, returned immediately to Berlin and resigned his post (1961). He accepted a post at the State Museum of Natural History in Stuttgart (then West Germany), where he headed a department for phylogenetic research.

Henning (1950, 1979) suggested that we should group and classify organisms strictly and only according to the historical sequences by which they descended from a common ancestor. Classifications, he claimed, should reflect the evolutionary history of a group. During evolution, characters evolve from an ancestral state (a *pleisiomorphous* state, in his terms) to an advanced, derived state (his *apomorphous* state). Within a group of related organisms in which a character is in more than one state, only one of these can be the ancestral state; and if so, then all others are the derived states, which have evolved within the group, from that ancestral state. Only by referring to shared derived features can we construct phylogenetic relations, evolutionary trees and hence natural classifications of a group. Hennig's Evolutionary Systematics thus differs from phenetics in that it seeks evolutionary relationships between groups of organisms.

Evolutionary systematics partitions character states into derived and ancestral ones. It further claims that we should recognize only groups that possess shared derived characters, and ignore ancestral characters. It thus excludes reliance on overall similarity. Hennig further claimed that every taxon should be monophyletic, namely, it should consist of the stem species, all its descendants and only its descendants; and that all sister groups should be accorded the same rank. To reconstruct the phylogeny of a group using character states that vary among its members, we first determine which state of each character was present in the common ancestor of the entire group. This is the ancestral character state for the group as a whole, and all other states of the character that arose later within the group are the "derived character states". The organisms or species that share derived character states form subsets within the group called clades (from the Greek *klados*, a branch). A clade corresponds to a unit of common evolutionary descent, and it includes all descendants of a particular ancestral lineage. The nested hierarchy of clades is presented as a branching diagram, called a cladogram.

To illustrate the differences between numerical phenetics and evolutionary systematics, an example in Fig. 8.11 presents three groups, A, B and C; and the character states of five binary characters, coded as 0s or 1s to present ancestral and derived states respectively.

In this example, the application of numeric phenetics leads to a different tree than that of evolutionary systematics, yet obviously, only one of these alternative trees can truly represent evolution. The phenetic constructed tree reflects the fact that A and B share more character states than either does with C. In contrast, the evolutionary constructed key shows B and C to be most closely related, even though it relies on just one derived character state, which is shared by B and C; no other pairs of groups share any derived features. (In this example, I am closely following Quickie 1996).

To evaluate which character state is ancestral and which state was derived later within a group, we consult an additional group of organisms (an outgroup) that is phylogenetically close but not within the group that we are studying. Any character state that we find both within the study group and in the outgroup is likely to be the ancestral state of the study group. As an example, we may consider as a character the dentition of amniotic vertebrates (reptiles, birds and mammals; again, from Quickie 1996). The presence versus absence of teeth in the jaws constitutes two different character states. Teeth are absent from birds but present in the other amniotes.

characters	taxa A	B	C
1	0	1	0
2	0	1	1
3	0	0	1
4	0	0	1
5	0	0	1

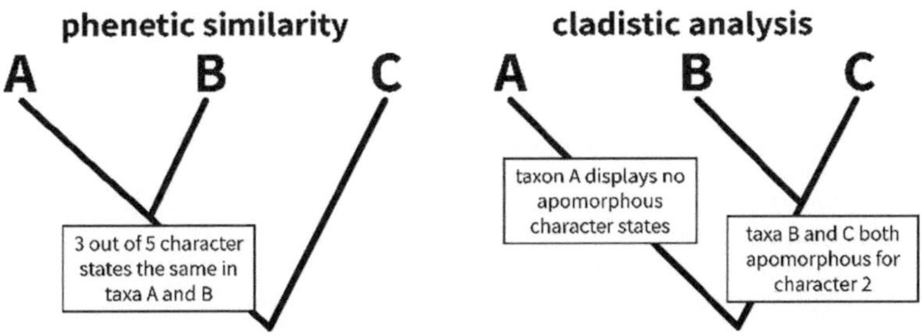

Fig. 8.11 A simple data matrix with three taxa and five characters, demonstrate how a tree constructed by overall phenetic similarity (a phenogram, left) can differ from one based on cladistics principles (a cladogram, right), which makes use of character states. (Modified after Quicke 1993 p. 6, Fig. 1.1.)

To evaluate which character state, presence or absence of teeth, characterized the direct ancestor of amniotes and which state was derived later within the amniotes, we seek a close outgroup. The amphibians are an appropriate outgroup to the amniotes for determining ancestor-derived variation in dentition of the amniotes. Teeth are usually present in amphibians. We infer, therefore, that the presence of teeth is ancestral for amniotes and the absence of teeth is the derived state. Teeth were secondarily lost in the ancestral lineage of all modern birds. To conclude this point, all character states in a study group, which are absent from its outgroup, are considered derived.

The phyletic approach is heavily dependent on the outgroup, and errors in determining which of the character states is ancestral and which is derived can produce mistakes. In our example, the outgroup is the amphibians, which usually bear teeth. Frogs, however, among amphibians, have lost their teeth over 20 times during their evolution, and only one frog, in more than 7000 species, has true teeth on both upper and lower jaws (*Gastrotheca guentheri*). This might cast a shade of doubt over frogs as a suitable outgroup.

Darwin's doctrine of evolution imposed upon classification the responsibility of displaying, in summary form, the evolutionary history of the animal kingdom. In line with his doctrine, modern taxonomy deals with samples of variable natural populations, and in the biological meaning of this variation. Faced as we now are with many more than a million different kinds of animals in different parts of the world, we must classify. Taxonomy today rests on far better foundations than it did in Darwin's day (Mayr and Ashlock 1991; Schuh 2000).

References

Barraclough, T. G. 2019. *The evolutionary biology of species*. Oxford University Press.

Capy, P., and P. Gibert. 2004. *Drosophila melanogaster, Drosophila simulans*: So similar yet so different. *Genetica* 120:5–16.

De Vries, H. 1901. *Die Mutationstheorie*. Leipzig: Veit.

De Vries, H. 1906. *Species and varieties, their origin by mutation*. 2nd ed. Chicago: Open Court.

Goldschmidt, R. 1940. *The material basis of evolution*. New Haven: Yale University Press. Third Printing, 1944.

Heather, J. M., and B. Chain. 2016. The sequence of sequencers: The history of sequencing DNA. *Genomics* 107:1–8.

Hennig, W. 1950. *Grundzüge einer Theorie der phylogenetischen Systematik*. Berlin: Deutscher Zentralverlag.

Hennig, W. 1979. *Phylogenetic systematics*. Translated from German by D. D. Davis and R. Zangerl. Chicago: Illinois University Press.

Marks, J. 2011. The rise and fall of DNA hybridization, ca. 1980–1995, or how I got interested in science studies. In *Workshop on "Mechanisms of fraud in biomedical research," organized by Christine Hauskeller and Helga Satzinger*. London: The Wellcome Trust.

Marks, J., C. W. Scmid, and V. M. Saric. 1988. DNA hybridization as a guide to phylogeny: Relations of the Hominoidea. *Journal of Human Evolution* 17:769–786.

Mayr, E. 1963. *Populations, species, and evolution*. Belknap: Harvard.

Mayr, E., and P. D. Ashlock. 1991. *Principles of systematic zoology*. New York: McGraw Hill.

Nordenskiold, E. 1929. *The history of biology. A survey*. New York: Tudor.
Nuttall, G. F. 1904. *Blood immunity and relationship*. Cambridge.
Poulton, E. R. 1908. *Essays on evolution, 1889–1907*. Oxford: Clarendon.
Quicke, D. L. J. 1993. *Principles and techniques of contemporary taxonomy*. London: Blackie Academic and Professional.
Robson, G. C. 1928. *The species problem. An introduction to the study of evolutionary divergence in natural populations*. London: Oliver & Boyd.
Schuh, R. T. 2000. *Biological systematics*. Ithaca: Cornell University Press.
Sibley, C. G., and J. E. Ahlquist. 1984. The phylogeny of the hominoid primates, as indicated by DNA-DNA hybridization. *Journal of Molecular Evolution* 20:2–15.
Sibley, C. G., and J. E. Ahlquist. 1987. DNA hybridization evidence of hominoid phylogeny: Results from an expanded data set. *Journal of Molecular Evolution* 26:99–12.
Sibley, C. G., and J. E. Ahlquist. 1990. *Phylogeny and classification of birds based on the data of DNA-DNA hybridization*. Yale University Press.
Sibley, C. G., J. E. Ahlquist, and B. L. Monroe. 1988. A classification of the living birds of the world based on DNA-DNA hybridization studies. *Auk* 105:409–423.
Sibley, C. G., J. A. Comstok, and J. E. Ahlquist. 1990. DNA hybridization evidence of hominoid phylogeny: A reanalysis of the data. *Journal of Molecular Evolution* 30:202–236.
Simpson, G. G. 1961. *Principles of animal taxonomy*. New York: Columbia University Press.
Sokal, R. R., and P. H. A. Sneath. 1973. *Principles of numerical taxonomy*. San Francisco: Freeman.
Sturtevant, A. H. 1920. Hybrids of *Drosophila simulans* with *melanogaster*. *Genetics* 5:488.
Werner, Y. L. 2016. *Reptile life in the land of Israel*. Frankfurt am Main: Chimaira.
Xu, X., and U. Arnason. 1996. A complete sequence of the mitochondrial genome of the western lowland gorilla. *Molecular Biology and Evolution* 13:691–698.

Chapter 9
After Darwin: Biogeography

Abstract Biogeography describes and explains distributions, today and in earlier times. Darwin's theory suggested that animal groups all evolved from one centre of creation. After him, one theory suggested that northern regions of the globe are cradles of successful, dominant modern groups. Cold climates force mammals to adapt to cold, arid, variable climatic conditions and favour active and enduring animals. These hardy-adapted animals disperse towards equatorial regions. Drops in sea levels enable northern faunas to disperse to southern continents, eventually attaining cosmopolitan distributions and displacing early, less adaptable remnants of ancient faunas that find refuge in the tropics. Another theory suggested that successful dominant groups arose in the Old World tropics and spread over much of the world, not polar faunas. A third theory claimed that a group with a narrow range is usually one of recent origin, and one with a wide range is an old one; consequently, the range of a group may indicate its age. To explain discontinuous distributions, another theory suggested that bridges had once crossed oceans and impassable mountains, and had thereby split ancestral, once broad distribution areas, into smaller areas. This theory depreciated the importance of long-distance dispersal.

Suggestions that Earth's continents have drifted apart, over time, began already four centuries ago. Later, naturalists suggested that Earth's internal parts are a fluid, the continents drifting upon it. Identical Carboniferous fossils in Europe and America suggested that these continents were then inter-connected, all southern continents once forming a single supercontinent. Others suggested that colliding continents thrust earth's surface upward, forming mountain ranges. Wegener independently developed the continental drift theory (1912), which modern plate tectonics supports: A huge southern supercontinent had existed 550 million years ago; it merged with a huge northern supercontinent 330 million years ago, forming one single giant supercontinent. 200 million years ago, this giant began breaking up, with continents splitting off; Africa and South America drifted apart 140 million years ago. Life's lineages evolved on each floating continent, thereby forming the various zoological realms. This explains the distribution of organisms in the Permian and Triassic. Among modern animals, it explains why marsupials are the only mammals native to Australia (except for bats) and why marsupials occur in both Australia and South America. It also explains the distribution of some insects. Some organisms

J. Heller, *Nature's Biodiversity*, https://doi.org/10.1007/978-3-031-89790-0_9

have thus achieved their modern disjunct distributions by the splitting of regions—Vicariance Biogeography.

Early biogeographic theory suggested that conditions are favourable in an area's centre, and unfavourable in marginal zones, where death exceeds reproduction. Excess reproduction near the centre and migration towards margins, maintain marginal populations and an equilibrium between these processes determining range limits. Modern theory applies to islands the fact that species undergo cycles of origin, spread, decline, retreat and extinction, due to processes of range expansion, diversification, adaptation, endemism and extinction. Insular faunas achieve saturation through colonization, with extinctions dynamically balancing immigration. The resulting equilibrium is a dynamic species richness, with turnover in species composition. Immigration rates of new species decrease with increasing numbers of species already present on an island, and extinction rates increase, as the number of species increases. The two processes balance at a point where the immigration rate equals extinction rates, and at this equilibrium, the number of species is constant. These factors determine island biodiversity. From island faunas to faunas in general, the theory predicts continual species turnovers, with ecological communities as open assemblages approaching a dynamic, steady-state species richness. In practice, some studies have found evidence for species equilibrium, on islands and in insular habitats, and for species turnover at points of equilibrium. Others have not.

The past two centuries have witnessed generalizations that correlate patterns of variation with biogeography, within and across species: Two species competing for the same limited resource cannot coexist. One species having even a slight advantage over another will dominate, leading to the extinction of the weaker competitor. Races of warm-blooded vertebrates from cooler climates are larger than those from warmer climates. Warm-blooded animals in warmer climates have larger appendages (tails, beaks and legs). Birds and mammals are darker in wet and warm environments. Body sizes are larger in deep-sea waters. Several environmental factors may cause similar responses in animals and may substitute one another in determining distribution. The establishment of a species in a habitat depends on a minimum presence of the factors necessary for the most sensitive developmental stage. Larger areas of land (or sea) usually contain more species of a group than smaller areas. Near the equator there is a rich diversity of life.

The biogeography of South America reflects a history of many millions of years. It was an island continent throughout much of the time, in which unique mammals evolved, including marsupials that preyed on rodents. Many extinct mammals were giant, heavily armoured grazers; others were carnivores or marine feeders of seagrasses. Among hoofed mammals, some resembled hares, others rhinoceroses and some were fast runners. There were large carnivorous flightless terror birds and other giant birds soared the skies. 40 million years ago, African caviomorph rodents rafted across the ocean, underwent explosive diversification and outcompeted native rodent-like mammals; some evolved into large, bison-size forms. African primates also reached the region by rafting. When the Panama Isthmus emerged, it enabled another biotic invasion, with migratory waves of terrestrial organisms occurring 26 million years ago. These migrations changed South America's predator guild from one of marsupials, terror birds and reptiles to one dominated by immigrant placental carnivores. Today, half the mammalian genera in the south are of northern origin. 12,000 years ago, Indians invaded and hunted many South American mammals to extinction. The Panama Isthmus also changed the marine fauna on the Atlantic side, which now had to adapt to an environment of lower productivity.

A large-scale migration of marine groups is occurring today: Species of tropical Indo-Pacific origin cross the Suez Canal and invade the Mediterranean. With 300 such migrating species since the canal opened a century and a half ago, this is one of the major biogeographic events of today. Though there is, as yet, no total replacement of native species by migrants, some Red Sea migrants are very abundant and some native species are only rarely encountered. Among sea snails and bivalves, about half the migrant species that ever reached the shores of the Mediterranean seem to have perished, and about one-third have successfully established themselves. Some of the colonizers are r-strategists, as compared to non-colonizers. Many Red Sea species have not invaded the Mediterranean.

This migration into the Mediterranean provides a unique opportunity to investigate large-scale processes of migration, invasion, colonization and outcompeting extinction when two faunas collide, processes that usually occur over long geological times.

9.1 Centres of Origin

Biogeography is a discipline of biology that describes and explains the distribution of animal and plant groups, today and in earlier times. As an integrative field of research, it unites concepts and information from ecology, evolution, taxonomy, geology, geography, palaeontology and climatology, and it refers both to local, regional and global levels. It is so close to the classification of organisms that I cannot imagine many (any?) taxonomists who are not interested in the distribution of at least his or her group. Macro-ecology, which in recent decades some have suggested as a new discipline, seeks to understand relationships between organisms and their environment on large spatial scales: how biodiversity and abiotic components of the Earth vary in space and time, and how they interact. It seeks processes that explain large-scale patterns of abundance, distribution and diversity, by considering questions of species richness, latitudinal gradients in species diversity, species-area curves, range size, endemism, body size and species abundance. In the questions they address, biogeography and macro-ecology are so similar that to me they seem to be the same discipline. Biogeography is a very dynamic scientific field, with excellent books covering its different aspects. This chapter is for the reader interested in its historic perspectives.

The discipline evolved slowly. By way of a telegraphic summary of preceding chapters, we recall Linnaeus, who believed that in some past, the land was merely a small island with all existing creatures. The island further emerged, its creatures began dispersing into new lands, and God placed different species in different parts of the world to prevent competition. Buffon suggested that a continuum via a land-bridge had once existed between faunas of the Old and New Worlds. As continents formed, sank and cooled, many animals went extinct. This released large quantities of molecules which, by spontaneous generation, built new faunas, better adapted to lower temperatures. In all places with similar climates, one finds similar plants and animals. Cuvier believed in drastic catastrophic events that wipe out existing faunas before new faunas arrive, but Lyell argued that also slight environmental changes, not only catastrophic events, provoke changes in animal ranges. He also claimed that there once were several centres of creation. Humboldt emphasized such environmental factors as temperature and humidity, as determining global vegetation patterns. Sclater, to account for populations of terrestrial species now separated by barriers of water, speculated that a now-sunken continent had once occupied much of the Indian Ocean.

We now come to Darwin, who sought evidence for natural selection in biogeography and devoted two chapters in his *Origin* (1859) to plant and animal distribution. He noted that environmental differences alone do not explain differences in the flora and fauna of separate regions. South America, Africa and Australia all have regions with similar climates at similar latitudes, but they have very different plants and animals. Furthermore, species in one area of a continent are often closer to species in other regions of that same continent than to species on other continents. He also noted that barriers to dispersal determine the differences between species of different regions, such as the few kilometres wide Panama Isthmus that formed a barrier between the coastal sea life of the Atlantic and Pacific sides of Central America because of which they have almost no species in common. He explained these differences by a combination of dispersal and descent with modification. We can thus understand how it is that at different latitudes, for instance, in South America, the inhabitants of the plains and mountains, of the forests, marshes and deserts, are linked together by affinity, and are likewise linked to extinct beings which formerly inhabited the same continent. Darwin went on to note that a volcanic island, formed a few hundred miles from a continent, might be colonized by a few species from that continent. These species may change and modify over time, but they still relate to species on the continent. Oceanic islands should have few inhabitants, but of these, a great number should be endemic or peculiar.

Wallace, who with Darwin had co-discovered the theory of natural selection, turned in later years to the study of biogeography. He discussed the animals of the Malay Archipelago (1869) and recognized that the faunas of New Guinea and Australia as sharply distinct and separate from those of Borneo and the Malay Archipelago. The boundary line separating these two faunal units is (what we today call) the Wallace Line. His *Geographical Distribution of Animals* (1876) was one of the first thorough presentations of the distribution of animals throughout the world. He amassed much information first-hand from his explorations in South America and in South-East Asia, but the bulk of his knowledge came from worldwide correspondence with other travellers and naturalists. He wisely noted that some animals reach some islands by stormy winds that carry them over oceans, whereas some islands preserve ancient faunal elements. He accepted, like Buffon and Lyell before him, that changes in climate modify animal distributions and, in this context, he paid much attention to the ice ages and their influence—he believed that climate changes raise and lower the snow line every 10,500 years. He confirmed that, in general, the different parts of the world support six greatly dissimilar assemblages of animals, in the zoological realms of the globe. Both Darwin and Wallace accepted geographic stability as granted, namely that the position and size of the continents had not changed over time. New species evolve in a limited area, a "centre of origin", from which they dispersed to other regions over the same pattern of world geography that we see today.

Lyell's theory of a moderate, ever so gradually changing globe was demolished on 26 August 1883 when the volcano on the small island of Krakatau in the Sumatran Straits erupted, killing 36,000 people (Thornton 1996). Hot ashes reached 80 km high, violent explosions were heard even in Mauritius (4800 km away), and airwaves circled the globe several times—the loudest sound ever recorded in modern history, it was heard on over 10% of Earth's area. Tsunamis reached heights of 46 metres near the island and they surrounded the globe three and a half times, gradually weakening. The eruption injected quantities of sulphur dioxide gas high into the air, which winds transported all over the planet; this led to a global increase in sulphuric acid concentration in high-level clouds, which increased their reflectivity. The clouds now reflected more incoming light from the Sun than usual and cooled the entire planet. Summer temperatures, in the year following the eruption, fell by 0.4 C in the northern hemisphere, and there were heavy rains; the suspended sulphur eventually fell to the ground, as acid precipitation. On Krakatau itself, where 70% of the island collapsed into an empty magma chamber, a thick layer of pumice covered the remaining part of the island, destroying all animal and plant life. The first researchers who reached Krakatau (May 1884) found just one living creature, a single spider. The huge amount of material deposited by the volcano also drastically altered the surrounding ocean floor, as 20 cubic kilometres of hardened tuff sank over 1100,000 square kilometres. Krakatau taught the scientific community that, though gradual environmental changes might sometimes determine some animal ranges, violent catastrophes might determine others (Thornton 1996).

The recolonization of Krakatau provided an opportunity to study dynamic aspects of biogeography regarding how plants and animals colonize new islands. It revealed different dynamics for seaborne, wind-borne and animal-borne dispersal, alongside the dominant role of succession in different habitats. These studies revealed that dispersal is the main process altering animal and plant distributions. Taxonomic groups originate in one confined area (a "centre of origin"), from where descendant populations spread.

After *The Origin* and its claim that species evolve and differentiate by modification from a common descent, the idea of several major centres of creation was no longer acceptable, and scientists could no longer believe that the lions of Africa, tigers of Asia and pumas of America had evolved in separate centres of creation. Clearly, they evolved from a single centre of creation. Was this centre of creation the same for all animals? If so, then where was it?

On this matter, one theory was that of the palaeontologist William Diller Matthew (1871–1930), curator at the American Museum of Natural History for over 30 years, and then director of the Museum of Palaeontology at the University of California. He insisted (1915) that during the past 65 million years (the Cenozoic era), both the sea basins and continents were as, and where, we find them today; and he rejected earlier theories, of submerged land bridges and continents to account for populations of terrestrial species now separated by barriers of water. Matthew argued that cyclical changes in global climate would result in corresponding changes in the sea level. He divided the continents of the globe into northern and southern ones (Europe, Asia and North America, versus South America, Africa, Australasia and Antarctica). Animals came to live where we find them today because of cyclic changes in the climate and sea level, along with a prevailing tendency for animals to disperse. Primitive and archaic types, such as marsupials, had once existed in the northern continents (as Cuvier had found), but today they abound mainly in the southern continents, as remnants of formerly cosmopolitan groups. The most "progressive and dominant types" (Mathew's words) all originate in the north, where cold climates accompanied by drops in sea level, especially in the interior of continents, would force terrestrial mammals to adapt to the cold, arid and highly variable climatic conditions. Such conditions also favour the evolution of higher, more active and more adaptable animals, with a sturdy built and endurance. From near the poles, where such adaptations first appear, adapted animals tend to disperse toward the equatorial regions. The terrestrial area of the continents becomes wider during drops in sea level, and this will enable an "expansive evolution" of the northern terrestrial faunae, which will eventually attain a cosmopolitan distribution. The northern animals, as they expand, will displace the early, less adaptable remnants of the more ancient fauna, first from the northern temperate regions and then from the southern landmasses. The early fauna will find its last refuge in the tropics of the southern continents. Northern areas of the globe are thus the cradle of the most successful, dominant modern genera and families. To support his theory, Mathew presented many animal groups in which fossils of the more primitive genera, today in the southern continents, once lived in the more northern regions (Fig. 9.1). His theories were not that far from those of Buffon, who had developed a theory whereby Earth was once a warm globe, in which life first began in (then warm) northern Eurasia, with heat-adapted animals. The earth was subject to overall cooling and eventually, with the cooling of the northern lands in which they originated, north-Eurasian animals migrated southwards, from the once warm poles to warmer (tropical) regions of the equator, where they live today.

Another biogeographic theory was that of John Christopher Willis FRS (1868–1958) an English botanist, who was director of the Royal Botanical Gardens

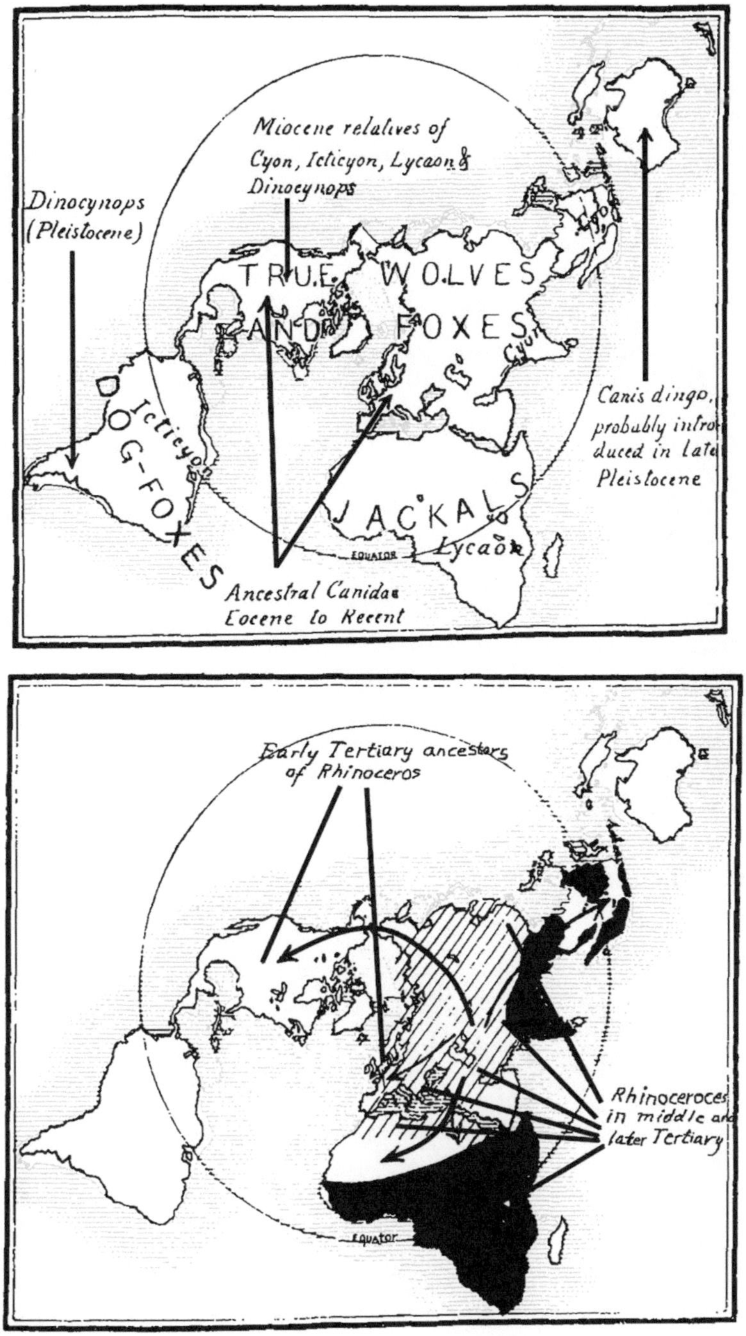

Fig. 9.1 Origins of animal groups in northern continents. (From Mathew 1915, slightly modified)

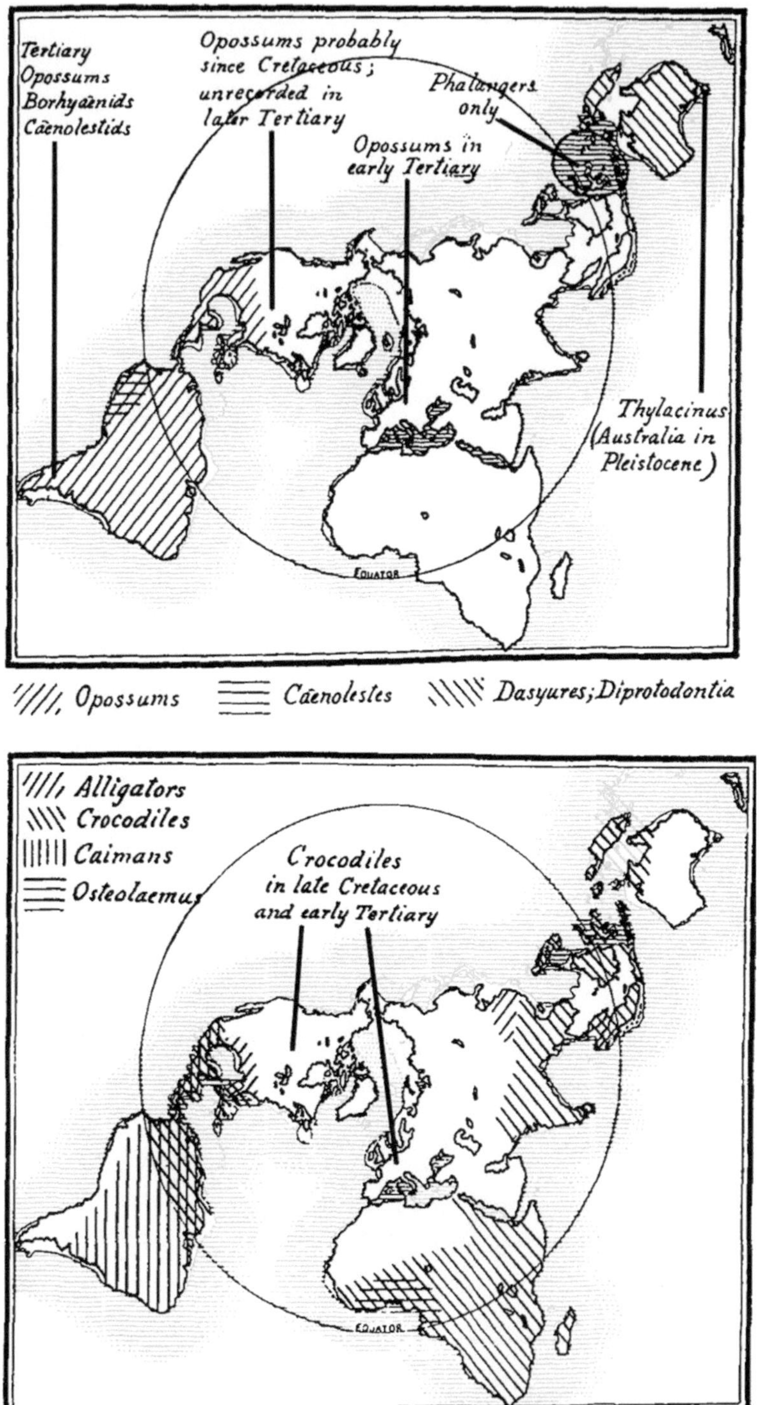

Fig. 9.1 (continued)

in Ceylon (now Sri Lanka) and later of the Botanical Gardens at Rio de Janeiro, Brazil. He claimed (1922) that the size of the area occupied by a group of closely allied plant species depends on its age, not on its peculiar adaptive needs. A group with a narrow range is usually one of recent origin, and vice versa, a group of species with a wide range is an old one. Time is dominant in determining distribution, and consequently, the range of a group of species may indicate its age. This Age-Area theory was criticized by many (e.g., Robson 1928) under the claim that factors other than time, such as hardiness, adaptability and migration, all share in determining the size of a plant's area. Plant groups may, of course, extend their distribution over time, but this does not make age the most important factor in determining range sizes. The Age-Area theory attracted some interest for about 20 years.

In 1957, Philip Darlington, Curator of Entomology at Harvard's Museum, pressed on with Mathew's theory. He suggested, like Mathew, that the main history of dispersal is the history of the cycles of movement of successions of dominant groups. Dominant groups are conspicuously successful ones, numerous in individuals, usually numerous in species, often (eventually) diverse in adaptations and often (eventually) widely and continuously distributed in unfavourable as well as favourable regions. The most dominant existing groups of land and freshwater vertebrates are cyprinids in fishes, the frog genus *Rana* in amphibians, colubrid snakes in reptiles, passerines in birds and rodents in mammals. Dominance reflects underlying qualities that make for initial success and lead to the evolution of numerous successful types, and dominance implies success in competition. The dominant groups are those that spread, indeed, the ability to spread is one of the attributes and evidence of dominance. Dominant animals spread because the qualities that make them dominant enable them to invade new areas and new habitats, become adapted to them and replace other animals there. Dominant, spreading animals over-run and destroy fewer dominant ones. The spread of successful dominant groups from the Old World tropics over much of the world, followed by zonation and differentiation according to climate and ocean barriers, and by retreat and replacement of old groups as new ones spread, has formed a broad global pattern of distribution. The pattern formed because evolution has tended to produce the most dominant animals in the largest and most favourable areas, which for most vertebrates are in the Old World tropics. Darlington, keeping Matthew's vision of dominance and cyclic replacement, thus merely reconsidered the area of origin. He shifted the principal source area for dominant groups to the Old World tropics, where, he believed, dominant terrestrial groups originated more frequently than elsewhere. Their species spread to parts of the world that Darlington now conceived to be peripheral: across the Bering land bridge to North America, and thence to Central and South America; across the Indonesian archipelagos to Australia and the Pacific archipelagos; and straight north into temperate Asia and south into southern Africa. We shall return to Darlington further on, in discussing island biogeography.

Some biogeographers sought to explain disjunct distributions. Leon Camille Marius Croizat (1894–1982), the son of French emigrants to Italy, studied law in Turin and then, at the age of 30, migrated to the United States, where he lived for several years as an artist, selling his artwork. During the 1930s, he worked at identifying plants, as part of an inventory conducted in the public parks of New York City. Eventually, at the age of 43, he became a technical assistant at the Harvard University Herbarium. In 1947, he moved to the Department of Agronomy at the Central University of Venezuela, and 4 years later, he became a Professor of Botany and Ecology at the University of the Andes. In 1953, he gave up all official academic positions to work full-time on biological research. At the age of 80, he took over as the first director of the *Jardin Botanico Xerofito*, a botanical garden that he had founded. A prolific writer, he published 300 scientific papers and seven books throughout his life. His book *Pangeography* (1958) proposed a method to reconstruct ancestral biotas of plants or animals: connect discontinuous distribution areas or collection localities of a particular group by straight lines (to be termed tracks) on a map. An individual track may cross an ocean or an impassable mountain. Next, superimpose several individual tracks of unrelated groups of organisms, and if they coincide, then consider the resulting summary lines as generalized tracks. These generalized tracks suggest the pre-existence of ancestral biotas, which tectonic and/or climate change had fragmented over time. Different ancestral biotic and geological fragments interrelate in space and time, constituting a composite area. Pan-biogeography thus explains all present distributions by the splitting of an ancestral, once broad distribution area, into smaller parts. Extinctions in a previously continuous range explain disjunct distributions. This approach depreciates the importance of long-distance dispersal.

Pan-biogeography, as the only process in biogeography, is somewhat problematic. If it is so hard for organisms to disperse across appreciable distances, then how did the common ancestor ever cover the whole area before the geographic splitting event happened? If the ancestor at some point could disperse so far, then why should we rule out this same possibility in its descendants? Furthermore, there are many readily observable cases of long-distance dispersal: wind-dispersed seeds, bird-dispersed seeds and ballooning spiders, to mention but a few. Clearly, long-distance dispersal does and can play a role in biogeography. Furthermore, the volcanic islands of Hawaii are a chain of islands formed by volcanoes rising above the ocean, thousands of kilometres distant from the nearest continents. If long-distance dispersal were not to be considered important, Hawaii would have no native plants and animals. In fact, it has a rich native flora and fauna whose closest relatives live, some in East Asia, some in Oceania and some in western North America.

9.2 Continental Drift

Darwin, in his theory of evolutionary diversification by common descent, suggested that during Earth's very long history, life had evolved, changed and diversified, spreading across the globe as circumstances permitted. Darwin's theory left a series of biogeographic questions open. Why do monkeys occur in both Africa and South America, yet not elsewhere? Why do marsupials occur in both South America and Australia, yet not elsewhere? Why do deer occur mainly in Eurasia, and not in

Africa? How could flightless birds have reached continents separated by very long stretches of deep oceans? Darwinian Theory could not explain these distribution patterns.

The mid-twentieth century brought the theory of continental drift into biogeography: the Earth's continents have moved relative to each other over geologic time, having drifted across the ocean bed. This theory began developing as early as a century after Columbus discovered America when a Belgian cartographer (Abraham Ortelius, 1596) suggested that earthquakes and floods had torn the Americas away from Europe and Africa. He added that "the vestiges of the rupture reveal themselves if someone brings forward a map of the world and considers carefully the coasts of the three [continents]".

Two centuries later, Benjamin Franklin, in a message to a French geologist (1782), described the way islands and continents might move about on the Earth's surface: Some part of it had sunk under the sea, and other parts, which had been under it, had risen above it. Such changes in the surface of the globe could not happen if the Earth were solid to the centre.

> I therefore imagine that the internal parts might be a fluid, denser than any of the solids we are acquainted with; which therefore might swim, in and upon that fluid. Thus, the surface of the Earth would be a shell, capable of being broken and disordered by the violent movements of the fluid on which it rested.

Franklin was thus describing the Earth as consisting of a thick, dense fluid inside, which causes the surface crust to be violently broken and mobile, and this is close to the modern theory of plate tectonics. Eight decades later, a French geographer and scientist (Antonio Snider-Pellegrini) proposed that during the Carboniferous, all continents were connected (Fig. 9.2). He based his theory on the fact that he had found identical plant fossils in both Europe and the United States.

Three decades later, Eduard Suess (Fig. 9.3), professor of palaeontology and geology at the University of Vienna, was studying *Glossopteris*, a fern of the Permian. Its seeds, much too large to be wind-borne, could not have blown across thousands of kilometres of the open sea, nor is it likely they had floated across vast oceans. Noticing that these fossils occur across several, now detached, landmasses (India, South America, Australia, Africa and Madagascar), Suess proposed (1885) that the southern continents had all once amalgamated into a single supercontinent. He named this ancient continent "Gondwanaland" (after the Gondwana region of India, the district in India where scientists first found *Glossopteris* remains; this is Sanskrit for "forest of the Gonds"). Suess did not propose continental drift. He believed that the oceans had flooded the existing spaces between those Gondwana lands. (Suess also postulated the early existence of the Tethys Ocean, which he named in 1893).

Roberto Mantovani proposed an early model of continental drift whereby thermal expansion and volcanism broke up a single large, ancient continent, forming new daughter continents that drifted away from each other because of further expansion of the rip-zones, where now the oceans lie (1889, 1909). Frank Bursley Taylor (1860–1938), an American geologist, proposed (1908) that the continents moved on

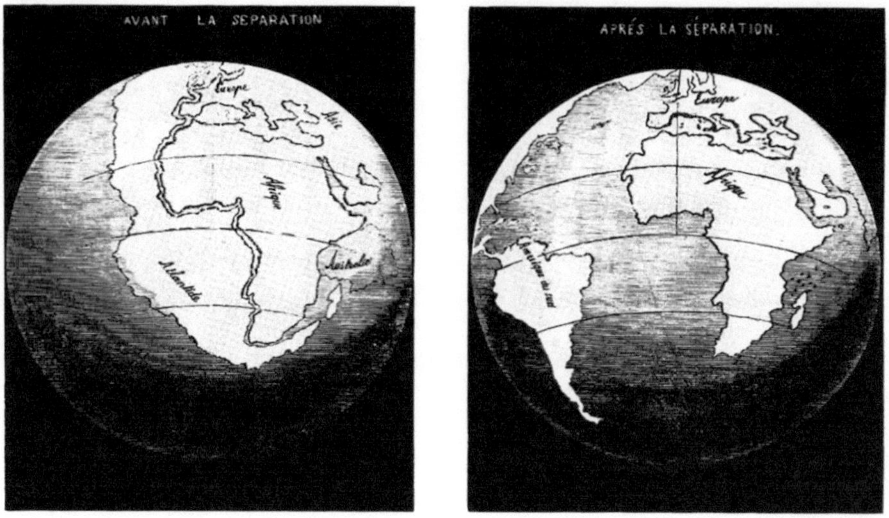

Fig. 9.2 Snider-Pellegrini's two maps (1858), depicting his interpretation of how America and Africa may once have fit together

Fig. 9.3 Suess

the Earth's surface, that a shallow region in the Atlantic marks where Africa and South America had once inter-joined, and that colliding continents could uplift mountains. Colossal lateral pressures, which thrust the earth's surface upward, had formed such mountain ranges as the Andes, Rockies, Alps and Himalayas. This occurred by their being dragged towards the equator by tidal forces during the hypothesized capture of the moon, resulting in a "general crustal creep" toward the equator. Earth's gravity captured the moon during the Cretaceous period, 100 million years ago, and the moon came so close to Earth that its tidal pull dragged the continents toward the equator. The continents, as they ploughed through the ocean floors towards the equator, wrinkled their equator-facing fronts, thereby producing

Fig. 9.4 Wegener during the 1912–1913 expedition to Greenland

the Himalayas and Alps. Fellow scientists of Taylor's time either ignored his theory or opposed it.

Three years later (1912), in Germany and unaware of Taylor's studies, Alfred Lothar Wegener (1880–1930, Fig. 9.4), who was a German polar researcher and meteorologist, independently developed the continental drift theory. He was born in Berlin into a clergyman's family and studied physics, meteorology and astronomy. Later, as an assistant at the *Aeronautisches Observatorium Lindenberg*, he pioneered the use of weather balloons to track air masses. To construct meteorological stations and record the polar climate, he participated in four expeditions to Greenland.

Around 1911, upon reading a paper that listed fossils of identical plants and animals on opposite sides of the Atlantic, he became interested in continental drift. He advocated that all continents were once a single landmass and that they had since drifted apart. The mechanisms causing the drift might, he supposed, be the centrifugal force of the Earth's rotation. He also speculated that the sea floor of the Atlantic keeps spreading from the mid-ocean ridges, thereby continuously tearing open and making space for fresh, relatively fluid and hot sima rising from depth (Fig. 9.5). When the World War I began (1914), the German army recruited Wegener as an infantry reserve officer but, after being wounded twice, he was declared unfit for active service and assigned to the army weather service. Travelling between various weather stations in Germany, the Balkans, the Western Front and the Baltic region, he found time, nevertheless, to publish the first version of his *Die Entstehung der Kontinente und Ozeane* (1915), in which he drew together evidence from various fields to advance the theory that there had once been a giant supercontinent. Interest in this small publication was, however, low because of wartime chaos. Later, revised editions during the 1920s presented further evidence. (The first English edition was published in 1924 as *The Origin of Continents and Oceans*, a translation of the third, 1922, German edition.) After the war, Wegener obtained a position as a meteorologist at a German Naval Observatory, and later he joined, first the University of Hamburg and eventually the University of Graz. In 1926, he presented his continental drift theory also at a symposium in New York City. The last edition (1929)

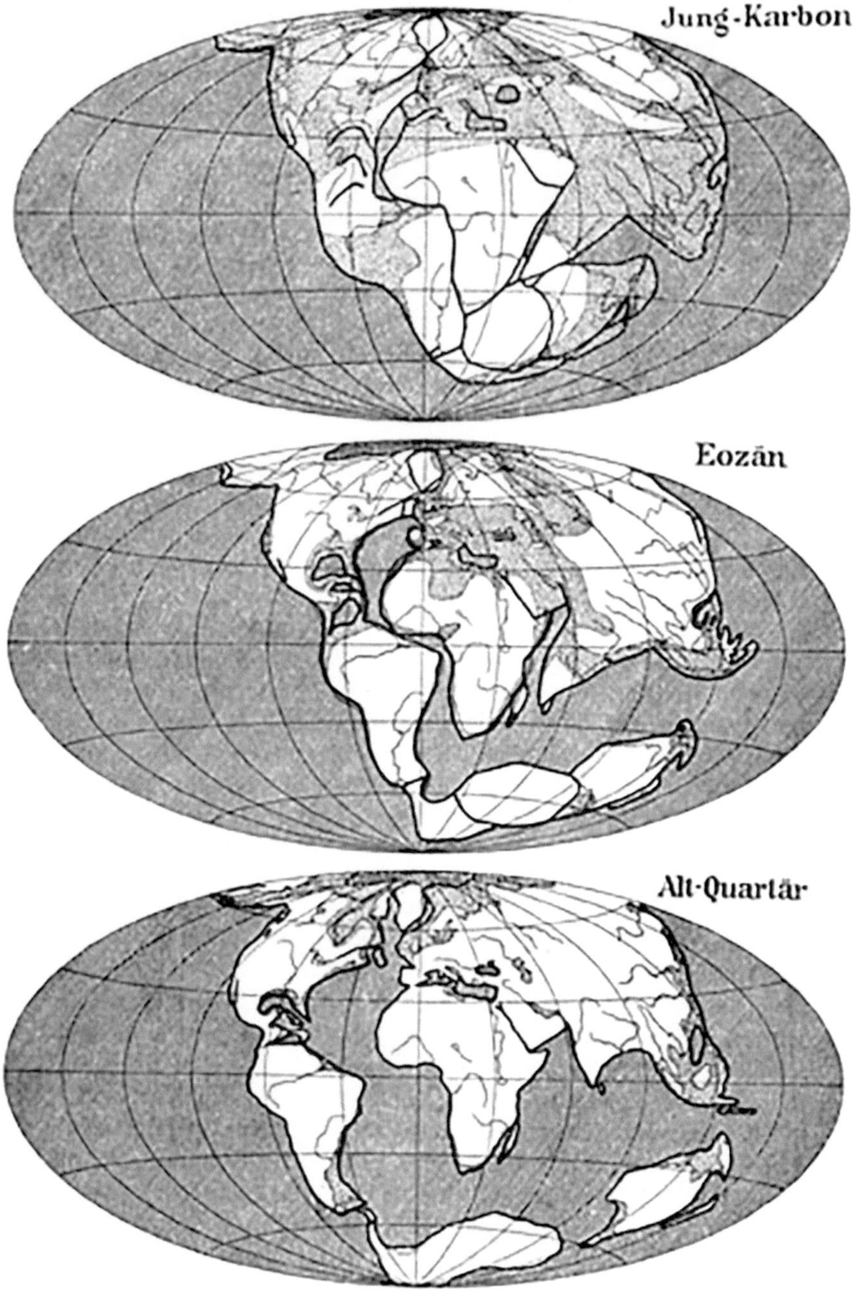

Fig. 9.5 Maps created by Wegener, showing Pangea and the continents drifting apart

revealed the significant observation that shallower oceans were geologically younger. Three years later, a fourth, further expanded edition appeared. Wegener died while returning from bringing food to a group of researchers camped in the middle of an icecap in Greenland, attempting to establish permanent stations that would report year-round Arctic weather observations.

Wegener was convinced that all of the Earth's continents were once part of an enormous, single landmass because he noticed that fossils of the ancient, one-metre-long reptile *Mesosaurus* occurred in the Perm (280–290 million years ago) of both southern Africa and South America. *Mesosaurus* could not have swum the Atlantic Ocean. The presence of *Mesosaurus* on these two, now-distant continents, suggested to him a single continuous landmass. Wegener also studied plant fossils from the frigid Arctic Archipelago of Svalbard, in Norway. He found that these plants were not of hardy species adapted to survive in the Arctic climate, but were tropical plants, which are adapted to a much warmer, more humid environment. The presence of these fossils suggested to him that Svalbard once had a tropical climate. Finally, Wegener studied the stratigraphy of different rocks and mountain ranges. He found that not only do the east coast of South America and the west coast of Africa fit together like pieces of a jigsaw puzzle, their rock layers fit just as clearly as well. South America and Africa were not the only continents with similar geology. Wegener discovered that the Appalachian Mountains of the eastern United States and the Caledonian Mountains of Scotland were also geologically related (Fig. 9.5).

Virtually all mainstream scientists of the day widely criticized and rejected Wegener's hypothesis. The criticism was so heavy that the American Association of Petroleum Geologists thought it worthy to organize a symposium, specifically opposing the continental drift hypothesis. Opponents argued that the good fit of the continents, the fossil evidence and the evidence from rocks were not sufficient proof and that the oceanic crust is too firm for the continents to plough through. Furthermore, Wegener suggested that the continents fit together 200 metre below sea level, at the level of the continental shelves where they match well, and not at the current shoreline. However, Schuchert, one of the participants in the symposium and one of Wegener's eminent critics, misunderstood that Wegener was suggesting the continents fit along the current coastline. He commented that during the vast time since the origin of the continents, sea waves had been continuously pounding against Africa and South America, and rivers have been bringing vast amounts of eroded material into the ocean. Wegener's claim (so he mistakenly believed) that the shorelines remained unchanged everywhere is therefore not reasonable. Furthermore, the sea level itself has not been unvarying, particularly during the Pleistocene, when millions of square miles of ice, made from water subtracted from the oceans, covered the lands and the sea level sank about 75 metres. Wegener was in the audience of Schuchert's lecture, but he did not respond, whether because of an inadequate command of English or because he considered this criticism irrelevant.

The major zoologist to join the fight against continental drift was the American palaeontologist George Gaylord Simpson (1902–1984). An expert on extinct mammals and their inter-continental migrations, he wrote a strong critique against continental drift (1943, 13 years after Wegener had died). The continental drift theory, he argued, allows for only one definitive break, or one definitive contact, of each connection. In the northern hemisphere, however, considerable intermigration had occurred throughout the Cenozoic, and this was because the continents were stable. He accepted that South America, Africa, Australia and India have some resemblances in flora and fauna that they do not share with northern continents. However, he declared, these resemblances are sometimes erroneous, some of them do not require land connections, and many are grossly exaggerated. Furthermore, he complained that some scientists classify evidence of groups for which fossils are nearly, or completely lacking, as palaeontological. "As far as I know, when fossil evidence has been found for a recent group of southern disjunctive distribution it has always turned out that this group also occurred in the north". Regarding the African fauna, Simpson argued that its Tertiary and Recent mammals resemble those of the Eastern Mediterranean and southern Asia and that it has no similarities with that of Australia, or of South America. Wegener (1924: 85–89) claimed that the Australian fauna represents two sources of origin. Its marsupial and monotreme mammals represent an archaic faunal element dating from the Jurassic to the Eocene, when Australia and South America were still attached (Antarctica intervening) but Australia was separated from other continents. The Australian placental mammals (dingo, rodents and bats) had reached Australia by an island route from Asia during the post-Pleistocene. To this two-source hypothesis of Wegener, Simpson replied that, though there are no fossil marsupials in Asia, this is likely to be an accident of non-discovery. As to the placental mammals of Australia, they had come from Asia already during the Pleistocene, as revealed by fossil evidence; indeed, the many endemic Australian genera and subfamilies suggest immigration as early as the early Pliocene. Both marsupial and placental mammals had thus entered Australia by an island route, from Asia. Simpson concluded that the Drifting Continents theory does not explain the past and present distributions of mammals, which are most consistent with fully stable continents throughout the entire mammalian history; the presumed evidence for drift is false or misinterpreted. He blamed those who supported Wegener's theory of looseness of thought, or method, of misrepresenting facts and of interpretations that are unreliable and unscientific. Indeed, "It must be almost unique in scientific history for a group of students, admittedly without special competence in a given field, thus to reject the all but unanimous verdict of those who do have such competence". In the 1940s, Simpson's critique of continental drift was the prevailing opinion.

A later critic of continental drift as applied to animals was Darlington (1957), whom we have already met, a few paragraphs above. He sought signs of drift in the present distribution of animals and, finding none, he insisted that modern animal distribution is, basically, a product of animal movement, not of land movement. Different kinds of animals, spreading and receding in many different ways, form an endless variety of different distribution patterns, which are determined by the animals' evolutionary ages, power of dispersal, relations to climate and relations to each other. The existing distributions of animals and plants are far too recent to be explained by tectonic movements in the Palaeozoic and Mesozoic. Nevertheless, believers in Gondwanan distribution try to use them. Their method is to try to fit existing plant and animal distributions to Gondwanan patterns of land and, when they find a fit, to claim it as evidence of a hypothesis that the ancient land existed. If a group of animals now occurs only in Africa and Brazil, they say that it dates from the hypothetical Afro-Brazilian continent. Their method, which is based on the assumption that animal distributions are more permanent than the land and that animals move less than continents, dictates its own conclusion. Consequently, if animals have not moved, the continents must have done so. However, animals do move. They spread and recede, sometimes over long distances in short times, and many factors determine their present distribution, besides the history of ancient lands.

Before proceeding, here is a paragraph in brackets about two of the above-mentioned eminent scientists, who so influentially and incorrectly opposed the continental drift theory. Charles Schuchert (1858–1942) was an American invertebrate palaeontologist who was a leader in the development of palaeography, the study of the distribution of lands and seas in the geological past. He served on the United States Geological Survey, was curator of the U.S. National Museum, and joined the Yale faculty. He was director of the Peabody Museum of Natural History at Yale University, first president of the Palaeontological Society and president of the Geological Society of America; the National Academy of Sciences awarded him the Mary Clark Thompson Medal. George Gaylord Simpson (1902–1984), one of the most influential palaeontologists of the twentieth century, was an expert on extinct mammals and their inter-continental migrations, who dispelled the myth that the evolution of the horse was a linear process culminating in the modern horse. He was a Professor of Zoology at Columbia University, Curator of the Department of Geology and Palaeontology at the American Museum of Natural History, Curator of the Museum of Comparative Zoology at Harvard University and Professor of Geosciences at the University of Arizona. He was awarded the Mary Clark Thompson Medal from the National Academy of Sciences, the Academy's Daniel Giraud Elliot Medal, the Linnaean Society of London's prestigious Darwin-Wallace Medal and the Royal Society's Darwin Medal "In recognition of his distinguished contributions to general evolutionary theory, based on a profound study of palaeontology, particularly of vertebrates". He also received the Golden Plate Award from the American Academy of Achievement. It is comforting to know that we ordinary people might be incorrect from time to time, just like very eminent scientists.

In his work, Wegener presented plentiful observational evidence in support of continental drift, but the mechanism remained a problem. In the early 1950s, however, the new science of paleo-magnetism was producing data in favour of Wegener's theory. By early 1953, samples taken from India showed that the country had previously been in the southern hemisphere, as predicted by Wegener. Scientists of the 1950s and 1960s, using echo sounders to map the ocean floor, found that ocean ridges stretch on for thousands of kilometres in long mountain chains, winding around the Earth's surface. They also observed that these ridges closely resembled terrestrial volcanic rift zones: At their crests, they had deep central valleys with steep faults on either side. This led to the conclusion that seafloor mid-ocean ridges were volcanoes. When these volcanoes erupted, they spewed out lava that cooled and solidified to become the new seafloor. As this lava cooled, magnetic particles within it aligned with Earth's magnetic field. Scientists further found that Earth's magnetic field had flip-flopped many times over its history, with the north and south poles exchanging places. Depending on when seafloor rocks formed their particles align in either one direction or the other, with either positive or negative magnetic anomalies. In the late 1960s, scientists found an alternating "striped" pattern of seafloor rocks. Rocks that formed when Earth's magnetic field was in one position alternated with rocks that formed when the field reversed. The stripes ran parallel to the mid-ocean ridges and extended out hundreds of kilometres on either side of them. The seafloor's magnetic signatures thus revealed that new ocean crust originated at the ridge crests and that it then spread outward, in both directions. In 1965, scientists combined the continental drift theory with the seafloor-spreading hypothesis, and proposed a new theory, that of plate tectonics. The Earth's crust consists of large, rigid plates that float atop an underlying rock layer that, under tremendous heat and pressure, behaves like a viscous liquid. The crustal plates have been sliding about on Earth's surface throughout its long history, sometimes widely separated, at other times agglomerated into large masses. (The earlier term, "Continental Drift", was no longer accurate, because the plates consist of both continental and oceanic crust, both of which drift over the Earth's surface.) Plate tectonics provides a unifying theory that explains fundamental processes that shape the face of Earth. In 1964, the Royal Society held a symposium on the subject. In later decades, with the advent of the Global Positioning System (GPS), it became possible to measure continental drift directly. It revealed that Wegener's estimate of the velocity of continental motion, 250 cm/year, was too high. The currently accepted rate for the separation of the Americas from Europe and Africa is about 2.5 centimetres per year. Today, we recognize continental drift as one of the major scientific revolutions of the twentieth century.

In brief, plate tectonics suggests that some 550 million years ago, Africa, South America, Australia, Antarctica, India and New Zealand formed a massive southern supercontinent that sat on the equator, Gondwanaland. It was the largest piece of continental crust in the Palaeozoic Era, covering an area of about 100,000,000 square kilometres, about one-fifth of the Earth's surface. It merged with another large supercontinent, Euramerica, about 330 million years ago, thereby forming a single giant supercontinent consisting of almost all the land masses on Earth (which we term Pangaea). Some 200 million years ago, the Central Atlantic opened, and the supercontinent Pangaea began breaking up, with today's continents splitting off, one by one. Some 140 million years ago, Africa and South America split through continental drift, as the southern Atlantic Ocean opened, ending any land link between these two continents. Over millions of years, Pangaea separated into pieces that moved away from one another. These pieces slowly assumed their positions as the continents we recognize today.

Scientists of the early decades of the twentieth century believed dispersal to be the only process determining plant and animal distributions, but those of later decades considered continental drift and offered a new outlook: Different parts of Earth have different histories of isolation and connection with other parts, and this influenced where animal groups could spread. As the continents drift about, their animal groups drift with them. In the Gondwanaland supercontinent, some animal groups inhabited only a single (present-day) continent, and others inhabited several continents, in which they spread and diversified. As the continents split off Gondwanaland, and as they split off from one another, they took their animal lineages with them. Continental drift and the time sequence of lineage origination, then, are what explain the existence of the zoological realms—animal lineages are unique both in time and space, and they originate not only at particular points in time but in particular regions of the globe. The distribution of biodiversity traces the geological history of the continents' movements. Life's lineages evolved on the surface of each floating continent, thereby forming the zoological realms. Of course, some organisms can easily disperse from their land of evolutionary origin and can reach new and distant regions, even if continents do not join. However, many aspects of distribution still bear the marks of ancient lineage originations and continents. This theory, that the modern distribution of organisms having low dispersal abilities is because they achieved their distributions while riding tectonic plates, is termed Vicariance Biogeography. It refers to patterns of disjunct distributions common to several members of a fauna, produced by the splitting up of an area. Croizat (1958) elegantly summarized this idea, that species can passively ride the continents as they split and drift across the surface of the Earth, in the famous sentence "Life and Earth evolve together", or in other words, geological barriers evolve together with biotas.

One of the best-known organisms of Gondwana was *Glossopteris*, a genus of an extinct Permian order of seed ferns, the Glossopteridales, that were the dominant trees of the southern flora during the Permian but disappeared at the end of the Permian. *Glossopteris* was a tree reaching a height of 30 metres, which grew in very wet, swampy soils. Its seeds, much too large to be wind-borne, could not have blown across thousands of miles of open sea, nor is it likely they had floated across vast oceans. We recall that Suess, noticing that *Glossopteris* occurs in India, South America, Australia, Africa and Madagascar, had proposed (1885) that the southern continents had all once amalgamated into a single supercontinent. *Glossopteris* was unknown from Antarctica at his time. Some 25 years after he had published his Gondwanaland theory, Robert Scott, in his ill-fated 1913 *Terra Nova* Expedition, discovered the first Antarctic specimens of *Glossopteris*. The expedition members abandoned much of their gear in an effort to reduce their load but kept 15–20 kilograms of *Glossopteris* fossils, and these were found alongside their bodies.

Another well-known organism of Gondwana was *Mesosaurus*, a reptile about 1 metre long, with webbed feet, a streamlined body and a long tail that may have supported a fin (Fig. 9.6). Its nostrils were located at the top, allowing the animal to breathe with only the upper part of its head breaking the surface, in a manner similar to a modern crocodile. It probably fed on crustaceans. Its thickened ribs enabled it to reach neutral buoyancy in the water. Many features suggest a mainly aquatic lifestyle. It lived in South Africa and South America (of today), a coastal animal that was unlikely to have crossed the Atlantic Ocean.

Fig. 9.6 *Mesosaurus.* (By Tuvia Kurz)

Lystrosaurus (Fig. 9.7) is an extinct genus of herbivorous therapsids from the Permian and Triassic (around 250 million years ago). It lived in what is now Antarctica, India, South Africa, China, Mongolia and European Russia, and ranged in size from that of a small dog to 2.5 metres long. It had only two teeth (a pair of tusk-like canines), and probably had a horny beak like those of turtles, for shearing off pieces of vegetation, which it then ground on a secondary palate. It was a heavily built herbivore that moved about with a semi-sprawling gait; its massive forelimbs were more robust than its hind limbs, and it may have been a powerful digger that nested in burrows. Dicynodonts, the group within which it is classified, were probably warm-blooded animals, as suggested by their highly vascularized bones, by their body proportions that were conducive to heat preservation, and by indirect evidence that they had hair. In the Early Triassic, they were by far the most common terrestrial vertebrates, accounting for as many as 95% of the total individuals in some fossil beds.

Cynognathus (Fig. 9.8) was an extinct genus of large-bodied therapsids that lived in the Middle Triassic. It was a 1.2 m long predator closely related to mammals and had a southern hemispheric distribution, in South Africa, South America and Antarctica. It was heavily built with a large head, wide jaws and sharp, differentiated teeth, evidence that it could effectively process its food before swallowing. The presence of a secondary palate in the mouth indicates that it would have been able

Fig. 9.7 *Lystrosaurus*. (By Tuvia Kurz)

to breathe and swallow simultaneously. Its hind limbs were directly beneath the body, but the forelimbs sprawled outwards in a more reptilian fashion. The possible lack of belly ribs in the stomach region suggests the presence of a diaphragm, an important muscle for mammalian breathing. Pits and canals on the bone of the snout indicate concentrations of nerves and blood vessels. In mammals, such structures suggest whiskers, which the animal uses as sensory organs. There is thus clear overall evidence of a Gondwanan distribution of organisms in the Permian and Triassic.

Can Gondwanaland also explain the distribution of modern animals found on different southern continents? In some cases, it can. It can explain why only marsupials live in Australia Fig. 9.9. The first marsupials came from Laurasia, a large landmass of the northern hemisphere, and spread into Gondwanaland at a time when there still was a land connection, from southernmost South America to Antarctica-with-Australia. Later on, by 125 million years ago, Gondwanaland had separated into two parts: South America with Africa and India, and Antarctica-with-Australia. The placental mammals diverged from the marsupials around 160 million years ago, and in dispersing southwards, they took nearly the same route as the marsupials had before them. By now, however, the connection between South America and the Antarctic/Australian landmass did not exist anymore. The more competitive placental mammals drove the marsupials of the northern lands to extinction, but they could not reach Australia. South America was then semi-isolated, and

Fig. 9.8 *Cynognathus.* (By Tuvia Kurz)

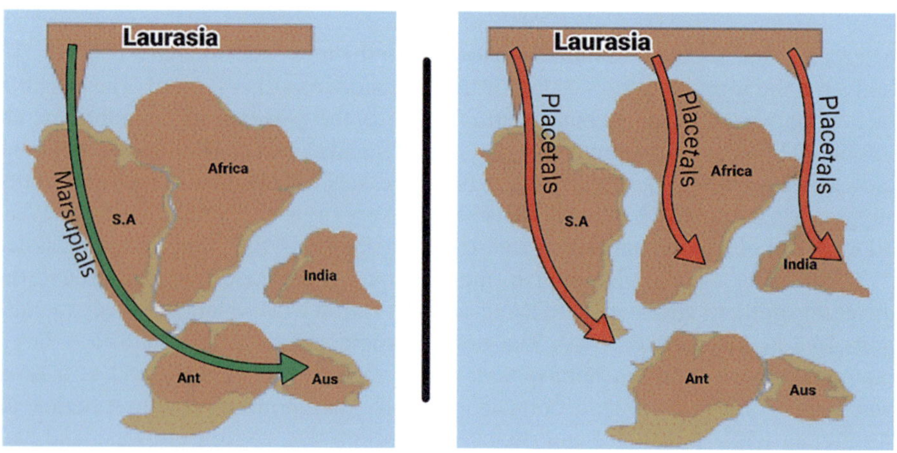

Fig. 9.9 Continental drift explains why marsupial mammals (left) reached Australia, and terrestrial placental mammals (right) did not

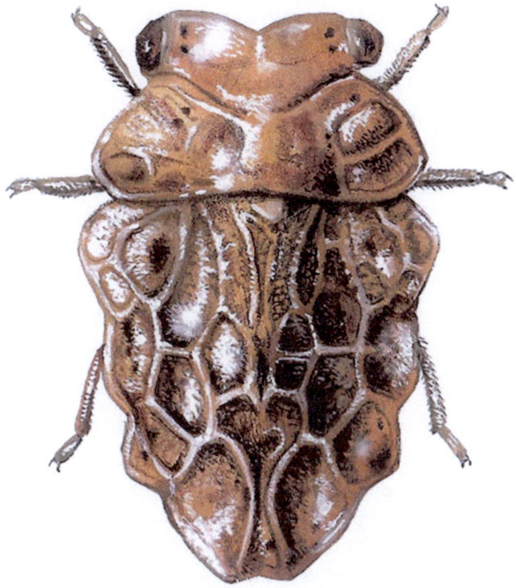

Fig. 9.10 A moss bug (Peloridiidae). (By Tuvia Kurz)

only a few of the placental mammals could invade it. We shall discuss the historical biogeography of South America in the next chapter.

Likewise, marsupials occur in Australia and South America because these two continents were once inter-connected in the supercontinent Pangea. Gradually, they drifted apart, with the marsupials in South America becoming opossums and those in Australia becoming kangaroos (so to say).

Another case of Gondwanan distribution is that of the moss bugs Peloridiidae (Fig. 9.10). These are cryptic, almost flightless insects, occurring in Australia, New Zealand, New Caledonia and Patagonia—a classical Gondwanan distribution. Molecular evidence demonstrates that they began to diversify in present-day Patagonia in the Jurassic, later expanding to Australia and New Zealand. About 100 million years ago, they split into three subgroups: Australia, Patagonia and New Zealand-with-Caledonia. About 80 million years ago, New Zealand broke off from the supercontinent and 40 million years ago, Australia broke off from Antarctica (Ye et al. 2018).

These bugs, with the marsupials, are two examples of continents drifting about and taking the ancestors of modern animal groups with them. Earth and life thus evolved together, the Gondwanaland break-up sketching the history of modern life. Gondwanaland explains why modern animal groups are where they are.

Other cases might only at first sight seem to represent a Gondwanan distribution, but careful observation reveals that this is not so. Africa, South America, Australia, India and New Zealand had undoubtedly all once formed the combined supercontinent of Gondwanaland. However, recent studies, using molecular clocks, reveal that many dispersal events occurred very long after the continents had separated: Long after continents had drifted apart, plants and animals were, somehow, hopping between them. Monkeys occur in both Africa and South America, seemingly because these two, once connected continents, gradually drifted apart; eventually, the South American monkeys evolved side-facing nostrils, prehensile tails, lack of trichromatic vision and usually non-opposing thumbs, features in which they differ from the monkeys and apes of Africa. However, the fossil record of monkeys in South America dates back to only 36 million years. Furthermore, monkeys appeared also in North America 21 million years ago. How, then, did monkeys travel around the world when there was no land for them to walk on?

They could have travelled on rafts. It is possible that the very ground beneath their feet had been swept away, leaving them stranded on a floating island. Such rafts are so far from our imagination that, to prove their (admittedly rare) existence, I cite Brewster (1924, in Darlington 1957: 15) who observed one incidentally while describing a nesting site of bitterns on a freshwater lake in Maine.

> … we came upon a small floating island … circular in shape, about twenty yards across and densely covered with bushes … with here and there a muddy or grassy space. Near its centre, a few green and vigorous young larch trees rose to a height of fifteen or twenty feet. Interlacing roots of trees, shrubs and grasses had combined with decaying vegetable matter, alluvial soil, and driftwood, to form a great, soggy raft, buoyant enough to rise and fall with the Lake, but so thin and flexible that big waves, rolling in under it, procured corresponding and very noticeable undulations of its surface. This was sufficiently elevated to appear comparatively dry and trodden on, when it gave beneath our steps, like that of a quaking bog, until the water came half-way to our knees. The island was then firmly anchored—probably by a few strong roots—to solid ground some five or six feet below, but we were told that not long before this it had come drifting across the flooded meadows from their western end. About ten years later, it again broke loose from its moorings and journeyed still father eastward, until stopped by a group of stubs.

Furthermore, a similar raft, with trees some 10 m high, was seen in the Atlantic, off the coast of North America, in 1892. It had drifted some 160 km (Powers 1911, in Darlington 1957: 15).

Back to our monkeys, we can imagine a group living near the western coast of Africa; a powerful storm battering the land, tears away a large chunk of ground and creates a large floating raft. The Atlantic Ocean was, at that time, a third less than its present 2800 kilometres wide, possibly only 2000 kilometres wide, and the journey from Africa to South America could then have been as short as about 14 days. With fresh water in the leaves, and with insects, fruits and other food items, an ancestral group of monkeys could have survived those two weeks. Once in South America, in a rainforest broadly similar to the one that they had left in Africa and with plentiful fruit, they dispersed and diversified throughout the new continent. Another possibility is that the individuals whose descendants would become the flat-nosed monkeys reached South America via a land bridge. The sea level dropped about 34 million years ago (during the Oligocene), and this perhaps produced a series of mid-Atlantic islands, as stepping-stones for dispersal. The Oligocene lowering of sea levels also allowed Asian mammals, such as rhinoceroses and ruminants, to enter Europe, and to drive endemic species to extinction. The earliest deer, giraffes, pigs and cattle appeared in the Oligocene in Eurasia (De Queiroz 2014).

The large flightless birds (ostrich, emu, cassowary, rhea, moa and kiwi, the Ratites lineage) and the side-necked terrapins (Pelomedusidae and Podocnemididae) are further cases of groups with Gondwanan distribution whose evolutionary history, as revealed by molecular studies, does not match the timing of the break-up of the Gondwanan continents. While continental drift does not explain all modern distributions on all southern continents, the theory behind it is valid and may explain other distributions.

For many decades, students of biogeography have viewed dispersal and vicariance as competing hypotheses in historical biogeography (Cox et al., 2010). However, the two processes are not mutually exclusive (Sanmartin 2012).

9.3 Seeking Biogeographic Generalizations

Eco-geographic rules describe and predict large-scale patterns of variation with biogeography, within and across species. These rules, formulated and developed over the past two centuries, have gained renewed attention due to climate change: Researchers seek to predict how species will respond to changing environmental conditions, and Mayr (1956) has already discussed geographical character gradients in animals, versus climatic adaptation. There is a battery of "eco-geographic" rules, of which I present several major ones. While using the term "rule" as it is widespread in the literature, I agree with Goldenberg et al. (2022) that "trend" is a more suitable term.

Competitive exclusion (Gause's rule) refers to the fact that two species competing for the same limited resource cannot coexist at constant population values. When one species has even the slightest advantage over another, the one with the advantage will dominate in the long term. This leads either to the extinction of the weaker competitor or to its behavioural shift towards a different niche. Georgy Gause (1910–1986), the son of a Russian professor of architecture, studied at the Moscow State University and worked at the University's Zoological Institute. He carried out laboratory competition experiments using two species of ciliates and found, while adding a constant flow of food, that one species drove the other to extinction, via exploitative resource competition. However, he was able to let the losing species survive by differing the environmental parameters of food, thus demonstrating that winning the competitive exclusion depends on environmental factors. He also studied competition between two species of yeast, finding that one species consistently outcompeted the other. He published his conclusion from this experimental work (1932) as the competitive exclusion principle: No two species with similar ecological niches can coexist in a stable equilibrium. When two species compete for exactly the same requirements, one will be slightly more efficient than the other and, as a result, it will reproduce at a higher rate. The fate of the less efficient species is local extinction.

Competitive exclusion is of major importance in biogeography because it explains why, so often, only one species of a genus inhabits a given area. It applies better to ground-dwelling animals (lizards, land snails and scorpions), which live in two dimensions, and to species within a genus that are similar in size or behaviour. It applies less well to flying, swimming and floating animals (birds, winged insects, fish and plankton).

Animals achieve a larger size in colder climates than in warmer ones (Bergmann's rule). Body size directs food requirements, and limits the shelter an animal can enter, and the performance of some organs, such as the vertebrate eye and ear, depends on the organ's absolute size. It also determines modes of locomotion, metabolism and thermal relations, and it is of extreme importance in male combats.

Bergmann, in 1847, suggested that within species and among closely related species of homoeotherms, animals often achieve a larger size in colder climates than in warmer ones and that this is linked to the energy budget of these animals. Based on the different rates of increase of volume and surface area with increasing size, he concluded that larger animals could maintain a constant temperature with less effort than smaller ones. Central to his suggestion was the connection between body size and the amount of heat that the animal needed to produce, to raise its temperature above that of the environment. The temperature budget and the need to keep a constant temperature were at the core of his reasoning. To give his suggestion biological significance, he compared animals of different sizes that were as similar as possible: different species within a single genus and in one case, different populations within a single species. One consequence of Bergmann's study was a global-level prediction of a size cline with temperature.

Some seven or eight decades after Bergmann's suggestion in the 1920s, scientists began formulating his ideas in terms of a general, well-defined "Bergmann's rule" but with a shift in emphasis. Rensch (1929) wrote that in closely related species, the larger forms always live in colder climates, and the smaller forms in warmer climates. He later (1936, 1938) reformulated Bergmann's rule as: "within a Rassenkreis [ring-species] of warm-blooded animals the races living in cooler climates are generally larger than the races living in warmer regions". Mayr (1956) adopted this definition of Rensch and wrote that Bergmann's rule states that "races of warm-blooded vertebrates from cooler climates tend to be larger than races of the same species from warmer climates". Mayr's definition has since become widely known as Bergmann's rule.

Modern studies reveal that Bergmann's rule does indeed apply in 70% of all mammals and birds investigated, with elks and muskrats being the best cases. However, this same tendency of larger animals towards the poles exists also in some ectothermic, cold-blooded animals (amphibians, reptiles, insects, fish and crabs), which need not follow the rule, as they are unable to generate internal heat by themselves and depend on external heat sources (Meiri and Dayan 2003; Meiri et al. 2007; Meiri 2010; Salewski and Watt 2017). Furthermore, the body size of Palearctic shrews (*Sorex cinereus*) decreases with increasing latitude, thus contradicting Bergmann's rule, and this trend might be due to food shortage during the cold northern winter (Yom Tov and Yom Tov 2005).

The poleward increase in body size could be because the minimum temperature threshold decreases as body mass increases (this is the temperature below which an endotherm must dedicate more metabolic energy to maintaining body heat). This would favour a larger body in cold environments, since the larger the animal, the less energy it will need to control its temperature. Another explanation: Animals with higher body mass can store more reserve substances and, in cold temperatures, this would allow them to go without food for longer periods. A larger animal can rely on its greater fat stores to provide the energy needed for survival, as well as being able to procreate for longer periods.

Bergmann's assumption of a constant rate of heat production per unit body mass is, however, problematic, in that as large mammals often have proportionally more body fat and are better insulated, this would influence their temperature budget and render the surface area/volume ratio invalid. Even more problematic is Bergmann's assumption that the volume of an animal is a measure of the production of heat, and his disregard for the possibility that heat production, per unit volume, may depend on other factors. Large endotherms lose more heat and expend more energy at all temperatures, compared to small endotherms, because metabolism and heat loss increase with body mass. Bergmann assumed that heat production and dissipation are associated with an animal's volume and surface area but if this assumption is not met, and larger animals need not produce less heat compared to smaller animals to increase body temperature, then Bergmann's (1847) rule is invalid, irrespective of size clines found in empirical studies (Geist 1987).

From mammals and birds to reptiles, the geographical size variation of most snakes and some lizards, counters Bergmann's rule in that, among related taxa, the larger ones live at warmer latitudes. Turtles as a group tend to obey Bergmann's rule, and this enables larger clutches (by larger individuals), where the climate would limit repeated clutches (Werner et al. 2016).

Warm-blooded animals in warmer climates have larger appendages (Allen's rule). Joel Asaph Allen (1838–1921) was curator of birds and mammals at Harvard's Museum of Comparative Zoology. He co-founded and became the first president of the American Ornithologists' Union, was chief editor of its journal *The Auk*, and became, eventually, the curator of birds and mammals at the American Museum of Natural History. He formulated the rule that, to facilitate more efficient heat exchange, warm-blooded animals in warmer climates have larger appendages (1877).

As animals can die from failure to maintain body temperature within critical limits, they employ two thermoregulatory methods to avoid lethal death: evaporative heat loss and dry heat exchange. Evaporative heat loss is intricately involved with water budgets and dehydration risk, whereas dry heat exchange strongly relates to energy budgets and morphology. An animal's ability to dissipate heat, via dry heat exchange, involves gradients in both body size and appendage size. Bergmann's rule, as we have seen, states that in cooler climates, the reduced body surface area/volume ratio of larger bodies enables animals to retain heat more effectively. Allen's rule focuses on the size of body appendages, such as tails, beaks and legs, rather than whole-body size, but is based on a similar argument (Ryding et al. 2021).

Appendages are important in the thermoregulation of endotherms as their surfaces provide efficient means to adjust heat dissipation and, consequently, larger appendages release more heat. This thermoregulatory role leads to geographic clines in morphology, whereby similar animals in warmer climates (lower latitudes) have larger appendages than those in cold climates. The use of appendages in thermoregulation is the proposed reason for the patterns seen in Allen's rule (Ryding et al. 2021).

Allen's rule has widespread support across various groups of birds: 58% of 110 bird species whose bill size was studied exhibit Allen's rule. The bird's bill is a highly vascularized surface not insulated by feathers and, as such, it is a site of significant heat exchange, and plays an exceptionally important role in body temperature regulation and heat exchange with the environment. In numerous groups, there is a positive association between bill size and environmental temperatures, both within and between species. Birds of the Paleogene had larger bills than birds of modern times, perhaps because of the drastically higher temperatures at the time, which, in the tropics, were on average, 9 °C higher than modern temperatures (Ryding et al. 2021).

Allen's rule is widespread also across various groups of mammals: In rabbits and in mice, heat exchange affects the size of ears and tails.

Birds and mammals are darker in wet and warm environments than in colder and drier areas (Gloger's rule). Bernhard Rensch coined the term in 1929 to honour Constantin Gloger, one of the first (1833) to describe correlations between climatic variation and animal pigmentation (see Mayr 1956). Rensch elaborated on this correlation by suggesting that colour variation associated with Gloger's rule results from differences in the deposition of melanin, the most common pigment in birds and mammals. Melanin comes in two main types: eumelanin, which yields dark-brown to black and various shades of grey; and pheo-melanin, which produces reddish to sandy yellowish colours (as in many foxes). Sometimes, both pigments mix, and they then produce intermediate shades. Rensch suggested that climate influences the deposition of both melanin types but in a different manner. Eumelanin deposition increases in response to humid, warm climates, pheomelanin increases in response to drier, cooler climates. As a result, the pale brown, reddish and yellowish (phaeomelanic) colours prevail under dry and warm conditions, in northern countries; whereas the heavily pigmented dark-brown and black (eumelanic) colours prevail under humid and warm regions, in the humid, warm tropics. Consequently, (the by now expanded) Gloger's rule predicts that the deposition of blackish eumelanins shall increase with humidity, whereas the brown-yellowish phaeomelanins shall prevail in dry and warm regions. Gloger further proposed that beyond melanin pigments, carotenoid-based colours (brilliant yellow, brilliant red) tend to be more intense towards the tropics and in warmer regions and, that this effect is more marked for yellow carotenoid-based colours than red carotenoid-based colours.

Gloger's rule thus comes in a simple version, which predicts that darker phenotypes tend to dwell near the equator, in warmer and wetter environments; and in a complex version, involving melanin production. It predicts that the deposition of eumelanins decreases at low temperatures and increases due to humidity, while pheomelanins increase in dry and hot places and decrease at low temperatures. I focus on the simple version, so as not to overburden the text.

The prediction of darker animals in wetter climates has received overwhelming support from a plethora of empirical studies. Many birds and mammals are indeed darker in humid areas, both within and across species. At the interspecific level of an entire avifauna (550 species of Australian land birds), darker coloured birds do indeed inhabit wetter regions and more densely vegetated habitats, thus supporting Gloger's rule for rainfall/vegetation. The vast majority of North American passerine species also support the rule. Turning to single species, the barn owl *Tyto alba* is darker in wetter areas worldwide, and in the black sparrow hawk *Accipiter melanoleucus*, a polymorphic species with dark and light morphs, the proportion of dark morphs in South Africa is higher in areas of greater rainfall during the breeding season, again consistent with Gloger's rule. Plumage colour variation in another bird, the variable antshrike *Thamnophilus caerulescens* of South America, also clearly supports Gloger's rule prediction. The darkest forms are montane, inhabiting the cool and rainy elevations high above sea level, whereas the brightest forms occupy the hot, dry lowlands. Gloger's rule applies also to primates, in all major clades (Marconides et al. 2020). Furthermore, Neotropical primates have darker eye masks and African monkeys nearer to the equator have darker faces, features that might be other instances of Gloger's rule. In leopards, as an extreme expression of Gloger's rule, all-black forms are most common in tropical and subtropical moist forests and their frequency is much lower in drier and more open habitats such as deserts, shrub lands, grasslands, savannahs and broadleaf forests. Likewise in jaguars, melanism is present in most forest habitats but is completely absent in grassland and savannah habitats. Gloger's rule applies also to rodents of Asia and furthermore, it might perhaps apply to insects too: The small fruit fly *Drosophila melanogaster* in sub-Saharan Africa is darker nearer the equator. In conclusion, the prediction that melanin levels increase along environmental gradients of dry to wet has been tested in many animal groups and in many environmental contexts, both within and between species. Clearly, humidity plays a predominant role in predicting melanistic colouration (Kamilar and Bradley 2011; Delhey 2019; Goldenberg et al. 2022).

Temperature effects on melanin are less clear and often contradict the predictions of Gloger's rule. When we separate humidity and temperature, temperature effects often have the opposite sign, in that darker colours are more likely in colder regions. Empirical evidence thus calls into question the component of Gloger's rule that predicts darker animals in hotter climates. The initial formulation of Gloger's rule involving both temperature and humidity might have developed from the intuitive perception that warm temperatures and high precipitation are always associated, as is true, for example in tropical lowland rainforests. However, many regions around the globe, such as tropical mountains, are cool and humid. Accordingly, recent research suggests reformulating Gloger's rule, exclusively in terms of humidity. This would bring Gloger's rule to formulate just one prediction, that animals should be darker in wetter climates.

Regarding the selective advantages of Gloger's rule, there are four adaptive suggestions (Kamilar and Bradley 2011, Delhey 2019, Goldenberg et al. 2022). One suggestion involves camouflage. Humid environments usually harbour dense vegetation with luxuriant growth and ample shade, soils dark due to a high content of organic matter and poor light conditions. They, therefore, tend to offer dark backgrounds and low light conditions, and these favour darker colouration for camouflage, greater concealment from (and of) predators. A match of this kind, between animal colour and environmental colour, affects predation probabilities. In contrast, drier areas tend to have light backgrounds. Background matching and camouflage might thus be an important driver of Gloger's rule. All-black jaguars can persist in the face of conspecific competition by non-melanistic conspecifics because of their superior camouflage in dense tropical forests (Mooring et al. 2020).

Another possible adaptive explanation concerns photo protection. Dark colouration has a photo-protective function because melanin provides protection from the effects of UV radiation. It is unclear to what extent such effects apply to pelage or plumage, which already shield the skin from most of the harmful radiation, but I note that the two diurnal rodents of Israel (a spiny mouse and a sand mouse) both have black skins, whereas other, nocturnal rodents, do not. The absorption of some amount of UV radiation is, of course, necessary, for vitamin D production.

A third adaptive explanation concerns parasites. Feathers in humid areas harbour a great load of feather-degrading bacteria (*Bacillus licheniformis*) that decompose keratin, of which feathers consist. These bacteria degrade pale feathers more quickly than darker feathers. As humid environments are more suitable for microbial growth, and as dark feathers are more difficult to break down, protection against feather-degrading bacteria could be a general explanation, for Gloger's rule: In warmer and more humid areas, birds encounter, both more virulent strains of feather-degrading bacteria and, in addition, superior conditions for bacterial growth. Consequently, birds would benefit from darker plumage.

A fourth adaptive explanation could be pleiotropy, namely the situation whereby a single gene can influence many traits. One key gene involved in melanin-based colouration, the melanocortin 1 receptor, has pleiotropic effects involving immunity, stress responses, basal metabolic rate and behavioural traits. This means that pale-to-dark colour variation along climatic gradients could simply be a side effect of selection on other traits. Many owls have two morphs, a pale grey one that generally lives in northern latitudes and a darker reddish morph, more common closer to the equator. In tawny owls, the grey morph survives better in harsh winters and the opposite is true in snow-poor winters. This could explain the geographic distribution of grey and rufous morphs. Grey morphs also have fewer blood parasites and develop stronger immune responses than rufous morphs. Links of this kind, between melanin-based colouration and behavioural and physiological traits, suggest that pleiotropic effects could perhaps underlie Gloger's rule (Kamilar and Bradley 2011; Delhey 2019; Goldenberg et al. 2022).

Gloger's rule, to conclude, might result from complementary selective pressures arising from many ecological factors, including camouflage, thermoregulation, parasite deterrence and resistance to feather abrasion.

Deep-sea gigantism is the tendency for deep-sea dwelling animals to be larger than their shallow-water relatives. Henry Nottidge Moseley FRS (1844–1891) was a British naturalist who sailed on the first global marine research expedition, covering over 120,000 km of the world's oceans on board HMS *Challenger*, a steam-assisted Royal Navy corvette (1872–1876). The Royal Society had converted it into a scientific vessel with trawls, dredges and great lengths of rope, with which to suspend equipment deep into the ocean. Mosely, one of the three naturalists on board, studied arthropods, corals and molluscs, and he observed that some crustaceans reach a gigantic size in the deep sea. He wrote (1880) of giant deep-sea sea-spiders (Pycnogonida) that they had dredged from the deep sea, of a gigantic ostracod, and of a gigantic isopod (Goldenberg et al. 2022). Today, we know that deep-sea gigantism occurs also in cnidarians and cephalopods.

The big red jellyfish *Tiburonia granrojo* lives down to 1500 m, across the Pacific Ocean. It reaches 76 cm in diameter, has thick fleshy oral arms, in place of the long tentacles found in most jellyfish, and dwells in close-to-freezing temperatures. The giant isopod *Bathynomus*, abundant in the cold, deep waters of the Atlantic, Pacific and Indian Oceans, is the largest isopod in the world. It reaches 50 cm in length, far larger than typical isopods that reach only 5 cm, and it dwells down to 2500 m and temperatures of 3 °C, where it scavenges dead whales, fish and squid; and preys on slow-moving animals such as sea cucumbers, perhaps even live fish. As food is scarce in the deep ocean and it must make do with whatever comes along, it can survive 5 years of famine. The female develops a brood pouch from which the young emerge almost fully developed, as miniatures of the adults. The giant ostracod (*Gigantocypris,* Cypridinidae) lives worldwide down to 3500 m and 2 °C. It is extremely large for ostracods, measuring up to 3.2 cm across. It has a pair of large eyes that, rather than using lenses to focus light onto a retina, use parabolic mirrors. These eyes are better at gathering light than those of any other animal, but they produce a poor resolution of the image. The giant ostracod uses them to find bioluminescent prey animals such as other ostracods, copepods, arrow worms and small fish. It swims by "rowing" with two featherlike antennae, while another pair of long, sensing antennae extend out in front of the swimming animal. It has a near-neutral buoyancy and swims smoothly (not in jerks) at a relatively high speed, indicating that it is an active predator. The female has a brood pouch, in which the embryos develop and the young resemble miniature adults. Adult males are uncommon.

Sea-spiders are arthropods (Pycnogonida) that dwell in oceans around the world. They have a small body and, usually, four pairs of walking legs (some have five or six pairs). Lacking gills, they absorb oxygen through their legs and transfer it by diffusion throughout the body. They either walk along the bottom with their stilt-like legs or swim just above it, in an umbrella-pulsing motion. They are mostly carnivores, that suck nutrients from sessile, soft-bodied invertebrates and their digestive tract has diverticula extending into the legs. Reproduction involves external fertilization, and only males care for the laid eggs and young. Most sea-spiders are small, about 1 cm, and dwell in shallow depths. Some, however, dwell in deep waters down to 7000 m and are quite large (*Colossendeis* reaches 40–50 cm).

Amphipoda, an order of malacostracan crustaceans with laterally compressed bodies, are mostly marine, but many groups live in freshwater, and some are terrestrial. Males are usually larger than females, and bear a brood pouch that holds the fertilized eggs, until the young hatch, directly into a juvenile form. Amphipods are typically less than 10 mm long but *Alicella gigantea*, of both the Atlantic and Pacific down to a depth of 5300 m, reaches 28–34 mm.

The colossal squid (*Mesonychoteuthis hamiltoni* (Cranchiidae) of the Southern Ocean, the largest squid species in terms of mass, is known from only a few specimens. Whereas squids typically weigh about 100–200 g and have a mantle length of about 30 cm, the colossal squid reaches at least 500 kilograms (perhaps 600–700 kilograms), and its total length is 9–10 m. Its vertical distribution appears to correlate with age in that the young dwell between 0 and 500 m, adolescents inhabit depths of 500–2000 m and adults live primarily in the bathypelagic regions of the open oceans. It is a slow-moving ambush predator with a very slow metabolic rate, and scientists have calculated (and speculated) that it needs only around 30 grams of prey daily, for an adult with a mass of 500 kilograms. Its eyes reach 30–40 cm in diameter, the largest eyes in the animal kingdom. This allows an increase in visual detection, to avoid predators such as the sperm whale.

Bigfin squids (*Magnapinna*, Magnapinnidae) are a group of rarely seen cephalopods, known only from larval and juvenile specimens. Since 1988, they have been recorded only ten times, throughout much of the Atlantic and Pacific, at depths that range from 2000 down to 4700 m. The bigfin's arms and tentacles reach 8 m long and are held perpendicular to the body, creating "elbows, and the fins are extremely large". They feed perhaps by dragging their arms and tentacles along the seafloor and grabbing edible organisms from the floor. Alternatively, they may perhaps passively, for prey to bump into their arms.

Deep-sea gigantism has been explained in terms of colder temperatures, food scarcity, reduced predation pressure and increased dissolved oxygen concentrations. Cold temperatures could promote delayed sexual maturity and an increased overall life span, both of which would lead to an increase in body size (crustaceans grow continuously throughout life). Food scarcity might also be a factor, since larger body size can improve foraging ability, for widely scattered resources. Dissolved oxygen levels may also play a role in deep-sea gigantism. The solubility of dissolved oxygen in the oceans increases with depth because of increasing pressure. There is a positive relationship between the body size of amphipods and increased oxygen levels in deeper waters, suggesting that larger animals may be favoured in deeper waters as they absorb more oxygen, allowing for adequate respiration. An additional possible influence is reduced predation pressure in deeper waters. Predation of brachiopods is nearly an order of magnitude less frequent at the greatest depths than in shallow waters (Goldberg et al. 2022, and references therein).

The *rule of Substitution* states that any external factor can substitute for another, as long as its impact on the organism is the same. For a hydrophilic organism requiring a great amount of water, this can be supplied in the form of precipitation, fog dripping, or atmospheric humidity. The organism is not concerned with how its requirements will be satisfied, as long as they are satisfied. Low air temperatures can also substitute for moisture, for evaporative water loss is much lower in cold than in warm air. Eduard August Rübel (1876–1960), a Swiss botanist, formulated this rule in 1935. He pointed out that the deciduous forests of Great Britain thrive, between 600 mm and 1000 mm yearly precipitation, in a fairly cool Atlantic climate; on the other hand, the closely related deciduous forests of the eastern United States require between 1100 mm and 1500 mm precipitation because of the much higher summer temperatures. He further noted that wind substitutes for aridity because wind dries out soil and plants alike and because evaporation is rapid in a windy climate, which affects birds and insects in Antarctica. Several essential environmental components may have the same response in an animal and may therefore act in substitution.

The *rule of Tolerance* maintains that the success of reproduction depends on the completeness of the complex conditions surrounding the animal. The failure of a species to breed in a particular habitat may be due to the deficiency or success of any one factor. Since the adaptations of each individual are limited, certain essential environmental components may limit the occurrence of individual animals in a locality. This concept is close to the idea that the establishment of a species in a habitat depends on a minimum presence of the factor necessary for the most sensitive development stage. We commonly call these factors and conditions limiting factors or conditions. Victor Shelford (1877–1968), an American zoologist from Illinois and the first president of the Ecological Society of America, first formulated the rule. He pointed out (Shelford 1911) that the same environmental factors govern local and geographic distribution.

The preceding paragraphs have suggested that within a single species throughout its range, geography may influence the size, appendage lengths and colour. However, geography may influence also the overall biodiversity of different types of organisms, such as all ants of a specific area, all butterflies, or all trees.

The species-area rule states that larger areas of land (or sea) usually contain more species of a type than smaller areas. One reason for this is habitat heterogeneity. In mainland habitats such as continents, physical features of weather, vegetation and soil often vary considerably, producing an environment that is more heterogeneous than a smaller area nested within this large area, and may therefore support a heterogeneous fauna. Another factor is speciation. More new species evolve in a large area than in a smaller area nested within it, and this increased speciation leads to more species in a larger area. A third factor is fragmentation: A larger area divides into more small habitats that are isolated from one another, than a smaller area nested within this large area.

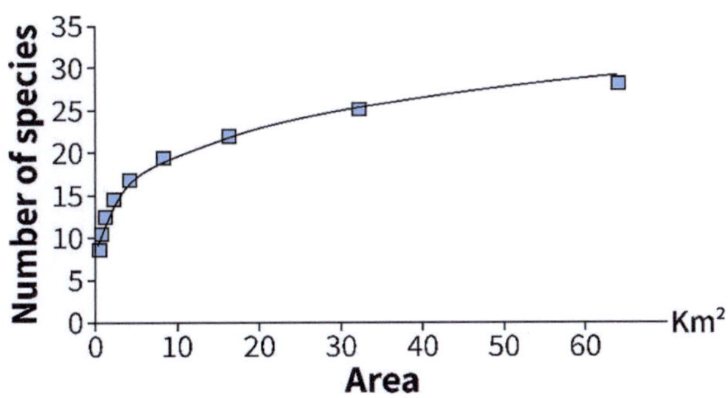

Fig. 9.11 The species-area curve formed by the eq. $S=CA^z$. We may consider each point on the curve as representing the area of (from left) a farm, village, county, district, province, region, state, country and continent

The graphic presentation of a species-area relationship yields a curve that is steep at first but gradually levels off (Fig. 9.11). This species-area curve expresses the relationship between areas of different sizes and the number of species found in those areas. If S is the number of species, A is the habitat area and z is the slope of the species-area relationship in log-log space, then the power function species-area relationship goes as:

$$S = CA^z$$

Here C is a constant that depends on the unit used for area measurement and equals the number of species that would exist if the habitat area were to be confined to one square unit.

The study of species-area relationships for mainland areas (contiguous habitats) often uses quadrats of successively larger size so that the area enclosed by each one includes the area enclosed by the smaller one (that is, the areas are nested). Consequently, species-area relationships for contiguous habitats will always increase as areas increase. Island species-area relationships differ, in that smaller islands do not nest within larger ones, and in that they do not reach the size of continents. Consequently, although larger islands tend to have more species, a smaller island may have more species than a larger one. Scientists often present species-area curves for terrestrial habitats of different sizes that are isolated from one another, such as woodlots in an agricultural landscape. Overall, species-area curves reflect regulating limits on species richness that scale with area, but multiple biogeographical processes can occur to achieve these limits.

The Latitudinal Biodiversity Gradient refers to a global pattern of species richness. One of the most broad-scale biogeographic patterns of life on Earth is that near the equator, in the tropics (namely at low latitudes), a rich diversity of species occurs, and this diversity gradually decreases towards the poles, towards high latitudes (Fig. 9.12). This global gradient in species richness is general to such an

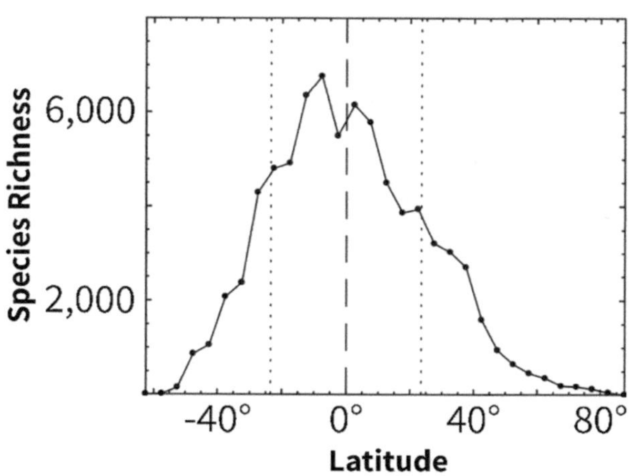

Fig. 9.12 Global and diversity: species richness by 5° latitudinal bands. (Modified from Economo et al. 2018, Fig. 2)

extent that it occurs in terrestrial, freshwater and marine habitats across many geographic regions and in a wide range of groups, from viruses via bacteria, plants and arthropods, to vertebrates (Gaston 2007; Dowle et al. 2013; Economo et al. 2018). The species richness of a group can vary by orders of magnitude between high and low latitudes, and it occurs in both local and regional assemblages. We term this increase the latitudinal biodiversity gradient. Wallace was one of the first to refer to it, when he, wrote (Wallace 1876) that "Animal life is, on the whole, far more abundant and more varied within the tropics than in any other part of the globe, and a great number of peculiar groups are found there which never extend into temperate regions". The question of why and how this pattern has arisen has thus persisted for more than 150 years.

Why does the gradient occur? There is no consensus. There is general agreement that variation in the number of species among regions results from variation in one of three basic phenomena underpinning the diversity of life on the earth: speciation, extinction and immigration rates; and regional biotas are the product of the interaction of these three processes. We do not understand, however, what factors drive these three, to form the latitudinal biodiversity gradient.

One environmental hypothesis suggests that in the tropics, high solar energy combined with steady year-round temperatures and with an abundance of water increases primary productivity. The greater the primary productivity, the more individuals that can be supported, and the more species there will be in a given area. Extinction rates are thus lower towards the equator because more populations can be sustained by the greater amount of available energy (and water) in the tropics; these lower extinction rates lead to more species in the tropics. A broadly complementary hypothesis suggests that the gradient exists because only a few species can tolerate the increasingly severe conditions of high latitudes, where the winters are colder than in the tropics. Climate stability might also be a reason for the gradient. Steady environments allow species to specialize in predictable resources, allowing them to have narrower niches, which would advance speciation, which seasonal environments may preclude. The tropics are more stable both seasonally and over geological timescales, and therefore their diversity is higher than that of other regions. Furthermore, the "timespan of life" in the tropics is year-round, whereas near the poles and in deserts, animals hibernate or aestivate throughout unfavourable seasons, and actively live only during very brief seasons that may last perhaps a tenth of that in the tropics. This approach thus, overall, suggests that environmental factors regulate species numbers, and that there are ecological limits to species numbers that vary with latitude, due to the influence of climate and/or available energy.

Another evolutionary hypothesis refers to the long evolutionary history of tropical habitats, some of which have existed for many millions of years, unaffected by glaciations and climatic shifts. Tropical rainforest species may have been able to evolve over long periods, during which they could have adapted to very many narrow, highly specialized niches. Towards the poles, on the other hand, species may have lacked sufficient time to colonize or recolonize areas since the Pleistocene glaciations, during which forests shifted further south, and this could express itself in low species richness. The latitudinal biodiversity gradient might thus be a consequence of a greater period of diversification in the tropics, through higher speciation rates combined perhaps with lower extinction rates. A somewhat similar hypothesis is that in the tropics, evolutionary rates are higher because the higher ambient temperatures cause higher mutation rates and/or faster physiological processes. This would cause higher speciation rates and hence higher diversity in the tropics. Diversity might be also determined by the evolutionary time under which ecosystems have existed under broadly unchanged conditions, and by evolutionary speed determined by the effects of environmental energy (temperature) on mutation rates, generation times and speed of selection. The tropics could thus be a "cradle of diversity" with high speciation rates, which also serve as a "museum of diversity" with low extinction rates (to use phrases of which students of biogeography are fond). Alternatively, ecosystems beyond the tropics could support as high a diversity as tropical regions, and diversification rates do not vary with latitude. Colder climates on Earth are, however, generally younger and have not had time to build-up diversity. Before the Eocene (56 to 34 million years ago), the Earth was much warmer than it is now and warm "mega-thermal" biomes, which today are limited to the tropics, covered much of the Earth's surface, including high-latitude areas, such as Europe. At the onset of the Oligocene (34 to 23 million years ago), climatic cooling opened up a vast expanse of cold-climate areas in which species had not yet diversified.

A third, biotic hypothesis, claims that species interactions such as competition, predation, mutualism and parasitism are stronger in the tropics; and these interactions promote species coexistence and specialization of species, leading to greater speciation. Intense predation could reduce the importance of competition, permit greater niche overlap and promote a higher richness of prey (Gaston 2007; Dowle et al. 2013; Economo et al. 2018).

There are many hypotheses to explain latitudinal differences in biodiversity, probably because it is difficult to disentangle the potential influences of the many abiotic and biotic variables that change with latitude. Rather than one single driving force, several contributory factors may well have generated the latitudinal biodiversity gradient.

9.4 Biogeography in Ecological Perspective: Islands

Island biogeography, a field within biogeography, examines factors affecting biodiversity of strongly isolated natural communities. Biologists originally investigated biodiversity in oceanic islands but the theory that they developed, for island biodiversity, has since turned out to be relevant to many isolated ecosystems: oases, high-altitude mountain peaks ("sky islands"), fragmented forests, isolated springs and even relict natural habitats isolated by human land development. An insular environment or "island", in this expanded context, is any area suitable for a specific ecosystem, surrounded by an expanse of unsuitable habitat (Cox et al. 2016).

In brief, the theory of island biogeography proposes that immigration and extinction determine the number of species in an insular environment. Back to the oceanic islands themselves, Darwin on his *Beagle* voyage in the Galapagos had found birds differing from island to island, and this had suggested to him that isolated populations might follow different evolutionary routes. Later decades were devoted to descriptive island biogeography, including the recolonization and floral succession of Krakatau after the volcanic outburst of 1883 (Thornton 1996).

Philip Jackson Darlington, Jr. (1904–1983) was an American entomologist. As a teenager, he wanted to collect specimens of beetles, and this became his ruling passion over the remainder of his life (they were Darwin's favourite group too). At the age of 28, he completed his Ph.D. at Harvard with a thesis on ground beetles (Carabidae). He then began a series of field expeditions to Australia and the West Indies, bringing back massive collections of insects, reptiles and other animals to Harvard's Museum of Comparative Zoology. He eventually became a curator of insects. He was also a Harvard University Professor of Zoology.

Shortly after Pearl Harbour, he enlisted in the Army Sanitary Corps Malaria Survey as a first lieutenant (retiring as a major in 1944). The army posted him to New Guinea, where he collected mosquitoes, ground beetles and other insects. On one occasion, alone in the jungle looking for specimens, he went out on a submerged log to sample water from the middle of a stagnant jungle pool, when a giant crocodile rose from the depths. While edging back toward shore, he slipped from the slimy log into the water. The crocodile rushed him, mouth gaping, huge teeth bared and spun him over and over, finally carrying him to the bottom. Thanks to a well-placed kick, the animal suddenly opened its jaws and Darlington swam free. He made for the shore, scrambled frantically up the bank and hitchhiked to a nearby hospital.

The bulk of Darlington's research was on the systematics, distribution and ecology of carabid beetles. From his thorough understanding of beetle distribution, he successfully challenged Darwin's theory that the loss of flight abilities in island insects is due to the wind sweeping away winged forms. He did this by demonstrating that small, low oceanic islands, fully exposed to strong winds, have predominantly fully winged faunas with very wide geographical ranges, whereas equally remote, high, forested islands tend to have a rich variety of flightless, endemic species. Darlington concluded (Darlington 1957) that evolution proceeds with the arrival of winged insects on mountainous islands, with subsequent specialization to the mountain habitats entailing a reduction of dispersal ability and loss of wings. He emphasized the great importance of wind as an agent of dispersal, noting that the effect of strong winds on smaller animals is far beyond anything a man can experience. The power of wind over small animals depends on the animals' ratio of body surface to weight. The surface varies as a square and weight as a cube. Decrease in size, therefore, decreases weight much more than surface. Very small animals have many times more surface for their weight than large ones do and are, therefore, much more at the mercy of the wind. Moreover, the force and carrying power increase (roughly) with the square of the wind's velocity. The effect of a wind of 100 miles per hour on an animal with the weight of a mouse is over 220 times as great as the effect of a 25-mile wind on a person. The effect of a 100-mile wind on, say a small frog is about 1000 times as great as the effect of the 25-mile wind on a person.

Moving on to factors that determine animal ranges, Darlington noted that the area occupied by a species is usually not uniform; it is more favourable in some parts than in others, and the ratio of reproduction to death often varies accordingly (Darlington 1957). Conditions are likely to be most favourable in the centre of the distribution area, and unfavourable in marginal zones where death exceeds reproduction. An excess of reproduction near the centre of the area, and a gradual migration of individuals towards the margins, maintain these marginal populations. The actual limits of a range will thus be determined, not by the limits of favourable ground but by a fluctuating equilibrium between the tendency to spread at the centre of the range and the tendency to recede at the margins. This makes a very delicate mechanism for the spreading and receding of a species. Any change, in either the species itself or the environment, which increases the rate of reproduction-to-death, will cause spreading. Any change that decreases the reproduction/death ratio will cause receding—extinction from marginal areas. Species thus fluctuate, both in their number of individuals and in their occupied area, spreading in favourable and receding in unfavourable times.

On the subject of dispersal, Darlington wrote that the simplest pattern of dispersal would be one in which groups of plants and animals just spread steadily in all possible directions from their place of origin. Plants and animals spread and recede in "dispersal cycles". They originate, rise to dominance, spread and then decline, retreat and become extinct. If an expanding group comes against a barrier that is difficult to cross, a new, independent cycle may begin across the barrier. This may happen when, more or less by accident, a member of an expanding group gets across (say) a water gap to a new land. It may happen also when a group originating in the tropics expands far enough northward so that some member of it because of its tolerance to cold rather than by accident, crosses a northern land bridge between the Old and New Worlds. We can expect such dispersal cycles in all groups of animals and plants, and the fossil record suggests them in many groups. One such group is the mastodons. They originated in Africa in the Oligocene and expanded to many parts of the World until, late in the Miocene, they extended northward far enough to cross into America; they then contracted, disappearing first in the Old World, then in America (Darlington 1957).

Many of Darlington's ideas, on dominance, on cycles of dispersal, on the balance between reproduction and death that might result in extinction, on cycles of dispersal and on the arrival at islands of generalist species that later on specialize, would inspire one of the next generation of entomologists, who would develop them further: Wilson.

Edward Osborne Wilson (1929–2021) had a passion for ants. When 23, he travelled to collect ants across Cuba and Mexico. Shortly later, Darlington, one of his influential mentors, suggested that he explore and collect ants in New Guinea and Melanesia. The prevailing theory of centres of origin, of dominant animals and of the successions of dynasties, enchanted Wilson. He sought to determine the biological nature of biogeographic dominance by extending this theory, from global scales down to more local scales, such as those of archipelagos and islands—a microcosm of the global cycle, as envisioned by Mathew and Darlington. Islands are more isolated (and less complex) than continents, so they provide better opportunities for developing principles that concern migrations and range expansions of terrestrial animals, over unfavourable obstacles.

Upon returning from his expedition and while studying his ants, Wilson described a cyclical pattern by which processes occurring in ecological time drive range expansion, diversification, adaptation and, over evolutionary time, their eventual extinction. Ants on islands reflect, in their distribution patterns, that ecological and evolutionary changes occur in several stages. Some species range continuously throughout an archipelago, with little geographic variation in their features. They are adapted to habitats along the beachfront of their source region, thus predisposing them to dispersal and pre-adapting, to colonization of islands fringed with similar habitats. Beachfronts are marginal, ephemeral habitats subject to succession, deterioration, or complete destruction from storms.

Later on, some of these colonizing populations might escape this beachfront fate by invading the interior rainforest habitats of an island and becoming isolated from their ancestral colonist populations. If evolutionary divergence then takes place, their descendant populations evolve into distinct endemic species that may disperse to other islands. Eventually, these broadly distributed (once) endemics contract in their distribution to habitats in which they specialize, forming relict species restricted to just one or two islands. The final step, for most species, is either extinction or specialization to extreme habitats. Wilson thus envisioned a “taxon cycle”.

Some years later Wilson developed a more comprehensive theory, of species replacement and of a balance of nature, which he termed “saturation”. Individual, insular faunas, approach upper limits, determined by the size of the islands—they are in either a saturated or near-saturated condition. New species can invade an island only if resident species go extinct, to make room for them. Saturation numbers of species vary with island area and isolation. Islands with large areas have more species, and Wilson attributed this species-area relationship to an inverse relationship, between-island area and extinctions. Species and genera disappear first from smaller islands and then, progressively, from larger islands. Interspecific interactions strongly alter species distributions among islands, where related species exclude one another unpredictably, thereby forming mosaic patterns. The entire form of the taxon cycle may vary in groups with different ecology and population structure, and among functionally different plant and animal groups, taxon cycles vary in a predictable manner. The taxon cycle maintains its headquarters in a given land mass indefinitely, expanding and contracting cyclically, or else it declines to extinction. The headquarters can shift from a larger to a smaller land mass (e.g. from New Guinea to Fiji) but insular endemics never colonize in the reverse direction. Small, local faunas vary in an unpredictable manner, introducing a random element (“faunal drift”) into the stages of the taxon cycle.

Turning from general theory to the real world, Wilson (1961) took the faunal dynamics that followed the catastrophic extinction of Krakatau. He noted that six widespread ant species had colonized the island by 1933, making the fauna approximately the same size as those on islands of comparable area. Island communities thus accumulate their faunas through rapid and recurrent colonization to achieve saturation, with extinctions eventually balancing immigration. Wilson now joined forces with MacArthur.

Robert Helmer MacArthur (1930–1972), born in Canada, was a mathematician and a devoted naturalist who, to watch birds, visited the tropics as often as he could. A professor, first at the University of Pennsylvania, and then at Princeton, he died young of renal cancer. In their first collaboration, MacArthur and Wilson sought to deal with equilibrium criteria (a term to replace "saturation"). They focused on patterns in species richness of bird communities and how they resulted from a balance or, as they put it, a "very dynamic equilibrium", in processes of immigration, extinction and speciation. When modelling dynamic equilibria and explaining geographic gradients in species richness (species-area and species-isolation relationships), they assumed that interspecific interactions, differences in immigration abilities and ecological dominance among species, could be ignored. Their 1963 model demonstrated not only how a dynamic equilibrium would explain species-area and species-isolation relationships, but also why species richness should be relatively stable, despite turnover in species composition. Furthermore, their model could estimate such properties as the degree of saturation of an island whose fauna was displaced from its equilibrium (say, due to hurricanes or volcanic eruptions), and the time it would take to again achieve equilibrium. The graphical form of their model allowed them to develop a set of falsifiable predictions regarding the biogeographic dynamics of insular communities: immigration, extinction and turnover rates should vary with area, isolation and climate of the island, and with the number of species in the source area. MacArthur and Wilson next wrote a full-scale book, *The Theory of Island Biogeography*, a general theory of biogeography based upon the dynamics of its fundamental processes—immigration, extinction, evolution and ecological interactions. A significant feature in this theory was their equilibrium model, which treats species as though they are equivalent with respect to their abilities to immigrate, survive, interact and evolve on islands. The book, with its species equilibrium theory, explored the many ramifications of a balance of species number on islands and on "habitat islands", namely any sharply demarcated habitat.

Island Biogeography Theory assumes that each species within an area (say an island), must have either emigrated from another island, or evolved from a local species. Within an island, many factors shape a species' birth, death, immigration and emigration rates. To avoid local extinction, the combined birth and immigration rates of a species must equal or surpass its death rates. These factors act in a dynamic fashion. Overall colonization and extinction rates determine the number of species on each island, and island size and proximity to the mainland will affect these numbers. These findings predict the establishment of a dynamic equilibrium, at which the number of species on an island stabilizes because of a balance between immigration and extinction rates (Fig. 9.13).

Species equilibrium, immigration to an island and species extinction rates on that island, are based on two assumptions.

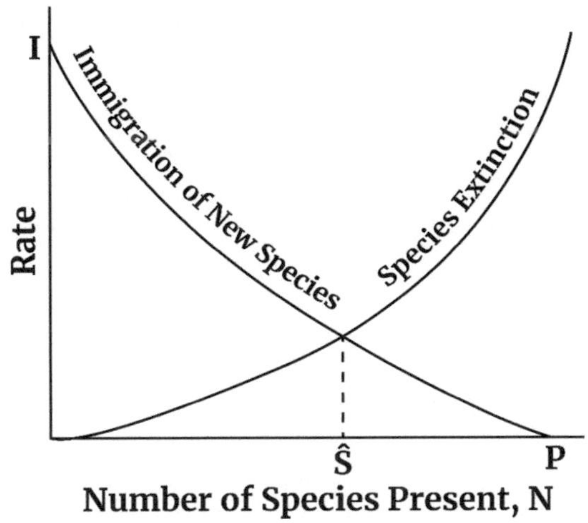

Fig. 9.13 The species equilibrium model. S is the number of species at equilibrium (saturation), and P is the number of species in the source pool. Rate curves are monotonic but nonlinear. The intercept of the dashed line on the ordinate is the turnover rate at equilibrium

1. The rate of immigration of new species decreases monotonically with increasing number of species already present. It reaches zero when all species in the source area are on the island (there are P of them).
2. The rate of extinction of species increases monotonically as the number of species increases (the more species there are, the more will become extinct).

The two curves that represent these two conflicting processes, immigration and extinction, are not linear. The immigration curve declines more rapidly at first and later slows down, because rapidly dispersing species would arrive sooner, leaving poorer and poorer dispersers as the only species not on the island, thereby reducing the rate of decline of the species immigration curve. Furthermore, species abundance generally results in a decrease in available resources and niches for new colonizers. The extinction curve, on the other hand, accelerates with increasing numbers of species, due to a larger number of potential competitive interactions among species and, also, due to decreasing population sizes, as the island fills up. As the number of species within an island increases, so too do the overall extinction rates, because increasing biodiversity increases the potential for competition or predation. The two curves intersect at a point where the rates of immigration and extinction are equal and at this equilibrium number, the number of species is constant. Accordingly, each island will support a limited number, both of individuals and species. The overall number of species may remain broadly constant because of this equilibrium, but the specific species composition of an island will constantly change, as new immigrants arrive and old residents become extinct. To reach balance, as the number of species on an island increases, the extinction rate increases and colonization rate decreases. These two factors together, determine the biodiversity of each island.

The point of saturation (dynamic equilibrium) will vary among islands. An island closer to the mainland will have a greater colonization rate than islands further away because colonization is more likely to occur over shorter distances, as it depends on the dispersal of organisms. Whereas birds can fly from one distant location to another and seeds can drift passively on the wind or float in the sea, many emigrating organisms (say, land snails or scorpions) cannot disperse to distant sites that easily. Another reason for between-island variation: Larger islands tend to support a greater number of species than smaller ones. This is so because larger areas can sustain a greater amount, of resources and niches, as well as larger populations. As the risk of extinction for each species decreases with more individuals, natural disasters are less likely to threaten larger populations. Since large islands near the mainland have the highest immigration rate and the lowest extinction rate, large islands shall support the greatest number of species. Smaller, more isolated islands shall support the lowest number. To conclude this point, the fauna at each island will be at dynamic equilibrium, as the number of species stabilizes because of a balance between immigration and extinction rates. Individual factors of each island, including its size, resources and species present, influence this balance (Fig. 9.14).

The theory of island biogeography predicts continual species turnover in ecological communities, rather than a stable species composition. Ecological communities are open assemblages of species that approach a steady state species richness that is dynamic, not static in its species composition. The concavity of the immigration and extinction rates (Figs. 9.13 and 9.14) makes late-arriving species experience lower successful immigration rates and higher extinction rates. Any species arriving late, regardless of whether it is a good colonizer or competitor, will exhibit the same rate of changes. Likewise, all species respond in an identical manner to variations in the size of the island and its distance from the mainland source area. It is noteworthy, however, that the immigration curve to an empty island is much faster than the loss of species through extinction from an "over-saturated" island. A colonizing event requires the arrival of only one individual of a given species, whereas extinction requires the death of all individual species on the island. Also noteworthy, when species have different relative abundances in the source area, then species no longer have equal probabilities of immigrating to the island.

The next step was to test experimentally the theory, that islands contain an equilibrium of species, as determined by the rate of immigration of new species and the rate of extinction of already-established species. This test required islands that lack species and, as this was not possible naturally, it required first to remove the fauna on an existing set of islands. Simberloff and Wilson selected seven islands in Florida

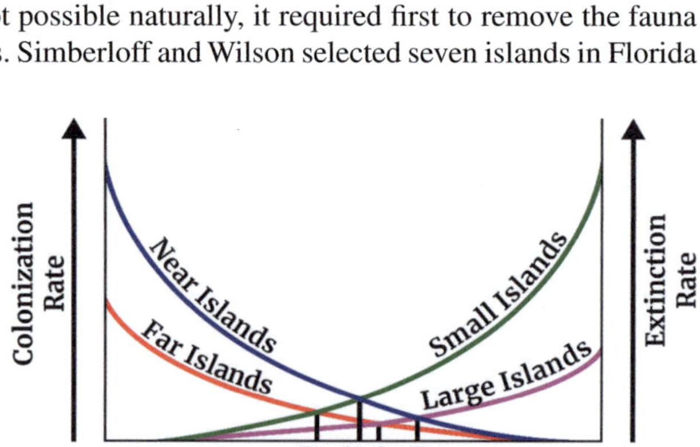

Fig. 9.14 Island biogeography: influence of island area, and of distance from mainland, on the equilibrium number of species

Bay, each of varying distances and direction from larger islands, ranging from only 11 to 18 metres in diameter, and consisting solely of red mangrove trees. They identified and counted the resident arthropods, which numbered between 20 and 50 species. They then erected tents that enclosed each island and fumigated the islands with methyl bromide, a chemical that kills arthropods but does not harm plants. After thus eradicating the islands' arthropods, they periodically monitored arthropod recolonization over a period of approximately 1 year. They found that arthropods rapidly recolonized the de-faunated islands and, after less than a year, all but the island farthest from the mainland had about the same number of species as before defaunation. However, though the islands returned to the same number of species, they did not necessarily return to the same species themselves: Species turnover rates were high. From these results, Simberloff and Wilson concluded that the islands supported an equilibrium species number, as suggested by the theory of island biogeography.

Since the studies of Simberloff and Wilson, many studies have sought evidence for species equilibrium on islands and on island-type habitats, by counting species numbers over several years; these have been summarized by Schoener. The number of bird species on islands off the California coast is largely constant and on islands in the Aegean, bird species counts did not change by more than one species over several years. On islands off Australia and New Zealand, the number of bird species (passerines) increased up to nine-fold the original values, and plants and ants of the Bahamas showed the opposite trend, in that the islands actually lost species. In the Bahamas, a hurricane with very high waves (Lili in 1996) was devastating for many small, low islands. After the hurricane, every lizard and web-spider individual originally inhabiting the islands was gone. For spiders, recolonization was rapid, and within 1 year the number of species was the same as before the hurricane struck; for lizards, however, recolonization was zero. This suggests that saturation closely depends on the organism. For a given rate of disturbance, organisms that are good dispersers are more likely to attain equilibrium. For spiders, highly agile organisms that disperse by ballooning, equilibrium recovers quickly. For lizards, which reach islands by rafting, achieving saturation may take a very long time, indeed a longer time than the next devastating hurricane. To conclude, saturation does indeed exist—but only occasionally.

The second assumption of the Species Equilibrium Model is that species lists vary in composition, even after attaining equilibrium (as well as on the way to equilibrium). In the newly gained equilibria number of arthropods of red mangrove islands, the species composition was substantially different from that determined shortly before the artificial de-faunation. As summarized by Schoener, there is a year-by-year turnover of 0.5–4.9% in birds of islands off the California coast; and on Krakatau, the turnover is 0.25–0.42% per year. On tropical islands other than Krakatau, on the Tres Marias Islands off Mexico and over a 72-year interval, only two immigrations occurred, Cocos Island had no turnover in 72 years, and One Tree Island (the Great Barrier Reef region) had no turnover during 6 years. A 7-year survey found no immigration and only one extinction for Guana Island in the Caribbean. The low turnover in tropical islands might perhaps be because tropical birds are more sedentary, and therefore they have lower migration rates. In spiders of the Bahamas, turnover does not involve the most abundant species, those that should often have the greatest food-web effects, and in any event, are of the most interest to the ecosystem. Often, the same spider species become extinct and then re-immigrate. Schoener concluded that turnover tends to be greater for small and distant islands and can be very low for tropical islands. In general, because of the strong relation between extinction rate and population size, species turnover comprises a low fraction of the total number of individuals in the biota.

The assumptions that close islands of the same area as distant islands have more species, and that large islands at the same distance as small islands have more species, are supported by numerous examples in the literature.

9.5 Biogeography in Historic Perspective: South America

In seeking historic perspectives of biogeography over a long stretch of time, South America offers one of the best possible views. After the late Mesozoic break-up of Gondwana, it spent much of the Cenozoic era as an island continent, and during this long "splendid isolation", unique mammals evolved that are found nowhere else on Earth. Some of these endemic mammals filled ecological niches similar to those of other mammals elsewhere, and converged on them in form; others were somewhat strange. (The Cenozoic is Earth's current geological era, representing the last 66 million years of Earth's history. Mammals, birds and flowering plants are the dominant groups of this era, in which the continents reached their current configuration.)

An extinct sister group to marsupials, the Sparassodonta, was an order of predatory mammals native to South America, that existed from the Eocene onwards (Fig. 9.15). A typical Sparassodont was *Thylacosmilus* of the Pliocene. It weighed

Fig. 9.15 *Thylacosmilus*, a sprassodont about the size of a jaguar. (By Tuvia Kurz)

80–120 kilograms, slightly less than a modern jaguar. Highly developed neck muscles powered the canines, which had open roots and grew constantly. The bite force of *Thylacosmilus* was much weaker than that of a leopard, indicating that its killing techniques differed from those of extant big cats. Its claws were not retractable, as in cats. The bones of the limbs were very robust and this, together with the fact that they walked with their toes and soles flat on the ground ("plantigrade feet"), show that it was not a runner. It probably ambushed its prey animals, grazers in open savannah-like grasslands or in sparsely forested areas.

Unlike placental mammals whose pelvis consists of only three pairs of bones, they had a fourth ("epi-pubic") pair of bones, as in marsupials; these bones were, however, highly reduced, and probably consisted of long cartilaginous elements. In their dentition, they replaced only the upper and lower third premolars throughout life, not other teeth. Their eyes faced almost completely to the sides, so they had very little binocular vision. They preyed on rodents and rodent-like mammals, raided seabird colonies, and some of them could crush bones like modern hyenas, wolverines and Tasmanian devils. In appearance, some of them closely resembled placental predators that had evolved separately, on other continents. Sparassodont diversity peaked in the mid-Miocene after which it began declining, and by about 3 million years ago, Sparassodonts were almost entirely extinct.

Moving on from marsupials to placental mammals, the Xenarthra include today's anteaters, armadillos and sloths. Among the many features that they share, and which most placentals lack, we note their heavily reduced and highly modified dentition. Their teeth have no functional enamel, are open-rooted and grow continuously. The front of the mouth usually has few or no teeth, and the teeth in the rear of the mouth all resemble one another. Xentharian ancestors fed on subterranean ants, which they would dig up. Several fossil xenarthrans evolved to chew plants and, to create grinding surfaces, their cheek teeth, lacking enamel, developed alternating harder and softer dentine. The vertebral joints along the xentharian spine usually have additional points of articulation, unlike any other mammal, that stiffen the spine, to suit a lifestyle that requires the digging up of food. They have single-colour vision suggesting that their ancestors had reduced eyesight, as in recent vertebrates living underground, or that they were nocturnal. They have the lowest metabolic rate of all marsupial and placental mammals. Xenarthrans are possibly the most primitive of placental mammals (Stehli and Webb 1985; Gaudin and Croft 2015).

Turning now from living to extinct xenarthrans, there were the glyptodonts: heavily armoured relatives of armadillos that evolved some 20 million years ago. They possessed a tortoise-like body armour consisting of bony deposits in their skin, somewhat like turtles. They could not withdraw their heads, but their armoured skin formed a bony cap on top of their skull. Their tails had a ring of bones, whether against predators for protection or against other members of their species, in aggressive combat. Many, such as *Glyptodon*, would weigh up to 2000 kilograms. These grazing herbivores had no incisor or canine teeth, but they had a number of cheek teeth that would have been able to grind up tough vegetation, such as grasses. Then, there were the Pampatheriidae ("Pampas beasts"), also related to armadillos. They attained a weight of up to 200 kg. Unlike glyptodonts, their armour shell had some flexibility. They were primarily grazers, with much of their diet consisting of coarse vegetation, unlike today's armadillos that feed on insects. They existed as a separate lineage since at least 45 million years ago (Eocene), finally disappearing about 12,000 years ago (Stehli and Webb 1985, Gaudin and Croft 2015).

Macroeuphractus, of the Late Miocene to Late Pliocene, was an armadillo that converted to a carnivorous lifestyle. It was a little over a metre in length, and weighed 30 (or perhaps 100) kilograms. It was fossorial, like most armadillos, probably digging out rodents and small notoungulates. Living in an epoch where sparassodonts and terror birds were in decline, it may have exploited this ecological vacancy.

Living sloths are arboreal, small-bodied and weigh less than 10 kilograms. In these, among other features, they differ from the diverse, extinct ground sloths, which reached massive sizes and heavy bodies. The thick bones of many extinct ground sloths gave their appendages tremendous power that, combined with their claws, provided a defence against predators. Some ground sloths had hearing abilities optimized for low frequencies, suggesting that perhaps they used infrasound for communication. The group includes *Megatherium*, the "great beast" described by Cuvier in 1789 (pp. XXX). Another ground sloth was *Thalassocnus*, which lived along the South American Pacific coast, from the Miocene to the Pliocene. Over 3 metres long, *Thalassocnus* (Fig. 9.16) was the only aquatic sloth, indeed the only aquatic xenarthan. It evolved several marine adaptations over the brief course of just 4 million years, one of which was bones that were dense and heavy, to counteract buoyancy and sink to the seafloor, as in modern manatees. Indeed, its bones were so compact that the cavity containing the bone marrow was nearly absent. In the head, to assist breathing while submerged, the nostrils shifted from the front to the top of the snout, as in seals. The head angled downwards, to aid in feeding off the sea floor, and some species had a short trunk, as in elephant seals. Possessing no adaptations for high-powered propulsion swimming, as in modern fishes and dolphins, they probably paddled like terrestrial mammals and walked across the seafloor rather than swimming through the water. Here, they would dig up seagrasses, using their strong lips (and a trunk?) to rip out grasses, somewhat as modern manatees do (Stehli and Webb 1985, Gaudin and Croft 2015).

We now come to South America's ancient, native hoofed mammals. All hoofed mammals share distant common ancestors. The South American ungulates (superorder Meridiungulata) are a sister group to the rhinoceros, horse and tapir (Perissodactyla), having diverged from them about 66 million years ago. One order of ungulates native to South America was the extinct Notoungulata, which lived from 60 million to 11,000 years ago. Some of them evolved ever-growing teeth, like rodents and rabbits. The group was highly diverse, many evolving along convergent lines, into forms that resembled mammals on other continents. *Pachyrukhos* (30 centimetres long, Late Oligocene to Mid-Miocene) converged into the form of rabbits and hares, with long hind feet and a short tail, and they probably also hopped. Its teeth suggest a diet of tough plants, perhaps also nuts, its large eyes suggest nocturnal activity, and its complex hearing apparatus suggests good hearing (Fig. 9.17).

Fig. 9.16 *Thalassocnus*, an extinct marine sloth. (Tuvia Kurz)

Homalodotherium of the Mid-Miocene was about 2 metres long, weighed 300 kilograms, and was partially bipedal. While rearing up on its hind legs, it would pull down tree branches, as do the giant panda and the gorilla. *Toxodon* lived from about 2 million to 11,000 years ago (Late Miocene to Holocene). Reaching 3 metres in body length, 1.5 metres in shoulder height, weighing up to 1400 kilograms, and broadly resembling a heavy, short-headed rhinoceros, it was among the largest members of its order. Its teeth grew throughout life, as in rodents. It lived in arid and semi-arid areas. Another ungulate group, the Litopterna, evolved toe reduction—some had only three toes, others had only a single hoof on each foot. Some were fast runners, converging on the adaptations of goats, antelopes and horses of other continents. Others, with long necks, muscular upper lips and short trunks, filled environmental roles similar to those of camels and giraffes. As browsers and grazers, they occupied ecological roles similar to those of hoofed mammals in Laurasia. *Macrauchenia* was a camel-like litoptern with a small proboscis. One-toed *Thoatherium* had legs like those of a horse, and horned *Trigodon* looked like a rhinoceros. The group reached its maximum diversity in the Late Miocene. Pyrotheria was another order of extinct meridiungulate mammals. These mastodon-like ungulates had the appearance of large, digitigrade, tapir-like mammals with short, slender limbs and five-toed feet with broad, flat phalanges. The dentition was complete

Fig. 9.17 *Pachyrukhos*, a hare-like notoungulate. (Tuvia Kurz)

Fig. 9.18 A terror bird. (Tuvia Kurz)

with strong, chisel-shaped incisors, strong sharp-pointed canines and low-crowned cheek teeth. They lived from the Palaeocene to the Oligocene.

Besides mammals, there were the fearsome "terror birds" (Phorusrhacidae), an extinct group of large carnivorous flightless birds that lived in South America from about 60 to 2 million years ago (Fig. 9.18). Reaching a height of 3 metres, they were the largest predators of their time. Their muscular yet highly flexible neck allowed them to strike with terrifying power and fast speed, causing fatal damage to the prey. The skull reached a length of 70 centimetres and the beak, exceeding 40 centimetres, curved in a hook shape as in eagles. Terror birds were extremely nimble and fast runners, reaching speeds of almost 50 kilometres per hour. All possessed a large, sharp beak, a powerful neck and sharp talons, and all were carnivorous, perhaps feeding on large prey. A terror bird would probably use its feet to injure prey by kicking it, and then hold the prey down, and it would then use its large beak to rip the flesh or slash vital organs. Terror birds were avian predators of the open plains.

The closest modern-day relatives of the terror birds are the 80-centimetre-tall seriemas, the sole living members of the small bird family Cariamidae, which is also the only surviving lineage of the order Cariamiformes. These are large, long-legged birds reaching a height of 70–90 centimetres, which live in grasslands, savannah, dry woodland and open forests of South America. They forage on foot and they run from danger, rather than fly (which they can, for short distances). Their long legs, necks, tails and short wings, reflect their way of life. They are among the largest ground-dwelling birds endemic to South America (rheas are much larger, of course). The seriema feeds on insects, lizards, snakes, frogs, young birds and rodents, often associating with grazing livestock, to take insects the animals disturb. Although capable of flying, it prefers to spend most of its time on the ground, taking flight only when necessary. Overnight, it takes shelter in the treetops, where it also builds its nests, about 1–5 metres above the ground. The adults reach the nest by foot rather than flying, through hops and the occasional flutter. Hatchlings reach maturity at five months. These are the closest living relatives of the gigantic, extinct terror birds.

Not all South American birds were ground birds. *Argentavis* (Fig. 9.19), which soared the skies over South America 6 million years ago, was one of the largest ever of flying birds. It had a wingspan of 6.5 metre, stood 1.8 metre high and weighed 70 kilograms—the heaviest flying bird ever. For comparison, the condor of today has a wingspan of up to only 3.3 metres, and weighs up to only 15 kilograms. It lived much like the modern condor, scanning large areas for carrion. The climate of the Andean foothills during the Late Miocene was warmer than today and thermal updrafts would have aided the bird in staying aloft. With meat requirements of 2.5 to 5 kilos per day, it probably preferred carrion scavenging but might also have killed small live prey.

Fig. 9.19 *Argentavis*. (Tuvia Kurz)

Beyond birds and on to the reptiles, some of South America's aquatic crocodilians (*Gryposuchus*, *Mourosuchus* and *Purussaurus*) reached monstrous sizes, with lengths up to 12 metres. They shared their habitat with one of the largest turtles of all time, the 3.3-metre long *Stupendemys*.

Invasions of South America started about 40 million ago (mid-Eocene), with West African caviomorph rodents crossing the ocean by rafting or by island hopping. Here, they underwent explosive diversification, outcompeting other animals in rodent-like niches, such as certain South American marsupials. Over time, some caviomorph rodents evolved into larger forms that, perhaps, competed with some of the native South American ungulates; this might have contributed to the gradual loss of diversity suffered by the latter, after the early Oligocene. The bison-sized *Phoberomys,* which lived in the ancient Orinoco River delta around 8 million years ago, was 3 metre long and weighed about 550 kilos, and *Josephoartigasia* of the Early Pliocene to Early Pleistocene was even larger, with a weight of 1000 kilograms (Fig. 9.20). These two giants relate to today's pacarana (*Dinomys*), a rare, slow-moving, nocturnal rodent living in tropical forests of the Amazon and adjacent foothills of the Andes. It weighs up to 15 kilograms and measures up to 80 centimetres in length (not including the tail). Today, caviomorph ecologies include burrowing, arboreal, running and aquatic forms, in environments that include grasslands, high mountains, forest edges and dense tropical forests.

Later, 36 million years ago, primates followed the rodents, again from Africa and again, in a rafting fashion similar to that of the rodents. All extant New World monkeys derive from radiation that, however, occurred long afterwards, 18 million years ago (Early Miocene; Schrago and Russo 2003, De Queiroz 2014). Tortoises, closely related to the African hinge-back tortoises, arrived in South America in the Oligocene. Their ability to float with heads up, and to survive up to six months without food or water, would have aided them in their oceanic dispersal. From the South American continent, tortoises then went on to colonize the West Indies and Galapagos Islands. Crocodiles (*Crocodylus*) rafted over, around 25 million years ago. Among lizards, geckos rafted over from Africa during both the Paleogene and

Fig. 9.20 *Josephoartigasia*, which weighed 1 tonne. (Tuvia Kurz)

Neogene, whereas skinks (*Mabuya*) crossed the Atlantic from Africa to South America, during the last 9 million years. Surprisingly, South America's burrowing, snake-like amphibians (amphisbenians) and blind snakes also rafted from Africa, as did the ancestors of the hoatzin, a weak-flying bird of South American rainforests.

Other invaders arrived from North America. The earliest was a relative of the raccoons that reached South America around seven million years ago, perhaps by island hopping. (Raccoons belong to the Procyonidae, a New World family of generally omnivorous carnivorans, comprising also ringtails, coatis, kinkajous and others). Hog-nosed skunks and sigmodontine rodents followed them into South America, whereas ground sloths invaded North America about 9 million years ago, as did terror birds some 5 million years ago. The invasion route used by these migrants was the Isthmus of Panama.

Thirty million years ago, a large ocean covered the area where Panama is today, and there was a large gap between the continents of North and South America. Beneath the surface, however, two drifting plates of the Earth's crust were slowly colliding, forcing one of them to slide under the other (the Caribbean under the Pacific Plate). The pressure and heat caused by this collision led to the formation of underwater volcanoes, some of which grew tall enough to break the surface of the ocean and form islands, as early as 15 million years ago. Over the next several million years, more and more volcanic islands filled in the area. Over time, large amounts of sand, soil and mud, which strong ocean currents had peeled away from both North and South America, poured through gaps between the newly forming islands. Little by little, over millions of years, the sediment deposits added to the islands, until they completely filled the gaps. Meanwhile, the movement of the two tectonic plates was also pushing up the sea floor, eventually forcing some areas above sea level. Eventually, a narrow strip of passable land emerged between North and South America, with water on either side. This narrow Isthmus of Panama, linking South America to North America, had a major impact on the biogeography of our world. It initiated one of the largest waves of biological migration between the two previously disconnected landmasses, waves that we term the Great American Biotic Interchange.

The Isthmus of Panama affected the marine animals of the area very differently from the terrestrial organisms. By severing the flow of water between the two oceans, the land bridge re-routed currents in both the Atlantic and Pacific oceans. It forced Atlantic currents northward and these, eventually, formed a new current, the Gulf Stream. With warm Caribbean waters flowing towards the northeast Atlantic, the climate of northwestern Europe grew warmer (winters there would have been as much as 10° C colder, without the transport of heat from the Gulf Stream). This and other changes helped establish the global ocean circulation pattern that we see today. It also set the now-separated populations on divergent evolutionary paths. Organisms on the Atlantic side had to adapt to an environment of lower productivity after the inflow of nutrient-rich water of deep Pacific origin was blocked. The Pacific coast of South America cooled, as the input of warm water from the Caribbean was cut off, and this may have caused the extinction of the marine sloths in the area. The Isthmus of Panama thus directly and indirectly influenced ocean and atmospheric circulation patterns, which regulated patterns of rainfall, which in turn sculpted landscapes, and changed the distribution of many animals.

When did the Isthmus form? Traditional studies suggest full closure by about 3.5 million years ago. A recent analysis of both molecular and fossil data, however, reveals that rather than a Pliocene, time-limited, single event, the formation of the Isthmus of Panama, and the Great American Biotic Interchange that enabled, were long and complex processes that began as early as the Oligocene-Miocene transition (Bacon et al. 2015). They reveal dramatic waves of dispersal and biotic turnover of terrestrial organisms, at approximately 20 and 6 million years ago, and corresponding events separating marine organisms in the Atlantic and Pacific oceans at about 23 and 7 million years ago. The direction of dispersal and their rates were symmetrical until 6 million years ago, when the northern migration of South American lineages increased. The cross-Isthmian migrations occurred in pulses, over a long time that probably correlated with the stepwise formation and emergence of the Isthmus of Panama, and in stark contrast to the assumption that a majority of events occurred at, or shortly after, a 3.5 million years ago closure.

There seems to have been an initial collision of the Panama Block and South America 25–23 million years ago with the development of extensive terrestrial (although not necessarily fully connected) landscapes in Panama by 20 million years ago. Pulses of dispersal of terrestrial organisms occurred over at least three periods during the last 30 million years. The underlying causes for these discrete migration pulses remain elusive but may be associated with landscape formation, volcanism, climate change and/or sea level fluctuations (Bacon et al. 2015). In summary, the Great American Biotic Interchange comprised several distinct pulses of migration. Marine and terrestrial clades exhibit similar dispersal pulses, occurring between about 23 and 20 million years ago and 8 and 6 million years ago.

This Great American Interchange involved the immigration into South America of such North American groups as the camelids, tapirs, deer, horses, gomphotheres, cougars, jaguars, sabre-tooths, canids, mustelids, procyonids and bears. Today, 50% of the mammalian genera in the south are of northern origin. It correspondingly involved the immigration into North America of such South American groups as ground sloths, terror birds, glyptodonts, pampatheres, capybaras and notoungulates. At first, the success of these two migrations was broadly symmetrical. Later on, however, the North American groups migrating southwards were more successful than the South American ones migrating northwards, establishing themselves in large numbers and diversifying considerably. The dispersal and subsequent explosive adaptive radiation of northern rodents throughout South America, leading to over 80 current genera, was vastly more successful, both spatially and by number of species, than any northward migration of South American mammals. Other North American groups that diversified conspicuously in South America were the canids and deer, both of which currently have three-four genera in North America, but six in South America.

The interchange considerably changed South America's terrestrial predator guild, from one composed almost entirely of marsupials, terror birds and reptiles, to one dominated by immigrant placental carnivores, such as dogs, bears, cats, weasels

Fig. 9.21 *Smilodon*. (Tuvia Kurz)

and skunks. Competitive exclusion from immigrating placental carnivores might have driven the native South American predator guild to extinction. One case might have been that of the invading sabre-tooth, *Smilodon*, with its upper canines extending down from the mouth when it is closed (Fig. 9.21). We recall South America's native sparassodont *Thylacosmilus*, of the Late Miocene to Pliocene, with its large canines and constantly growing open roots. New-coming *Smilodon* might have outcompeted *Thylacosmilus*, driving it to extinction. *Smilodon* might have been a social predator, like the lions of today. Regardless of how this turnover in the predator guild happened, it is clear that the newcomers benefitted from it. Dogs and cats underwent adaptive radiation in South America after dispersing there, indeed, the greatest modern diversity of canids in the world is in South America.

Another case may have been that of the invading camelids, tapirs, deer, horses and elephant-sized gomphotheres. We recall that one of South America's native ungulate groups, the litopterns, which displayed toe reduction, evolved adaptations for fast running and occupied a variety of ecological roles similar to those of hoofed mammals in Laurasia. New-coming camelids, tapirs, deer and horses may have outcompeted the litopterns, driving them to the brink of extinction. Likewise, large cats may have replaced the borhyaenids. Counterparts of analogous groups thus replaced one another. These successful migrants might have led to the extinction of a large proportion of the South American fauna. This successful diversification did not apply to all habitats, thus, one South American niche that the North American invaders did not dominate was for massive herbivores. After the interchange, South America was home to about 25 species of herbivores weighing more than 1000 kilograms. There were the Old Timers: the ground sloths, glyptodonts and toxodontids. Alongside them were the newcomers: the gomphotheres and camelids of North American origin. Native South American forms made up about 75% of this guild.

Interestingly, this outcompeting of native fauna by an invading immigrant one, which we have offered for South America, did not occur in North America. South American groups invading North America included opossums, armadillos and their relatives, pampatheres, glyptodonts, several ground sloths, New World porcupines, capybaras, notoungulates and bats, alongside terror birds and hummingbirds. However, none of these groups diversified and indeed, no migrant from the south ever drove to extinction any of the native northern fauna. Furthermore, after initially establishing themselves, many of these South Americans went extinct.

This leads us to a big "Why" question. Why did the North American migrants in South America succeed and why did the South American migrants in North America fail? Why did the North American mammals prove superior? Some authors believe that it was the history of inter-continental connections, which played into the hands of the northern invaders. During the last 66 million years of Earth's history (the Cenozoic), the Bering Straits repeatedly connected North America to Eurasia, enabling recurring animal migrations back and forth, and thus uniting the faunas of the two continents. Eurasia, in turn, connected to Africa, and this further contributed to the species that made their way to North America. South America, however, connected only to Antarctica and Australia, two small and less hospitable continents, and only in the early Cenozoic. Moreover, this land connection does not seem to have carried much traffic in the direction of South America: no mammals other than marsupials and, perhaps, a few relatives of the platypus (monotremes) had ever migrated by this route. This meant that northern hemisphere animal groups arose within a land area six times greater than was available to South American groups. North American groups were thus products of a larger, more competitive arena, where evolution would have proceeded more rapidly. They tended to be more efficient and more brainy, generally able to outrun and outwit their South American counterparts, who were products of an evolutionary backwater. In the cases of ungulates and their predators, the invaders replaced the South American forms because of these advantages. The greater eventual success of South America's African immigrants compared to its native early Cenozoic mammal fauna further illustrates this phenomenon, since the former evolved over a greater land area; their ancestors migrated from Eurasia to Africa, two significantly larger continents, before finding their way to South America.

North American mammals had lived a more competitive life, in a larger continent with more species, than the isolated southern mammals. The "arms race" of competition had moved further in the north. When the North American mammals invaded the south, they had been prepared by 50 million years or so of more demanding competition, which enabled them to over-run the southern mammals. The South American fauna, in conclusion, often could not compete for resources as well as the North American groups already occupying the same ecological niches, and those that did succeed in establishing themselves for some time, often did not survive for long. To me, this argument merely explains away the question, it does not elucidate it.

At the end of the Pleistocene epoch, about 12,000 years ago, Native Americans invaded and occupied the New World, bringing to extinction many of the successful participants of the Great American Interchange, as well as other species that had not migrated. All the pampatheres, glyptodonts, ground sloths, equids, proboscids, giant short-faced bears, dire wolves and machairodonts of both continents disappeared. The last of the South American notoungulates and litopterns died out, as well as North America's giant beavers, lions, cheetahs and many of its antilocaprid, bovid, cervid, tapirid and tayassuid ungulates. Some groups disappeared over most or all of their original range but survived in their adopted homes, such as South American tapirs, camelids and tremarctine bears (cougars and jaguars may have been temporarily reduced to South American refuges also). Others, such as capybaras, survived in their original range but died out in areas to which they had migrated. This extinction pulse eliminated all South American migrants to North America larger than about 15 kg, and all native South American mammals larger than about 65 kg. In contrast, the largest surviving native North American mammal, the wood bison, can exceed 900 kg, and the largest surviving Nearctic migrant to South America, Baird's tapir, can reach 400 kg. Similar mega-faunal extinctions have occurred on other recently populated landmasses (Australia, Japan, Madagascar, New Zealand and many smaller islands around the world, such as Cyprus, Crete and New Caledonia), at different times that correspond closely to the first arrival of humans at each location.

These extinction pulses invariably swept rapidly over the full extent of a contiguous land mass, regardless of whether it was an island or a hemisphere-spanning set of connected continents. On sizeable islands far enough offshore from newly occupied territory to escape immediate human colonization, megafaunal species sometimes survived for many thousands of years after they or related species became extinct on the mainland. These included giant kangaroos in Tasmania; giant *Chelonides* tortoises of the Galapagos Islands; giant *Dipsochelys* tortoises of Seychelles; giant meiolaniid turtles on Lord Howe Island, New Caledonia and Vanatu; ground sloths on the Antilles; Steller's sea-cows off the Commander Islands; and woolly mammoths on Wrangel Island and Saint Paul Island.

South America was subject to one wave of migrating mammals after another, and from a historic perspective, its modern fauna is today highly multilayered. Another great biotic migration is occurring today and in front of our eyes, in the Mediterranean Sea.

9.6 Dynamic Biogeography: The Great Modern Migration into the Mediterranean

In today's globalized world, introductions of non-indigenous species occur at unprecedented rates and this is probably the dominant biogeographic event of our times. A very large-scale migration of marine species is occurring today. Animal species of tropical Indo-Pacific origin, coming from the Red Sea, cross the Suez Canal and invade the Mediterranean Sea; more uncommonly, animals move in the opposite direction: Mediterranean species of temperate Atlantic origin cross the canal, and migrate into the Red Sea. The canal (Fig. 9.22), 160 kilometres long, 15 metres deep and 200–300 metres wide, opened in 1869 to provide a direct trade route from Europe to India and the Far East, and in doing so, it exposed each sea to invasion of organisms from the other sea. Over the past 150 years, this has become one of the major biogeographic events of our times: Some 300 marine species have migrated from the Red Sea into the Eastern Mediterranean, which is undergoing an unparalleled species composition transformation (Belmaker et al. 2012). We term this ongoing event Lessepsian Migration in recognition of Ferdinand de Lesseps, the French diplomat, engineer and spirit behind the construction of the Suez Canal.

The surface of the Red Sea is higher than the Eastern Mediterranean by about 1.2 metres, and therefore Red Sea waters pour, through the canal, into the Mediterranean. Accordingly, most migrations between the two seas are invasions of Red Sea species into the Mediterranean, and only a few migrations occur in the opposite direction, colonizing the Red Sea from the Mediterranean ("anti-Lessepsian" migrants). The Red Sea is a tropical water body in which temperatures do not fall below 18°C in winter, and cool temperatures constitute a physiological constraint to fish of tropical origin. Tropical species require summer temperatures high enough for the reproductive process and development of eggs, and minimum winter temperatures above their lethal limits. Currently, low winter temperatures seem to be a factor that prevents the spread of the bulk of Red Sea animals into the Mediterranean Sea, particularly reef-building corals (Galil 2007; Galil et al. 2021).

The first marine Lessepsian migrant was recorded (1891) just 14 years after the canal had opened. Today, with 300 recorded migrants, about 10% of all fauna in the Levant basin (the southeast basin of the Mediterranean) is of Indo-Pacific origin. There is a clear east-west gradient in the distribution of migrant species in the Mediterranean. In fishes, for example, there are 54 migrant species along the

Fig. 9.22 A troopship sailing to India through the Suez Canal, a few years after the canal opened. (My collection)

Levantine coast but only 30 on the southern coast of Turkey, only four along the southern coast of Italy, and none in the western basin of the Mediterranean. Furthermore, among fishes, the year of the first record of an individual migrant species in the Mediterranean broadly correlates with this species' current abundance along the Mediterranean coast: species that appeared earlier are currently more abundant. The vast majority of species recorded only since 1970, are rare today (Galil 2007, Galil et al. 2021).

The study of fish is advantageous in that, due to their commercial importance, constant information on quantitative and qualitative changes in migrant and native populations is available, through the local fisheries. The migrant species constitute only 13% of the fish fauna in the Levant, but almost half of the catch of trawl fishery on the Mediterranean coast of Israel consists of migrant species (Fig. 9.23).

Do the migrants outcompete the locals? In goatfishes, there are two migrant species and two native Mediterranean species. The first record in the Mediterranean of the Red Sea gold-band goatfish (Fig. 9.24) was in the 1930s. By the late 1940s, it made up about 10% of the total goatfish catches off the Israeli coast; in 1954–1955, its percentage increased to over 80% of the catch and it began replacing the native red mullet, in commercial fisheries. Today it inhabits the shallow waters almost exclusively, whereas the native mullet dwells in deep, cool waters, of 70 metres and below. As the feeding habits of the migrant and native species are similar, they may

Fig. 9.23 *Mullus barbatus*, native to the Mediterranean (above); and *Upeneus moluccensis*, a Lessepsian migrator (below). (Tuvia Kurz)

Fig. 9.24 Lessepsian immigration: migrating species (left) and native Mediterranean species they are displacing (right). (Tuvia Kurz)

well have partitioned the niche along bathymetric axes, the colonizer occupying a shallower sector. However, as we lack data on fishing depths from before the 1930s, the period before the invasion, it is difficult to determine whether the colonizers displaced the local species, or whether the latter occupied the same bathymetric niche prior to the migration. In general, we lack direct tests of competition leading to niche limitation or displacement, between migrants and native species.

From fishes to molluscs (Fig. 9.24), an update (to 2021) list of all alien species along the Mediterranean coast of Israel reveals about 140 migrants. Our best knowledge of the colonization fate of these species concerns the rocky intertidal. A small Red Sea mussel (*Brachidontes pharaonis*) is one of the earliest Red Sea invaders into the Mediterranean (1876). Today, it is common in the Levant, in dense clusters on mid-littoral and infra-littoral rocks, piers and debris; and has spread as far west as Sicily. In the early 1970s, it was still about 250 times rarer than the native mussel *Mytilaster minimus*, which then formed dense beds on intertidal rocky ledges along the Israeli coastline. By the late 1990s, however, a rapid shift in dominance occurred, with dense populations of *Brachidontes* on rocky platforms where mussel beds were absent in the past. Today, this migrant completely covers many rocks, and one only rarely encounters the native species. The Red Sea limpet *Cellana rota*, commonly found in the Red Sea and Suez Canal, was first collected in the Mediterranean in 1961. By 2000, it had spread along the southeast Levantine coast, where it occupied sites less exposed to waves in the intertidal zone, whereas the native Mediterranean limpet, *Patella caerulea*, inhabited rocks exposed to the surf. Today, the Red Sea limpet dominates the upper rocky littoral, where it has replaced the native limpet, at many sites. The Red Sea spiny oyster, *Spondylus spinosus*, has completely outcompeted the native congener, within a decade of its first record in Haifa Bay. Similarly, the Red Sea jewel box oyster *Chama pacifica* outnumbers its native congener. The native Mediterranean creepers *Cerithium vulgatum* and *C. lividulum*, common and abundant in shallow water along the coast of Israel until the 1970s, have been supplanted by the Red Sea creepers *Cerithium scabridum* and *Rhinoclavis kochi*.

Of the 90 sea snail species of the Red Sea that have invaded the Mediterranean, the champions of success on the list—if we measure success in terms of biodiversity—are the creepers (Cerithioidea) with 14 species in 8 genera (Galil et al. 2021). In general, most creepers bury into soft sand and mud rich in organic matter, and many of the 200 odd genera have invaded freshwaters. The genital system of both sexes is primitive in that the male does not have a penis and not a closed sperm duct. Similarly, the female's ovules cross the mantle cavity floor in an open groove rather than in a closed oviduct. In the creepers of Florida, there are two types of development. The first and most common is indirect development involving many eggs, rapid cleavage, an encapsulated period that is brief and the emergence of free-swimming veligers. The second pattern involves direct development, fewer and larger eggs and a longer period of encapsulation, with the young hatching only when metamorphosed as small crawling snails. Another migrant creeper, *Planaxis*, lives in the rocky intertidal and is widely distributed in the Indian Ocean, where some populations consist of both males and females and others consist of females only. Reproduction in the all-female populations may occur without a male (parthenogenesis). The extent to which planktonic versus direct development, and parthenogenesis versus sexual reproduction, have contributed to the successful colonization and population establishment of Red Sea creepers in the Mediterranean, remain open questions.

H. Mienis, curator of molluscs at the Hebrew University, has placed all Lessepsian migrant molluscs on a scale, ranging from a single record to rare, local, common, or abundant (in Galil et al. 2021). Of the 90 migrant snail species, he placed 45 as known from only a single record or as rare; 23 as common and 4 as abundant (I leave out the intermediate group of "local"). Thus, since the canal opened about a century and a half ago, about half the gastropod migrant species that ever reached the shores of Israel might well have perished, and about one-third have successfully established themselves. In bivalves, with their 46 migrant species, Mienis placed 23 as single records or rare; and 13 as common or abundant (again, I leave out the "local" category). Thus, about half the bivalve migrant species that ever reached the shores of Israel might have perished, and about 10%–20% have successfully established themselves, in the Levant basin of the Mediterranean Sea (Galil et al. 2021).

It is noteworthy that in the Red Sea, a number of fish families have several species, of whom only a few have established populations in the Mediterranean. There are fifteen squirrelfish species in the Red Sea but only one colonizer in the Mediterranean; there are 14 goatfish species in the Red Sea but only two have migrated to the Mediterranean (Golani 1998; Golani et al. 2019).

What makes a species a successful migrant? Can we distinguish a potential colonizer? Is there a correlation to successful colonization in genetic variability, breadth of trophic niche, life history strategy, or relative importance, in the fish assemblage Red Sea? The Red Sea—Mediterranean migration provides a unique opportunity to investigate large-scale processes of migration, invasion, colonization and outcompeting extinction when two faunas collide, processes that usually occur over long geological time scales. Among fishes, alien species occupy a larger range of environments in their native ranges than do species that are not invaders, with those Red Sea species that naturally experience high maximum temperatures in their native range having a high probability of becoming alien (Belmaker et al. 2012). Upon reaching the Mediterranean, tropical invading fish species may spread far beyond their native niches, and this makes it difficult to forecast species invasion and their response to environmental changes (Paravicini et al. 2015).

References

Bacon, C. D., D. Silvestro, C. Jarmillo, B. Tilston Smith, P. Chkrabarty, and A. Antonelli. 2015. Biological evidence supports an early and complex emergence of the isthmus of Panama. *PNAS* 112:6110–6115.

Belmaker, J., V. Paravicini, and M. Kulbicki. 2012. Ecological traits and environmental affinity explain Red Sea fish introduction into the Mediterranean. *Global Change Biology* 19:1373–1382.

Bergmann, C. 1847. Ueber die Verhältnisse der Wärmeökonomie der Thiere zu ihrer Grösse. *Göttinger Studien* 1:595–708.

Cox, C. B., et al. 2010. *Biogeography: An Ecological and Evolutionary Approach.*

Cox, C. B., R. J. Ladle, and P. D. Moore. 2016. Biogeography. In *An ecological and evolutionary approach*, 10th ed. Wiley.

Croizat, L. C. M. 1958. *Panbiogeography or an introductory synthesis of zoogeography, phytogeography, geology; with notes on evolution, systematics, ecology, anthropology, etc*. Published by the author, Caracas.

Darlington, P. J. J. R. 1957. *Zoogeography: The geographical distribution of animals.* New York: Wiley.

De Queiroz, A. 2014. *The monkey's voyage, how improbable journeys shaped the history of life.* Basic Books.

Delhey, K. 2019. A review of Gloger's rule, an ecogeographical rule of colour: Definitions, interpretations and evidence. *Biological Reviews of the Cambridge Philosophical Society* 94 (1708): https://doi.org/10.1111/brv.12503.

Dowle, E. J., M. Morgan-Richards, and S. A. Trewick. 2013. Molecular evolution and the latitudinal biodiversity gradient. *Heredity* 110:501–510.

Economo, E. P., N. Narula, N. R. Friedman, M. Weiser, and B. Guenard. 2018. Macroecology and macroevolution of the latitudinal diversity gradient in ants. *Nature Communications* 9:1778.

Galil, B. S. 2007. Seeing red: Alien species along the Mediterranean coast of Israel. *Aquatic Invasions* 2:281–312.

Galil, B. S., H. K. Mienis, R. Hoffman, and M. Goren. 2021. Non-indigenous species along the Israeli Mediterranean coast: tally, policy, outlook. *Hydrobiologia* 848:2011–2029. Supplement 1.

Gaston, K. J. 2007. Quick guide latitudinal gradient in species richness. *Current Biology* 17:R574.

Gaudin, T. J., and D. A. Croft. 2015. Paleogene Xenarthra and the evolution of south American mammals. *Journal of Mammalogy* 96:622–634.

Geist, V. 1987. Bergmann's rule is invalid. *Canadian Journal of Zoology* 65:1035–1038. https://doi.org/10.1139/z87-164.

Golani, D. 1998. Impact of Red Sea fish migrants through the Suez Canal on the aquatic environment of the eastern Mediterranean. *Bulletin of the Yale School of Forestry Enviromental Studies* 103:375–387.

Golani, D., R. Fricke, and B. Appelbaum-Golani. 2019. The zoogeographic distribution patterns of Red Sea fish species: Is there a correlation between wide distribution and success in colonization of the Mediterranean?. *Front Mar. Sci. Conference Abstract: XVI European Congress of Ichthyology.*

Goldenberg, J., K. Bisschop, L. D'Alba, and M. D. Shawkey. 2022. The link between body size, colouration and thermoregulation, and their integration into ecogeographical rules: A critical appraisal in light of climate change. *Oikos*. https://doi.org/10.1111/oik.09152.

Kamilar, J. M., and B. Bradley. 2011. Interspecific variation in primate coat color supports Gloger's rule. *Journal of Biogeography* 38:2270–2277.

Marconides, R. S., K. F. Stryjewski, and R. T. Brumfield. 2020. Testing the simple and complex versions of Gloger's rule in the variable Antshrike (*Thamnophilus caerulescens*, Thamnophilidae). *The Auk* 137. https://doi.org/10.1093/auk/ukaa026.

Mathew, W. D. 1915. Climate and evolution. *Annals of the New York Academy of Sciences* 24:171–318.

Mayr, E. 1956. Geographical character gradients and climatic adaptation. *Evolution* 10:105–108.

Meiri, S. 2010. Bergmann's rule—What's in a name? *Global Ecology and Biogeography* 20:203–207.

Meiri, S., and T. Dayan. 2003. On the validity of Bergmann's rule. *Journal of Biogeography* 30:331–351.

Meiri, S., Y. Yom Tov, and E. Geffen. 2007. What determines conformity to Bergmann's rule? *Global Ecology and Biogeography* 16:788–794.

Mooring, M. S., A. A. Eppert, and R. T. Botts. 2020. Natural selection of Melanism in Costa Rican Jaguar and Oncilla: A test of Gloger's rule and the temporal segregation hypothesis research. *Tropical Conservation Science*. https://doi.org/10.1177/1940082920910364.

Paravicini, V., E. Azzuro, M. Kulbicki, and J. Belmaker. 2015. Niche shifts can impair the ability to predict invasion risk in the marine realm: An illustration using Mediterranean fish invaders. *Ecology Letters* 18:246–253.

Rensch, B. 1929. *Das Prinzip geographischer Rassenkreise und das Problem der Artbildung*. Berlin: Gebrüder Bornträger.

Robson, G. C. 1928. *The species problem: An introduction to the study of evolutionary divergence in natural populations*. Geological Magazine.

Ryding, S., M. Klaasen, G. J. Tattersal, J. L. Gardner, and R. E. Symonds. 2021. Shape shifting: Changing animal morphologies as a response to climatic warming. *TREE*1–13. https://doi.org/10.1016/j.tree.2021.07.006.

Salewski, V., and C. Watt. 2017. Bergmann's rule: A biophysiological rule examined in birds. *Oikos* 126:161–172. https://doi.org/10.1111/oik.03698.

Sanmartin, I. 2012. Historical biogeography: Evolution in time and space. *Evolution: Education and Outreach* 5:555–568.

Schrago, C. G., and C. A. M. Russo. 2003. Timing the origin of New World monkeys. *Molecular Biology and Evolution* 20:1620–1625.

Shelford, V. E. 1911. Physiological animal geography. *Journal of Morphology* 22:551–618.

Stehli, F. G., and S. D. Webb. 1985. *The great American biotic interchange*. New York: Plenum Press. Google Scholar.

Thornton, I. 1996. *Krakatau. The destruction and reassembly of an island ecosystem*. Cambridge, MA: Harvard University Press.

Wallace, A. R. 1876. *The geographical distribution of animals. With a study of the relations of living and extinct faunas as elucidating the past changes of the earth's surface*. New York: Harper & Brothers.

Werner, Y. L., N. Korolker, G. Sion, and B. Gocemen. 2016. Bergman's and Rench's rules and the spur-thighed tortoise (*Testudo graeca*). *Biological Journal of the Linnean Society* 117:796–811.

Wegener, A. 1924. *The Origin of Continents and Oceans*.

Willis, J. C. 1922. *Age and area: A study of geographical distribution and origin of species*. Cambridge.

Wilson, E. O. 1961. The Nature of the Taxon Cycle in the Melanesian Ant Fauna. *The American Naturalist* 95.

Ye, Z., J. Damgaard, D. Burckhardt, G. Gibbs, J. Yuan, H. Yang, and W. Bu. 2018. Phylogeny and historical biogeography of Gondwanan moss-bugs (Insecta: Hemiptera: Coleorrhyncha: Peloridiidae). *Cladistics*. https://doi.org/10.1111/cla.12237.

Yom Tov, Y., and J. Yom Tov. 2005. Global warming, Bergmann's rule and body size in the masked shrew *Sorex cinereus* Kerr in Alaska. *Journal of Animal Ecology* 74:803–808.

MIX
Papier aus verantwortungsvollen Quellen
Paper from responsible sources
FSC® C105338

If you have any concerns about our products,
you can contact us on
ProductSafety@springernature.com

In case Publisher is established outside the EU,
the EU authorized representative is:
Springer Nature Customer Service Center GmbH
Europaplatz 3, 69115 Heidelberg, Germany

Printed by Libri Plureos GmbH
in Hamburg, Germany